AF595354

Corrosion and Protection of Materials

Corrosion and Protection of Materials

Editor

Marina Cabrini

MDPI • Basel • Beijing • Wuhan • Barcelona • Belgrade • Manchester • Tokyo • Cluj • Tianjin

Editor
Marina Cabrini
University of Bergamo
Italy

Editorial Office
MDPI
St. Alban-Anlage 66
4052 Basel, Switzerland

This is a reprint of articles from the Special Issue published online in the open access journal *Materials* (ISSN 1996-1944) (available at: https://www.mdpi.com/journal/materials/special_issues/corrosion_protection_materials).

For citation purposes, cite each article independently as indicated on the article page online and as indicated below:

LastName, A.A.; LastName, B.B.; LastName, C.C. Article Title. *Journal Name* **Year**, *Volume Number*, Page Range.

ISBN 978-3-0365-0290-8 (Hbk)
ISBN 978-3-0365-0291-5 (PDF)

Cover image courtesy of Marina Cabrini.

© 2021 by the authors. Articles in this book are Open Access and distributed under the Creative Commons Attribution (CC BY) license, which allows users to download, copy and build upon published articles, as long as the author and publisher are properly credited, which ensures maximum dissemination and a wider impact of our publications.
The book as a whole is distributed by MDPI under the terms and conditions of the Creative Commons license CC BY-NC-ND.

Contents

About the Editor

Marina Cabrini has been a professor of Science and Technology of Materials at the Mechanical Engineering Faculty of the University of Bergamo since 2001. Her courses are "Metallic Materials" and "Polymer, Composites and Ceramics" in Mechanical Engineering, "Biomaterials" for Technology Engineering for Health and "Electrochemistry and Electrochemical Technologies" for the PhD student in Engineering and Applied Science of the University of Bergamo.

Her research activity is on electrochemistry and corrosion, primarily focused on the environmentally assisted cracking of traditional and innovative steels for the oil and gas industry. She has conducted some research on biomaterials, on the electrochemical characterization of the kinetic of passivity film formation on rebar in concrete, on corrosion inhibitors for chloride-contaminated concrete and on the corrosion evaluation of carbon steel in CCTS (carbon capture transport and storage).

Nowadays, she is working on the corrosion behavior of aluminum, titanium and nickel alloys obtained by means of direct metal laser sintering in collaboration with Polytechnic of Turin, on the corrosion and stress corrosion cracking of aluminum alloys (AA7075 and AA 2024) welded by means of friction stir welding, the effect of loading on the hydrogen diffusion and embrittlement of low alloyed high strength steels and the corrosion in oil and gas and geothermal systems.

Preface to "Corrosion and Protection of Materials"

The corrosion phenomenon is the deterioration of metallic structures caused by the reaction of the metal with the environment. Several recent case studies pointed the attention towards the importance of the security and monitoring of facilities and infrastructures. The first estimate of the economic damages caused by corrosion and the protection of materials was carried out in the 1970s by the British Government, with the conclusion that the amount of expenses for the restoration of the damaged structures was around 3% of the gross domestic product (GPD), and that 90% of this figure would have been easily saved if the technical notions of protection and choice of materials already acquired had been applied correctly. A similar study commissioned published on the web site of the NACE (National Association of Corrosion Engineering) indicates a surprisingly identical value: 3.4% of GDP (2013)! It is impressive how in forty years the economic importance of the corrosion of materials has not decreased at all, on the contrary—considering the growth of GDP, the data from an economic point of view are certainly alarming. On the basis of these considerations, it could be inferred that over more than forty years, not much has been done for the prevention of corrosion, but obviously this consideration is superficial. It is much more realistic to think that the conditions in which materials operate are in many plants more and more exasperated (for example, think of the increase in the depth of exploitation of oil fields), as there is a constant search for more performing materials, with high weight/performance ratios in order to reduce energy consumption, not to mention the increased sensitivity towards the environment that led to the abandonment of some anticorrosion processes and/or methods (an example for all the use of chromates as corrosion inhibitors or anticorrosive pigments) and the research on environmentally friendly systems. In addition, new alloys, with different microstructures, like high entropy alloys, ultra-fine grains metals, or new technologies of production and installation, like additive manufacturing, innovative coating or friction stir welding, need to improve the consolidated knowledge on the corrosion mechanisms and protection methods of metallic materials. At the base of each correct design must therefore be an increased sensitivity about the increase in its useful life, both to reduce the energy impact of production plants and disposal, and to reduce the possibility of contamination of the environment with the fluids of the process or with the same substances used for corrosion protection. The aim of this Special Issue was to draw a panorama of the actual research on the corrosion and protection of traditional and innovative materials in natural environments or in the industrial context. This Special Issue contains thirty papers by 140 authors of 16 countries, with the geographical distribution observed in Figure 1, testifying to the importance of studying corrosion problems all over the world. Several different items are presented. The first papers of the volume are on alloys of new technologies, and it is opened by an invited review on the corrosion of additively manufactured aluminum alloys; the other two papers are, respectively, on alloy 625 obtained by additive manufacturing and on CrFeCoNiNbx high-entropy alloys. The following seven papers are mainly on the mechanism of corrosion, including two interesting reviews, one on the alternate current corrosion on carbon steel under cathodic protection and the other on the electrochemical polishing of austenitic stainless steels. Studies on the mechanism of corrosion of pure iron, martensitic and austenitic stainless steels, nickel–aluminum–bronze, and Al–Zn alloy are presented in the following papers. Microbiological corrosion, biocide treatments and corrosion inhibitors are always topical in industrial practice. These topics are dealt with in the following five papers, with particular attention to the environmental impact of the chemicals studied, which must necessarily be eco-friendly. Many studies are underway regarding the development of anticorrosive coatings, the most current trend is aimed at ceramic and metal coatings, as demonstrated by six works published in this Special Issue. Stress corrosion cracking and hydrogen embrittlement are widely studied, but they remain of strong actuality. This item in the Special Issue is reported on in five papers regarding the traditional stress corrosion phenomena in sour environments or in pipeline steels, high-strength steels wires, as well as aluminum or magnesium alloys. Finally, the last four works present interesting aspects related to monitoring the corrosion of structures and the effects of corrosion on the residual mechanical resistance of civil structures, such as bridges or reinforced concrete. These arguments are of vital importance given the increase in the average life of infrastructures and their natural deterioration which makes the problem of their control and recovery of primary importance. Personally, I am honored to have had the opportunity to be the guest editor of this volume, and I want to thank MPDI for giving me the opportunity; thanks you above all to Andy Zhang for his precious contribution in the realization of this Special Issue of *Materials*.

Figure 1. Geografical distribution of the authors of the papers present in this special issue.

Marina Cabrini
Editor

Review

Corrosion and Corrosion Protection of Additively Manufactured Aluminium Alloys—A Critical Review

Reynier I. Revilla *, Donovan Verkens, Tim Rubben and Iris De Graeve

Research group of Electrochemical and Surface Engineering, Department of Materials and Chemistry (MACH), Vrije Universiteit Brussel, Pleinlaan 2, 1050 Brussels, Belgium; Donovan.Verkens@vub.be (D.V.); tim.rubben@vub.be (T.R.); Iris.De.Graeve@vub.be (I.D.G.)

* Correspondence: rrevilla@vub.be

Received: 28 September 2020; Accepted: 22 October 2020; Published: 28 October 2020

Abstract: Metal additive manufacturing (MAM), also known as metal 3D printing, is a rapidly growing industry based on the fabrication of complex metal parts with improved functionalities. During MAM, metal parts are produced in a layer by layer fashion using 3D computer-aided design models. The advantages of using this technology include the reduction of materials waste, high efficiency for small production runs, near net shape manufacturing, ease of change or revision of versions of a product, support of lattice structures, and rapid prototyping. Numerous metals and alloys can nowadays be processed by additive manufacturing techniques. Among them, Al-based alloys are of great interest in the automotive and aeronautic industry due to their relatively high strength and stiffness to weight ratio, good wear and corrosion resistance, and recycling potential. The special conditions associated with the MAM processes are known to produce in these materials a fine microstructure with unique directional growth features far from equilibrium. This distinctive microstructure, together with other special features and microstructural defects originating from the additive manufacturing process, is known to greatly influence the corrosion behaviour of these materials. Several works have already been conducted in this direction. However, several issues concerning the corrosion and corrosion protection of these materials are still not well understood. This work reviews the main studies to date investigating the corrosion aspects of additively manufactured aluminium alloys. It also provides a summary and outlook of relevant directions to be explored in future research.

Keywords: metal additive manufacturing; aluminium alloys; corrosion behaviour; microstructure; corrosion protection

---

1. Introduction

Metal additive manufacturing (MAM), commonly known as metal 3D-printing, is a process by which complex multifunctional metal parts are produced in a layer by layer fashion using 3D computer-aided design (CAD) models [1–6]. Several MAM techniques are available. They can be separated into two main groups: direct energy deposition (DED) methods and powder bed fusion (PBF) technology [7]. During direct energy deposition, focused thermal energy is used to fuse materials by melting as they are being deposited; while during powder bed fusion, thermal energy selectively fuses regions of a powder bed [7]. DED processes such as wire arc additive manufacturing (WAAM) and laser metal deposition (LMD) can generally be used on existing parts of arbitrary geometry with a relatively high deposition rate; however, the shape complexity is limited. This makes DED processes the preferred methodology for repairing or improving existing parts [8]. On the other hand, on PBF methods such as selective laser melting (SLM), selective laser sintering (SLS), and electron beam melting (EBM), the dimension of the pieces is limited and the starting substrate has to be a flat surface. However, they generally allow the fabrication of pieces with extremely high structural complexity at a relatively high level of precision. Among the several MAM processes, those utilizing a metal powder feedstock and a laser source to achieve the metal fusion are the most widely used [1–6]. From those, SLM is regarded as the most used and studied MAM method. This is not only because it allows a higher level of precision compared to other MAM techniques, but also because (in contrast to SLS) it allows the full melting of the material, and therefore the production of solid and dense metal parts in a single process (without the need to use binders and/or other post-process furnace operations).

Additive manufacturing is considered one of the enabling technologies for Industry 4.0 [9]. In particular, MAM presented a global market valued at € 2.02 billion in 2019 [10]. This included systems, materials, and services. MAM allows the near-net shape manufacturing of geometrically complex parts such as lattice structures and 3D structures with undercuts or cavities, which is why this technology has found numerous applications in industries such as medical implants, energy, aerospace, and automotive. As an example in aerospace applications, MAM has made possible the re-design of complex fuel injector nozzles (commonly requiring the assembly of more than 20 parts) in a single operation [11,12], as well as the re-design of several other complex engineered parts, resulting in considerable cost and weight reduction. In aerospace, as well as in the automotive industry, MAM is also actively used in prototyping and the fabrication of custom tooling.

Nowadays, a great number of metals and alloys can be processed by additive manufacturing techniques, depending mainly on the availability of the raw materials as metal powders or metal wires [5]. Amongst these, aluminium alloys are of great interest for applications requiring high strength and stiffness to weight ratio, good wear and corrosion resistance, and recycling potential, which is why they are attracting increasing attention of the automotive and aerospace industries [12,13]. The most common Al-based alloys processed by additive manufacturing (AM) either for commercial use or for research purposes are [5,14,15]: AlSi12, AlSi10Mg, AlSi7Mg0.6, AlSi9Cu3, AlSi5Cu3Mg, AA1050, AA2017, AA2024, AA2219, AA6061, AA7020, AA7050, AA7075, and AA5083; next to proprietary industrial alloys like Scalmalloy. From those, Al-Si alloys, and more specifically AlSi10Mg (followed by AlSi12), are undoubtedly the most investigated and commercially used additively manufactured Al-based alloys. These materials, particularly relevant for light-weight and high-strength applications, are widely used for aluminium casting due to the proximity to the eutectic composition (~12.5% Si) [16]. Therefore, they are relatively easy to process by laser applications, which are known to lead to a small solidification range [17]. Additionally, minor additions of magnesium (0.3–0.5 wt.% Mg) are known to induce hardening of the alloy by forming Mg_2Si precipitates upon natural or artificial ageing treatments [18]. However, the actual formation of these precipitates on additively manufactured Al-Si parts is still a topic of discussion.

Due to the special conditions of the MAM processes (namely that the metal powder used is already pre-alloyed, and the melting occurs in small pools that rapidly solidify), a very fine microstructure with unique directional growth features far from equilibrium is achieved [1]. This distinctive microstructure, together with other special features and microstructural defects originating from the additive manufacturing process is known to greatly influence the corrosion behaviour of these materials. Sander et al. [19] and Kong et al. [20] reviewed the impact of these special features and defects on the corrosion performance of additively manufactured metals. These works also consider the main corrosion issues of several additively manufactured materials, including some studies on Al alloys. Zhang et al. [17] presented a review of Al-based alloys summarizing the microstructural characteristics and mechanical properties. Chen et al. [21] reviewed the main research studying the corrosion behaviour of selective laser melted Al alloys, classified/structured by the used technique.

The present work reviews the main studies to date investigating the corrosion aspects of additively manufactured aluminium alloys from a general perspective. Some of the defects and intrinsic issues from the additive manufacturing process that can influence the corrosion performance of these materials are discussed. Additionally, the most relevant studies and results concerning the effect of microstructure, heat treatments, Si content, surface roughness, and surface treatments on the corrosion behaviour of Al-based alloys are identified and discussed. Moreover, the main aspects concerning the behaviour of these materials to corrosion protection surface treatments such as anodizing are summarized and discussed for the first time. A summary and outlook of relevant directions to be explored in future works are also presented. The majority of the works investigating additively manufactured Al alloys was conducted on SLM materials, and a few studies were realized using SLS specimens. The microstructural characteristics and corrosion behaviours of the materials fabricated using those two methods are remarkably similar. Therefore, the use of "AM Al alloys" in this work refers, in general, to SLM and/or SLS Al alloys.

2. Influence of Defects in AM Specimens on Corrosion Resistance

Metal additive manufacturing presents, without a doubt, great potential to become a key manufacturing technology in several industries and daily life applications. One of the conditions still needed to achieve this goal is better control and understanding of the numerous macro- and micro-structural defects introduced by the special manufacturing conditions of these processes. These defects can greatly influence the materials' performance against corrosion [22]. Further investigations are needed to better understand the effect of features such as porosity, surface roughness, and residual stresses, among others, on the corrosion resistance of these materials.

Metal additive manufacturing is, in general, a very complex process with a great number of operation parameters involved (for instance the laser/beam power, scanning speed, the spot size of the power source, hatch distance, layer thickness, powder flow speed in the case of DED processes, etc.). The effect of each process parameter in the final printed part cannot be seen independently of the rest. Instead, the energy density (E), which indicates the energy input into the material (see Equation (1)) and is equal to the laser/beam power (P) divided by the scanning speed (v), the layer thickness (d), and the hatch distance (h), was the quantity considered to evaluate the effect of process parameters on the properties of the final parts [23]. Previous studies have shown that the energy input is the determining parameter for defining and optimizing the "process window" during additive manufacturing [23]. A low energy input per unit length results in droplet formation and a bad wetting to the previous layer. On the other hand, a relatively high energy input per unit length causes distortions and irregularities due to big melt pool volumes and the balling effect. The specific process parameters and, therefore, the energy density used, are usually optimized per alloy type in each AM system to obtain high-density parts with the lowest possible level of porosity. Nevertheless, confirming the fact that MAM is a complex process, other researchers believe that the energy density cannot be the sole parameter analysed during the process optimization [24]. Other parameters such as the scanning strategy, laser spot size, and materials' properties (i.e., thermal conductivity and reflectivity) should also be considered. Moreover, the researchers suggest that the laser power has a greater influence than the rest of the parameters, and therefore should also be considered independently [24]. Furthermore, Leung et al. [25] also demonstrated that the oxidation state of the metal powder could have a great influence on the formation of defects in the printed parts.

$$E = \frac{P}{v \times h \times d} \tag{1}$$

An incorrect choice of process parameters could lead to high levels of porosity within additively manufactured metal parts. Pores in additive manufacturing can be classified into two groups: trapped-gas pores and lack-of-fusion pores. Trapped-gas pores, as its name says, are the result of gas trapped inside the powder particles during gas atomization, or inside the printed part during the actual manufacturing. These pores have a rather spherical shape as can be seen in Figure 1a. Lack-of-fusion pores have, on the other hand, an irregular shape (see Figure 1b) and are much larger than trapped-gas pores. Lack-of-fusion pores appear when there is no complete adherence of the current melt to the surrounding part (when powder particles are only partially molten) [26]. These two types of pores possess intrinsically different characteristics, and therefore, they could affect/impact corrosion in different ways and/or to a different extent.

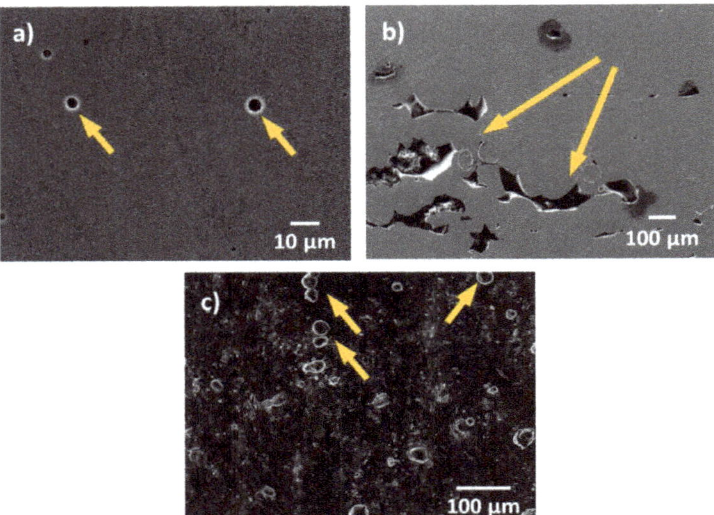

Figure 1. Secondary electron image representing defects that can be present in additively manufactured metal parts: (**a**) trapped-gas pores; (**b**) lack-of-fusion pores; (**c**) unmolten powder.

In general, close-to-the-surface pores can compromise the passivation properties of the native surface oxides, and therefore the pitting resistance of the material. This was demonstrated in previous studies for stainless steel [27,28], in which the researchers have shown that the level of porosity and the size of the pores play an important role in the resistance to pitting corrosion. However, no work has been conducted to date to systematically study the effect of porosity (dimension, type, and extent of porosity) on the corrosion behaviour of AM Al-based alloys. Therefore, more focus should be given to this issue in future studies. Moreover, pores might also play an important role in the susceptibility of the printed parts to stress corrosion cracking (SCC), since pores are intrinsically stress concentrators during mechanical loading. Once corrosive species reach the pores, occluded cell conditions could rapidly built-up and promote a premature failure of the printed part. This is a topic that needs more attention in the future.

As in the case of porosity, unmolten, or partially molten powder (see Figure 1c) results from incorrect choices of process parameters. Unmolten powder on the surface of the printed part can compromise the passive behaviour of the material by introducing defects in the native oxide film and greatly increases the surface roughness. There are currently several processes available (such as abrasive blast, shoot peening, electrochemical polishing, etc.) that are used to improve the surface finishing and reduce the roughness of AM metal parts [29,30]. These methods greatly affect the surface state and therefore, the surface energy, which influences the nucleation and growth of the native oxide film. This can, consequently, affect the barrier properties of the native oxide, and hence the passivity of the material.

Another intrinsic feature of additively manufactured metal parts that could compromise their physical integrity is the existence of residual stresses. Residual stress formation in metal additive manufacturing is caused by the high thermal gradients and cooling rates associated with these processes [31]. Particularly high tensile residual stresses are very typical near the surface [31]. These residual stresses can affect the materials' susceptibility to stress corrosion cracking. Previous studies have shown the formation of micro-cracks in as-built AM Al-Si alloys after exposure to corrosive media (see Figure 2) [32,33]. The researchers associated this cracking with the combined effect of the local disruption of the Si network, the selective dissolution of the Al matrix around the disrupted zones, and the existence of residual stresses from the MAM process [33]. Additionally, surface residual stresses could, in general, influence the surface energy and therefore, the nucleation and growth of the native oxide film, affecting as a consequence the passivity and pitting resistance of the material. Nevertheless, no relevant work has been published so far on the study of the effect of residual stresses on the corrosion behaviour of additively manufactured metals.

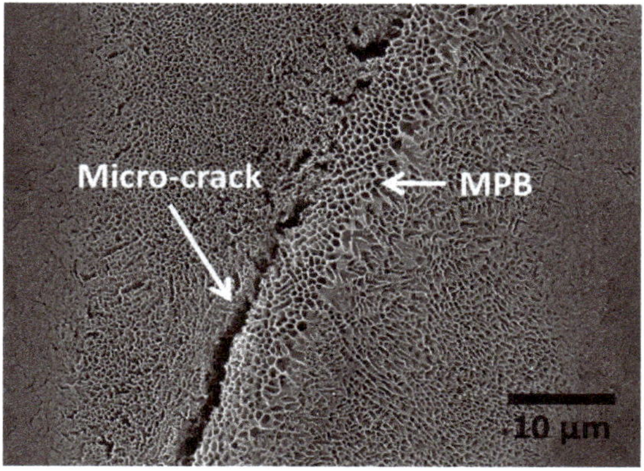

Figure 2. Secondary electron microscopy image of the surface of an as-built additive manufacturing (AM) Al-Si (AlSi10Mg) after immersion for 48 h in 0.1 M NaCl. The formation of micro-cracks in the heat affected zone next to the MPB can be seen. MPB—melt pool border.

3. Al-Si Alloys

3.1. Effect of Microstructure on Corrosion Behavior

Several studies characterizing the microstructure of as-built additively manufactured Al-Si alloys have been conducted in recent years [33–43]. As-built specimens exhibit a fine distribution of Si, forming a three-dimensional network that encloses the primary face-centred cubic α-Al in very small cells (see Figure 3). The size of these cells varies over the melt pool due to the thermal gradient created by the moving heat source. Finer cells are formed towards the middle of the melt pools (MP), while coarser cells are present in the melt pool borders (MPBs) [36]. A marked anisotropy has been described in past studies concerning the shape of these cells. These cells are known to present an approximately round shape in the plane parallel to the building platform (XY), whereas in the plane perpendicular to the building platform (XZ) more elongated cells have been observed [36,42]. A heat-affected zone (HAZ) located right outside the borders of the melt pools has also been identified (see Figure 3). This HAZ, in which the silicon network is partly broken and discontinuous, has been associated with overheating of the underlying layer during the scanning of a newly deposited layer [36].

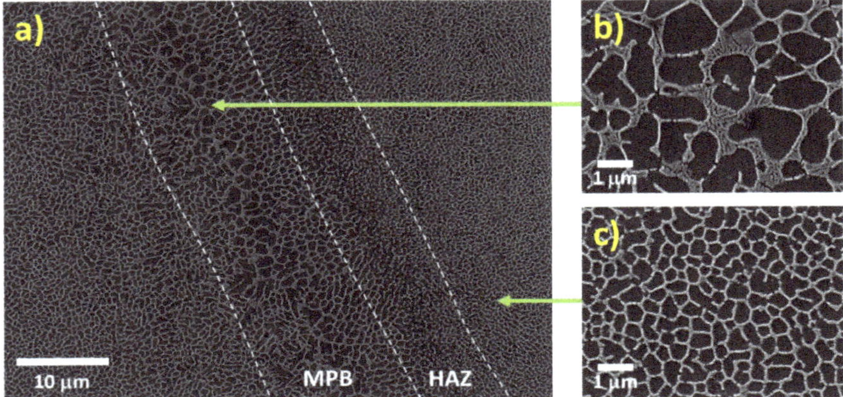

Figure 3. (**a**) Secondary electron image representing the microstructure of additively manufactured Al-Si alloys (AlSi10Mg). The surface parallel to the building platform is represented here. Similar features can be observed in the surface perpendicular to the building platform, but the shape of the cells is more elongated in that case. (**b**) Higher magnification image of a zone in the melt pool border. (**c**) Higher magnification image of a zone within the melt pool. MPB—melt pool border; HAZ—heat-affected zone. (Adapted from reference [43]).

A great number of works have already shown that the corrosion behaviour of additively manufactured Al-Si alloys is greatly influenced by this special and unique microstructure created during the additive manufacturing process.

3.1.1. Influence of the Melt Pool Borders on Corrosion

Among the different details of the microstructural features of as-built AM Al-Si alloys, the borders of the melt pools have been pointed out to be a key element in the corrosion of these materials. Several studies have shown that after potentiodynamic polarization tests and/or immersion in NaCl solution a particularly pronounced selective corrosion of the α-Al cells in the melt pool borders is observed (see Figure 4) [32,33,44–57]. This selective corrosion attack along the borders of the melt pools has been reported for polished and ground specimens [32,33,44–56], but also for as-built materials presenting a low surface roughness [57], for which the microstructural features are believed to control the electrochemical performance of the material.

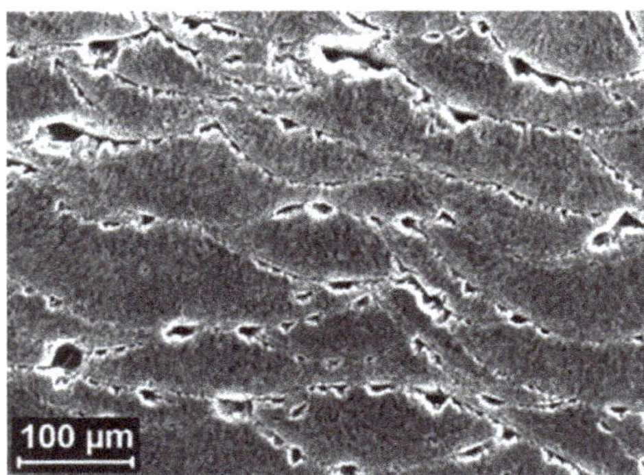

Figure 4. Secondary electron microscopy image of the surface of an as-built AM Al-Si (AlSi10Mg) corroded in NaCl solution. (Adapted from reference [47]).

The initiation and further propagation of the corrosion attack in the Al cells along the MPBs are caused by the higher driving force for micro-galvanic corrosion between the α-Al and the Si phase in these regions compared to other areas within the melt pools [32,45,54]. Due to the presence of a relatively larger microstructure at the MPBs (see Figure 5), a greater Volta-potential difference has been reported between the Al and the Si phase compared to regions within the MPs (Figure 5c). Nevertheless, Kubacki et al. [51] do not support the assumption of a galvanic couple between the Al and Si phase accelerating corrosion in simulated atmospheric conditions at the melt pool borders. The researchers believe that under these conditions the cathodic kinetics on the Si phase is not fast enough to support the active dissolution of Al. Instead, the selective/pronounced corrosion attack at the borders of the melt pools is attributed to the high discontinuities of the Si network around the MPBs (in the heat-affected zone).

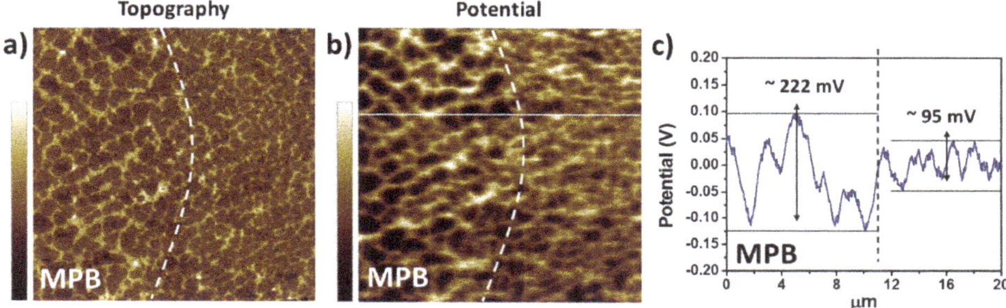

Figure 5. (**a**) Topography and (**b**) surface potential map (obtained by scanning Kelvin probe force microscopy—SKPFM) of an area on the surface of a polished AM AlSi10Mg specimen in which a melt pool border is visible (delimited by the discontinuous line). Scan size: 20 × 20 μm^2. Colour bar: (**a**) 30 nm range, (**b**) 210 mV range. (**c**) Surface potential profile of the line represented in (**b**). (Adapted from reference [32]).

The morphology of the corrosion attack in as-built AM Al-Si alloys has been described by several researchers to be rather superficial due to the existence of a considerably connected Si network that holds the corrosion from penetrating deeper into the material [32,33,51,54,58]. However, several works have found a large penetration of the corrosion attack along the melt pool borders [46,50–53,55]. Moreover, the formation of micro-cracks during corrosion has been reported in regions next to the MPBs, where heat-affected zones exist [32,33,54,55,59]. As mentioned above, Revilla et al. [33] associated the formation of these micro-cracks to the synergistic effect of the selective dissolution of the α-Al cells along the MPBs, next to which the disrupted heat-affected zones are found, and the existence of internal and superficial residual stresses from the MAM process. Rafiezad et al. [49] also confirmed through intergranular corrosion test an accelerated preferential attack combined with the formation of micro-cracks along the melt pool borders for specimens fabricated using recycled powder. Moreover, while conducting electrochemical tests in NaCl solution, Girelli et al. [59] found that the MPBs/HAZ represent preferential paths for exfoliation-like corrosion to occur.

The schematic in Figure 6 shows the proposed evolution of the corrosion attack in as-built AM Al-Si alloys [33]. Because of the greater Volta-potential difference between the α-Al and the Si phase at the melt pool borders (as measured by scanning Kelvin probe force microscopy SKPFM [32,45,54]—see Figure 5), a higher driving force for galvanic corrosion provokes the initiation of the corrosion attack in these regions. The corrosion spreads superficially to adjacent zones, including the neighbouring heat-affected zones. This corrosion is partially contained by the connected portion of the Si network. However, due to the disruption in the heat-affected zone, the corrosion can propagate further into the material along the MPBs.

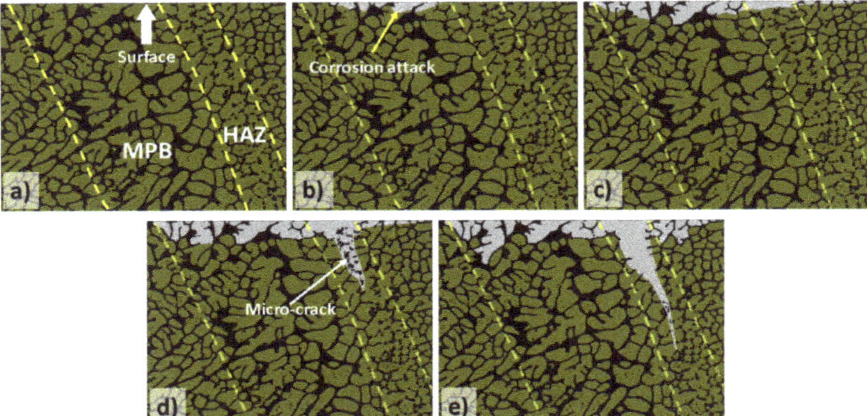

Figure 6. Schematic diagram representing the process of corrosion initiation and corrosion propagation for as-built additively manufactured Al-Si alloys. The illustrations portray the corrosion process seen from a cross-sectional perspective, with the top side of the images representing the surface exposed to the corrosive medium. The Al phase is represented with green, while black portrays the Si phase. A melt pool border (MPB), where a coarser microstructure is seen, and a heat-affected zone (HAZ), characterized by discontinuities in the Si network, can be observed in the images (**a**). The corrosion attack initiates at the MPB, where there is a larger potential difference between the Al and the Si phase, and therefore, a larger driving force for galvanic corrosion (**b**). Because of the partial containment of the corrosion by the Si network, the attack spreads superficially to adjacent zones, reaching the HAZ (**c**). Due to the disruption of the Si network in the HAZ, combined with the existence of internal stresses from the MAM process, micro-cracks are formed along the HAZ (**d**). In a later stage of the corrosion, additional lateral spreading of the attack occurs, accompanied by further propagation of the crack and corrosive medium through the HAZ (**e**) [33].

3.1.2. Anisotropy

As already mentioned, the special microstructure of additively manufactured Al-Si alloys is the result of the exceptionally high cooling rates and direction of the thermal gradients [36]. The particular directional distribution of thermal gradients during the fusion and solidification of the different layers of the material produces an intrinsically anisotropic microstructure [36,42]. The small α-Al cells surrounded by the Si network are rather rounded in the plane parallel to the building platform, while in the plane perpendicular to the building platform these cells have an elongated shape [36,42]. Additionally, in the surfaces parallel to the building platform (XY plane), elongated laser tracks are easily identified, while in the surfaces perpendicular to the building platform (XZ plane) a scale-like feature of melt pool borders is generally seen [32,33,44–48].

Concerning the influence of this anisotropy on the corrosion behaviour/resistance of these materials, various and contradictory results have been reported in the literature. For AlSi10Mg prepared by SLM, Cabrini et al. [48,50] concluded from potentiodynamic polarization experiments in diluted Harrison's solution that the surface of the XZ plane presents a slightly lower pitting corrosion resistance than the XY plane. The researchers associated this behaviour with the higher density of melt pool borders found in the XZ plane compared to that of the XY plane. However, in a later work, the same researchers concluded through a statistical approach that the building direction does not significantly influence the corrosion resistance of the analysed surface [56]. Moreover, Revilla et al. [32,33,54] found no difference in the electrochemical behaviour of the different planes during potentiodynamic polarization tests in aerated NaCl solution.

On the other hand, Cabrini et al. [46] demonstrated in another work by conducting an intergranular corrosion test that the corrosion in as-built specimens propagates mainly along the MPBs. Therefore, the penetration depth of the corrosion attack is highly influenced by the anisotropy of the melt pools. A more penetrative corrosion attack was seen on the surface parallel to the building platform compared to the surface perpendicular to the building platform. This could demonstrate that even if similar electrochemical behaviour is seen during potentiodynamic polarization tests, great anisotropy can still exist on the morphology of the corrosion attack and corrosion penetration depth.

For SLM AlSi7Mg0.6 and AlSi12, no considerable difference was reported between the behaviours of the two different planes during potentiodynamic polarization tests in NaCl solution [33]. Nevertheless, a clear distinction in the corrosion resistance of the different planes was shown by Chen et al. [42] for SLM Al-Si12. The researchers presented experimental data (open circuit potential, potentiodynamic polarization, and electrochemical impedance spectroscopy measurements) supporting a better corrosion resistance of the XZ plane compared to the XY plane. This is the opposite trend as that reported for AlSi10Mg by Cabrini et al. [48,50]. Chen et al. [42] associated this behaviour with the depth of the Al/Si cells in each plane. The small and round cells were seen in the XY plane are deep since they are elongated in the perpendicular direction. According to the researchers, this could lead to the growth and even deposition of corrosion products, which could extrude and crack the Si shells, exposing the underlying Al substrate to further corrosion attack. On the other hand, the cells seen in the XZ plane are shallow, which could limit/prevent the deposition of corrosion products. Moreover, Chen et al. [42] reported a clear difference in the weight loss rate during corrosion in NaCl solution for the different planes. These results are shown in Figure 7. A reduced weight loss rate is seen for the XZ plane compared to the XY plane, possibly due to the same reason as that given for the difference in the electrochemical behaviour.

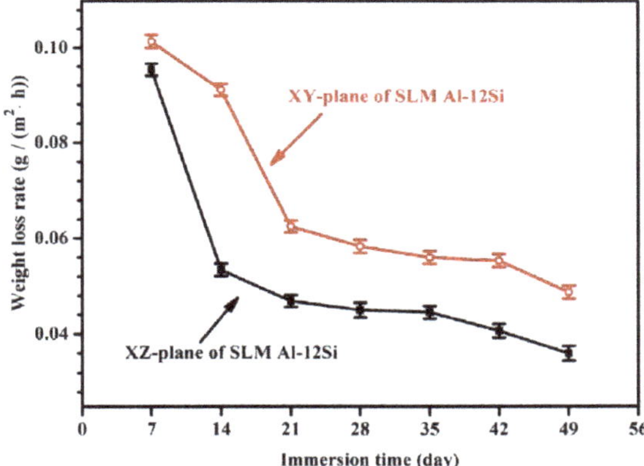

Figure 7. The weight loss rate for the different surface planes (parallel to the building platform—XY—and perpendicular to the building platform—XZ) of as-built and mechanically polished SLM AlSi12 after immersion in 3.5 wt.% NaCl solution at room temperature [42].

Due to the limited and contradictory results found in literature concerning the effect of the anisotropic microstructure on the corrosion behaviour, further research is needed to better understand this issue. It is also important to keep in mind that different alloys (even from the same alloy family) might display intrinsically different behaviours.

3.1.3. Comparison with Conventional Alloys

While the microstructure of as-built AM Al-Si alloys presents a very fine distribution of alloying elements into a three-dimensional network, the traditional cast alloy is characterized by the presence of large Si crystals [32,44] and for the case of cast AlSi10Mg, other intermetallic particles such as AlFeSi and Mg_2Si, among others, are also present [44]. These great differences in microstructure could result in differences in the corrosion behaviour of these materials. The corrosion resistance of AM Al-Si alloys has also been compared to that of conventionally manufactured alloys [32,44,51,59–63]. The results obtained for AlSi10Mg and AlSi12 are somewhat varied and in some cases contradictory.

AlSi10Mg

Table 1 summarizes, in a comparative way, the main results from potentiodynamic polarization tests obtained for AlSi10Mg. While some works report a similar corrosion current density for cast and AM AlSi10Mg [32,51,61], others claim a higher value for the cast alloy compared to the additively manufactured specimen [44,59,60]. Besides, while some studies show a similar value of corrosion potential for cast and AM AlSi10Mg [32,61], others report a nobler value of corrosion potential for AM AlSi10Mg in comparison with its cast counterpart [44,59,60]. On the contrary, Kubacki et al. [51] measured a lower value of corrosion potential for the additively manufactured material compared to the cast alloy. Nevertheless, the AM material presented a much higher value of pitting potential and a much wider passive region than the cast alloy [51].

Table 1. Comparison between Cast and as-built AM AlSi10Mg from results obtained in potentiodynamic polarization tests. Only results acquired in polished/ground samples were considered. E_{corr}, I_{corr}, and E_{pit} refer to corrosion potential, corrosion current density, and pitting potential, respectively.

Reference	Electrolyte	E_{corr}	I_{corr}	E_{pit}	Corrosion Rate
[44]	Aerated 3.5 wt.% NaCl	AM > Cast	Cast > AM		Cast > AM
[51]	Diluted Harrison's solution	Cast > AM	Cast ~ AM	AM > Cast	
[59]	Aerated 3.5 wt.% NaCl	AM > Cast	Cast > AM		Cast > AM
[60]	Aerated 3.5 wt.% NaCl	AM > Cast	Cast > AM	Cast ~ AM	
[32]	Aerated 0.1 M NaCl	Cast ~ AM	Cast ~ AM		
[61]	Aerated 3.5 wt.%NaCl	Cast ~ AM	Cast ~ AM		Cast ~ AM

While the corrosion attack in AM AlSi10Mg is mainly localized along the melt pool borders [32,33,44–57], the corrosion of the cast alloy is characterized by a severe localized corrosion of the α-Al matrix in the periphery of the large Si precipitates, as well as around the Mg_2Si and Fe-containing intermetallics [32,44,51]. In general, several researchers claim that the corrosion resistance of additively manufactured AlSi10Mg is higher than that of the cast alloy of approximately the same chemical composition [44,60,61]. Girelli et al. [59] and Leon et al. [61] found a slightly lower mass loss during corrosion in aerated 3.5 wt.% NaCl for the AM specimen compared to the cast alloy. However, Kubacki et al. [51] concluded that while both the cast and AM AlSi10Mg presented similar surface damage and median corrosion depth after immersion in a modified G85-A2 (cyclic acidified salt fog/spray test) cycle, the additively manufactured material presented a greater maximum damage depth compared to the cast alloy.

AlSi12

The AlSi12 alloy has been, in general, less studied than AlSi10Mg. Only a few works have compared the corrosion resistance of the additively manufactured specimens with that of the conventional cast alloy [62,63]. Yang et al. [63] reported a higher corrosion resistance for the AM material compared to the cast alloy. Their experimental data in NaCl solution showed not only a better electrochemical performance during open circuit potential and potentiodynamic polarization tests for the AM AlSi12 sample than for the cast alloy, but also a much lower weight loss during immersion in the same solution for the AM sample compared to the cast material [63]. On the other hand, other researchers reported that the cast and the AM AlSi12 alloy display similar corrosion behaviour in the HNO_3 solution [62]. Weight loss measurements after immersion in 0.1 M HNO_3 showed an almost perfect overlap between the curves of the cast and additively manufactured AlSi12 alloy [62].

3.2. Effect of Heat Treatment on Microstructure and Corrosion

As mentioned above, the unique processing conditions during MAM cause rapid solidification of melt pools due to the extremely high cooling rates reached (~10^5 K/s) [39,58]. This leads to a very fine cellular microstructure far from equilibrium, consisting of a primary Al-rich phase [39,64] supersaturated in Si (up to 11 wt.%) [52] with residual Si segregated at the cellular boundaries [39]. The extremely fine Si network gives a large total interfacial energy, resulting in a high driving force for Si coarsening [38]. As thermal gradients and grain growth rates decrease towards the melt pool borders, coarser cellular structures can be found there [36]. In addition, the fast cooling rates can introduce residual thermal stresses that can lead to dimensional inaccuracy and distortion [55]. A variety of heat treatments can be performed to act on these factors. Due to the presence of Mg in most Al-Si alloys, artificial aging (150–180 °C) [65] can be performed to induce age hardening through Mg_2Si precipitation [34,52,55]. The increase in the solid solubility of Si in Al due to the rapid solidification can enhance the effect of solution hardening and strengthening [66]. Partial annealing treatments can be performed to release residual stress [50,52–55,58,67,68]. Alternatively, the building platform can also be heated during manufacturing [34,52].

Fiocchi et al. [69] performed Differential Scanning Calorimetry (DSC) experiments to determine the phase transformations occurring in an AlSi10Mg alloy prepared by selective laser melting. Several heating rates (2, 5, 10, 20, and 30 °C/min) were applied for a temperature range of 0 to 500 °C. Two exothermic transformations were identified for the SLM material. The first peak was found to be common to SLM material, unmolten powder, and cast alloy of similar composition. It was determined to result from the alloying elements and was identified as Mg_2Si precipitation. The second peak was found to be unique to the SLM material. It was ascribed to the rupture and spheroidization of the Si network, which is characteristic of SLM parts. The isothermal endpoint temperatures of these transformations were calculated using second-degree polynomial regressions, resulting in 263 °C for Mg_2Si precipitation and 294 °C for Si rupture and spheroidization [69]. Similar DSC experiments on SLM AlSi10Mg alloy were performed by Rafieazad et al. [55] for a temperature range of 0 to 550 °C. They determined the peak values for the two exothermal transformations at several heating rates (2, 5, 10, and 20 °C/min) and calculated the isothermal temperatures using second-degree polynomial regressions. These were determined to be 232.9 and 273.2 °C. Similar to the work of Fiocchi et al. [69], these were interpreted as being respectively the Mg_2Si β phase precipitation and the Si phase precipitation via solid-state diffusion [55].

3.2.1. Effect on Microstructure

Stress Release Heat Treatments

Several studies have focused on the effect of stress-relieving heat treatments on the microstructure. Cabrini et al. [52] studied the effect of stress-release at 200, 300, 400, and 500 °C for 2 h. Similarly, Rubben et al. [54] investigated the effect of stress-release at 250 and 300 °C for 2 h, while Rafieazad et al. [55] investigated heat treatments at 200, 300, and 350 °C for 3 h. Other papers discuss annealing with similar conditions [68] but without the expressed aim of stress release.

Cabrini et al. [52] observed that stress release heat treatments up to 300 °C for 2 h do not cause a significant change in the characteristic melt pool macrostructure. It should be noted however that although melt pool tracks are retained, significant microstructure alteration can still occur. For the heat treatment of 200 °C for 3 h, no microstructure change was noticed [55]. This is attributed to the temperature being significantly lower than the minimum required for Si interdiffusion in Al [55]. After a stress release at 250 °C for 2 h, a limited breakup of the Si network was observed [54] (see Figure 8b). The limited nature of this breakup can be explained by considering that the temperature of stress release was below the peak temperature value of Si network coarsening [55], resulting in slow kinetics. More significant changes occurred after stress release at 300 °C for 2 h. The morphology of the Si phase changed completely, going from a continuous silicon network towards separate Si precipitates (see Figure 8c) [52,54,68]. Despite this, larger microstructures were still observed at MPBs. As a result, the melt pool macrostructure was retained. After stress release at 400 °C for 2 h, the MPBs became harder to distinguish [52]. Rafieazad et al. [55] reported a similar diminishing of MPBs after 3 h at 350 °C. In both cases, an α-Al matrix with rounded silicon particles was revealed at higher magnification [52,55,68]. When the temperature was increased to 500 °C, the characteristic melt pool macrostructure completely disappeared. Silicon particle coalescence occurred, leading to much larger silicon particles [52].

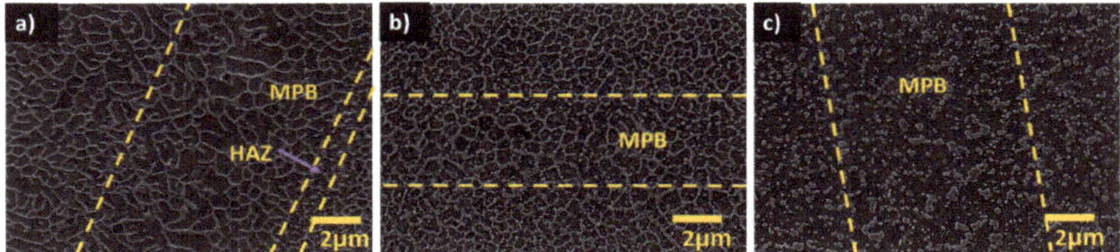

Figure 8. Secondary electron image showing the microstructure of (**a**) an as-built AM Al-Si sample (AlSi10Mg), as well as two stress-released specimens: one at 250 °C for 2 h (**b**), and another at 300 °C for 2 h (**c**). MPB refers to the melt pool border, while HAZ stands for the heat-affected zone. (Adapted from reference [54]).

Heating of Building Platform

Heating the building platform during manufacturing can be used as an alternative to the stress release heat treatment. Residual stresses can be reduced by increasing the building platform temperature without affecting the mechanical strength [52]. Cabrini et al. [52] studied the effect of two different building platform temperatures, 35 and 100 °C. The temperature change did not cause a noticeable change in the macrostructure as both show similar melt pools and heat-affected zones. At higher magnification, some small changes were noticed in the melt pools. Limited formation of the eutectic-like phase by silicon particles was more visible for specimens built at 100 °C. The small changes in microstructure explain the lack of change in mechanical strength. A higher building temperature of 300 °C was investigated by Brandl et al. [34]. No significant change in macrostructure was obtained. It should be noted that the used magnification was not high enough to be able to observe variations in the cellular network.

Solution Treatment—T6

Several researchers [34,38,65] have focused on the T6 heat treatment, consisting of a solution treatment (450–550 °C) [38,58] followed by quenching and artificial aging (150–180 °C) [65]. Both Li et al. [38] and Gu et al. [58] solution treated SLM produced AlSi10Mg material at 450, 500, and 550 °C for 2 h, followed by water quenching. In both cases, results showed that the microstructure became coarser with increasing solution temperatures (see Figure 9). After solution treatment at 450 °C for 2 h (see Figure 9d), Si precipitates were small (<1 μm) and uniformly distributed in the Al matrix. When the temperature was further increased to 500 and 550 °C (see Figure 9e,f), particle coalescence and Oswald ripening occurred, resulting in a further increase in particle size (2–4 μm) [38]. The volume fraction of Si increased, proving that Si precipitates during heat treatment [58]. In the work of Li et al. [38], half of the specimens additionally underwent artificial aging at 180 °C for 12 h after the solution treatment. Further coarsening occurred with the addition of the artificial aging procedure (see Figure 9g–i), despite the temperature being seemingly too low to induce Si coarsening [38,55]. This is in contrast to the work of Aboulkhair et al. [65], where a solution treatment at 520 °C was applied for 4 h. The addition of 6 h of artificial aging at 160 °C did not seem to alter particle size or density.

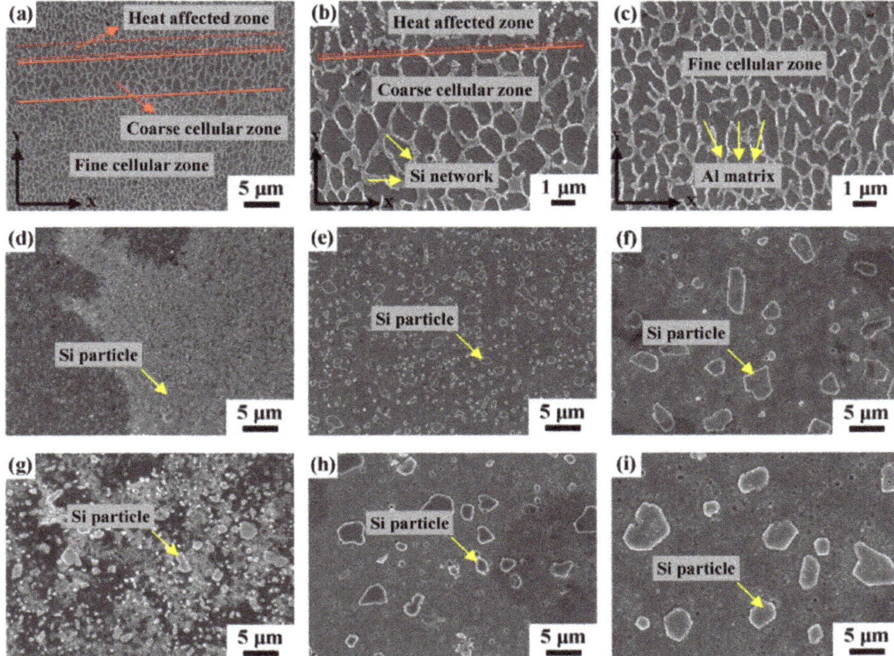

Figure 9. Secondary electron micrographs showing the microstructure of as-built and heat-treated AM AlSi10Mg. (**a**–**c**) Represent the microstructure of the as-built samples in different zones and at different magnifications. (**d**–**i**) Represent the microstructure of the samples after different heat treatments: (**d**) 450 °C for 2 h; (**e**) 500 °C for 2 h; (**f**) 550 °C for 2 h; (**g**) 450 °C for 2 h + 180 °C for 12 h; (**h**) 500 °C for 2 h + 180 °C for 12 h; (**i**) 550 °C for 2 h + 180 °C for 12 h. (Adapted from reference [38]).

Artificial Aging Heat Treatments

In most studies, artificial aging (AA) heat treatments are performed after a solution treatment (e.g., T6 heat treatment). Kempen et al. [23] noted that this is not ideal for SLM parts as it undoes the fine microstructure giving the superior mechanical properties. The effect of AA heat treatment on the microstructure of as produced SLM material was studied by Rubben et al. [54]. No significant microstructure evolution was observed compared to the untreated specimens. An explanation for this can be found in the works of Fiocchi et al. [69] and Rafieazad et al. [55]. The temperature of the heat treatment is significantly below the peak temperature (273 °C [55]) of precipitation, coarsening, and spheroidization of Si particles.

3.2.2. Effect on Corrosion

Stress Release Heat Treatments

Several studies have focused on the effect of stress-relieving heat treatments on corrosion behaviour. Cabrini et al. [52] studied the effect of stress release for 2 h at 200, 300, 400, and 500 °C. They performed intergranular corrosion tests in 30 g/L NaCl + 10 mL/L of HCl at room temperature for 24 h. Another study by Rubben et al. [54] focused on the effect of stress release for 2 h at 250 and 300 °C. To achieve this, they performed open circuit potential, potentiodynamic polarization, and immersion experiments in 0.1 M NaCl.

On untreated specimens, selective corrosion attacks occur at the melt pool borders [52,54]. Galvanic coupling between α-Al and Si causes the dissolution of Al (as already explained above). Selective corrosion attacks at MPBs were still detected after performing stress release up to 300 °C for 2 h [52,54]. This is despite significant microstructure evolution, with the microstructure changing from a continuous silicon network to separate Si precipitates, with larger precipitates formed at the MPBs. Revilla et al. [32] and Rubben et al. [54] performed SKPFM measurements on specimens before [32,54] and after stress release at 300 °C for 2 h [54]. For both cases, measurements showed Volta potential differences between the primary aluminium and the more noble Si, with larger Volta potential differences at the melt pool borders compared to the interior of the melt pools. This indicated that the larger microstructure found at the MPBs is more prone to galvanic corrosion, explaining the selective attack at the MPBs before and after heat treatment.

While preferential corrosion at MPBs was noticed for these types of specimens, Rubben et al. [54] illustrated that the underlying mechanism may change depending on the temperature of stress release. This was linked to a corresponding change in microstructure (as explained in the previous section). When the Si network is still intact (which is the case for untreated specimens, and heat-treated at 200 °C), the corrosion is mostly superficial with microcrack formation at the heat-affected zone next to the MPBs. This allows corrosion to locally spread deeper in the material. When the Si network has broken up (for instance after stress release at 300 °C for 2 h), the spread of corrosion is no longer stopped by the presence of an Si network, leading to a more broadly penetrating attack around MPBs. This behaviour is summarized in Figure 10. Depending on how discontinuous the silicon network is, both mechanisms may be active at the same time [54].

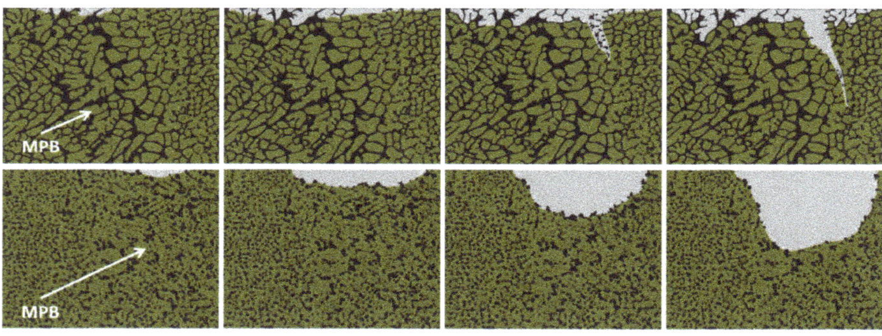

Figure 10. Schematic diagram representing the evolution of the corrosion attack for two separate cases: (**top**) When a connected Si network is present, as is the case for as-built AM Al-Si specimens, and (**bottom**) when the Si is broken up in separate precipitates, as is the case for heat-treated AM Al-Si parts at relatively high temperatures (~300 °C or higher). The illustrations portray the corrosion process seen from a cross-sectional perspective, with the top side of the images representing the surface exposed to the corrosive medium. The Al phase is represented with green, while black portrays the Si phase. For as-built as well as heat-treated specimens, the corrosion initiates at the melt pool borders (MPB), due to the higher driving force for galvanic corrosion in these regions. For as-built specimens corrosion spreads superficially, accompanied by the formation of micro-cracks along the heat-affected zones. On the other hand, a dip and relatively wide penetration of the corrosion attack characterizes heat-treated specimens due to the presence of separate Si precipitates. (Adapted from reference [54]).

After stress release at 400 or 500 °C for 2 h, no more selective attacks at MPBs were detected after intergranular corrosion tests [52]. Instead, more general corrosion morphologies were noticed. This was linked to a significant modification of the microstructure, showing an α-Al matrix and rounded precipitates of Si. This leads to the disappearance of the characteristic melt pool macrostructure. As there is no longer a larger microstructure at the MPBs, no preferential attack occurs in those cases.

Heating of Building Platform

Intergranular corrosion tests were performed by Cabrini et al. [52] on specimens fabricated with a building platform temperature of either 35 or 100 °C. Specimens were immersed for 24 h in 30 g/L NaCl + 10 mL/L of HCl at room temperature. The corrosion morphology showed a superficial attack, with a deeper attack for specimens built at 35 °C. This change can be attributed to either the small change in microstructure (see the previous section) or to the reduction of thermal stresses. The latter is supported by the works of Revilla et al. [32,33,54], where it was postulated that the presence of residual stresses combined with the exposure of the silicon network after corrosion might lead to the development of micro-cracks, allowing the spread of corrosion further in-depth.

Solution Treatment—T6

Some research has been performed on the effect of solution heat treatments on the corrosion behaviour of SLM AlSi10Mg. In one study, Gu et al. [68] solution treated SLM produced AlSi10Mg material at 300 and 400 °C for 2 h under an argon atmosphere, followed by water quenching. In another study by Gu et al. [58], SLM produced AlSi10Mg material was solution treated at 450, 500, and 550 °C for 2 h under an argon atmosphere, followed by water quenching. In both studies, experiments were conducted in 3.5 wt.% NaCl solution.

In untreated specimens, the continuous Si network impedes contact between the underlying metal matrix and the electrolyte, restricting the transfer of Al^{3+} ions towards the electrolyte [58]. After heat treatment, the continuous Si network devolved into separate Si precipitates (see the previous section). Higher temperature heat treatments, coinciding with the growth of the Si particles, resulted in higher corrosion current densities and lower corrosion potentials, film resistances, and charge transfer resistances [58,68]. This was attributed to the change in the morphology of the Si phase. Larger separate Si precipitates restrict the formation of a compact oxide layer [58,68] while Si precipitating from the Al matrix leads to an increasing cathode area and a more active matrix [58]. Further explanations can be found by looking at the work of Revilla et al. [32]. SKPFM measurements showed that the Volta potential difference between the Si phase and the matrix increases with the size of the particles due to the influence of the surrounding matrix. As the Si particles grow in size with increasing temperature, a higher driving force for galvanic corrosion is obtained.

Artificial Aging Heat Treatments

Rubben et al. [54] studied the effect of AA on the corrosion properties of as produced SLM specimens. Open circuit potential (OCP), linear sweep voltammetry (LSV), and immersion experiments in 0.1 M NaCl showed a similar behaviour, resulting from the lack of significant changes in microstructure compared to the untreated specimens (see the previous section). This shows that performing the artificial aging without prior solution treatment is not only critical for retaining the mechanical properties [36] but also for retaining the corrosion resistance of SLM material.

3.3. *Effect of Si Content on Microstructure and Corrosion*

Even though several studies have been dedicated to investigating the corrosion behaviour of AM AlSi10Mg [32,33,44–61,68,70], and a few other works have studied the corrosion performance of AlSi12 alloy [33,42,62,63], very limited research explores the influence of Si on the microstructure and corrosion behaviour of these materials. A recent comparative study carried out on polished as-built AlSi7Mg0.6, AlSi10Mg, and AlSi12 demonstrated that the general appearance concerning the structure of melt pools for all of these additively manufactured specimens was approximately the same [33]. All the specimens, independently on the amount of Si in the alloy, presented a fine cellular structure within the melt pools, while at the borders coarser cells were detected. Additionally, heat-affected zones, in which idiomorphic Si particles and a relatively more discontinuous Si network are present, were identified next to the MPBs for all of these specimens. Nevertheless, higher resolution SEM analysis revealed that even though the size of the Al cells was independent of the amount of Si in the alloy, the connectivity of the Si network was highly affected by the Si content. A much thinner and partially broken Si network was observed for the as-built AlSi7Mg0.6 specimen, while a thicker and much more connected Si network was seen in the as-built AlSi12 sample. By increasing the amount of Si, a greater level of connectivity is observed in the silicon network of the AM specimens. Similarly, a relatively higher level of connectivity was seen in the silicon network within the HAZ for higher amounts of Si [33].

The corrosion studies on these three AM specimens revealed that, even though the corrosion attacks seemed rather superficial for all the samples, the corrosion resistance in NaCl solution is slightly influenced by the specific alloying content: materials with higher amounts of Si (and lower Mg content) showed higher resistance against corrosion [33]. The corrosion resistance of the materials showed the following relationship: AlSi12 > AlSi10Mg > AlSi7Mg0.6. The relatively low resistance against corrosion for the AlSi7Mg0.6 specimen could be due to the higher amount of Mg present in this material compared to the others, since the presence of Mg in Al-Si alloys has repeatedly been shown to greatly influence their corrosion behaviour through the formation of Mg_2Si precipitates [71–73]. Furthermore, earlier research has shown that the corrosion resistance of Al-Si alloys decreases by increasing the content of Mg [73]. Nevertheless, it is important to notice that (to the best of our knowledge) the actual formation of Mg_2Si particles in additively manufactured Al-Si alloys has not yet been confirmed, due possibly to the extremely high cooling and solidification rates resulting in a very fine (out-of-equilibrium) distribution of alloying elements with almost no time for the formation of the usual precipitates. We believe that the possible formation of Mg- and Fe-containing precipitates in these materials should be more systematically studied in future works.

Moreover, due to the relatively high connectivity of the Si network in the as-built AM AlSi12 material, no micro-cracks were formed after corrosion; while for the case of AlSi7Mg0.6 and AlSi10Mg several micro-cracks were seen at heat-affected-zones next to the MPBs [33]. This selective cracking seems to be caused by the combination of several factors: the relatively larger disruption of the Si network in the HAZ, the selective dissolution of the Al matrix around the MPBs/HAZ due to the presence of a corrosive medium, and the existence of residual internal stresses from the MAM process [33].

Cabrini et al. also conducted a comparative study using AlSi7Mg0.6 and AlSi10Mg to investigate the combined effect of Si content and heat treatment on the susceptibility of these materials to corrosion attack [46]. They demonstrated that in the case of as-built (untreated) specimens, higher Si content in the material resulted in a lower penetration depth of the corrosion attack. As mentioned above, this could be due to the higher level of connectivity achieved in the Si network for higher amounts of Si in the alloy [33]. On the other hand, after heat treatments, the Si network (partially) breaks into isolated Si particles, resulting in greater penetration depth of the corrosion attacks for samples with the highest Si content. For higher amounts of Si, larger Si particles will be formed after heat treatment. These isolated Si particles present a more cathodic potential than the Al matrix, being the main cause for the localized corrosion in these materials (due to galvanic coupling effect). For larger Si content and Si particles formed after heat treatment, larger will be the extent of this corrosion attack, which will then not be contained/stopped by the (partially) broken Si network.

3.4. *Effect of Surface Roughness on Corrosion*

In order to study the effect of surface roughness on the corrosion behaviour of additively manufactured Al-Si alloys, several works compare the corrosion performance of as-built AM specimens with that of polished or ground samples [44,50,53,60,70]. In most of these cases, it was shown that the corrosion resistance in NaCl solution [44,50,53,60], as well as the low cycle corrosion fatigue life span [70], is improved after mechanical polishing. The reduced corrosion resistance of as-produced

SLM samples was ascribed to the excessive number of cavities and other surface defects generated during the SLM process at the external sample surface. These surface defects will induce localized corrosion in the form of pits, which could then be considered as crack initiation sites. The stimulated crack initiation and propagation of the unpolished samples thus resulted in a relatively accelerated corrosion fatigue failure [70]. Nevertheless, while studying the effect of surface finishing on the corrosion properties of AlSi10Mg prepared by direct laser sintering, Fathi et al. [60] concluded that during the initial stage of immersion in NaCl solution, SLS-prepared samples had a higher corrosion resistance than ground specimens. Ground specimens were characterised by a high selective attack mainly at the heat-affected zones next to the MPBs, while only a minor attack on the surface of as produced samples was seen. The authors believe that this is caused by a less protective passive film on the ground samples. However, for long immersion times, a change in corrosion behaviour was seen. It was reported that the ground SLS samples now had the highest corrosion resistance. This might indicate that during the initial stage of immersion in a corrosive medium, the internal microstructural features (such as MPBs), which are more pronounced for polished/ground specimens, play a more dominant role in the corrosion process.

An improved surface quality of AM samples is, however, not always possible to obtain by post-printing operations, like mechanical polishing/grinding due to the complex shape of the final printed parts. Therefore, methods such as shot peening and sandblasting (among others) are sometimes applied in order to improve the surface quality of the as-built parts. Some works have been dedicated to studying the effect of such methods on the corrosion behaviour of these materials. The results demonstrate that shot-peened and sandblasted materials present a slightly higher resistance against corrosion than as-built parts due to the reduction of surface roughness and superficial defects [50,60]. However, a much better corrosion performance is obtained always for the polished or ground samples. Fathi et al. [60] propose that it is of the utmost importance to perform a post-grinding operation, in order to substantially improve the corrosion properties of the printed parts. Cabrini et al. [48] further showed an improvement of the corrosion resistance of as-produced parts by bright dipping in a phosphoric/nitric acid bath, attributed to the removal of the oxide film formed during printing. Moreover, they noticed that the corrosion resistance worsened after a bright dipping of mechanically polished parts, caused by the silicon enrichment at the surface during etching.

Furthermore, the surface roughness can also be improved by changing the printing parameters. Calignano et al. [74] studied the effect of different printing parameters on the surface roughness of AlSi10Mg DMLS produced samples. They found that the scanning speed had the greatest influence on the surface roughness, followed by the hatching distance. Fathi et al. [57] studied the corrosion properties of as-produced SLS samples with improved surface quality. The reduced surface roughness was obtained by adjusting the beam offset and reducing the scanning speed and hatching distance, as was suggested by Calignano et al. [74]. They reported that by changing the printing parameters, like the hatch distance, not only does the surface quality change for the better, but the degree of overlap between melt pools changes as well. Furthermore, this study showed that by reducing the hatch distance a material with periodically large and small melt pools is obtained. This overlap between melt pools was shown to play an important role in the solidification behaviour: for higher overlap, the higher the solidification rate will be. The authors claimed that in those cases not the surface roughness, but the resulting microstructure has the largest effect on the corrosion properties of the as-produced parts. Moreover, the material with the finest microstructure, originating from the higher degree of melt pool overlap, showed the best corrosion properties. This was attributed to the fact that the coarser the microstructures of the Al dendrites and the Si particles are, the higher the galvanic coupling and galvanic corrosion [57].

3.5. Corrosion Protection

Even though several studies have demonstrated that the special conditions during MAM have a great influence on the microstructure and corrosion behaviour of these materials, not many works have been dedicated to investigating the influence of their special microstructure on the mechanisms of corrosion protection. To date, only a few studies explore the impact of microstructure, heat treatments, Si content, and defects on the anodizing behaviour of Al–Si alloys prepared by selective laser melting. Anodizing of aluminium and aluminium alloys is done to improve their corrosion protection. Galvanostatic anodizing of SLM produced Al-Si alloys in H_2SO_4 electrolyte was studied by Revilla et al. [41,43,67,75]. They concluded that it is possible to anodize SLM produced Al–Si alloys. However, it was shown that the characteristic microstructure of SLM produced parts will have a significant impact on the voltage-time response and on the formed anodic oxide film. Furthermore, they showed a rather significantly different anodizing behaviour when compared to the cast alloy of similar chemical composition. The voltage-time response of the cast alloy shows the typical steady-state growth regime, while the AM material, on the other hand, showed a continued increase in voltage until eventually a steady-state was reached. Due to the fine distribution and high connectivity of the Si network in as-built AM Al-Si alloys, the moving oxide front will be obstructed to a larger extent resulting in the formation of a thinner oxide film. Moreover, a significantly lower oxide growth rate was seen for the SLM alloy compared to the cast alloy. This lower oxide growth rate was attributed to a larger fraction of the anodic charge consumed by Si oxidation in the SLM alloy. The effect of the Si distribution on the anodizing behaviour of Al-Si alloys was also studied [43]. It was shown that the melt pool borders in SLM produced parts have a eutectic Al-Si structure with alternating lamellae consisting of Al and Si. This lamellae structure was attributed to the lower cooling rate in these melt pool borders compared to the centre of the melt pools. Furthermore, it was reported that in these melt pool borders the Al content was slightly lower compared to the average value in the whole sample, and the Si content slightly higher [43]. This slightly higher Si content and eutectic Al-Si lamellar structure in the melt pool borders was suggested to cause the relatively thinner oxide film formed at these melt pool borders. An XPS analysis reported that in the anodic film of the cast alloy only a superficial layer of the Si precipitates is oxidized. However, in the SLM samples, most of the Si in the anodic film is oxidized; this was suggested to cause a severe reduction of the anodizing efficiency compared to the cast alloy. It was shown that the anodic film formed on the cast alloy has a much higher roughness than the anodic film formed on the AM specimen. This roughness difference was attributed to the difference in aluminium cell size between the cast alloy and the SLM alloy [43].

Moreover, a great anisotropy was also seen during the galvanostatic anodizing of as-built AM Al-Si samples [41,75]. The voltage–time response curves of the AM material showed a different response depending on the orientation of the surface that was anodized. The voltage–time response of a surface parallel to the building platform (XY surface) was shown to give a higher steady-state potential value than that of a surface with an orientation perpendicular to the building platform (XZ surface). This asymmetric anodizing behaviour was found to be independent of the Si content in the alloy. The authors proposed that this asymmetry could be related to the difference in the size of aluminium cells in the different planes [41], or differences in the density of melt pool borders encountered by the anodizing front [75].

A unique pore structure was seen for the SLM alloys. A branched-like pore structure was seen throughout the whole anodic film. This pore structure was attributed to the fine distribution of the silicon phase in an almost continuous network encapsulating the aluminium in small cells [41]. Revilla et al. [75] further reported that the Si content in the alloy has a significant effect on the pore structure of the anodic film, i.e., the higher the Si content the higher the voltage response will be. As a consequence of the higher voltage response, wider pores with a greater inter-pore distance are obtained. Furthermore, the pore density is shown to decrease with Si content.

Rubben et al. [67] studied the effect of several heat treatments on the anodizing behaviour of SLM produced AlSi10Mg. It was shown that anodizing behaviour depends strongly on the morphology of the Si phase, which can be highly affected by the heat treatment applied. For the non-heat treated and the artificially aged (at 170 °C for 6 h) samples, the Si phase consists out of a rather continuous network, consequently, most of the Si encountered by the anodizing front will be oxidized. For the stress released heat-treated samples (at 250 and 300 °C for 2 h), the Si network is broken up into separated Si precipitates [67]. As a consequence, only a fraction of the Si phase will be oxidized. It was shown that the difference in Si phase morphology can give rise to significantly different anodizing behaviour. Furthermore, they reported that voltage–time response during anodizing varied strongly depending on the heat treatment applied. The steady-state anodizing potential was reported to increase with the anodized Si fractions [67]. Additionally, the anodizing efficiency was shown to decrease with a higher anodized Si fraction.

Finally, Revilla et al. [75] investigated the influence of internal pores, resulting from the MAM process, on anodizing behaviour. They showed that when the moving oxide front encounters such a pore, the sides of the pores get anodized. Furthermore, the sides of the anodized internal pores showed cracks, probably caused by the volume expansion of anodic layers advancing in opposite directions [75]. Even though the cracks observed were relatively small (1–2 μm), they could compromise the physical integrity of the final piece. Additionally, these cracks could facilitate the access of corrosive media to the metal matrix.

Other studies are needed in the future to further understand the anodizing behaviour of these materials under other anodizing conditions such as: potentiostatic anodizing regime, different electrolytes, and study the effect of surface roughness. Moreover, the effect of microstructure and microstructural defects on other surface treatments aimed at the protection of the materials against corrosion should also be investigated.

4. Other Al Alloys

Besides the Al-Si alloys discussed in previous sections, only very few other additively manufactured Al-based alloys have been considered so far for corrosion studies. The only other cases are, to the best of our knowledge, AA2024 and AA7075 [76,77]. For AA2024, the main alloying agent added is Cu, while for AA7075 the primary alloying element is Zn. The specifications concerning the chemical composition of these alloys are given in Table 2. AA2024 is a heat-treatable alloy widely used in aircraft structures due to its high strength to weight ratio, as well as good fatigue resistance. AA7075 has high specific strength, low density, and good thermal properties, for which it is widely used in transport applications, including marine, automotive, and aviation, as well as in-mold tool manufacturing.

Table 2. Specification of chemical composition for AA2024 and AA7075.

Material	Al	Si	Fe	Cu	Mn	Mg	Cr	Zn	Ti
AA2024	Balance	<0.5	<0.5	**3.8–4.9**	0.3–0.9	1.2–1.8	<0.1	<0.25	<0.15
AA7075	Balance	<0.4	<0.5	1.2–2	<0.3	2.1–2.9	0.18–0.28	**5.1–6.1**	<0.2

Previous work on polished wrought and additively manufactured AA2024 revealed that due to the special conditions associated with the selective laser melting process (i.e., highly localized melting and subsequent rapid solidification), the microstructure of AM AA2024 was characterized by a refined particle size [76]. The traditional micrometre-sized constituent particles and the S-phase Al_2CuMg present in wrought AA2024-T3 were instead replaced by nm-sized particles, which were mainly determined to be θ-phase Al_2Cu. This highly refined microstructure was shown to greatly impact the corrosion behaviour of this material according to results obtained in polished samples [76]. While the anodic current was found to increase rapidly for potentials above the corrosion potential during anodic polarization of wrought AA2024-T3, the additively manufactured AA2024 materials presented a passive-like window. This could be due to the finer microstructure found in the additively manufactured material compared to the wrought specimen, which resulted in a thicker and more stable native oxide film; as well as the absence of S-phase precipitates in AM AA2024. It was also shown that the corrosion rate of the Al matrix in 0.01 M NaCl for the AM AA2024 material was about five times lower than that measured in wrought AA2024-T3. Moreover, the dissolution ratio between the alloying elements (Cu and Mg) and that of Al was about ten times higher for the AM AA2024 sample than for the wrought AA2024-T3. As the authors stated, the dissolution of these alloying elements at the early stages of corrosion is beneficial since it reduces the number of cathodic sites that would be detrimental for subsequent localized corrosion.

The relation between microstructure and corrosion behaviour has also been studied for AA7075 materials [77]. Gharbi et al. [77] demonstrated that while coarse (up to approximately 15 µm in size) second phase Al-Cu-Fe(-Si) particles exist in the traditional wrought AA7075-T6 material, finely distributed nm-sized Mg-Zn-Cu(-Al) (ν-phase) and Mg_2Si (β-phase) characterize the microstructure of additively manufactured AA7075 prepared by SLM. This microstructure was also shown to vary significantly after solutionising and subsequent artificial ageing. The existing ν-phase in as-built specimens was dissolved after solutionising, while the subsequent aging resulted in the formation of $MgZn_2$ precipitates. Concerning their corrosion behaviour, the authors demonstrated that, depending on the post-heat-treatment, AM AA7075 presented a higher corrosion resistance compared to wrought AA7075-T6 [77]. While as-built AM AA7075 and wrought AA7075-T6 presented a highly active behaviour during anodic polarization in NaCl solution, the heat-treated specimens presented a passive-like behaviour. The immersion of the samples in a NaCl solution revealed that pits formed in as-built AM AA7075 were notably smaller compared to those formed in wrought AA7075-T6, possibly due to the fine microstructural features and the absence of large second phase particles in the AM material compared to its wrought counterpart. Moreover, while pitting occurred predominantly along the melt pool borders in the case of polished as-built AM AA7075, pits were uniformly distributed on the Al matrix for the solutionised AM material. Due to the formation of $MgZn_2$ precipitates in the solutionised aged samples, large pits were formed along grain boundaries [77].

To enhance the metallurgical state of the AA7075 alloy, and therefore, make it more suitable for processing by additive manufacturing, the minor addition of transition elements such as Sc, Zr, Ti, B, Fe, and Ni has been explored [78]. The addition of these elements influenced the resulting microstructure of AM AA7075. Therefore, further studies should focus on investigating the effect of this microstructure and additional alloying elements on the corrosion resistance of this material.

In general, as for AM Al-Si alloys, the special conditions during additive manufacturing (i.e., highly localized melting and solidification accompanied by extremely high cooling rates) seem to promote a high refinement of the microstructure of these materials and/or the annihilation of certain precipitates/phases. This is generally beneficial in terms of corrosion resistance since it greatly reduces the possible cathodic sites and the galvanic interactions that can promote localized corrosion. Nevertheless, there is a need for further investigations of the corrosion behaviour of other AM Al-based alloys, or Al alloys prepared by MAM techniques other than selective laser melting.

5. Summary and Outlook

1. Due to the special conditions associated with the metal additive manufacturing processes, numerous macro- and micro-structural defects can exist within the printed parts. Defects such as porosity, remaining unmolten powder, high surface roughness, and residual stresses, can greatly influence the corrosion performance of these materials. Even though the existence of some of these defects can be limited by carefully tuning the process parameters, they cannot be fully avoided due to the large number of parameters involved and the high complexity of the process. In general, more research should be conducted in order to better understand the influence of these defects and the combined impact of process parameters on the corrosion resistance of these materials.
2. Aside from Al-Si alloys (and specially AlSi10Mg alloy), very limited research has been published concerning the corrosion resistance of other additively manufactured Al alloy parts. More focus should be given to other types of 3D-printed Al-based alloys. Moreover, further research is needed to study the microstructure and corrosion behavior of additively manufactured parts prepared using other metal additive manufacturing methods besides selective laser melting/sintering.

3. In general, most of the published works on additively manufactured Al-based alloys agree that their corrosion resistance is similar to or higher than that of the same alloys fabricated using traditional manufacturing methods.
4. Due to the special conditions associated with the metal additive manufacturing processes (for instance: the use of pre-alloyed metal powder and highly localized (re)melting combined with high cooling rates), a unique microstructure formed by a fine Si network that encloses the α-Al in small cells of varying sizes across the melt pools characterizes the printed Al-Si alloys specimens. A great number of works have already shown that the corrosion behavior of these materials is greatly influenced by this special and unique microstructure. The borders of the melt pools (where larger cells are found) have been identified as vulnerable sites for corrosion to initiate and further propagate. However, in general, the existence of a rather connected Si network for the case of as-built specimens prevents/holds the corrosion from penetrating deeper into the material.
5. The specific (out of equilibrium) microstructure obtained by SLM processing makes the materials susceptible to certain heat treatments (such as stress release and solution T6 heat treatments). Depending on the temperature of the heat treatment, these can disrupt the silicon network in Al-Si alloys, leading to the formation of large and separate Si precipitates. Aside from negatively impacting the mechanical properties, this also affects the corrosion resistance. The disruption of the Si network allows corrosion to penetrate deeper into the material, while larger Si precipitates disrupt the formation of a compact passive layer and form larger cathodic sites that increase the driving force for galvanic corrosion. Additionally, the presence of Mg in most Al-Si alloys makes the material susceptible to precipitation hardening through the formation of Mg_2Si precipitates. However, the presence of these precipitates can greatly affect the pitting corrosion process in these materials. An alternative to stress release heat treatment is the heating of the building platform during production. In this way, alternation of the microstructure and subsequent change in corrosion behavior can be prevented. However, more research on different building platform temperatures and their effect on the microstructure and corrosion behavior is required. Moreover, further work should focus on adapting existing heat treatments to the specific microstructure of the additively manufactured materials.
6. For additively manufactured Al-Si alloys, increasing the amount of Si, greatly increases the level of connectivity of the silicon network, which reduces the penetration depth of the corrosion attack. However, in these alloys, the presence of small amounts of Mg and Fe cannot be ignored during the analysis of the corrosion behavior.
7. It has been shown that the choice of printing parameters can greatly affect the surface roughness of the printed parts, and thus also their corrosion resistance. The scanning speed, followed by the hatching distance, was reported to have the largest effect. Altering the manufacturing parameters to reduce the surface roughness could also affect the microstructure. In general, an improvement of the corrosion resistance has been reported for samples that underwent post-printing operations to reduce their surface roughness (such as shot peening, sandblasting, and bright dipping) compared to as-produced samples. However, further work is needed to better understand the performance of other post-printing processes (for instance electro-polishing), as well as their effect on the corrosion behavior of additively manufactured Al-based alloys.
8. In general, a limited number of studies have been conducted to understand the mechanisms of corrosion protection of additively manufactured metal parts. Studies on the galvanostatic anodizing of 3D-printed Al-Si alloys demonstrated that the special microstructure of these materials greatly influences the typical voltage-time behavior as well as the characteristics of the anodic oxide layer. A significantly lower oxide growth rate was seen for additively manufactured Al-Si parts compared to cast alloy, attributed to the larger fraction of the anodic charge consumed by the oxidation of the Si phase. The porous anodic oxide film, thinner around the melt pool borders, was characterized by the formation of branched-like pores. The anodizing behavior as well as the anodic oxide layer of additively manufactured Al-Si alloys is greatly

affected by defects and special features typical from the metal additive manufacturing process. The intrinsic anisotropy, melt pool borders, and internal pores are among those features affecting the anodization process. More research is still needed to better understand the impact of other defects such as unmolten powder and residual stresses on the anodizing behavior of these materials, as well as to further investigate the properties of the formed anodic oxides. Moreover, additional research is required to understand the anodizing behaviour of these materials under other anodizing conditions such as potentiostatic anodizing regime and different electrolytes.

Author Contributions: Conceptualization, R.I.R.; writing—original draft preparation, R.I.R., D.V., and T.R.; writing—review and editing, I.D.G., D.V., T.R., and R.I.R. All authors have read and agreed to the published version of the manuscript.

Funding: This research received no external funding.

Conflicts of Interest: The authors declare no conflict of interest.

References

1. Murr, L.E.; Gaytan, S.M.; Ramirez, D.A.; Martinez, E.; Hernandez, J.; Amato, K.N.; Shindo, P.W.; Medina, F.R.; Wicker, R.B. Metal Fabrication by Additive Manufacturing Using Laser and Electron Beam Melting Technologies. *J. Mater. Sci. Technol.* **2012**, *28*, 1–14. [CrossRef]
2. Murr, L.; Martinez, E.; Amato, K.N.; Gaytan, S.M.; Hernandez, J.; Ramirez, D.A.; Shindo, P.W.; Medina, F.; Wicker, R.B. Fabrication of Metal and Alloy Components by Additive Manufacturing: Examples of 3D Materials Science. *J. Mater. Res. Technol.* **2012**, *1*, 42–54. [CrossRef]
3. Herzog, D.; Seyda, V.; Wycisk, E.; Emmelmann, C. Additive manufacturing of metals. *Acta Mater.* **2016**, *117*, 371–392. [CrossRef]
4. Sames, W.J.; List, F.A.; Pannala, S.; Dehoff, R.R.; Babu, S.S. The metallurgy and processing science of metal additive manufacturing. *Int. Mater. Rev.* **2016**, *61*, 315–360. [CrossRef]
5. Gorsse, S.; Hutchinson, C.; Goune, M.; Banerjee, R. Additive manufacturing of metals: A brief review of the characteristic microstructures and properties of steels, Ti-6Al-4V and high-entropy alloys. *Sci. Technol. Adv. Mater.* **2017**, *18*, 584–610. [CrossRef] [PubMed]
6. Gu, D.; Meiners, W.; Wissenbach, K.; Poprawe, R. Laser additive manufacturing of metallic components: Materials, processes and mechanisms. *Int. Mater. Rev.* **2012**, *57*, 133–164. [CrossRef]
7. ISO/ASTM52900-15. *Standard Terminology for Additive Manufacturing—General Principles—Terminology*; ASTM International: West Conshohocken, PA, USA, 2015.
8. Sun, G.; Shen, X.; Wang, Z.; Zhan, M.; Yao, S.; Zhou, R.; Ni, Z. Laser metal deposition as repair technology for 316L stainless steel: Influence of feeding powder compositions on microstructure and mechanical properties. *Opt. Laser Technol.* **2019**, *109*, 71–83. [CrossRef]
9. Dilberoglu, U.M.; Gharehpapagh, B.; Yaman, U.; Dolen, M. The Role of Additive Manufacturing in the Era of Industry 4.0. *Procedia Manuf.* **2017**, *11*, 545–554. [CrossRef]
10. Additive Manufacturing Report. Available online: https://additive-manufacturing-report.com/additive-manufacturing-market/ (accessed on 16 October 2020).
11. DebRoy, T.; Wei, H.L.; Zuback, J.S.; Mukherjee, T.; Elmer, J.W.; Milewski, J.O.; Beese, A.M.; Wilson-Heid, A.; De, A.; Zhang, W. Additive manufacturing of metallic components—Process, structure and properties. *Prog. Mater. Sci.* **2018**, *92*, 112–224. [CrossRef]
12. *Additive Manufacturing for the Aerospace Industry*; Elsevier: Amsterdam, The Netherlands, 2019.
13. Miller, W.; Zhuang, L.; Bottema, J.; Wittebrood, A.; De Smet, P.; Haszler, A.; Vieregge, A. Recent development in aluminium alloys for the automotive industry. *Mater. Sci. Eng. A* **2000**, *280*, 37–49. [CrossRef]
14. Available Materials for Metal Additive Manufacturing: Characteristics & Applications, Farinia Group. Available online: https://www.farinia.com/additive-manufacturing/3dmaterials/characteristics-and-applications-ofavailable-metals-for-additive-manufacturing (accessed on 7 September 2020).
15. Aboulkhair, N.T.; Simonelli, M.; Parry, L.; Ashcroft, I.; Tuck, C.; Hague, R. 3D printing of Aluminium alloys: Additive Manufacturing of Aluminium alloys using selective laser melting. *Prog. Mater. Sci.* **2019**, *106*, 100578. [CrossRef]

16. Vojtech, D.; Serak, J.; Ekrt, O. Improving the casting properties of high-strength aluminium alloys. *Mater. Technol.* **2004**, *38*, 99–102.
17. Zhang, J.; Song, B.; Wei, Q.; Bourell, D.; Shi, Y. A review of selective laser melting of aluminum alloys: Processing, microstructure, property and developing trends. *J. Mater. Sci. Technol.* **2019**, *35*, 270–284. [CrossRef]
18. *ASM Handbook*; ASM International: Materials Park, OH, USA, 1991; Volume 4.
19. Sander, G.; Tan, J.; Balan, P.; Gharbi, O.; Feenstra, D.; Singer, L.; Thomas, S.; Kelly, R.; Scully, J.; Birbilis, N. Corrosion of Additively Manufactured Alloys: A Review. *Corrosion* **2018**, *74*, 1318–1350. [CrossRef]
20. Kong, D.; Dong, C.; Ni, X.; Li, X. Corrosion of metallic materials fabricated by selective laser melting. *NPJ Mater. Degrad.* **2019**, *3*, 24. [CrossRef]
21. Chen, H.; Zhang, C.; Jia, D.; Wellmann, D.; Liu, W. Corrosion Behaviors of Selective Laser Melted Aluminum Alloys: A Review. *Metals* **2020**, *10*, 102. [CrossRef]
22. Ornek, C. Additive manufacturing–a general corrosion perspective. *Corr. Eng. Sci. Technol.* **2018**, *53*, 531–535. [CrossRef]
23. Kempen, K.; Thijs, L.; Van Humbeeck, J.; Kruth, J.-P. Processing AlSi10Mg by selective laser melting: Parameter optimisation and material characterisation. *Mater. Sci. Technol.* **2014**, *31*, 917–923. [CrossRef]
24. Prashanth, K.G.; Scudino, S.; Maity, T.; Das, J.; Eckert, J. Is the energy density a reliable parameter for materials synthesis by selective laser melting? *Mater. Res. Lett.* **2017**, *5*, 386–390. [CrossRef]
25. Leung, C.L.A.; Marussi, S.; Towrie, M.; Atwood, R.C.; Withers, P.J.; Lee, P.D. The effect of powder oxidation on defect formation in laser additive manufacturing. *Acta Mater.* **2019**, *166*, 294–305. [CrossRef]
26. Coeck, S.; Bisht, M.; Plas, J.; Verbist, F. Prediction of lack of fusion porosity in selective laser malting based on melt pool monitoring data. *Addit. Manuf.* **2019**, *25*, 347–356.
27. Sander, G.; Thomas, S.; Cruz, V.; Jurg, M.; Birbilis, N.; Gao, X.; Brameld, M.; Hutchinson, C.R. On the Corrosion and Metastable Pitting Characteristics of 316L Stainless Steel Produced by Selective Laser Melting. *J. Electrochem. Soc.* **2017**, *164*, C250–C257. [CrossRef]
28. Schaller, R.; Taylor, J.; Rodelas, J.; Schindelholz, E. Corrosion Properties of Powder Bed Fusion Additively Manufactured 17-4 PH Stainless Steel. *Corrosion* **2017**, *73*, 796–807. [CrossRef]
29. Kumbhar, N.N.; Mulay, A.V. Post Processing Methods used to Improve Surface Finish of Products which are Manufactured by Additive Manufacturing Technologies: A Review. *J. Inst. Eng. Ser. C* **2016**, *99*, 481–487. [CrossRef]
30. Scherillo, F. Chemical surface finishing of AlSi10Mg components made by additive manufacturing. *Manuf. Lett.* **2019**, *19*, 5–9. [CrossRef]
31. Li, C.; Liu, Z.; Fang, X.; Guo, Y. Residual Stress in Metal Additive Manufacturing. *Procedia CIRP* **2018**, *71*, 348–353. [CrossRef]
32. Revilla, R.I.; Liang, J.; Godet, S.; De Graeve, I. Local Corrosion Behavior of Additive Manufactured AlSiMg Alloy Assessed by SEM and SKPFM. *J. Electrochem. Soc.* **2016**, *164*, C27–C35. [CrossRef]
33. Revilla, R.I.; De Graeve, I. Influence of Si content on the microstructure and corrosion behaviour of additive manufactured Al-Si alloys. *J. Electrochem. Soc.* **2018**, *165*, C926–C932. [CrossRef]
34. Brandl, E.; Heckenberger, U.; Holzinger, V.; Buchbinder, D. Additive manufactured AlSi10Mg samples using Selective Laser Melting (SLM): Microstructure, high cycle fatigue, and fracture behavior. *Mater. Des.* **2012**, *34*, 159–169. [CrossRef]
35. Manfredi, D.; Calignano, F.; Krishnan, M.; Canali, R.; Ambrosio, E.P.; Atzeni, E. From Powders to Dense Metal Parts: Characterization of a Commercial AlSiMg Alloy Processed through Direct Metal Laser Sintering. *Materials* **2013**, *6*, 856–869. [CrossRef]
36. Thijs, L.; Kempen, K.; Kruth, J.-P.; Van Humbeeck, J. Fine-structured aluminium products with controllable texture by selective laser melting of pre-alloyed AlSi10Mg powder. *Acta Mater.* **2013**, *61*, 1809–1819. [CrossRef]
37. Yan, C.; Hao, L.; Hussein, A.; Young, P.; Huang, J.; Zhu, W. Microstructure and mechanical properties of aluminium alloy cellular lattice structures manufactured by direct metal laser sintering. *Mater. Sci. Eng. A* **2015**, *628*, 238–246. [CrossRef]
38. Li, W.; Li, S.; Liu, J.; Zhang, A.; Zhou, Y.; Wei, Q.; Yan, C.; Shi, Y. Effect of heat treatment on AlSi10Mg alloy fabricated by selective laser melting: Microstructure evolution, mechanical properties and fracture mechanism. *Mater. Sci. Eng. A* **2016**, *663*, 116–125. [CrossRef]

39. Prashanth, K.; Scudino, S.; Klauss, H.; Surreddi, K.; Löber, L.; Wang, Z.; Chaubey, A.; Kühn, U.; Eckert, J. Microstructure and mechanical properties of Al–12Si produced by selective laser melting: Effect of heat treatment. *Mater. Sci. Eng. A* **2014**, *590*, 153–160. [CrossRef]
40. Kimura, T.; Nakamoto, T.; Mizuno, M.; Araki, H. Effect of silicon content on densification, mechanical and thermal properties of Al-xSi binary alloys fabricated using selective laser melting. *Mater. Sci. Eng. A* **2017**, *682*, 593–602. [CrossRef]
41. Revilla, R.I.; Verkens, D.; Couturiaux, G.; Malet, L.; Thijs, L.; Godet, S.; De Graeve, I. Galvanostatic Anodizing of Additive Manufactured Al-Si10-Mg Alloy. *J. Electrochem. Soc.* **2017**, *164*, C1027–C1034. [CrossRef]
42. Chen, Y.; Zhang, J.; Gu, X.; Dai, N.; Qin, P.; Zhang, L.-C. Distinction of corrosion resistance of selective laser melted Al-12Si alloy on different planes. *J. Alloys Compd.* **2018**, *747*, 648–658. [CrossRef]
43. Revilla, R.I.; Terryn, H.; De Graeve, I. Role of Si in the Anodizing Behavior of Al-Si Alloys: Additive Manufactured and Cast Al-Si10-Mg. *J. Electrochem. Soc.* **2018**, *165*, C532–C541. [CrossRef]
44. Fathi, P.; Mohammadi, M.; Duan, X.; Nasiri, A. A comparative study on corrosion and microstructure of direct metal laser sintered AlSi10Mg_200C and die cast A360.1 aluminum. *J. Mater. Process. Technol.* **2018**, *259*, 1–14. [CrossRef]
45. Cabrini, M.; Lorenzi, S.; Pastore, T.; Testa, C.; Manfredi, D.; Lorusso, M.; Calignano, F.; Pavese, M.; Andreatta, F. Corrosion behavior of AlSi10Mg alloy produced by laser powder bed fusion under chloride exposure. *Corros. Sci.* **2019**, *152*, 101–108. [CrossRef]
46. Cabrini, M.; Lorenzi, S.; Pastore, T. Corrosion behavior of aluminum-silicon alloys obtained by Direct Metal Laser Sintering. In Proceedings of the EUROCORR 2017—The Annual Congress of the European Federation of Corrosion, 20th International Corrosion Congress and Process Safety Congress 2017, Prague, Czech Republic, 3–7 September 2017.
47. Cabrini, M.; Lorenzi, S.; Pastore, T.; Testa, C.; Manfredi, D.; Cattano, G.; Calignano, F. Corrosion resistance in chloride solution of the AlSi10Mg alloy obtained by means of LPBF. *Surf. Interface Anal.* **2018**, *51*, 1159–1164. [CrossRef]
48. Cabrini, M.; Lorenzi, S.; Pastore, T.; Pellegrini, S.; Pavese, M.; Fino, P.; Ambrosio, E.P.; Calignano, F.; Manfredi, D. Corrosion resistance of direct metal laser sintering AlSiMg alloy. *Surf. Interface Anal.* **2016**, *48*, 818–826. [CrossRef]
49. Rafieazad, M.; Chatterjee, A.; Nasiri, A.M. Effects of Recycled Powder on Solidification Defects, Microstructure, and Corrosion Properties of DMLS Fabricated AlSi10Mg. *JOM* **2019**, *71*, 3241–3252. [CrossRef]
50. Cabrini, M.; Lorenzi, S.; Pastore, T.; Pellegrini, S.; Manfredi, D.G.; Fino, P.; Biamino, S.; Badini, C.F. Evaluation of corrosion resistance of Al–10Si–Mg alloy obtained by means of Direct Metal Laser Sintering. *J. Mater. Process. Technol.* **2016**, *231*, 326–335. [CrossRef]
51. Kubacki, G.; Brownhill, J.P.; Kelly, R.G. Comparison of Atmospheric Corrosion of Additively Manufactured and Cast Al-10Si-Mg over a Range of Heat Treatments. *Corrosion* **2019**, *75*, 1527–1540. [CrossRef]
52. Cabrini, M.; Calignano, F.; Fino, P.; Lorenzi, S.; Lorusso, M.; Manfredi, D.; Testa, C.; Pastore, T. Corrosion Behavior of Heat-Treated AlSi10Mg Manufactured by Laser Powder Bed Fusion. *Materials* **2018**, *11*, 1051. [CrossRef]
53. Cabrini, M.; Lorenzi, S.; Pastore, T.; Pellegrini, S.; Ambrosio, E.P.; Calignano, F.; Manfredi, D.G.; Pavese, M.; Fino, P. Effect of heat treatment on corrosion resistance of DMLS AlSi10Mg alloy. *Electrochim. Acta* **2016**, *206*, 346–355. [CrossRef]
54. Rubben, T.; Revilla, R.I.; De Graeve, I. Influence of heat treatments on the corrosion mechanism of additive manufactured AlSi10Mg. *Corros. Sci.* **2019**, *147*, 406–415. [CrossRef]
55. Rafieazad, M.; Mohammadi, M.; Nasiri, A. On microstructure and early stage corrosion performance of heat treated direct metal laser sintered AlSi10Mg. *Addit. Manuf.* **2019**, *28*, 107–119. [CrossRef]
56. Cabrini, M.; Lorenzi, S.; Testa, C.; Pastore, T.; Manfredi, D.; Lorusso, M.; Calignano, F.; Fino, P. Statistical approach for electrochemical evaluation of the effect of heat treatments on the corrosion resistance of AlSi10Mg alloy by laser powder bed fusion. *Electrochim. Acta* **2019**, *305*, 459–466. [CrossRef]
57. Fathi, P.; Rafieazad, M.; Duan, X.; Mohammadi, M.; Nasiri, A. On microstructure and corrosion behaviour of AlSi10Mg alloy with low surface roughness fabricated by direct metal laser sintering. *Corros. Sci.* **2019**, *157*, 126–145. [CrossRef]
58. Gu, X.-H.; Zhang, J.-X.; Fan, X.-L.; Zhang, L.-C. Corrosion Behavior of Selective Laser Melted AlSi10Mg Alloy in NaCl Solution and Its Dependence on Heat Treatment. *Acta Met. Sin.* **2019**, *33*, 327–337. [CrossRef]

59. Girelli, L.; Tocci, M.; Conte, M.; Giovanardi, R.; Veronesi, P.; Gelfi, M.; Pola, A. Effect of the T6 heat treatment on corrosion behavior of additive manufactured and gravity cast AlSi10Mg alloy. *Mater. Corros.* **2019**, *70*, 1808–1816. [CrossRef]
60. Fathi, P.; Mohammadi, M.; Duan, X.; Nasiri, A. Effects of Surface Finishing Procedures on Corrosion Behavior of DMLS-AlSi10Mg_200C Alloy Versus Die-Cast A360.1 Aluminum. *JOM* **2019**, *71*, 1748–1759. [CrossRef]
61. Leon, A.; Shirizly, A.; Aghion, E. Corrosion Behavior of AlSi10Mg Alloy Produced by Additive Manufacturing (AM) vs. Its Counterpart Gravity Cast Alloy. *Metals* **2016**, *6*, 148. [CrossRef]
62. Prashanth, K.; Debalina, B.; Wang, Z.; Gostin, P.; Gebert, A.; Calin, M.; Kühn, U.; Kamaraj, M.; Scudino, S.; Eckert, J. Tribological and corrosion properties of Al–12Si produced by selective laser melting. *J. Mater. Res.* **2014**, *29*, 2044–2054. [CrossRef]
63. Yang, Y.; Chen, Y.; Zhang, J.; Gu, X.; Qin, P.; Dai, N.; Li, X.; Kruth, J.-P.; Zhang, L.-C. Improved corrosion behavior of ultrafine-grained eutectic Al-12Si alloy produced by selective laser melting. *Mater. Des.* **2018**, *146*, 239–248. [CrossRef]
64. Yang, P.; Deibler, L.A.; Bradley, D.R.; Stefan, D.K.; Carroll, J.D. Microstructure evolution and thermal properties of an additively manufactured, solution treatable AlSi10Mg part. *J. Mater. Res.* **2018**, *33*, 4040–4052.
65. Aboulkhair, N.T.; Tuck, C.; Aschroft, I.A.; Maskery, I.; Everitt, N.M. On the Precipitation Hardening of Selective Laser Melted AlSi10Mg. *Met. Mater. Trans. A* **2015**, *46*, 3337–3341. [CrossRef]
66. Yang, P.; Rodriguez, M.A.; Deibler, L.A.; Jared, B.H.; Griego, J.; Kilgo, A.; Allen, A.; Stefan, D.K. Effect of thermal annealing on microstructure evolution and mechanical behavior of an additive manufactured AlSi10Mg part. *J. Mater. Res.* **2018**, *33*, 1701–1712. [CrossRef]
67. Rubben, T.; Revilla, R.I.; De Graeve, I. Effect of Heat Treatments on the Anodizing Behavior of Additive Manufactured AlSi10Mg. *J. Electrochem. Soc.* **2019**, *166*, C42–C48. [CrossRef]
68. Gu, X.; Zhang, J.; Fan, X.; Dai, N.; Xiao, Y.; Zhang, L.-C. Abnormal corrosion behavior of selective laser melted AlSi10Mg alloy induced by heat treatment at 300 °C. *J. Alloys Compd.* **2019**, *803*, 314–324. [CrossRef]
69. Fiocchi, J.; Tuissi, A.; Bassani, P.; Biffi, C.A. Low temperature annealing dedicated to AlSi10Mg selective laser melting products. *J. Alloys Compd.* **2017**, *695*, 3402–3409. [CrossRef]
70. Leon, A.; Aghion, E. Effect of surface roughness on corrosion fatigue performance of AlSi10Mg alloy produced by Selective Laser Melting (SLM). *Mater. Charact.* **2017**, *131*, 188–194. [CrossRef]
71. Gupta, R.K.; Sukiman, N.L.; Fleming, K.M.; Gibson, M.A.; Birbilis, N. Electrochemical Behavior and Localized Corrosion Associated with Mg2Si Particles in Al and Mg Alloys. *ECS Electrochem. Lett.* **2012**, *1*, C1–C3.
72. Zeng, F.-L.; Wei, Z.-L.; Li, J.; Li, C.-X.; Tan, X.; Zhang, Z.; Zheng, Z.-Q. Corrosion mechanism associated with Mg2Si and Si particles in Al–Mg–Si alloys. *Trans. Nonferr. Met. Soc. China* **2011**, *21*, 2559–2567. [CrossRef]
73. Ahlatci, H. Production and corrosion behaviours of the Al–12Si–XMg alloys containing in situ Mg2Si particles. *J. Alloys Compd.* **2010**, *503*, 122–126. [CrossRef]
74. Calignano, F.; Manfredi, D.; Ambrosio, E.P.; Iuliano, L.; Fino, P. Influence of process parameters on surface roughness of aluminum parts produced by DMLS. *Int. J. Adv. Manuf. Technol.* **2012**, *67*, 2743–2751. [CrossRef]
75. Revilla, R.I.; Rojas, Y.; De Graeve, I. On the Impact of Si Content and Porosity Artifacts on the Anodizing Behavior of Additve Manufactured Al-Si Alloys. *J. Electrochem. Soc.* **2019**, *166*, C530–C537. [CrossRef]
76. Gharbi, O.; Jiang, D.; Feenstra, D.; Kairy, S.; Wu, Y.; Hutchinson, C.; Birbilis, N. On the corrosion of additively manufactured aluminium alloy AA2024 prepared by selective laser melting. *Corros. Sci.* **2018**, *143*, 93–106.
77. Gharbi, O.; Kairy, S.K.; De Lima, P.R.; Jiang, D.; Nicklaus, J.; Birbilis, N. Microstructure and corrosion evolution of additively manufactured aluminium alloy AA7075 as a function of ageing. *NPJ Mater. Degrad.* **2019**, *3*, 1–11. [CrossRef]
78. Khalil, A.; Loginova, I.; Pozdnyakov, A.V.; Mosleh, A.; Solonin, A.N. Evaluation of the Microstructure and Mechanical Properties of a New Modified Cast and Laser-Melted AA7075 Alloy. *Materials* **2019**, *12*, 3430.

Publisher's Note: MDPI stays neutral with regard to jurisdictional claims in published maps and institutional affiliations.

© 2020 by the authors. Licensee MDPI, Basel, Switzerland. This article is an open access article distributed under the terms and conditions of the Creative Commons Attribution (CC BY) license (http://creativecommons.org/licenses/by/4.0/).

Article

Microstructure and Selective Corrosion of Alloy 625 Obtained by Means of Laser Powder Bed Fusion

Marina Cabrini [1,2,*], Sergio Lorenzi [1,2], Cristian Testa [1,2], Fabio Brevi [3], Sara Biamino [4], Paolo Fino [4], Diego Manfredi [5], Giulio Marchese [4], Flaviana Calignano [6] and Tommaso Pastore [1,2]

1 Consorzio INSTM, via G. Giusti, 9, 50121 Firenze, Italy; sergio.lorenzi@unibg.it (S.L.); cristian.testa@unibg.it (C.T.); tommaso.pastore@unibg.it (T.P.)

2 University of Bergamo, Department of Engineering and Applied Sciences, Viale Marconi 5, 24044 Dalmine, Italy

3 OMB Valves SpA, Via Europa, 7, 24069 Cenate Sotto, Italy; fabio.brevi@ombvalves.com

4 Department of Applied Science and Technology, Politecnico di Torino, Corso Duca degli Abruzzi 24, 10129 Torino, Italy; sara.biamino@polito.it (S.B.); paolo.fino@poilito.it (P.F.); giulio.marchese@polito.it (G.M.)

5 Center for Sustainable Futures Technologies-CSFT@POLITO, Istituto Italiano di Tecnologia, Via Livorno 60, 10144 Torino, Italy; diego.manfredi@iit.it

6 Department of Management and Production Engineering, Politecnico di Torino, Corso Duca degli Abruzzi 24, 10129 Torino, Italy; flaviana.calignano@polito.it

* Correspondence: marina.cabrini@unibg.it; Tel.: +390352052318

Received: 8 April 2019; Accepted: 27 May 2019; Published: 29 May 2019

Abstract: The effect of microstructure on the susceptibility to selective corrosion of Alloy 625 produced by laser powder bed fusion (LPBF) process was investigated through intergranular corrosion tests according to ASTM G28 standard. The effect of heat treatment on selective corrosion susceptibility was also evaluated. The behavior was compared to commercial hot-worked, heat treated Grade 1 Alloy 625. The morphology of attack after boiling ferric sulfate-sulfuric acid test according to ASTM G28 standard is less penetrating for LPBF 625 alloy compared to hot-worked and heat-treated alloy both in as-built condition and after heat treatment. The different attack morphology can be ascribed to the oversaturation of the alloying elements in the nickel austenitic matrix obtained due to the very high cooling rate. On as-built specimens, a shallow selective attack of the border of the melt pools was observed, which disappeared after the heat treatment. The results confirmed similar intergranular corrosion susceptibility, but different corrosion morphologies were detected. The results are discussed in relation to the unique microstructures of LPBF manufactured alloys.

Keywords: additive manufacturing; corrosion; alloy 625; selective corrosion; oil and gas; materials qualification

1. Introduction

Alloy 625 (UNS N06625) is one of the most used material in chemical and oil and gas industries for its corrosion resistance in either oxidant or non-oxidant acids. It is also used in the aerospace industry thanks to the resistance to thermal oxidation [1,2]. In the oil and gas industry, high aggressiveness of the environment requires the use of materials with outstanding corrosion resistance to ensure the reliability of assets, reduce the risk of breakage, prevent consequences on the environment and guarantee the safety of people. The continuous growth of demand and the depletion of reservoir lead the oil and natural gas extraction industry to exploit deeper wells, with extreme conditions (High Pressure High Temperature Wells—HPHT). The use of traditional alloys with low both mechanical and corrosion resistance is not suitable under such hypothesis and international standards i.e., NACE MR 0175/ISO 15156 defines such environment as "sour" even in presence of few ppm of H_2S in the formation gas due to the high operating pressures. Materials resistant to sulfide stress corrosion cracking (SSCC) are required in these environments and high temperatures, high fraction of water in the formation and presence of CO_2 and H_2S implies the use of general and localized corrosion resistant alloys (Corrosion Resistance Alloys—CRA). Inconel 625® (Alloy 625) is a nickel-based superalloy strengthened by solid-solution hardening of Nb and Mo in a Ni-Cr matrix [3]. Alloy 625 has a combination of high yield strength, fatigue strength, and excellent corrosion resistance in aggressive environments, it has found widespread applications in the aerospace, marine, and nuclear industries where complex shapes are often required. The alloy shows general corrosion rates less than 0.5 mm/year in concentrated non-oxidizing organic and inorganic acids [4,5]. The resistance to pitting is very high, much higher than the traditional austenitic stainless steels. Thanks to the high nickel content, it is immune to chloride stress corrosion (SCC, Stress Corrosion Cracking). Alloy 625 is precipitation-hardening alloy, due to the formation of very fine phases, which strengthen the austenitic matrix, because of aging treatments between 550–750 °C. The formation of these phases or their decomposition modifies both the yield strength and ductility of the alloy [6]. Furthermore, the formation of carbides and secondary phases can also affect the corrosion resistance. In particular, the alloy can become susceptible to intergranular corrosion if subjected to improper solubilization treatment [7].

In the last years, a very promising additive manufacturing (AM) technique has been developed in order to find alternative processes based on Laser Powder Bed Fusion (LPBF) of metal powders—layer by layer—by using a laser following the 3D CAD model. This production method has advantages in terms of cost reduction and lead time, because it eliminates swarf—typical of traditional subtractive machining—welding operations, and assembly phases. Lot of efforts has been devoted to AM of Alloy 625 alloy [8–14]. Alloy 625 is sensitive to precipitation of intermetallic phases such as Ni_3M phases, Ni_2(Cr, Mo) Laves phase as well as MC primary carbides and M_6C and $M_{23}C_6$ secondary carbides. Large amount of literature exists on the heat-treatment-induced phase evolution of wrought Alloy 625 [3,15,16]. The effect of post-weld heat treatment has been also studied by several groups [17–19]. As AM techniques are processes characterized by melting and solidification, microstructures are expected to be different due to different thermal gradients. Precipitation of phases that occurs in AM Alloy 625 are not present at all in wrought material or they only appear after tens to hundreds of hours [2].

Despite huge amount of data and scientific papers regarding the mechanical, physical and microstructural characterization of AM of nickel-based alloys, on the authors knowledge there are no works on the corrosion behavior of the alloy obtained by means of LPBF. On the contrary several works underline the effect of the different microstructures obtained by means of LPBF on the corrosion behavior of Ti6Vl4V alloy [20–22], on AISI 316L [23–25], high strength steels [26] on Cr/Co alloys [27,28] and AlSi10Mg [29–34].

The relation between microstructure and corrosion resistance is of fundamental importance for the qualification of materials selected for hostile environments. The paper is devoted to the study of susceptibility to intergranular corrosion of Alloy 625 produced by LPBF in as-built condition and after typical heat treatment usually recommended for oil and gas service. The behavior of LPBF alloy has been compared to hot-worked, heat treated commercial Alloy 625.

2. Materials and Methods

Commercial gas atomized Alloy 625 powder was used for LPBF process and hot-worked bar with 16 mm diameter was considered for comparison purposes. The bar was furnished in annealed condition, after heat treatment at 980 °C for 32 min and water quenching (Grade 1, according to ASTM B446 standard). The chemical composition of LPBF and hot-worked alloys as well as microstructures are shown in Figure 1. The microstructure was revealed by grinding with silicon carbides emery papers up to 4000 grit, polishing with diamond paste up to 1 μm, and then etching with Kalling's N°2 reagent or electrolytic oxalic acid according to ASTM A262—Practice A. As-built LPBF specimens were etched by Mixed Acid Solution (15 mL HCl, 10 mL H_3COOH and 10 mL HNO_3, Carlo Erba RPA reagents, Cornaredo, Milan, Italy) (Figure 1). The microstructures of LPBF specimens were observed on planes parallel or perpendicular to the building direction, whereas hot-worked specimens were only examined on transverse plane.

[% wt.]	C	Si	Mn	P	S	Cr	Mo	Ni **	Nb	Ti	Al	Co	Ta	Fe	Nb+Ta
Hot worked *	0.036	0.25	0.19	0.007	0.0010	21.6	8.26	61.92	3.66	0.243	0.199	0.020	0.010	3.11	3.67
LPBF	0.010	0.08	0.03	<0.001	0.002	22.4	8.2	64.59	3.73	0.18	<0.01	0.17	0.13	0.45	3.86

* Heat composition
** Balance

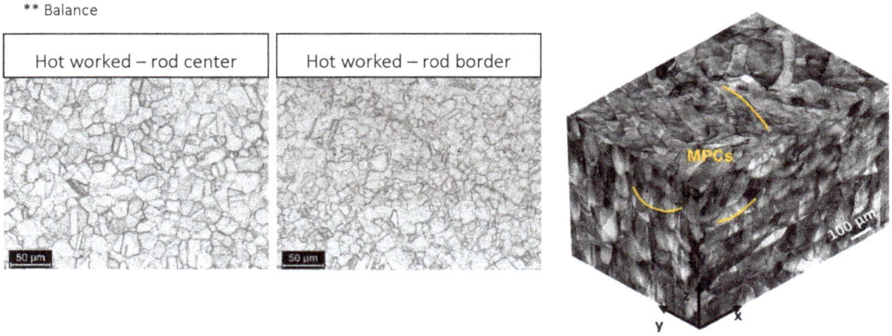

Figure 1. Chemical composition and 3D optical image composite of as-built IN625 sample showing the melt pool contours (MPCs) along the building direction (z-axis) and perpendicular to the building direction (x-y) plane; mixed acids reagent was used.

2.1. Specimens

Cubic specimens of 15 mm side were printed by means of an EOSINT M270 (EOS Gmbh, Krailling, Germany) Dual Mode version working under argon atmosphere. The device operates at 195 W laser power, 1200 mm/s scan speed, 0.09 mm hatching distance as well as 0.02 mm of layer thickness. The scanning strategy involves the rotation of laser beam of 67° between consecutive layers based on the EOS strategy. The combination of the process parameters and scanning strategy allows the production of parts with a residual porosity less than 0.1%. More in-depth information about optimization procedure and building parameters were previously published [14,35]. Specimens in as-built condition and after post-processing annealing heat treatment at 980 °C for 32 min were considered. Five millimeters high cylindrical specimens were obtained by cutting the hot-worked bar.

2.2. Corrosion Tests

Boiling ferric sulfate/sulfuric acid corrosion tests were carried out according to ASTM G28 standard, method A. Before testing, the surface of specimens was grinded up to 1200 grit by silicon carbides emery paper. Two faces of LPBF specimens were ground and polished up to 1 μm diamond paste in order to permit observations of corrosion morphology on main planes, i.e., XY plane and XZ plane parallel and perpendicular to the building plane, respectively. One face was polished for observing morphology on plane transversal to rolling direction on disk specimens. Afterwards, the specimens were degreased in acetone in ultrasonic bath, rinsed in water and dried before their immersion in the boiling test solution. Before and after tests, the specimens were weighed by means of analytic scale to measure the weight loss and corrosion rate.

3. Results and Discussion

Table 1 shows the results of 120 h weight loss tests in boiling ferric sulfate/sulfuric acid solutions (ASTM G28 Method A). The highest corrosion rate was observed on hot-worked alloys. The LPBF specimens showed about the half and the third the corrosion rate measured on hot-worked specimens in as-built conditions and after heat treatment at 980 °C for 32 min, respectively. The corrosion morphology of these specimens is flat and even. On the contrary, hot-worked alloy showed penetrating corrosion attack, more severe compared to LPBF specimens.

Table 1. Results of 120 h weight loss test in boiling ferric sulphate/sulfuric acid solution (ASTM G28).

Metallurgical Condition	Specimens	Exposed Surface [cm^2]	Initial Mass * [g]	Final Mass * [g]	Mass Variation [g]	Corrosion Rate [mdd]
Hot worked bar/Grade 1 annealing	HW1	6.05	6.73923	6.63074	0.10849	359
	HW2	6.13	7.16643	7.06459	0.10184	332
LPBF as built	LPBF NHT1	13.50	28.23744	28.12392	0.11352	168.2
	LPBF NHT2	13.35	27.03907	26.92816	0.11091	166.2
LPBF/Grade 1 annealing	LPBF HT1	13.06	28.24230	28.16770	0.0746	114.2
	LPBF HT2	13.12	28.34171	28.26959	0.07212	109.9

* Average of 3 measurements.

Figures 2 and 3 show the corrosion morphology of the specimens after 120 h exposure tests in boiling ferric sulphate/sulfuric acid solution according to ASTM G28 standard. The building direction of LPBF specimens and the rolling direction of the hot-worked bar is represented by the yellow arrows. Figure 2 shows the images of the metallographic sections of the alloys taken at low and high magnification. HW specimens showed penetrating intergranular attack mainly located at the center of the specimen (Figure 2a,d).

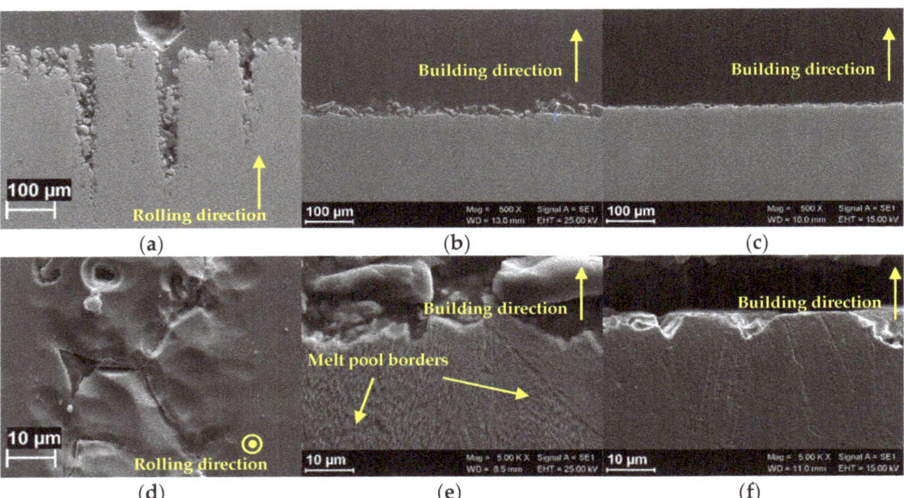

Figure 2. SEM images of Alloy 625 after intergranular corrosion tests. Metallographic section of commercial rolled alloy along the rolling direction (**a**) detail of the attack on the surface perpendicular to the rolling direction (**d**). Metallographic section of LPBF alloys along the building direction in as produced conditions (**b**,**c**) and after heat treatment (**e**,**f**).

Figure 3. SEM images of the polished surface of the specimens after the tests of intergranular corrosion. (**a**) hot-worked bar, (**b**) LPBF as-built, (**c**) LPBF heat treated.

The intergranular attack produced also grain dropping, as evidenced by the analysis of the surface after exposure (Figure 3a). The high value of the corrosion rate is practically due to grain dropping. The morphology of the corrosion attack on LPBF as-built is far more even (Figure 2b,e), but the attack follows different paths. The morphological analysis emphasizes the unique melt pool microstructure-typical of LPBF processing—formed by overlapped laser scan traces (Figure 3b). The melt pool structure is well-defined and penetrating attacks along the borders can be clearly detected. Finally, the attack takes place in form of shallow attacks on the heat treated LPBF specimens (Figure 2c,f) and the melt pool borders are no more evident. Equiaxed grain microstructure is evidenced by the corrosion attack (Figure 3c).

The differences in the morphology of attack is strictly related to the microstructure of the alloy, which depends upon the manufacturing process and the heat treatments. HW alloy shows a deep intergranular attack very close to the axis of the bar, where strong segregation of several precipitates at the grain boundary occurred. This is well evidenced by the presence of certain amount of both macro and micro-precipitates (Figure 4). The EDS analysis carried out on these precipitates confirmed the presence of Nb-rich phases, mainly MC carbides.

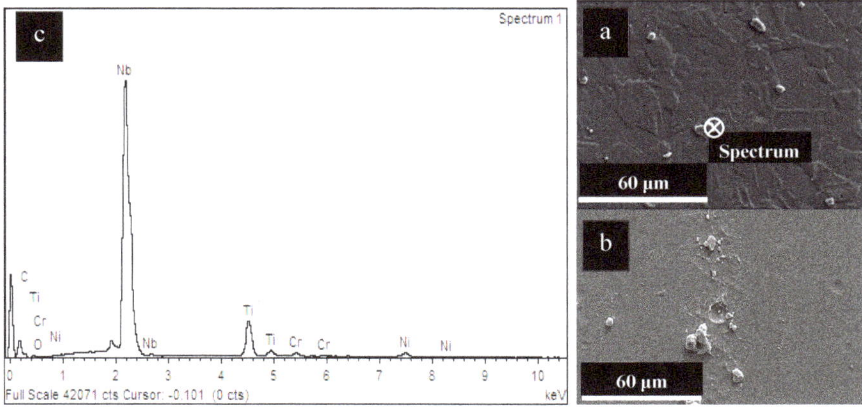

Figure 4. Microstructure of HW specimens (**a**,**b**) (Kalling's attack) and EDS spectrum of macro-precipitates (**c**) taken in the point evidenced in (**a**).

Owing Kim and Lee, the formation of carbides can be a factor contributing to the degradation of corrosion resistance of Alloy 625, because intergranular areas of carbides are thermodynamically more unstable and react more readily than other intergranular ones [36]. Cortial et al. studied the effect of heat treatment on the corrosion resistance of forged and welded nickel superalloys [37,38]. They concluded that hardening between 650 and 800 °C results in dislocation gliding of matrix, γ''

precipitation and maximum of intergranular precipitation of fine $M_{23}C_6$ carbides. Tensile strength and intergranular corrosion become maxima, impact strength and reduction of area become minima in the field of $M_{23}C_6$ and M_6C intergranular precipitation between 700 and 950 °C. Furthermore, the carbides have a higher Cr concentration than that in the base metal and they deplete the surrounding matrix in chrome, thus decreasing its corrosion resistance. However, the precipitates founded in this works were mainly enriched in Nb. The role of Nb rich precipitates in the corrosion of Alloy 625 alloy was underlined by Tawancy et al. [39]. Authors made tests in boiling 10% nitric acid on Alloy 626 annealed and aged at different time and temperature. They concluded that Ni_3Nb phase formed during aging is preferentially corroded owing the lower chromium content compared to the surrounding matrix. The intense intergranular corrosion attack observed on the hot-worked and heat treated reference alloy can be then ascribed both to the precipitation of MC carbides and Ni_3Nb phase at the grain boundary which is favored at the center of the bar due to the production process.

The intergranular attack is not so penetrating in as-built LPBF Alloy 625 alloy, and it is not detectable at all on the LPBF alloy after the heat treatment. The microstructure of the as-built samples obtained in this work shows columnar grains (CGs) developing epitaxially, thus crossing several melt pools along the building direction (z-axis) (Figure 5). The marks underline the different melt pool borders (MPCs) along both the building direction (z-axis) and perpendicular to the building platform (x-y plane). The melt pools are not all placed straight along the building direction due to the scan strategy which involves laser beam rotation of 67° between two consecutive layers [14,40,41].

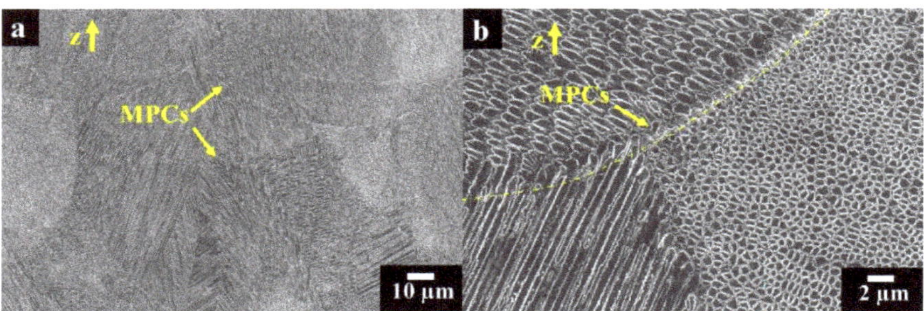

Figure 5. FESEM images at different magnification (**a**,**b**) exhibiting melt pools with columnar and cellular primary dendrites for as-built IN625 sample; mixed acids reagent.

During the LPBF process, the solidification rate is different between the upper and lower zones of the melt pool. In the lower part of the layer the cooling speed is very high, while the upper part cools more slowly.

Figure 5a reveals the FESEM (Field Emission Scanning Electron Microscope) image of as-built Alloy 625 samples along the building direction. The melt pools are made up of very fine primary dendritic structures which have both cellular and columnar shapes, as shown in Figure 5b. The cellular dendrites structure is due to the alteration of columnar dendrite structures caused by very high cooling rates, as previously reported [42].

At higher magnification (Figure 6), very fine bright precipitates along the dendritic areas indicated by arrows 1 and bright elongated phases along the inter-dendritic regions pointed out by arrows 2 can be observed. Previous TEM investigations performed on as-built Alloy 625 (produced with the same process parameters) [43] reported a relatively high concentration of Nb and C for these precipitates, suggesting the early stage formation of fine MC carbides. EDS results revealed enrichment in Nb and Mo within the inter-dendritic areas, indicated possible segregation of Nb and Mo due to their elevated tendency to segregate, as confirmed by the literature [44,45].

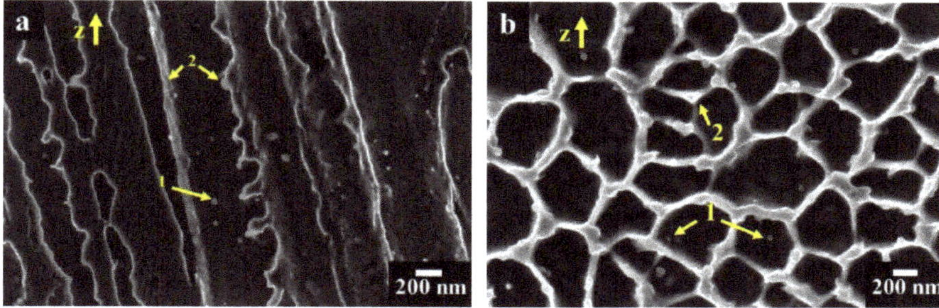

Figure 6. FESEM images of as-built IN625 sample at high magnification (**a**) columnar dendritic structures with nanoprecipitates (1) and segregation of Nb and Mo within the inter-dendritic zones (2); (**b**) cellular dendritic structures with nanoprecipitates (1) and segregation of Nb and Mo within the inter-dendritic areas (2).

These precipitates did not influence the corrosion resistance of the alloys owing to their very small sizes. On the other hand, the attack seems to be preferentially located along the border of the melt pools.

Dinda et al. hypothesized the precipitation of intermetallic γ'' [Ni_3Nb] and δ [Ni_3Nb] phases in the heat affected zone at the border of melt pool [9], taking into account also the reticular parameters modifications in the Ni matrix. Anam et al. reported the presence of γ'' [Ni_3Nb] precipitates at the border of melt pool [46]. A similar attack morphology was observed also for aluminum alloys [29,32,34]. However, the preferential attack at the border of the melt pool does not penetrate in depth for Alloy 625.

The heat treatment at 980 °C for 32 min dissolved the melt pool microstructure obtained by the LPBF process (Figure 7a). The specimens presented columnar grains along the building direction, coupled with the formation of fine carbides as displayed in Figure 7b,c. The carbides had dimensions from nanometric size (around 40 nm) as well as larger size roughly around 500 nm, as can be seen in Figure 7c,b, respectively. In order to detect a chemical composition variation, the EDS analysis was performed on the largest carbides.

EDS scan line (Figure 7d) revealed that the carbides are enriched in Nb and depleted of Cr suggesting the formation of MC carbides, which is in accordance with the TTT (Time Temperature Transformation) diagram of Alloy 625 [47].

The very low susceptibility to intergranular corrosion of the heat-treated specimens could be attributed to the absence of macroscopic precipitation as the size and distribution of the precipitate are coherent with the rounded pits on the surface of the specimen at the end of the intergranular corrosion test (Figure 7a). The precipitation of second phases is not continuous. This promotes a quite even not penetrating corrosion attack that does not penetrate significantly in depth along the grain boundaries, without any dropping. Thus, very low loss of weight was measured due to this fact.

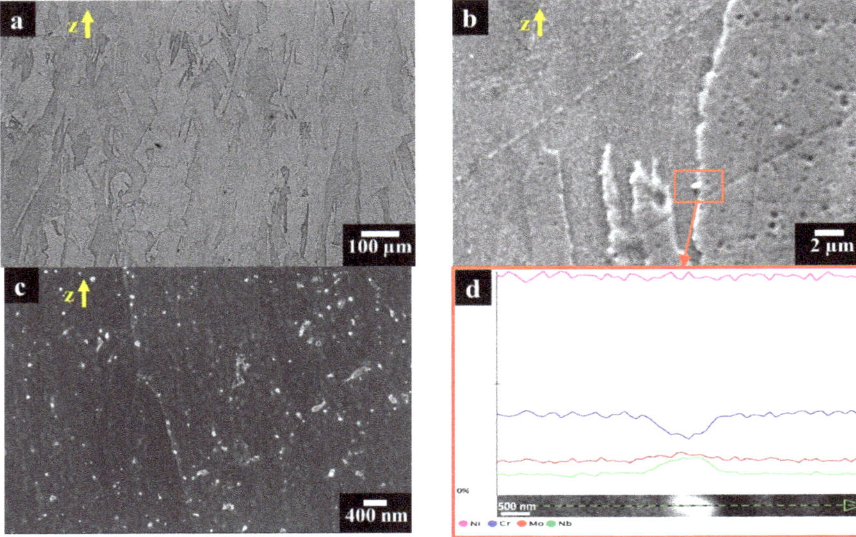

Figure 7. (**a**) Optical microscopy image of LPBF Alloy 625 annealed at 980 °C for 32 min and water quenched showing columnar grains along the building direction; (**b**,**c**) FESEM images of IN625 sample annealed at 980 °C for 32 min and water quenched revealing fine precipitates throughout the material, with the largest located along the grain boundaries; (**d**) EDS scan line (at higher magnification) on a carbides showing the enrichment in Nb and the depletion of Cr with respect to the austenitic matrix.

4. Conclusions

The susceptibility to selective corrosion behavior of Alloy 625 obtained by means of LPBF was studied. The behavior was compared to commercial hot-worked, heat treated Grade 1 Alloy 625. The morphology of attack after boiling ferric sulfate-sulfuric acid test according to ASTM G28 standard is less penetrating for LPBF 625 alloy compared to hot-worked and heat treated alloy both in as-built condition and after heat treatment at 980 °C for 32 min. Furthermore, lower weight loss of LPBF alloy were measured.

The different attack morphology can be ascribed to the oversaturation of the alloying elements in the nickel austenitic matrix obtained due to the very high cooling rate. The precipitates of second phases along and inside the micro-dendrites are too small to promote preferential dissolution paths. On as-built specimens, a shallow selective attack of the border of the melt pools was observed. The Grade 1 heat treatment at 980 °C for 32 min followed by water quenching caused the melt pools microstructure to disappear and prevent the in-depth penetration of the selective attack.

Author Contributions: Methodology, S.L., C.T., D.M., G.M., F.C.; investigation, S.L., C.T., D.M., G.M.; data curation, S.B., P.F., M.C.; all the authors contributed to writing the original draft of the paper. writing—review & editing T.P., P.F., M.C., S.B.; supervision, T.P., S.B.; project administration, T.P.; funding acquisition, T.P., F.B.

Funding: The work was realized with the contribution and reliance on the Collaboration Agreement of INSTM and the Lombardy Region signed on 24th September 2015.

Conflicts of Interest: The authors declare no conflict of interest.

References

1. Eiselstein, H.; Tillack, D. The Invention and Definition of Alloy 625. *Superalloys* **1991**. [CrossRef]
2. Radavich, J.F.; Fort, A. Effects of Long Time Exposure in Alloy 625 at 1200°F, 1400°F and 1600°F. *Superalloys* **1994**, *718*, 635–647.

3. Rai, S.K.; Kumar, A.; Shankar, V.; Jayakumar, T.; Rao, K.B.S.; Raj, B. Characterization of microstructures in Inconel 625 using X-ray diffraction peak broadening and lattice parameter measurements. *Scr. Mater.* **2004**, *51*, 59–63. [CrossRef]
4. Kolts, J.; Wu, J.B.C.; Manning, P.E.; Asphahani, A.I. Highly Alloyed Austenitic Material for Corrosion Resistance. *Corros. Rev.* **1986**, *6*, 279–326.
5. *Metals Handbook*, 9th ed.; ASM International: Metals Park, Geauga, OH, USA, 1987.
6. Shankar, V.; Bhanu Sankara Rao, K.; Mannan, S. Microstructure and mechanical properties of Inconel 625 superalloy. *J. Nucl. Mater.* **2001**, *288*, 222–232. [CrossRef]
7. Harris, J.A.; Scarberry, R.C. Effect of metallurgical reactions in INCONEL nickel-chromium-molybdenum alloy 625 on corrosion resistance in nitric acid. *JOM* **1971**, *23*, 45–49. [CrossRef]
8. Murr, L.E.; Martinez, E.; Amato, K.N.; Gaytan, S.M.; Hernandez, J.; Ramirez, D.A.; Shindo, P.W.; Medina, F.; Wicker, R.B. Fabrication of metal and alloy components by additive manufacturing: Examples of 3D materials science. *J. Mater. Res. Technol.* **2012**, *1*, 42–54. [CrossRef]
9. Dinda, G.P.; Dasgupta, A.K.; Mazumder, J. Laser aided direct metal deposition of Inconel 625 superalloy: Microstructural evolution and thermal stability. *Mater. Sci. Eng. A* **2009**, *509*, 98–104. [CrossRef]
10. Paul, C.P.; Ganesh, P.; Mishra, S.K.; Bhargava, P.; Negi, J.; Nath, A.K. Investigating laser rapid manufacturing for Inconel-625 components. *Opt. Laser Technol.* **2007**, *39*, 800–805. [CrossRef]
11. Witkin, D.B.; Adams, P.; Albright, T. Microstructural evolution and mechanical behavior of nickel-based superalloy 625 made by selective laser melting. In Proceedings of the SPIE, San Francisco, CA, USA, 16 March 2015.
12. Yadroitsev, I.; Gusarov, A.; Yadroitsava, I.; Smurov, I. Single track formation in selective laser melting of metal powders. *J. Mater. Process. Technol.* **2010**, *210*, 1624–1631. [CrossRef]
13. Keller, T.; Lindwall, G.; Ghosh, S.; Ma, L.; Lane, B.M.; Zhang, F.; Kattner, U.R.; Lass, E.A.; Heigel, J.C.; Idell, Y.; Williams, M.E.; Allen, A.J.; Guyer, J.E.; Levine, L.E. Application of finite element, phase-field, and CALPHAD-based methods to additive manufacturing of Ni-based superalloys. *Acta Mater.* **2017**, *139*, 244–253. [CrossRef]
14. Marchese, G.; Garmendia Colera, X.; Calignano, F.; Lorusso, M.; Biamino, S.; Minetola, P.; Manfredi, D. Characterization and Comparison of Inconel 625 Processed by Selective Laser Melting and Laser Metal Deposition. *Adv. Eng. Mater.* **2017**, *19*, 1600635. [CrossRef]
15. Suave, L.M.; Cormier, J.; Villechaise, P.; Soula, A.; Hervier, Z.; Bertheau, D.; Laigo, J. Microstructural evolutions during thermal aging of alloy 625: Impact of temperature and forming process. *Metall. Mater. Trans. A Phys. Metall. Mater. Sci.* **2014**, *45*, 2963–2982. [CrossRef]
16. Petrzak, P.; Kowalski, K.; Blicharski, M. Analysis of phase transformations in Inconel 625 alloy during annealing. *Acta Phys. Pol. A* **2016**, *130*, 1043–1044. [CrossRef]
17. Dupont, J.N.; Robino, C.V.; Marder, A.R.; Notis, M.R. Solidification of Nb-bearing superalloys: Part II. Pseudoternary solidification surfaces. *Metall. Mater. Trans. A Phys. Metall. Mater. Sci.* **1998**, *29*, 2797–2806. [CrossRef]
18. Xu, F.; Lv, Y.; Liu, Y.; Xu, B.; He, P. Effect of heat treatment on microstructure and mechanical properties of Inconel 625 alloy fabricated by pulsed plasma arc deposition. *Phys. Procedia* **2013**, *50*, 48–54. [CrossRef]
19. Li, S.; Wei, Q.; Shi, Y.; Zhu, Z.; Zhang, D. Microstructure Characteristics of Inconel 625 Superalloy Manufactured by Selective Laser Melting. *J. Mater. Sci. Technol.* **2015**, *31*, 946–952. [CrossRef]
20. Dai, N.; Zhang, L.-C.; Zhang, J.; Chen, Q.; Wu, M. Corrosion behavior of selective laser melted Ti-6Al-4V alloy in NaCl solution. *Corros. Sci.* **2016**, *102*, 484–489. [CrossRef]
21. Yang, J.; Yang, H.; Yu, H.; Wang, Z.; Zeng, X. Corrosion Behavior of Additive Manufactured Ti-6Al-4V Alloy in NaCl Solution. *Metall. Mater. Trans. A* **2017**, *48*, 3583–3593. [CrossRef]
22. Liu, L.; He, M.; Xu, X.; Zhao, C.; Gan, Y.; Lin, J.; Luo, J.; Lin, J. Preliminary study on the corrosion resistance, antibacterial activity and cytotoxicity of selective-laser-melted Ti6Al4V-xCu alloys. *Mater. Sci. Eng. C* **2017**, *72*, 631–640. [CrossRef]
23. Ziętala, M.; Durejko, T.; Polański, M.; Kunce, I.; Płociński, T.; Zieliński, W.; Łazińska, M.; Stępniowski, W.; Czujko, T.; Kurzydłowski, K.J.; et al. The microstructure, mechanical properties and corrosion resistance of 316 L stainless steel fabricated using laser engineered net shaping. *Mater. Sci. Eng. A* **2016**, *677*, 1–10. [CrossRef]

24. Trelewicz, J.R.; Halada, G.P.; Donaldson, O.K.; Manogharan, G. Microstructure and Corrosion Resistance of Laser Additively Manufactured 316L Stainless Steel. *JOM* **2016**, *68*, 850–859. [CrossRef]
25. Sander, G.; Thomas, S.; Cruz, V.; Jurg, M.; Birbilis, N.; Gao, X.; Brameld, M.; Hutchinson, C.R. On The Corrosion and Metastable Pitting Characteristics of 316L Stainless Steel Produced by Selective Laser Melting. *J. Electrochem. Soc.* **2017**, *164*, C250–C257. [CrossRef]
26. Schmidt, D.P.; Jelis, E.; Clemente, M.P.; Ravindra, N.M. Corrosion of 3D Printed Steel. Available online: https://www.researchgate.net/publication/283346384_CORROSION_OF_3D_PRINTED_STEEL (accessed on 28 May 2019).
27. Tuna, S.H.; Özçiçek Pekmez, N.; Kürkçüoğlu, I. Corrosion resistance assessment of Co-Cr alloy frameworks fabricated by CAD/CAM milling, laser sintering, and casting methods. *J. Prosthet. Dent.* **2015**, *114*, 725–734. [CrossRef]
28. Puskar, T.; Jevremovic, D.; Williams, R.J.; Eggbeer, D.; Vukelic, D.; Budak, I. A comparative analysis of the corrosive effect of artificial saliva of variable pH on DMLS and cast Co-Cr-Mo dental alloy. *Materials* **2014**, *7*, 6486–6501. [CrossRef]
29. Cabrini, M.; Lorenzi, S.; Pastore, T.; Pellegrini, S.; Ambrosio, E.P.; Calignano, F.; Manfredi, D.; Pavese, M.; Fino, P. Effect of heat treatment on corrosion resistance of DMLS AlSi10Mg alloy. *Electrochim. Acta* **2016**, *206*, 346–355. [CrossRef]
30. Cabrini, M.; Lorenzi, S.; Pastore, T.; Pellegrini, S.; Testa, C.; Manfredi, D.; Calignano, F.; Fino, P.; Cattano, G. Evaluation of corrosion resistance of aluminum silicon alloys manufactured by LPBF | Valutazione della resistenza a corrosione di leghe di alluminio-silicio ottenute per LPBF. *Metall. Ital.* **2017**, *109*, 123–126.
31. Cabrini, M.; Lorenzi, S.; Pastore, T.; Testa, C.; Manfredi, D.; Lorusso, M.; Calignano, F.; Pavese, M.; Andreatta, F. Corrosion behavior of AlSi10Mg alloy produced by laser powder bed fusion under chloride exposure. *Corros. Sci.* **2019**, *152*, 101–108. [CrossRef]
32. Cabrini, M.; Calignano, F.; Fino, P.; Lorenzi, S.; Lorusso, M.; Manfredi, D.; Testa, C.; Pastore, T. Corrosion Behavior of Heat-Treated AlSi10Mg Manufactured by Laser Powder Bed Fusion. *Materials* **2018**, *11*, 1051. [CrossRef]
33. Cabrini, M.; Lorenzi, S.; Pastore, T.; Testa, C.; Manfredi, D.; Cattano, G.; Calignano, F. Corrosion resistance in chloride solution of the AlSi10Mg alloy obtained by means of LPBF. *Surf. Interface Anal.* **2016**, *48*, 818–826. [CrossRef]
34. Cabrini, M.; Lorenzi, S.; Testa, C.; Pastore, T.; Manfredi, D.; Lorusso, M.; Calignano, F.; Fino, P. Statistical approach for electrochemical evaluation of the effect of heat treatments on the corrosion resistance of AlSi10Mg alloy by laser powder bed fusion. *Electrochim. Acta* **2019**, *305*, 459–466. [CrossRef]
35. Marchese, G.; Lorusso, M.; Calignano, F.; Ambrosio, E.P.; Manfredi, D.; Pavese, M.; Biamino, S.; Ugues, D.; Fino, P. Inconel 625 by direct metal laser sintering: Effects of the process parameters and heat treatments on microstructure and hardness. In Proceedings of the 13th International Symposium of Superalloys, Seven Springs, PA, USA, 11–15 September 2016.
36. Kim, J.S.; Lee, H.W. A study on effect of intergranular corrosion by heat input on Inconel 625 overlay weld metal. *Int. J. Electrochem. Sci.* **2015**, *10*, 6454–6464.
37. Corrieu, J.; Vernot-Loier, C.; Cortial, F. Influence of Heat Treatments on Corrosion Behavior of Alloy 625 Forged Rod. *Superalloys* **1994**. [CrossRef]
38. Cortial, F.; Corrieu, J.M; Vernot-Loier, C. Influence of heat treatments on microstructure, mechanical properties, and corrosion resistance of weld alloy 625. *Metall. Mater. Trans. A* **1995**, *26*, 1273–1286. [CrossRef]
39. Tawancy, H.M.; Allam, I.M.; Abbas, N.M. Effect of Ni_3Nb precipitation on the corrosion resistance of Inconel alloy 625. *J. Mater. Sci. Lett.* **1990**, *9*, 343–347. [CrossRef]
40. Manfredi, D.; Calignano, F.; Krishnan, M.; Canali, R.; Ambrosio, E.P.; Atzeni, E. From powders to dense metal parts: Characterization of a commercial AlSi10Mg alloy processed through direct metal laser sintering. *Materials* **2013**, *6*, 856–869. [CrossRef]
41. Kreitcberg, A.; Brailovski, V.; Turenne, S. Elevated temperature mechanical behavior of IN625 alloy processed by laser powder-bed fusion. *Mater. Sci. Eng. A* **2017**, *700*, 540–553. [CrossRef]
42. Antonsson, T.; Fredriksson, H. The effect of cooling rate on the solidification of INCONEL 718. *Metall. Mater. Trans. B-Process Metall. Mater. Process. Sci.* **2005**, *36*, 85–96. [CrossRef]

43. Marchese, G.; Lorusso, M.; Parizia, S.; Bassini, E.; Lee, J.W.; Calignano, F.; Manfredi, D.; Terner, M.; Hong, H.U.; Ugues, D.; Lombardi, M.; Biamino, S. Influence of heat treatments on microstructure evolution and mechanical properties of Inconel 625 processed by laser powder bed fusion. *Mater. Sci. Eng. A* **2018**, *729*, 64–95. [CrossRef]
44. Harrison, N.J.; Todd, I.; Mumtaz, K. Reduction of micro-cracking in nickel superalloys processed by Selective Laser Melting: A fundamental alloy design approach. *Acta Mater.* **2015**, *94*, 59–68. [CrossRef]
45. Tobergte, D.R.; Curtis, S. Modeling and Experimental validation of Nickel-based super alloy (Inconel 625) made using Selective Laser Melting. *Solid Free. Fabr.* **2013**. [CrossRef]
46. Anam, M.A.; Dilip, J.J.S.; Pal, D.; Stucker, B. Effect of scan pattern on the microstructural evolution of Inconel 625 during selective laser melting. In Proceedings of the 25th Annual International Solid Freeform Fabrication Symposium, Austin, TX, USA, 4–6 August 2014.
47. Floreen, S.; Fuchs, G.E.; Yang, W.J. The Metallurgy of Alloy 625. In *Superalloys 718, 625, 706 and Various Derivatives*; Minerals, Metals and Materials Society: Pittsburgh, PA, USA, 1994; pp. 13–39.

© 2019 by the authors. Licensee MDPI, Basel, Switzerland. This article is an open access article distributed under the terms and conditions of the Creative Commons Attribution (CC BY) license (http://creativecommons.org/licenses/by/4.0/).

Article

The Effect of Nb-Content on the Microstructures and Corrosion Properties of $CrFeCoNiNb_x$ High-Entropy Alloys

Chun-Huei Tsau *, Chen-Yu Yeh and Meng-Chi Tsai

Institute of Nanomaterials, Chinese Culture University, Taipei 111, Taiwan; qazwsx61515@gmail.com (C.-Y.Y.); asd99586@yahoo.com.tw (M.-C.T.)

* Correspondence: chtsau@staff.pccu.edu.tw

Received: 10 October 2019; Accepted: 7 November 2019; Published: 11 November 2019

Abstract: This work studied the effect of niobium-content on the microstructures, hardness, and corrosion properties of $CrFeCoNiNb_x$ alloys. Results indicated that the microstructures of these alloys changed from granular structures (CrFeCoNi alloy) to the hypereutectic structures ($CrFeCoNiNb_{0.2}$ and $CrFeCoNi_{0.4}$ alloys), and then to the hypoeutectic microstructures ($CrFeCoNiNb_{0.6}$ and CrFeCoNi alloys). The lattice constants of the major two phases in these alloys, fcc and Laves phases (hcp), increased with the increasing of Nb-content because of solid-solution strengthening. The hardness of $CrFeCoNiNb_x$ alloys also had the same tendency. Adding niobium would slightly lessen the corrosion resistance of $CrFeCoNiNb_x$ alloys in 1 M deaerated sulfuric acid and 1 M deaerated sodium chloride solutions, but the $CrFeCoNiNb_x$ alloys still had better corrosion resistance in comparison with commercial 304 stainless steel. In these dual-phased $CrFeCoNiNb_x$ alloys, the fcc phase was more severely corroded than the Laves phase after polarization tests in these two solutions.

Keywords: $CrFeCoNiNb_x$ alloys; microstructure; hardness; corrosion property

---

1. Introduction

The high-entropy alloy concept has been announced for many years [1], and many important results have been published recently [2–4]. This concept creates new alloys from many elements with different atomic radiuses and crystal structures. The final structures of the high-entropy alloys are thus not dominated by any one of the elements because the molar fraction of each element is not high enough. The high-entropy alloy concept has four important core effects, which are: High entropy, lattice distortion, sluggish diffusion, and cocktail effects [4,5]. The high-entropy alloy concept is thus used to develop new alloys with special properties because of these four core effects. Corrosion resistance is also an important property for structural application of high-entropy alloys.

Many studies of the corrosion behaviors of high-entropy alloys have been published. $Al_xCrFe_{1.5}MnNi_{0.5}$ alloy possesses good workability and high-temperature oxidation resistance [6]. The pitting resistance of $Al_{0.3}CrFe_{1.5}MnNi_{0.5}$ alloy can be enhanced after anodic treatment [7] because a stable passive film can be formed on the surface. The corrosion resistance of $Fe_{73.5}Cu_1Nb_xSi_{13.5}B_{10}$ alloy can be improved after adding a suitable amount of Nb [8]. Moreover, $FeCoNiCrCu_x$ alloy possesses a good corrosion resistance in a 3.5% sodium chloride solution [9]. Similar results are also observed in $Al_{7.5}Cr_{22.5}Fe_{35}Mn_{20}Ni_{15}$ and $Al_{0.5}CoCrFeNi$ alloys [10,11]. These literatures proved that the high-entropy alloy concept is a good method to improve the corrosion resistance of structural materials.

Referring to our previous studies, CrFeCoNi alloy has an fcc granular structure, and some hcp Cr-rich precipitates are in the matrix [12]; it has a good corrosion resistance in 1 M deaerated sulfuric acid and 1 M deaerated sodium chloride solutions. However, the CrFeCoNi alloy is very soft, which limits its application. The hardness of CrFeCoNi alloy could be enhanced after the adding of

Nb [13]. The corrosion resistance of CrFeCoNiNb alloy slightly decreases in comparison with that of CrFeCoNi alloy, but it is still better than that of commercial 304 stainless steel. The microstructures and mechanical properties of $CrFeCoNiNb_x$ alloys are also investigated because of their high performance for commercial applications [14–16]. This study investigates the microstructures, hardness, and corrosion behaviors of $CrFeCoNiNb_x$ (x = 0–1) alloys.

2. Experimental Section

The experimental $CrFeCoNiNb_x$ alloys were prepared by an arc melter in an Ar atmosphere with a partial pressure of 200 torr, and the purities of Cr, Fe, Co, Ni, and Nb were all higher than 99.9%. The nominal compositions of $CrFeCoNiNb_x$ alloys are listed in Table 1. The microstructures of $CrFeCoNiNb_x$ alloys for observation were metallographically prepared and chemically etched in aqua regia (25% HNO_3 and 75% HCl in volume fraction). The microstructures and chemical compositions of the alloys were done using a field emission scanning electron microscope with an energy dispersive spectrometer (SEM/EDS, JEOL JSM-6335, JEOL Ltd., Tokyo, Japan) operated at 15 kV. The phases in the alloys were identified by an X-ray diffractometer (XRD, Rigaku ME510-FM2, Rigaku Ltd., Tokyo, Japan) with Cu–Kα (with a wavelength of 1.541 Å) radiation operated at 30 kV. The microstructures and corresponding selection area diffraction patterns (SAD) of the alloys were observed by a transmission electron microscope (TEM, JEOL JEM-2000EX II, JEOL Ltd., Tokyo, Japan) operated at 200 kV. The TEM specimens were electrochemically prepared by a model 110 digital Fischione twin-jet electropolisher at a potential of 30 V (Fischione Instruments Co., Pittsburgh, PA, USA), and the solution was 10 vol.% perchloric acid and 90 vol.% methanol. The hardness of the alloys was measured using a Matsuzawa Seiki MV1 Vicker's hardness tester (Matsuzawa Co., Akita, Japan) under a load of 294 N (30 kgf), and more than five points were averaged for each alloy.

Table 1. Nominal compositions of $CrFeCoNiNb_x$ alloys.

Alloys (at.%)	Compositions (wt.%)				
	Cr	Fe	Co	Ni	Nb
CrFeCoNi	23.06	24.77	26.13	26.04	N/A
$CrFeCoNiNb_{0.2}$	21.31	22.88	24.14	24.05	7.62
$CrFeCoNiNb_{0.4}$	19.80	21.26	22.44	22.35	14.15
$CrFeCoNiNb_{0.6}$	18.49	19.86	20.95	20.88	19.82
CrFeCoNiNb	16.33	17.54	18.51	18.44	29.18

Polarization curves of the alloys were tested in a potentiostat/galvanostat with a three-electrode system (Autolab PGSTAT302N, Metrohm Autolab B.V., Utrecht, the Netherlands). The exposed surface of each as-cast $CrFeCoNiNb_x$ alloy for polarization testing was fixed at 0.1964 cm^2 (with a diameter of 0.5 cm). The reference electrode was a saturated silver chloride electrode (Ag/AgCl), and the counter electrode was a Pt wire. The potential of a saturated silver chloride electrode (SSE, V_{SSE}) is 222 mV higher than that of the standard hydrogen electrode (SHE, V_{SHE}) at 25 °C [17]. The specimens were all mechanically polished using 1200 SiC grit paper. Test solutions were prepared from reagent-grade sulfuric acid (H_2SO_4) and sodium chloride (NaCl) that were dissolved in distilled water, and the concentrations of these two solutions were fixed at 1 M. To eliminate any effect of dissolved oxygen, the solutions were deaerated by bubbling nitrogen gas with a flow rate of 10 sccm/min through them before and during the polarization experiments. The immersion time before the experiment was fixed at 900 s for stabilizing the open circuit potential (E_{OC}). The scanning rate was fixed at 1 mV/s, and the scanning range was from E_{OC} −1 V to E_{OC} +2 V in a single pass. The polarization data were also compared with those of commercial 304 stainless steel (304SS), and the composition of 304SS was 71.61% Fe, 18.11% Cr, 8.24% Ni, 1.12% Mn, 0.75% Si, 0.05% Co, 0.02% Mo, 0.05% C, 0.03% P, and 0.02% S by weight.

3. Results and Discussion

The as-cast SEM microstructures of $CrFeCoNiNb_x$ alloys are shown in Figure 1. Figure 1a shows the microstructure of as-cast CrFeCoNi alloy. The CrFeCoNi alloy had a granular microstructure with an average grain size of several hundred micrometers; some hcp-structured Cr-rich precipitates in the fcc matrix were reported in our previous study [12]. However, the microstructures of as-cast $CrFeCoNiNb_x$ alloys became a dendritic microstructure after adding niobium. The dendrites were a single phase, and the interdendrites showed a eutectic structure. The dendrites of these alloys had different morphologies after etching. The dendrites of $CrFeCoNiNb_{0.2}$ and $CrFeCoNiNb_{0.4}$ alloys had significant imprints after etching, as shown in Figure 1b,c, but those of $CrFeCoNiNb_{0.6}$ and CrFeCoNiNb alloys exhibited almost smooth surfaces after etching, as shown in Figure 1d,e. These differences were caused by the different phases of the dendrites in these alloys. The dendrites of $CrFeCoNiNb_{0.2}$ and $CrFeCoNiNb_{0.4}$ alloys were an fcc phase, and the dendrites of $CrFeCoNiNb_{0.6}$ and CrFeCoNiNb alloys were a Laves phase (hcp structure). The interdendrites of $CrFeCoNiNb_x$ alloys had two major phases which were fcc and hcp phases (Laves phase). These were identified and described below.

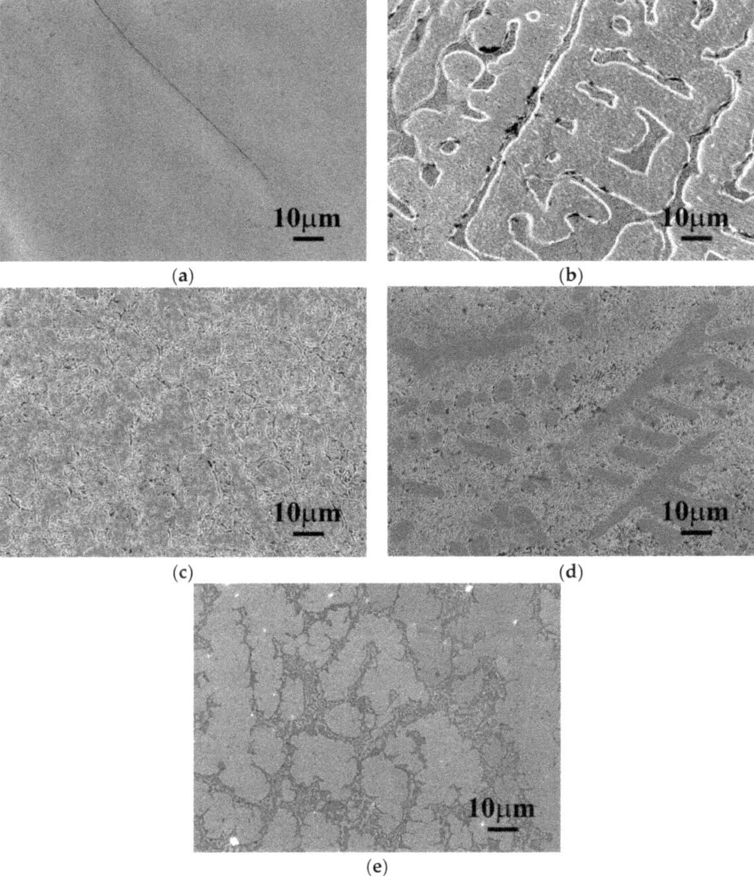

Figure 1. The SEM microstructures of as-cast $CrFeCoNiNb_x$ alloys, (**a**) CrFeCoNi; (**b**) $CrFeCoNiNb_{0.2}$; (**c**) $CrFeCoNiNb_{0.4}$; (**d**) $CrFeCoNiNb_{0.6}$; and (**e**) CrFeCoNiNb alloys.

Figure 2 shows the XRD patterns of $CrFeCoNiNb_x$ alloys. There are two phases which existed in the $CrFeCoNiNb_x$ alloys, which are fcc and Laves phases. The intensities of Laves phases in the $CrFeCoNiNb_x$ alloys increase with increasing Nb-content. Table 2 lists the lattice constants of fcc and Laves phases analyzed from the XRD patterns. The lattice constants of fcc and Laves phases increased with increasing Nb-content in the $CrFeCoNiNb_x$ alloys. This result was caused by the larger atomic radius of niobium. The atomic radiuses of Cr, Fe, Co, Ni, and Nb are 0.128, 0.124, 0.125, 0.125, and 0.143 nm, respectively [18]. The radiuses of Cr, Fe, Co, and Ni are very close, but niobium has a much larger atomic radius. Therefore, the lattice constants of fcc and Laves phases increased when the solid-solution amount of Nb-content increased in these alloys.

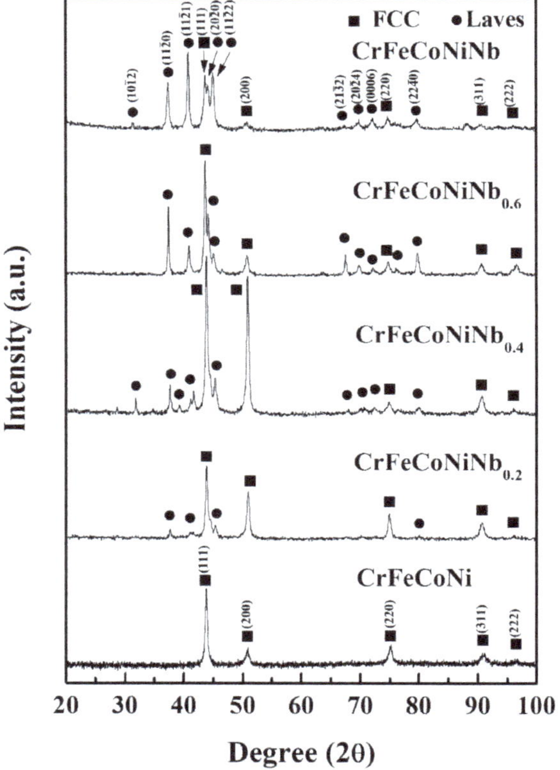

Figure 2. XRD patterns of $CrFeCoNiNb_x$ alloys.

Table 2. Lattice constants of fcc and Laves phases in as-cast $CrFeCoNiNb_x$ alloys.

Alloys	fcc (Å)	Laves Phase (hcp) (Å)
CrFeCoNi	3.577	N/A
$CrFeCoNiNb_{0.2}$	3.578	$a = 4.773$; $c = 7.818$
$CrFeCoNiNb_{0.4}$	3.581	$a = 4.773$; $c = 7.831$
$CrFeCoNiNb_{0.6}$	3.590	$a = 4.798$; $c = 7.841$
CrFeCoNiNb	3.590	$a = 4.802$; $c = 7.848$

Figure 3 shows the TEM bright field (BF) images to examine the phases of the $CrFrCoNiNb_x$ alloys, and the inserts are their corresponding select area diffraction patterns (SADs). Figure 3a is an image of the dendrite of $CrFrCoNiNb_{0.4}$ alloy. The corresponding SAD proves that the dendrite is an fcc phase. Figure 3b is an image of the interdendrite of $CrFrCoNiNb_{0.4}$ alloy, and the corresponding SAD was taken from the Laves phase in the interdendrite. Figure 3c is an image of the dendrite of $CrFrCoNiNb_{0.6}$ alloy. The corresponding SAD proves that the dendrite is a Laves phase (hcp structure). Figure 3d shows the image of the interdendrite of $CrFrCoNiNb_{0.6}$ alloy, and the corresponding SAD was taken from the fcc phase in the interdendrite of $CrFrCoNiNb_{0.6}$ alloy. Therefore, Figure 3 indicates that the $CrFeCoNiNb_{0.4}$ alloy is a hypoeutectic alloy and that the dendrites of this alloy are an fcc phase; the $CrFeCoNiNb_{0.6}$ alloy is a hypereutectic alloy, and the dendrites of the alloy are the Laves phase. In all of these alloys, the interdendrites are a dual-phased (fcc and Laves phase) eutectic structure. Chanda and Das [14] also pointed out that the Nb-content of eutectic $CrFeCoNiNb_x$ alloy should be located between 0.5 and 0.55.

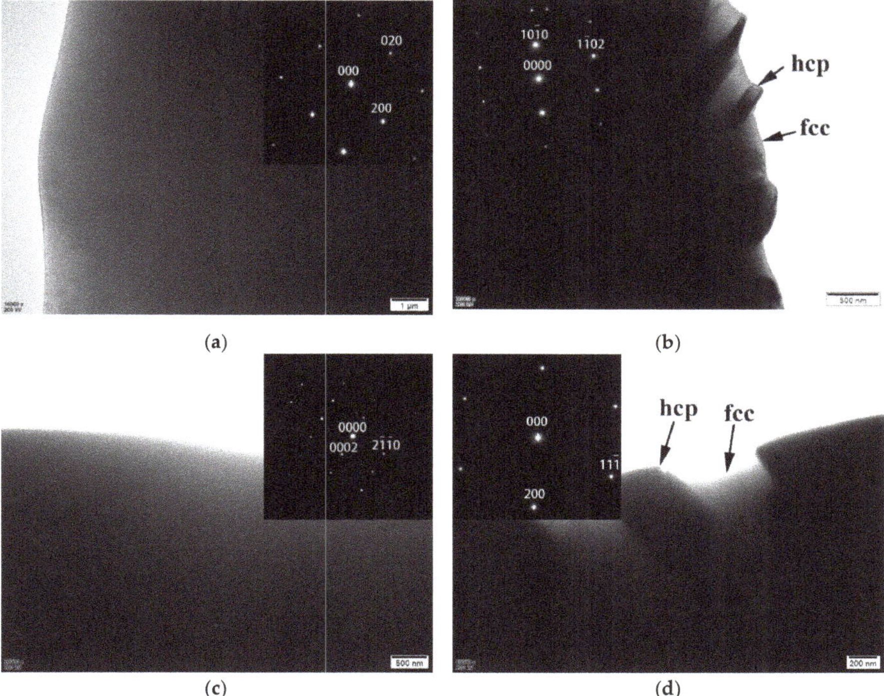

Figure 3. TEM bright field (BF) images and the corresponding selection area diffraction patterns (SAD) of $CrFeCoNiNb_x$ alloys: (**a**) the dendrite of $CrFeCoNiNb_{0.4}$ alloy, and the corresponding SAD taken from [001] of fcc phase; (**b**) the Laves phase in the interdendrite of $CrFeCoNiNb_{0.4}$ alloy, and the corresponding SAD taken from [$\bar{2}4\bar{2}3$] of Laves phase; (**c**) the Laves phase in the dendrite of $CrFeCoNiNb_{0.6}$ alloy, and the corresponding SAD taken from [01$\bar{1}$0] of Laves phase; and (**d**) the fcc phase in the interdendrite of $CrFeCoNiNb_{0.6}$ alloy, and the corresponding SAD taken from [011] of fcc phase.

Table 3 lists the chemical compositions of the $CrFeCoNiNb_x$ alloys which were analyzed by SEM/EDS. The fcc phase had higher Cr-, Fe-, Co-, and Ni-contents, but less Nb-content than the average compositions of the alloys. On the contrary, the Laves phase of $CrFeCoNiNb_x$ alloys had higher Nb-content and less Cr-, Fe-, Co-, and Ni-contents than the average.

Table 3. Chemical compositions of the $CrFeCoNiNb_x$ alloys analyzed by SEM/EDS.

Alloys	Compositions (Atomic Percent)				
	Cr	Fe	Co	Ni	Nb
CrFeCoNi					
Overall	23.1	24.8	26.1	26.0	N/A
Cr-rich precipitate	53.8	16.1	15.9	14.2	N/A
$CrFeCoNiNb_{0.2}$					
Overall	23.4	23.5	23.3	22.3	7.4
fcc	24.2	26.5	24.3	23.2	1.8
hcp	19.4	18.2	23.1	21.2	18.1
$CrFeCoNiNb_{0.4}$					
Overall	22.7	21.6	22.3	21.6	11.8
fcc	25.2	24.3	22.2	24.9	3.5
hcp	21.3	19.3	19.8	19.1	20.5
$CrFeCoNiNb_{0.6}$					
Overall	21.4	20.6	21.5	19.3	17.2
fcc	25.1	24.4	21.8	23.0	5.7
hcp	16.5	18.4	22.3	15.9	26.9
CrFeCoNiNb					
Overall	19.8	19.8	19.0	19.4	22.0
fcc	26.7	24.3	18.4	25.5	5.1
hcp	17.1	18.1	19.7	14.8	30.3
precipitate	16.7	16.7	19.1	15.7	31.8

Figure 4 plots the overall hardness of $CrFeCoNiNb_x$ alloys as a function of x-ratio. The hardness of $CrFeCoNiNb_x$ alloys almost linearly increases with increasing Nb-content. The two phases, fcc and Laves phases, in these $CrFeCoNiNb_x$ alloys were two solid-solution phases, because no superlattice peak was observed from the XRD patterns and no superlattice spot was found from the TEM SADs of the $CrFeCoNiNb_x$ alloys. Therefore, the effect of solid-solution strengthening was caused by the larger-radius element Nb in the fcc and Laves phases.

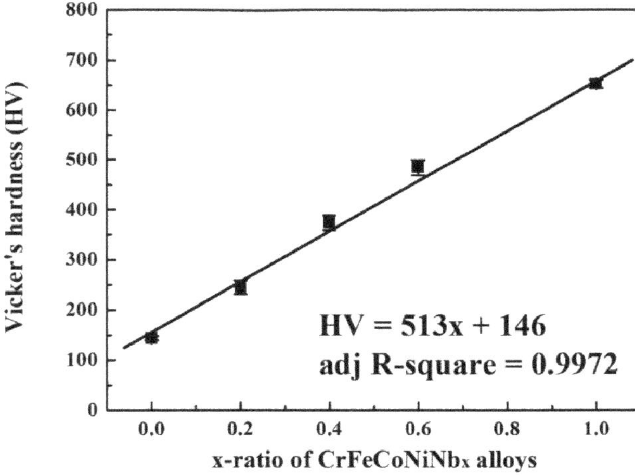

Figure 4. Plot of the overall hardness of $CrFeCoNiNb_x$ alloys as a function of x-ratio.

The polarization curves of $CrFeCoNiNb_x$ alloys and 304SS in 1 M deaerated H_2SO_4 solution are shown in Figure 5, and Table 4 lists the polarization data of the alloys. The values of free corrosion potential (E_{corr}) of these alloys are close. Increasing the Nb-content of $CrFeCoNiNb_x$ alloys resulted in slightly increasing the E_{corr} of the alloys. The free corrosion current density (i_{corr}) varied randomly in these alloys, because the microstructures and the fractions of the phases in the alloys changed significantly. The $CrFeCoNiNb_{0.6}$ alloy had the lowest i_{corr} among these alloys. In addition, increasing the Nb-content of $CrFeCoNiNb_x$ alloys resulted in decreasing the anodic peaks of these alloys. The 304 stainless steel had the largest anodic peak among these alloys, and the CrFeCoNiNb alloy had the smallest anodic peak. The largest current density (i_{crit}) and the potential (E_{pp}) of the anodic peak of each alloy are listed in Table 4. Therefore, the CrFeCoNiNb alloy entered the passivation region more easily than the other alloy in this solution. Moreover, adding Nb could stabilize the passivation region of $CrFeCoNiNb_x$ alloys. The passivation regions of these alloys broke down at potentials of about 1.2 V_{SHE} because of an oxygen evolution reaction [19]. All of the passivation regions of $CrFeCoNiNb_x$ alloys were broader than that of 304SS.

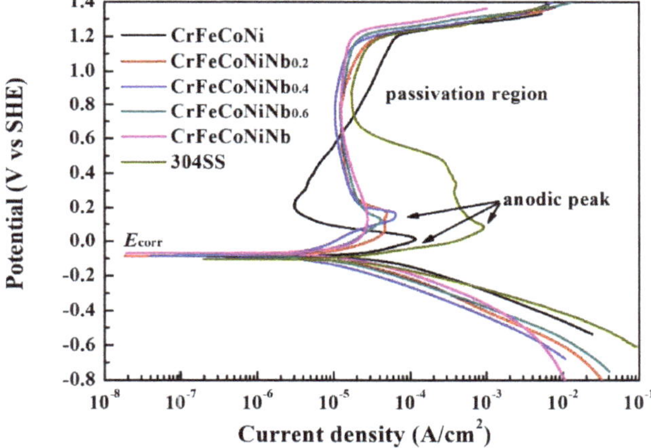

Figure 5. Polarization curves of $CrFeCoNiNb_x$ alloys and 304SS in 1 M deaerated H_2SO_4 solution at 30 °C.

Table 4. Polarization data of the alloys in 1 M deaerated H_2SO_4 solution at 30 °C.

Alloys	E_{corr} (V_{SHE})	i_{corr} (A/cm^2)	E_{pp} (V_{SHE})	i_{crit} (A/cm^2)
CrFeCoNi	−0.086	35.1	0.014	120
$CrFeCoNiNb_{0.2}$	−0.084	15.8	0.082	47.1
$CrFeCoNiNb_{0.4}$	−0.082	52.1	0.183	64.4
$CrFeCoNiNb_{0.6}$	−0.078	10.1	0.122	41.9
CrFeCoNiNb	−0.068	22.3	0.132	28.0
304SS	−0.101	30.0	0.082	930

The morphologies of $CrFeCoNiNb_x$ alloys after polarization tests in 1 M deaerated H_2SO_4 solution at 30 °C are shown in Figure 6. The morphology of CrFeCoNi alloy indicated that it was uniformly and slightly corroded after the polarization test, as shown in Figure 6a. Figure 6b shows the micrograph of $CrFeCoNiNb_{0.2}$ alloy after the test. The fcc-dendrites of $CrFeCoNiNb_{0.2}$ alloy were concave, which meant that the dendrites were more severely corroded than the interdendrites. This implied that the fcc phase was more active than the hcp phase; the fcc phase thus preferred to be corroded more than the Laves phase. Figure 6c clearly indicates this phenomenon: The fcc dendrites of $CrFeCoNiNb_{0.4}$ alloy were more severely corroded and almost vanished. However, the dendrites of $CrFeCoNiNb_{0.6}$ alloy changed to the Laves phase, and the dendrites almost kept their original shapes;

the matrix of interdendrites (fcc phase) was corroded after the polarization test, as shown in Figure 6d. The CrFeCoNiNb alloy also had a similar result, shown in Figure 6e.

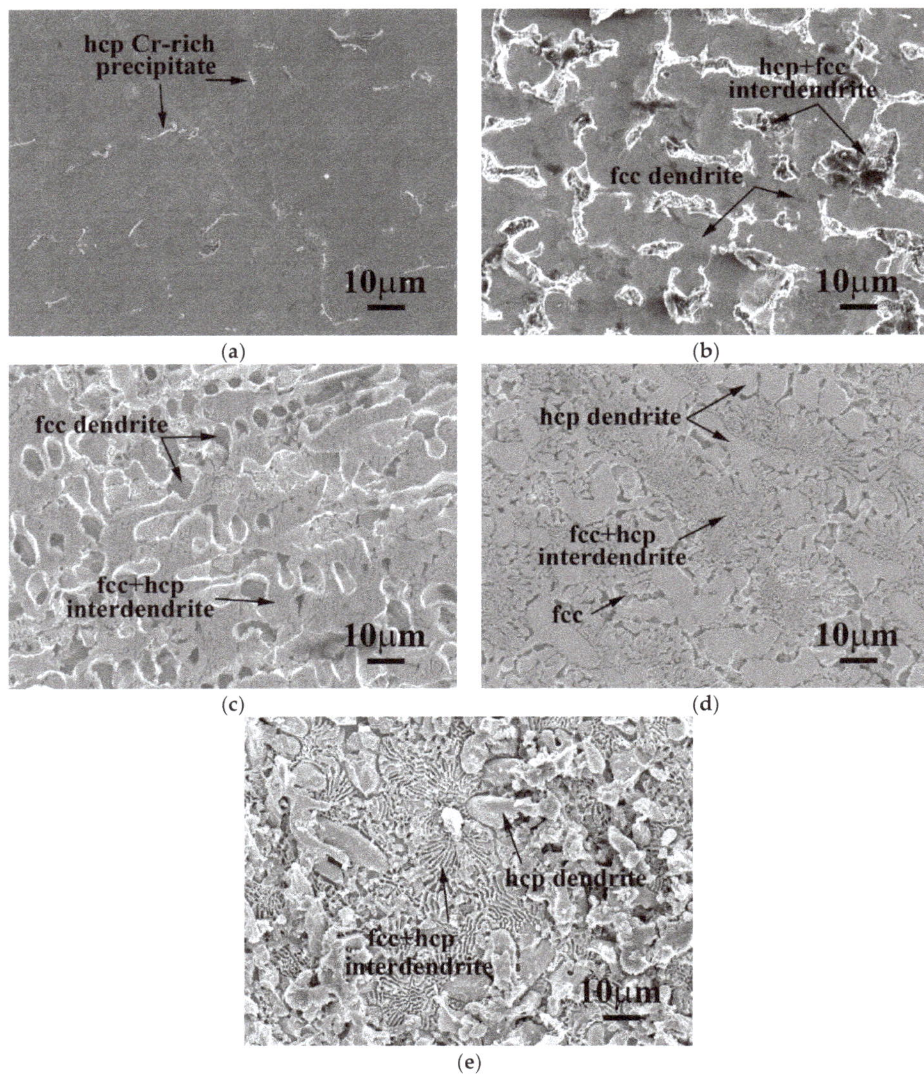

Figure 6. SEM micrographs of the alloys after polarization test in 1 M deaerated H_2SO_4 solution at 30 °C (**a**) CrFeCoNi; (**b**) $CrFeCoNiNb_{0.2}$; (**c**) $CrFeCoNiNb_{0.4}$; (**d**) $CrFeCoNiNb_{0.6}$; and (**e**) CrFeCoNiNb alloys.

The polarization curves of $CrFeCoNiNb_x$ alloys and 304SS in 1 M deaerated NaCl solution are shown in Figure 7, and Table 5 lists the polarization data of the alloys. The values of E_{corr} of $CrFeCoNiNb_x$ alloys were very close; they are nobler than that of 304SS, but more active than that of CrFeCoNi alloy. The values of i_{corr} of $CrFeCoNiNb_x$ alloys were all higher than that of CrFeCoNi alloy. The i_{corr} of $CrFeCoNiNb_{0.6}$ alloy had the highest value. The $CrFeCoNiNb_{0.2}$ and CrFeCoNiNb alloys had smaller i_{corr}, and their values were almost the same as that of 304SS. The primary anodic peak of CrFeCoNi alloy was significant, but increasing the Nb-content of $CrFeCoNiNb_x$ alloys would diminish the anodic peaks of these alloys. Therefore, the CrFeCoNiNb alloy had no anodic peak. Moreover, adding Nb could expand the passivation regions of $CrFeCoNiNb_x$ alloys.

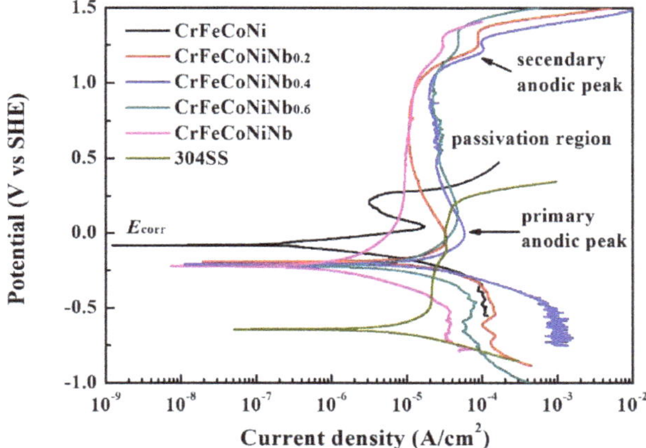

Figure 7. Polarization curves of $CrFeCoNiNb_x$ alloys and 304SS in 1 M deaerated NaCl solution at 30 °C.

Table 5. Polarization data of the alloys in 1 M deaerated NaCl solution at 30 °C.

Alloys	E_{corr} (V_{SHE})	i_{corr} ($\mu A/cm^2$)	E_{pp} (V_{SHE})	i_{crit} ($\mu A/cm^2$)
CrFeCoNi	−0.081	0.28	0.049	17.3
$CrFeCoNiNb_{0.2}$	−0.192	12.3	−0.086	33.2
$CrFeCoNiNb_{0.4}$	−0.207	21.1	−0.006	58.1
$CrFeCoNiNb_{0.6}$	−0.212	72.5	0.152	46.7
CrFeCoNiNb	−0.221	12.0	N/A	N/A
304SS	−0.638	12.9	N/A	N/A

The morphologies of $CrFeCoNiNb_x$ alloys after polarization tests in 1 M deaerated NaCl solution at 30 °C are shown in Figure 8. The micrograph of CrFeCoNi alloy after the polarization test indicated that it was only slightly corroded, as shown in Figure 8a. However, the fcc dendrites of $CrFeCoNiNb_{0.2}$ and $CrFeCoNiNb_{0.4}$ alloys were severely corroded after the tests, as shown in Figure 8b,c. The dendrites (Laves phase) of $CrFeCoNiNb_{0.6}$ and CrFeCoNiNb alloys almost kept their original shapes, but the matrices of the interdendrites (fcc phase) were corroded after the tests, shown in Figure 8d,e. Therefore, the fcc phase was more active than the Laves phase in the $CrFeCoNiNb_x$ alloys, and the fcc phase thus preferred to corrode more than Laves phase in 1 M deaerated H_2SO_4 and 1 M deaerated NaCl solutions.

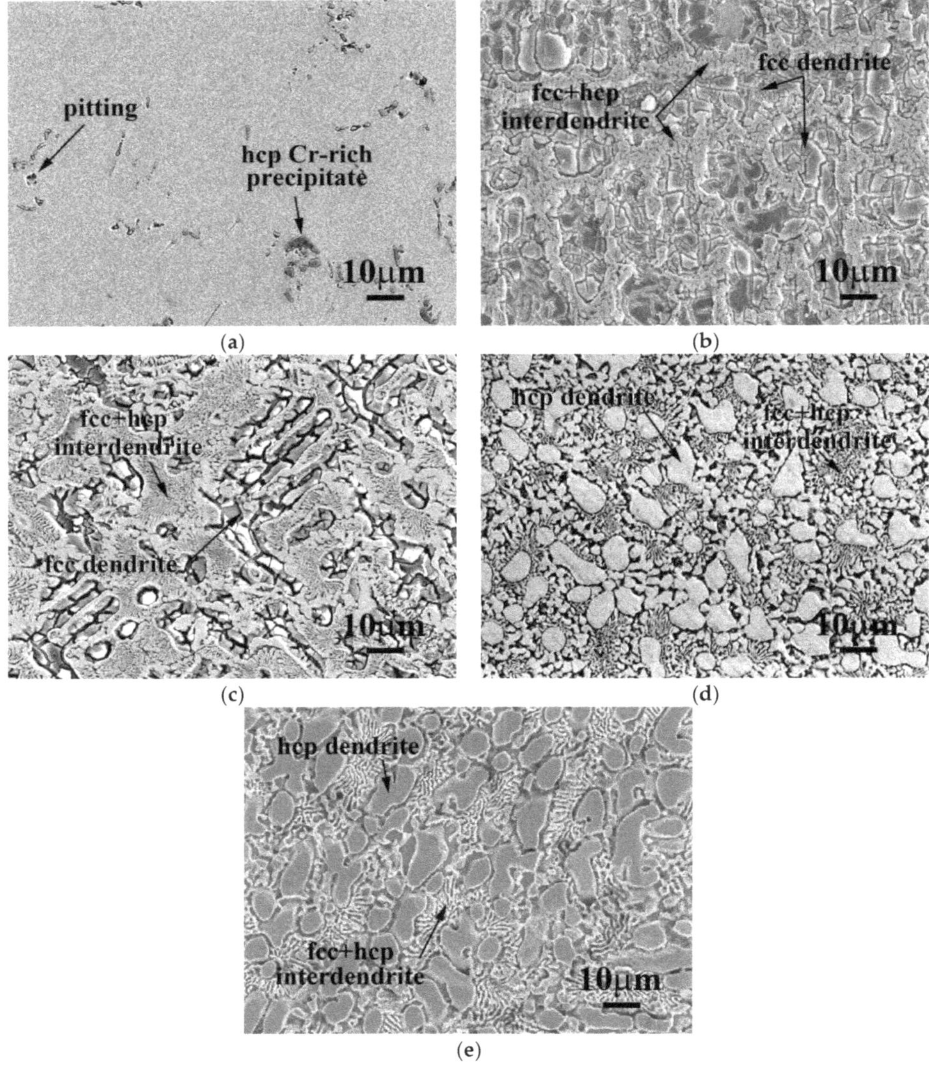

Figure 8. SEM micrographs of the alloys after polarization tests in 1 M deaerated NaCl solution at 30 °C (**a**) CrFeCoNi; (**b**) $CrFeCoNiNb_{0.2}$; (**c**) $CrFeCoNiNb_{0.4}$; (**d**) $CrFeCoNiNb_{0.6}$; and (**e**) CrFeCoNiNb alloys.

4. Conclusions

1. The CrFeCoNi alloy had an fcc granular structure. The microstructures became dual-phased dendritic microstructures after adding niobium. The $CrFeCoNiNb_{0.2}$ and $CrFeCoNiNb_{0.4}$ alloys were hypereutectic alloys; their dendrites were fcc phases, and the interdendrites were eutectic structures of fcc and Laves phases (hcp). The $CrFeCoNiNb_{0.6}$ and CrFeCoNiNb alloys were hypoeutectic alloys; their dendrites were a Laves phase, and the interdendrites were still eutectic structures of fcc and Laves phases.
2. The lattice constants of fcc and Laves phases in these $CrFeCoNiNb_x$ alloys increased with increasing Nb-content due to solid-solution strengthening. Increasing Nb-content also resulted

in increasing the hardness of $CrFeCoNiNb_x$ alloys. The hardness increased from HV144 of CrFeCoNi alloy to HV652 of CrFeCoNiNb alloy.

3. The corrosion resistance of $CrFeCoNiNb_x$ alloys slightly decreased after adding niobium because of their dual-phased dendritic microstructures. In addition, adding niobium into $CrFeCoNiNb_x$ alloys could stabilize and expand the passivation regions of these alloys in these two solutions. The fcc phase of each $CrFeCoNiNb_x$ alloy was more severely corroded than the Laves phase after polarization tests in 1 M deaerated H_2SO_4 and 1 M deaerated NaCl solutions.
4. The $CrFeCoNiNb_x$ alloys possessed good corrosion resistance, and their hardness changed with Nb-content. This work thus provides a method to design a $CrFeCoNiNb_x$ alloy with suitable hardness and good corrosion resistance for different commercial applications.

Author Contributions: C.-H.T. conceived and designed the experiments; C.-Y.Y. and M.-C.T. performed the experiments; all of the authors analyzed the data; C.-Y.Y. and M.-C.T. contributed reagents/materials/analysis tools; C.-H.T. wrote the paper. All of the authors have read and approved the final manuscript.

Funding: We are grateful to the Ministry of Science and Technology of Republic of China for its financial support under the project MOST 106-2221-E-034-008.

Conflicts of Interest: The authors declare no conflict of interest. The funding sponsors had no role in the design of the study; in the collection, analyses, or interpretation of data; in the writing of the manuscript, and in the decision to publish the results.

References

1. Yeh, J.W.; Chen, S.K.; Lin, S.J.; Gan, J.Y.; Chin, T.S.; Shun, T.T.; Tsau, C.H.; Chang, S.Y. Nanostructured high-entropy alloys with multiple principal elements: Novel alloy design concepts and outcomes. *Adv. Eng. Mater.* **2004**, *6*, 299–303. [CrossRef]
2. Murty, B.S.; Yeh, J.W.; Ranganathan, S. *High-Entropy Alloys*; Butterworth-Heinemann Co.: London, UK, 2014, pp. 13–36.
3. George, E.P.; Raabe, D.; Ritchie, R.O. High-entropy alloys. *Nat. Rev. Mater.* **2019**, *4*, 515–534. [CrossRef]
4. Gao, M.C.; Yeh, J.W.; Liaw, P.K.; Zhang, Y. *High-Entropy Alloys, Fundamentals and Applications*; Springer Berlin, Germany, 2016; pp. 1–19.
5. Tsai, M.H.; Yeh, J.W. High-Entropy Alloys: A Critical Review. *Mater. Res. Lett.* **2014**, *2*, 107–123. [CrossRef]
6. Chen, S.T.; Tang, W.Y.; Kuo, Y.F.; Chen, S.Y.; Tsau, C.H.; Shun, T.T.; Yeh, J.W. Microstructure and properties of age-hardenable $Al_xCrFe_{1.5}MnNi_{0.5}$ alloys. *Mater. Sci. Eng. A* **2010**, *547*, 5818–5825. [CrossRef]
7. Lee, C.P.; Chen, Y.Y.; Hsu, C.Y.; Yeh, J.W.; Shih, H.C. Enhancing pitting corrosion resistance of $Al_xCrFe_{1.5}MnNi_{0.5}$ high-entropy alloys by anodic treatment in sulfuric acid. *Thin Solid Films* **2008**, *517*, 1301–1305. [CrossRef]
8. Mariano, N.A.; Souza, C.A.C.; May, J.E.; Kuri, S.E. Influence of Nb content on the corrosion resistance and saturation magnetic density of FeCuNbSiB alloys. *Mater. Sci. Eng. A* **2003**, *354*, 1–5. [CrossRef]
9. Hsu, Y.J.; Chiang, W.C.; Wu, J.K. Corrosion behavior of $FeCoNiCrCu_x$ high-entropy alloys in 3.5% sodium chloride solution. *Mater. Chem. Phys.* **2005**, *92*, 112–117. [CrossRef]
10. Tsau, C.H.; Lee, P.Y. Microstructures of $Al_{7.5}Cr_{22.5}Fe_{35}Mn_{20}Ni_{15}$ High-Entropy Alloy and Its Polarization Behaviors in Sulfuric Acid, Nitric Acid and Hydrochloric Acid Solutions. *Entropy* **2016**, *18*, 288. [CrossRef]
11. Lin, C.M.; Tsai, H.L. Evolution of microstructure, hardness, and corrosion properties of high-entropy $Al_{0.5}CoCrFeNi$ alloy. *Intermetallics* **2011**, *19*, 288–294. [CrossRef]
12. Tsau, C.H.; Lin, S.X.; Fang, C.H. Microstructures and corrosion behaviors of FeCoNi and CrFeCoNi equimolar alloys. *Mater. Chem. Phys.* **2017**, *186*, 534–540. [CrossRef]
13. Tsau, C.H.; Tsai, M.C. The effects of Mo and Nb on the microstructures and properties of CrFeCoNi(Nb,Mo) alloys. *Entropy* **2018**, *20*, 648. [CrossRef]
14. Chanda, B.; Das, J. Composition Dependence on the Evolution of Nanoeutectic in CoCrFeNiNbx ($0.45 \leq x \leq 0.65$) High Entropy Alloys. *Adv. Eng. Mater.* **2017**, *1700908*, 1–9.
15. Jiang, H.; Han, K.; Li, D.; Cao, Z. Microstructure and Mechanical Properties Investigation of the $CoCrFeNiNb_x$ High Entropy Alloy Coatings. *Crystals* **2018**, *8*, 409. [CrossRef]

16. Zhang, M.; Zhang, L.; Liaw, P.K.; Li, G. Effect of Nb content on thermal stability, mechanical and corrosion behaviors of hypoeutectic $CoCrFeNiNb_X$ high-entropy alloys. *J. Mater. Res.* **2018**, *33*, 3276–3286. [CrossRef]
17. Revie, R.W.; Uhlig, H.H. *Corrosion and Corrosion Control*, 4th ed.; John Wiley & Sons Co.: Hoboken, NJ, USA, 2008; pp. 31–63.
18. Smith, W.F. *Foundations of Materials Science and Engineering*, 3rd ed.; McGraw-Hill Co.: New York, NY, USA, 2004; pp. 877–878.
19. Callister, W.D., Jr. *Fundamentals of Materials Science and Engineering: An Integrated Approach*, 2nd ed.; John Wiley & Sons, Inc.: New York, NY, USA, 2005; p. 674.

© 2019 by the authors. Licensee MDPI, Basel, Switzerland. This article is an open access article distributed under the terms and conditions of the Creative Commons Attribution (CC BY) license (http://creativecommons.org/licenses/by/4.0/).

Article

Corrosion Initiation and Propagation on Carburized Martensitic Stainless Steel Surfaces Studied via Advanced Scanning Probe Microscopy

Armen Kvryan [1], Corey M. Efaw [1], Kari A. Higginbotham [1], Olivia O. Maryon [1], Paul H. Davis [1], Elton Graugnard [1], Hitesh K. Trivedi [2] and Michael F. Hurley [1,*]

[1] Micron School of Materials Science & Engineering, Boise State University, Boise, ID 83725-2090, USA; armenkvryan@u.boisestate.edu (A.K.); coreyefaw@u.boisestate.edu (C.M.E.); karilivingston@u.boisestate.edu (K.A.H.); oliviamaryon@u.boisestate.edu (O.O.M.) pauldavis2@boisestate.edu (P.H.D.); eltongraugnard@boisestate.edu (E.G.)

[2] UES, Inc., 4401 Dayton Xenia Rd, Dayton, OH 45432, USA; hitesh.trivedi.ctr@us.af.mil

* Correspondence: mikehurley@boisestate.edu

Received: 11 February 2019; Accepted: 13 March 2019; Published: 21 March 2019

Abstract: Historically, high carbon steels have been used in mechanical applications because their high surface hardness contributes to excellent wear performance. However, in aggressive environments, current bearing steels exhibit insufficient corrosion resistance. Martensitic stainless steels are attractive for bearing applications due to their high corrosion resistance and ability to be surface hardened via carburizing heat treatments. Here three different carburizing heat treatments were applied to UNS S42670: a high-temperature temper (HTT), a low-temperature temper (LTT), and carbo-nitriding (CN). Magnetic force microscopy showed differences in magnetic domains between the matrix and carbides, while scanning Kelvin probe force microscopy (SKPFM) revealed a 90–200 mV Volta potential difference between the two phases. Corrosion progression was monitored on the nanoscale via SKPFM and in situ atomic force microscopy (AFM), revealing different corrosion modes among heat treatments that predicted bulk corrosion behavior in electrochemical testing. HTT outperforms LTT and CN in wear testing and thus is recommended for non-corrosive aerospace applications, whereas CN is recommended for corrosion-prone applications as it exhibits exceptional corrosion resistance. The results reported here support the use of scanning probe microscopy for predicting bulk corrosion behavior by measuring nanoscale surface differences in properties between carbides and the surrounding matrix.

Keywords: corrosion; bearing steels; martensitic stainless steel; aerospace; atomic force microscopy (AFM); scanning Kelvin probe microscopy (SKPFM); nanoscale; electrochemistry; wear; Pyrowear 675/AMS 5930

1. Introduction

The performance of advanced gas turbine engines is currently limited by degradation of the mechanical components, in particular, rolling bearing elements, such as the raceway [1]. This is because aerospace engine bearings are subject to extreme operating conditions, including elevated temperatures, high speeds, vibratory stresses, rolling contact fatigue, and complex lubricant and environment interactions [2]. Accordingly, both high hardness and high toughness are critical requirements for aerospace bearing materials, yet achieving both in a single material is challenging. M50, a through-hardened carbon steel, was developed for aircraft engine bearing applications and has become the standard bearing steel used in the United States due to its ability to perform well at high temperatures while maintaining relatively high fracture toughness compared to earlier generation carbon steels, such as AISI 52100 (UNS G52986) [1,3,4]. In the case of sea-based or coastal aircraft operations however, open turbine engine systems can limit the ability of ester-based lubricants to provide wear and corrosion protection, as the surrounding environment introduces water and marine aerosols into the engine during both storage and operation [5]. The presence of water in the lubricant can then serve to initiate aqueous corrosion during engine cycling and downtime [5]. Consequently, current aero-engine performance is limited by corrosion-enhanced wear of the metallic bearings and drive components, which leads to increased maintenance and premature failure [1,6–8]. Thus, there has been significant research effort to develop alternative bearing steels to M50 that exhibit enhanced corrosion resistance to support increased engine performance [3,4,7–10].

Martensitic stainless steels (MSSs) were developed for use in applications where high wear resistance and toughness is required whilst maintaining high corrosion resistance. These properties, combined with their potential for high hardness upon heat treatment [1,11–15], have led to MSSs being implemented in many demanding applications, including bearings, molds, nuclear reactors, hydroelectric engines, and petrochemical steam and gas turbines and buckets [1,11–20]. To improve surface wear resistance while maintaining the corrosion resistance of the core, MSSs can instead be surface treated (carburized), with carbon incorporated into the sample surface at elevated temperatures to form hard carbides with alloying elements such as chromium or vanadium [1,21–23].

Highly corrosion-resistant MSSs (e.g., Cronidur 30 or XD15NW) include additions of alloying elements (and/or nitrogen) and can have poor adhesive and wear performance [24]. While not as corrosion resistant, UNS S42670 or AMS 59030B (referred to herein as P675) are relatively cost-efficient MSSs with high corrosion resistance (equivalent to 440C steel) and bulk fracture toughness (higher than M50) [25]. P675 was specifically engineered for aerospace bearing applications in advanced gas-turbine engines, where conventional bearing steels (e.g., M50 and 440C) are adversely affected by corrosion in aggressive environments and/or do not have sufficient high temperature wear performance [8]. Although P675 shows improvement in corrosion resistance relative to conventional bearing steels, higher surface hardness would lead to a longer wear lifetime in-service. Accordingly, secondary surface processing has been targeted as a way to increase the hardness and wear resistance of P675 [7,9,10,26]. Such surface treatments impart a graded microstructure that extends ~1000 μm below the metal surface. Optimized wear properties are obtained by balancing the surface hardness and core ductility of composite microstructures across the gradient region. However, the increased surface hardness typically comes at the expense of corrosion resistance, as the formation of carbides on the surface locally depletes corrosion-resistant elements (e.g., chromium, vanadium, molybdenum) from the surrounding matrix [7,20,22,23,27,28].

The corrosion performance of various P675 surface treatments has been previously assessed through accelerated DC and AC electrochemical testing in aqueous solutions [7,9,10]. These investigations provided a ranking of corrosion performance, showing that the final tempering temperature and processing atmosphere had a considerable influence on both the overall corrosion rate and damage morphology. Compared to M50, surface hardened P675 can be significantly more corrosion-resistant, and higher processing temperatures typically increased susceptibility to general corrosion damage, while lower temperatures exhibited more localized corrosion relative to untreated P675 [7]. The influence of processing on P675 wear performance for the same steels in non-corrosive wear testing has also been reported, where higher processing temperatures (HTT) yielded longer bearing lifetimes compared to low-temperature temper (LTT) [29,30]. However, there remains a need for research into the interdependency between simultaneously balancing corrosion resistance and surface hardness for bearing applications, since wear resistance (i.e., bearing performance) in corrosive environments is ultimately limited by corrosion [11].

Investigation of surface electronic properties can provide information to aid in the prediction of corrosion initiation sites [31]. Recently, scanning Kelvin probe force microscopy (SKPFM) has been used to investigate the role of nano- and micro-scale surface features on corrosion behavior [19,32–42]. Additionally, magnetic force microscopy (MFM) [43–45] has been used to similarly provide insight into the magnetic behavior of alloy surfaces. SKPFM permits measurement with nanoscale resolution of Volta potential differences (VPDs), which are related to the electronic work function (EWF), while MFM provides information regarding the magnitude and orientation of the magnetic moments of surface domains. Likewise, in situ atomic force microscopy (AFM) has been used to monitor morphological changes during corrosion in electrolyte solution and link them to the electrochemical behavior of the material [19,46–49]. The current work presents the first application of such techniques to investigate corrosion behavior of MSS P675 with various surface treatments. Since corrosion is the most common precursor to wear damage during aero-engine operation [8], the time to onset and rate of corrosion can directly control maintenance requirements and operational costs. Initiation and propagation are critical considerations because they determine both wear behavior as well as the lifetime of the part or engine [8,50,51]. The focus of this study is to understand the effects of heat treatment processing parameters on corrosion evolution in P675 by utilizing a combination of scanning probe microscopy (SPM) techniques and accelerated corrosion testing, thereby linking surface microstructural differences (on the nanoscale) with observed macroscale surface corrosion behavior and wear performance.

2. Materials and Methods

2.1. Materials

The nominal bulk composition of P675 (UNS S42670, the MSS studied here) prior to heat treatment is shown in Table 1 [29]. To increase surface hardness, P675 samples were carburized, followed by quenching and tempering, to harden the outer layer or case. Samples were cylindrical (9.5 mm diameter × 12 mm height) with post-treatment case depths of 750–1250 μm radially inward [9]. Samples differed in the final tempering temperature and carburization atmosphere: high-temperature tempering (HTT) at 496 °C, low-temperature tempering (LTT) at 315 °C, and carbo-nitrided (CN) where the case was obtained through a carburizing cycle followed by nitriding cycle during heat treating. Further details on the processing routes are discussed in previous works [9,10,29,30]. Prior to SPM characterization, samples were mechanically ground with SiC paper (to 2000 grit) in deionized (DI) water, followed by sequential polishing to 0.02 μm with a colloidal silica aqueous slurry. After polishing, samples were rinsed with ethanol and sonicated for 1 min in ethanol to remove any polishing residue.

Table 1. Nominal composition (wt.%) of P675 alloy (remainder is Fe). Adapted from Trivedi, et al. [29].

Steel	C	Mn	Cr	Mo	Si	Ni	S	V	Co
P675 (AMS 5930B)	0.07	0.75	13	2	0.4	2.5	0.010	0.6	6.5

2.2. Electron Microscopy

A field emission scanning electron microscope (SEM, FEI Teneo, Hillsboro, USA) coupled to an energy-dispersive X-ray spectrometer (EDS, 80 mm^2 Energy+, Oxford Instruments, Abingdon, UK) was utilized to characterize the surface microstructure and corrosion morphology of all samples, as well as construct elemental composition maps of the heat-treated surfaces. SEM analyses were conducted in both secondary electron (SE) and backscattered electron (BSE) imaging modes using 10–20 keV accelerating voltages.

2.3. Scanning Probe Microscopy

2.3.1. Ex situ Scanning Probe Microscopy (SPM)

Ex situ AFM, MFM, and SKPFM were performed under an inert argon atmosphere containing <0.1 ppm H_2O and O_2 using a Bruker Dimension Icon AFM housed in an MBraun glovebox (MBraun, Stratham, USA). Prior to imaging, previously polished and sonicated samples were cleaned with HPLC/spectrophotometric grade ethanol (Sigma-Aldrich, 200 proof, St. Louis, USA) using lint-free wipes (Kimtech). Following ethanol cleaning, compressed ultra-high purity nitrogen gas (Norco UHP, 99.999%) was used to dry the surface of the steel and remove any remaining surface particulates before introducing the samples into the glovebox antechamber.

Both MFM and SKPFM were performed using a dual-pass lift mode implementation in which the first pass over each scan line acquires surface topography. Upon completing the first pass, the probe then lifts off the surface to a user-defined height above the surface. This height (i.e., tip-sample separation, 100 nm in this study) remains constant throughout the second pass as the electromagnetic property of interest (i.e., Volta potential difference in the case of SKPFM or magnetic moment in the case of MFM) is measured. Surface topography was mapped using either intermittent contact (tapping) mode in the case of MFM imaging or PeakForce tapping mode (Bruker Nano, Santa Barbara, USA), which employs rapid force curve acquisition with a user-defined force setpoint (typically 2 nN here), in the case of AFM and SKPFM. In MFM, the magnetic force gradient between a magnetized Co-Cr coated AFM probe (Bruker MESP, $k = 2.8$ N/m, $f_0 = 75$ kHz, $\mu = 1 \times 10^{-13}$ EMU, where 1 EMU = 1 erg G^{-1}) and the surface of the material was observed during the lift mode pass. For consistency, all MFM imaging reported herein was performed with the same MESP probe, which was magnetized immediately prior to imaging with its magnetic axis perpendicular to the sample surface. In SKPFM, the Volta potential difference (VPD) between a conductive probe (Bruker PFQNE-AL, $k = 0.8$ N/m, $f_0 = 300$ kHz) and the surface was quantified by application of a DC bias to null the tip-sample electric force gradient arising from the difference in Volta potential between the probe and sample surface. VPD maps were acquired utilizing frequency modulation SKPFM [31], as described in detail elsewhere [37,38]. These VPD maps were used to predict the corrosion behavior of the samples by suggesting the cathodic and anodic sites and the relative driving force for galvanic corrosion.

SKPFM was also used to observe corrosion initiation and propagation mechanisms by carrying out intermittent imaging at well-defined intervals throughout the corrosion process. While all such imaging was carried out within the controlled environment (<0.1 ppm H_2O and O_2) of the argon-filled glovebox, corrosion was initiated and allowed to proceed outside the glovebox, where samples were sequentially soaked for prescribed amounts of time in a 1 M NaCl solution prepared from reagent grade NaCl (Sigma Aldrich, St. Louis, USA) and deionized (DI) water. After each time increment, samples were rinsed with DI water to remove any adhered salt, dried with UHP nitrogen, and cleaned with ultrapure ethanol using lint-free wipes. The samples were then reintroduced into the glovebox and imaged via dual-pass SKPFM. Repeated nanoscale imaging at specific recurrent locations with micron-scale positional accuracy was made possible by fiduciary marks created with a diamond tip indenter. Testing and imaging were performed ~500 µm away from the fiduciary mark to ensure results obtained were not influenced by the indent.

2.3.2. In Situ SPM

To capture images of corrosion initiation and propagation in real time, in situ PeakForce tapping (topographical) AFM was also performed. In contrast to the ex situ (i.e., glovebox) SPM imaging, samples for in situ AFM imaging were mounted in a fluid cell and immersed in a 0.1 M NaCl solution under ambient atmosphere. The NaCl concentration was chosen such that it would initiate corrosion on samples at an appropriate timescale to reveal changes in topography concurrent with corrosion propagation and progression. Silicon nitride probes with a nominal tip radius of 20 nm (Bruker ScanAsyst-Fluid, $k = 0.7$ N/m, $f_0 = 150$ kHz) were used for repetitive imaging (0.5 Hz scan rate) of 10×10 µm^2 areas at 512×512 pixel resolution, corresponding to a refresh rate of ~8.5 min to capture each image. Due to differences in time between initial immersion of each sample and the initial image capturing (driven by optimization of imaging parameters), the specific timing of subsequent images is not exact between samples. The total amount of time each sample had been exposed to the corrosive salt solution was documented at both the start and end of captured images.

2.3.3. Image Processing

SPM image processing and quantitative analysis were conducted using NanoScope Analysis 1.90 (Bruker). All topographical images were processed with a first order flatten filter to remove sample tip and tilt as well as any individual line-to-line offsets. The images for HTT at 116 and 135 min required a 2nd order flatten procedure to account for the deposited debris. To quantify the findings from SKPFM mapping, a threshold technique was implemented (see example image in Figure 1 below) that utilized a user-determined cut-off potential based on the distribution of Volta potentials observed in the corresponding data histogram (512 bins). From the resulting thresholded data, the average Volta potential (with corresponding standard deviation) was calculated for each of the two phases present on the surface (i.e., matrix and carbides, identities confirmed through SEM/EDS characterization) [52]. Figure 1a shows a representative SKPFM Volta potential map for HTT P675. Figure 1b shows the matrix in dark brown with the carbides (data in blue) excluded, while the light brown areas visible in Figure 1c correspond to the carbides (with the matrix excluded and indicated by the dark blue areas). Using this method, an average VPD between the matrix and carbides was calculated for each SKPFM image.

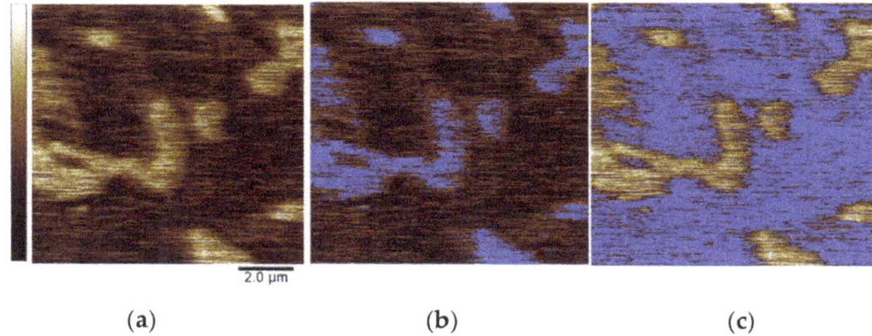

(a) (b) (c)

Figure 1. Representative 10 × 10 µm^2 scanning Kelvin probe microscopy (SKPFM) images of P675-high-temperature temper (HTT). Dark brown corresponds to the softer matrix phase, which is lower in height following polishing than the harder, lighter brown carbides. Images show (**a**) the original Volta potential image (600 mV full-scale range) and subsequent implementation of thresholding cutoffs (blue) to calculate average Volta potential differences (VPDs) for the (**b**) matrix and (**c**) carbides.

2.4. Electrochemical Corrosion Testing

Electrochemical cyclic polarization testing was used to characterize corrosion behavior for each type of heat-treated steel. Sample preparation details can be found in a previous publication, thus the sample testing area was defined by masking off the sample such that only a circular area (diameter ~6.6 mm) test area was in contact with the electrolyte solution [7]. Testing was conducted in 0.01 M NaCl electrolyte solution with a potentiostat (SP-300, Bio-Logic, Seyssinet-Pariset, France) used to control and monitor a three-electrode system in a modified flat cell. A saturated calomel electrode (SCE) served as the reference electrode and a platinum mesh as the counter electrode. Following sample immersion, open circuit potential (OCP) was monitored for 30 min. The sample was then polarized at a scan rate of 0.5 mV/s from 100 mV below OCP to 600 mV above OCP or when pitting had stabilized, followed by a reverse scan back to OCP.

3. Results

3.1. Surface Composition

The carburizing and carbo-nitriding heat treatment processes performed on MSS P675 resulted in the development of well distributed metal-carbon precipitates (carbides) ranging in size from approximately 10 nm to 2 µm in diameter (Figure 2a), surrounded by the martensitic matrix at the sample surface. In addition to the surface, the carbides are present diminishingly, approximately 1000 μm radially inward into each of the samples (data not shown). Sample surfaces were analyzed via EDS (Figure 2b) to resolve carbide chemistry and determine alloying elements that segregated from the matrix to form these carbides during heat treatment. Carbides resulting from all three surface treatments were found to be predominantly carbon- and chromium-rich with lesser amounts of vanadium, molybdenum and/or manganese, while the surrounding matrix showed primarily iron, cobalt, and nickel. In previous work done on P675, X-ray diffraction (XRD) and electron beam backscattered diffraction (EBSD) determined M_7C_3 (orthorhombic) and $M_{23}C_6$ (face-centered cubic) to be the primary carbides formed in P675 (M represents the metal in the carbide), with $M_{23}C_6$ precipitating after M_7C_3, and chromium being the primary metal constituent present in the carbides [27,53]. HTT contains a greater population of $M_{23}C_6$ carbides than LTT and CN due to its higher tempering temperature (i.e., increased kinetics). By stoichiometry, the HTT carbides contain more chromium than the carbides of the other two surface treated steels despite all having the same bulk composition before heat treating. The large amount of chromium present in the bulk (pre-heat treatment) P675 alloy (Table 1), coupled with the presence of molybdenum, should yield a magnetic MSS [54–56]. However, EDS analysis (Figure 2a,b), performed on the bulk surface of each steel, showed that the chromium and molybdenum primarily segregated within the carbides following heat treatment (thereby increasing the likelihood of magnetic carbides). EDS was performed on the bulk steel and not on the individual carbides due to inconsistent results obtained since large interaction volumes (by the EDS) penetrated both the carbide and surrounding matrix. In contrast, nickel, in the presence of iron and carbon acts as an austenite stabilizer and thus promotes a non-magnetic austenitic (fcc) structure [57]. MFM was therefore utilized to observe how the secondary processing performed on these steels affected the magnetic properties of the surface.

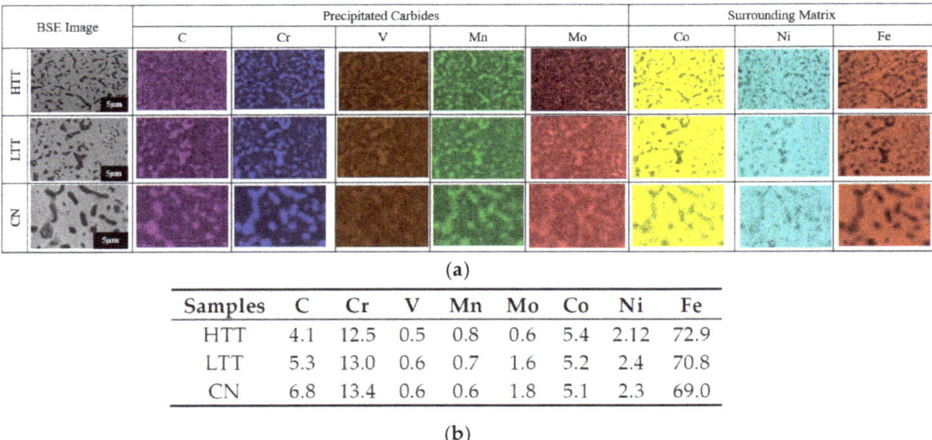

(a)

Samples	C	Cr	V	Mn	Mo	Co	Ni	Fe
HTT	4.1	12.5	0.5	0.8	0.6	5.4	2.12	72.9
LTT	5.3	13.0	0.6	0.7	1.6	5.2	2.4	70.8
CN	6.8	13.4	0.6	0.6	1.8	5.1	2.3	69.0

(b)

Figure 2. (**a**) Grayscale backscattered electron (BSE) images (left column) of the three different P675 surface-treated samples (carbides appear darker than surrounding matrix) with corresponding colored energy-dispersive X-ray spectrometer (EDS) compositional maps highlighting the principal components of the carbides (middle columns) and bulk matrix (right columns) for the HTT, low-temperature tempered (LTT), and carbo-nitrided (CN) samples (images for each row share the same micron bar). (**b**) Elemental composition in wt.% (determined via EDS) for the surface of each steel (not individual carbides).

3.2. Scanning Probe Microscopy

3.2.1. Magnetic Force Microscopy (MFM)

MFM was utilized to map variations in the magnetic moment projections (surface normal direction) on the surface of the steels (Figure 3). In Figure 3, purple regions are identified as carbides since these coincide with regions that are raised in topography and visually similar to carbides seen in SEM/EDS analysis (see Figure 2). Topographical relief of the carbides was expected due to differential polishing rates during sample prep, resulting in the harder carbides slightly protruding above the surrounding matrix. MFM results indicated that the carbides and the matrix both exhibit out-of-plane magnetic domains (i.e., positive magnetic direction, non-parallel to surface), but with varying magnitude, carbides being noticeably larger than the matrix as expected from the enhanced chromium concentration (see Figure 2). Within the matrix, nanoscale variations in magnetic domain were also evident. In HTT these were larger and more elongated those on either HTT or CN. CN had the finest distribution of different magnetic domain regions able to be resolved within the matrix.

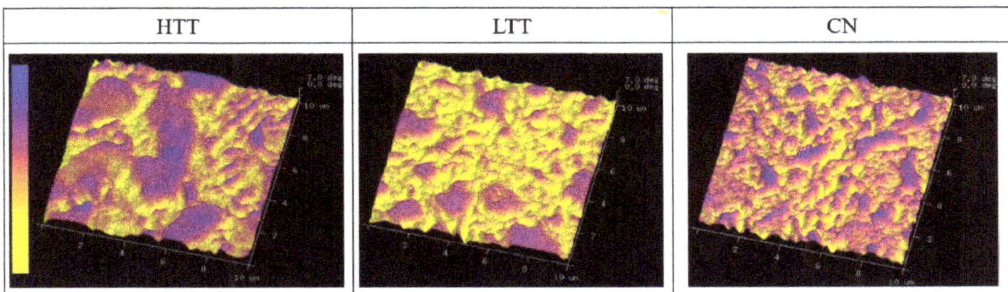

Figure 3. 3D magnetic response maps with changes in height representative of differences in magnetism. Color scale ranges are 7 degrees (0° = yellow, +7° = blue) for magnetic response.

3.2.2. Inert Environment Scanning Kelvin Probe Microscopy (SKPFM)

Freshly polished, cleaned, and dried samples underwent ex situ SPM imaging in an inert atmosphere glovebox. Images were acquired using sequentially larger scan areas of 10×10 μm^2, 20×20 μm^2, and 90×90 μm^2, with contrast between carbides and the surrounding matrix observed in both Volta potential and topography (Figure 4). Numerical VPD results were calculated per the method described earlier and compiled for comparison (see Figure 5, error bars are indicative of one standard deviation). The measured VPD of the carbides ranged from 60 to 200 mV greater than the steel matrix, depending on the scan size analyzed, with HTT possessing the highest difference and CN the lowest. The relative magnitudes of the carbide-matrix VPDs remained consistent regardless of scan size, suggesting even the smallest imaging areas chosen (10×10 μm^2) were large enough to be representative of the sample while also providing the highest spatial resolution of VPD variations.

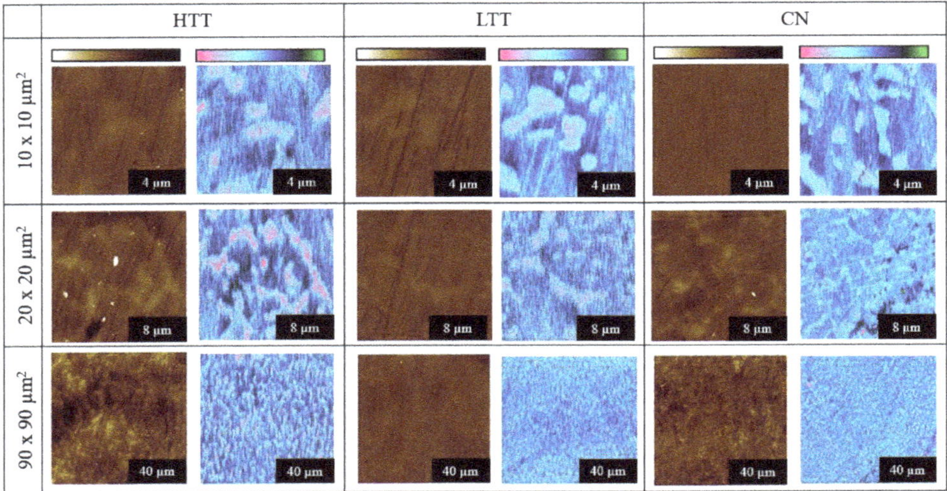

Figure 4. High-resolution atomic force microscopy (AFM) topography (dark brown to white color scale, 100 nm full scale) and SKPFM Volta potential (green to pink color scale, 600 mV full scale) images over different size scan areas showing the different sizes and shapes of carbides distributed throughout the three sample types.

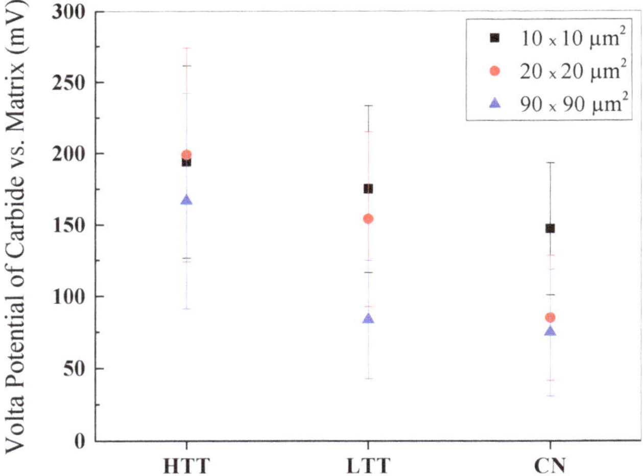

Figure 5. Plot of measured VPDs (with standard deviation error bars) of carbide precipitates versus the surrounding matrix for the three P675 surface-treated steels as a function of scan area.

3.2.3. Intermittent SKPFM

Intermittent ex situ SKPFM was performed to track the evolution of the surfaces resulting from sequential sustained exposure to corrosive conditions. Samples were placed in a corrosive salt solution and the VPD maps were obtained at intervals of 0, 1, 2, 10, and 15 cumulative minutes of exposure to 1 M NaCl solution (Figure 6). Qualitative differences in both appearance (surface topography and morphology) and carbide-matrix VPD over time were observed for the steels. The HTT sample showed the formation of particulates on the surface and degraded uniformly with time, leading to a progressively lower variation in surface VPD. In contrast, the CN sample showed little change in

VPD or topography on the surface, indicating corrosion reaction kinetics were much slower despite the distinct VPD contrast between the carbides and matrix. LTT exhibited behavior somewhere in between the other two steels. Initially, salt deposits on the LTT surface obscured the steel topography and VPD variations. However, with increasing time LTT appeared similar to CN, as evidenced by the relatively large contrast in topography and VPD apparent by the 15 min mark (see Figure 6).

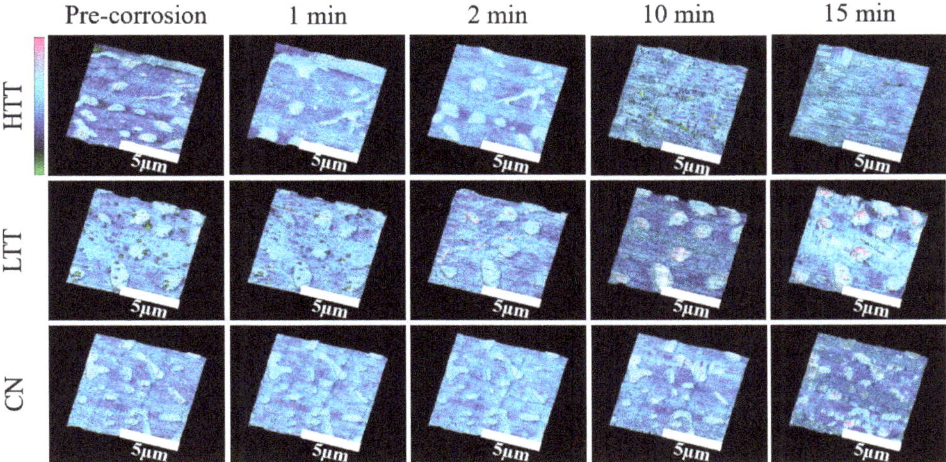

Figure 6. SKPFM Volta potential maps (green to pink color scale, 400 mV full scale) overlaid on the evolving 3D topography (30 nm full scale) of the three heat-treated MSSs as a function of immersion time in 1 M NaCl solution.

Figure 7 presents VPD maps (left column) and plots of Volta potential versus location (middle and right columns) for cross sections of different carbide/matrix interfaces as a function of exposure time. As can be seen in the top row of Figure 7, the VPD between the HTT carbides and the surrounding matrix decreased with exposure time, while the VPDs of the LTT (Figure 7(b1,b2)) and CN (Figure 7(c1,c2)) carbides remained relatively constant throughout testing. For HTT, corrosion proceeded simultaneously both along grain and carbide boundaries as well as within the matrix. Corrosion products evolved and settled on both the matrix and surface carbides, where cathodic activity was supporting anodic dissolution of the matrix. With this production and deposition of corrosion products, the VPD between carbides with a native oxide and matrix decreased on the HTT surface until there was very little difference observed between the two, as seen in Figure 7(a1,a2). Conversely, the LTT and CN samples underwent typical localized corrosion (see Figure 6), wherein highly localized attack adjacent to grain boundaries/carbides was seen, as evidenced by particulates settling on or near the carbide-matrix interface. As time in solution progressed, the VPD between the carbides and steel matrix remained essentially unchanged throughout the duration of testing, with matrix attack relatively shallow. Therefore, there are notable differences in the initiation of corrosion mechanisms between different heat treated samples.

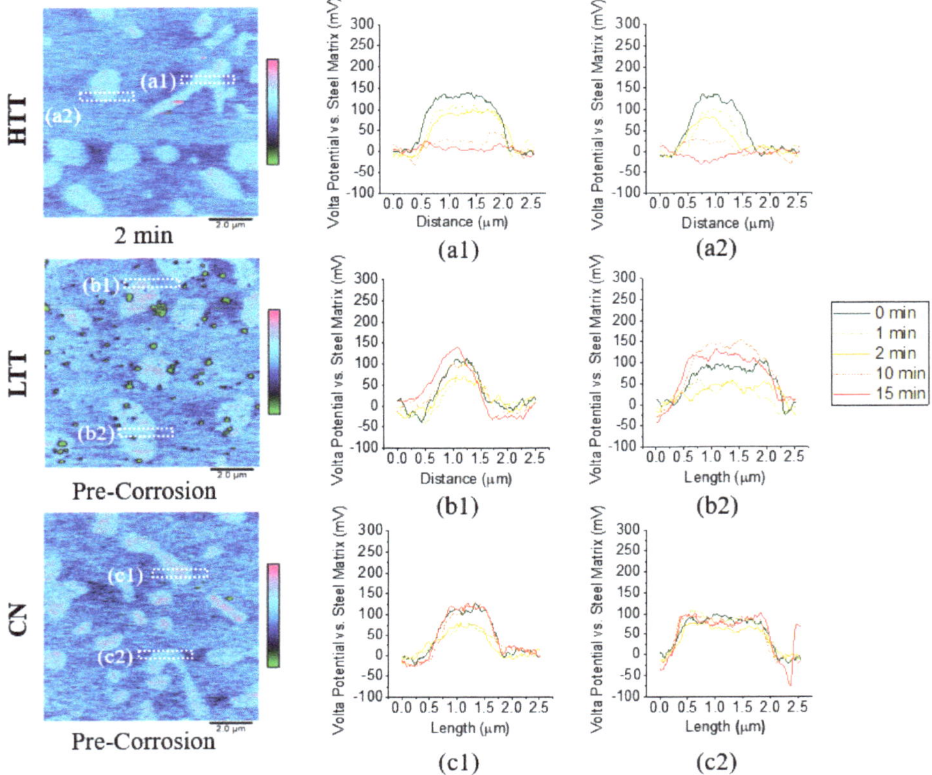

Figure 7. SKPFM Volta potential maps ((**a**,**b**,**c**) 600 mV full scale, exposure time given below each image) for each of the three heat-treated MSSs with time-dependent Volta potential profiles (**a1**–**c2**) across two representative carbides plotted as a function of duration of exposure to 1 M NaCl solution. The location of the carbide represented by each profile is indicated by the corresponding dotted box in the exemplary SKPFM maps at left.

3.2.4. In Situ Atomic Force Microscopy (AFM)

To observe the progression of corrosion in real time while samples were immersed in 0.1 M NaCl solution, in situ AFM was employed to monitor topographical changes over time. Figure 8 shows the results for the three heat-treated P675 steels with no applied bias voltage. (Variations in exposure time across samples are due to differences in corrosion rate and the time necessary to implement optimized imaging parameters.) For HTT, corrosion activity rapidly progressed and large surface deposits (~1–2 μm wide) appeared on the surface after ~107 min (Figure 8). EDS analysis indicated these large features to be iron-rich corrosion products with NaCl (analysis not shown). Despite the deposited particles, distinct localized corrosion was not seen on the HTT sample. As testing progressed, corrosion reactions proceeded, depositing corrosion product particulates on the surface (see Figure 8–HTT 116 & 134 min). In comparison, highly localized corrosion was evident at the carbide-matrix interfaces in both the CN and LTT samples. CN showed the greatest segregation of corrosion between matrix attack and the unaffected carbides, as indicated by near complete but shallow etching attack along carbide boundaries (Figure 8). LTT appeared to behave somewhere in the middle of these two extremes, with particle build-up similar to HTT seen initially, but eventually, these particles cleared to reveal evidence of localized corrosion propagation in the matrix adjacent to some of the carbides, similar to CN.

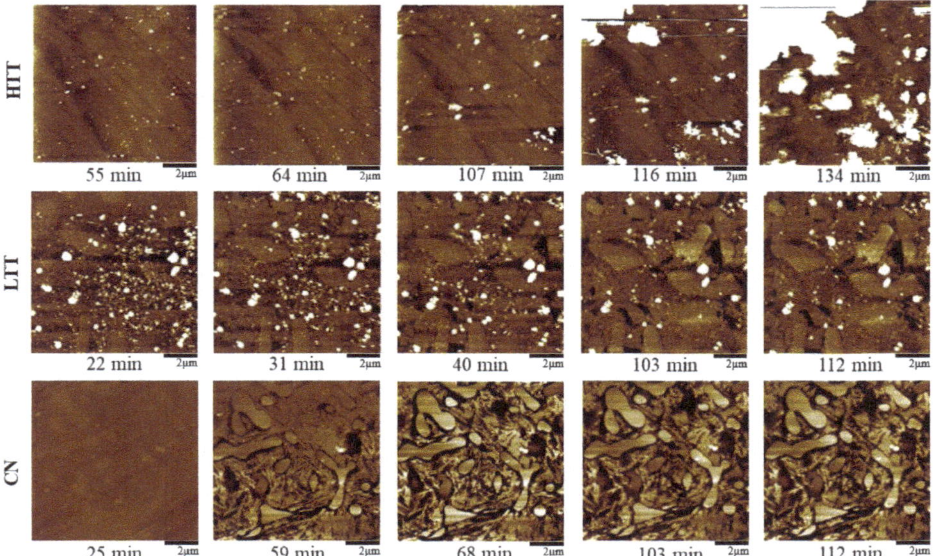

Figure 8. Time-lapse in situ AFM topography maps (160 nm full scale) for each of the heat-treated MSSs in 0.1 M NaCl solution, with approximate exposure time at the end of each scan indicated below the corresponding map (image time was ~8.5 min).

Time-dependent line profile analysis of selected carbide particles was conducted on each of the samples (Figure 9), confirming the qualitative observations arising from the images presented in Figure 8. HTT showed an increase in surface contrast of the carbides, up to 50 nm, with corresponding slight, uniform changes in the height of the surrounding matrix. For LTT, height changes across the carbide/matrix interface initially (44 min) showed ~100 nm deep attack immediately adjacent to the carbides (Figure 9). Then at longer times (112 min), the height of the carbides increased, accompanied by shallower apparent depth of attack in the adjacent matrix area. These changes are likely associated with the production and deposition of insoluble corrosion products. CN exhibited the sharpest contrast in topography by the end of exposure to salt solution, with the carbide surface height increasing by ~25 nm relative to the adjacent bulk matrix, with matrix attack limited to ~75 nm deep and only extending approximately 0.5 μm away from the carbide interface. The depth of attack also decreased from 103 min to 112 min, indicating slight corrosion product deposition within the highly localized area of matrix attack.

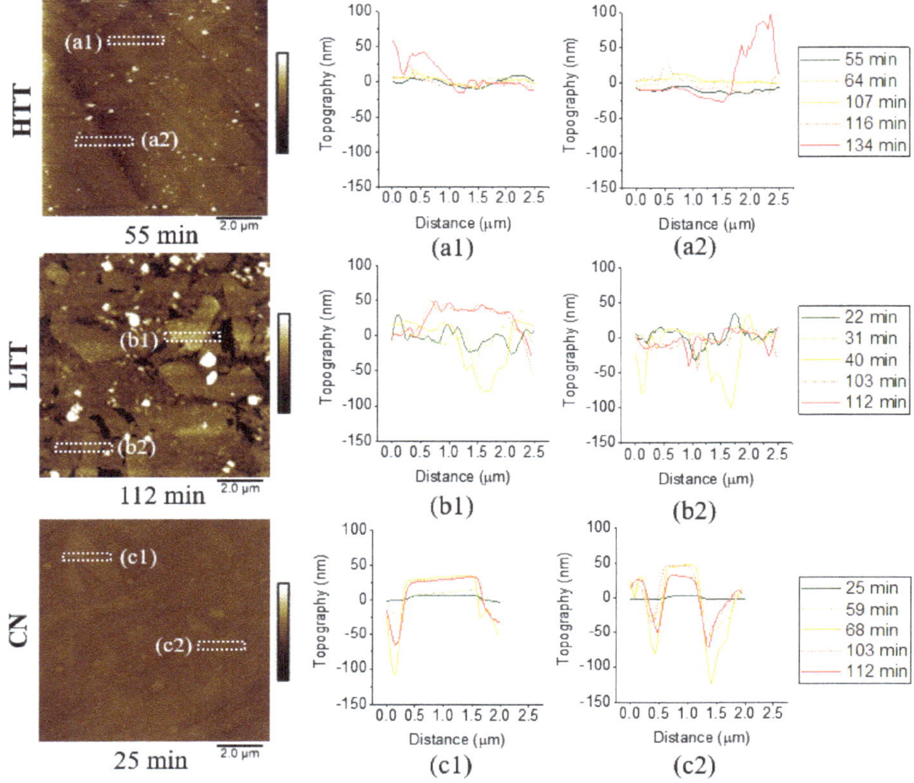

Figure 9. Topography maps ((**a**,**b**,**c**), 160 nm full scale, exposure time indicated below corresponding map) for each of the three heat-treated MSSs with height profiles across selected carbide-matrix interfaces shown as a function of exposure time to 0.1M NaCl solution (**a1**–**c2**). Location of each profile is indicated by the corresponding box in the exemplary topography maps presented at left for each of the three heat-treated steels.

Post-testing SEM imaging was conducted on the same sample surfaces (Figure 10) to record surface morphological differences following the in situ AFM testing. HTT exhibited a distinctively different surface morphology compared to LTT and CN, characterized by the presence of large, fluffy appearing salt-laden corrosion deposits. Beneath these deposits and surrounding the carbides, the entire matrix surface area was uniformly corroded with no indication of matrix passivity. In contrast, both the LTT and CN carbide boundaries were attacked, with NaCl particles present along the grain boundaries and carbide-matrix separation and subsequent grain separation (Figure 10). LTT showed some attack along carbide boundaries as well as some generalized attack as indicated by roughening of the entire surface due to corrosion product deposition. CN displayed much more localized attack at the carbide boundaries than LTT (dotted oval in the right panel of Figure 10), and narrow "valleys" on the order of ~0.5 μm wide were observed around the CN carbides, confirming observations in Figure 8. Furthermore, unlike LTT or HTT, CN did not show evidence of adhered or deposited corrosion products. Tracing the representative "line of attack" for the CN sample in Figure 10 reveals a grain undergoing intergranular attack, indicative of microgalvanic corrosion between the noble carbides and the active matrix.

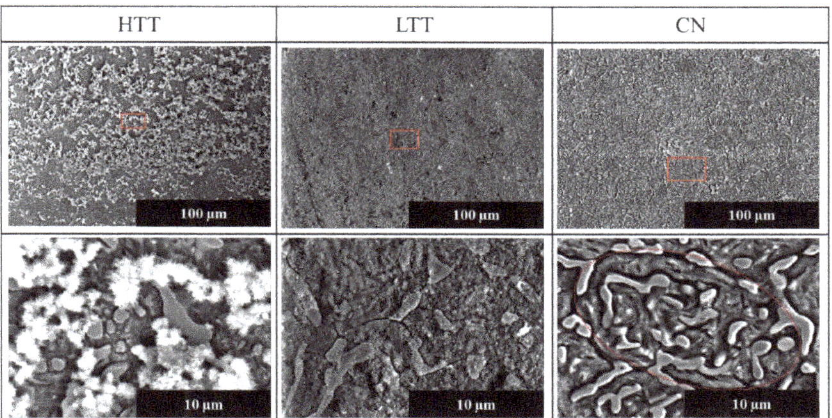

Figure 10. SE SEM images of the sample surfaces following in situ AFM testing. Red squares in the images in the top panels indicate areas of magnified images below. Dotted red oval area in magnified CN image indicates the "line of attack" (see discussion).

3.3. Electrochemical Corrosion Testing

To elucidate the corrosion pitting and repassivation behavior of the samples, cyclic potentiodynamic polarization (CPP) scans were conducted on each of the samples to explore the effects of the different heat treatments. Figure 11 shows the resultant polarization curves, along with macro images of the sample surfaces post-electrochemical testing. Testing indicated that HTT had the lowest OCP (−400 mV), followed by LTT (−200 mV) and CN (−80 mV), respectively. This ranking is in agreement with previous studies that ranked corrosion resistance for these same steels (i.e., corrosion rate determined via electrochemical methods) [7,9]. The LTT and CN samples exhibited a rapid change in potential over a minimal increase in current density (Figure 11a, green boxed areas), indicative of typical passive behavior. The breakdown potential of the LTT and CN samples occurred at 40 mV and 95 mV, respectively. Conversely, the HTT sample showed active corrosion behavior as demonstrated by linear growth of the current density over the potential sweep. However, pits were initially observed on the HTT surface (−200 mV), but did not grow and as the anodic overpotential continued to increase. The post-corrosion images in Figure 11b show the difference in corrosion morphology for each sample following CPP testing. For HTT, the entire test area darkened due to corrosion product formation (Figure 11b), engulfing the initially isolated areas of pitting. Arrows in Figure 11b indicate the four pits that first formed on the HTT sample before the entire test area underwent generalized corrosion. As expected from previous work [7], LTT and CN showed a distinctly different morphology of corrosion attack, with corrosion limited to only several dispersed pits on the surface of the sample. Compared to HTT, LTT showed limited regions of depassivation emanating from corrosion pits, evidenced by regions of minor surface darkening. In contrast, corrosion attack on CN displayed only highly localized, isolated pits (Figure 11b) with no visual evidence of any other associated areas of depassivation.

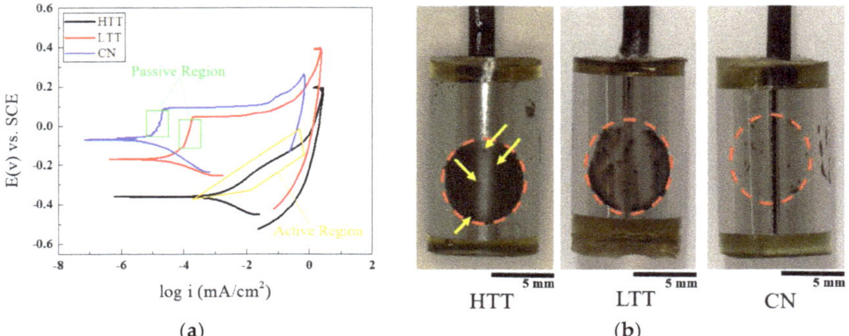

Figure 11. (**a**) Cyclic potentiodynamic polarization (CPP) scans (0.01 M NaCl electrolyte) for all three surface treatment samples. Passive regions for LTT and CN are indicated by green squares. (**b**) Images of the samples post-testing (after the area masking tape was removed) with dotted red circles indicating the test location on each sample surface. All samples display some isolated pitting. However, due to the difficulty in clearly seeing the pits on the HTT sample (which, in contrast to the other samples, underwent generalized corrosion attack), yellow arrows indicate the location of the pits present on the HTT sample.

4. Discussion

4.1. Nanoscale Origins of Corrosion Initiation

Determining the nanoscale contributions to a material's bulk corrosion rate is inherently difficult due to the complexity and multitude of variables that influence its behavior in a corrosive environment. Corrosion is a spontaneous process driven by thermodynamics [58,59]. In a microgalvanic couple, the difference in electrode potential of the anode and cathode regions on the surface correlates with the magnitude of negative free energy change (thermodynamic propensity) for local corrosion to occur. SKPFM Volta potential (VPD) mapping is the highest spatial resolution method available to directly measure the relative thermodynamic propensity for corrosion between nanoscale heterogeneities in a material. For the MSSs considered in this study, the relatively high Cr composition of the carbides suggests they are likely noble in comparison to the matrix based on the galvanic series [60]. Hence, a larger VPD between carbides and the matrix will lead to a greater drive (i.e., increased microgalvanic full-cell potential) for corrosion of the matrix. Among the steels studied, HTT consistently exhibited the largest VPD between the carbides and the matrix (200 mV), while LTT (150 mV) and CN (90 mV) were considerably less (Figure 5). The relative magnitudes of these VPDs can likely be attributed to carbide chemistry, as HTT carbide composition is predominantly $M(Cr)_{23}C_6$ compared to predominantly $M(Cr)_7C_3$ compounds for LTT and CN. An interesting finding of this study is that for each of the surface treatments considered, the bulk OCP values measured inversely corresponded with the magnitude of the VPD between the carbide and matrix phases (Figure 5). HTT had the greatest carbide/matrix VPD and least noble OCP (−400 mV), CN had the lowest VPD and most noble OCP (−80 mV), and LTT was intermediate with a carbide/matrix OCP of −200mV. This observation demonstrates how local SKPFM measurements of the relative microgalvanic couple potential contribute to the bulk OCP observed on each of the different surface-treated MSSs investigated. In addition, variations in chromium enrichment of the carbides subsequently influenced both the VPD and degree of passivity of the surrounding chromium-depleted matrix. The steepest VPD gradients measured were across the carbide/matrix interface (Figure 7), and so SKPFM measurements also provided a technique to predict and locate expected points of microgalvanic corrosion initiation on the surface.

4.2. Corrosion Propagation

SKPFM measures VPDs on the surface, which are influenced by the presence of oxide layers. With MSSs, passivating chromium oxide layers are readily formed and act as a kinetic barrier to corrosion, which complicates any correlation of thermodynamic propensity derived from SKPFM measurements. However, for the steels considered herein, since the bulk composition is the same, data obtained from SKPFM also provided information on the spatial variations in surface properties that influence corrosion propagation. Intermittent SKPFM testing was conducted to monitor shifts in microgalvanic couples' VPD over time due to corrosion activity. For HTT, the VPD between the carbides and the matrix decreased with time (Figure 7). As a result, as the duration of corrosion propagation increased, the VPD between carbides and the matrix approached 0 mV for HTT, resulting in a more thermodynamically homogenous surface. In contrast, for LTT and CN, the initial VPD between the carbides and matrix phase was smaller, but remained nearly constant throughout testing, with only minor evidence of the corrosion activity apparent on the surface (Figures 6 and 7). This behavioral difference can be attributed to differences in the passive oxide layer performance, and is also reflected in the VPD measurements, which are highly influenced by the presence of surface oxides. Previous work by Schmutz and Frankel showed similar behavior on aluminum alloys and indicates that the shift in VPD observed on HTT following active corrosion was caused by oxide growth at cathodic sites and the generation and deposition of corrosion products at active sites creating a more homogenous surface [51]. For carburized MSSs, the magnitude of VPD surface variation measured by SKPFM pre-corrosion provided an indication of the how the VPD evolved as a result of exposure to corrosion conditions: smaller initial VPD between the carbides and matrix phase indicated more robust passivity during corrosion, as seen in CN and LTT steels. For HTT, the higher initial VPD between the carbides and matrix indicated a greater susceptibility to depassivation and more uniform corrosion activity during propagation. These findings were validated with bulk electrochemical testing (Figure 11), where CPP testing showed that LTT and CN had a more protective oxide layer as indicated by the presence of a passive region in the CPP scan. Moreover, during intermittent SKPFM testing the VPD on HTT evolved rapidly and HTT exhibited active corrosion behavior throughout CPP testing.

While the bulk amount of chromium present at the surface is the same for all steels considered, the spatial distribution is different among the three surface treatments, leading to distinctly different corrosion properties and behavior. Relative to LTT and CN, HTT tended to corrode more uniformly and had a higher VPD between carbides and matrix. HTT was more prone to depassivation compared to LTT despite both having identical bulk chemical composition and same carburization cycle (carburized in single furnace load). The different carbide-matrix VPDs among the samples influences or indicates how local solution chemistry likely evolves during active corrosion on MSSs. This suggests that for HTT, as pitting progressed, the local solution chemistry, most likely due to higher sensitization during tempering cycle, was sufficiently aggressive to cause widespread depassivation. Conversely, with LTT and CN samples, the VPD between carbides was smaller and pitting was unable to transition to more widespread corrosion, suggesting local solution chemistry evolution did not support auto-catalytic depassivation as corrosion propagated. Here the lower VPD observed for LTT and CN indicated the matrix phases exhibited more robust passivity than the matrix of HTT. The in situ SKPFM VPD measurements correlate with the observed corrosion morphology of the steels. That is, the measured carbide-matrix VPD for each steel is inversely proportional to the extent of general (uniform) corrosion resistance of the steel. The efforts in this paper show that SKPFM is able to effectively predict bulk corrosion behavior of different surface treatments by observing and measuring nanoscale surface VPD differences between carbides and the underlying matrix.

4.3. SPM Characterization and Implications on Wear

MFM provides a method to characterize local variations in magnetic properties that contribute to the bulk magnetic properties. For all steels studied, the carbides showed variable shades of purple/blue in the MFM maps (~1–3° phase shift), indicating slightly different magnetic properties within the individual phases (Figure 3), likely due to different carbide compositions in terms of the relative amounts of chromium and molybdenum, which influence the magnetic properties of phases [61–64]. Sample CN had a much less homogenous matrix that showed considerable variation in magnetic properties and is likely an effect of the relatively high surface retained austenite (18–22%) found within the matrix phases compared to LTT (10–13%) and HTT (1–2%) [9,30]. The bulk magnetism of the steels will change with tempering temperature and heat treatment process, following changes to the microstructural phases formed [9,30]. Further work is currently underway to investigate the implications of local magnetism and magnetic domains on resulting wear and corrosion mechanisms.

Similarly, the ability to resolve nanoscale variations in the resistance to deformation (elastic modulus) on a material's surface could help improve prediction of the wear behavior. The PeakForce tapping mode employed here measured differences in the elastic modulus distribution, as determined via the Derjaguin-Muller-Toporov (DMT) model [65], for the carbide and the matrices of the steels simultaneously with topography (see exemplary Figure 12). As seen in the CN image presented in Figure 12, carbides had a higher relative modulus than the matrix, suggesting potential sites for development of micro-cracking and fracture would likely lie at the interface between carbides and matrix where local modulus variation was greatest. Further work is underway to determine how these local differences in recorded modulus correlate to a material's ability to handle loads/stress in bearing applications.

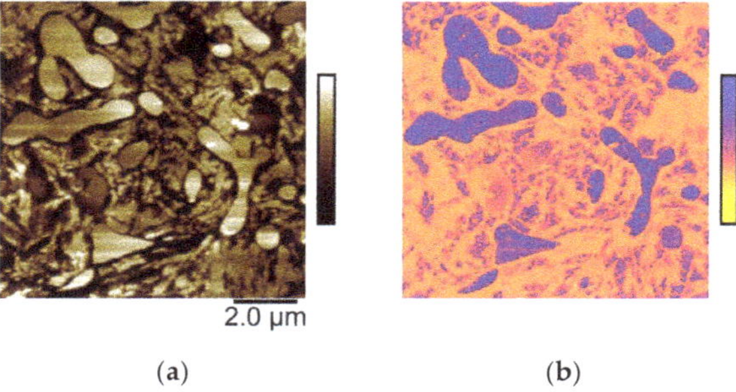

(**a**) (**b**)

Figure 12. (**a**) CN topography (160 nm full scale), and (**b**) DMT Modulus (1.5 GPa full scale). Images are representative of 103–112 min submersion in 0.1M NaCl solution.

In service, the uniform degradation seen on HTT could be effectively monitored conventionally via visual inspection, detection of wear debris, or thickness monitors installed on bearing raceways. For CN and LTT, current methods of monitoring engine health are less effective since significantly lower amounts of reaction products are generated from highly localized corrosion. Localized corrosion may not be detected until it has led to significant wear damage. Bearing steel developers should, therefore, be cautious with heat treatments that yield a surface similar to CN which, although highly corrosion resistant, the passive surface will inevitably be compromised in wear applications. Small areas of highly localized corrosion pits lead to surface crater development which can potentially lead to highly undesirable and unpredictable failure via spalling. LTT behavior was intermediate between the two other surface treatments, with some localized attack on grain/carbide boundaries as well as some evidence of wider depassivation. In corrosive environments, the overall wear lifetime may be controlled by resistance to corrosion initiation, in which case LTT and CN could provide greater benefit than HTT. Previously conducted wear studies are in agreement with the recommendations given, and the results of this study provide nanoscale insight to help understand why HTT outperformed both CN and LTT during rolling contact fatigue testing even though it had significantly lower corrosion resistance [29,30]. Based on this work, P675 HTT would be recommended over the other two tempering procedures for use in aerospace bearings where corrosion is not a primary concern. However, when the bearing assembly is prone to corrosion attack, HTT is not recommended due to its overall low corrosion resistance [7] which would lead to premature failure via degradation of the material. In this case, CN is recommended for bearing use due to its high resistance to both corrosion onset and propagation [7].

5. Conclusions

P675 carburizable martensitic stainless steel (UNS S42670) samples were processed using two different heat treatment methods (carburizing and carbo-nitriding (CN)) and two tempering temperatures (HTT and LTT). Following, the research conducted in this paper highlights the viability of SKPFM to effectively predict bulk corrosion behavior by measuring nanoscale surface differences in VPDs between carbides and the surrounding matrix, thereby providing insight into bulk observations by using information obtained at the nanoscale. More generally, SPM can be used to evaluate the potential efficacy of different steels and/or surface treatments for use in corrosive environments.

- MFM imaging distinguished local differences in magnetic properties where precipitated carbides exhibited a larger magnetic moment than the matrix, likely due to the presence of chromium relative to the chromium-depleted matrix.
- SKPFM VPD measurements in an inert environment showed HTT as the thermodynamically most favorable to experience microgalvanic corrosion between the chromium-rich precipitated carbides and the surrounding martensitic matrix, with a measured carbide-matrix VPD of 200 mV, while LTT (150 mV) and CN (90 mV) were less.
- Intermittent SKPFM showed the HTT sample behaved differently during corrosion than the LTT and CN samples; by the end of the testing period, there was minimal VPD between the HTT carbides and the surrounding matrix, whereas the carbides present in the LTT and CN samples retained their relative nobility throughout testing.
- Corrosion propagation was also monitored in real time via in situ AFM and revealed that HTT underwent the most rapid spread of corrosion attack across the sample, while LTT and CN were less affected and showed much more localized, intergranular attack and adjacent to carbides.
- Bulk electrochemical testing results agreed with in situ AFM results, with LTT and CN showing distinct passive regions as compared to HTT, confirming the nanoscale differences in corrosion behavior observed between the steel heat treatments investigated.

Author Contributions: All authors contributed to writing the manuscript. Conceptualization, A.K., C.M.E., K.A.H., P.H.D., M.F.H.; Investigation, A.K., C.M.E., K.A.H., O.O.M.; Project Administration, P.H.D., E.G.; M.F.H.; Visualization, A.K., C.M.E., K.A.H., O.O.M., P.H.D., M.F.H.; Resources, P.H.D., E.G., H.K.T., M.F.H.; Funding Acquisition, H.K.T., E.G., P.H.D., M.F.H.; Validation, H.K.T.; Supervision, M.F.H.

Funding: This research was funded by Micron School of Materials Science and Engineering, NSF MRI #1727026

Conflicts of Interest: The authors declare no conflicts of interest. The funders had no role in the design of the study; in the collection, analyses, or interpretation of data; in the writing of the manuscript, or in the decision to publish the results.

References

1. Bhadeshia, H.K.D.H. Steels for bearings. *Prog. Mater. Sci.* **2012**, *57*, 268–435. [CrossRef]
2. Davies, D.P. Gear materials in helicopter transmissions. *Metals Mater.* **1986**, *2*, 342–348.
3. Zaretsky, E.V. Rolling bearing steels—A technical and historical perspective. *Mater. Sci. Technol.* **2012**, *28*, 58–69. [CrossRef]
4. Zaretsky, E.V. Bearing and gear steels for aerospace applications. *NASA Technical Memorandum 102529*; 1990. Available online: https://ntrs.nasa.gov/archive/nasa/casi.ntrs.nasa.gov/19900011075.pdf (accessed on 16 March 2019).
5. Hurley, M.F.; Marx, B.M.; Allahar, K.N.; Smith, C.P.; Chin, H.A.; Ogden, W.P.; Butt, D.P. Corrosion Assessment and Characterization of Aerospace-Bearing Steels in Seawater and Ester-Based Lubricants. *Corrosion* **2012**, *68*, 645–661. [CrossRef]

6. Popgoshev, D.; Valori, R. *Rolling Contact Fatigue Evaluation of Advanced Bearing Steels STP 771*; ASTM International: West Conshohocken, PA, USA, 1982.
7. Kvryan, A.; Faulkner, E.; Lysne, D.; Carter, N.; Acharya, S.; Rafla, V.; Trivedi, H.K.; Hurley, M.F. Electrochemical Corrosion Test Methods for Rapid Assessment of Aerospace Bearing Steel Performance. In *Bearing Steel Technologies:11th Volume, Progress in Steel Technologies and Bearing Steeel Quality Assurance, ASTM STP1600*; ASTM International: West Conshohocken, PA, USA, 2017; pp. 466–486.
8. Zaretsky, E.V.; Branzai, E.V. Rolling Bearing Service Life Based on Probable Cause for Removal—A Tutorial. *Tribol. Trans.* **2017**, *60*, 300–312. [CrossRef]
9. Trivedi, H.; Otto, F.; McCoy, B. Low Temperature Plasma Nitriding of Pyrowear 675. In *Bearing steel Technologies: 10th Volume, Advances in Steel Technologies for Rolling Bearings, STP 1580*; ASTM International: West Conshocken, PA, USA, 2015; Volume 10, pp. 444–464.
10. Trivedi, H.; Monahan, R. Low Temperature Plasma Nitriding of Pyrowear 675. In *Bearing steel Technologies: 10th Volume, Advances in Steel Technologies for Rolling Bearings, STP 1580*; ASTM International: West Conshocken, PA, USA, 2015; Volume 10, pp. 444–464.
11. Dalmau, A.; Richard, C.; Igual–Muñoz, A. Degradation mechanisms in martensitic stainless steels: Wear, corrosion and tribocorrosion appraisal. *Tribol. Int.* **2018**, *121*, 167–179. [CrossRef]
12. Ma, X.; Wang, L.; Liu, C.; Subramanian, S. Microstructure and properties of 13Cr5Ni1Mo0. 025Nb0. 09V0. 06N super martensitic stainless steel. *Mater. Sci. Eng. A* **2012**, *539*, 271–279. [CrossRef]
13. Thibault, D.; Bocher, P.; Thomas, M. Residual stress and microstructure in welds of 13% Cr–4% Ni martensitic stainless steel. *J. Mater. Process. Technol.* **2009**, *209*, 2195–2202. [CrossRef]
14. Li, C.; Bell, T. Corrosion properties of plasma nitrided AISI 410 martensitic stainless steel in 3.5% NaCl and 1% HCl aqueous solutions. *Corros. Sci.* **2006**, *48*, 2036–2049. [CrossRef]
15. Smallman, R.E.; Ngan, A.H.W. *Physical Metallurgy and Advanced Materials*, 7th ed.; Elsevier Ltd.: Burlington, MA, USA, 2007.
16. Klueh, R.; Nelson, A. Ferritic/martensitic steels for next-generation reactors. *J. Nuclear Mater.* **2007**, *371*, 37–52. [CrossRef]
17. Zinkle, S.J. Advanced materials for fusion technology. *Fusion Eng. Des.* **2005**, *74*, 31–40. [CrossRef]
18. Puli, R.; Ram, G.J. Microstructures and properties of friction surfaced coatings in AISI 440C martensitic stainless steel. *Surf. Coat. Technol.* **2012**, *207*, 310–318. [CrossRef]
19. Anantha, K.H.; Örnek, C.; Ejnermark, S.; Medvedeva, A.; Sjöström, J.; Pan, J. Correlative Microstructure Analysis and In Situ Corrosion Study of AISI 420 Martensitic Stainless Steel for Plastic Molding Applications. *J. Electrochem. Soc.* **2017**, *164*, C85–C93. [CrossRef]
20. Klecka, M.A. *Microstructure-Property Relationships and Constitutive Response of Plastically Graded Case Hardened Steels*; University of Florida: Gainesville, FL, USA, 2011.
21. Schneider, J.M.; Chatterjee, M. *Introduction to Surface Hardening of Steels*; ASM International: West Conshohocken, PA, USA, 2013; Volume 4A, pp. 389–398.
22. Godec, M.; Batic, B.S.; Mandrino, D.; Nagode, A.; Leskovsek, V.; Skapin, S.D.; Jenko, M. Characterization of the carbides and the martensite phase in powder-metallurgy high-speed steel. *Mater. Charact.* **2010**, *61*, 452–458. [CrossRef]
23. Callister, W.D.; Rethwisch, D.G. *Materials Science and Engineering: An Introduction*; Wiley: New York, NY, USA, 2007; Volume 7.
24. Johnson, M.; Laritz, J.; Rhoads, M. *Thin Dense Chrome Bearing Insertion Program: Pyrowear 675 and Cronidur Wear Testing*; GE Advanced Engineering Technologies Dept: Cincinnati, OH, USA, 1998.
25. Wert, D.E. Development of a carburizing stainless steel alloy. *Adv. Mater. Process.* **1994**, *145*, 89–91.
26. Trivedi, H.K.; Wedeven, V.; Black, W. Effect of Silicon Nitride Ball on Adhesive Wear of Martensitic Stainless Steel Pyrowear 675 and AISI M-50 Races with Type II Ester Oil. *Tribol. Trans.* **2016**, *59*, 363–374. [CrossRef]
27. Hetzner, D.W.; Van Geertruyden, W. Crystallography and metallography of carbides in high alloy steels. *Mater. Charact.* **2008**, *59*, 825–841. [CrossRef]
28. Goldschmidt, H. The structure of carbides in alloy steels. Part 2-Carbide formation in high-speed steels. *J. Iron Steel Inst.* **1952**, *170*, 189.

29. Trivedi, H.; Rosado, L.; Gerardi, D.; Givan, G.; McCoy, B. Fatigue Life Performance of Hybrid Angular Contact Pyrowear 675 Bearings. In *Bearing Steel Technologies: 11th Volume, Progress in Steel Technologies and Bearing Steel Quality Assurance, ASTM STP 1600*; ASTM International: West Conshohocken, PA, USA, 2017; pp. 275–295.
30. Kirsch, M.; Trivedi, H. Microstructural Changes in Aerospace Bearing Steels under Accelerated Rolling Contact Fatigue Life Testing. In *Bearing Steel Technologies: 11th Volume, Progress in Steel Technologies and Bearing Steel Quality Assurance, ASTM STP 1600*; ASTM International: West Conshohocken, PA, USA, 2017; pp. 92–107.
31. Reynaud-Laporte, I.; Vayer, M.; Kauffmann, J.-P.; Erre, R. An electrochemical-AFM study of the initiation of the pitting corrosion of a martensitic stainless steel. *Microsc. Microanal. Microstr.* **1997**, *8*, 175–185. [CrossRef]
32. Leblanc, P.P.; Frankel, G. Investigation of filiform corrosion of epoxy-coated 1045 carbon steel by scanning Kelvin probe force microscopy. *J. Electrochem. Soc.* **2004**, *151*, B105–B113. [CrossRef]
33. Schmutz, P.; Frankel, G.S. Characterization of AA2024-T3 by scanning Kelvin probe force microscopy. *J. Electrochem. Soc.* **1998**, *145*, 2285–2295. [CrossRef]
34. Campestrini, P.; van Westing, E.P.M.; van Rooijen, H.W.; de Wit, J.H.W. Relation between microstructural aspects of AA2024 and its corrosion behaviour investigated using AFM scanning potential technique. *Corros. Sci.* **2000**, *42*, 1853–1861. [CrossRef]
35. de Wit, J.H.W. Local potential measurements with the SKPFM on aluminium alloys. *Electrochim. Acta* **2004**, *49*, 2841–2850. [CrossRef]
36. Larignon, C.; Alexis, J.; Andrieu, E.; Lacroix, L.; Odemer, G.; Blanc, C. Combined Kelvin probe force microscopy and secondary ion mass spectrometry for hydrogen detection in corroded 2024 aluminium alloy. *Electrochim. Acta* **2013**, *110*, 484–490. [CrossRef]
37. Hurley, M.F.; Efaw, C.M.; Davis, P.H.; Croteau, J.R.; Graugnard, E.; Birbilis, N. Volta Potentials Measured by Scanning Kelvin Probe Force Microscopy as Relevant to Corrosion of Magnesium Alloys. *Corrosion* **2015**, *71*, 160–170. [CrossRef]
38. Kvryan, A.; Livingston, K.; Efaw, C.M.; Knori, K.; Jaques, B.J.; Davis, P.H.; Butt, D.P.; Hurley, M.F. Microgalvanic Corrosion Behavior of Cu-Ag Active Braze Alloys Investigated with SKPFM. *Metals* **2016**, *6*, 91. [CrossRef]
39. Guillaumin, V.; Schmutz, P.; Frankel, G.S. Characterization of Corrosion Interfaces by the Scanning Kelvin Probe Force Microscopy Technique. *Electrochem. Soc.* **2001**, *148*, 163–173. [CrossRef]
40. Andreatta, F.; Terryn, H.; De Wit, J. Corrosion behaviour of different tempers of AA7075 aluminium alloy. *Electrochim. Acta* **2004**, *49*, 2851–2862. [CrossRef]
41. Örnek, C.; Leygraf, C.; Pan, J. Passive film characterisation of duplex stainless steel using scanning Kelvin prove force microscopy in combination with electrochemical measurements. *NPJ Mater. Degrad.* **2019**, *3*, 8. [CrossRef]
42. Anantha, K.H.; Örnek, C.; Ejnermark, S.; Medvedeva, A.; Sjöström, J.; Pan, J. In Situ AFM Study of Localized Corrosion Processes of Tempered AISI 420 Martensitic Stainless Steel: Effect of Secondary Hardening. *J. Electrochem. Soc.* **2017**, *164*, C810–C818. [CrossRef]
43. Ramírez-Salgado, J.; Domínguez-Aguilar, M.; Castro-Domínguez, B.; Hernández-Hernández, P.; Newman, R. Detection of secondary phases in duplex stainless steel by magnetic force microscopy and scanning Kelvin probe force microscopy. *Mater. Charact.* **2013**, *86*, 250–262. [CrossRef]
44. Sathirachinda, N.; Pettersson, R.; Pan, J.S. Depletion effects at phase boundaries in 2205 duplex stainless steel characterized with SKPFM and TEM/EDS. *Corros. Sci.* **2009**, *51*, 1850–1860. [CrossRef]
45. Femenia, M.; Canalias, C.; Pan, J.; Leygraf, C. Scanning Kelvin probe force microscopy and magnetic force microscopy for characterization of duplex stainless steels. *J. Electrochem. Soc.* **2003**, *150*, B274–B281. [CrossRef]
46. Birbilis, N.; Meyer, K.; Muddle, B.; Lynch, S. In situ measurement of corrosion on the nanoscale. *Corros. Sci.* **2009**, *51*, 1569–1572. [CrossRef]
47. Davoodi, A.; Pan, J.; Leygraf, C.; Norgren, S. In situ investigation of localized corrosion of aluminum alloys in chloride solution using integrated EC-AFM/SECM techniques. *Electrochem. Solid-State Lett.* **2005**, *8*, B21–B24. [CrossRef]
48. Guo, L.; Li, M.; Shi, X.L.; Yan, Y.; Li, X.; Qiao, L. Effect of annealing temperature on the corrosion behavior of duplex stainless steel studied by in situ techniques. *Corros. Sci.* **2011**, *53*, 3733–3741. [CrossRef]

49. Park, J.; Kalnaus, S.; Han, S.; Lee, Y.K.; Less, G.B.; Dudney, N.J.; Daniel, C.; Sastry, A.M. In situ atomic force microscopy studies on lithium (de) intercalation-induced morphology changes in LixCoO2 micro-machined thin film electrodes. *J. Power Sources* **2013**, *222*, 417–425. [CrossRef]
50. Leblanc, P.; Frankel, G.S. A study of corrosion and pitting initiation of AA2024-T3 using atomic force microscopy. *J. Electrochem. Soc.* **2002**, *149*, B239–B247. [CrossRef]
51. Schmutz, P.; Frankel, G.S. Corrosion study of AA2024-T3 by scanning Kelvin probe force microscopy and in situ atomic force microscopy scratching. *J. Electrochem. Soc.* **1998**, *145*, 2295–2306. [CrossRef]
52. Efaw, C.M.; da Silva, T.; Davis, P.H.; Li, L.; Graugnard, E.; Hurley, M.F. Toward Improving Ambient Volta Potential Measurements with SKPFM for Corrosion Studies. *J. Electrochem. Soc.* **2019**, *166*, C3018–C3027. [CrossRef]
53. Klecka, M.A.; Subhash, G.; Arakere, N.K. Microstructure-Property Relationships in M50-NiL and P675 Case-Hardened Bearing Steels. *Tribol. Trans.* **2013**, *56*, 1046–1059. [CrossRef]
54. Lind, M.A. *The Infrared Reflectivity of Chromium and Chromium-Aluminum Alloys*; Iowa State University: Ames, IA, USA, 1972.
55. Fawcett, E.; Alberts, H.; Galkin, V.Y.; Noakes, D.; Yakhmi, J. Spin-density-wave antiferromagnetism in chromium alloys. *Rev. Modern Phys.* **1994**, *66*, 25. [CrossRef]
56. Iga, A.; Tawara, Y. Magnetic properties of molybdenum-and wolfram-Modified Mn3B4. *J. Phys. Soc. Jpn.* **1968**, *24*, 28–35. [CrossRef]
57. Mouritz, A.P. *Introduction to Aerospace Materials*; Woodhead Publishing: Philadelphia, PA, USA, 2012.
58. Ahmad, Z. *Principles of Corrosion Engineering and Corrosion Control*; Elsevier: Burlington, MA, USA, 2006.
59. Lauter, V.; Lauter, H.; Glavic, A.; Toperverg, B. *Reference Module in Materials Science and Materials Engineering*; Elsevier: Burlington, MA, USA, 2016.
60. Forman, C.M.; Verchot, E.A. *Practical Galvanic Series*; RS-TR-67-11; Report No. RS-TR-67-11; Army Missile Command Redstone Arsenal AL Systems Research Directorate: Redstone Arsenal, AL, USA, 1967.
61. Sathirachinda, N.; Gubner, R.; Pan, J.; Kivisäkk, U. Characterization of phases in duplex stainless steel by magnetic force microscopy/scanning Kelvin probe force microscopy. *Electrochem. Solid-State Lett.* **2008**, *11*, C41–C45. [CrossRef]
62. Mészáros, I.; Szabo, P. Complex magnetic and microstructural investigation of duplex stainless steel. *Non-destruct. Test. Eval. Int.* **2005**, *38*, 517–521. [CrossRef]
63. Tavares, S.; Da Silva, M.; Neto, J. Magnetic property changes during embrittlement of a duplex stainless steel. *J. Alloys Compd.* **2000**, *313*, 168–173. [CrossRef]
64. Tavares, S.; Pedrosa, P.; Teodosio, J.; Da Silva, M.; Neto, J.; Pairis, S. Magnetic properties of the UNS S39205 duplex stainless steel. *J. Alloys Compd.* **2003**, *351*, 283–288. [CrossRef]
65. Derjaguin, B.V.; Muller, V.M.; Toporov, Y.P. Effect of contact deformations on the adhesion of particles. *J. Colloid Interface Sci.* **1975**, *53*, 314–326. [CrossRef]

© 2019 by the authors. Licensee MDPI, Basel, Switzerland. This article is an open access article distributed under the terms and conditions of the Creative Commons Attribution (CC BY) license (http://creativecommons.org/licenses/by/4.0/).

Article

Effect of Cast Defects on the Corrosion Behavior and Mechanism of UNS C95810 Alloy in Artificial Seawater

Xu Zhao [1], Yuhong Qi [1,*], Jintao Wang [1], Zhanping Zhang [1], Jing Zhu [2], Linlin Quan [2] and Dachuan He [2]

1 Department of Materials Science and Engineering, Dalian Maritime University, Dalian 116026, China; zx1988@dlmu.edu.cn (X.Z.); 18342299941@163.com (J.W.); zzp@dlmu.edu.cn (Z.Z.)

2 Dalian Marine Propeller Co., Ltd., Dalian 116021, China; jingzhu2009@yeah.net (J.Z.); quanlinlin@163.com (L.Q.); xyz@dmpp.cn (D.H.)

* Correspondence: yuhong_qi@dlmu.edu.cn

Received: 21 March 2020; Accepted: 8 April 2020; Published: 10 April 2020

Abstract: To study the effect of cast defects on the corrosion behavior and mechanism of the UNS C95810 alloy in seawater, an investigation was conducted by weight loss determination, scanning electron microscopy (SEM), confocal laser scanning microscopy (CLSM), X-ray diffraction (XRD) and electrochemical testing of the specimen with and without cast defects on the surface. The results show that the corrosion rate of the alloy with cast defects is higher than that of the alloy without cast defects, but the defects do not change the composition of the resulting corrosion products. The defects increase the complexity of the alloy microstructure and the tendency toward galvanic corrosion, which reduce the corrosion potential from −3.83 to −86.31 mV and increase the corrosion current density from 0.228 to 0.23 $\mu A \cdot cm^{-2}$.

Keywords: nickel-aluminum bronze; cast defects; electrochemical corrosion; mechanism; microstructure

1. Introduction

Due to its good mechanical properties, corrosion fatigue resistance, corrosion resistance and cavitation resistance, Ni-Al bronze (NAB) has been widely used for marine components, such as ship propellers, pumps and valves [1–3]. NAB parts used in seawater are inevitably subject to corrosion by the seawater. To date, there are many studies on NAB corrosion behavior in seawater [1,3–8]. The results showed that an α phase corroded while a κ phase did not corrode in near-neutral seawater and 3.5% NaCl solution [1,8,9]. Neodo et al. [7] studied the corrosion behavior of NAB in a 3.5% NaCl solution with different pH values. It was found that the κ phase is preferentially corroded when the pH<4 but the α phase dissolves first when the pH > 4. The results of Weill-Couly et al. [10] showed that the corrosion resistance of NAB in seawater was related to its microstructures. The phase transformation of a small amount of $\beta' \rightarrow \alpha + \kappa$ could effectively improve the corrosion resistance of the alloy in seawater. The excellent corrosion resistance of NAB is derived from a corrosion product film produced on the surface [4,11–14]. Schüssler et al. [4] proposed that the corrosion product film produced on the surface of NAB contained an inner layer of Al_2O_3 and an outer layer of Cu_2O; furthermore, a thickness of 800–1000 nm could reduce the corrosion rate by 20–30 times. If time is sufficient, corrosion products such as $Cu_2(OH)_3Cl$ will also form on the outer layer of the film [12,13].

The as-cast NAB pieces used for ship propellers are very large, weighing as much as 200 t. Therefore, cast defects are unavoidable. Large and concentrated defects can be repaired in various ways to reduce the adverse effects of defects. For example, Sabbaghzadeh et al. [15] repaired NAB by fusion welding, and the galvanic corrosion current between the welded and the as-cast microstructures was only a few nA. Hanke et al. [16] applied a friction surface treatment for as-cast NAB, resulting in refinement and homogenization of the surface microstructure, which repaired cast defects. Ni et al. [17] used friction stir processing for as-cast Ni-Al bronze and compared the weight loss of the treated NAB and as-cast NAB in 3.5% NaCl solution. It was found that the weight loss of the treated NAB was significantly smaller than that of as-cast NAB, improving corrosion performance. That is, welding can improve the corrosion resistance of copper alloys with large cast defects in seawater.

However, for small and dispersed defects, they are not only unavoidable but also difficult to repair. Therefore, NAB propellers that are put into use generally have small cast defects, and their effect on the corrosion behavior of NAB in practical applications should be considered. Unfortunately, there is little literature available on this issue. Therefore, the effect of small cast defects on the corrosion behavior of the UNS C95810 alloy will be investigated in this paper by comparing the corrosion rate, morphology, product composition and electrochemical corrosion parameter of the specimen with and without defects.

2. Experiment

2.1. Materials and Corrosive Medium

The experimental material was UNS C95810 and its chemical composition is shown in Table 1.

Table 1. Composition of the UNS C95810.

Elements	Cu	Mn	Fe	Al	Ni	Zn, Sn, Pb, C, Si, P, Sb
UNS C95810 (wt.%)	79.9	1.39	4.8	9.24	4.43	Trace

The measured yield strength, tensile strength and elongation were 266 MPa, 679 MPa and 29.5%, respectively. Specimens with and without defects were taken from a UNS C95810 on a large as-cast sample, and they were cut out (of $10 \times 10 \times 10$ mm^3) and observed under optical microscopy to determine whether there was a defect on the surface of each sample. For simplicity, the specimens with and without defects are represented by S-Defect and S-Cast, respectively. To effectively control the uncertainties of the test, artificial seawater was used as a corrosive medium, and its main chemical composition from ASTM D1141 was listed in Table 2.

Table 2. Chemical components of artificial seawater [18].

Components	NaCl	$MgCl_2$	Na_2SO_4	$CaCl_2$	KCl	$NaHCO_3$	KBr	H_3BO_3	$SrCl_2$	NaF
Mass Concentration (g/L)	24.53	5.2	4.09	1.16	0.695	0.201	0.101	0.027	0.025	0.003

2.2. Preparation of Test Specimens

2.2.1. Specimens for Electrochemical Tests

Each of the specimens was mounted in plastic tubes by a two-component epoxy resin (WSR618, Nantong Xingchen Synthetic Material Co., Ltd, Nantong, China) with a Cu wire welded on their back, leaving an area of 0.38 cm^2 to be in contact with the artificial seawater. The study surface of each specimen was ground to 1000 grit by silicon carbide abrasive paper and polished. Then, the sample was cleaned with absolute ethanol to remove organics, embedded polishing media, etc.

2.2.2. Specimens for the Corrosion Test and Microstructure Analysis

Each specimen used for the weight loss measurements, scanning electron microscopy (SEM, Supra-55-sapphire, Carl Zeiss AG, Jena, Germany), confocal laser scanning microscopy (CLSM, OLS4000, Olympus, Tokyo, Japan) and X-ray diffraction (XRD, D/MAX-Ultima+, Rigaku, Tokyo, Japan) was mounted in plastic tubes by a two-component epoxy resin, leaving an area of 1 cm^2 to be in contact with the artificial seawater. The test surface of each specimen was ground to 1000 grit by silicon carbide abrasive paper and polished. Then, the sample was cleaned with absolute ethanol to remove organics, embedded polishing media, etc.

2.3. Measurement and Characterization

2.3.1. Weight Loss

The specimens prepared in 2.2.2 were ultrasonically cleaned in ethanol, dried with blowing air and weighed as m_1. They were then immersed in artificial seawater. After immersion for 72 h, 168 h, 432 h, 720 h and 960 h, the specimens were removed, rinsed with deionized water, dried, and weighed as m_2. Subsequently, they were immersed in HCl solution (containing 500 mL of commercially available 32% HCl and 500 mL of deionized water) for 2 min to remove the corrosion products and weighed as m_3 after ultrasonically cleaning and drying. The weight loss of a fresh sample after immersion in this solution for 2 min was less than 0.1 mg to ensure the base metal was not attacked vigorously by the acid. The weight loss of the specimen and the weight of the corrosion products were noted as m_1–m_3 and m_2–m_3, respectively. The weight loss rate = $\frac{m_1-m_3}{S \cdot t}$, where S is the surface area of the sample in contact with artificial seawater and t is the immersion time.

2.3.2. Microstructure Analysis and Corrosion Morphology Observation

The surface morphology of the specimen prepared in as described in Section 2.2.2 was observed by SEM, and the defects and their surrounding as-cast microstructures were analyzed by an energy dispersive spectrometer (EDS, Ultim Extreme, OXFORD Instruments, Oxford, UK) before and after immersion in artificial seawater. After immersion for 72 h and 240 h, they were removed, rinsed with deionized water, dried, and observed by CLSM. After immersion in artificial seawater for 720 h, the washed and dried specimens were subjected to SEM observation and EDS analysis.

2.3.3. XRD

The specimens prepared in 2.2.2 were tested using an X-ray diffractometer (D/MAX-Ultima$^+$, Rigaku, Tokyo, Japan) with Co Kα radiation. The diffraction angle range was from 10° to 100°. The specimens that had been tested by XRD were immersed in artificial seawater. After immersion for 240 h and 720 h, they were removed, rinsed with deionized water, dried, and tested by XRD.

2.3.4. Electrochemical Test

Electrochemical measurements were conducted in a typical three-electrode system. The working electrode was the surface of the specimen with an area of 0.38 cm^2 prepared in 2.2.1. A platinum net and a saturated calomel electrode served as the counter and reference electrode, respectively. The test medium was 100 mL of artificial seawater. The polarization curves were recorded at a sweep rate of 1 mV/s from -500 mV to 500 mV vs. the open circuit potential using an IM6ex Electrochemical Workstation (IM6ex, ZAHNER, Kronach, Germany). The voltage perturbation was 10 mV. To avoid any inconsistency of the experiment, two or three repeated tests were carried out for each sample. The electrochemical corrosion parameters in this paper were then obtained by fitting a linear polarization region of approximately 10 mV around the resting potential [19].

3. Result

3.1. Weight Loss

The test results of S-Cast and S-Defect weight loss in artificial seawater are shown in Figures 1–3. The results show that the weight loss and film weight of S-Cast and S-Defect generally show a gradual increase trend with corrosion time but their weight loss rates gradually decrease with corrosion time.

Moreover, the weight loss, film weight and weight loss rate of S-Defect are larger than that of S-Cast. Therefore, the defects accelerate the corrosion of UNS C95810. Defects induce a discontinuous corrosion product film on NAB in seawater, which results in a decrease in the protective effect of the film on the substrate and an increase in the corrosion rate and the thickness of the film.

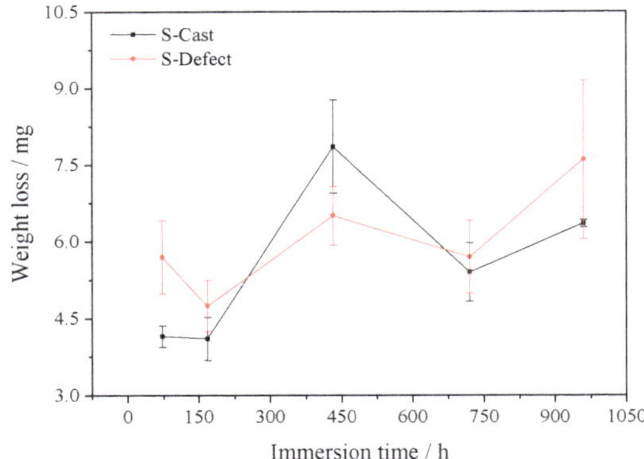

Figure 1. Weight loss of S-Cast and S-Defect immersed for different time periods in artificial seawater.

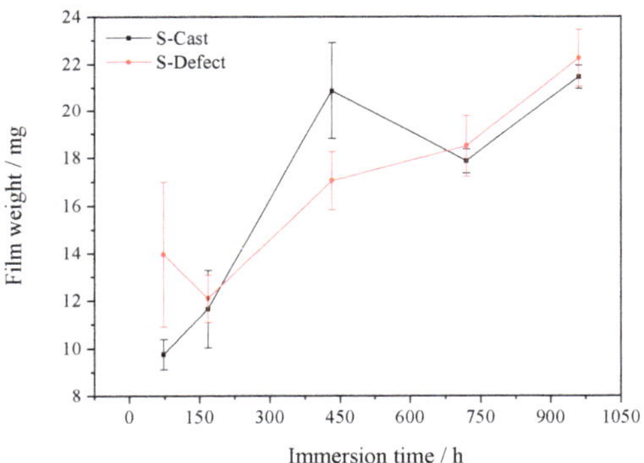

Figure 2. Film weight of S-Cast and S-Defect immersed for different time periods in artificial seawater.

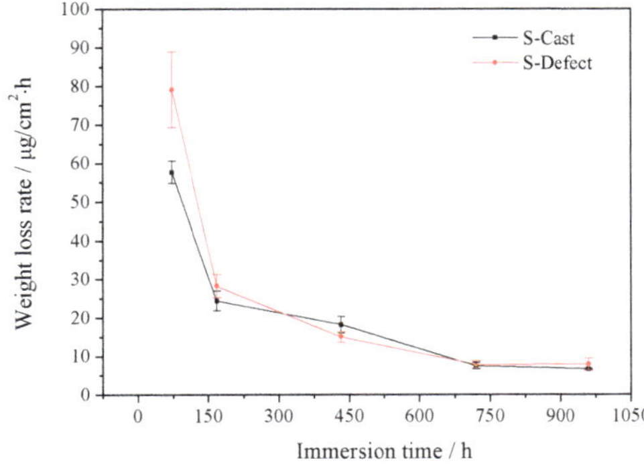

Figure 3. Weight loss rate of S-Cast and S-Defect immersed for different time periods in artificial seawater.

3.2. Microstructure and Corrosion Morphology

The normal microstructure of UNS C95810 is mainly composed of a α phase matrix and a dispersed κ phase (Figure 4).

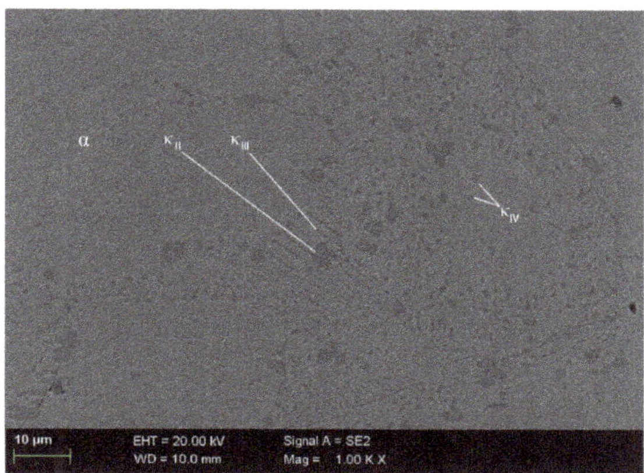

Figure 4. The microstructure of S-Cast before corrosion.

According to the composition and shape of the κ phase, it can be divided into κ_{II}, κ_{III} and κ_{IV}. The phase κ_{II} is an intermetallic compound based on Fe_3Al, which is flower-like or spherical with a size of 5 to 10 µm and distributed at the boundary of the α phase [9]. The κ_{III} phase is a lamellar intermetallic compound based on NiAl [9]. The κ_{IV} phase is an intermetallic compound based on Fe_3Al and is distributed in the α phase with a size smaller than 2 µm [9]. The size of the α phase is generally more than 100 µm [20]. To determine the chemical composition of the defects, EDS tests of selected area and selected points (Figure 5) are carried out. The results show that there is only copper in selected point while there are many oxygen and few chlorine elements in alloys in selected areas, as shown in Table 3.

Table 3. The composition of the studied specimens by SEM with EDS.

	Content of Element (atom.%)						
Specimen	O	Al	Si	S	Cl	Fe	Cu
Selected point in Figure 5							100
Selected area in Figure 5	59.29	13.39	4.09	1.68	1.59	1.64	18.32

It can be inferred that the microstructures in defects mainly include pure copper, an oxide of copper and aluminum and the κ phase.

S-Cast and S-Defect differ in their corrosion processes. S-Cast is preferentially corroded at the boundary of the αphase and some locations within the α phase, and the corrosion near the boundary of the α phase is more serious (Figure 6a,b).

As the corrosion progresses, the α phase also begins to undergo significant corrosion (Figure 6c,d). When the corrosion reaches 30 days, obvious corrosion products have formed on the surface of the alloy (Figure 6e). S-Defect is preferentially corroded at the defects, and the corrosion near the defects is more serious (Figure 7a,b).

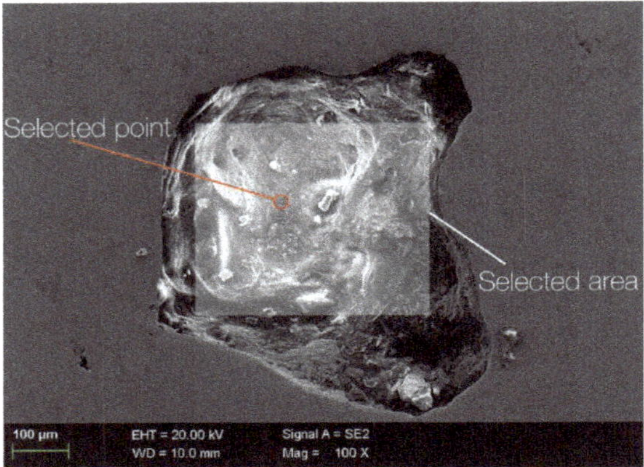

Figure 5. Surface and point scan of S-Defect before corrosion.

After the corrosion occurs at the defects and their vicinity, the corrosion at the αphase boundary and some locations within the αphase gradually becomes apparent (Figure 7a–d). When the corrosion reaches 30 days, a large area of severe corrosion is occurred at the defects compared to other locations (Figure 7e). In addition to affecting the location of preferential corrosion, defects increase the complexity of UNS C95810 microstructures, leading to an increase in tendency toward galvanic corrosion, thus accelerating the corrosion of UNS C95810 (Figure 6 vs. Figure 7). To determine whether there is a difference between the corrosion product composition of the defects and that of the surrounding as-cast microstructures, EDS analysis is performed on the position shown in Figure 7e. The results (Table 4) show that there is almost no composition difference between the corrosion product of the defects and that of the surrounding as-cast microstructures.

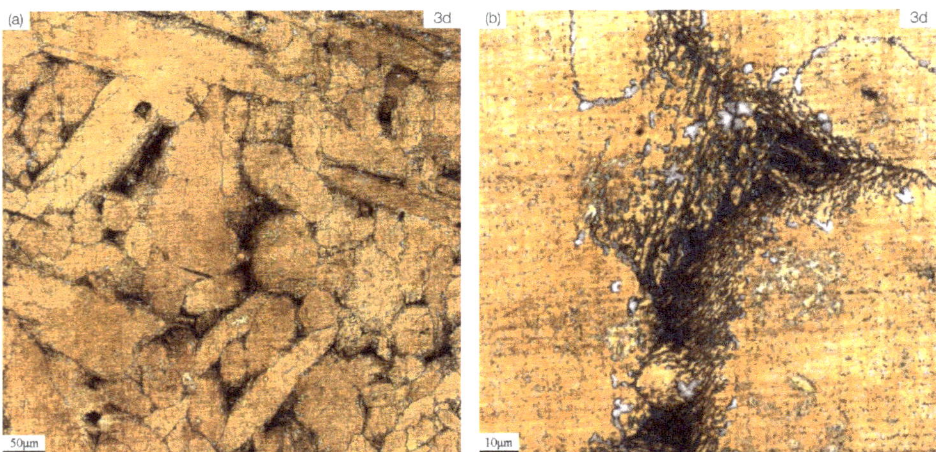

Figure 6. *Cont.*

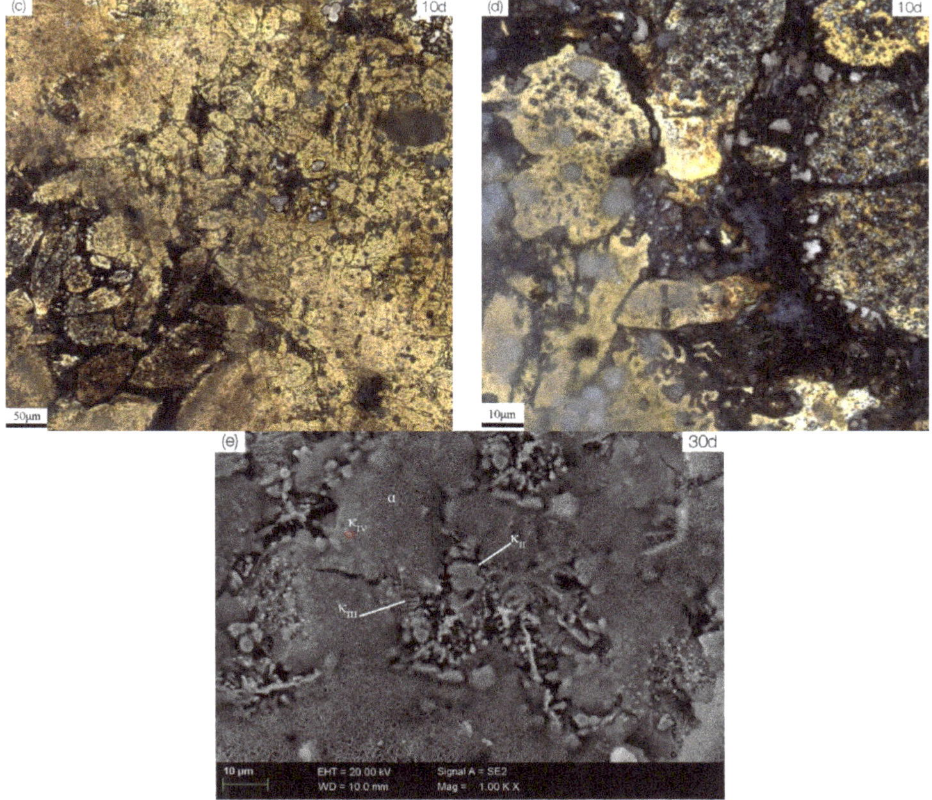

Figure 6. The corrosion morphology of for S-Cast immersed in artificial seawater at different times: (**a**) 3 d; (**b**) local magnification at 3 d; (**c**) 10 d; (**d**) local magnification at 10 d; (**e**) 30 d.

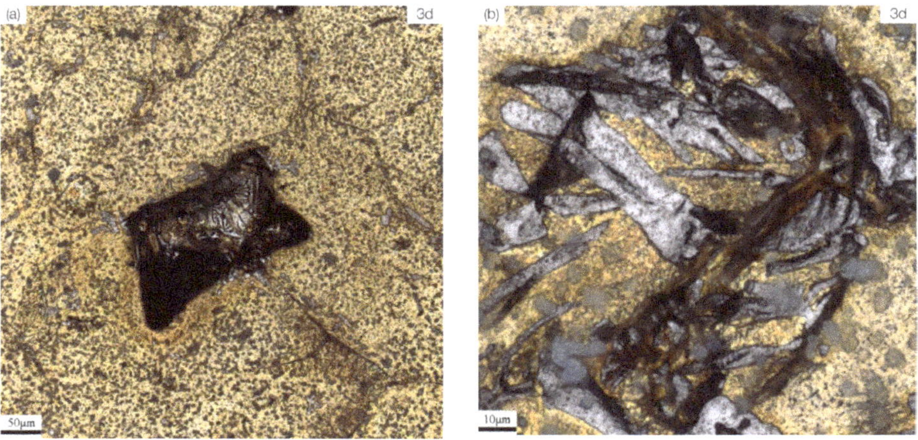

Figure 7. *Cont.*

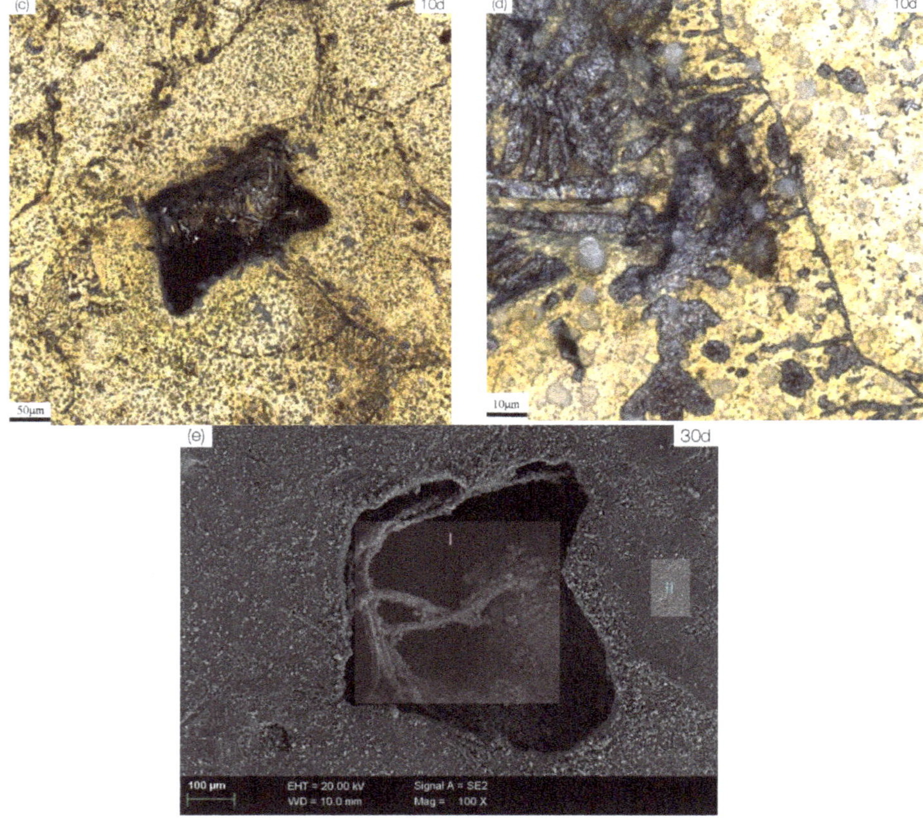

Figure 7. The corrosion morphology of for S-Defect immersed in artificial seawater at different times: (**a**) 3 d; (**b**) local magnification at 3 d; (**c**) 10 d; (**d**) local magnification at 10d; (**e**) 30 d.

Table 4. The composition of the studied position by SEM with EDS.

	Content of Element (atom.%)								
Position	O	Mg	Al	S	Cl	Ca	Fe	Ni	Cu
Zone I in Figure 7e	78.79		2.99	1.30	3.43	0.85	0.85		11.79
Zone II in Figure 7e	72.01	2.31	6.62	1.38	3.67	0.29	0.47	1.13	12.12

3.3. Corrosion Product Film

The XRD results of S-Defect and S-Cast for different immersion periods are shown in Figures 8 and 9. Cu_2O, $Cu(OH,Cl)_2$ and $Cu_2(OH)_3Cl$ are detected in the XRD results of S-Defect and S-Cast. Cu_2O changes little with corrosion while $Cu(OH,Cl)_2$ and $Cu_2(OH)_3Cl$ vary greatly with corrosion (Figure 8). $Cu(OH,Cl)_2$ is more abundant in the corrosion products after 10 days than that of $Cu_2(OH)_3Cl$ (Figure 8). Therefore, it is considered that during the corrosion process, Cu dissolves to Cu^+, and then it reacts with Cl^- to form $CuCl_2^-$ [13,20].

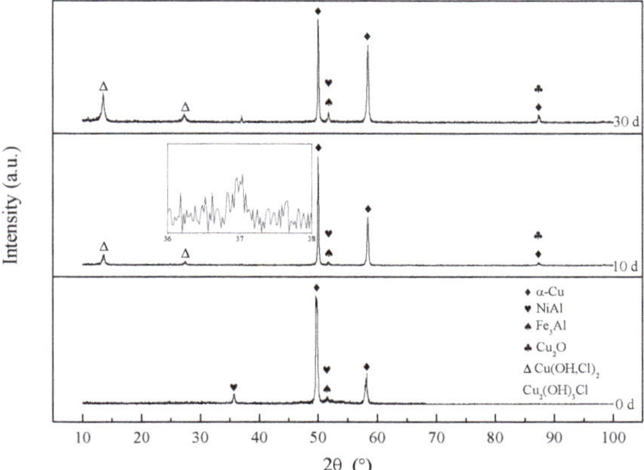

Figure 8. The XRD patterns of S-Defect immersed for different time periods in artificial seawater.

Over time, it converts to Cu_2O [20,21]. As the amount of Cu_2O increases, it will react with Cl^- to form more stable phases, such as $Cu_2(OH)_3$ Cl [20–22]. This is similar to the experimental results of Song [21] and Du [22]. Song [21] proposed that $Cu_2(OH)_3Cl$ was located on the outermost layer of a NAB corrosion product film. Du [22] proposed that $Cu(OH)_2$ and $CuCl_2$ were located between Cu_2O and $Cu_2(OH)_3Cl$. It is believed that since Al is more active than Cu, it first dissolves to form Al_2O_3, and then Cu dissolves to form CuO and Cu_2O. $Cu(OH)_2$, $CuCl_2$ and $Cu_2(OH)_3Cl$ are formed on the basis of products formed by the dissolution of Cu. Therefore, they should be located on the outer layers of the Al_2O_3, CuO and Cu_2O. The XRD results in this study indicate that the Al_2O_3, Cu_2O, CuO, $Cu(OH)_2$, $CuCl_2$ and $Cu_2(OH)_3Cl$ phases were detected on the corrosion product film with increasing corrosive time. In summary, the corrosion product film of UNS C95810 has a three-layer structure.

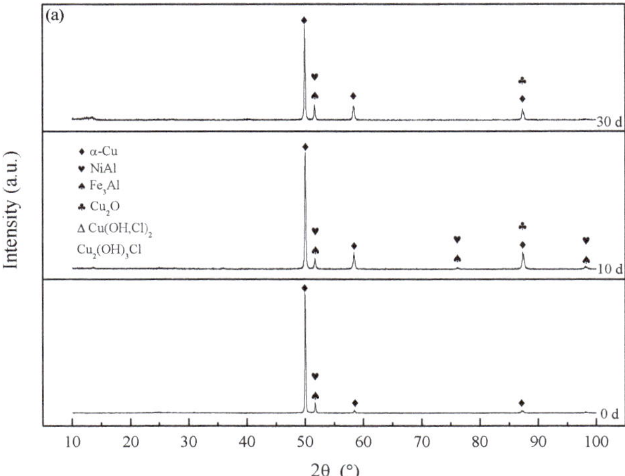

Figure 9. *Cont.*

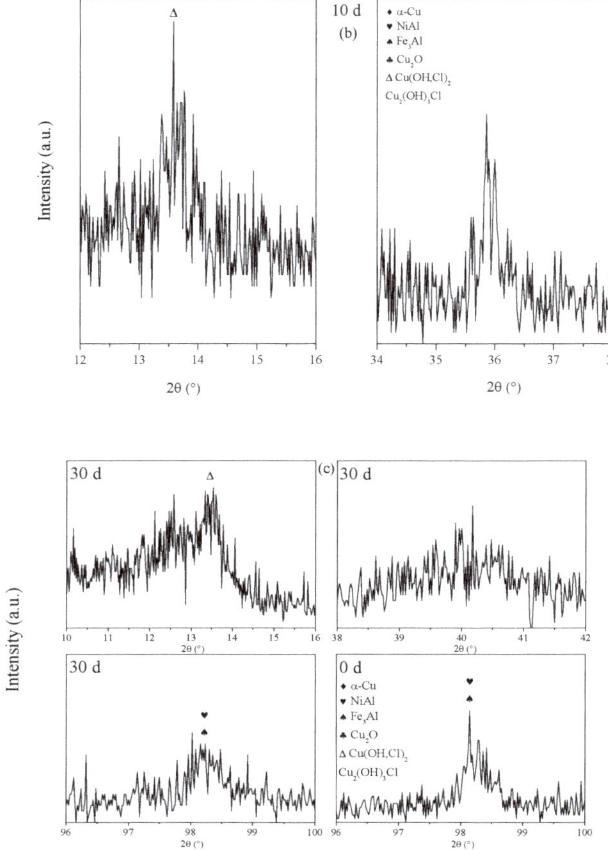

Figure 9. The XRD patterns of (**a**) S-Cast immersed for different time periods in artificial seawater and (**b**,**c**) S-Cast partial magnification.

3.4. Electrochemistry

The polarization curves of S-Cast and S-Defect in artificial seawater are shown in Figure 10.

The electrochemical corrosion parameters obtained by fitting a linear polarization region of approximately 10 mV around the resting potential are shown in Table 5.

The corrosion potential and corrosion current density of S-Defect are −86.31 mV and 0.23 μA·cm^{-2}, respectively, while those of S-Cast are −3.83 mV and 0.228 μA·cm^{-2}, respectively. Generally, defects decrease the corrosion potential and increase the corrosion current density. Defects increase the tendency toward galvanic corrosion of UNS C95810, which leads to the promotion of the corrosion driving force, thus accelerating corrosion. Therefore, S-Defect has a lower corrosion potential and a larger corrosion current density than those of S-Cast.

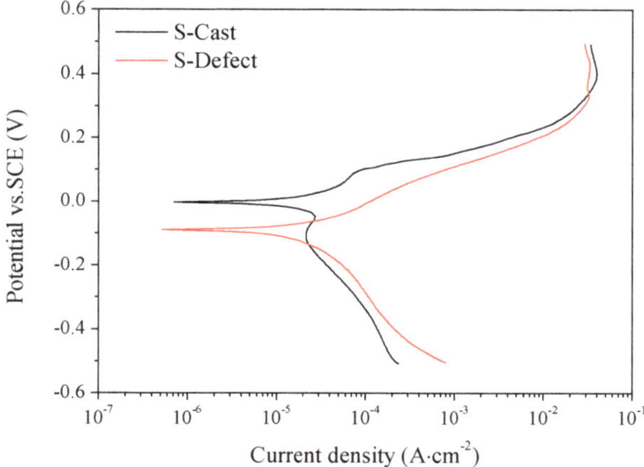

Figure 10. The polarization curve of S-Cast and S-Defect.

Table 5. Corrosion parameters of S-Cast and S-Defect.

Sample	Mean (Standard Deviation)		
	Potential vs. SCE (mV)	**Current Density (μA·cm^{-2})**	**Rp (kΩ·cm^2)**
S-Cast	-3.83 (1.93)	0.228 (0.12)	18.9 (2.12)
S-Defect	-86.31 (3.62)	0.23 (0.22)	14.9 (2.89)

4. Discussion

4.1. Corrosion Rate of UNS C95810

Defects can affect the diffusion of Cl^- and the corrosion driving force, thus accelerating the corrosion of NAB. The presence of defects causes the passivation film formed on NAB to become uneven. Generally, the electric field strength in the passivation film (ε) obeys the following formula: $\varepsilon = \frac{V}{l}$, where V is the potential difference between the two sides of the passivation film and l is the thickness of the passivation film [19]. Because the passivation film at the defects is thinner than that at other locations [19], the field strength at the defects is higher than that at other locations. This causes Cl^- to diffuse more easily to the defects and to enrich there (Figure 11 and Table 6), resulting in faster destruction of the passivation film.

Table 6. The composition of S-Defect immersed in artificial seawater for 3 h.

Content of Element (atom.%)							
Specimen	**O**	**Al**	**Mn**	**Cl**	**Fe**	**Ni**	**Cu**
Position I	73.30	3.33	-	2.13	2.90	1.44	16.90
Position II	-	15.07	1.41	-	2.65	2.33	78.54
Position III	-	29.27	2.6	-	46.35	12.96	8.82

Once the passivation film is broken, the substrate is directly in contact with the corrosive medium, which accelerates the corrosion rate. Therefore, the presence of defects affects the diffusion of Cl^- and the corrosion process.

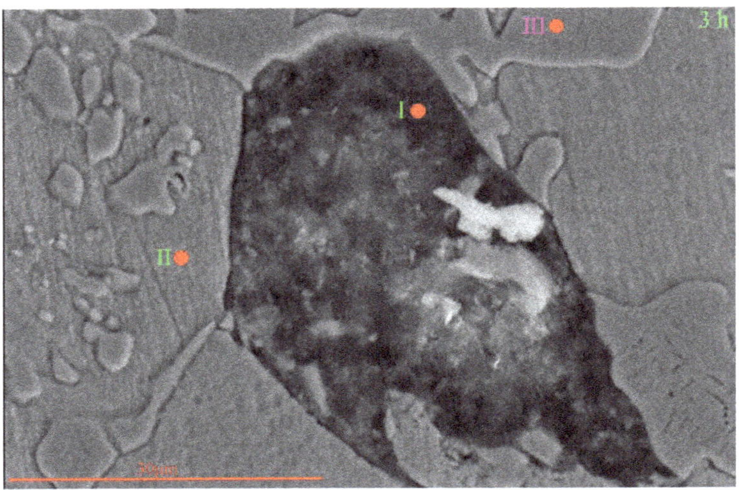

Figure 11. SEM image of S-Defect immersed in artificial seawater for 3 h.

4.2. Corrosion Product Film Structure of S-Defect

The results of the XRD test show that the X-ray diffraction pattern of both the S-Cast and S-Defect samples are basically the same, as shown in Figures 8 and 9, but the intensity of some phases in the corrosion products is different. Therefore, they have similar corrosion product film structures but different corrosion product film thicknesses (Figures 2, 8 and 9). It had been reported that a corrosion product film with Cu_2O in the outer layer and Al_2O_3 in the inner layer was formed on a NAB [4] and a ternary Cu-Al-Ni alloy [23,24]. Cu_2O and Al_2O_3 were formed by the following reactions [12,13,22,24,25]:

$$Al(s) + 4Cl^- \rightarrow AlCl_4{}^- + 3e^- \quad (1)$$

$$AlCl_4{}^- + 2H_2O \rightarrow Al_2O_3(s) + 4Cl^- + 3H^+ \quad (2)$$

$$Cu(s) \rightarrow Cu^+ + e^- \quad (3)$$

$$Cu^+ + Cl^- \rightarrow CuCl(s) \quad (4)$$

$$CuCl(s) + Cl^- \rightarrow CuCl_2{}^- \quad (5)$$

$$2CuCl_2{}^- + H_2O \rightarrow Cu_2O(s) + 4Cl^- + 2H^+ \quad (6)$$

Cu_2O is a p-type semiconductor with cation vacancies. It can accept foreign ions, such as Ni and Fe ions, to incorporate into inner cation vacancies, further improving the protection from the film [14,26,27]. Based on previous research [4,14,21,23,24,26,27], Cu_2O should be the main corrosion product in the inner layer of the film for both S-Cast and S-Defect. Fe and Ni is enriched in the inner layer by incorporating into the lattice of Cu_2O. The microstructures of the S-Cast and S-Defect are complex, resulting in different Cu dissolution rates in different phases. Because Al_2O_3 cannot form simultaneously on different phases, the discontinuous Al_2O_3 layer on the surface is not impermeable to the passage of cuprous cations [20]. As a result, the inner layer of the corrosion product film consists of Al_2O_3 and Cu_2O with the incorporation of Fe and Ni [20]. In the XRD tests, $Cu(OH,Cl)_2$ and $Cu_2(OH)_3Cl$ are detected in the corrosion products of S-Cast and S-Defect. They could form by the following reactions [13,25,26,28]:

$$Cu^+ + \frac{1}{2}O_2 + e^- \rightarrow CuO(s) \quad (7)$$

$$CuO(s) + H_2O \rightarrow Cu(OH)_2(s) \quad (8)$$

$$Cu_2O(s) + Cl^- + 2H_2O \rightarrow Cu_2(OH)_3Cl(s) + H^+ + 2e^- \quad (9)$$

According to Song's research [21] on a NAB corrosion product film, $Cu_2(OH)_3Cl$ was located in the outer layer of the NAB corrosion product film. In the structure of the NAB corrosion product film proposed by Du [22], $Cu(OH)_2$ or CuO was located between Cu_2O and $Cu_2(OH)_3Cl$. Therefore, combined with the above analysis and the results of the XRD test, it can be concluded that the structure of the S-Defect corrosion product film is as follows (Figure 12): the inner layer is Al_2O_3 and Cu_2O with the incorporation of Fe and Ni, the middle layer is $Cu(OH,Cl)_2$ or CuO, and the outer layer is $Cu_2(OH)_3Cl$.

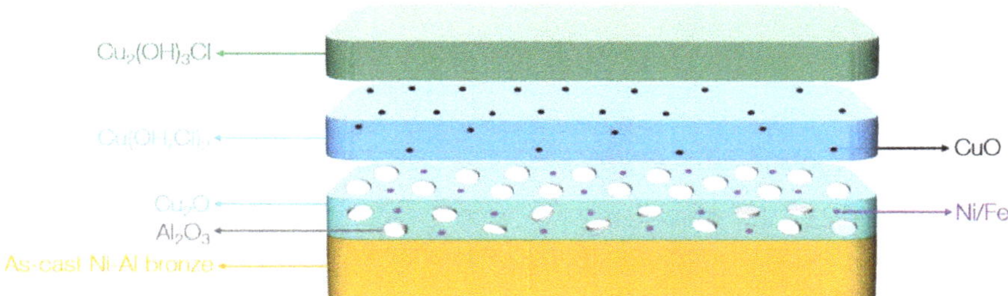

Figure 12. The corrosion product film structure of S-Defect immersed in artificial seawater for a long time.

4.3. Microstructure and Corrosion Behavior of UNS C95810

The stability of the phase and its corrosion products can affect the corrosion behavior, and the stability of the phase can be expressed with a work function. There is a corresponding relationship between work function and Volta potential: $V_{CPD} = \left(\Phi_{tip} - \Phi_{sample}\right)/e$, where V_{CPD} is the Volta potential, Φ_{tip} and Φ_{sample} are the work functions of the tip and the sample, respectively, and e is the value of the electronic charge [29]. The value of Φ_{tip} is constant. Thus, higher V_{CPD} results in lower Φ_{sample}. The work function is defined as the minimum energy required for an electron to escape from the surface of a solid. A lower work function of a material represents more likely occurrences of corrosion [30]. Therefore, a substance with a higher Volta potential is more susceptible to corrosion. From the measurements of the Volta potential of an as-cast NAB by Song, it can be seen that the Volta potential of the κ_{IV} phase was higher than that of the α phase but lower than that of the κ_{II} phase [21]. Hence, the stability of the phases from low to high should be κ_{II}, κ_{IV} and α. However, in near-neutral artificial seawater, Al_2O_3 is formed on the surface of the Al-rich κ phase and its stability is much higher than that of Cu_2O, so the κ phase as the cathode phase will not continue to corrode but accelerates the corrosion rate of the surrounding microstructures [31]. Since the κ_{IV} phase and κ_{II} phase are located in the α phase and the boundary of the α phase, corrosion preferentially occurs in the αphase boundary and some locations in the α phase. The Volta potential between the κ_{II} phase and α phase is 60–80 mV, while that between the κ_{IV} phase and α phase is 20–40 mV [21]. The areas with a higher Volta potential have a stronger driving force for corrosion. Therefore, the corrosion in the α phase boundary is more prone to occur than that in the α phase.

The existence of defects accelerates the corrosion process to a certain extent, mainly due to the following reasons. First, the defect microstructures mainly include oxides of copper and aluminum, along with pure copper. This is a more complex phase composition than the as-cast microstructures. Such a phase composition makes it easier for the inside of defects and between the defects and the surrounding microstructures to form a galvanic corrosion, thereby accelerating the corrosion of the defects and the microstructures near the defects. Second, defects not only make the microstructures of as-cast NAB more complex but also cause the inhomogeneity of the corrosion product film, resulting in greater stress in the growth process of the film [21]. The increase in growth stress can cause the film to become loose or even crack, thereby causing the corrosive medium to more easily enter the film and corrode the substrate. Finally, the corrosion potential of UNS C95810 decreases due to defects. The lower corrosion potential provides a more powerful driving force for corrosion, thereby accelerating the corrosion of UNS C95810.

4.4. Corrosion Mechanism of UNS C95810

According to the experimental results obtained in this paper and related literature [21], the stability of the phases in NAB from low to high is κ_{II}, κ_{IV}, κ_{III} and. Since the κ phase is an Al-rich phase and Al is a more active metal than Cu, the κ phase will preferentially corrode and form Al_2O_3 on the surface when NAB is in contact with seawater. Al_2O_3 is a more stable oxide than that of Cu_2O and CuO, so the κ phase covered by Al_2O_3 will no longer corrode as an anode phase but will act as a cathode phase in galvanic corrosion to accelerate the corrosion of the surrounding microstructures. The above results in the formation of Cu_2O, $Cu(OH,Cl)_2$ and $Cu_2(OH)_3Cl$ on the surface of the surrounding microstructures. The Volta potential between the κ_{II}, κ_{III}, κ_{IV} phases and the α phase is 60–80 mV, 10–30 mV and 20–40 mV, respectively [21]. Therefore, the corrosion near the κ_{II} phase is the fastest and the most serious, followed by that near the κ_{IV} phase, and finally that near the κ_{III} phase. In summary, the specific corrosion behavior of S-Cast can be summarized as Figure 13.

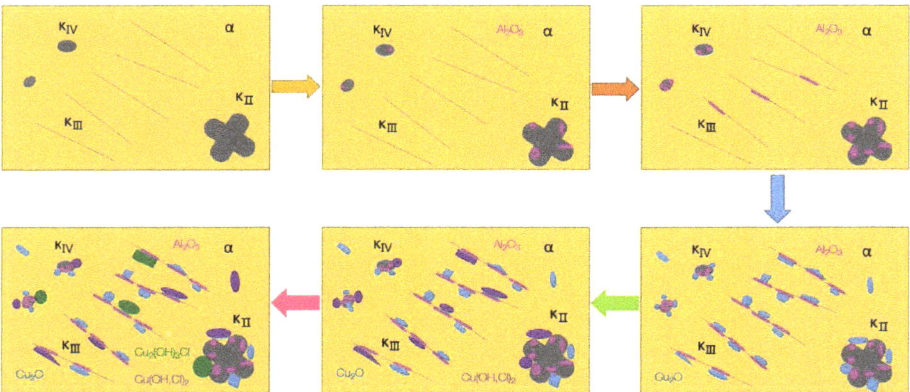

Figure 13. Process mechanism of corrosion behavior of S-Cast immersed in artificial seawater.

For S-Defect, in addition to the corrosion characteristics of S-Cast, it has some additional characteristics due to the presence of defects. The presence of defects not only promotes the formation of galvanic corrosion between the internal microstructures of the defects but also promotes that between the defects and the surrounding microstructures, thereby greatly increasing the tendency toward galvanic corrosion. Therefore, the corrosion behavior of S-Defect can be obtained by combining the corrosion process of S-Cast with an additional galvanic corrosion process introduced due to the existence of defects. Its corrosion mechanism can be summarized as shown in Figure 14.

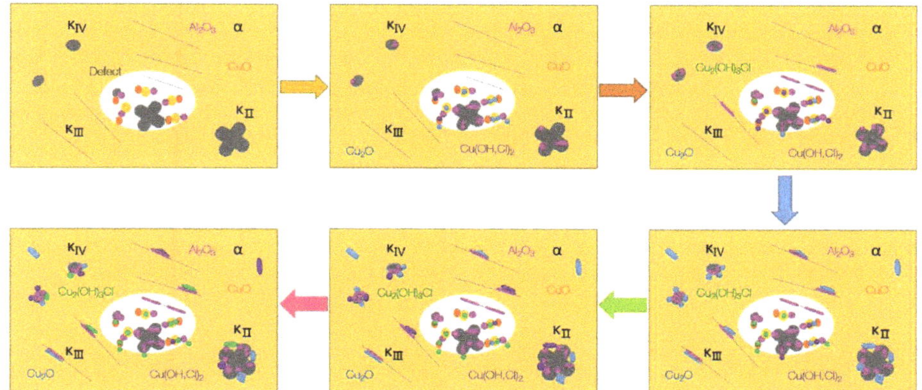

Figure 14. Process mechanism of corrosion behavior of S-Defect immersed in artificial seawater.

5. Conclusions

In this paper, the effects of cast defects on the corrosion behavior and mechanism of the UNS C95810 in artificial seawater were studied. The conclusions are as follows:

1) The microstructure of the UNS C95810 alloy consists of an α phase and a κ phase. Except for these cast phases, there are oxides of copper and aluminum, along with pure copper in defects.
2) Compared to those of S-Cast, S-Defect has a thicker corrosion product film, more negative corrosion potential and greater corrosion current density.
3) The corrosion product film of both S-Defect and S-cast consists of three layers. The inner layer is Al_2O_3 and Cu_2O with the incorporation of Fe and Ni, the middle layer is $Cu(OH,Cl)_2$ or CuO, and the outer layer is $Cu_2(OH)_3Cl$. The defects do not affect the phase composition of the corrosion products, but they have an effect on the amount of the specific phase in the corrosion products.

Author Contributions: Conceptualization, X.Z., Y.Q. and Z.Z.; methodology, X.Z., Y.Q. and Z.Z.; software, X.Z.; validation, J.W.; formal analysis, X.Z. and Z.Z.; investigation, X.Z. and Z.Z.; resources, J.Z., L.Q., D.H., Y.Q. and Z.Z.; data curation, X.Z.; writing—original draft preparation, X.Z.; writing—review and editing, X.Z., Z.Z. and Y.Q.; visualization, X.Z. and J.W.; supervision, Z.Z. and Y.Q.; project administration, Z.Z. and Y.Q.; funding acquisition, Z.Z. and Y.Q. All authors have read and agreed to the published version of the manuscript.

Funding: This research was funded by Project of Equipment Pre-research Field Fund, grant number 61409220304 and Equipment Pre-research Sharing Technology Project, grant number 41404010306,41423060314.

Acknowledgments: Thanks to Zeng Kai for his help in the experimental part of this article.

Conflicts of Interest: The authors declare no conflict of interest.

References

1. Culpan, E.A.; Rose, G. Corrosion Behaviour of Cast Nickel Aluminium Bronze in Sea Water. *Br. Corros. J.* **2014**, *14*, 160–166. [CrossRef]
2. Al-Hashem, A.; Caceres, P.G.; Shalaby, H.M.; Riad, W.T. Cavitation Corrosion Behavior of Cast Nickel-Aluminum Bronze in Seawater. *Corrosion* **1995**, *51*, 331–342. [CrossRef]
3. Wharton, J.A.; Barik, R.C.; Kear, G.; Wood, R.J.K.; Stokes, K.R.; Walsh, F.C. The corrosion of nickel–aluminium bronze in seawater. *Corros. Sci.* **2005**, *47*, 3336–3367. [CrossRef]
4. Schüssler, A.; Exner, H.E. The corrosion of nickel-aluminium bronzes in seawater—I. Protective layer formation and the passivation mechanism. *Corros. Sci.* **1993**, *34*, 1793–1802. [CrossRef]
5. Schüssler, A.; Exner, H.E. The corrosion of nickel-aluminium bronzes in seawater—II. The corrosion mechanism in the presence of sulphide pollution. *Corros. Sci.* **1993**, *34*, 1803–1811. [CrossRef]

6. Ferrara, R.J.; Caton, T.E. Review of dealloying of cast aluminum bronze and nickel-aluminum bronze alloys in sea water service. *Mater. Perform.* **1982**, *21*, 30–34.
7. Neodo, S.; Carugo, D.; Wharton, J.A.; Stokes, K.R. Electrochemical behaviour of nickel–aluminium bronze in chloride media: Influence of pH and benzotriazole. *J. Electroanal. Chem.* **2013**, *695*, 38–46. [CrossRef]
8. Lorimer, G.W.; Hasan, F.; Iqbal, J.; Ridley, N. Observation of microstructure and corrosion behaviour of some aluminium bronzes. *Br. Corros. J.* **2013**, *21*, 244–248. [CrossRef]
9. Hasan, F.; Jahanafrooz, A.; Lorimer, G.W.; Ridley, N. The morphology, crystallography, and chemistry of phases in as-cast nickel-aluminum bronze. *Metall. Trans. A* **1982**, *13*, 1337–1345. [CrossRef]
10. Weill-Couly, P.; Arnaud, D. Effect of the Composition and Structure of Aluminum Bronzes on their Behavior in Service. *Fonderie* **1973**, 123–135.
11. Tuthill, A.H. Guidelines for the use of copper alloys in seawater. *Mater. Perform.* **1987**, *26*, 12–22.
12. Ateya, B.G.; Ashour, E.A.; Sayed, S.M. Corrosion of Î±Al bronze in saline water. *J. Electrochem. Soc.* **1994**, *141*, 71–77. [CrossRef]
13. Kato, C.; Castle, J.E.; Ateya, B.G.; Pickering, H.W. On the mechanism of corrosion of Cu-9.4Ni-1.7Fe alloy in air saturated aqueous NaCl solution 2. Composition of the protective surface layer. *J. Electrochem. Soc.* **1980**, *127*, 1897–1903. [CrossRef]
14. North, R.F.; Pryor, M.J. The influence of corrosion product structure on the corrosion rate of Cu-Ni alloys. *Corros. Sci.* **1970**, *10*, 297–311. [CrossRef]
15. Sabbaghzadeh, B.; Parvizi, R.; Davoodi, A.; Moayed, M.H. Corrosion evaluation of multi-pass welded nickel–aluminum bronze alloy in 3.5% sodium chloride solution: A restorative application of gas tungsten arc welding process. *Mater. Des.* **2014**, *58*, 346–356. [CrossRef]
16. Hanke, S.; Fischer, A.; Beyer, M.; Santos, J.D. Cavitation erosion of NiAl-bronze layers generated by friction surfacing. *Wear* **2011**, *273*, 32–37. [CrossRef]
17. Ni, D.R.; Xiao, B.L.; Ma, Z.Y.; Qiao, Y.X.; Zheng, Y.G. Corrosion properties of friction–stir processed cast NiAl bronze. *Corros. Sci.* **2010**, *52*, 1610–1617. [CrossRef]
18. Wang, J.; Yan, F.; Xue, Q. Tribological behaviors of some polymeric materials in sea water. *Sci. Bull.* **2009**, *54*, 4541–4548. [CrossRef]
19. Cao, C.N. *Principles of Electrochemistry of Corrosion*, 3rd ed.; Chemical Industry Press: Beijing, China, 2008.
20. Song, Q.N.; Zheng, Y.G.; Ni, D.R.; Ma, Z.Y. Characterization of the Corrosion Product Films Formed on the As-Cast and Friction-Stir Processed Ni-Al Bronze in a 3.5 wt% NaCl Solution. *Corrosion* **2015**, *71*, 606–614. [CrossRef]
21. Song, Q.N. Effect of Friction Stir Processing on the Cavitation Erosion and Corrosion Behaviors of As-Cast Ni-Al Bronze. Ph.D. Thesis, University of Chinese Academy of Sciences, Beijing, China, 2015.
22. Du, C.Y. Research on Corrosion Resistance of the Nickel-Aluminum Bronze and the Surface Treatment on It. Master's Thesis, Jiangsu University of Science and Technology, Zhenjiang, China, 2014.
23. Badawy, W.A.; Al-Kharafi, F.M.; El-Azab, A.S. Electrochemical behaviour and corrosion inhibition of Al, Al-6061 and Al–Cu in neutral aqueous solutions. *Corros. Sci.* **1999**, *41*, 709–727. [CrossRef]
24. Nady, H.; Helal, N.H.; El-Rabiee, M.M.; Badawy, W.A. The role of Ni content on the stability of Cu–Al–Ni ternary alloy in neutral chloride solutions. *Mater. Chem. Phys.* **2012**, *134*, 945–950. [CrossRef]
25. Mansfeld, F.; Liu, G.; Xiao, H.; Tsai, C.H.; Little, B.J. The corrosion behavior of copper alloys, stainless steels and titanium in seawater. *Corros. Sci.* **1994**, *36*, 2063–2095. [CrossRef]
26. Popplewell, J.M.; Hart, R.J.; Ford, J.A. The effect of iron on the corrosion characteristics of 90-10 cupro nickel in quiescent 3·4%NaCl solution. *Corros. Sci.* **1973**, *13*, 295–309. [CrossRef]
27. Milošev, I.; Metikoš-Huković, M. The behaviour of Cu-xNi (x = 10 to 40 wt%) alloys in alkaline solutions containing chloride ions. *Electrochim. Acta* **1997**, *42*, 1537–1548. [CrossRef]
28. Lenglet, M.; Lopitaux, J.; Leygraf, C.; Odnevall, I.; Carballeira, M.; Noualhaguet, J.C.; Guinement, J.; Gautier, J.; Boissel, J. Analysis of corrosion products formed on copper in Cl2/H2S/NO2 exposure. *J. Electrochem. Soc.* **1995**, *142*, 3690–3696. [CrossRef]
29. Li, M.; Guo, L.Q.; Qiao, L.J.; Bai, Y. The mechanism of hydrogen-induced pitting corrosion in duplex stainless steel studied by SKPFM. *Corros. Sci.* **2012**, *60*, 76–81. [CrossRef]

30. Li, W.; Li, D.Y. Variations of work function and corrosion behaviors of deformed copper surfaces. *Appl. Surf. Sci.* **2005**, *240*, 388–395. [CrossRef]
31. Wharton, J.A.; Stokes, K.R. The influence of nickel–aluminium bronze microstructure and crevice solution on the initiation of crevice corrosion. *Electrochim. Acta* **2008**, *53*, 2463–2473. [CrossRef]

© 2020 by the authors. Licensee MDPI, Basel, Switzerland. This article is an open access article distributed under the terms and conditions of the Creative Commons Attribution (CC BY) license (http://creativecommons.org/licenses/by/4.0/).

Article

Effects of Different Ions and Temperature on Corrosion Behavior of Pure Iron in Anoxic Simulated Groundwater

Teng Li [1,*], Guokai Huang [2], Yanpeng Feng [2], Miao Yang [2], Lingyu Wang [1], Daqing Cui [1,3] and Xian Zhang [4,*]

1 Department of Radiochemistry, China Institute of Atomic Energy, Beijing 102413, China; linyuwang2955@163.com (L.W.); daqing.cui@studsvik.se (D.C.)
2 School of Chemistry, Chemical Engineering and Life Sciences, Wuhan University of Technology, Wuhan 430070, China; guokaihuang@whut.edu.cn (G.H.); fengyanpeng1996@gmail.com (Y.F.); yangmiao@whut.edu.cn (M.Y.)
3 Department of Materials and Environmental Chemistry, Stockholm University, SE-106 91 Stockholm, Sweden
4 The State Key Laboratory of Refractories and Metallurgy, Hubei Province Key Laboratory of Systems Science in Metallurgical Process, Collaborative Innovation Center for Advanced Steels, Wuhan University of Science and Technology, Wuhan 430081, China
* Correspondence: liteng@ciae.ac.cn (T.L.); xianzhang@wust.edu.cn (X.Z.)

Received: 22 May 2020; Accepted: 12 June 2020; Published: 15 June 2020

Abstract: As a typical material of the insert in high-level radioactive waste (HLW) geological disposal canisters, iron-based materials will directly contact with groundwater after the failure of a metallic canister, acting as a chemical barrier to prevent HLW leaking into groundwater. In this paper, anoxic groundwater was simulated by mixing 10 mM NaCl and 2 mM $NaHCO_3$ purged by Ar gas (containing 0.3% CO_2) with different added ions (Ca^{2+}, $CO_3{}^{2-}$ and $SiO_3{}^{2-}$) and operation temperatures (25, 40 and 60 °C). An electrochemical measurement, immersion tests and surface characterization were carried out to study the corrosion behavior of pure iron in the simulated groundwater. The effects of Ca^{2+} on the corrosion behavior of iron is negligible, however, Cl^- plays an important role in accelerating the corrosion activity with the increased concentration and temperature. With increased concentrations of $CO_3{}^{2-}$ and $SiO_3{}^{2-}$, the corrosion resistance of iron is largely improved, which is attributed to the formation of a uniform passivation film. The independent effects of temperature on the corrosion behavior of iron are resulted from the repeated passivation–dissolution processes in the formation of the passivation film, resulting from the synergistic effects of $CO_3{}^{2-}$/$SiO_3{}^{2-}$ and Cl^-. The formation of ferric silicate is dominant in the passivation film with the addition of $SiO_3{}^{2-}$, which effectively protects the iron surface from corrosion.

Keywords: pure iron; groundwater; corrosion behavior; ions; temperature

1. Introduction

The metallic canister is the first barrier to prevent high-level radioactive waste (HLW) from leakage in different countries. Although with the same concept on selecting the candidate materials for the canister, i.e., with good mechanical properties and corrosion resistance, the relevant countries have made relatively different choices. To be specific, a Cu canister with a nodular cast iron insert is adopted in Sweden and Finland [1,2]; a Cu coating on a welded steel vessel is being designed in Canada [3,4]; a stainless steel canister with a glass or ceramic waste form is being planned in the US [5]; and a stainless steel with cast iron is selected by France [6]. Iron-based materials are considered not only because of their mechanical performance but also their nature as reductants [1].

Pure iron has been proved to be able to reduce highly dissolved U(VI), Se(IV) and Tc(VII) to insoluble UO_2, $FeSe_2$ and TcO_2 in an anoxic solution [1,7–11] since the airborne O_2 in the deep geological repository will have been consumed completely by the iron minerals before the canister failure [12,13]. In case of canister failure, the intrusion of underground water together with a radiation effect may lead to the dissolution of toxic U(VI). However, based on the nature of pure iron, it can be predicted that the toxic dissolved U(VI) from HLW can be reduced to insoluble UO_2, thereby not entering the biosphere. Considering this, the service behavior and development of corrosion products of iron-based materials in a groundwater environment may affect the reduction and inhibition of HLW. Therefore, it is of utmost significance to study the corrosion behavior of pure iron in a groundwater environment under anoxic conditions.

The lifetime of a canister is affected by environmental factors such as dissolved oxygen, composition of groundwater and temperature in the repository. Based on the surveys in different sites for repository [14], it demonstrates that the main cations are Na^+, K^+, Ca^{2+} and Mg^{2+}, while the main anions are HCO_3^-/CO_3^{2-}, Cl^- and SO_4^{2-}. At the beginning period of the geological disposal, the temperature rises in a short period of time due to the release of residual heat from the high-level radioactive waste, and then decreases gradually. N.R. Smart and co-workers [15] reported that the corrosion rate of low-carbon steel in an anoxic simulated groundwater solution of Sweden at 30–85 °C is below 0.1 μm/y. C. Liu et al. [16] measured the average corrosion rate of low-carbon steel in aerobic and unsaturated bentonite after irradiation and thermal aging treatment at 90 °C. F. A. Martin et al. [17] studied the corrosion behavior of low-alloy steel in anoxic simulated groundwater using the electrochemical impedance method at 90 °C and the results showed that the corrosion rate decreased with the increase in reaction time. The research on the candidate canister materials for a deep geological repository mainly focused on iron-based materials, such as low-carbon steel, low-alloy steel and nodular cast iron [18–22], while as the basic of iron-based materials, pure iron has rarely been studied regarding its corrosion behavior in a deep geological disposal repository.

Pure iron can be oxidized by water ferric iron via ferrous iron even in an anoxic solution without strong oxidants, i.e., O_2, H_2O_2 and so on [23]. Consistent with the anoxic corrosion of pure iron, the solution pH and corrosion products increase while Eh decreases. In addition, co-existing ions in the solution have different effects on the corrosion of pure iron under anoxic conditions. To be specific, cations like As(V), Se(VI) can be reduced by pure iron nanoparticles [24], while anions like nitrate and sulfate may lower the reduction rate of Se(VI) [25]. Therefore, it is important to investigate the effects of other anions, like Cl^-, HCO_3^-, CO_3^{2-} and SiO_3^{2-}, which are typical in underground water, on the anoxic corrosion of pure iron.

In this paper, the corrosion behavior of pure iron in anoxic simulated groundwater was studied. Specifically, the anoxic simulated groundwater (SG) solution was simulated by using 10 mM NaCl and 2 mM $NaHCO_3$ [1], with different concentrations of $CaCl_2$, Na_2CO_3 and Na_2SiO_3, respectively. The corrosion behavior of pure iron coupons in these simulated solutions was studied via an immersion test, electrochemical test and surface characterization. The effect of temperature was also evaluated by varying the temperature from 25 to 60 °C. Based on the results, a corrosion mechanism of pure iron affected by different ions and temperatures in anoxic simulated groundwater is proposed.

2. Experimental

2.1. Materials

The chemical composition of the commercial iron plate (> 99.99%) is shown in Table 1. For the immersion experiment, the iron plate was cut into specimens with dimension of 20 mm × 10 mm × 2 mm. Each sample was mechanically grounded to 2000 grit by silicon carbide papers, washed and dried with ethanol and then stored in a desiccator. For the electrochemical experiment, the iron plate was cut into specimens with dimension of 10 mm× 10 mm× 2 mm to make electrodes. The prepared working electrodes were grounded to 2000 grit and polished to a mirror surface, after which they were washed and dried with ethanol and then stored in a desiccator.

Table 1. Chemical composition of pure iron.

Material	Composition (wt.%)					
	C	S	P	Si	Mn	Fe
Pure iron	0.003	0.002	0.002	0.002	0.001	balance

2.2. Simulated Groundwater Solution

The simulated groundwater includes the major component of an SG solution (10 mM NaCl + 2 mM $NaHCO_3$) and different ions (Ca^{2+}, $CO_3{}^{2-}$, $SiO_3{}^{2-}$). The concentrations of ions were set as 0, 1, 5, 10 and 20 mM, respectively, by adding $CaCl_2$, Na_2CO_3 and Na_2SiO_3 solution. The details of the chemical composition of the simulated groundwater for the different groups are shown in Table 2.

Table 2. Chemical composition of simulated groundwater.

Solution	Chemicals Concentration (mM)				
	NaCl	$NaHCO_3$	$CaCl_2$	Na_2CO_3	Na_2SiO_3
SG	10	2	0	0	0
SG + $CaCl_2$	10	2	1	0	0
	10	2	5	0	0
	10	2	10	0	0
	10	2	20	0	0
SG + Na_2CO_3	10	2	0	1	0
	10	2	0	5	0
	10	2	0	10	0
	10	2	0	20	0
SG + Na_2SiO_3	10	2	0	0	1
	10	2	0	0	5
	10	2	0	0	10
	10	2	0	0	20

2.3. Electrochemical Measurement

In order to figure out the corrosion behavior of pure iron in the anoxic simulated groundwater, the measurements of the potentiodynamic polarization curve and electrochemical impedance spectroscopy (EIS) of the iron samples were carried out using a Zahner electrochemical workstation, Germany, with varied ions concentrations (see Table 2) and temperatures (25, 40 and 60 °C). An electrochemical cell of three electrodes (the iron sample acted as the working electrode, a saturated Hg/HgCl electrode worked as the reference electrode and a platinum foil served as the counter electrode) was set up, followed by purging the respective solution with argon gas (containing 0.3% CO_2) for 20 min to obtain the H_2- and O_2-free solution. Then, the open-circuit potential was tested for 20 min and when the potential was stable, EIS was tested with a direct current potential of 0 V relative to the open-circuit, alternate current (AC) amplitude of 10 mV, initial frequency of 10,000 Hz and termination frequency of 0.01 Hz. Finally, the potentiodynamic polarization curve was tested with a potential range of −0.3~+0.8/1.2 V relative to the open-circuit potential. The potential sweep rate is 0.5 mV/s.

2.4. Immersion Test

Immersion tests of iron samples were carried out in 20 mM SG and SG + $CaCl_2$/Na_2CO_3/Na_2SiO_3 solution at 25, 40 and 60 °C for 4 weeks, which were replaced each week. Parallel experiments were performed and three samples were taken out after 4 weeks. Two parallel samples used for the determination of the corrosion rate were firstly washed by a rust removal solution (500 mL HCl, 500 mL deionized water and 10 g hexamethylenetetramine), followed by drying and weighing.

The rest of the sample was directly stored in a desiccator for the characterization of the corrosion product and passivation film. The solutions (pH 8.2) were maintained anoxic by purging Ar gas (containing 0.3% CO_2) prior to the experiment and during the replacement of the solutions.

2.5. Surface Characterization

Scanning electron microscopy with energy dispersive spectroscopy (SEM/EDS) was conducted to analyze the surface morphologies and elemental compositions of the iron surfaces after the immersion tests, by using a FEI Quanta FEG 250 SEM equipped with an Oxford Inca X-act 2000 EDS system (FEI, Hillsboro, OR, USA). An accelerating voltage of 15 kV and a working distance of 10 mm were used to analyze secondary electron images.

Confocal Raman microscopy (CRM, Renishaw, London, UK) measurements were carried out to analyze the formation of the corrosion products after the immersion tests using a Renishaw inVia Qontor CRM system equipped with a laser source with the wavelength of 532 nm. The scan range was 0–1200 cm^{-1} with a spectral resolution of 1 cm^{-1}.

X-ray photoelectron spectroscopy (XPS, Kratos, Manchester, UK) measurements were performed to characterize the formation of the passivation film after the immersion tests by using a Kratos AXIS Ultra DLD spectrometer, with a monochromated Al K-α X-ray source (hv = 1486.69 eV) at the power of 150 W. The working voltage was set as 15 kV and the transmission current was set as 10 mA. The chemical state assessment was achieved by curve-fitting the spectra using the XPSpeak software (XPSpeak4.1, Hong Kong, China).

3. Result and Discussion

3.1. Potentiodynamic Polarization Curves

The electrochemical corrosion behavior of pure iron in different groundwater solutions and at different temperatures was investigated by potentiodynamic polarization measurements. The polarization curves measured at 40 °C are shown in Figure 1, indicating distinguishing characteristics in different simulated groundwater solutions. By means of the Tafel extrapolation method, electrochemical parameters obtained from the fitted curves are presented in Table 3, showing the corrosion current density and corrosion potential. The variation in the corrosion current density with varying concentrations/temperatures is displayed in Figure 2.

Table 3. Electrochemical parameters obtained from the fitted curves.

40 °C	C/mM i_0/E_0	0	1	5	10	20
$CaCl_2$	$i_0/A\cdot cm^{-2}$	4.40×10^{-6}	8.25×10^{-6}	2.33×10^{-5}	3.09×10^{-5}	3.76×10^{-5}
	E_0/V	−0.543	−0.542	−0.542	−0.540	−0.536
Na_2CO_3	$i_0/A\cdot cm^{-2}$	4.40×10^{-6}	7.19×10^{-6}	7.91×10^{-6}	1.33×10^{-6}	3.27×10^{-6}
	E_0/V	−0.543	−0.537	−0.610	−0.642	−0.161
Na_2SiO_3	$i_0/A\cdot cm^{-2}$	4.40×10^{-6}	1.56×10^{-5}	4.19×10^{-6}	5.40×10^{-6}	3.93×10^{-6}
	E_0/V	−0.543	−0.621	−0.634	−0.655	−0.377

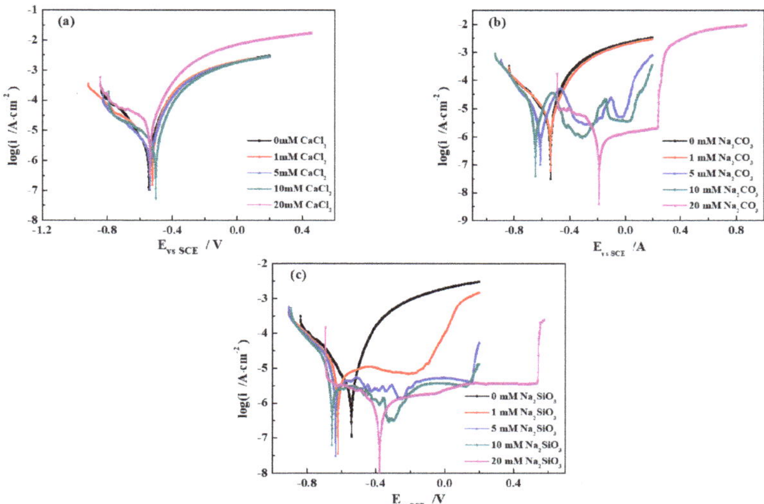

Figure 1. Polarization curves of pure iron in different simulated groundwaters at 40 °C. (**a**) SG + $CaCl_2$; (**b**) SG + Na_2CO_3; (**c**) SG + Na_2SiO_3.

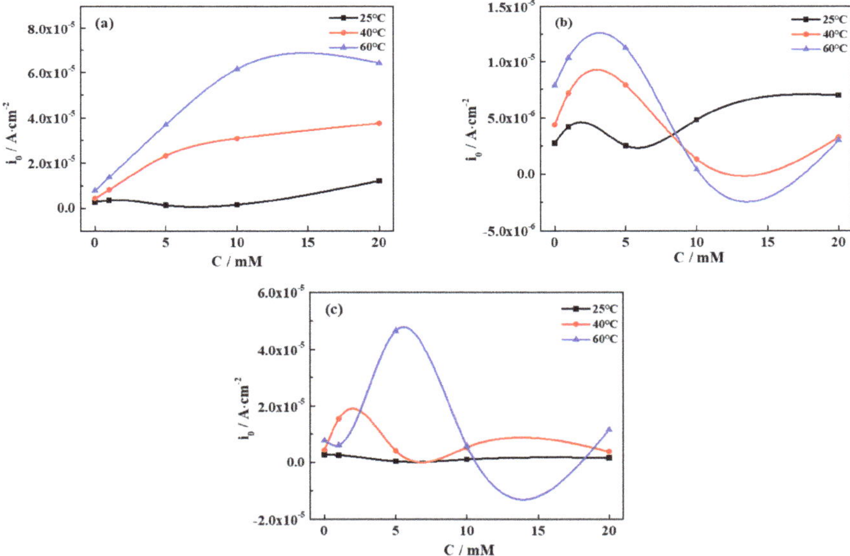

Figure 2. The variation in the corrosion current density with concentration/temperature in different simulated groundwaters. (**a**) SG + $CaCl_2$; (**b**) SG + Na_2CO_3; (**c**) SG + Na_2SiO_3.

In the SG + $CaCl_2$ solution (Figure 1a), the curves are very similar, characterized as active dissolution control in both the anodic and cathodic areas. The anodic polarization curves appear in the active dissolution region, but are without occurrence in the passivation region [26,27]. Both the SG and $CaCl_2$ solutions contain active Cl^-, which easily destroys the protective oxide film formed on the iron surface and triggers the dissolution of the iron, further accelerating the electrochemical corrosion activity.

Seen from Table 3 and Figure 2a, it is evident that the corrosion current density demonstrates a rising trend with increased concentrations of $CaCl_2$ and temperature, suggesting a higher corrosion activity of iron. The increase in concentration and temperature accelerates the activity of Cl^- and promotes the dissolution of the corrosion products, leading to the increase in the corrosion rate [21]. The corrosion rate of iron in the SG + $CaCl_2$ solution is higher than that in other environments, indicating Cl^- plays dominant roles in the electrochemical processes.

In the SG + Na_2CO_3 solution (Figure 1b), the polarization curves in the 0 and 1 mM Na_2CO_3 solutions are very similar to Figure 1a, indicating Cl^- plays a major role in the kinetic control of active dissolution. However, with increased concentrations of Na_2CO_3 (5 and 10 mM), the polarization curves display an obvious inflexion after −0.55 V and then appear as fluctuating transition and passivation regions. It suggests that a discontinuous passivation film gradually forms on the iron surface. When the concentration of Na_2CO_3 reaches 20 mM, the corrosion potential positively shifts to −0.2 V and the obviously different curves show very stable passivation and over passivation regions, but without a transition passivation region. It is evident that the high concentration of ${CO_3}^{2-}$ is related to the stable passivation region in the anodic polarization curve, indicating the formation of a continuous passivation film on the iron surface.

Seen from Table 3 and Figure 2b, with increased concentrations of Na_2CO_3, the overall trend of the corrosion current density firstly rises and then falls, and gradually becomes stable. The effects of active Cl^- competes with the formation of the passivation film, so the fluctuated corrosion rates are dependent on the synergetic effect. However, the effects of temperature on the corrosion current density are not consistent, which is dependent on the concentration of the solution. In the low-concentration solution, the corrosion current density increases with the increased temperature, which is in relation to the enhanced activity of Cl^-. The discontinuous passivation film formed in the low-concentration solution is easily destroyed by Cl^-, resulting in an increased corrosion rate of iron. However, a high concentration of ${CO_3}^{2-}$ promotes the formation of the passive film, further preventing the penetration of Cl^-. The decreased corrosion current density attributes to the barrier effects of the passivation film, resulting in a lower corrosion rate with the increased temperature.

In the SG + Na_2SiO_3 solution (Figure 1c), the characteristics of the polarization curves are very similar to the Na_2CO_3 solution. With increased concentrations of Na_2SiO_3, the passivation region becomes more and more stable. When the concentration of Na_2SiO_3 reaches 20 mM, the wide potential range of the passivation region infers the formation of a denser and more uniform passive film than that formed in the Na_2CO_3 solution.

Seen from Table 3 and Figure 2c, similar to the Na_2CO_3 solution, the variation in the corrosion current density is fluctuating and not stable with the increased concentrations of Na_2SiO_3, indicating the competition of synergetic effects between the active Cl^- and the passivation film. The effects of temperature on the corrosion current density are also dependent on the concentration. On the one hand, a high temperature promotes ${SiO_3}^{2-}$ to form a dense and uniform passivation film. On the other hand, a high temperature accelerates Cl^- to destroy the passivation film.

3.2. Electrochemical Impedance Spectroscopy

Electrochemical impedance measurements were performed to characterize the barrier effect of the oxide film formed on the pure iron in different groundwater solutions. The EIS spectra measured at 40 °C are shown in Figure 3, displaying different characteristics in different groundwater solutions. Electrochemical parameters obtained from the fitted EIS spectra are presented in Table 4, based on a different equivalent circuit (Figure 4). In order to illuminate the effect of different ions and temperature on the corrosion resistance of iron in simulated groundwater, the variation in the polarization resistance with varying concentrations/temperatures are displayed in Figure 5.

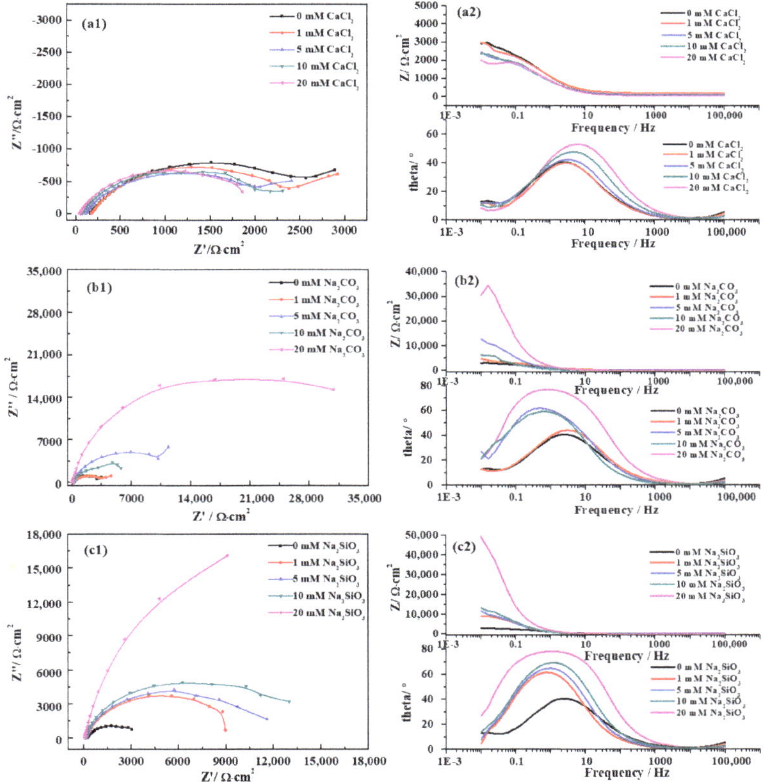

Figure 3. EIS Nyquist (1) and Bode (2) spectra of pure iron in different simulated groundwaters at 40 °C. (**a**) SG + $CaCl_2$; (**b**) SG + Na_2CO_3; (**c**) SG + Na_2SiO_3.

Table 4. Electrochemical parameters of equivalent circuit of electrochemical impedance spectroscopy (EIS).

Solution	C/mM	R_s/Ω	R_{ct}/Ω	R_f/Ω	R_p/Ω	C_r/S·sec^n	C_{dl}/S·sec^n
$CaCl_2$	0 mM	169.0	2889.0	–	2889.0	–	0.00024
	1 mM	176.8	2653.0	–	2653.0	–	0.000023
	5 mM	108.6	226.4	–	226.4	–	0.00030
	10 mM	72.3	227.1	–	227.1	–	0.00026
	20 mM	44.9	1977.0	–	1977.0	–	0.00022
Na_2CO_3	1 mM	174.8	3535.0	1277.0	4812	0.00017	0.0085
	5 mM	115.9	12.7	13,280.0	13,292.7	0.00021	0.000031
	10 mM	68.9	13.5	8780.0	8793.5	0.000052	0.00044
	20 mM	54.9	89.6	40,060.0	40,149.6	0.000063	0.000043
Na_2SiO_3	1 mM	168.7	8164.0	275.1	8439.1	0.00020	0.0020
	5 mM	275.1	9494.0	289.1	9783.1	0.00021	0.021
	10 mM	85.1	0.3	11,400.0	11,400.3	0.00023	0.00019
	20 mM	51.6	28.3	51,770.0	51,798.3	0.00016	0.00010

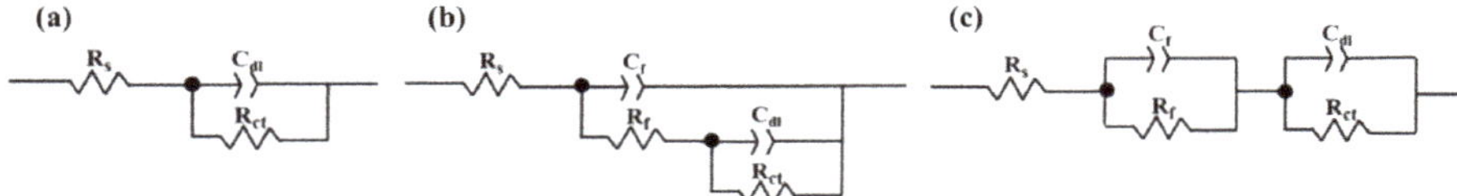

Figure 4. Equivalent electric circuits used to extract parameters from the EIS spectra. (**a**) SG + $CaCl_2$; (**b**) SG + Na_2CO_3; (**c**) SG + Na_2SiO_3. (R_s: electrolyte resistance; R_{ct}: charge-transfer resistance; R_f: film resistance; C_{dl}: double layer capacitance; C_f: film capacitance).

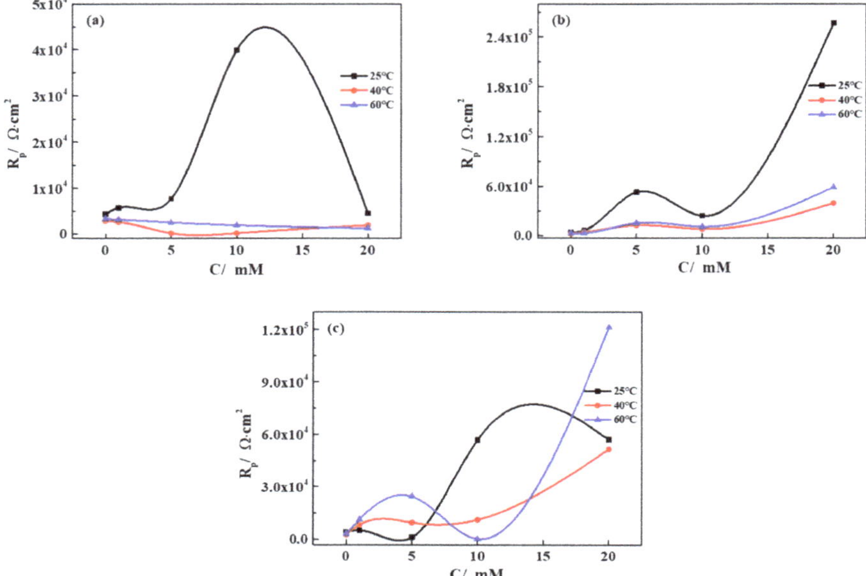

Figure 5. The variation in the polarization resistance of iron with concentration/temperature in different simulated groundwaters. (**a**) SG + $CaCl_2$; (**b**) SG + Na_2CO_3; (**c**) SG + Na_2SiO_3.

In the SG + $CaCl_2$ solution (Figure 3a), the Nyquist spectra show only one capacitive loop and the diameter of the loop decreases with the increased concentration of $CaCl_2$. From the Bode plot, it can be seen that the major process has a capacitive slope below 10 Hz. With the increased concentration, the resistance of pure iron at low frequency decreased from about 3000 to 2000 $\Omega\cdot cm^2$. A time constant is evident at low frequency, probably assigned to a charge transfer process [28]. The equivalent circuit used for modeling the Nyquist plots consisted of electrolyte resistance (R_s), charge-transfer resistance (R_{ct}) and double layer capacitance (C_{dl}) (Figure 5a). The polarization resistance (R_p) equals to R_{ct}.

Seen from Table 4 and Figure 4a, the polarization resistance is basically conserved with the increased concentration of $CaCl_2$ (10 mM at 25 °C is the only exception), indicating the corrosion resistance of iron is independent from the concentration. Similar to the polarization results, the effects of the formation of the passivation film compete with the active Cl^-, which is not beneficial to enhance corrosion resistance. However, the effects of the increased temperature on polarization resistance were generally decreasing, indicating that the high temperature accelerated the activity of Cl^- and promoted the dissolution of the corrosion products, resulting in a lower corrosion resistance. The corrosion resistance of iron in the SG + $CaCl_2$ solution was much lower than that in other environments, indicating Cl^- plays dominant roles in the electrochemical processes.

In the SG + Na_2CO_3 solution (Figure 3b), a capacitive loop appears in the Nyquist spectra. Different from the $CaCl_2$ solution, the diameter of the loop increases with the increased concentration of Na_2CO_3. The Bode plot indicates that the major process has a capacitive slope below 1 Hz. With the increased concentration, the resistance of pure iron at low frequency increased from about 3000 to 34,000 $\Omega \cdot cm^2$. A time constant is also evident at low frequency, probably assigned to a charge transfer process. However, with the increased concentration, the characteristics of the peaks of the phase angle changed from symmetry to asymmetry, suggesting the possible existence of two overlapped peaks [29,30]. Hence, two time constants were required for a proper fitting, indicating two electrochemical reactions occurred on the iron surface. Based on the results from the polarization curves, the high concentration of CO_3^{2-} promoted the formation of the passive film, so Cl^- plays an important role in the dissolution of both the passivation film and iron matrix. The equivalent circuit used for modeling the Nyquist plots consists of electrolyte resistance (R_s), film resistance for passivation film formed on the iron surface (R_f), film capacitance (C_f), charge-transfer resistance (R_{ct}) and double layer capacitance (C_{dl}) (Figure 5b). The polarization resistance (R_p) is calculated by adding R_f and R_{ct}.

Seen from Table 4 and Figure 5b, it is indicated that raising the concentration of Na_2CO_3 increases both R_f and R_p of pure iron. It is obvious that CO_3^{2-} promotes the formation of the passive film, further preventing the penetration of Cl^-. The passivation film becomes uniform and dense with the increased concentration, demonstrating higher barrier effects on the surface. Even though the corrosion rate is fluctuant, shown in the polarization curves, the corrosion resistance of iron largely enhances with the increased concentration. However, the polarization resistance decreased with the increased temperature, which is related to the reduced barrier effects of the passivation film. The high temperature easily accelerates the activity of Cl^-, leading to the dissolution of the passivation film, further lowering the corrosion resistance.

In the SG + Na_2SiO_3 solution (Figure 3c), the characteristics of the EIS spectra are similar to the Na_2CO_3 solution. With the increased concentration, the resistance of pure iron at low frequency increased from about 3000 to 50,000 $\Omega \cdot cm^2$, which is even higher than the resistance in the Na_2CO_3 solution. Due to the asymmetry peaks of the phase angles, two time constants were also required for a proper fitting. The equivalent circuit used for modeling the Nyquist plots consists of the same components used for the Na_2CO_3 solution, but two series of networks were used to address the effect of a dense and uniform passivation film.

Seen from Table 4 and Figure 5c, the variation in the polarization resistance shows a fluctuating but increasing trend in the SG + Na_2SiO_3 solution (20 mM at 25 °C is the only exception). Similar to the Na_2CO_3 solution, with the increased concentration, an increased R_f evidently indicates the formation of a uniform and dense passivation film, which supplies a high barrier effect on the surface. Even though the corrosion rate is fluctuant with the concentration, shown in the polarization curves, the corrosion resistance is enhanced, showing an even larger value compared with Figure 5b. Hence, SiO_3^{2-} promotes the formation of a denser and more uniform passivation film compared with CO_3^{2-}. However, the effects of temperature on the corrosion current density are also dependent on the concentration, similar to the results from the polarization curves. The formation of the passivation film competes with the active Cl^-, so the corrosion resistance is dependent on the synergetic effect. The synergetic effects of CO_3^{2-}/SiO_3^{2-} and Cl^- contribute to a dynamic process in the electrochemical reactions, leading to repeated passivation–dissolution processes for the formation of the passivation film.

3.3. Corrosion Rate Obtained from Immersion Tests

In order to investigate the corrosion rate of iron samples, immersion tests were carried out in simulated groundwater with the addition of 20 mM ions. After four weeks of exposure, the corrosion products were removed by a mixed solution of 500 mL HCl, 500 mL deionized H_2O and 10 g $C_6H_{12}N_4$. The visual appearances of the corroded samples' surfaces are shown in Figure 6. The surfaces after immersion in the SG solution and the SG + $CaCl_2$ solution are completely covered with a thick layer of corrosion products. In the SG + Na_2CO_3 solution, the surfaces are partially covered with corrosion products, mainly located at the side and edge of the samples. Besides corrosion products, other parts of the surfaces are not corroded. However, the visual appearances of the surfaces after immersion in the SG + Na_2CO_3 solution are quite different, showing barely an iron matrix covered by a thin layer of passivation film.

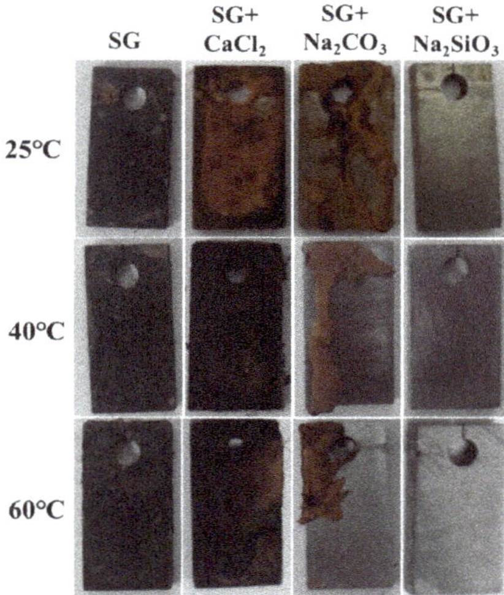

Figure 6. Visual appearance of iron after four weeks immersion in 20 mM SG and SG + $CaCl_2$/Na_2CO_3/Na_2SiO_3 solutions.

The corrosion rate of the iron samples after the immersion test was calculated by the weight loss method. The calculation formula of the corrosion rate is as follows [31]:

$$CR\ (mm/y) = \frac{87600 \times (M - M_1)}{S \times T \times \rho} \tag{1}$$

where CR represents the corrosion rate (mm/y, corrosion depth per year), M represents the weight of the sample before immersion (g), M_1 represents the weight of the sample after four weeks of immersion (g), S represents the total surface area of the sample (cm^2), T represents the immersion period (H) and d represents the density of the sample (g/m^3). The results are expressed as an average corrosion rate, which was calculated by two samples.

The weight loss corrosion rates of the pure iron in different simulated groundwaters are compared in Figure 7. It seems that the general corrosion rates of iron in the SG/SG + $CaCl_2$/SG + Na_2CO_3 solution are much higher than that in the SG + Na_2SiO_3 solution, consistent with the electrochemical results. In the SG + $CaCl_2$ solution, the corrosion rates increased with the increased temperature, showing an accelerated corrosion activity of iron. In the SG + Na_2CO_3/Na_2SiO_3 solution, the corrosion rates are also fluctuant with the increased temperature. In agreement with the electrochemical results, it is indicated that Cl^- competes with the passivation film, contributing synergetic effects to the corrosion rate. The corrosion rates of iron with the addition of Na_2SiO_3 at 25 and 60 °C are approximately zero, which indicates that the barrier effects of the passivation film largely protect the iron matrix.

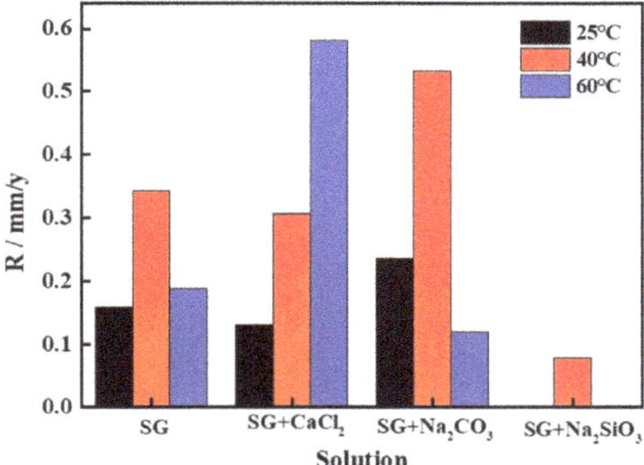

Figure 7. Weight loss corrosion rates of iron after immersion tests in different simulated groundwaters.

3.4. Characterization of Corrosion Products and Passivation Film

In order to analyze the morphology and composition of the corrosion products and passivation film, SEM/EDS, CRM and XPS measurements were performed on the iron surfaces after the immersion tests.

Seen from Figure 8a–f, the characteristics of the corrosion products change with different ions and temperatures in the SG and SG + $CaCl_2$ solutions. In the SG solution, granular corrosion products are partially distributed on the iron surface at 25 °C. Corrosion products gradually accumulate together, and it was found that flower-like clusters appear with the increased temperature. In the SG + $CaCl_2$ solution, a loose layer of corrosion products was found, consisting of flower-like and rod-like clusters. With the increased temperature, the corrosion products clusters become larger and larger, showing a distinguished lamellar structure.

Figure 8. Surface morphology of iron after 4 weeks immersion in 20 mM SG and SG + $CaCl_2$/ Na_2CO_3/Na_2SiO_3 solutions (**a**–**l**).

The morphology of the passivation film formed in the SG + Na_2CO_3/Na_2SiO_3 solution is displayed in Figure 8g–l. Different from the formation of a large number of corrosion products in the SG and SG + $CaCl_2$ solutions, only a small amount of corrosion products are partially distributed on the iron surface. With the increased temperature, the amount of corrosion products further decreases and scratches on the iron surface could be observed, indicating the substrate is protected by a thin film of the passivation layer.

When comparing the distribution of the corrosion products formed with the different ions, it can be seen that adding $CO_3{}^{2-}$ and $SiO_3{}^{2-}$ significantly reduces the amount of corrosion products, showing the characteristics of the passivation film. When comparing the different temperatures in the SG and SG + $CaCl_2$ solutions, it can be seen that the corrosion products clusters significantly accumulated and grew in size, indicating that a high temperature will promote the electrochemical reactions and accelerate the corrosion behavior of iron. However, with the increased temperature in the SG + Na_2CO_3/Na_2SiO_3 solution, the increased barrier effect of the passivation film protects the substrate, further reducing the amount of corrosion products.

The corresponding EDS analyses of the corrosion products are displayed in Figure 9. The corrosion products formed in the SG and SG + $CaCl_2$ solutions are composed of Fe and O, inferring different types of iron oxide. Ca and C are also detected on the iron surface after immersion in the SG + $CaCl_2$ solution. Although Ca^{2+} may react with $HCO_3{}^-$ to form calcium-rich products, it seemed the products failed to protect the iron surface due to the strong destroying ability of Cl^-. In the SG + Na_2CO_3/Na_2SiO_3 solution, the main components of the passivation film are composed of Fe, O, C and Si. Carbonate-rich and silicate-rich passivation films possibly form on the iron surface during electrochemical reactions, which prevent the penetration of Cl^-, resulting in a protective effect to lower the corrosion rate.

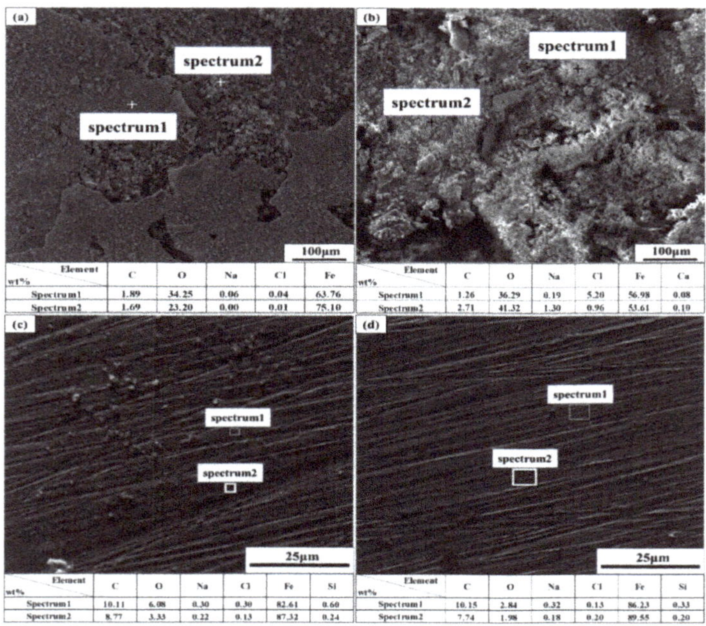

(a)

Element / wt%	C	O	Na	Cl	Fe
Spectrum1	1.89	34.25	0.06	0.04	63.76
Spectrum2	1.69	23.20	0.00	0.01	75.10

(b)

Element / wt%	C	O	Na	Cl	Fe	Ca
Spectrum1	1.26	36.29	0.19	5.20	56.98	0.08
Spectrum2	2.71	41.32	1.30	0.96	53.61	0.10

(c)

Element / wt%	C	O	Na	Cl	Fe	Si
Spectrum1	10.11	6.08	0.30	0.30	82.61	0.60
Spectrum2	8.77	3.33	0.22	0.13	87.32	0.24

(d)

Element / wt%	C	O	Na	Cl	Fe	Si
Spectrum1	10.15	2.84	0.32	0.13	86.23	0.33
Spectrum2	7.74	1.98	0.18	0.20	89.55	0.20

Figure 9. EDS analysis of iron after 4 weeks immersion in different simulated groundwaters. (**a**) SG solution. (**b**) SG + 20 mM $CaCl_2$; (**c**) SG + 20 mM Na_2CO_3; (**d**) SG + 20 mM Na_2SiO_3.

CRM measurements after the immersion tests were conducted to identify the distribution of the corrosion products on the iron surfaces in the SG and SG + $CaCl_2$ solutions. The Raman spectra in

Figure 10 reveal that the formations of corrosion products are very similar without or with the addition of $CaCl_2$. In Figure 10a, it is obvious that the corrosion products mainly consist of lepidocrocite (γ-FeOOH; associated with the bands at 248, 376, 525 and 644 cm^{-1}) at 25 °C. When the temperature increases to 40 and 60 °C, besides the formation of lepidocrocite, the corrosion products are mainly composed of goethite (α-FeOOH; associated with the bands at 300, 379 and 533 cm^{-1}) and maghemite (γ-Fe_2O_3; associated with band at 716 cm^{-1}) [32–34]. In the SG + $CaCl_2$ solutions, Figure 10b demonstrates a similar trend of corrosion products formation, which is mainly composed of γ-FeOOH, α-FeOOH and γ-Fe_2O_3. However, the passivation film formed in the SG + Na_2CO_3/Na_2SiO_3 solution is too thin to be detected by the CRM measurements.

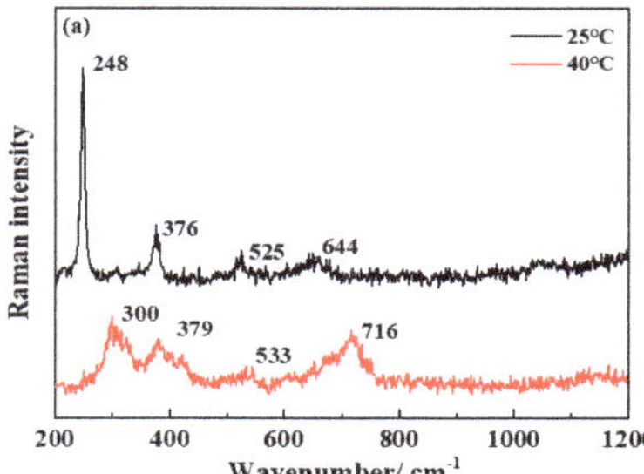

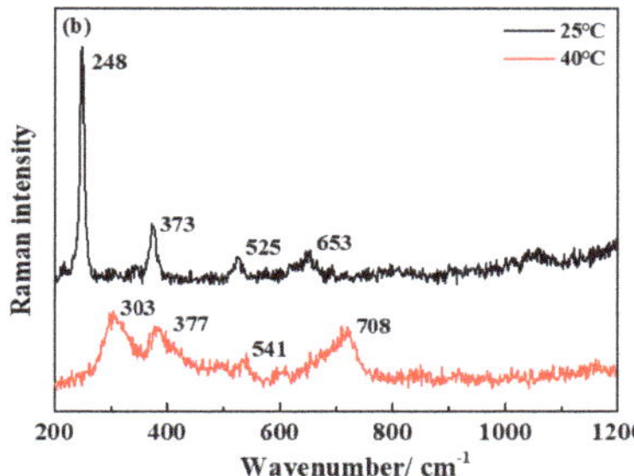

Figure 10. Raman spectra obtained on iron surfaces after 4 weeks immersion in the SG (**a**) and SG + 20 mM $CaCl_2$ (**b**) solutions.

High-resolution XPS spectra were performed to achieve detailed information about the passivation films formed in the SG + Na_2CO_3/Na_2SiO_3 solution. From the patterns shown in Figure 11a, the C 1s core levels were deconvoluted into three different components of C-C (284.8 eV), C-O-C (285.7 eV) and O-C=O (289.1 eV) bonds, indicating the possible existence of CO_3^{2-} and CO_2 [35]. The Si 2p core levels, shown in Figure 11b, were deconvoluted into two different components. The component with a binding energy around 102.6 eV is reported for silicon bonded to oxygen in the silica compound (SiO_2), where the component with a lower binding energy (101.9 eV) is assigned to the silicate-based species [36,37]. The formation of SiO_2 is attributed to the oxidation of Si in the iron substrate. The silicate content is significantly higher for the passivation film formed in the SG + Na_2CO_3 solution, whereas SiO_2 predominates for the film formed in the SG + Na_2SiO_3 solution. From Figure 11c, the complex Fe2p core level structure is composed of three components. Two pronounced satellite peaks with a binding energy around 711.4 and 725.0 eV correspond to nonstoichiometric FeOOH and Fe_2O_3, respectively, which is possibly connected to the small amount of corrosion products (Figure 8g,j). Another weak satellite peak at 720.0 eV is assigned to the Fe^{3+} species [36,38], indicating the possible existence of ferric carbonate and silicate. The O 1s spectra (Figure 11d) can be deconvoluted into two components of SiO_2 (531.6 eV) and Fe-O bonds (530.3 eV) [38]. Based on the XPS information, it is evident that the passivation film formed in SG + Na_2CO_3/Na_2SiO_3 is mainly composed of SiO_2, ferric carbonate and silicate, where ferric silicate is dominant with the addition of SiO_3^{2-}.

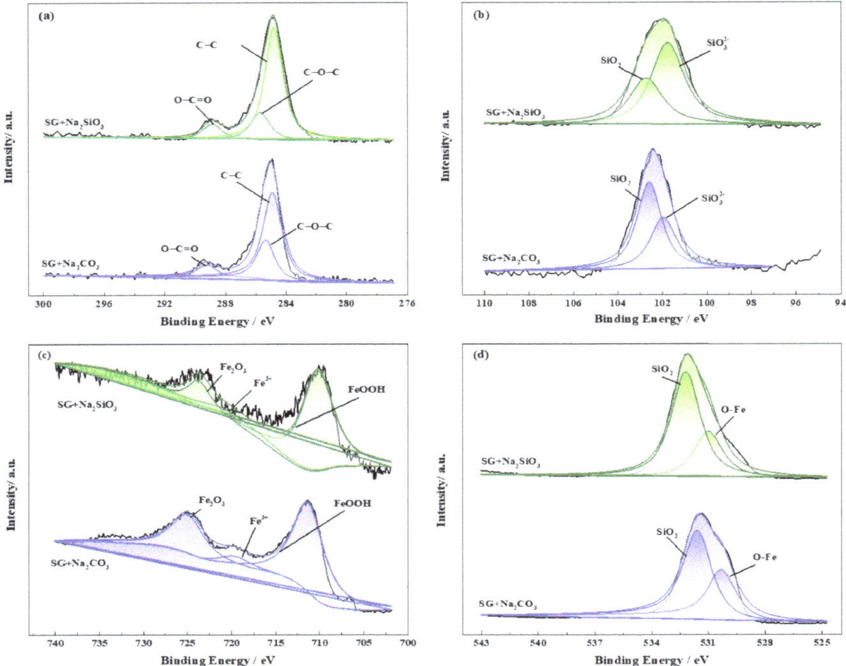

Figure 11. XPS spectra obtained on iron surfaces after 4 weeks immersion in the SG + 20 mM Na_2CO_3/Na_2SiO_3 solution. (**a**) C 1s; (**b**) Si 2p; (**c**) Fe 2p; (**d**) O 1s.

3.5. Mechanism of Corrosion Behavior of Iron in Simulated Groundwater

The mechanism of the corrosion behavior of iron in simulated groundwater is elucidated in Figure 12. The effects of different ions and temperatures on the corrosion behavior are discussed based on the electrochemical results, weight loss corrosion rates and surface characterization.

Seen from Figure 12a,b, both the SG and SG + $CaCl_2$ solutions contain active Cl^-, which easily destroys the protective oxide film formed on the iron surface and triggers the dissolution of iron, further accelerating the electrochemical corrosion processes. The corrosion products formed on the iron surface mainly consist of γ-FeOOH, α-FeOOH and γ-Fe_2O_3. The increased concentration and temperature accelerate the activity of Cl^- and promote the dissolution of the corrosion products, leading to the increase in the corrosion rate. Although Ca^{2+} may react with OH^- to form calcium-rich products, it seems the products are easily destroyed by Cl^-. Hence, the effects of Ca^{2+} on the corrosion behavior of iron is negligible, however, the enhanced effects of chloride ions with temperature lead to more severe corrosion.

In summary, the anodic and cathodic reactions during the corrosion process of pure iron in the anoxic SG and SG + $CaCl_2$ solutions occurred as follows [23,24,39]:

Anodic reaction:

$$Fe \longrightarrow Fe^{2+} + 2e^- \tag{2}$$

Cathodic reaction:

$$2H_2O + 2e^- \longrightarrow 2OH^- + H_2 \tag{3}$$

Overall reactions:

$$Fe + 2H_2O \longrightarrow Fe(OH)_2 + H_2 \tag{4}$$

$$2Fe + 3H_2O \longrightarrow Fe_2O_3 + 3H_2 \quad (5)$$

$$2Fe(OH)_2 + 2H_2O \longrightarrow 2Fe(OH)_3 + H_2 \quad (6)$$

$$Fe(OH)_3 \longrightarrow FeOOH + H_2O \quad (7)$$

$$Ca^{2+} + 2OH^- \longrightarrow Ca(OH)_2 \quad (8)$$

In the SG + Na_2CO_3/Na_2SiO_3 solution, shown in Figure 12c,d, additional $CO_3{}^{2-}$ and $SiO_3{}^{2-}$ play important roles in the formation of the passivation film, reducing the formation of corrosion products on the iron surface with the increased concentration. However, the effects of temperature on the corrosion rates are not consistent. Generally, in the low-concentration solution, Cl^- plays important roles in enhancing the corrosion activity, so the corrosion rates increase with the increased temperature. A uniform passivation film gradually forms with the increased concentration of $CO_3{}^{2-}$ and $SiO_3{}^{2-}$, composed of protective SiO_2, ferric carbonate and silicate. In the high-concentration solution, on one hand, the passivation film plays an important role in enhancing the barrier effects. On the other hand, Cl^- accelerates the dissolution processes with the increased temperature. The synergistic effects lead to repeated passivation–dissolution processes in the formation of the passivation film, leading to a fluctuant corrosion rate varying with the temperature.

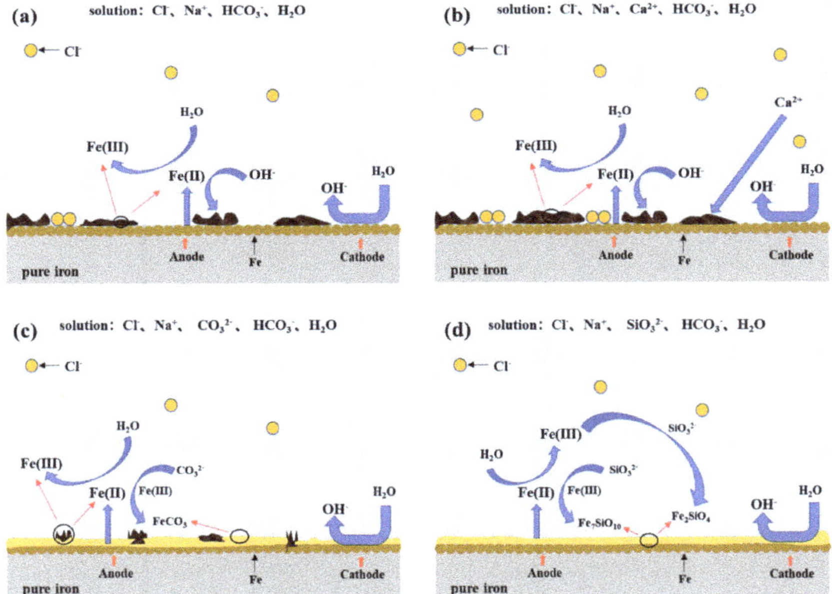

Figure 12. The mechanistic scheme of corrosion behavior of iron in simulated groundwaters. (**a**) SG; (**b**) SG + $CaCl_2$; (**c**) SG + Na_2CO_3; (**d**) SG + Na_2SiO_3.

In summary, the anodic and cathodic reactions during the corrosion process of pure iron in the anoxic SG + Na_2CO_3/Na_2SiO_3 solution occurred as follows:

Anodic reaction [23,40]:

$$Fe \longrightarrow Fe^{2+} + 2e^- \quad (9)$$

Cathodic reaction:

$$2H_2O + 2e^- \longrightarrow 2OH^- + H_2 \quad (10)$$

Overall reactions:

$$Fe^{2+} + CO_3^{2-} \longrightarrow FeCO_3 \quad (11)$$

$$2Fe^{2+} + SiO_3^{2-} + 2OH^- \longrightarrow Fe_2SiO_4 + H_2O \quad (12)$$

$$5Fe^{2+} + 2Fe^{3+} + SiO_3^{2-} + 14OH^- \longrightarrow Fe_7SiO_{10} + 7H_2O \quad (13)$$

Hence, the effects of CO_3^{2-} and SiO_3^{2-} are beneficial to form a passivation film in anoxic simulated groundwater. In the Na_2SiO_3 solution, the formation of ferric silicate predominates, resulting in a more uniform and denser passivation film compared with that formed in the Na_2CO_3 solution. The penetration of Cl^- is largely prevented, improving the protective effects of the iron surfaces.

4. Conclusions

The effects of Ca^{2+}, $CO_3{}^{2-}$ and $SiO_3{}^{2-}$ on the corrosion behavior of pure iron in SG solutions were investigated by varying the concentration of added ions and reaction temperatures. The conclusions are as follows:

1. The corrosion rate of pure iron in the SG + $CaCl_2$ solution increases with the increasing concentration and temperature. The effects of Ca^{2-} on the corrosion behavior of iron is negligible, however, Cl^- plays important roles in the corrosion processes. The increased concentration and temperature accelerate the activity of Cl^- and promote the dissolution of corrosion products, leading to more severe corrosion behavior. The formation of the corrosion products is mainly consisted of γ-FeOOH, α-FeOOH and γ-Fe_2O_3 in the SG and SG + $CaCl_2$ solutions;
2. The corrosion resistance of iron is largely improved by adding $CO_3{}^{2-}$ and $SiO_3{}^{2-}$ in the SG + Na_2CO_3/Na_2SiO_3 solution. A uniform passivation film gradually forms with the increased concentration, playing important roles in increasing the barrier effects of the iron surface and decreasing the formation of corrosion products. The passivation film is mainly composed of SiO_2, ferric carbonate and silicate;
3. The effects of temperature on the corrosion behavior of iron in the SG + Na_2CO_3/Na_2SiO_3 solution are not consistent. Cl^- is dominant in enhancing the corrosion activity in the low-concentration solution, so the corrosion rates increase with the increased temperature. In the high-concentration solution, the synergistic effects of $CO_3{}^{2-}$/$SiO_3{}^{2-}$ and Cl^- contribute to the synergistic effects in the formation of the passivation film, leading to the cycles of passivation–dissolution processes. Hence, the corrosion rates are fluctuant with the varying temperature;
4. The effects of $CO_3{}^{2-}$ and $SiO_3{}^{2-}$ are beneficial to iron in simulated groundwater. The formation of ferric silicate is dominant with the addition of $SiO_3{}^{2-}$, resulting in a more uniform and denser passivation film than the film formed with the addition of $CO_3{}^{2-}$. The penetration of Cl^- is effectively prevented and the corrosion resistance of iron is largely improved.

Author Contributions: Conceptualization, T.L., M.Y. and X.Z.; data curation, T.L. and G.H.; formal analysis, T.L., G.H. and X.Z.; investigation, G.H. and Y.F.; methodology, T.L. and X.Z.; project administration, L.W. and D.C.; resources, L.W. and D.C.; software, G.H. and Y.F.; validation, M.Y.; writing—original draft, T.L. and G.H.; writing—review and editing, T.L. and X.Z. All authors have read and agreed to the published version of the manuscript.

Funding: This research was funded by the Nuclear Material Innovation Foundation of China, grant number No. ICNM-2019-ZH-05.

Conflicts of Interest: The authors declare no conflict of interest.

References

1. Cui, D.; Spahiu, K. The reduction of U(VI) on corroded iron under anoxic conditions. *Radiochim. Acta* **2002**, *90*, 623–628. [CrossRef]
2. Johnson, L.; Snellman, N.M.; Pastina, B. Safety assessment for a KBS3H spent nuclear fuel repository at Olkiluoto. *Atlas Clin. Urol.* **2007**, *358*, 127–140.

3. Keech, P.G.; Vo, P.; Ramamurthy, S.; Chen, J.; Jacklin, R.; Shoesmith, D.W. Design and development of copper coatings for long term storage of used nuclear fuel. *Corros. Eng. Sci. Technol.* **2014**, *49*, 425–430. [CrossRef]
4. Ibrahim, B.; Zagidulin, D.; Behazin, M.; Ramamurthy, S.; Wren, J.C.; Shoesmith, D.W. The corrosion of copper in irradiated and unirradiated humid air. *Corros. Sci.* **2018**, *141*, 53–62. [CrossRef]
5. Guo, X.; Gin, S.; Lei, P.; Yao, T.; Liu, H.; Schreiber, D.K.; Ngo, D.; Viswanathan, G.; Li, T.; Kim, S.H.; et al. Self-accelerated corrosion of nuclear waste forms at material interfaces. *Nat. Mater.* **2020**. [CrossRef] [PubMed]
6. Gaudin, A.; Gaboreau, S.; Tinseau, E.; Bartier, D.; Petit, S.; Grauby, O.; Foct, F.; Beaufort, D. Mineralogical reactions in the Tournemire argillite after in-situ interaction with steels. *Appl. Clay Sci.* **2009**, *43*, 196–207. [CrossRef]
7. Cui, D.; Low, J.; Rondinella, V.V.; Spahiu, K. Hydrogen catalytic effects of nanostructured alloy particles in spent fuel on radionuclide immobilization. *Appl. Catal. B Environ.* **2010**, *94*, 173–178. [CrossRef]
8. Yang, H.; Cui, D.; Grolimund, D.; Rondinella, V.V.; Brütsch, R.; Amme, M.; Kutahyali, C.; Wiss, A.T.; Puranen, A.; Spahiu, K. Reductive precipitation of neptunium on iron surfaces under anaerobic conditions. *J. Nucl. Mater.* **2017**, *496*, 109–116. [CrossRef]
9. Odorowski, M.; Jegou, C.; De Windt, L.; Broudic, V.; Jouan, G.; Peuget, S.; Martin, C. Effect of metallic iron on the oxidative dissolution of UO_2 doped with a radioactive alpha emitter in synthetic Callovian-Oxfordian groundwater. *Geochim. Cosmochim. Acta* **2017**, *219*, 1–21. [CrossRef]
10. He, J.; Ma, B.; Kang, M.; Wang, C.; Nie, Z.; Liu, C. Migration of ^{75}Se(IV) in crushed Beishan granite: Effects of the iron content. *J. Hazard. Mater.* **2017**, *324*, 564–572. [CrossRef]
11. Cui, D.; Rondinella, V.V.; Fortner, J.A.; Kropf, A.J.; Eriksson, L.; Wronkiewicz, D.J.; Spahiu, K. Characterization of alloy particles extracted from spent nuclear fuel. *J. Nucl. Mater.* **2012**, *420*, 328–333. [CrossRef]
12. Björkbacka, Å.; Hosseinpour, S.; Johnson, M.; Leygraf, C.; Jonsson, M. Radiation induced corrosion of copper for spent nuclear fuel storage. *Radiat. Phys. Chem.* **2013**, *92*, 80–86. [CrossRef]
13. Cui, D.; Ranebo, Y.; Low, J.; Rondinella, V.V.; Pan, J.; Spahiu, K. Immobilization of radionuclides on iron canister material at simulated near-field conditions. In Proceedings of the Materials Research Society Symposium, Boston, MA, USA, 2–4 December 2009; Volume 1124, pp. 111–116.
14. Rosborg, B.; Werme, L. The Swedish nuclear waste program and the long-term corrosion behaviour of copper. *J. Nucl. Mater.* **2008**, *379*, 142–153. [CrossRef]
15. Smart, N.R.; Blackwood, D.J.; Werme, L. Anaerobic corrosion of carbon steel and cast iron in artificial groundwaters: Part 2-Gas generation. *Corrosion* **2002**, *58*, 627–637. [CrossRef]
16. Liu, C.; Wang, J.; Zhang, Z.; Han, E.H. Studies on corrosion behaviour of low carbon steel canister with and without γ-irradiation in China's HLW disposal repository. *Corros. Eng. Sci. Technol.* **2017**, *52*, 136–140. [CrossRef]
17. Martin, F.A.; Perrin, S.; Bataillon, C. Evaluating the corrosion rate of low alloyed steel in Callovo-Oxfordian clay: Towards a complementary EIS, gravimetric and structural study. In Proceedings of the Materials Research Society Symposium, Boston, MA, USA, 25–30 November 2012.
18. Standish, T.E.; Braithwaite, L.J.; Shoesmith, D.W.; Noël, J.J. Influence of area ratio and chloride concentration on the galvanic coupling of copper and carbon steel. *J. Electrochem. Soc.* **2019**. [CrossRef]
19. Zhang, Q.; Zheng, M.; Huang, Y.; Kunte, H.J.; Wang, X.; Liu, Y.; Zheng, C. Long term corrosion estimation of carbon steel, titanium and its alloy in backfill material of compacted bentonite for nuclear waste repository. *Sci. Rep.* **2019**, *9*, 3195. [CrossRef] [PubMed]
20. Cheshire, M.C.; Caporuscio, F.A.; Jové Colón, C.F.; Norskog, K.E. Fe-saponite growth on low-carbon and stainless steel in hydrothermal-bentonite experiments. *J. Nucl. Mater.* **2018**. [CrossRef]
21. Lu, Y.; Dong, J.; Ke, W. Effects of Cl^- ions on the corrosion behaviour of low alloy steel in deaerated bicarbonate solutions. *J. Mater. Sci. Technol.* **2016**. [CrossRef]
22. Liu, C.; Wang, J.; Zhang, Z.; Han, E.-H.; Liu, W.; Liang, D.; Yang, Z. Characterization of corrosion behavior of irradiated X65 low carbon steel in aerobic and unsaturated gaomiaozi bentonite. *Acta Metall. Sin. Engl. Lett.* **2019**, *032*, 506–516. [CrossRef]
23. Crane, R.A.; Scott, T.B. Nanoscale zero-valent iron: Future prospects for an emerging water treatment technology. *J. Hazard. Mater.* **2012**, *211–212*, 112–125. [CrossRef] [PubMed]

24. Tuček, J.; Prucek, R.; Kolařík, J.; Zoppellaro, G.; Petr, M.; Filip, J.; Sharma, V.K.; Zbořil, R. Zero-valent iron nanoparticles reduce arsenites and arsenates to As(0) firmly embedded in core-shell superstructure: Challenging strategy of arsenic treatment under anoxic conditions. *ACS Sustain. Chem. Eng.* **2017**, *5*, 3027–3038. [CrossRef]
25. Das, S.; Lindsay, M.B.; Hendry, M.J. Selenate removal by zero-valent iron under anoxic conditions effects. *Environ. Earth Sci.* **2019**, *78*, 528. [CrossRef]
26. Ni, Q.; Xia, X.; Zhang, J.; Dai, N.; Fan, Y. Electrochemical and SVET studies on the typical polarity reversal of Cu–304 stainless steel galvanic couple in Cl^--containing solution with different pH. *Electrochim. Acta* **2017**, *247*, 207–215. [CrossRef]
27. Li, J.; Wu, J.; Wang, Z.; Zhang, S.; Wu, X.; Huang, Y.; Li, X. The effect of nanosized NbC precipitates on electrochemical corrosion behavior of high-strength low-alloy steel in 3.5%NaCl solution. *Int. J. Hydrogen Energy* **2017**, *42*, 22175–22184. [CrossRef]
28. Kazum, O.; Mathan, B.; Beladi, H.; Timokhina, I.; Hodgson, P.; Khoddam, S. Aqueous corrosion performance of nanostructured bainitic steel. *Mater. Des.* **2014**, *54*, 67–71. [CrossRef]
29. Marques, A.G.; Simões, A.M. EIS and SVET assessment of corrosion resistance of thin Zn-55% Al-rich primers: Effect of immersion and of controlled deformation. *Electrochim. Acta* **2014**, *148*, 153–163. [CrossRef]
30. Moreto, J.A.; Marino, C.E.B.; Filho, W.W.B.; Rocha, L.A.; Fernandes, J.C.S. SVET, SKP and EIS study of the corrosion behaviour of high strength Al and Al–Li alloys used in aircraft fabrication. *Corros. Sci.* **2014**, *84*, 30–41. [CrossRef]
31. Xu, Q.; Gao, K.; Lv, W.; Pang, X. Effects of alloyed Cr and Cu on the corrosion behavior of low-alloy steel in a simulated groundwater solution. *Corros. Sci.* **2016**, *102*, 114–124. [CrossRef]
32. Cui, Y.; Liu, S.; Smith, K.; Yu, K.; Hu, H.; Jiang, W.; Li, Y. Characterization of corrosion scale formed on stainless steel delivery pipe for reclaimed water treatment. *Water Res.* **2016**, *88*, 816–825. [CrossRef]
33. Thompson, S.P.; Day, S.J.; Parker, J.E.; Evans, A.; Tang, C.C. Fine-grained amorphous calcium silicate CaSiO 3 from vacuum dried sol-gel—Production, characterisation and thermal behaviour. *J. Non Cryst. Solids* **2012**, *358*, 885–892. [CrossRef]
34. Edwards, H.G.M.; Villar, S.E.J.; Jehlicka, J.; Munshi, T. FT-Raman spectroscopic study of calcium-rich and magnesium-rich carbonate minerals. *Spectrochim. Acta Part A Mol. Biomol. Spectrosc.* **2005**, *61*, 2273–2280. [CrossRef] [PubMed]
35. Fang, L.; Liu, Z.; Zhou, C.; Guo, Y.; Feng, Y.; Yang, M. Degradation mechanism of methylene blue by H_2O_2 and synthesized carbon nanodots/graphitic carbon nitride/Fe(II) composite. *J. Phys. Chem. C* **2019**, *123*, 26921–26931. [CrossRef]
36. De Oliveira, L.A.; Correa, O.V.; Dos Santos, D.J.; Páez, A.A.Z.; De Oliveira, M.C.L.; Antunes, R.A. Effect of silicate-based films on the corrosion behavior of the API 5L X80 pipeline steel. *Corros. Sci.* **2018**, *139*, 21–34. [CrossRef]
37. Lopez-Garrity, O.; Frankel, G.S. Corrosion inhibition of aa2024-t3 by sodium silicate. *Electrochim. Acta* **2014**, *130*, 9–21. [CrossRef]
38. Wang, D.; Zhu, Q.; Su, Y.; Li, J.; Wang, A.; Xing, Z. Preparation of MgAlFe-LDHs as a deicer corrosion inhibitor to reduce corrosion of chloride ions in deicing salts. *Ecotoxicol. Environ. Saf.* **2019**, *174*, 164–174. [CrossRef]
39. Gibson, B.D.; Blowes, D.W.; Lindsay, M.B.J.; Ptacek, C.J. Mechanistic investigations of Se(VI) treatment in anoxic groundwater using granular iron and organic carbon: An EXAFS study. *J. Hazard. Mater.* **2012**, *241–242*, 92–100. [CrossRef]
40. Wang, C.; Chen, J.; Hu, B.; Liu, Z.; Wang, C.; Han, J.; Su, M.; Li, Y.; Li, C. Modified chitosan-oligosaccharide and sodium silicate as efficient sustainable inhibitor for carbon steel against chloride-induced corrosion. *J. Clean. Prod.* **2019**, *238*, 117823. [CrossRef]

© 2020 by the authors. Licensee MDPI, Basel, Switzerland. This article is an open access article distributed under the terms and conditions of the Creative Commons Attribution (CC BY) license (http://creativecommons.org/licenses/by/4.0/).

Review

AC Corrosion of Carbon Steel under Cathodic Protection Condition: Assessment, Criteria and Mechanism. A Review

Andrea Brenna *, Silvia Beretta and Marco Ormellese

Dipartimento di Chimica, Materiali e Ingegneria Chimica "Giulio Natta", Politecnico di Milano, I-20131 Milan, Italy; silvia.beretta@polimi.it (S.B.); marco.ormellese@polimi.it (M.O.)

* Correspondence: andrea.brenna@polimi.it

Received: 21 January 2020; Accepted: 6 May 2020; Published: 7 May 2020

Abstract: Cathodic protection (CP), in combination with an insulating coating, is a preventative system to control corrosion of buried carbon steel pipes. The corrosion protection of coating defects is achieved by means of a cathodic polarization below the protection potential, namely −0.85 V vs. CSE (CSE, copper-copper sulfate reference electrode) for carbon steel in aerated soil. The presence of alternating current (AC) interference, induced by high-voltage power lines (HVPL) or AC-electrified railways, may represent a corrosion threat for coated carbon steel structures, although the potential protection criterion is matched. Nowadays, the protection criteria in the presence of AC, as well as AC corrosion mechanisms in CP condition, are still controversial and discussed. This paper deals with a narrative literature review, which includes selected journal articles, conference proceedings and grey literature, on the assessment, acceptable criteria and corrosion mechanism of carbon steel structures in CP condition with AC interference. The study shows that the assessment of AC corrosion likelihood should be based on the measurement of AC and DC (direct current) related parameters, namely AC voltage, AC and DC densities and potential measurements. Threshold values of the mentioned parameters are discussed. Overprotection ($E_{\text{IR-free}} < -1.2$ V vs. CSE) is the most dangerous condition in the presence of AC: the combination of strong alkalization close to the coating defect due to the high CP current density and the action of AC interference provokes localized corrosion of carbon steel.

Keywords: alternating current; cathodic protection; carbon steel; pipeline; AC interference corrosion; AC corrosion assessment; protection criteria; corrosion mechanism

---

1. Introduction

Cathodic protection (CP), in combination with an insulating coating, is a well-known electrochemical technique that reduces (or halts) the external corrosion rate of buried carbon steel pipes used to transport liquid or gas. In CP condition, the corrosion rate is reduced below 0.01 mm·a^{-1}, which is the maximum acceptable value fixed by CP standards [1,2]. Carbon steel in aerated soil, i.e., where oxygen reduction is the controlling cathodic process, operates in CP condition if the IR-free potential (excluding the ohmic drop contribution in soil) is more negative than −0.85 V vs. CSE ($Cu/CuSO_4$ reference electrode, +0.318 V vs. standard hydrogen electrode, SHE) [1,2].

The presence of alternating current (AC) interference on buried pipelines in free corrosion or under CP condition can lead to severe localized corrosion through the pipe thickness. In the case of AC interference, the sources of electrical disturbance are the high-voltage power lines (HVPL) or the AC-electrified railways (fed by a high voltage line at 50 or 60 Hz), the receptor is the pipeline that runs parallel to the interference source and the coupling mechanism occurs mainly via a resistive (or conductive) and inductive (or electromagnetic) mechanism [3].

The resistive coupling is primarily a concern when there is a fault or an unbalanced condition on the power line and large currents and voltages are conveyed to the earth during the HVPL short circuit. Although the interference time is short, it represents a hazard to the operators and to the buried pipe corresponding to the coating defects. The inductive coupling occurs when AC flowing in phase wires produces an electromagnetic field inducing alternating currents and voltages to the pipeline, which shares the way with the power line. The induced AC voltage depends on the length of parallelism with the power line, and it is inversely proportional to the distance between the HVPL and the pipeline; the AC density, i.e., the current for unit surface, is a function of AC voltage, the coating defect dimension and soil resistivity.

Nowadays, there is agreement that corrosion induced by AC interference can occur even on carbon steel structures that fully respect the CP criterion and that AC corrosion is less than that provoked by the equivalent direct current, i.e., considering the same current density. In the presence of AC interference, the CP criteria reported by international standards [1,2] are not sufficient to prove that steel is protected from corrosion. In the past 50 years, great effort has been made in order to propose criteria to assess AC corrosion likelihood and to understand the mechanism by which AC causes corrosion. This effort brought about the international standard ISO 18086 (Corrosion of metals and alloys—Determination of AC corrosion—Protection criteria) [4] that replaced in 2017 the EN 15280 standard (Evaluation of a.c. corrosion likelihood of buried pipelines applicable to cathodically protected pipelines). The ISO 18086 standard provides monitoring procedures, mitigation measures and information to deal with long-term AC interference. Nevertheless, some aspects related to the phenomenon were not fully understood and the protection criteria as well as the corrosion mechanism have been debated for a long time.

This paper deals with a narrative literature review, which includes a deep analysis of the AC corrosion phenomenon, in particular the assessment of AC interference, and the evaluation of AC corrosion likelihood and interference levels, the corrosion mechanism.

2. Assessment of AC Corrosion Likelihood

The assessment of AC interference likelihood on a buried pipeline should include several parameters related to both the interference source and the interfered structure. During the design phase, the evaluation of AC interference on a buried structure can be carried out by mathematical/electrical modelling, e.g., according to EN 50443 (Effects of electromagnetic interference on pipelines caused by high voltage AC electric traction systems and/or high voltage AC power supply systems) [5] or IEEE Guide for Safety in AC Substation Grounding [6]. These approaches aim to evaluate the tolerable AC voltage based on parameters, such as the electrical configuration of the AC power line, the distance between the AC source (power line or traction system) and the pipeline, the insulation properties of the coating as well as soil resistivity. In the case of existing structures, field measurements can be used as an alternative to calculation. According to calculations or field measurements, relevant mitigation measures should be installed to decrease the AC corrosion probability. Nevertheless, not only electrical parameters are involved in the AC corrosion mechanism and an electrochemical approach is required for an understanding of the mechanism, in particular in the presence of cathodic protection.

According to ISO 18086 [4], the assessment of AC corrosion should be performed by evaluation of some or all of the following parameters:

- AC voltage, V_{AC};
- AC density, i_{AC};
- DC density, i_{DC};
- AC/DC densities ratio, i_{AC}/i_{DC};
- DC potential (IR-free potential, $E_{IR\text{-}free}$, and ON potential, E_{ON});
- Soil resistivity, ρ.

2.1. AC Voltage

The measurement of the AC voltage, V_{AC}, on a pipeline is carried out with respect to a reference electrode located at remote position, i.e., where the AC voltage gradient does not change and is close to zero. The AC voltage gradient is measured by means of two-reference electrodes spaced 1 to 5 m transverse to the pipeline.

According to [4], acceptable AC voltages on the pipeline in CP condition are lower than 15 V r.m.s. measured as an average over a representative time (e.g., 24 h). According to NACE SP0177 (Mitigation of Alternating Current and Lightning Effects on Metallic Structures and Corrosion Control Systems) [7], the maximum AC voltage is set at 15 V with respect to local earth (approximately 1 m); this threshold is mostly driven by safety considerations (shock hazard).

In a recent work, Tang et al. [8] investigate the effects of several parameters on the electric field distribution of AC interference, such as the unbalanced current magnitude, soil and coating resistivity and the distance between the power line and the pipeline. By means of numerical simulation, the authors conclude that the reference electrode should be placed farther from the pipeline route with the increase of mitigation wire length, soil resistivity and the distance between the power line and the structure; conversely, the earth remote position is closer to the pipe by increasing coating resistivity if mitigation is applied [8].

2.2. AC Density

AC and DC densities on a coating defect control both the AC interference and the CP level, respectively. Contrary to the AC voltage measurement, the AC density, i_{AC}, cannot be readily determined. The numerical approach considers the calculation of AC density from the AC voltage, the soil resistance, ρ, and the diameter, ϕ, of a circular coating defect, according to the following equation, as reported in [8–12]:

$$i_{AC} = \frac{8 \cdot V_{AC}}{\rho \cdot \pi \cdot \phi} \quad (1)$$

Considering a maximum AC voltage of 15 V measured on a circular coating defect of 1 cm^2 and a medium soil resistivity of 100 $\Omega \cdot m$, the expected AC density threshold is about 30 $A \cdot m^{-2}$. Nevertheless, this calculation is generally not possible since the coating defect area is not known. Moreover, the application of CP can significantly change the electrolyte composition in proximity to the coating defect and consequently the local soil resistivity. The formula is valid when the coating defect size is larger than the coating thickness, although rigorous calculations are available [13]. The current density can only be estimated by means of coupons or probes. According to ISO 18086 [4], the measurement of AC density has to be carried out on a 1 cm^2 coupon surface area connected to the structure.

The definition of a critical threshold of AC density over which AC corrosion could occur is still controversial and large data variability is found. Compared to DC interference corrosion, AC corrosion of carbon steel is lower considering the same current density. Since the sixties of the last century [14,15], the effect of AC density was determined in terms of "equivalent DC density", defined as the percent ratio of the weight loss caused by AC to the expected weight loss due to the same DC density. The values of equivalent DC density are in the range between 0.1% and 0.3% with AC density up to 600 $A \cdot m^{-2}$ [14,15]. Using laboratory tests on carbon steel in soil-simulating solution (1200 $mg \cdot dm^{-3}$ sulphates, 200 $mg \cdot dm^{-3}$ chlorides), Goidanich et al. [16] reported that AC corrosion efficiency (defined similarly to the "equivalent DC density") is lower than 1% when AC density ranges from 50 to 500 $A \cdot m^{-2}$, but it increases up to 4% for AC density lower than 50 $A \cdot m^{-2}$ (Figure 1). In 2010, Fu and Cheng [17] reported comparable results.

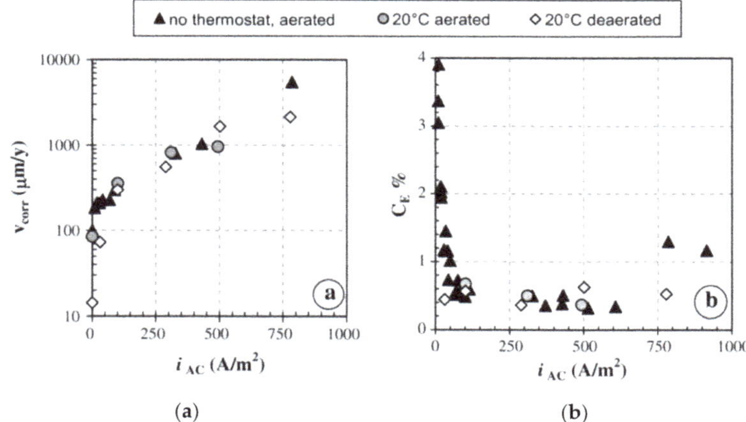

Figure 1. Weight loss tests on carbon steel in free corrosion condition exposed to a soil-simulating solution: (**a**) corrosion rates vs. AC density (i_{AC}) and (**b**) AC corrosion efficiency vs. AC density (i_{AC}) as reported by Goidanich et al. [16].

Gummow et al. [18] stated that the corrosion rate increases with increased AC density greater than 20 A·m^{-2} and becomes significant at AC densities greater than 100 A·m^{-2}, regardless of the magnitude of CP density. Based on laboratory tests, Pourbaix et al. [19] reported that AC corrosion is associated with the IR-free potential oscillation during interference but it is not related to a critical value of the AC density. As reported by Yunovich and Thompson [20], steel corrosion can significantly increase in the presence of 20 A·m^{-2} AC density: the measured corrosion rate at 20 A·m^{-2} AC density is nearly two times higher than that of the control specimen in free corrosion condition and decreases with the application of CP. This last consideration introduces the need of the additional consideration of the cathodic DC density.

2.3. AC/DC Current Density Ratio

For carbon steel structures under CP conditions, AC corrosion likelihood should be evaluated also considering the level of DC polarization, by means of the IR-free potential or DC density. The latter can be measured by means of a corrosion coupon or probes with a known surface area, e.g., 1 cm^2. In order to assess AC corrosion conditions, it is better to refer to the AC density-DC density ratio (i_{AC}/i_{DC}), which is dimensionless. Nevertheless, use of only the i_{AC}/i_{DC} ratio could be misleading in the assessment of AC corrosion likelihood, i.e., different AC corrosion conditions can be represented by the same i_{AC}/i_{DC} ratio. For instance, an i_{AC}/i_{DC} ratio equals to 10 results from an interference condition of 30 A·m^{-2} AC density in the presence of 3 A·m^{-2} DC density or 3 A·m^{-2} AC density with 0.3 A·m^{-2} DC density. Although the ratio between the current densities is equal in the two conditions, they represent a different corrosion risk, i.e., 3 A·m^{-2} AC density is not recognized as a threat, dissimilarly from 30 A·m^{-2}. As discussed in Paragraph 3, several authors [21–27] investigated the effect of i_{AC}/i_{DC} ratio on corrosion rate, proposing different threshold limits. ISO 18086 standard [4] reports that AC corrosion can be mitigated by maintaining the i_{AC}/i_{DC} ratio less than 3 over a representative time (e.g., 24 h) and it is valid for DC density greater than 10 A·m^{-2} (severe over-protection condition) and AC density over 30 A·m^{-2}.

2.4. AC Frequency

There is full agreement that the corrosion rate decreases by increasing the frequency of the AC signal. The effect of frequency has been investigated on mild steel, nickel and copper-nickel alloys [28–35]. AC can cause severe corrosion at the industrial frequencies of 50 or 60 Hz, while the effect decreases at frequencies higher than 150 Hz.

Fernandes et al. [28] and the other authors [29–32] proposed a kinetic interpretation: by increasing the frequency, the time between the anodic and cathodic half-cycles becomes shorter and the metallic ions dissolved during the anodic period would be available for the subsequent deposition in the cathodic cycle.

Guo et al. [34,35] reported that at an AC density of 50 $A \cdot m^{-2}$, the corrosion rate of X60 steel decreases from 1.2 $mm \cdot a^{-1}$ at 10 Hz frequency to about 0.6 $mm \cdot a^{-1}$ at 50 Hz. In parallel, the free corrosion potential increases to about 50 mV. Yunovich and Thompson [36] proposed an electrical circuit in order to simulate the behavior of a steel specimen exposed to soil varying the frequency of the AC signal. The model shows that the corrosion current in the circuit decreases with increasing frequency and is approximately 0.3% of the total current at 60 Hz frequency, in agreement with the results using weight loss tests, as reported in Figure 1.

2.5. Soil Resistivity and Chemical Composition

According to Equation (1), the AC density corresponding to a coating defect depends on the alternating voltage and on the spread resistance, which is the ohmic resistance through a coating defect (or a corrosion coupon) to earth. The ISO 18086 standard [4] reports an empirical relation between soil resistivity and AC corrosion risk:

- $\rho < 25\ \Omega \cdot m$: very high risk;
- $25\ \Omega \cdot m < \rho < 100\ \Omega \cdot m$: high risk;
- $100\ \Omega \cdot m < \rho < 300\ \Omega \cdot m$: medium risk;
- $\rho > 300\ \Omega \cdot m$: low risk.

Soil resistivity close to a coating defect is significantly affected by the electrochemical reactions at the metal-to-electrolyte interface, due to the application of the CP current. In CP condition, oxygen reduction ($O_2 + 2H_2O + 4e^- \rightarrow 4OH^-$) and, at lower potential, hydrogen evolution ($2H^+ + 2e^- \rightarrow H_2$), cause a growth of alkalinity at the metal surface. The pH value can increase over 10 and up to 12–13 at very high cathodic current densities. The local soil chemical composition can play a crucial role in the AC corrosion assessment, as documented in [27,37]. Earth-alkaline ions (as Ca^{2+} and Mg^{2+}), moved towards the metal surface by the CP electric field, form slightly soluble hydroxides; the pH increase, shifting the carbonate-bicarbonate chemical equilibrium, favors the growth of a scale of calcium and magnesium carbonate that increases the spread resistance. Otherwise, alkaline cations (as Na^+, K^+ or Li^+) form not-scaling hydroxides. Büchler et al. [37] reported a reduction of AC density due to the growth of chalk layers on the surface in the presence of calcium ions.

Recently, Xiao et al. [27] reported that the spread resistance of a X70 steel specimen at constant CP potential and different AC densities is higher in the presence of calcium and magnesium ions. Moreover, the corrosion rate of the specimens exposed to higher content of Na^+ was greater than that in the presence of earth-alkaline ions at the same potential and AC density (100 $A \cdot m^{-2}$, 300 $A \cdot m^{-2}$).

2.6. Effect on DC Potential (Free Corrosion Condition)

2.6.1. Negative Shift of Potential

There is general agreement that the free corrosion potential of carbon steel, i.e., without cathodic protection, decreases as the AC density increases. This has been documented from the sixties of the last century. Bolzoni et al. [38] reported laboratory tests on the influence of AC interference on carbon steel corrosion in free corrosion condition in different environments (sulfate and chloride aqueous solutions, with or without oxygen, simulating soil conditions and seawater). AC was overlapped to the specimens ranging from 10 to 6000 $A \cdot m^{-2}$. The free corrosion potential of carbon steel in chloride and sulfate solutions decreases as AC density increases. At AC densities below 100 $A \cdot m^{-2}$, the DC potential variation was low (about 50 mV); above 100 A·m^{-2}, the effect was higher (100–200 mV). In chloride solutions, the DC potential variation is less significant at high AC density (higher than 1000 A·m^{-2}). Results were confirmed in [39]: except for carbon steel in soil-simulating solution (1200 ppm SO_4^{2-} (Na_2SO_4) + 200 ppm Cl^- ($CaCl_2 \cdot 2H_2O$) [16]), the corrosion potential of galvanized steel, copper and carbon steel in different environmental conditions decreases with increasing AC density. Authors investigated the effects of AC density on anodic and cathodic overvoltages: AC has a significant effect on the kinetics parameters, with a decrease of overvoltages and increase of exchange current density of anodic and cathodic processes [39,40]. Nevertheless, in these papers some inconsistencies were observed between the experimental tests and the results expected from mathematical models based on the asymmetry of anodic and cathodic reactions. Li et al. [41] and Wang et al. [42] have measured a lowering of free corrosion potential on X70 and X80 carbon steel samples at various AC densities in simulated soil solution and 3.5% sodium chloride solution. Potential shift in both environments is about 0.2 V at 300 A·m^{-2} AC density. Moreover, authors investigated the kinetic effect of AC interference on anodic and cathodic overvoltages, measuring a variation of anodic and cathodic Tafel slope and their ratio in the presence of AC. Zhang et al. [43] proposed a nonlinear model (an electrical equivalent model considering the anodic and cathodic reactions under activation control) for the investigation of AC interference effect on corrosion potential and corrosion rate. Results show that the expected variation of the free corrosion potential depends on the AC peak potential, as expected, and on the ratio of the anodic and cathodic Tafel slope ($r = \beta a/\beta c$). When $r = 1$ (symmetry of anodic and cathodic overvoltages), no DC potential variations are predicted by the model. For $r > 1$, a positive (anodic) potential shift is expected, while for $r < 1$ the DC potential lowers as AC peak potential increases. The latter covers the electrochemical condition of active carbon steel in soil or waters where the cathodic processes (oxygen reduction and/or hydrogen evolution) have a higher Tafel slope than that of steel dissolution. These data are consistent with the observations made in [44] and in previous works [45–47].

2.6.2. Positive Shift of Potential

In 2012, He et al. [24] report that the average corrosion potential of X65 steel in loam soil moves to more positive values by increasing AC density from 5 to 150 A·m^{-2}. At 150 A·m^{-2}, a positive shift of about 200 mV has been measured with respect to the condition without interference. Xu et al. [48] examined the effect of AC (60 Hz frequency) on 16Mn steel potential in a simulated soil solution by means of real-time AC/DC signal acquisition. AC moves corrosion potential negatively at an AC density lower than 400 A·m^{-2}, while at higher AC density, the DC potential variation with respect to the absence of AC interference is positive. In a recent work, Wu et al. [49] reported polarization curves of X70 steel tested at AC densities up to 100 A·m^{-2} in simulated seawater. The presence of AC has a strong effect on the polarization curves with a general shift toward higher current density and a positive (anodic) variation of the zero-current potential, i.e., the free corrosion potential. Nevertheless, as the AC density was raised from 10 to 100 A·m^{-2}, the corrosion current density and the free corrosion potential roughly remained constant.

2.7. *Effect on DC Potential (Cathodic Protection Condition)*

The potential measurement is affected by AC interference, even if CP is applied. Several authors have investigated in the last decades the effect of AC on IR-free potential. Bolzoni et al. [38] investigated the influence of AC interference on carbon steel in CP condition in different environments (sulfate and chloride aqueous solutions, with or without oxygen). In the presence of cathodic polarization, the potential trend depends on DC density: at 0.1 A·m^{-2}, the DC potential is lowered after AC application; conversely, at 1 and 10 A·m^{-2} DC density, the DC potential increases as the AC density increases. In 2008 [50], and later in 2010 [23], Ormellese et al. reported the measurements of IR-free potential of carbon steel specimens exposed for about four months to a soil-simulating solution. DC and AC density were in the range 0.1–10 and 10–500 A·m^{-2}, respectively. The increment of potential is not significant at 10 A·m^{-2} AC, while it is about 0.1 and 0.2 V at 100 and 200 A·m^{-2}, respectively. The effect of AC density on IR-free potential is more pronounced at high DC density [23].

Xu et al. [51,52] investigated the effects of AC on the CP potential reading of a 16Mn pipeline steel in a simulated soil solution. At −0.85 V vs. SCE (maintained in a galvanostatic way, SCE—saturated calomel electrode), AC moves DC potential negatively. Furthermore, the higher the AC density, the more negative the DC potential is. Conversely at −1 V vs. SCE, AC shifts potential in the positive direction. Similar observations were reported by Kuang et al. [53,54]: the DC potential of X65 steel in near-neutral pH bicarbonate solution is shifted negatively by AC at −0.85 V vs. CSE, but positively shifted by AC under the CP of −1 V vs. CSE (Figure 2). Nevertheless, differently to what can be expected, the potential variation reported is higher at smaller AC density (Figure 2b). When the applied CP level was −0.925 V vs. CSE (data not shown), the DC potential becomes more positive at low AC densities of 10 and 50 A·m^{-2}, while it decreases with 100 A·m^{-2} AC density. Recently, Wang et al. [55] reported similar conclusions for X70 steel in near-neutral bicarbonate solution: at −0.775 V vs. SCE, the DC potential is shifted negatively consequently to the application of AC, while at −0.95 V vs. SCE and −1.2 V vs. SCE, an increase of DC potential is measured (Figure 3). In this case, the potential variation is proportional to AC density.

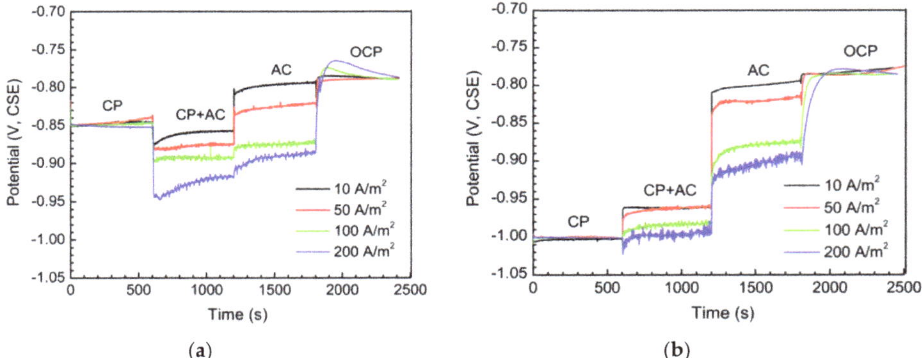

Figure 2. Effect of AC density on DC potential of X65 steel in near-neutral pH bicarbonate solution in cathodic protection condition: (**a**) −0.85 V vs. CSE, (**b**) −1 V vs. CSE [54].

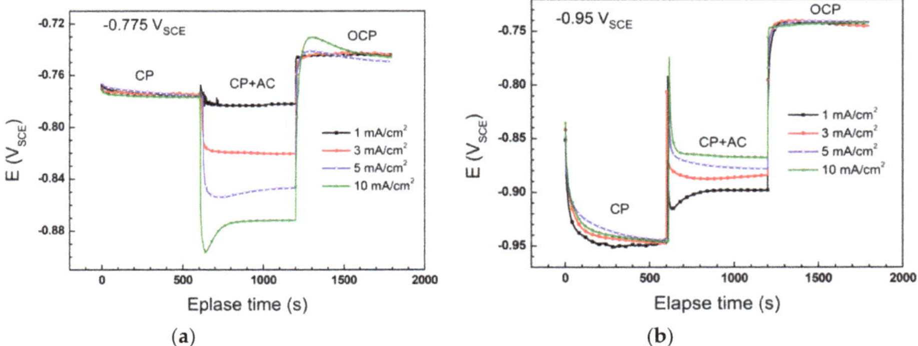

Figure 3. Effect of AC density on DC potential of X70 steel in near-neutral pH bicarbonate solution in cathodic protection condition: (**a**) −0.775 V vs. saturated calomel electrode (SCE) (−0.850 V vs. CSE) (**b**) −0.95 V vs. SCE (−1.02 V vs. CSE). 1 mA·cm^{-2} corresponds to 10 A·m^{-2} [55].

Generally, it can be concluded that there is good agreement on the increase of DC potential in the presence of AC, although a negative shift is measured at small CP current density.

3. Acceptable AC Interference Levels—Protection Criteria

There is large agreement that corrosion can occur on AC interfered carbon steel structures that fully match the CP potential criterion ($E < E_{prot}$) defined by ISO 15589-1 [2]. Much effort has been made in the last decades in order to define acceptable AC interference levels for carbon steel under CP condition. Kajiyama et al. [56–59] proposed a CP criterion based on the ratio between DC and AC densities, measured by means of corrosion coupons. The criterion can be summarized as follows (Figure 4):

- if $0.1\ \text{A·m}^{-2} \leq i_{DC} < 1\ \text{A·m}^{-2}$, then $i_{AC}/i_{DC} < 25$,
- if $1\ \text{A·m}^{-2} \leq i_{DC} \leq 20\ \text{A·m}^{-2}$, then $i_{AC} < 70\ \text{A·m}^{-2}$.

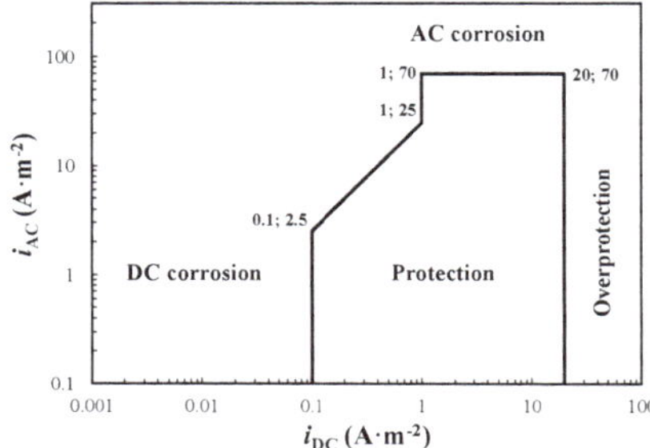

Figure 4. Acceptable AC interference levels based on current densities criterion according to Kajiyama et al. (adapted from [56]).

Accordingly, the maximum AC density depends on the CP level: at higher DC densities (i.e., more negative potential), a higher AC density can be tolerated. Even if the criterion has been applied successfully to some case studied [58], some authors recognized a greater AC corrosion risk at higher DC density differently to this criterion, as discussed later. Moreover, this criterion does not consider directly the value of the measured potential.

In 2012, He et al. [24] reported a similar approach based on current densities: the AC density threshold increases linearly with CP current density. The criterion (Figure 5) suggests there is not corrosion risk if $i_{AC} < 10 + 100{\cdot}i_{DC}$ (with $i_{DC} \geq 0.01\ \text{A·m}^{-2}$). Comparing the two criteria, the latter (Figure 5) is less conservative at a DC density lower than 1 A·m^{-2} and does not take into account AC corrosion at greater DC densities. For instance, at 0.1 A·m^{-2} DC density, the maximum allowed AC density is 2.5 and 20 A·m^{-2}, considering the corrosion criterion of Figures 4 and 5, respectively.

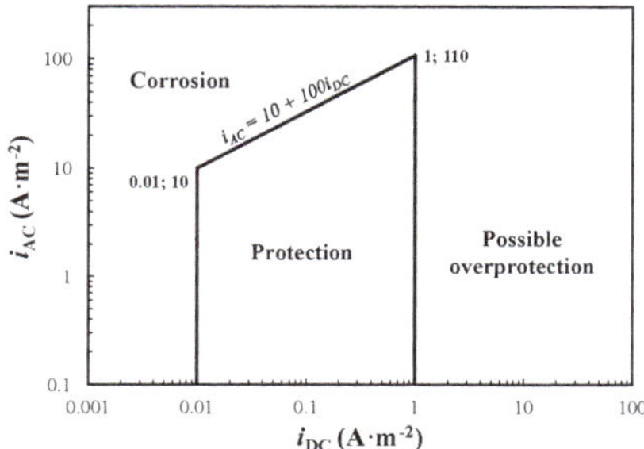

Figure 5. Acceptable AC interference levels based on current densities criterion according to He et al. (adapted from [24]).

The effect of potential has been investigated by Ormellese et al. in [50] and later in [23,26]. The authors propose corrosion maps based on corrosion rate data evaluated using a weight loss test of carbon steel specimens exposed to a soil simulation environment, with varying AC interference and CP levels. Two AC corrosion risk regions are defined, low and high, for corrosion rates lower or greater than 10 μm·a^{-1}, respectively. Corrosion risk increases by increasing the i_{AC}/i_{DC} ratio (Figure 6a): corrosion protection is achieved up to a maximum value of the i_{AC}/i_{DC} ratio, which decreases as the IR-free potential becomes more negative. Differently from the criteria discussed previously, in overprotection condition a few A·m^{-2} of AC density (ranging from 5 to 20 A·m^{-2}, depending on potential) provokes corrosion of overprotected carbon steel. The authors proposed the following criterion (Figure 6b):

- if 0.1 A·m^{-2} ≤ i_{DC} < 1 A·m^{-2}, then i_{AC} < 30 A·m^{-2},
- if 1 A·m^{-2} ≤ i_{DC} ≤ 10 A·m^{-2}, then i_{AC} < 10 A·m^{-2}.

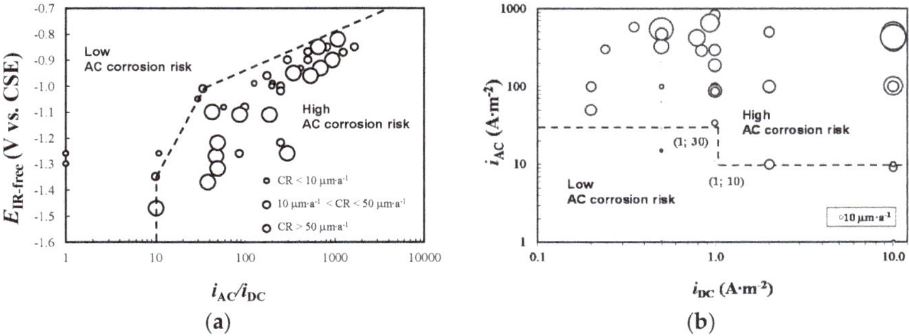

Figure 6. AC corrosion protection criterion as proposed by Ormellese et al.: (**a**) AC corrosion risk diagram based on IR-free potential and current densities ratio, (**b**) protection criterion based on AC and DC (i.e., cathodic protection—CP) current densities.

The −0.85 V vs. CSE criterion is not always safe in the presence of AC interference; in overprotection condition ($E_{IR\text{-}free}$ < −1.2 V vs. CSE), severe AC corrosion could occur.

Fu et al. [60] proposed a criterion based on both potential and current densities: potentials more positive than −0.95 V vs. CSE are considered not safe in the presence of AC. The maximum acceptable AC density increases as the potential becomes more negative: at −0.95 V vs. CSE, the maximum AC density is 20 A·m^{-2}, while at −1.05 V vs. CSE, the threshold increases up to 100 A·m^{-2}.

Büchler in 2012 [61] and previously in 2009 [22] investigated new protection criteria based on laboratory and field investigations. For current density average values measured by means of on-site coupons, one of the criteria below must be met (Figure 7):

- $i_{AC} < 30\ A{\cdot}m^{-2}$;
- $i_{DC} < 1\ A{\cdot}m^{-2}$;
- $i_{AC}/i_{DC} < 3$.

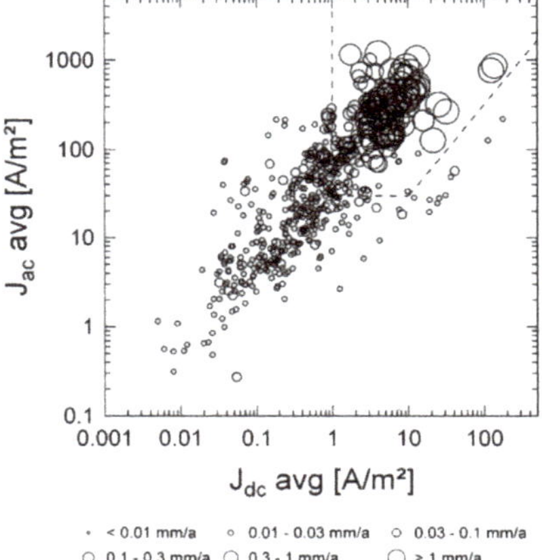

Figure 7. Protection criterion based on AC and DC densities as proposed by Büchler [61]. The region confined by the dotted lines corresponds to very severe AC corrosion and must be avoided.

Nevertheless, the author [61] stated that the measurement with coupons is fraught with problems, since the obtained results are affected by coupon geometry and local soil conditions. The use of ON potential, E_{ON}, and AC voltage (V_{AC}, indicated by the authors as U_{AC}) is therefore suggested (Figure 8). For ON potential, one of the following criteria must be met:

- average $V_{AC} < 15$ V and average E_{ON} more positive than −1.2 V vs. CSE;
- $V_{AC} < 3{\cdot}(|E_{ON}| - 1.2)$ where E_{ON} is in V vs. CSE and $E_{ON} < -1.2$ V vs. CSE.

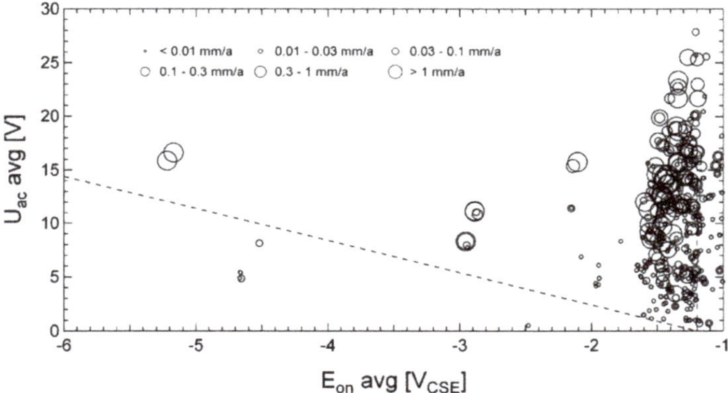

Figure 8. Protection criterion based on AC voltage and ON potential, as proposed by Büchler [61]. The region below the dotted lines corresponds to acceptable AC interference conditions.

In 2015, these efforts brought to the AC corrosion protection criteria of the ISO 18086 standard [4] for buried carbon steel in CP condition:

- As a first step, the AC voltage on the pipeline should be decreased below 15 V r.m.s. The AC voltage is measured as an average over a representative time (e.g., 24 h) with respect to a reference electrode located in remote position;
- As a second step, AC corrosion mitigation is achieved by matching the CP protection potentials defined in ISO 15589-1 [2] and — maintaining the AC density (i_{AC}) lower than 30 A·m^{-2} on a 1 cm^2 coupon or probe over a representative time (e.g., 24 h), or — maintaining the average cathodic current density lower than 1 A·m^{-2} on a 1 cm^2 coupon or probe over a representative time (e.g., 24 h), if AC density is higher than 30 A·m^{-2}, or — maintaining the ratio between AC and DC densities (i_{AC}/i_{DC}) less than 3 over a representative time (e.g., 24 h).

In Annex E (informative), the standard reports other criteria that have been used in the presence of AC. These criteria have been derived from Büchler's work [61] and are based on AC voltage, ON potential and current densities, as discussed. The protection criteria reported in the ISO 18086 standard are shown in Figure 9.

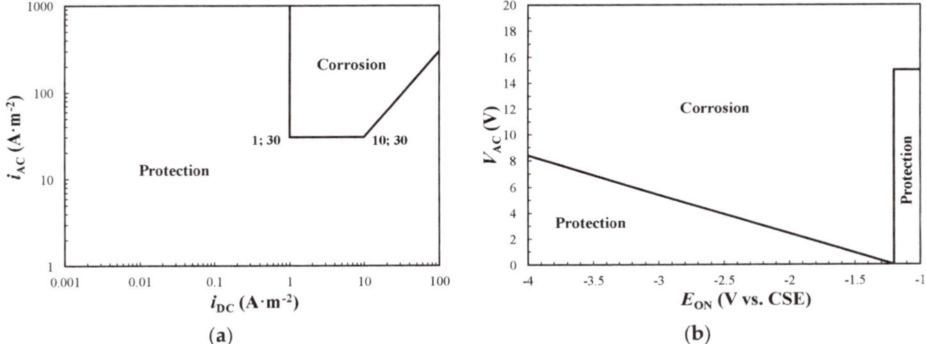

Figure 9. Graphical representation of the AC protection criteria reported on ISO 18086 [4]: (**a**) i_{AC} vs. i_{DC}, (**b**) V_{AC} vs. E_{ON}.

In 2018, Junker et al. [62] reported results of laboratory and field tests varying AC and DC levels and soil chemistry. Both laboratory and field data confirmed very high AC corrosion rates under excessive CP (> 1 A·m^{-2}) and AC interference higher than 30 A·m^{-2}. Moreover, they recognized in the spread resistance of a coating defect a highly dynamic parameter under AC and DC influence. The investigations illustrate that the chemical environment alters the AC and DC density limits for AC corrosion, however the present limits of ISO 18086 constitute a safe strategy in most environments.

Figure 10 reports the AC corrosion rate measured on corrosion coupons with varying AC and DC density, as reported by Nielsen [63]. Data are overlapped to the AC protection criteria reported in ISO 18086.

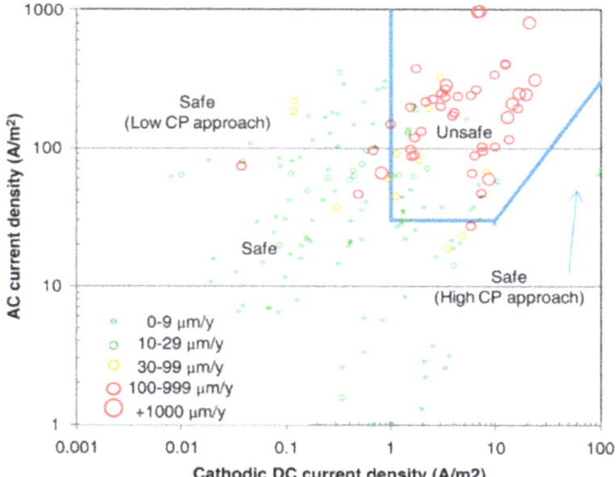

Figure 10. AC corrosion rate measured on corrosion coupons as reported by Nielsen [63]. Data are compared with the AC protection criteria reported in ISO 18086.

It can be concluded that overprotection (namely $E_{IR\text{-}free}$ lowers than −1.2 V vs. CSE [2] or i_{DC} > 1 A·m^{-2}) is the most dangerous condition in the presence of AC interference. At "high" CP levels, the maximum tolerable AC density is 30 A·m^{-2}. Below 1 A·m^{-2} DC density, the AC corrosion likelihood decreases. Nevertheless, some doubts are revealed regarding the inexistence of an AC density threshold at "low" CP condition (i_{DC} < 1 A·m^{-2}).

4. AC Corrosion Mechanism

Numerous theories and models have been proposed on the AC corrosion mechanism of carbon steel. Some models referring to carbon steel structures in free corrosion conditions as well as in the presence of CP are discussed hereafter.

4.1. *Effect of Anodic and Cathodic AC Half-Wave on Metal Dissolution*

Büchler et al. [22,61] proposed a corrosion mechanism based on thermodynamic and kinetic considerations on the reactions involved during AC interference on cathodically protected carbon steel. When an AC voltage is present, current will flow through the coating defects exposed to soil. If the pH value is sufficiently high (above 10, as in CP condition), during the anodic half-cycle, steel oxidation occurs, promoting the formation of a passive film. During the cathodic half-wave, the passive film is electrochemically destroyed and converted in porous rust. In the successive anodic cycle, the passive film is reformed under the non-protective rust layer. Moreover, the Fe^{2+} present in the rust layer is oxidized to Fe^{3+} ($Fe^{2+} \rightarrow Fe^{3+} + e^-$). In the subsequent cathodic cycle, the dissolution of the passive film will increase the volume of porous rust. Hence, every AC cycle results in the oxidation of the metal with a significant metal loss in the long term (Figure 11). A simplified description of this mechanism is reported also in the ISO 18086 standard [4].

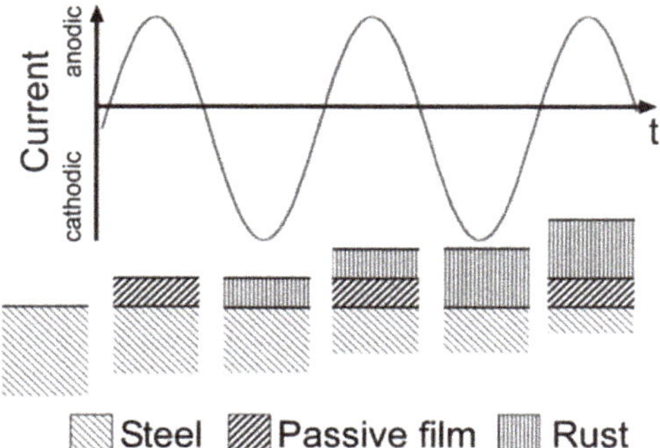

Figure 11. Schematic representation of AC corrosion mechanism according to Büchler et al. [61].

4.2. The Alkalization Mechanism and the Effect of Spread Resistance

Nielsen et al. [64–66] proposed the "alkalization model" of AC-induced corrosion of carbon steel under CP condition. The model is based on the combination of two effects: (1) the alkalization of the metal-to-electrolyte interface of overprotected carbon steel, and (2) potential oscillations across the immunity, the passive and the high-pH corrosion domain of the iron potential-pH diagram.

As known, the presence of a cathodic current on the metal surface under CP condition is beneficial because it promotes the reduction (or zeroing) of the corrosion rate and the increase of alkalinity due to the accumulation of hydroxyl ions (OH^-) close to coating defect. The pH increase is proportional to the cathodic current density and depends on the diffusion and electrical migration of ions towards and from the metal surface. In overprotection condition, namely IR-free potential more negative than −1.2 V vs. CSE, the high cathodic current density (in the order of a few $A{\cdot}m^{-2}$) can promote a significant increase of pH up to 13 or higher. According to the Pourbaix diagram, at elevated pH the potential oscillations caused by AC interference could lead to periodic entry in the high-pH corrosion region with formation of dissolved $HFeO_2^-$ ions. The authors report the presence of an "incubation time" defined as the period to reach a critical pH (close to 14) at the metal-to-soil interface, with a significant lowering of the spread resistance and increase of AC density due to depolarization effects of the AC (Figure 12). Then, potential oscillations could cause corrosion due to different time constants associated to iron dissolution (fast) and the formation of a passive film (slower). Accordingly, AC corrosion of carbon steel in CP condition cannot be reduced by adding a surplus of cathodic current, as in the case of DC corrosion phenomena, but by avoiding high DC densities and the overprotection condition, in agreement with the protection criteria of the ISO standard [4].

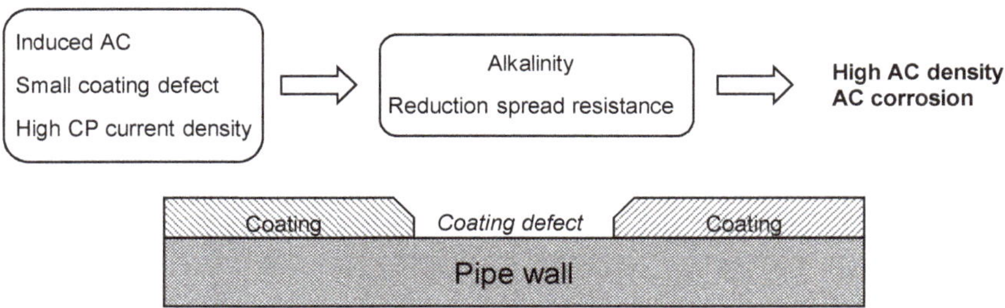

Figure 12. Schematic representation of the AC corrosion mechanism of carbon steel under CP condition, as proposed by Nielsen et al. [63–66].

This "vicious circle" is supported by the data shown in Figure 13a–f [63], which illustrates experimental results in a laboratory soil box environment in purified quartz sand. At a constant

AC voltage (15 V), six different levels of CP (ON potential) were applied for some weeks to monitor the corrosion rate of an electrical resistance probe and various electrical parameters. At a fixed AC voltage, the corrosion rate depends strongly on the CP level (ON potential), and therefore, AC voltage alone cannot be considered a valid indicator of AC corrosion risk. Despite the constant AC voltage, the AC density varies from about 100 A·m^{-2} at low CP levels up to 500 A·m^{-2} at higher CP levels. Figure 13e shows corrosion rate as a function of cathodic current density: the corrosion rate increases with increasing CP current density, in agreement with the proposed mechanism. At the same time, the spread resistance is strongly influenced by the DC density with the consequent increase of AC density and corrosion rate (Figure 13f).

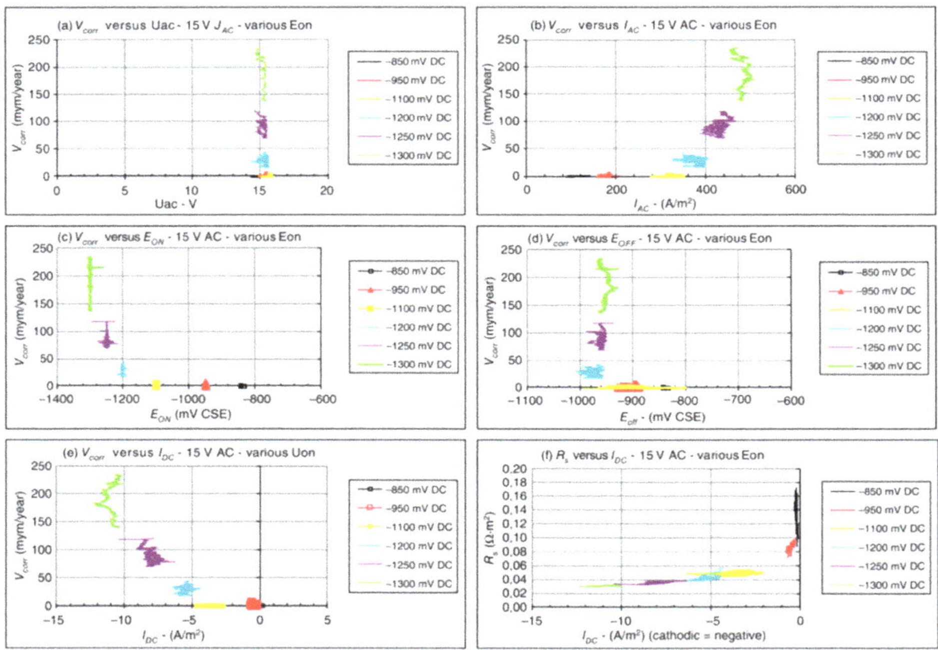

Figure 13. Example of corrosion rate vs. various electrical parameters, as reported by Nielsen [63].

The crucial role of soil chemical composition, pH and spread resistance was carefully investigated by Junker et al. [62,67–69]. The spread resistance is identified as a key parameter, controlling the current densities, and is highly influenced by the formation of calcareous deposits or corrosion products. At high CP density, AC corrosion of an ER probe element is associated with strong alkalization of the electrolyte and consequently dense calcareous deposit formation. The calcareous deposit dramatically increases the spread resistance and reduces the AC and DC densities. Corrosion decreases, but only as long as the calcareous deposit is stable and fully covering the surface. Due to brittle fracture or a 'flake of' mechanism of the scale (probably provoked by hydrogen evolution at low potentials), the spread resistance suddenly decreases, causing an increase of current densities and AC corrosion. The cathodic reactions on the re-exposed probe surface will restart the alkalization and precipitation of calcareous deposits and with time (days) the corrosion stops again. This causes a cyclic variation of spread resistance, current densities and corrosion rate. The detailed chemical investigation of stone hard soil formed on cathodically protected pipeline under AC interference is reported in [68].

The effect of spread resistance on AC corrosion has been also documented by Nielsen and Cohn [70] with the help of an electrical equivalent circuit analysis that represents the impedances existing between the pipe and remote earth. AC and DC sources impose a DC and AC voltage on the pipe: the simulated AC source is the HVTL (High Voltage Transmission Line), whereas the DC source represents the CP system. The authors consider static and dynamic elements. Static elements, namely spread resistance and charge transfer resistance, are defined as elements where impedance is frequency independent. Conversely, dynamic elements are frequency dependent: these include the double layer capacitance and diffusion elements. Because of the low impedance of the capacitance, the spread resistance is the dominant impedance element at 50–60 Hz frequency and plays a key role in controlling the AC corrosion process.

4.3. Effect of AC on Anodic and Cathodic Overvoltages

The increase of corrosion rate in the presence of AC has been explained by some authors using the effect of AC on anodic and cathodic overvoltages [39,46,49,65,71,72]. Goidanich et al. [39] investigated by means of laboratory tests the influence of AC on kinetic characteristics of carbon steel, galvanized steel, copper and zinc under different experimental conditions. Results showed that AC has a significant influence on kinetic parameters, such as Tafel slope and exchange current density, and on corrosion and equilibrium potential. The authors proposed a "mixed" corrosion mechanism, with a general decrease of overvoltages and increase of exchange current density in the presence of AC. This effect could be related to the local rise in temperature, associated with high AC densities, as reported in [16]. In a recent work, Wu et al. [49] reported polarization curves of X70 steel tested with varying AC densities (up to 100 A·m^{-2}) in simulated seawater (Figure 14). AC shifts toward higher current density in the polarization curves, promoting both the anodic and cathodic processes.

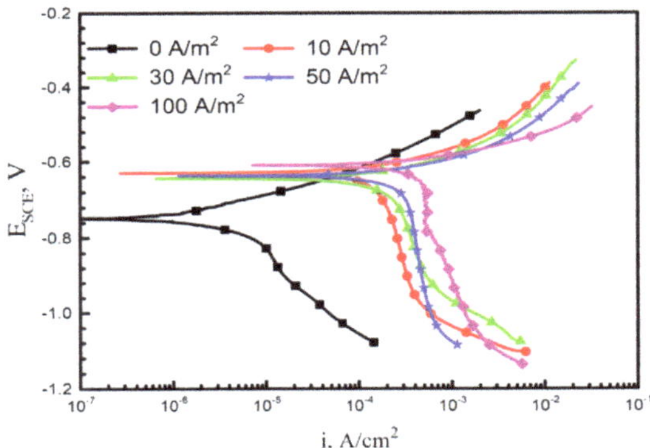

Figure 14. Polarization curves on X70 steel tested at various AC densities in simulated seawater [49].

As discussed previously, some authors [43–47] proposed theoretical models based on the fundamental thermodynamic and kinetic laws of corrosion in order to investigate the effect of AC interference on DC potential and corrosion rate. Accordingly, the sensitivity of the corroding system is influenced by the ratio of the anodic-to-cathodic Tafel slope ($r = \beta_a/\beta_c$). The effect of r on DC potential variation has already been discussed in Section 2.6. Considering a corrosion system under activation control, Lalvani and Lin [45] proposed an analytical solution for the relationship between corrosion rate and voltage peak amplitude. In a revised model [46], the authors introduced the effect of the double layer capacity, without considering the resistance of the electrolyte. The model indicates that corrosion current increases with voltage peak for all values of r, while the potential shift depends strongly on the anodic and cathodic characteristic curves, i.e., on the ratio between the anodic and cathodic Tafel slope. Potentiodynamic polarization curves were obtained using the revised model; nevertheless, these approaches predict that corrosion current and corrosion potential are independent of the frequency of the AC signal, differently to what is observed.

The model was improved in 2008 [43,44]; the authors considered three elements in an electrical equivalent circuit of a metal subjected to an induced AC voltage: the polarization impedance, the double layer capacitance and the electrolyte resistance. The model shows that the corrosion current increases as the frequency of the AC signal decreases (Figure 15a), in agreement with experimental observations, and by increasing the peak potential. Moreover, the model shows that corrosion current increases by decreasing the DC corrosion potential. For instance, by decreasing the DC corrosion potential from −0.6 to −0.7 V vs. SCE at a peak potential of 1.25 V vs. SCE, the corrosion rate increases several orders of magnitude (Figure 15b); an increase of DC corrosion potential from −0.2 to 0.0 V does not result in a further reduction of corrosion current [43].

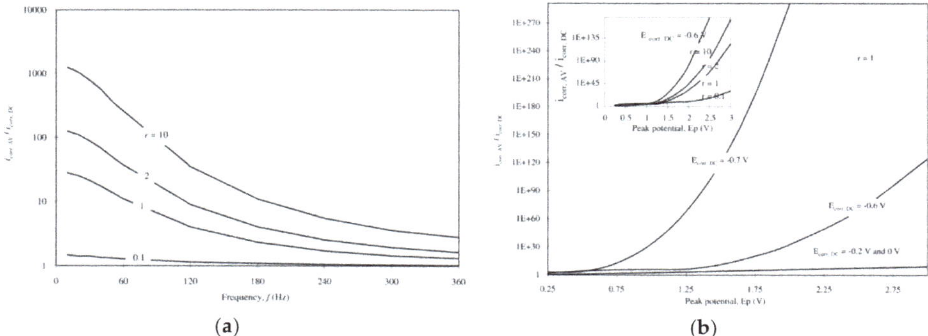

(**a**) (**b**)

Figure 15. A nonlinear model for corrosion of metals subjected to AC corrosion [43]: (**a**) dimensionless corrosion current vs. frequency, (**b**) dimensionless corrosion current vs. peak potential and DC corrosion potential. Legend: $E_{corr,DC}$ = corrosion potential in the absence of AC (V vs. SCE); E_p = AC peak potential (V vs. SCE); $r = \beta_a/\beta_c$.

Recently, Ibrahim et al. [73–75] proposed a theoretical approach (Part 1, 2, 3) to evaluate the effect of double layer capacitance and electrolyte resistance on corrosion current density and potential shift. The authors stated that corrosion rate enhancement is due to the faradaic rectification as a consequence of the nonlinear current-potential relationship [73].

4.4. Breakdown of the Passive Film and High-pH Corrosion

Recently, Brenna et al. [76,77] proposed a two-step AC corrosion mechanism. In the first step, AC causes the weakening of the passive film formed under CP condition on carbon steel, due to electromechanical stresses. Electrostriction appears to be a convincing explanation of the passive film breakdown mechanism, because of the presence of high alternating electric field (in the order of 10^6 V· cm^{-1} [78]) across the passive film [77]. The effect of AC on passive condition has been documented for carbon steel in alkaline solution or concrete and for stainless steel in neutral solution [79–85]: AC causes localized corrosion of passive metals with a decrease of corrosion resistance. Mechanical failure of the film can result from high electromechanical stresses (electrostriction pressure) generated by the presence of an electric field across the film and by the interfacial tension, which is not negligible as a result of the thin thickness of the oxide.

After film breakdown, high-pH chemical corrosion (i.e., potential independent) occurs in the overprotection condition because of the high cathodic current density supplied to the metal. According to the Pourbaix diagram of iron, high-pH corrosion can occur with formation of di-hypo ferrite ions ($HFeO_2^-$). This mechanism can also explain the unexpected corrosion of the metal below its equilibrium potential. Indeed, CP electrons are involved in the cathodic process, which depends on the potential assumed by the metal. Thus, below the equilibrium potential, the applied cathodic current makes electrons available to the metal; therefore, no anodic electrochemical reactions can take place.

Consequently, if no oxidation reaction takes place, such as iron ion production, this can occur only through chemical corrosion, which is not potential dependent, or, in other words, is not influenced by the electrons made available otherwise [86].

5. AC Corrosion Monitoring

According to ISO 18086 [4], AC voltage should be measured with respect to remote earth at test posts during the general and detailed assessment of CP effectiveness, as defined in ISO 15589-1. Additional measurements should be carried out at sites where AC corrosion risk is suspected, e.g., areas where the soil resistivity is low (lower than 25 Ω·m), areas with highest AC interference levels, areas where AC corrosion has previously taken place and areas where local DC polarization conditions can favor AC corrosion, as high levels of CP.

The measurement of AC and DC densities should be carried out by means of coupons or probes installed in the same soil or backfill as the pipeline itself. The measurements with respect to the criteria defined in ISO 18086 have to be carried out on a 1 cm^2 coupon surface area. Coupon or probe currents can be measured by the voltage drop across a series resistor; the value of the resistance should be sufficiently low to avoid disturbance of the system. For field measurements, a typical value is 10 Ω for a 1 cm^2 coupon.

For corrosion rate measurements, weight loss measurements, perforation measurements or electrical resistance (ER) measurements can be applied. Weight loss measurements require installation of pre-weighed coupons. After some time of operation (months to years), the coupon is excavated and weight loss rate is determined. The main drawback of this measurement is that the coupon provides no information until the excavation. Perforation measurements are made on special perforation probes: a signal is generated when the corrosion process has perforated the wall thickness of the coupon. In this case, information about the maximum (localized) corrosion depth is available without excavating the coupon; the primary disadvantage is that this information is not available until the coupon is perforated. Electrical resistance measurements require the installation of electrical resistance probes (ER probes). Corrosion is detected by the increase of the electrical resistance of the coupon when corrosion progressively decreases its thickness [64,87,88].

6. Conclusions

This paper deals with a narrative literature review on AC corrosion assessment, protection criteria and corrosion mechanisms for buried carbon steel structures in CP condition. Main conclusions can be summarized as follows:

- The assessment of AC corrosion likelihood should be based on the measurement of AC and DC related parameters. The AC interference level is evaluated by AC remote voltage and AC density, while the CP level is assessed by DC density and potential measurements;
- AC and DC densities should be measured by means of a corrosion coupon (1 cm^2 area) connected to the structure in CP condition; IR-free potential is considered more accurate than ON potential, because it does not contain the ohmic drop contribution in soil;
- There is general agreement that the DC potential of carbon steel in CP condition increases in the presence of AC interference, although a negative shift is measured at small DC density. Conversely, in free corrosion condition, i.e., without CP, the potential decreases as the AC density increases;
- Overprotection (namely $E_{\text{IR-free}} < -1.2$ V vs. CSE) is the most dangerous condition in the presence of AC interference. At "high" CP levels, the maximum tolerable AC density is 30 $A{\cdot}m^{-2}$. Below 1 $A{\cdot}m^{-2}$ DC density, the AC corrosion likelihood decreases. Nevertheless, some doubts are revealed regarding the inexistence of the criterion reported in the ISO 18086 standard of an AC density threshold at "low" CP condition ($i_{DC} < 1$ $A{\cdot}m^{-2}$);
- The higher AC corrosion likelihood at high CP levels could be explained by a corrosion mechanism that involves both the AC and DC levels:

- ○ At high DC density, a strong alkalization of the electrolyte close to the coating defect occurs with formation of a passive film and deposits (as a calcareous deposit) on carbon steel. Soil chemical composition, pH and spread resistance at the coating defects seem to have a crucial role, controlling the local AC and DC densities;
- ○ AC interference provokes a weakening of the passive condition due to an effect on anodic and cathodic overvoltages; moreover, the scale formed in CP condition is not stable in the presence of AC due to potential oscillations that could break the protective layer;
- ○ High-pH corrosion occurs with localized corrosion attacks; chemical corrosion (i.e., potential independent) with formation of di-hypo ferrite ions ($HFeO_2^-$) is a possible explanation for the occurrence of corrosion at low potentials.

Author Contributions: A.B. conceptualized and prepared the manuscript; M.O. and S.B. contributed to literature analysis and to the organization of the manuscript. All authors have read and agreed to the published version of the manuscript.

Funding: This research received no external funding.

Conflicts of Interest: The authors declare no conflicts of interest.

References

1. *General Principles of Cathodic Protection of Buried or Immersed Onshore Metallic Structures*; EN 12954; British Standards Institution: London, UK, 29 August 2019.
2. *Petroleum, Petrochemical and Natural Gas Industries. Cathodic Protection of Pipeline Systems. Part 1: On-land Pipelines*; ISO 15589-1; ISO - International Organization for Standardization: Geneva, Switzerland, 1 March 2015.
3. CIGRE Working Group 36.02. *Guide on the Influence of High Voltage AC Power Systems on Metallic Pipelines*; CIGRE Technical Brochure no. 095; CIGRE: Paris, France, 1995.
4. *Corrosion of Metals and Alloys. Determination of AC Corrosion. Protection Criteria*; ISO 18086; ISO - International Organization for Standardization: Geneva, Switzerland, 2019.
5. *Effects of Electromagnetic Interference on Pipelines Caused by High Voltage a.c. Electric Traction Systems and/or High Voltage a.c. Power Supply Systems*; EN 50443; CEN-CENELEC - European Committee for Electrotechnical Standardization: Brussels, Belgium, 1 January 2011.
6. IEEE. *IEEE Guide for Safety in AC Substation Grounding*; IEEE Std. 80-2000; IEEE - The Institute of Electrical and Electronics Engineers, Inc.: New York, NY, USA, 2000.
7. *Mitigation of Alternating Current and Lightning Effects on Metallic Structures and Corrosion Control Systems*; NACE SP0177; NACE International: Houston, TX, USA, 22 June 2007.
8. Tang, D.; Du, Y.; Lu, M.; Chen, S.; Jiang, Z.; Dong, L. Study on location of reference electrode for measurement of induced alternating current voltage on pipeline. *Int. Trans. Electr. Energy Syst.* **2013**, *25*, 99–119. [CrossRef]
9. NACE International Task Group 327. *AC corrosion State-of-the-Art: Corrosion Rate, Mechanism, and Mitigation Requirements*; NACE Technical Committee Report 35110; NACE International: Houston, TX, USA, 2010.
10. Gummow, R.A.; Segall, S.M.; Fieltsch, W. Pipeline ac mitigation misconceptions. In Proceedings of the Northern Area Western Conference, Calgary, AB, Canada, 15–18 February 2010; p. 14.
11. Panossian, Z.; Filho, S.E.A.; De Almeida, N.L.; Filho, M.L.P.; De L. Silva, D.; Laurino, E.W.; De L. Oliver, J.H.; De S. Pimenta, G.; De C. Albertini, J.Á. Effect of alternating current by high power lines voltage and electric transmission systems in pipelines corrosion. In Proceedings of the Corrosion 2009 Conference & Expo, Atlanta, GA, USA, 22–26 March 2009; p. 41.
12. Wakelin, R.G.; Gummow, R.A.; Segall, S.M. AC corrosion—Case histories, test procedures, & mitigation. In Proceedings of the Corrosion/98, San Diego, CA, USA, 22–27 March 1998; p. 14.
13. Ouadah, M.; Touhami, O.; Ibtiouen, R.; Zergoug, M. Method for diagnosis of the effect of AC on the X70 pipeline due to an inductive coupling caused by HVPL. *IET Sci. Meas. Technol.* **2017**, *11*, 766–772. [CrossRef]
14. Williams, J.F. Corrosion of metals under the influence of alternating current. *Mater. Prot.* **1966**, *5*, 52–53.
15. Pookote, S.R.; Chin, D.-T. Effect of alternating current on the underground corrosion of steels. *Mater. Performance* **1978**, *17*, 9–15.

16. Goidanich, S.; Lazzari, L.; Ormellese, M. AC corrosion. Part 2: Parameters influencing corrosion rate. *Corros. Sci.* **2010**, *52*, 916–922. [CrossRef]
17. Fu, A.; Cheng, Y.F. Effects of alternating current on corrosion of a coated pipeline steel in a chloride-containing carbonate/bicarbonate solution. *Corros. Sci.* **2010**, *52*, 612–619. [CrossRef]
18. Gummow, R.A.; Wakelin, R.G.; Segall, S.M. AC corrosion—A new challenge to pipeline integrity. In Proceedings of the Corrosion/98, San Diego, CA, USA, 22–27 March 1998; p. 18.
19. Pourbaix, A.; Carpentiers, P.; Gregoor, R. Detection and assessment of AC induced corrosion. *Mater. Performance* **2000**, *39*, 34–37.
20. Yunovich, M.; Thompson, N.G. AC corrosion: Corrosion rate and mitigation requirements. In Proceedings of the Corrosion/2004, New Orleans, LA, USA, 22 March–1 April 2004; p. 18.
21. Ragault, I. AC corrosion induced by V.H.V. electrical lines on polyethylene coated steel gas pipelines. In Proceedings of the Corrosion/98, San Diego, CA, USA, 22–27 March 1998; p. 14.
22. Büchler, M.; Schöneich, H.-G. Investigation of Alternating Current Corrosion of Cathodically Protected Pipelines: Development of a Detection Method, Mitigation Measures, and a Model for the Mechanism. *Corrosion* **2009**, *65*, 578–586. [CrossRef]
23. Ormellese, M.; Lazzari, L.; Brenna, A.; Trombetta, A. Proposal of CP criterion in the presence of AC-interference. In Proceedings of the Corrosion 2010 Conference & Expo, San Antonio, TX, USA, 14–18 March 2010; p. 17.
24. He, X.; Jiang, G.; Qiu, Y.; Zhang, G.; Jin, X.; Xiang, Z.; Zhang, Z.; Tang, H. Study of criterion for assuring the effectiveness of cathodic protection of buried steel pipelines being interfered with alternative current. *Mater. Corros.* **2011**, *63*, 534–543. [CrossRef]
25. Ding, H.; Li, J.; Wang, S.; Yang, Y. Experimental study on stray current corrosion of coated pipeline steel. *J. Nat. Gas Sci. Eng.* **2015**, *27*, 1555–1561. [CrossRef]
26. Ormellese, M.; Lazzari, L.; Brenna, A. AC corrosion of cathodically protected buried pipelines: Critical interference values and protection criteria. In Proceedings of the Corrosion 2015 Conference & Expo, Dallas, TX, USA, 15–19 March 2015; p. 11.
27. Xiao, Y.; Du, Y.; Tang, D.; Ou, L.; Sun, H.; Lu, Y. Study on the influence of environmental factors on AC corrosion behavior and its mechanism. *Mater. Corros.* **2017**, *69*, 601–613. [CrossRef]
28. Fernandes, S.Z.; Mehendale, S.G.; Venkatachalam, S. Influence of frequency of alternating current on the electrochemical dissolution of mild steel and nickel. *J. Appl. Electrochem.* **1980**, *10*, 649–654. [CrossRef]
29. Lalvani, S.; Zhang, G. The corrosion of carbon steel in a chloride environment due to periodic voltage modulation: Part I. *Corros. Sci.* **1995**, *37*, 1567–1582. [CrossRef]
30. Qiu, W.W.; Pagano, M.; Zhang, G.; Lalvani, S. A periodic voltage modulation effect on the corrosion oF Cu-Ni Alloy. *Corros. Sci.* **1995**, *37*, 97–110. [CrossRef]
31. Lalvani, S.B.; Kang, J.-C.; Mandich, N.V. The corrosion of Cu-Ni alloy in a chloride solution subjected to periodic voltage modulation: Part I. *Corros. Sci.* **1998**, *40*, 69–89. [CrossRef]
32. Lalvani, S.B.; Kang, J.-C.; Mandich, N.V. The corrosion of Cu-Ni alloy in a chloride solution subjected to periodic voltage modulation: Part II. *Corros. Sci.* **1998**, *40*, 201–214. [CrossRef]
33. Song, H.; Kim, Y.; Lee, S.; Kho, Y.; Park, Y. Competition of AC and DC current in AC corrosion under cathodic protection. In Proceedings of the Corrosion/2002, Denver, CO, USA, 7–11 April 2002; p. 12.
34. Guo, Y.-B.; Liu, C.; Wang, D.-G.; Liu, S.-H. Effects of alternating current interference on corrosion of X60 pipeline steel. *Pet. Sci.* **2015**, *12*, 316–324. [CrossRef]
35. Guo, Y.; Meng, T.; Wang, D.; Tan, H.; He, R. Experimental research on the corrosion of X series pipeline steels under alternating current interference. *Eng. Fail. Anal.* **2017**, *78*, 87–98. [CrossRef]
36. Yunovich, M.; Thompson, N.G. AC Corrosion: Mechanism and Proposed Model. In Proceedings of the 2004 International Pipeline Conference, Calgary, AB, Canada, 4–8 October 2004; Volumes 1–3, pp. 183–195.
37. Büchler, M.; Voûte, C.-H.; Schöneich, H.-G.; Stalder, F. Characteristics of potential measurements in the field of ac corrosion. In Proceedings of the CeoCor International Congress and Technical Exhibition, Giardini Naxos, Italy, 13–16 May 2003.
38. Bolzoni, F.; Goidanich, S.; Lazzari, L.; Ormellese, M. Laboratory test results of AC interference on polarized steel. In Proceedings of the Corrosion/2003, San Diego, CA, USA, 16–20 April 2003; p. 13.
39. Goidanich, S.; Lazzari, L.; Ormellese, M. AC corrosion—Part 1: Effects on overpotentials of anodic and cathodic processes. *Corros. Sci.* **2010**, *52*, 491–497. [CrossRef]

40. Bolzoni, F.; Goidanich, S.; Lazzari, L.; Ormellese, M.; Pedeferri, M. Laboratory testing on the influence of alternated current on steel corrosion. In Proceedings of the Corrosion/2004, New Orleans, LA, USA, 22 March–1 April 2004; p. 11.
41. Li, Y. Effects of Stray AC Interference on Corrosion Behavior of X70 Pipeline Steel in a Simulated Marine Soil Solution. *Int. J. Electrochem. Sci.* **2017**, *12*, 1829–1845. [CrossRef]
42. Wang, X. Corrosion Behavior of X70 and X80 Pipeline Steels in Simulated Soil Solution. *Int. J. Electrochem. Sci.* **2018**, *13*, 6436–6450. [CrossRef]
43. Zhang, R.; Vairavanathan, P.R.; Lalvani, S. Perturbation method analysis of AC-induced corrosion. *Corros. Sci.* **2008**, *50*, 1664–1671. [CrossRef]
44. Xiao, H.; Lalvani, S.B. A Linear Model of Alternating Voltage-Induced Corrosion. *J. Electrochem. Soc.* **2008**, *155*, C69. [CrossRef]
45. Lalvani, S.; Lin, X. A theoretical approach for predicting AC-induced corrosion. *Corros. Sci.* **1994**, *36*, 1039–1046. [CrossRef]
46. Lalvani, S.; Lin, X. A revised model for predicting corrosion of materials induced by alternating voltages. *Corros. Sci.* **1996**, *38*, 1709–1719. [CrossRef]
47. Bosch, R.; Bogaerts, W. A theoretical study of AC-induced corrosion considering diffusion phenomena. *Corros. Sci.* **1998**, *40*, 323–336. [CrossRef]
48. Xu, L.; Su, X.; Yin, Z.; Tang, Y.; Cheng, Y.F. Development of a real-time AC/DC data acquisition technique for studies of AC corrosion of pipelines. *Corros. Sci.* **2012**, *61*, 215–223. [CrossRef]
49. Wu, W.; Pan, Y.; Liu, Z.; Du, C.; Li, X. Electrochemical and Stress Corrosion Mechanism of Submarine Pipeline in Simulated Seawater in Presence of Different Alternating Current Densities. *Materials* **2018**, *11*, 1074. [CrossRef]
50. Ormellese, M.; Lazzari, L.; Goidanich, S.; Sesia, V. CP criteria assessment in the presence of AC interference. In Proceedings of the Corrosion 2008 Conference & Expo, New Orleans, LA, USA, 16–20 March 2008; p. 10.
51. Xu, L.; Su, X.; Cheng, Y.F. Effect of alternating current on cathodic protection on pipelines. *Corros. Sci.* **2013**, *66*, 263–268. [CrossRef]
52. Xu, L.; Cheng, Y.F. A Real-Time AC/DC Measurement Technique for Investigation of AC Corrosion of Pipelines and Its Effect on the Cathodic Protection Effectiveness. In Proceedings of the Corrosion 2013 Conference & Expo, Orlando, FL, USA, 17–21 March 2013; p. 10.
53. Kuang, D.; Cheng, Y.F. Effect of alternating current interference on coating disbondment and cathodic protection shielding on pipelines. *Corros. Eng. Sci. Technol.* **2015**, *50*, 211–217. [CrossRef]
54. Kuang, D.; Cheng, Y.F. Effects of alternating current interference on cathodic protection potential and its effectiveness for corrosion protection of pipelines. *Corros. Eng. Sci. Technol.* **2016**, *52*, 1–7. [CrossRef]
55. Wang, L.; Cheng, L.; Li, J.; Zhu, Z.; Bai, S.; Cui, Z. Combined Effect of Alternating Current Interference and Cathodic Protection on Pitting Corrosion and Stress Corrosion Cracking Behavior of X70 Pipeline Steel in Near-Neutral pH Environment. *Materials* **2018**, *11*, 465. [CrossRef] [PubMed]
56. Hosokawa, Y.; Kajiyama, F.; Nakamura, Y. New CP criteria for elimination of the risks of ac corrosion and overprotection on cathodically protected pipelines. In Proceedings of the Corrosion/2002, Denver, CO, USA, 7–11 April 2002; p. 12.
57. Hosokawa, Y.; Kajiyama, F. New CP maintenance concept for buried steel pipelines—Current density-based CP criteria, and on-line surveillance system for CP rectifiers. In Proceedings of the Corrosion/2004, New Orleans, LA, USA, 22 March–1 April 2004; p. 15.
58. Hosokawa, Y.; Kajiyama, F. Case studies on the assessment of AC and DC interference using steel coupons with respect to current density CP criteria. In Proceedings of the Corrosion 2006 Conference & Expo, Orlando, FL, USA, 10–14 September 2006; p. 15.
59. Kajiyama, F.; Nakamura, Y. Development of an advanced instrumentation for assessing the AC corrosion risk of buried pipelines. In Proceedings of the Corrosion 2010 Conference & Expo, San Antonio, TX, USA, 14–18 March 2010; p. 13.
60. Fu, A.Q.; Cheng, Y.F. Effect of alternating current on corrosion and effectiveness of cathodic protection of pipelines. *Can. Met. Q.* **2012**, *51*, 81–90. [CrossRef]
61. Büchler, M. Alternating current corrosion of cathodically protected pipelines: Discussion of the involved processes and their consequences on the critical interference values. *Mater. Corros.* **2012**, *63*, 1181–1187. [CrossRef]

62. Junker, A.; Nielsen, L.V.; Heinrich, C.; Møller, P. Laboratory and field investigation of the effect of the chemical environment on AC corrosion. In Proceedings of the Corrosion 2018 Conference & Expo, Phoenix, AZ, USA, 15–19 April 2018; p. 15.
63. Nielsen, L.V. AC Corrosion. In *Oil and Gas Pipelines*; John Wiley & Sons: Hoboken, NJ, USA, 2015; pp. 363–386.
64. Nielsen, L.V.; Nielsen, K.V.; Baumgarten, B.; Breuning-Madsen, H.; Cohn, P.; Rosenberg, H. AC-induced corrosion in pipelines: Detection, characterization, and mitigation. In Proceedings of the Corrosion/2004, New Orleans, LA, USA, 22 March–1 April 2004; p. 16.
65. Nielsen, L.V. Role of alkalization in AC induced corrosion of pipelines and consequences hereof in relation to CP requirements. In Proceedings of the Corrosion/2005, Houston, TX, USA, 3–7 April 2005; p. 11.
66. Nielsen, L.V.; Baumgarten, B.; Cohn, P. On-site measurements of AC induced corrosion: Effect of AC and DC parameters—A report from the Danish activities. In Proceedings of the CeoCor International Congress and Technical Exhibition, Dresden, Germany, 15–16 June 2004.
67. Junker, A.; Møller, P.; Nielsen, L.V. Effect of chemical environment and pH on AC corrosion of cathodically protected structures. In Proceedings of the Corrosion 2017 Conference & Expo, New Orleans, LA, USA, 26–30 March 2017; p. 14.
68. Junker, A.; Belmonte, L.J.; Kioupis, N.; Nielsen, L.V.; Møller, P. Investigation of stone-hard-soil formation from AC corrosion of cathodically protected pipeline. *Mater. Corros.* **2018**, *69*, 1170–1179. [CrossRef]
69. Santana-Diaz, E.; Nielsen, L.V.; Junker-Holst, A. A critical review of parameters for meaningful AC corrosion modelling. In Proceedings of the Corrosion 2018 Conference & Expo, Phoenix, AZ, USA, 15–19 April 2018; p. 15.
70. Nielsen, L.V.; Cohn, P. AC corrosion and electrical equivalent diagrams. In Proceedings of the CeoCor International Congress and Technical Exhibition, Bruxelles, Belgium, 9–12 May 2000.
71. Wang, X.; Wang, Z.; Chen, Y.; Song, X.; Xu, C. Research on the Corrosion Behavior of X70 Pipeline Steel Under Coupling Effect of AC + DC and Stress. *J. Mater. Eng. Perform.* **2019**, *28*, 1958–1968. [CrossRef]
72. Cui, Y.; Shen, T.; Ding, Q. Study on the Influence of AC Stray Current on X80 Steel under Stripped Coating by Electrochemical Method. *Int. J. Corros.* **2019**, *2019*, 1–8. [CrossRef]
73. Ibrahim, I.; Tribollet, B.; Takenouti, H.; Meyer, M. AC-Induced Corrosion of Underground Steel Pipelines. Faradaic Rectification under Cathodic Protection: I. Theoretical Approach with Negligible Electrolyte Resistance. *J. Braz. Chem. Soc.* **2014**, *26*, 196–208. [CrossRef]
74. Ibrahim, I.; Meyer, M.; Takenouti, H.; Tribollet, B. AC Induced Corrosion of Underground Steel Pipelines. Faradaic Rectification under Cathodic Protection: II. Theoretical Approach with Electrolyte Resistance and Double Layer Capacitance for Bi-Tafelian Corrosion Mechanism. *J. Braz. Chem. Soc.* **2015**, *27*, 605–615. [CrossRef]
75. Ibrahim, I.; Meyer, M.; Takenouti, H.; Tribollet, B. AC Induced Corrosion of Underground Steel Pipelines under Cathodic Protection: III. Theoretical Approach with Electrolyte Resistance and Double Layer Capacitance for Mixed Corrosion Kinetics. *J. Braz. Chem. Soc.* **2017**, *28*, 1483–1493. [CrossRef]
76. Brenna, A.; Diamanti, M.V.; Lazzari, L.; Ormellese, M. A proposal of AC corrosion mechanism in cathodic protection. In Proceedings of the 2011 NSTI Nanotechnology Conference and Expo, Boston, MA, USA, 13–16 June 2011; pp. 553–556.
77. Brenna, A.; Ormellese, M.; Lazzari, L. Electro-mechanical breakdown mechanism of passive film in AC-related corrosion of carbon steel under cathodic protection condition. *Corrosion* **2016**, *72*, 1055–1063. [CrossRef]
78. Sato, N. A theory for breakdown of anodic oxide films on metals. *Electrochimica Acta* **1971**, *16*, 1683–1692. [CrossRef]
79. Bertolini, L.; Carsana, M.; Pedeferri, P. Corrosion behaviour of steel in concrete in the presence of stray current. *Corros. Sci.* **2007**, *49*, 1056–1068. [CrossRef]
80. Ormellese, M.; Brenna, A.; Lazzari, L. Effects of AC-interference on passive metals corrosion. In Proceedings of the Corrosion 2011 Conference & Expo, Houston, TX, USA, 13–17 March 2011; p. 12.
81. Zhu, M.; Du, C.-W.; Li, X.; Liu, Z.-Y.; Wu, X.-G. Effect of AC on corrosion behavior of X80 pipeline steel in high pH solution. *Mater. Corros.* **2014**, *66*, 486–493. [CrossRef]
82. Brenna, A.; Beretta, S.; Bolzoni, F.M.; Pedeferri, M.; Ormellese, M. Effects of AC-interference on chloride-induced corrosion of reinforced concrete. *Constr. Build. Mater.* **2017**, *137*, 76–84. [CrossRef]
83. Zhu, M. Corrosion Behavior of X65 and X80 Pipeline Steels under AC Interference Condition in High pH Solution. *Int. J. Electrochem. Sci.* **2018**, *13*, 10669–10678. [CrossRef]

84. Zhu, M. Synergistic Effect of AC and Cl- on Stress Corrosion Cracking Behavior of X80 Pipeline Steel in Alkaline Environment. *Int. J. Electrochem. Sci.* **2018**, *13*, 10527–10538. [CrossRef]
85. Wang, H.; Du, C.; Liu, Z.; Wang, L.; Ding, D. Effect of Alternating Current on the Cathodic Protection and Interface Structure of X80 Steel. *Materials* **2017**, *10*, 851. [CrossRef]
86. Drazic, D.; Popic, J. Anomalous dissolution of metals and chemical corrosion. *J. Serbian Chem. Soc.* **2005**, *70*, 489–511. [CrossRef]
87. Nielsen, L.V.; Nielsen, K.V. Differential ER-technology for measuring degree of accumulated corrosion as well as instant corrosion rate. In Proceedings of the Corrosion/2003, San Diego, CA, USA, 16–20 April 2003; p. 13.
88. Nielsen, L.V.; Galsgaard, F. Sensor technology for on-line monitoring of ac-induced corrosion along pipelines. In Proceedings of the Corrosion/2005, Houston, TX, USA, 3–7 April 2005; p. 12.

© 2020 by the authors. Licensee MDPI, Basel, Switzerland. This article is an open access article distributed under the terms and conditions of the Creative Commons Attribution (CC BY) license (http://creativecommons.org/licenses/by/4.0/).

Article

Crystallographic Evaluation of Susceptibility to Intergranular Corrosion in Austenitic Stainless Steel with Various Degrees of Sensitization

Tomoyuki Fujii *, Takaya Furumoto, Keiichiro Tohgo and Yoshinobu Shimamura

Department of Mechanical Engineering, Shizuoka University, Hamamatsu 4328561, Japan; furumoto.takaya.15@shizuoka.ac.jp (T.F.); tohgo.keiichiro@shizuoka.ac.jp (K.T.); shimamura.yoshinobu@shizuoka.ac.jp (Y.S.)

* Correspondence: fujii.tomoyuki@shizuoka.ac.jp; Tel.: +81-53-478-1029

Received: 17 December 2019; Accepted: 26 January 2020; Published: 30 January 2020

Abstract: This study investigated the susceptibility to intergranular corrosion (IGC) in austenitic stainless steel with various degrees of sensitization (DOSs) from a microstructural viewpoint based on the coincidence site lattice (CSL) model. IGC testing was conducted using oxalic acid and type 304 stainless steel specimens with electrochemical potentiokinetic reactivation (EPR) ratios that varied from 3 to 30%. As a measure of IGC susceptibility, the width of the corroded groove was used. The relationship between IGC susceptibility, grain boundaries (GB) structure, and EPR ratio of the specimens was evaluated. As a result, the IGC susceptibility cannot be characterized using the Σ value, irrespective of the DOS of the specimen. The IGC susceptibility increases with increasing unit cell area of CSL boundaries, which is a measure of the stability of the CSL boundaries, and then levels off. The relationship between the IGC susceptibility and unit cell area is sigmoidal, irrespective of the DOS of the specimen. The sigmoid curve shifts rightward and the upper bound of IGC susceptibility decreases with decreasing DOS of the specimen.

Keywords: intergranular corrosion; sensitization; austenitic stainless steel; grain boundary; grain boundary structure

---

1. Introduction

Austenitic stainless steel exhibits superior mechanical properties and corrosion resistance to aqueous, gaseous, and high-temperature environments, and hence is used for pipes in chemical plants and coolant pipes in light water reactors. Machines and structures made of steel are usually assembled by welding, and it is well-known that sensitization of stainless steel occurs during welding. At high temperature, precipitation of chromium carbide at grain boundaries (GBs) occurs in stainless steel, which forms chromium depletion zones in the vicinity of the GBs, resulting in a loss of resistance to intergranular corrosion (IGC) and intergranular stress corrosion cracking [1].

In an effort to improve the resistance of stainless steel to IGC, the IGC susceptibility has been investigated from electrochemical and microstructural viewpoints. Xin et al. [2] investigated the IGC susceptibility in TIG-welded 316LN stainless steel by the double-loop electrochemical potentiokinetic reactivation (DL-EPR) technique, and found that a welded area exhibits a slight IGC susceptibility, although the base metal exhibits superior IGC resistance. These results indicate that chromium depletion zones were formed near GBs during welding and the IGC resistance decreased even in high-corrosion resistant 316LN stainless steel. Aquino et al. [3] demonstrated the same results for the effect of welding on IGC susceptibility in supermartensitic stainless steel. Iacoviello et al. [4] and Morshed-Behbahani et al. [5] investigated the influence of the heat treatment period on IGC susceptibility in stainless steel using the DL-EPR technique and found that the formation zone of chromium carbide precipitation followed by chromium depletion depends on the heat treatment period, resulting in a change in corrosion behavior. Suresh et al. [6] discussed the IGC resistance of type 304L stainless steel in simulated groundwater based on the carbon content in the steel. They found that high-angle GBs in 304L stainless steel containing more than 0.02% carbon exhibited high IGC susceptibility.

From a microstructural/crystallographic viewpoint, atomic force microscopy (AFM), scanning transmission electron microscopy (STEM), and electron backscattered diffraction (EBSD) techniques have been used to examine the formation of chromium carbide precipitation followed by chromium depletion, and the susceptibility of GBs to IGC. Murr and coworkers [7–12] conducted comprehensive studies on the effects of GB misorientation, carbon content, deformation, etc., on the formation behavior of chromium carbide precipitation at GBs in austenitic stainless steel. Liu et al. [13] examined the geometry of corroded grooves via AFM and discussed the relationship between the degree of sensitization (DOS) and the width and depth of the grooves. Bruemmer et al. [14] investigated chromium depletion zones at GBs by energy dispersive X-ray spectroscopy in a TEM and discussed the relationship between the DOS and chromium-depleted zone size. Bi et al. [15] examined the relationship between chromium depletion formation sites and GB structure and found that chromium depletion did not tend to occur at low-energy GBs. Note that GBs have been classified based on their misorientation, namely the difference in crystallographic orientation between adjacent two grains, and the Σ value based on the coincidence site lattice (CSL) model [16]. Srinivasan et al. [17] and Qi et al. [18] discussed the relationship between IGC susceptibility and microstructural features such as grain size, GB structure, and microscopic plastic strain. An et al. [19] investigated the relationship between IGC behavior and GB structure measured by 3D-orientation microscopy consisting of EBSD and serial sectioning techniques, and classified the GB structure into GBs with high and low IGC resistance based on the GB structure. Recently, Haruna et al. [20] pointed out that it is insufficient to characterize the IGC susceptibility of Σ3 boundaries in sensitized austenitic stainless steel determined by EBSD analysis. Fujii et al. [21] also concluded the same result and proposed a parameter (the unit cell area of CSL boundaries) reasonably to characterize the susceptibility of CSL boundaries to IGC. Note that the unit cell area will be explained in Section 2.2 in detail. Figure 1a,b shows the relationships between IGC susceptibility (IGC width) and Σ value and between IGC susceptibility and the unit cell area, respectively. Although the relationship between IGC susceptibility and Σ value exhibited large scatter, the relationship between IGC susceptibility and unit cell area was characterized by a sigmoid curve. The unit cell area is determined based solely on the arrangement of atoms at CSL boundaries, and it is expected that the area is a simple measure that can characterize the IGC susceptibility of CSL boundaries. However, while they demonstrated the IGC susceptibility of type 304 stainless steel with one level of DOS, the applicability of the technique to other DOSs remained to be clarified.

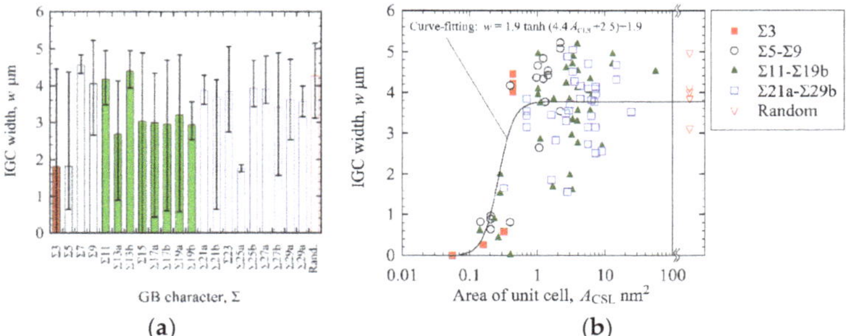

Figure 1. Intergranular corrosion (IGC) susceptibility of grain boundaries (GBs) in sensitized stainless steel (electrochemical potentiokinetic reactivation (EPR) ratio of 29.3%) [21]. (**a**) IGC width as a function of Σ value; (**b**) IGC width as a function of the area of unit cell of coincidence site lattice (CSL) boundaries.

This study aims to elucidate the susceptibility of GBs to IGC in type 304 austenitic stainless steel specimens with various DOSs. IGC testing was conducted under a constant current in oxalic acid, and then a corroded groove formed at the GBs was observed. The crystal orientations of the steel were analyzed via EBSD, and the relationships between the groove size and crystallographic parameters, such as the misorientation, Σ value, and unit cell area of CSL boundaries, were investigated. Especially, the applicability of the evaluation technique using the unit cell area was examined focusing on CSL boundaries.

2. Experimental

This section outlines the procedures of specimen preparation and IGC testing used in this study. These procedures are similar to those used in our previous study [21]. See the previous study for details on the testing procedures.

2.1. Material

Type 304 austenitic stainless steel was used. Table 1 shows the chemical composition of the steel used. The steel was heated at 1100 °C for 1 h for solutionizing. The average grain diameter of the steel was 119 μm, measured based on Japanese Industry Standard (JIS) G 0551.

Table 1. Chemical composition of type 304 austenitic stainless steel (mass %).

C	Si	Mn	P	S	Ni	Cr	Fe
0.06	0.47	0.87	0.03	0.003	8.05	18.16	Bal.

Specimens of 14 × 14 × 5 mm were machined from heat-treated bulk material, as shown in Figure 2. To change the DOS of the specimens, additional heat treatments were applied to the solutionized steel. Figure 3 shows the heating schedule (700 °C for 2 h and 500 °C for 24 h, furnace cooling) used in the previous studies [21–26]. The electrochemical potentiokinetic reactivation (EPR) ratio of the steel after the full heating schedule as a measure of DOS was 29.3%, measured by the EPR technique (JIS G 0580). In this study, the heating schedule was terminated after various heating periods to decrease the EPR ratio of the specimen. After terminating the heating schedule, each specimen was furnace-cooled. The EPR ratios of the heat-treated specimens are shown in Figure 3, demonstrating that the EPR ratio was lower for a shorter heating time.

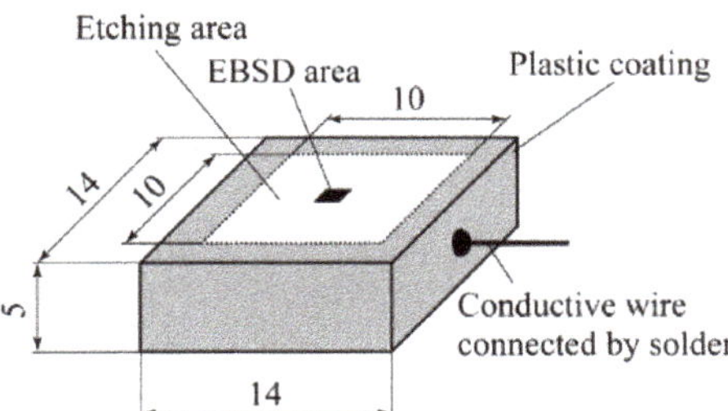

Figure 2. Dimensions of specimens (in units of mm) and preparation for corrosion testing.

After the heat treatment, the surface of each specimen was ground with emery paper up to grade #2000 and polished with diamond paste of 3 m grain diameter and colloidal silica (OP-U, Struers, Nagoya, Japan). Then, the surface was ion-milled by argon gas (EM RES101, Leica Microsystems, Wetzlar, Germany) to obtain a smooth and clean surface for EBSD analysis.

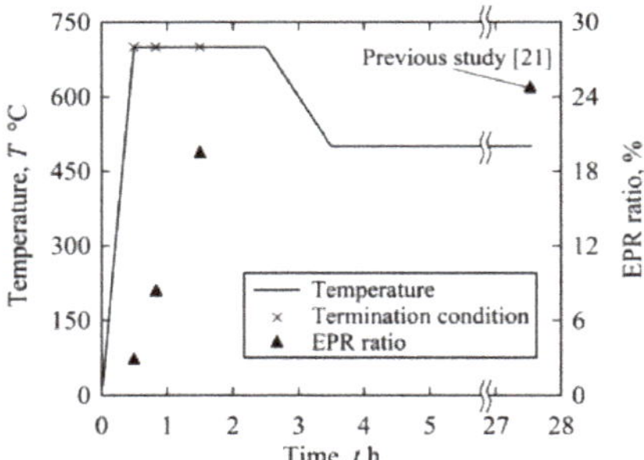

Figure 3. Heating schedule for sensitization and change in EPR ratio of a specimen.

2.2. Crystallography

Prior to IGC testing, the crystal orientations of a 1 × 1 mm central area in the etching area (black area in Figure 2) were measured with an EBSD system consisting of an orientation imaging microscopy (TSL Solutions, Sagamihara, Kanagawa, Japan). The measurement data were acquired and analyzed by OIM data correction and OIM analysis (TSL Solutions, Sagamihara, Kanagawa, Japan), respectively. Table 2 lists the EBSD measurement conditions. GBs were classified as low-angle (misorientation angle ranging from 5 to 15°) or high-angle (misorientation angle more than 15°), and the high-angle GBs were subclassified as CSL if they exhibited a Σ3–Σ29 structure and random boundaries otherwise. The tolerance angle was used based on the Brandon criterion [27] to determine the CSL boundaries.

Table 2. Conditions of electron backscattered diffraction (EBSD) measurement and analysis using OIM.

Grid Shape	Hexagonal
Measurement interval (m)	3.5
Grain tolerance angle (°)	5
Minimum grain size (pixel)	2
Clean up	Grain dilation, single iteration

In common EBSD analysis, the structure of a GB at a specimen surface is identified only by the relative orientation relationship between the two adjacent grains. However, Haruna et al. [20] and Fujii et al. [21] pointed out that the analysis yielded an ambiguous result for the GB structure. Each CSL boundary actually exhibits several GB planes (rotation angles and axes). For example, a Σ3 boundary can be formed by the planes {1 1 0}, {1 1 1}, {2 1 0}, {2 1 1}, or {3 1 1}, as shown in Figure 4. We proposed a technique to identify the actual CSL boundary using not only the relative orientation between two adjacent grains, but also the GB trace angle, and applied this technique in this study. To characterize the IGC susceptibility of the CSL boundaries, we used the area of the CSL unit cell, A_{CSL}. Figure 5 shows the arrangement of atoms on a Σ3 boundary and the technique for determining the value of A_{CSL}. The solid circles in Figure 5 denote sharing atoms on the boundary, and the value of A_{CSL} is defined by the area of the smallest repeating square. The value of A_{CSL} is considered to be a measure of the stability of the CSL boundary because the larger the area, the smaller the defects on the CSL boundary. The lattice constant a of the steel used was set to 0.359 nm in this calculation [28].

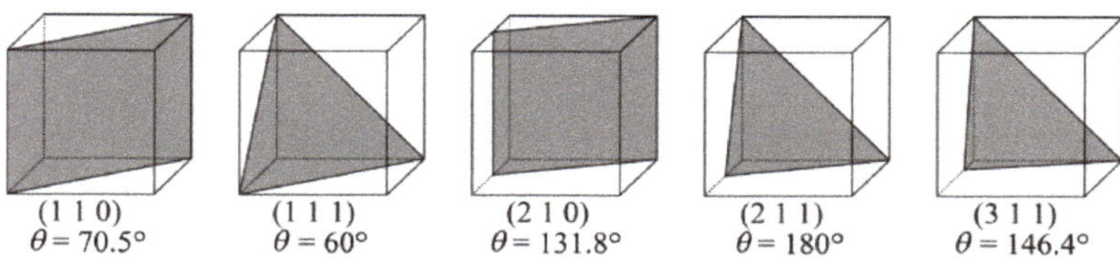

Figure 4. Schematic of GB planes and rotation angles for a Σ3 boundary. θ is a misorientation between adjacent grains.

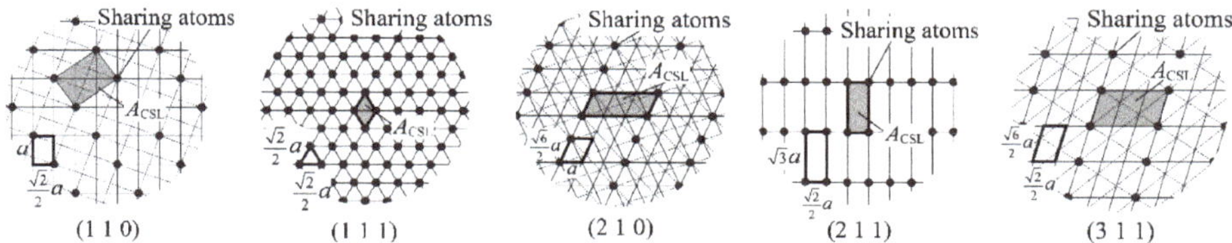

Figure 5. Five kinds of GB structure with atomic configurations for a Σ3 boundary and definition of the area of the CSL unit cell, A_{CSL} [21]. a is the lattice constant of the steel.

2.3. Electrochemical Method for Intergranular IGC Testing

After EBSD analysis, a conductive wire was connected to the specimen using solder (63/37 Sn-Pb, SE-0ST16, Taiyo Electric Ind. Co., Ltd, Fukuyama, Hiroshima, Japan) the melting temperature of 183–184 °C) and liquid flux (BS-45, Taiyo Electric Ind. Co., Ltd, Fukuyama, Hiroshima, Japan), and all the surfaces except for a 10 × 10 mm etching area were coated with plastic. IGC testing was then conducted based on the oxalic acid etch test for stainless steel (JIS G 0571). The surface was etched in 10% oxalic acid under 1 A/cm^2 for 90 s with a power source (POLIPOWER, Struers, Nagoya, Japan). After testing, the etched surface was observed under a high-resolution optical microscope (OM) (MS-Z420, Asahi Kogaku Manuf Co. Ltd., Tokyo, Japan). The width of the corroded groove was measured at the midpoint of each GB by the OM, which was regarded as a measure of the susceptibility of the GB to IGC for the sake of simplicity. The GBs for which the corroded groove width was measured were arbitrarily selected in the etched area for each GB class for each specimen.

3. Results and Discussion

3.1. IGC Behavior

Figures 6 and 7 show the crystal orientation maps and IGC profiles of specimens with high DOS (EPR ratio of 19.4%) and low DOS (EPR ratio of 2.8%), respectively. The black lines in Figures 6a and 7a denote GBs with misorientations of more than 5°. In the high DOS specimen, it was confirmed that IGC occurred at some GBs and not at others. In the low DOS specimen, the trend in IGC occurrence was almost the same. The width of IGC in the high DOS specimen seemed to be thicker than that in the low DOS specimen. The relationship between GB structure, EPR ratio, and susceptibility to IGC is discussed in the following sections.

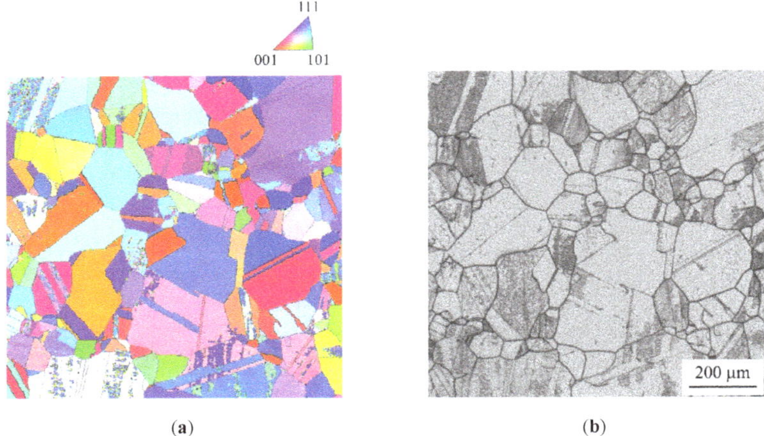

Figure 6. Microstructure of a specimen with high degrees of sensitization (DOS) (EPR of 19.4%). (**a**) Crystal orientation map of specimen surface before etching; (**b**) optical microscope (OM) image of specimen surface etched with oxalic acid corresponding to (a).

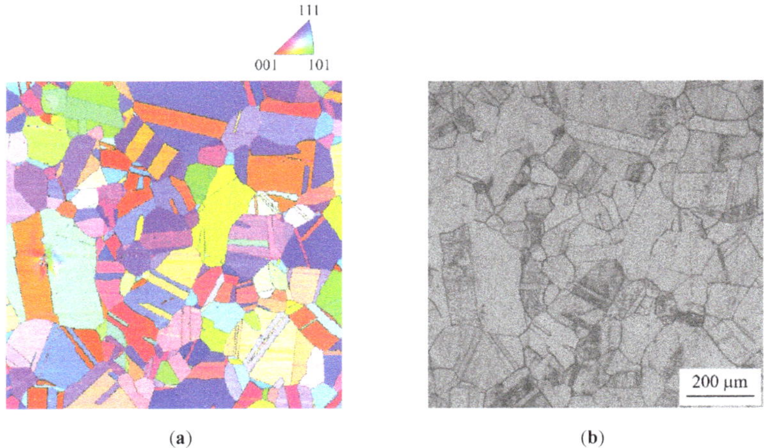

Figure 7. Microstructure of a specimen with low DOS (EPR of 2.8%). (**a**) Crystal orientation map of specimen surface before etching; (**b**) OM image of specimen surface etched with oxalic acid corresponding to (**a**).

3.2. IGC at Low-Angle GBs

Figure 8 shows the relationship between IGC width and misorientation in low-angle GBs. In high DOS specimens (EPR ratios of 29.3% and 19.4%), IGC did not occur at GBs with a misorientation of less than approximately 10°. For GBs with misorientations of 10–15°, IGC occurred at some of the GBs and not at others, and the IGC width was constant for the corroded GBs. Hence, for high DOS specimens, GBs with misorientations <10° do not exhibit IGC susceptibility and GBs with misorientations of 10–15° may or may not exhibit IGC susceptibility. The trend for the relationship between misorientation and IGC width is the same for low DOS specimens (EPR ratios of 8.3% and 2.8%). However, the critical misorientation shifts rightward and the widths of corroded GBs decrease with decreasing EPR ratio. Moreover, the number of GBs at which no IGC occurred seems to increase with decreasing EPR ratio.

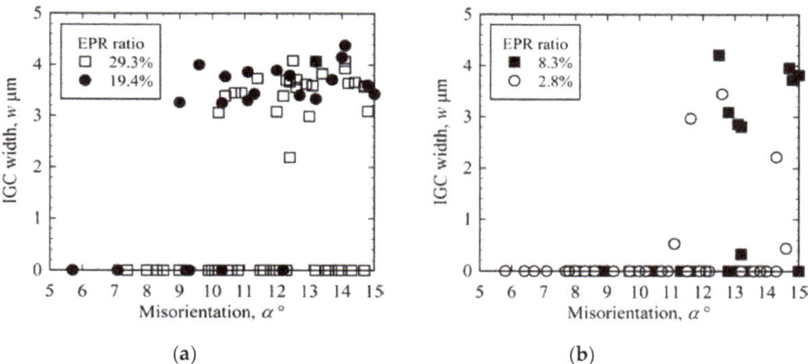

Figure 8. IGC width as a function of misorientation in low-angle GBs. (**a**) High DOS specimens; (**b**) low DOS specimens.

The GBs with misorientations of less than a critical value exhibit little IGC susceptibility and those with misorientations beyond a critical value may or may not exhibit IGC susceptibility, irrespective of the DOS. It is well-known that low-angle GBs have a periodic arrangement of dislocations and the structural stability of low-angle GBs decreases with increasing misorientation between adjacent grains [29]. A chromium depletion zone tends to be formed at GBs with lower structural stability, and the GBs with lower misorientation exhibit little IGC susceptibility. Some of the GBs beyond the critical value exhibit IGC susceptibility, while the others exhibit little IGC susceptibility. This could be explained by the following two reasons. First, the structural stability of GBs is affected not only by the misorientation, but also by the geometric relationship between the GB planes [30]. The formation of chromium carbide followed by a chromium-depleted zone at GBs would depend on their structural stability, resulting in the differences in IGC susceptibility between the GBs. Second, we measured the IGC width at the midpoint of each GB, though the width along the corroded GBs varied. Figure 9 shows an example of a corroded groove at a low-angle GB with a misorientation of 7.1°. Although the IGC width at the midpoint of the GB between Grains A and B in the figure is null, the IGC is several micrometers wide near both ends. It is known that chromium carbide is formed insularly along a GB [31] and a chromium depletion zone would grow from spots along a GB. Hence, as the IGC occurs at the segmentally formed chromium depletion zone along a GB, the IGC susceptibility will vary along a single GB. It would be difficult therefore to characterize in detail the IGC susceptibility distribution along such a GB using a single parameter. Further study related to a measure to evaluate the IGC susceptibility should be conducted.

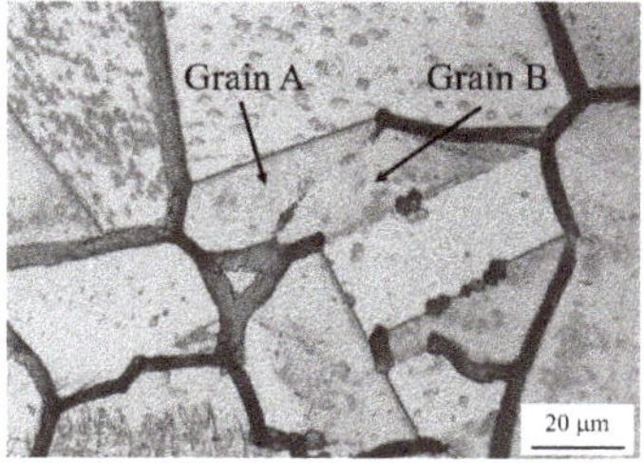

Figure 9. Example of a corroded groove segment at a low-angle GBs in a specimen with an EPR ratio of 19.4%. The misorientation of Grains A and B is 7.1°.

3.3. IGC at High-Angle GBs

Figure 10 shows the relationship between IGC width and GB character (Σ value) for high-angle GBs. The scatter bars denote the maximum and minimum widths measured. For the bars with no scatter bar, only one GB was detected in the EBSD-analyzed area. Note that some of the CSL boundaries were not detected in the EBSD-analyzed area, which are denoted as "no data" in this figure.

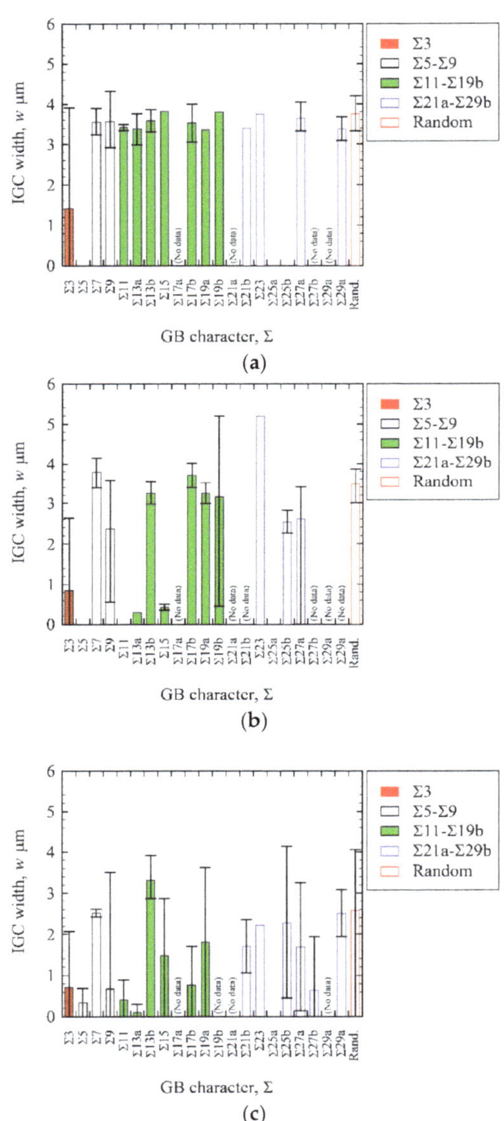

Figure 10. IGC width as a function of value in high-angle GBs. (**a**) EPR ratio of 19.4%; (**b**) EPR ratio of 8.3%; (**c**) EPR ratio of 2.8%.

For the high DOS specimens (EPR ratio: 19.4%), the IGC width drastically fluctuates with increasing Σ value. Additionally, the IGC width for each Σ value exhibits a large scatter. For the low DOS specimens (EPR ratios: 8.3% and 2.8%), the trends are almost the same: the IGC width is unrelated to the Σ value and the IGC width exhibits a large scatter.

Figure 11 shows the IGC width w as a function of A_{CSL}. The IGC widths of the random boundaries are plotted on the right-hand side in the figure because the random boundaries do not have a regular periodic structure and the unit cell cannot be defined. The relationship between IGC width and A_{CSL} takes on a sigmoid shape in every case irrespective of the EPR ratio. When A_{CSL} is small, no IGC is observed. The IGC width increases with increasing A_{CSL} and then levels off. To investigate the influence of the EPR ratio on IGC susceptibility, the trendlines of the relationship between IGC width and unit cell area for all specimens are drawn in Figure 12. Each trendline was obtained by curve-fitting using a hyperbolic tangent function:

$$w = a\tanh(bA_{CSL} + c) + d \tag{1}$$

The parameters of a, b, c, and, d are obtained using a generalized reduced gradient method to fit the experimental results of each condition. The trendline shifts rightward and the IGC width decreases with decreasing EPR ratio.

The stability is generally evaluated using the Σ value based on the CSL model. The Σ value is defined as the reciprocal of the density of coincidence sites, and it is expected that the stability of CSL boundaries decreases with increasing Σ value. Hence, CSL boundaries with low Σ value may be not sensitized, resulting in a low IGC susceptibility of the boundaries. From Figure 10, however, the IGC width seems not to be dependent on the Σ value, irrespective of the EPR ratio of the specimen From Figures 11 and 12, on the other hand, the IGC width increases with increasing A_{CSL}, irrespective of EPR ratio. Consequently, the IGC susceptibility of sensitized stainless steel cannot be characterized using the Σ value, but can be characterized using A_{CSL}. Figure 13 schematically illustrates the influence of the DOS on IGC susceptibility of high-angle GBs. The relationship between IGC susceptibility and A_{CSL} is sigmoid, irrespective of the DOS, and the relationship shifts rightward with decreasing DOS. Romero and Murr [7] found that chromium carbide was formed first at random boundaries, then at incoherent twin boundaries, and then at coherent twin boundaries in type 304 austenitic stainless steel. Suresh et al. [6] and Bi et al. [15] also found that chromium carbide tends to be distributed along high angle random boundaries. These results mean that chromium carbide (followed by a chromium depletion zone) is easily formed at unstable GBs, and the IGC susceptibility increases with decreasing GB structural stability. This study quantitatively expresses this trend of IGC susceptibility using A_{CSL} (GB structural stability). In addition, the IGC susceptibility of GBs with large A_{CSL} decreases with decreasing DOS. When a specimen is exposed at a high temperature for a shorter period, the chromium depletion zone formation changes and, as for stable GBs such as CSL boundaries with low A_{CSL}, a smaller chromium depletion zone is formed. For relatively unstable GBs such as random boundaries and CSL boundaries with high A_{CSL}, the chromium depletion zone is formed in a narrower region in the vicinity of the GBs. As a result, GBs with a larger A_{CSL} exhibit less IGC susceptibility, resulting in the low DOS of the specimen.

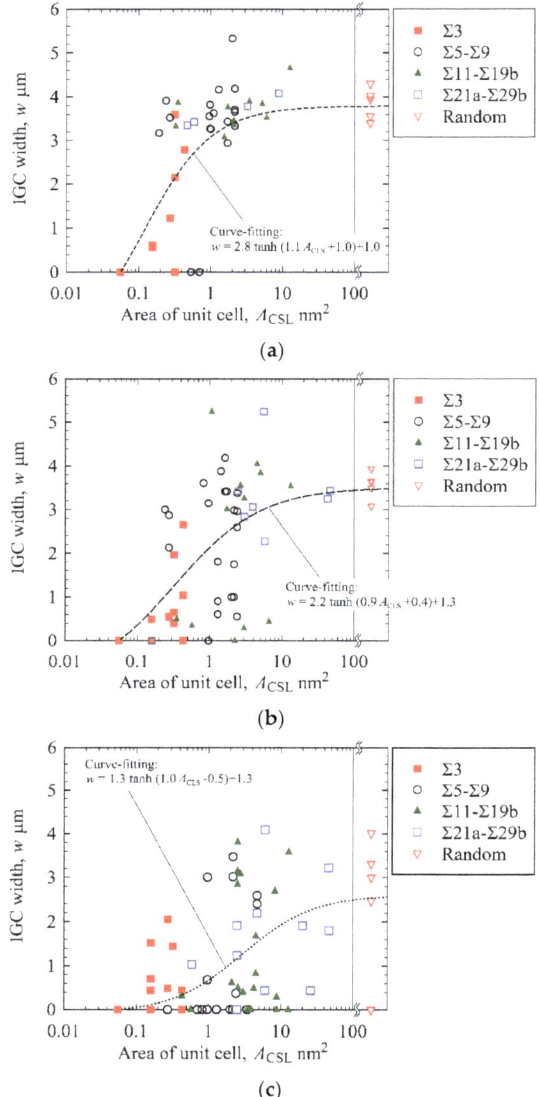

Figure 11. Relationship between IGC width and A_{CSL} at GBs. (**a**) EPR ratio of 19.4%; (**b**) EPR ratio of 8.3%; (**c**) EPR ratio of 2.8%.

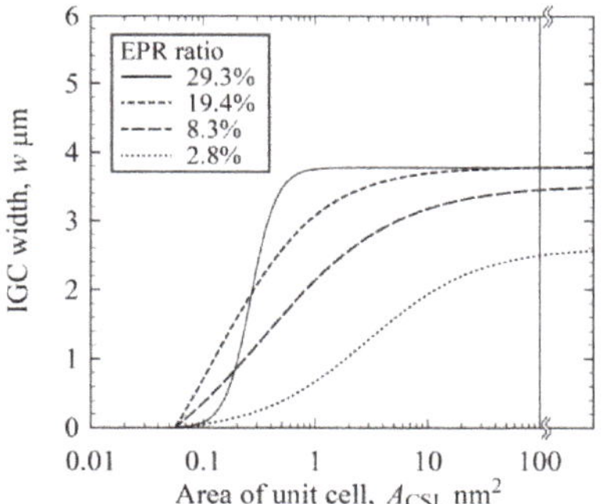

Figure 12. Trendlines of the relationship between IGC width and A_{CSL} for all specimens.

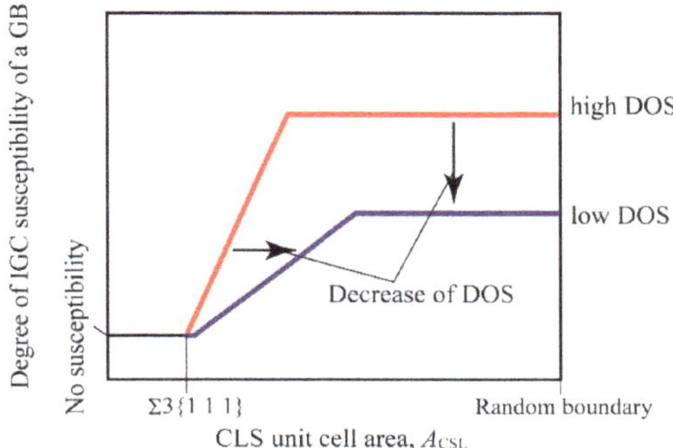

Figure 13. Schematic illustration of the influence of DOS on IGC susceptibility at high-angle GBs. The blue and red lines denote the IGC susceptibility of GBs in high and low DOS specimens, respectively.

The use of A_{CSL} is not valid for low-angle GBs and random boundaries because it cannot be defined. A candidate for evaluating the structural stability and IGC susceptibility, irrespective of GB structure, is the GB energy [32–34]. The energy can be determined based on the type and arrangement of atoms by numerical methods such as molecular dynamics. However, it is difficult to calculate the energy for each GB in actual conditions. Although some experimental techniques were proposed to measure the GB energy [35,36], the energy could not be determined based on the nondestructive operation: a specimen should be annealed at a high temperature, etc. Hence, it is very difficult to experimentally and/or analytically measure the GB energy of GBs of a material subjected to corrosion. As mentioned before, the unit cell area A_{CSL} is simply defined by atom arrangement on CSL boundaries and is easily determined by specimen surface observation via SEM and crystal orientation measurement via EBSD. Although limited to the evaluation of IGC susceptibility at CSL boundaries, this evaluation technique using A_{CSL} is a very effective approach. Further investigation of a unified parameter to characterize the susceptibility of low-angle and high-angle GBs is needed.

4. Conclusions

In this study, IGC testing was conducted using oxalic acid on thermally sensitized type 304 stainless steel specimens with varied EPR ratios. The susceptibility of GBs to IGC was investigated based on the GB structure, such as their misorientation and Σ value. For low-angle GBs (misorientations ranging from 5° to 15°), GBs with misorientations of less than 10° exhibit little IGC susceptibility, while some GBs with misorientations of 10–15° exhibit IGC susceptibility for high DOS specimens. In high-angle

GBs (misorientations >15°), the IGC susceptibility cannot be characterized by the Σ value, but can be characterized by the unit cell area of CSL boundaries, irrespective of the EPR ratio of the steel. The IGC susceptibility increases with increasing unit cell area, and then levels off. The IGC susceptibility of the GBs increases with increasing EPR ratio. To easily characterize the IGC susceptibility of CSL boundaries, the unit cell area of A_{CSL}, which is simply determined based on atom arrangement at CSL boundaries, is very useful.

Author Contributions: Conceptualization, T.F. (Tomoyuki Fujii) and K.T.; validation, T.F. (Tomoyuki Fujii), K.T. and Y.S.; formal analysis, T.F. (Tomoyuki Fujii), K.T. and Y.S.; investigation, T.F. (Tomoyuki Fujii) and T.F. (Takaya Furumoto); resources, T.F. (Tomoyuki Fujii); data curation, T.F. (Tomoyuki Fujii) and T.F. (Takaya Furumoto); writing—original draft preparation, T.F. (Tomoyuki Fujii); visualization, T.F. (Tomoyuki Fujii) and T.F. (Takaya Furumoto); supervision, T.F. (Tomoyuki Fujii); project administration, T.F. (Tomoyuki Fujii); funding acquisition, T.F. (Tomoyuki Fujii). All authors have read and agreed to the published version of the manuscript.

Funding: This work was supported by JSPS KAKENHI Grant Number 17H04899.

Conflicts of Interest: The authors declare no conflict of interest.

References

1. Gaudett, M.A.; Scully, J.R. Distributions of Cr Depletion Levels in Sensitized AISI 304 Stainless Steel and Its Implications Concerning Intergranular Corrosion Phenomena. *J. Electrochem. Soc.* **1993**, *140*, 3425. [CrossRef]
2. Xin, J.; Song, Y.; Fang, C.; Wei, J.; Huang, C.; Wang, S. Evaluation of inter-granular corrosion susceptibility in 316LN austenitic stainless steel weldments. *Fusion Eng. Des.* **2018**, *133*, 70–76. [CrossRef]
3. Aquino, J.M.; Della Rovere, C.; Kuri, S. Intergranular corrosion susceptibility in supermartensitic stainless steel weldments. *Corros. Sci.* **2009**, *51*, 2316–2323. [CrossRef]
4. Iacoviello, F.; Di Cocco, V.; D'Agostino, L. Analysis of the intergranular corrosion susceptibility in stainless steel by means of potentiostatic reactivation tests. *Procedia Struct. Integr.* **2017**, *3*, 269–275. [CrossRef]
5. Morshed-Behbahani, K.; Najafisayar, P.; Pakshir, M.; Shahsavari, M. An electrochemical study on the effect of stabilization and sensitization heat treatments on the intergranular corrosion behaviour of AISI 321H austenitic stainless steel. *Corros. Sci.* **2018**, *138*, 28–41. [CrossRef]
6. Suresh, G.; Parida, P.K.; Bandi, S.; Ningshen, S. Effect of carbon content on the low temperature sensitization of 304L SS and its corrosion resistance in simulated ground water. *Mater. Chem. Phys.* **2019**, *226*, 184–194. [CrossRef]
7. Romero, R.J.; Murr, L.E. Fundamental and comparative studies of M23C6 carbide growth associated with twin boundaries and grain boundaries in 304 stainless steel. *Microstruct. Sci.* **1995**, *22*, 367–380.
8. Trillo, E.A.; Murr, L.E. A TEM investigation of M23C6 carbide precipitation behaviour on varying grain boundary misorientations in 304 stainless steels. *J. Mater. Sci.* **1998**, *33*, 1263–1271. [CrossRef]
9. Trillo, E.; Murr, L. Effects of carbon content, deformation, and interfacial energetics on carbide precipitation and corrosion sensitization in 304 stainless steel. *Acta Mater.* **1998**, *47*, 235–245. [CrossRef]
10. Almanza, E.; Murr, L. A comparison of sensitization kinetics in 304 and 316 stainless steels. *J. Mater. Sci.* **2000**, *35*, 3181–3188. [CrossRef]
11. Ramírez, L.; Almanza, E.; Murr, L. Effect of uniaxial deformation to 50% on the sensitization process in 316 stainless steel. *Mater. Charact.* **2004**, *53*, 79–82. [CrossRef]
12. Alvarez, C.J.; Almanza, E.; Murr, L.E. Evaluation of the sensitization process in 304 stainless steel strained 50% by cold-rolling. *J. Mater. Sci.* **2005**, *40*, 2965–2969. [CrossRef]
13. Liu, L.; Mitamura, T.; Terasawa, M.; Takatani, Y.; Fukumoto, S.; Tsubakino, H. AFM Observations of Intergranular Corrosion of 304 Stainless Steels. *J. Soc. Mater. Sci. Jpn.* **2000**, *49*, 222–226. [CrossRef]
14. Bruemmer, S.M.; Chariot, L.A.; Arey, B.W. Sensitization Development in Austenitic Stainless Steel: Correlation between STEM-EDS and EPR Measurements. *Corrosion* **1988**, *44*, 328–333. [CrossRef]
15. Bi, H.Y.; Kokawa, H.; Wang, Z.J.; Shimada, M.; Sato, Y.S. Suppression of chromium depletion by grain boundary structural change during twin-induced grain boundary engineering of 304 stainless steel. *Scr. Mater.* **2003**, *49*, 219–223.
16. Shvindlerman, L.; Straumal, B. Regions of existence of special and non-special grain boundaries. *Acta Met.* **1985**, *33*, 1735–1749. [CrossRef]

17. Srinivasan, N.; Kain, V.; Birbilis, N.; Krishna, K.M.; Shekhawat, S.; Samajdar, I. Near boundary gradient zone and sensitization control in austenitic stainless steel. *Corros. Sci.* **2015**, *100*, 544–555. [CrossRef]
18. Qi, J.; Huang, B.; Wang, Z.; Ding, H.; Xi, J.; Fu, W. Dependence of corrosion resistance on grain boundary characteristics in a high nitrogen CrMn austenitic stainless steel. *J. Mater. Sci. Technol.* **2017**, *33*, 1621–1628. [CrossRef]
19. An, D.; Griffiths, T.; Konijnenberg, P.; Mandal, S.; Wang, Z.; Zaefferer, S. Correlating the five parameter grain boundary character distribution and the intergranular corrosion behaviour of a stainless steel using 3D orientation microscopy based on mechanical polishing serial sectioning. *Acta Mater.* **2018**, *156*, 297–309. [CrossRef]
20. Haruna, T.; Shimizu, T.; Kamaya, M. Correlation between Grain Boundary Character and Susceptibility of Sensitization for SUS304 Stainless Steel Foil with Course-Grained Structure. *J. Jpn. Inst. Met.* **2012**, *76*, 314–320. [CrossRef]
21. Fujii, T.; Tohgo, K.; Mori, Y.; Shimamura, Y. Crystallography of intergranular corrosion in sensitized austenitic stainless steel. *Mater. Charact.* **2018**, *144*, 219–226. [CrossRef]
22. Fujii, T.; Tohgo, K.; Ishizuka, N.; Shimamura, Y.; Takanashi, M.; Itabashi, Y.; Nakayama, G.; Sakakibara, Y.; Hirano, T. Fracture Mechanics Study on Stress Corrosion Cracking Behavior under Corrosive Environment. *J. Solid Mech. Mater. Eng.* **2013**, *7*, 341–356. [CrossRef]
23. Fujii, T.; Tohgo, K.; Kenmochi, A.; Shimamura, Y. Experimental and numerical investigation of stress corrosion cracking of sensitized type 304 stainless steel under high-temperature and high-purity water. *Corros. Sci.* **2015**, *97*, 139–149. [CrossRef]
24. Fujii, T.; Tohgo, K.; Kawamori, M.; Shimamura, Y. Characterization of stress corrosion crack growth in austenitic stainless steel under variable loading in low- and high-scale yielding conditions. *Eng. Fract. Mech.* **2019**, *205*, 94–107. [CrossRef]
25. Fujii, T.; Tohgo, K.; Mori, Y.; Miura, Y.; Shimamura, Y. Crystallographic and mechanical investigation of intergranular stress corrosion crack initiation in austenitic stainless steel. *Mater. Sci. Eng. A* **2019**, *751*, 160–170. [CrossRef]
26. Fujii, T.; Yamakawa, R.; Tohgo, K.; Shimamura, Y. Strain-based approach to investigate intergranular stress corrosion crack initiation on a smooth surface of austenitic stainless steel. *Mater. Sci. Eng. A* **2019**, *756*, 518–527. [CrossRef]
27. Brandon, D. The structure of high-angle grain boundaries. *Acta Met.* **1966**, *14*, 1479–1484. [CrossRef]
28. JCPDS International Centre for Diffraction Data, # 33-0397, Powder Diffraction File, Sets 33-34, Inorganic Volume. **1989**, *142*. Available online: http://www.icdd.com/pdj/ (accessed on 30 January 2020).
29. Read, W.T.; Shockley, W. Dislocation Models of Crystal Grain Boundaries. *Phys. Rev.* **1950**, *78*, 275–289 [CrossRef]
30. Tschopp, M.A.; Coleman, S.P.; McDowell, D.L. Symmetric and asymmetric tilt grain boundary structure and energy in Cu and Al (and transferability to other fcc metals). *Integrating Mater. Manuf. Innov.* **2015**, *4*, 176–189. [CrossRef]
31. Parvathavarthini, N.; Mulki, S.; Dayal, R.; Samajdar, I.; Mani, K.; Raj, B. Sensitization control in AISI 316L(N) austenitic stainless steel: Defining the role of the nature of grain boundary. *Corros. Sci.* **2009**, *51*, 2144–2150. [CrossRef]
32. Olmsted, D.L.; Foiles, S.M.; Holm, E.A. Survey of computed grain boundary properties in face-centered cubic metals: I. Grain boundary energy. *Acta Mater.* **2009**, *57*, 3694–3703. [CrossRef]
33. Rohrer, G.S. Grain boundary energy anisotropy: A review. *J. Mater. Sci.* **2011**, *46*, 5881–5895. [CrossRef]
34. Han, J.; Vitek, V.; Srolovitz, D.J. Grain-boundary metastability and its statistical properties. *Acta Mater.* **2016**, *104*, 259–273. [CrossRef]
35. Inman, M.C.; Tipler, H.R. Interfacial energy and composition in metals and alloys. *Met. Rev.* **1963**, *8*, 105–166. [CrossRef]
36. Gjostein, N.; Rhines, F. Absolute interfacial energies of [001] tilt and twist grain boundaries in copper. *Acta Met.* **1959**, *7*, 319–330. [CrossRef]

© 2020 by the authors. Licensee MDPI, Basel, Switzerland. This article is an open access article distributed under the terms and conditions of the Creative Commons Attribution (CC BY) license (http://creativecommons.org/licenses/by/4.0/).

Review

Electrochemical Polishing of Austenitic Stainless Steels

Edyta Łyczkowska-Widłak [1,*], Paweł Lochyński [1] and Ginter Nawrat [2]

[1] Institute of Environmental Engineering, Wrocław University of Environmental and Life Sciences, 50-363 Wrocław, Poland; pawel.lochynski@upwr.edu.pl
[2] Department of Inorganic Chemistry, Silesian University of Technology, Analytical Chemistry and Electrochemistry, 44-100 Gliwice, Poland; Ginter.Nawrat@polsl.pl
* Correspondence: edyta.lyczkowska@gmail.com

Received: 16 April 2020; Accepted: 28 May 2020; Published: 4 June 2020

Abstract: Improvement of the corrosion resistance capability, surface roughness, shining of stainless-steel surface elements after electrochemical polishing (EP) is one of the most important process characteristics. In this paper, the mechanism, obtained parameters, and results were studied on electropolishing of stainless-steel samples based on the review of the literature. The effects of the EP process parameters, especially current density, temperature, time, and the baths used were presented and compared among different studies. The samples made of stainless steel presented in the articles were analysed in terms of, among other things, surface roughness, resistance to corrosion, microhardness, and chemical composition. All results showed that the EP process greatly improved the analysed properties of the stainless-steel surface elements.

Keywords: electropolishing; mechanism; stainless steel; corrosion; surface roughness; microhardness

---

1. Introduction

Metals and their alloys, when immersed in a suitable bath and in adequate electrolysis conditions, may undergo anodic dissolution so that their surface becomes glossy and smooth [1]. The typical reactions, anodic (1) and cathodic (2), in the electrochemical polishing process are [2]:

$$M^0 - ze^- \rightarrow M^{z+} \quad (1)$$

$$2H^+ + 2e^- \rightarrow H_2 \quad (2)$$

Anodic dissolution of metals and their transmission to the solution in the form of simple or complex hydrated ions may be described with the following equations:

$$Me + xH_2O - ze^- \rightarrow Me^{z+} \cdot xH_2O \quad (3)$$

$$Me + xH_2O + yA^- - ze^- \rightarrow [MeA_y]^{z-y} \cdot xH_2O \quad (4)$$

The effects of electropolishing the surface of metal elements are:

- Macropolishing, i.e., removing the peaks of a height of approximately 100 μm which smoothens the surface;
- Micropolishing, i.e., removing the peaks of a height of approximately 10 μm which makes the surface glossy;
- Passivation which means the formation of a passive, oxide layer on the surface of metal [3–5].

Thus, during the electrochemical polishing process, the geometry of the surface is subject to changes in the micrometric range, which results in smoothing the surface. In the range below 10 μm, this results in a highly ordered, single-direction reflection of the rays of light that gives the surface mirror-like properties.

In the electropolishing process, the surface is smoothened and lustred without using any mechanical tools, so that the superficial layer is protected against structural changes, while the crystalline structure of the core of the processed element is revealed. Moreover, the process results in uniform passivation of the whole surface of the polished element that protects it against corrosion, together with spots that are hard to access and gives the element a decorative appearance.

Electrochemical polishing was probably first applied in industry after World War II to process various types of carbon and alloy steels [6,7]. The best technical and economic results were obtained for stainless steels, for which mechanical polishing did not lead to the same results as electropolishing. Austenitic chromium and nickel steels (e.g., AISI 304) and chromium–nickel–molybdenum (e.g., AISI 316) are widely used in heavy industry, food processing, the aviation, the automotive industry, the space industry, jewellery making as well as in sport and biomedical engineering [8–10]. However, not all types of metallic materials may be polished with comparable results. Good electrochemical polishing results are obtained for metals or alloys of a homogeneous and fine-grained structure which are also free from non-metallic intrusions.

The main advantages of electropolishing include:

- Giving an aesthetic appearance;
- Obtaining good anti-corrosion properties due to the creation of the passive oxide layer containing chromium oxide Cr_2O_3, nickel oxide NiO, molybdenum oxide Mo_2O_3, and iron oxide Fe_2O_3 [11–13];
- Facilitating washing and cleaning of elements subjected to electrochemical polishing (the removal of dirt and bacteria is easier) [14];
- Removing microstresses in the superficial layer caused by processing and restoring uniform micro-hardness of the native material [15,16];
- The possibility to polish surfaces and spots that are inaccessible to mechanical polishing [17,18].

The possibility to polish soft metals and alloys and elements of a delicate, openwork structure is no less important.

Electropolishing produces a surface free from deformations without interfering with the crystalline network structure and superficial stresses. The degree of smoothening is higher the more homogeneous the alloy structure. The literature on the subject includes studies that present the influence of the polishing method on fatigue. The results of these studies demonstrated that low cycle fatigue (LCF) of electrochemical polished samples was 87% higher than that of samples polished mechanically using abrasive materials [19].

The fact that surfaces subjected to electrochemical polishing are easy to clean makes it possible to achieve a high quality of surface cleanliness and sterility. This is an important requirement in the pharmaceutical and food processing industries, because metal elements that come into direct contact with food must not cause contamination or any changes in the taste or colour of the food product.

These advantages of electropolishing support the use of this method in finishing surgical instruments and implants as well. Studies on the electrochemical polishing process conducted by Nawrat [20] led to implementing the electropolishing and chemical passivation of implants made from AISI 316L steel (e.g., screws, plates, and compressive-distractive apparati) at the "Mikromed" facility in Dąbrowa Górnicza. Pursuant to the PN-EN ISO 14630 standard, implants made from AISI 316L steel after the electrochemical polishing process were characterised by an average roughness coefficient $Ra < 0.16$ μm and very good corrosion resistance [20–22].

2. Mechanism of the Electrochemical Polishing Process

The first patent on electrolytic polishing were published by Spitalsky in 1910 [23], who worked on polishing silver in cyanide solutions.

The mechanism of the electrochemical polishing process was then analysed by numerous scientists. The first description of this process was presented by Jacguet [24,25] in the 1930s. According to Jacguet articles [24,25], the factor that has the greatest influence on surface smoothening during electropolishing is the emergence of a highly viscous layer on the anode [26]. The layer is formed on the surface of the anode as a result of the polarisation of the processed material. The anodic diffusion layer is flat on the side facing the solution and the cathode, while on the side adjacent to the anode it takes the form of the anodic surface. The layer is characterised by high electric resistance. However, this layer is thinner on micro-peaks than on micro-valleys. As a result, the peaks of roughness are dissolved first, as higher density current passes through them [26].

The process of anodic dissolution takes place in the conditions of threshold current. Due to the low electric conductivity and high viscosity of the solution forming the anodic layer that fosters the sustainable conditions of the process, the peaks of surface roughness are dissolved as a result of the differences in current density between peaks and indentations.

According to another hypothesis, also based on the presence of a highly viscous layer, the acceptors of the metal dissolution process plays an important role. The acceptors of the anodic process may be, for example, particles of water and anions that are present in the bath. Water causes hydration of metal ions, enabling them to detach from the surface, so the diffusion of acceptors towards the surface of electrochemical polished elements is decisive for the anodic dissolution process. During electropolishing, the concentration of acceptors in the vicinity of the anode is low, and the gradient of their concentration increases at micro-peaks, so that they are the first to be subjected to anodic dissolution. On the other hand, in micro-indentations the dissolution process is slightly delayed until the metal ions close to micro-peaks are released from the metal surface [27].

The manner of transporting metal ions by the said viscous layer was also studied by Datt and Landolt [28–30]. These authors explained the mechanism of the electrochemical polishing process by the emergence of a layer of high density, low water content, and high concentration of anodic metal ions (i.e., the aforementioned viscous layer) on the surface of the anode. According to them [28], a film emerges on the surface of the anode, consisting of dissolution products and the bath of increased density and high viscosity. It is subject to continuous transformations, as it decomposes and rebuilds itself. Considering the decomposition of this layer and the following reconstruction, one may claim that it is a dynamic system. According to the authors, the film in question is a layer saturated with salts, through which diffusion and migration transport of ions occurs in the absence of convection. Hypothetically, one may assume that in the event of the absence of concentration and temperature gradients and in a perfectly mixed bath, only migration transport of metal ions would occur.

Electrochemical polishing (EP) is also defined as a diffusion–adsorption process that also assumes the existence of a thin layer on the metal–electrolyte border that it created from the products of anodic dissolution in the suitable bath characterised by high density and viscosity. The mechanism of electropolishing based on the presence of a viscous layer in which the migration and diffusion of anodic dissolution processes occur in the direction from the anode and process acceptor (i.e., water particles and anions that are present in the anodic layer) towards the anode was also confirmed by Hryniewicz [31]. During electropolishing of stainless steel, in the electrolyte containing phosphoric acid (V) and sulphuric acid (VI), water is bound in the hydration zone of ions. Metal ions are transported into the electrolyte in the form of hydrated ions and complexes. Water deficit, which is further reinforced by electrolytic decomposition of water and the emission of oxygen, results in the conditions that foster the emergence of insoluble chemical bonds that form a thin film in the form of the so-called viscous layer. In the opinion of the author, the formation of passive layers may further hinder the transport of ions. High resistance of the electrolyte solution results in high heat emission, which fosters convection. It is also enhanced by changes in the solution density. Bagdach and Ciszewski [3,32] also describe a mechanism where a viscous layer, which is thin yet thicker than the height of micro-roughnesses, emerges on the anodic surface. This layer is characterised by higher electric resistance and increased concentration of dissolution products compared to the rest of the electrolytic solution. As a result, the current density on peaks increases, so that the dissolution rate is higher on peaks and lower in indentations. Strong adhesion of the viscous layer to the surface contributes to the stability of the described conditions near the surface of the anode.

A summary of the models of electrochemical polishing based on the existence of the viscous layer was presented by Lin and Hu [33,34]. The authors logically combined the hypotheses discussed above, taking into account the molecular interactions between water particles that are assigned metal ions and the presence of acids and additives that increase the viscosity of the solution. The authors divided the layer that was formed between the anodic surface (substrate E) and the depth of the solution (layer A) into three zones (B, C, and D, Figure 1).

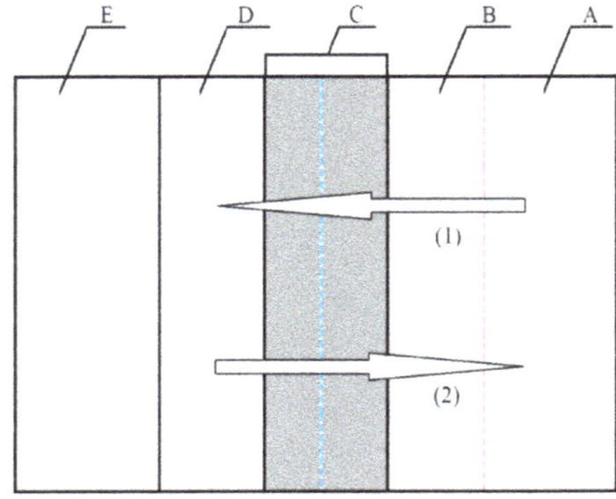

Figure 1. Model of anodic dissolution in the EP process (1) direction of acceptor movement, (2) direction of movement of metal ions [33,34].

B refers to the zone which is a thin layer formed on metal ions and acceptors that were previously adsorbed on the anode. Zone C is a layer consisting of salt solution that contains the bath and metal ions from the D layer and acceptors from layer B. The D layer consists of accumulated metal ions that emerge as a result of dissolution of metal or alloy substrate (E). It is usually a super-saturated solution. Mass transport in the electrochemical polishing process depends on the gradient of concentrations that exist in layers B and C. Metal ions migrate and diffuse from the D layer towards the depth of the solution (layer A) by electrostatic forces and thanks to the gradient of concentration. The dissolution of substrate (E) and the transport of the mass of dissolved metal ions are dominated by the micro-dimensions of the B, C, and D layers that depend on the solution used for electrochemical polishing (Figure 1). The diffusion of metal ions (E) to the bath (A) depends on the microstructure of the C and D zones. In this case, the D layer is a layer of metal salts that emerge as a result of the deposition of phosphates (V) and sulphates (VI) that are products of the reaction with bath components—phosphoric acid (V) and sulphuric acid (VI). Layer C is a liquid layer that contains concentrated metal salts that are highly concentrated, up to saturation. If the electropolishing process takes place according to the acceptor mechanism, then the diffusion of acceptors towards the anode and the migration of complexed metal ions from the anode should occur simultaneously in layer B. Hence, the microstructure of layer B is formed by the mass transport connected with these processes. The D layer disappears, while layers B and C are liquid, with a relatively concentrated layer of metal salts deposited on substrate E. This mechanism applies to chromium steel electrochemical polished in acids with a low content of water as the acceptor. When baths containing glycerine are used, polymer structures or aggregates may emerge in layer B. This results from the presence of three functional groups $-OH^-$ in the particle of glycerine. These groups determine the existence of strong hydrogen bonds between them. As a result, the effect of hindering the diffusion of acceptors towards the substrate (E) and the migration of ions towards the depth of the solution (A) are reinforced. Pursuant to this mechanism, covering the substrate surface (E) by weak or strong polymer aggregate structures (e.g., particles of glycerine) influences the speed of steel dissolution.

The current convection used in the diffusion layer into the depth used is possible. However, this is a dynamic system in which the near electrode layer is constantly reproducing on the anode side and convection on the solution side. In the diffusion layer there is still a layer of adsorbed polar particles that adsorb onto the oxide layer, e.g., amines or multi-alcohols. The oxide film forms on the anode and due to the large shortage of water molecules in the layer diffusion, available in the solvation process of metal ions formed on the anode, which are cations and they change in the electric field towards the cathode, that is, into the depth of action. In these conditions, oxides or hydrated metal oxides accumulate, which are moved by moving the anodes far away from the thermodynamic state balance of the dissolution of metals:

$$Me - ze^- + nH_2O \rightarrow Me^{z+} \cdot nH_2O \tag{5}$$

which is characteristic of the metal etching process. It is also worth mentioning the role of adsorbing substances on the anode surface, such as amines and multi-alcohols, which shift potential after adsorption on the anodes far from the state of thermodynamic equilibrium of the reaction:

$$Me - ze^- \rightarrow Me^{z+} \tag{6}$$

The basic condition in the process of electrochemical polishing is a higher displacement of anode potential, far from thermodynamic conditions. Then the digestion process is not decided by active centres on the anode, such as the location of the metal atom in the anode metal crystals, and begins to play a role in the topography of the surface.

In the electrochemical polishing process, the anode is usually covered with an oxide film. There is only one case where no oxide layer is formed on the anode surface in the electropolishing process, i.e., the case of silver electropolishing in cyanide solutions. A salt film is then formed, which is fully the same as the oxide film in the required electropolishing.

Electrochemical polishing solutions require the ability to transfer electrical charges, which occurs in the results of anion and cation analysis in the electric field. Anions of oxygen data as well as anions of organic compounds migrating to the anode surface are capable of forming metal oxides as a result of the anode reaction. Only halogen anions cannot form metal oxides. Anhydrous solutions, such as perchlorate solution made from acetic anhydride, and concentrated chloric acid (VII) also contain chlorate (VII), ClO_4^- and acetate anions, $CH3COO^-$ which may be involved in the formation of oxide oxygen. Acetic anhydride can react with metal anodes with metal oxide and metal acetate. For structures forming bivalent Me^{2+} ions, a hypothetical reaction can be used:

$$(CH_3COO)_2O + 3Me - 2e^- \rightarrow MeO + Me^{2+} + (CH_3COO)_2Me \tag{7}$$

This reaction is much more complicated, because it results in high-molecular complex substances that have the nature of surface-active substances capable of adsorbing on the anode surface.

The presence of micro-peaks and micro-indentations of the surface during the polishing process is attributed to local differences in the rate of dissolution of the structural elements of stainless steel. The higher the difference in the rate of dissolution of specific structural components, the more uneven and rough the surface will be after electropolishing [35].

If the solution contains substances that form complex compounds with ions of the dissolved metal, then the adsorption and diffusion of the emerging complex compounds may play a significant role in the mechanism of the electrochemical polishing process. As complex compounds usually have high molar mass and a complex structure, their mobility is usually low. Polishing alloy steels in phosphate(V)–sulphate(VI) baths involves the emergence of ions $[Fe(PO_4)_3]^{6-}$, $[Fe(HPO_4)_3]^{3-}$, and if the bath contains oxalic acid: $[Fe(C_2O_4)_3]^{3-}$ [20].

Another known hypothesis is based on the presence of an oxide layer that emerges on the surface of the electropolishing anode in the conditions of anodic passivation. This layer is labile in a strongly acidic environment and it is in a state of dynamic balance as a result of two contradictory processes: the electrochemical formation of the oxide and its chemical degradation. The oxide layer conducts current and is characterised by variable thickness that depends on the duration and the shape of the anodic surface. As water particles and anions may access spots located on peaks of the surface more easily, the rate of chemical reactions increases which, in turn, leads to decreasing thickness of the oxide layer and increased speed of the electrochemical reaction. It should be emphasised that the emergence of oxides that form the passive layer is facilitated on the borders of grains and in spots of network defects of the metal. As a result, the structure of the processed metal or alloy influences the smoothness of the electrochemical polished element [36].

The course of the electrochemical polishing process may also be analysed in terms of the energy needs of the phase of ionisation of atoms of the dissolved metal. The energy required for the ionisation of a superficial atom during anodic dissolution is the lowest on peaks and the highest in surface indentations.

The high number of various hypotheses concerning the mechanism (Figure 2) of smoothening and shining the surface in electropolishing results from the complexity of the process.

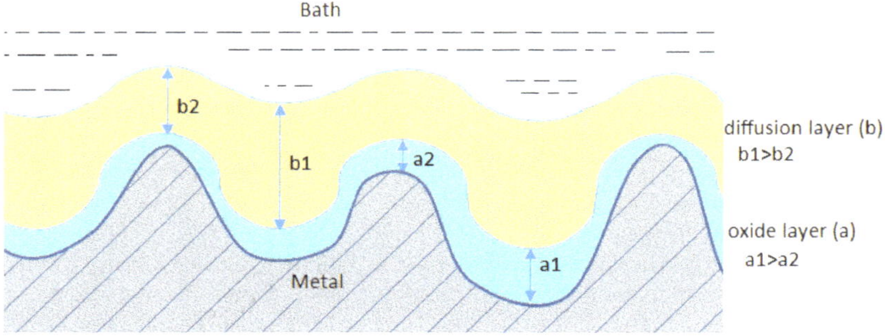

Figure 2. Mechanisms of the electropolishing processes.

However, apart from the diffusion layer, also known as the viscous layer, that emerges near the anode, an oxide layer usually also emerges directly on the surface of the anode. A common phenomenon resulting from the simultaneous formation of both these layers is the process of diffusion of water particles and the migration of anions towards the surface of the anode. These substances are necessary in electrochemical formation and chemical dissolution of the oxides that constitute the passive layer. Due to the competitiveness of both these processes, even small differences in the rate of migration and diffusion of the acceptors of the dissolution process may lead to significant differences in the rate of dissolution of individual elements of the anode surface structure.

3. Description of the Technological Electrochemical Polishing Process

The important factors that influence the appearance of the polished surface are current density, the composition and concentration of electrolyte, mixing and bath temperature [37,38]. Current density is the ratio of the total current I (A) to the total surface S (dm^2). Excessively high current density may cause intensive emission of oxygen, which may result in the emergence of surface defects. The type of current, including whether it is direct current or pulsating, also influences the quality of the process [3]. The composition and concentration of electrolyte, current density, and mixing rate are complementary. The conductivity of the bath depends on the concentration of its components and, as a consequence, on the density of the electrolyte. Increased temperature significantly influences the conductivity of the solution and the decrease in the clamp voltage of electrolysis. If the viscosity of the solution increases (as a result of an addition of lactic or oxalic acid or glycerine), the content of adsorbed particles on the surface of the electrochemical polished metal may increase, eventually resulting in decreased roughness.

During electropolishing, the processed element is immersed in the bath and connected to the positive pole of a direct current source (the anode). Inside the tank, cathodes are placed, which are connected to the negative terminal of the current source. The cathodes and anodes placed in electrolyte create an electric cell. In industry, tanks made from steel sheet metal layered with lead are usually used, or alternately tanks made from stainless steel and cathodes made from lead or stainless steel. The electric couplings of the hooks used for electrochemical polishing steel elements are typically made from copper wire. The electrochemical polishing process requires access to proper ventilation.

The best results of the electropolishing process are obtained after performing three basic stages of the procedure (Figure 3):

1. Preparing the surface (removing dirt that may interfere with the electrochemical polishing process);
2. Electrochemical polishing (softening sharp edges and electrochemical polishing);
3. Final processing (rinsing and removing remains of the bath, drying the metal surface).

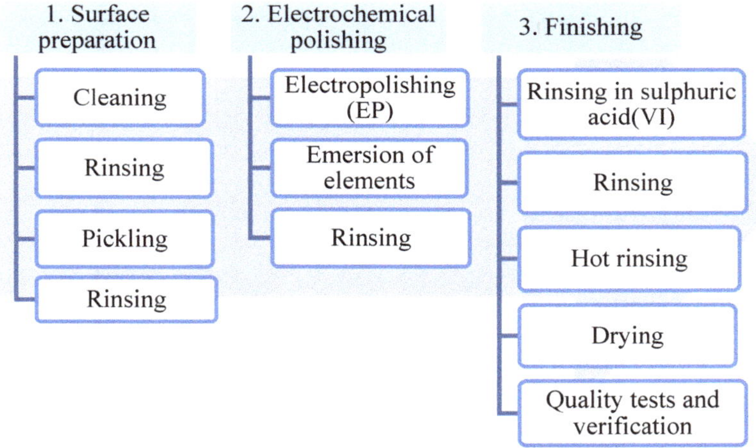

Figure 3. Technological diagram of a typical electrochemical polishing process.

Figure 3 presents a technological diagram of a typical electrochemical polishing process.

Rinsing stainless-steel elements with diluted sulphuric acid (VI) additionally helps to clean the surface of the element of contaminants that may emerge from electrochemical polishing. After each stage, the processed elements are rinsed with demineralised water.

4. Baths and Parameters of the Electrochemical Polishing Process of Stainless Steel

The electrolyte plays several important roles:

- It is the medium in which chemical processes take place;
- It enables the transport of electric load in the solution;
- Eemoves the products of anodic dissolution from the processing zone.

The proper selection of components of the bath used for electropolishing on a laboratory and industrial scale is an important issue [39,40]. In order for the electrochemical processing to bring the desired technological and economical results, it is necessary to select such bath composition that will be adequate to the chemical composition and structure of the processed material. As attempts to determine the appropriate composition of the electrolyte are very time-consuming, baths, which are of great practical importance, are usually protected by patents.

Additionally, the scope of application of certain electrolytes is limited by their corrosive influence on the equipment and the processed element as well as harmful impact on humans and the environment [41].

In order for the thin, viscous layer to be created from the products of anodic dissolution on the anodic surface in the electrolyte it is required to maintain a low concentration of iron and chromium ions. For this purpose, after preparing the bath, it is "worked" to obtain a specific, low concentration of these ions in the solution.

During electropolishing, the concentration of metal ions in the electrolyte increases and its viscosity changes. The content of dissolved iron in the bath influences the specific weight, acidity and viscosity of the solution and the current efficiency. In practice, it is known that at 3% weight (approximately 70 g Fe/dm^3) content of this metal in electrolyte, the electrochemical polishing process is stopped. In such event, the bath should be regenerated or replaced [31]. The durability of electrochemical polishing baths is usually determined by the value of current load that has passed through a volume unit of the bath since the moment when it was prepared. The durability of baths, e.g., for electrochemical polishing carbon steel is not high and it ranges from 100 to 180 Ah/dm^3. Bath maintenance usually consists in removing 30–50% of the volume of the solution after processing approximately 80 Ah/dm^3 and then supplementing it with a fresh solution [3]. Baths for polishing chromium and nickel steel are more durable. Their maintenance consists in regular supplementing of the glossing additives (e.g., triethanolamine) in half the amount specified in the formula.

The main components of solutions used for electrochemical polishing stainless steel are sulphuric acid (VI) and orthophosphoric acid (V) [42]. Sulphuric acid (VI) ensures that the electric conductivity of the bath is appropriately high, while phosphoric acid (V) is responsible for the properties of the anodic layer on the polished metal surface [43]. These acids contribute to the smoothening of the surface during electrochemical polishing and they are used to reduce the contamination of the bath (for regeneration). During electropolishing of stainless steel, the components of the steel are dissolved: iron, chromium and nickel on the anode:

$$Fe \rightarrow Fe^{3+} + 3e^- \tag{8}$$

$$Cr \rightarrow Cr^{3+} + 3e^- \tag{9}$$

$$Ni \rightarrow Ni^{2+} + 2e^- \tag{10}$$

$$2H_2O \rightarrow O_2 + 4H^+ + 4e^- \tag{11}$$

Hydrogen is emitted on the cathode:

$$2H_3O^+ + 2e^- \rightarrow 2H_2O + H_2 \tag{12}$$

The review of literature on the surface processing of metals revealed that elements made from chromium and nickel steel may be electrochemical polished with baths that contain sulphuric acid (VI), orthophosphoric acid (V) [44], triethanolamine [45–47] and ethylene glycol, oxalic acid, and acetanilide [21]. Stainless steel may also be electrochemical polished in a solution consisting of sulphuric acid (VI) and citric acid, as well as of sulphuric acid (VI), phosphoric acid (V) and lactic acid [48,49]. Some other additives include: 2-amino-2-methyl-1-propanol [50], ethanolamine, diethanolamine and triethanolamine [5,51], glycerine [34], and natrium phosphate [52]. Until recently, chromic acid (VI) was also used. It was added in form of chromic acid anhydride (VI), i.e., chromium oxide (VI) [53–56]. Currently, it is used more rarely due to the fact that chromium (VI) compounds are harmful for the environment. Abbot [57], in his studies, proposed a bath consisting of ionic liquids of ethylene glycol and choline chloride instead of baths made from water solutions of acids. ChCl:EG mixtures can be a potential alternative to sulphuric/phosphoric acid-based electrolytes for electrochemical polishing stainless steel. The mechanism of electropolishing in aqueous acid solutions is different from that in the ChCl:EG mixture. The removal of the oxide is slower and more potential dependent than in aqueous solutions. When the oxide has been removed from the surface, the metal atoms are oxidised and the treatment process is controlled by mass transport limitations. When electrolytic polishing stainless steels on an industrial scale, only aqueous solutions are used, which are concentrated solutions of phosphoric and sulfuric acids. Optionally with the addition of organic substances such as glycerin, triethanolamine.

Eliaz, Nissan, and Sojitra [58,59] presented the process of electrochemical polishing implants made from 316L steel in solutions that are less harmful for the environment, i.e., in baths consisting of sulphuric acid (VI) 96% (50% vol.) and phosphoric acid (V) 85%, (50% vol.). Haïdopoulos and Zhao [60,61] conducted research on a bath consisting of glycerine 99% (47–50% vol.), phosphoric acid (V) 85% (35–42% vol.) and distilled water. These baths were used to electropolish stents made from 316L steel. Before electrochemical electropolishing, the samples were pickled in baths consisting of nitric acid and hydrofluoric acid. The process parameters for the first bath were: temperature 75 °C, current density 0.44 A/cm^2, duration 3 min. The second bath worked at a temperature of 90–95 °C, the current density was 1.2 A/cm^2, and time 1–10 min. Sample baths and parameters of the electrochemical polishing process of stainless steel are presented in Table 1.

Table 1. Sample baths and parameters of the electrochemical polishing process of stainless steel.

Material	Bath	Parameters			Source
		j (A/dm^2)	t (min)	T (°C)	
Steel AISI 304, 316	Sulphuric acid (VI) (35 wt.%), orthophosphoric acid (V) (51 wt.%), triethanolamine 99%. (3 wt.%), H_2O (11 wt.%)	20	12	55	[17]
	Sulphuric acid (VI) 96% (40% vol.), orthophosphoric acid (V) 85% (60% vol.), additives: ethylene glycol 99%—200 g/dm^3, oxalic acid—200 g/dm^3, acetanilide—200 g/dm^3.	35–50	1–50	60	[21]
	Sulphuric acid (VI) 96% (50% vol.), orthophosphoric acid (V) 85%, (50% vol.)	15	1–3	40–75	[58,59]
	Orthophosphoric acid (V) 85% (35% vol.), glycerine 99% (50% vol.), distilled water (15% vol.)	75	1–10	60–95	[60,61]
	Base solution: orthophosphoric acid 85%: sulphuric acid (VI) 97% at a ratio from 2:1 to 3:2; (75% vol.), glycerine 99% (25% vol.)	50	1–10	30–90	[34]

The surface roughness of stents electrochemical polished in the first bath, after pickling in a solution consisting of hydrofluoric acid and nitric acid, was approx. Ra = 250 nm [59], while the roughness of stents that had been electrochemical polished in the second solution was Ra = 120.52 ± 25.65 nm. After pickling in hydrofluoric and nitric acid was Ra = 126.07 ± 37.13 nm [61]. Acid pickling in a bath consisting of hydrofluoric acid and nitric acid was carried out to remove slag and metal oxides attached after other types of surface processing (for example, laser micromachining) before electrochemical polishing. Electropolishing stents in the first solution resulted in a surface roughness Ra = 14.77 nm for the selected working parameters, while roughness after electrochemical polishing in the second bath was Ra = 13.13 nm. Weight loss for the first bath after pickling was 7.43%, while after the electrochemical polishing process 17.99%. In the second case, weight loss after pickling was 7.7% and 16.7% after electrochemical polishing.

5. Electrochemical Polishing and Other Methods of Surface Processing

Finishing is the final phase of the whole surface processing operation. Its aim is to obtain a suitably high quality of the processed element, to comply with the required technological specifications concerning the accuracy of dimensions and shape, and surface roughness. Mechanical polishing and electrochemical polishing are applied in order to obtain an appropriately smooth surface (e.g., polished surfaces of metal mirrors, ornamental elements, medical instruments, and implants).

Based on the literature, it was determined that the selection of an adequate surface processing method depends on the size and internal structure of the elements. In many cases, electrochemical polishing is preferred for the use of objects with complex shapes, because mechanical polishing is very difficult. For example [60,61], electropolishing leads to smoothing of samples of complex shapes (laser cut stents and endovascular stents) and is comparable to commercially available stents. Electropolishing does not deform the microstructure of stents; it improves it to a homogeneous structure and smooth, defect-free and contamination-free surface. Lukas Löber [62] in his studies presented a manner of surface processing of elements made from 316L steel of the dimensions of 10 × 10 × 10 mm. In order to reduce roughness, the processed elements manufactured in the SLM (Selective Laser Melting) technology were subjected to the following mechanical processing: blast cleaning, grinding, mechanical polishing as well as electrochemical polishing and electro-plasmatic polishing (Table 2).

Table 2. Results of roughness measurements for elements from 316L steel manufactured in the SLM (Selective Laser Melting) technology, after selected processing methods [62].

Processing Method	Abrasive Grit	*Ra* (μm)
SLM		15.03
Grinding	P 80	2.22
	P 240	1.15
	P 300	0.52
	P 500	0.43
Blast cleaning with glass	50–150 mm	8.85
Electrochemical polishing after SLM		15.03
Grinding and electrochemical polishing	P 80	9.28
	P 240	1.46
	P 500	0.64
Mechanical polishing and electrochemical polishing	-	0.21
Mechanical polishing and electro plasmatic polishing	-	0.12

For example, test results for processing elements manufactured from 316L steel in SLM technology, discussed in the study [62], are presented in Table 2. Positive effects were obtained after combining mechanical polishing with electropolishing: the roughness parameter *Ra* decreased from 15.03 μm to 0.21 μm and after mechanical polishing combined with electro plasmatic polishing, after which the roughness was 0.12 μm. The author of the quoted paper also mentioned the possibility to process elements after the SLM process by means of chemical pickling. Unfortunately, the level of surface roughness obtained after only mechanical initial processing (e.g., grinding) was unsatisfactory.

Table 3, based on reviewing the literature, contains the parameters of the electrochemical polishing process presented in the subject literature:

Table 3. Sample parameters of the electrochemical polishing (EP) processes, bath compositions and test results.

EP Process Parameters	EP Bath	Type of Analyses/Material	Test Results	Source
T = 40, 50, 58, 67 °C, U = 2–4 V, I = 0.04 A, j = 0.15 A/cm^2, $t \leq$ 10 min, S = 0.27 cm^2, $q \leq$ 24.7 mAh/cm^2.	H_2SO_4 96%, H_3PO_4 85% (50% *v/v*).	Surface roughness/316LVM.	Mean roughness: AFM: SS: Ra = 18.4 ± 4.5 nm, after EP: Ra = 2.1 ± 0.8 nm. Profile meter: SS: Ra = 183.1 ± 80.6 nm, after EP: Ra = 76.2 ± 67.9 nm.	[58]
U = 10–12 V, I = 1.2 A, T = 90–95 °C, t = 1 min, S—surface of one stent.	H_3PO_4 85% (42 wt.%), glycerine (47 wt.%), H_2O (11 wt.%).	Surface roughness/316L. stents: length 15 mm, diameter 1.6 mm, wall thickness 95 µm.	Mean roughness: (a) central area of the sample: SS: Ra = 120.52 ± 26.65 nm, pickled: Ra = 126.07 ± 37.13 nm, relaxed: Ra = 142.71 ± 26.20 nm, after EP: Ra = 13.13 ± 1.56 nm, (b) laser cutting area: SS: Ra = 491.26 ± 52.46 nm, pickled: Ra = 268.67 ± 27.7 nm, relaxed: Ra = 302.90 ± 23.33 nm, after EP: Ra = 15.01 ± 1.79 nm.	[61]
U = 9.5 V, I = 0.44 A, T = 75 °C, t = 3 min, S—surface of one stent.	H_2SO_4 96%, H_3PO_4 85% (50% *v/v*).	Surface roughness, corrosion resistance/316LVM, stents: length 16 mm, diameter 1.720 mm, wall thickness 110–115 µm. Corrosion tests were conducted in Phosphate Buffer Saline solution (PBS) at pH 7.4 and 37 ± 1 °C constant temperature. Open circuit potential was measured for 1000 s with respect to saturated calomel electrode (SCE).	Processing result: 78.10% roughness reduction: SS: Ra = 250 nm, after EP: Ra = 14.77 nm, SS: Rq = 97.56 nm, after EP: Rq = 4.895 nm, SS: Rp = 208.7 nm after EP: Rp = 32.49 nm, SS: Rv = 130 nm, after EP: Rv = 8.921 nm. Corrosion potential: -after pickling: $Ecor$ = −326 mV, -after laser cutting: $Ecor$ = −259 mV, -after EP: $Ecor$ = −173 mV.	[59]
I = 6.8 A, $j \leq$ 40 A/dm^2, T = 55 ± 1 °C, t = 4, 6 min, S = 20 cm^2, $q \leq$ 40 mAh/cm^2.	1—H_3PO_4 (51 wt.%), H_2SO_4 (35 wt.%), H_2O (14 wt.%), 2—H_3PO_4 (51 wt.%), H_2SO_4 (35 wt.%), glycerine (3 wt.%), H_2O (11 wt.%). 3—H_3PO_4 (51 wt.%), H_2SO_4 (35 wt.%), triethanolamine (3 wt.%), H_2O (11 wt.%). Other organic additives: triethylamine, ethanoloamine, diethanolamine, butyldiglycol.	Surface roughness, corrosion resistance, gloss/samples from 304 steel, dimensions: 90 × 25 × 1.5 mm.	Bath 1: pickled: Ra = 0.35–0.38 µm, after EP: (q = 0.02 Ah/cm^2) Ra = 0.12 µm, after EP: (q = 0.03 Ah/cm^2) Ra = 0.10 µm, after EP: (q = 0.04 Ah/cm^2) Ra = 0.12 µm. Gloss: (t = 6 min, j = 30 A/dm^2, q = 0.03 Ah/cm^2) = 890 ± 10 GU. Bath 2: pickled: Ra = 0.36–0.40 µm, after EP: (q = 0.02 Ah/cm^2) Ra = 0.14 µm, after EP: (q = 0.03 Ah/cm^2) Ra = 0.10 µm, after EP: (q = 0.04 Ah/cm^2) Ra = 0.09 µm. Gloss: (t = 6 min, j = 30 A/dm^2, q = 0.03 Ah/cm^2) = 680 ± 10 GU. Bath 3: pickled: Ra = 0.36–0.38 µm, after EP: (q = 0.02 Ah/cm^2) Ra = 0.095 µm, after EP: (q = 0.03 Ah/cm^2) Ra = 0.09 µm, after EP: (q = 0.04 Ah/cm^2) Ra = 0.079 µm. Gloss: (t = 6 min, j = 30 A/dm^2, q = 0.03 Ah/cm^2) = 900 ± 20 GU. Recommended processing parameters: bath 3, j = 30 A/dm^2, T = 55 °C, t = 6 min.	[63]

Table 3. *Cont.*

EP Process Parameters	EP Bath	Type of Analyses/Material	Test Results	Source
$U = 2.5; 4; 10$ V, $T = 65–70$ °C, $t = 3$ min, $S = 3.33$ cm^2.	H_3PO_4 85% (60% v/v), H_2SO_4 95–97% (20% v/v), glycerine 99.5 % (10% v/v), H_2O (10% v/v).	Surface roughness, corrosion, chemical composition of surface, blood cells adhesion/316L, diameter 12.7 mm, wall thickness 2 mm. Corrosion tests were conducted in 0.16 M NaCl.	SS: $Ra = 188 \pm 9$ nm, after EP: ($U = 2.5$ V) $Ra = 107 \pm 6$ nm, after EP: ($U = 4$ V) $Ra = 77 \pm 4$ nm, after EP: ($U = 10$ V) $Ra = 97 \pm 11$ nm. Corrosion potential: SS: $OCP = 0.34$ V, after EP: ($U = 2.5$ V) $OCP = 0.29$ V, after EP: ($U = 4$ V) $OCP = 0.18$ V, Adhesion-decrease: after EP: ($U = 2.5$ V) by 71%, after EP: ($U = 4$ V) by 89%, after EP: ($U = 10$ V) by 93%.	[64]
$U = 10$ V, $T = 60$ °C, (EPO—oxygen evolution potential), $T = 25$ °C (EPBO—oxygen evolution plateau), $t = 5$ min.	1—H_3PO_4 85%, H_2SO_4 93%-7:3 vol. (EPO), 2—100 mL CH_3OH, 300 mL H_2SO_4 93% (EPBO).	Surface roughness, corrosion resistance, chemical composition of the surface, cell adhesion/316L Corrosion tests were conducted in Phosphate Buffer Saline solution (PBS) at pH 37 °C constant temperature. Calomel electrode (SCE) was used as the reference electrode and graphite rod was used as a counter electrode.	SS: $Ra = 33.51 \pm 5.54$ nm, after EPO: $Ra = 11.50 \pm 1.61$ nm, after EPBO: $Ra = 6.07 \pm 0.73$ nm, SS: $Rq = 48.28 \pm 8.65$ nm, after EPO: $Rq = 14.79 \pm 1.90$ nm, after EPBO: $Rq = 7.91 \pm 0.73$ nm. Corrosion current density: SS: $jcorr = 56.1$ mA/m^2, after EPO: $jcorr = 23.7$ mA/m^2, after EPBO: $jcorr = 12.9$ mA/m^2, Corrosion potential: SS: $Ecor = -410$ mV, after EPO: $Ecor = -353$ mV, after EPBO: $Ecor = -288$ V, Linear corrosion rate: SS: $Vp = 0.0475$ mm/y, after EPO: $Vp = 0.0182$ mm/y, after EPBO: $Vp = 0.0123$ mm/y, Polarisation resistance: SS: $Rs = 93.40$·Ωcm^2, after EPO: $Rs = 104.2$ Ω·cm^2, after EPBO: $Rs = 100.8$ Ω·cm^2, Chemical composition of the surface oxide layer (XPS): SS: Fe2p = 13.89 wt.%, after EPO: Fe2p = 10.51 wt.%, after EPBO: Fe2p = 11.5 wt.%, SS: Cr2p = 1.86 wt.%, after EPO: Cr2p = 7.55%, after EPBO: Cr2p = 8.55 wt.%, after EPO: Cr2p3 = 5.65 wt.%, after EPBO: Cr2p3 = 1.23 wt.%, SS: O = 43.92 wt.%, after EPO: O = 36.51 wt.%, after EPBO: O = 38.75 wt.%.	[65]
$U = 5$ V, $I = 0.6$ A, $j = 25$ A/dm^2, $T = 50$ °C, $t = 20$ min, $S = 2.4$ cm^2, $q \leq 83$ mAh/cm^2.	H_3PO_4 (60% v/v), H_2SO_4 (40% v/v).	Surface roughness, corrosion resistance, chemical composition of the surface, cell adhesion/316L, 10 × 10 × 1 mm. Corrosion tests were conducted in ringer solution at 37 °C constant temperature. Silver chloride electrode (Ag/AgCl) and platinum counter electrode were used as reference electrode and a working electrode.	SS: AFM $Sa = 161.34 \pm 57.15$ nm, after EP: $Sa = 5.05 \pm 0.28$ nm, after EP + chemical processing: $Sa = 0.96 \pm 0.29$%, SS: $Srms = 206.58 \pm 70.06$ nm, after EP: $Srms = 8.43 \pm 0.40$ nm, after EP + chemical processing: $Srms = 1.71 \pm 0.78$%, Corrosion current density: SS: $jkor = 0.921$ µA/cm^2, after EP: $jkor = 0.61$ µA/cm^2, after EP + chemical processing: $jkor = 0.0066$ µA/cm^2, Corrosion potential: SS: $Ecor = -343$ mV, after EP: $Ecor = -292$ mV, after EP + chemical processing: $Ecor = -16$ mV, XPS: SS: C = 80.2 at.%, after EP: C = 40.8 at.%, after EP + chemical processing: C = 37 at.%, SS: O = 19.8.2 at.%, after EP: O = 49 at.%, after EP + chemical processing: O = 44.2 at.%, Fe: after EP: Fe = 3.4 at.%, after EP + chemical processing: Fe = 3.2 at.%, Cr: after EP: Cr = 4.2 at.%, after EP + chemical processing: Cr = 6.0 at.%, P: after EP: P= 2.7 at.%, after EP + chemical processing: P = 9.2 at.%, Cell viability: approx. 80% (samples subjected to electrochemical and chemical processing).	[66]

Table 3. *Cont.*

EP Process Parameters	EP Bath	Type of Analyses/Material	Test Results	Source
T = 20–90 °C, I = 1.5 A, j = 0.75 A/cm^2, t = 1–6 min, S = 2 cm^2 (immersed surface) $q \leq 75$ mAh/cm^2.	glycerine 99% (50% v/v), H_3PO_4 85% (35% v/v), H_2O (15% v/v).	Surface roughness, chemical composition of the surface/316, sample dimensions: 15 × 10 × 1.5 mm.	AFM: SS: Ra = 2.2 ± 0.15 nm, ground: Ra = 2.3 ± 0.2 nm, Mechanically polished: Ra = 0.04 ± 0.01 nm, after EP: Ra = 0.07 ± 0.02 nm.Chemical composition–increase at.% Cr from 5.7% to after EP 10% (for T = 20 °C) and to 11.5% (for T > 20 °C). Recommended parameters: t = 3 min, T = 90 °C.	[60]
For baths 1 and 2 *EP*: I = 4 A, j = 20 A/dm^2, T = 55 °C, t = 1–20 min, For bath 3 *EP*: T = 90 °C, t = 120 min, S = 20 cm^2, $q \leq 66.7$ mAh/cm^2.	1: - H_2SO_4 96% (35% v/v), H_3PO_4 85% (60.5% v/v), triethanolamine 99% (4.5% v/v). 2: - H_2SO_4 96% (40% v/v), H_3PO_4 85% 60% v/v, glycol 99% 200 g/dm^3, oxalic acid 200 g/dm^3, acetanilide 200 g/dm^3. 3: - glycerine 99% (50% v/v), H_3PO_4 85% (35% v/v), H_2O (15% v/v).	Surface roughness, chemical composition of the bath/316L, sample dimensions: 80 × 20 × 2 mm.	SS: Ra = 0.17–0.20 µm, Ra for bath 1 after EP: Ra = 0.063–0.065 µm, Ra for bath 2 after EP: Ra = 0.091–0.096 µm, Ra for bath 3 after EP (at 90 °C): Ra = 0.070 µm. Recommended parameters for baths 1 and 2: t = 15–20 min, T = 55 °C. Recommended parameters for bath 3: T = 90 °C.	[67]
j = 20 A/dm^2, T = 55 °C, t = 12 min.	H_3PO_4 (51 wt.%), H_2SO_4 (35 wt.%), triethanolamine (3 wt.%), H_2O (wt.% 11).	Surface roughness/bath contamination/304.	SS: Ra = 0.17–0.25 µm, pickled: Ra = 0.37–0.43 µm, after EP in industrial baths I and II: Ra = 0.14–0.20 µm (where iron content: 35–55 g Fe/dm^3), after EP in industrial bath III: Ra = 0.17–0.28 µm (where iron content: 67–75 g Fe/dm^3), after EP in laboratory bath with low iron content (3–6 g Fe/dm^3): Ra < 0.11 µm.	[17]
j = 0.5–3.0 A/cm^2, T = 50–95 °C, t = 3, 6, 9 min.	H_2SO_4: H_3PO_4 (vol.): 5:5, 4:6, 3:7, H_2O, glycerine.	Surface roughness, corrosion resistance/316L Corrosion tests were conducted in $FeCl_3$ solution (workpiece was put 2 cm deep). The container was sealed, and the temperature was controlled at 50 °C for 72 h.	$Rmax$ = 0.8 µm, Ra = 0.08 µm (for recommended parameters), Recommended parameters: j = 1 A/cm^2, T = 85 °C, t = 3–5 min, Bath composition: H_2SO_4; H_3PO_4—vol. 4:6, addition of 10% H_2O.	[68]
j = 0.8 A/cm^2, T = 40 °C, t = 420 s.	H_3PO_4 (64 wt.%), H_2SO_4 (13 wt.%), H_2O (23 wt.%).	Surface roughness, corrosion resistance/316L For corrosion tests the saturated calomel electrode (SCE) was applied as the reference electrode and platinum foil as a counter electrode. Experiments were performed in physiological solution represented by 0.9 % NaCl solution.	SS: Ra = 0.256 µm, after EP: Ra = 0.078 µm, SS: Rq = 0.363 µm, after EP: Rq = 0.096 µm, SS: Rz = 2.290 µm, after EP: Rz = 0.474 µm.	[69]
j = 10, 29, 48, 67 A/dm^2, T = 35, 45, 55, 65 °C, t = 3, 14, 25, 36 min.	1—H_2SO_4 15%, H_3PO_4 63%, H_2O 22%,2—H_2SO_4 35%, H_3PO_4 45%, H_2O 17%, CrO_3 3%,3—H_2SO_4 35%, H_3PO_4 45%, H_2O 20%.	Surface roughness, chemical composition of the bath/316L.	Maximum Ra ranges after EP: 80–90% for: Recommended parameters: j = 48 A/dm^2, T = 35 °C, t = 25 min, ΔRa best in bath 3, worst in bath 1. Addition of CrO_3 in bath 2 did not influence the electrochemical polishing result.	[70]

Table 3. *Cont.*

EP Process Parameters	EP Bath	Type of Analyses/Material	Test Results	Source
j = 30–50 A/dm^2, T = 55 °C, t = 5–20 min, S = 4 cm^2, $q \leq$ 167 mAh/cm^2.	H_3PO_4 (51 wt.%), H_2SO_4 (35 wt.%), triethanolamine (3 wt.%), H_2O (11 wt.%).	Corrosion resistance, chemical composition of the surface/304 During the corrosion tests, a tested electrode was 304 steel, reference saturated calomel electrode (SCE) was the electrode and the counter electrode was a platinum electrode.	SS: *Epit* = 0.30 V, pickled: *Epit* = 0.38 V, after EP (for t = 6 min, j = 30 A/dm^2): *Epit* = 0.57 V, after EP (for t = 5 min, 20 min, j = 50 A/dm^2): *Epit* (for 5 min) = 0.60 V, *Epit* (for 20 min) = 0.49 V, at.% ΣFe (FeO, Fe_2O_3, Fe_3O_4, FeOOH): SS: at.% ΣFe = 6.1%, pickled: at.% ΣFe = 11.9%, after EP: at.% ΣFe = 14.5%, at.% ΣCr (CrO_2, Cr_2O_3, CrO_3, $Cr(OH)_3$): SS: at.% ΣCr = 18.0%, pickled: at.% ΣCr = 17.9%, after EP: at.% ΣCr = 28.6%.	[71,72]
U = 25 V, T = 30 °C, t = 20 s.	$HClO_4$ 70% (20% v/v), CH_3COOH 98% (80% v/v).	Corrosion resistance, chemical composition of the surface/316L, sample dimensions: 10 × 10 × 1 Corrosion tests were conducted at 30 °C in a solution with composition water, i.e., 1000 mg/L of B as H_3BO_3 and 2 mg/L of Li as LiOH. The reference electrode was a saturated calomel electrode (SCE), and a counter electrode was a platinum plate.	XPS analysis: after EP: at.% Cr (hydroxide) ≈ 35%, at.% Cr (oxide) ≈ 7%, mechanical processing: after CPS (polishing with colloidal silicate) at.% Cr (oxide) ≈ 30%, after EP: at.% Fe (hydroxide) ≈11%, at.% Fe (oxide) ≈ 25%, mechanical processing: after CPS: at.% Fe (oxide) ≈ 28%, corrosion: after EP: *Rs* = 8736 Ω·cm^2, after CPS: *Rs* = 8840 Ω·cm^2, after EP: *C1* = 78.2 µF/cm^2, after CPS *C1* = 42.5 µF/cm^2, after EP: *R1* = 0.41 Ω·cm^2, after CPS *R1* = 0.97 Ω·cm^2, after EP: *C2* = 63.8 µF/cm^2, after CPS *C2* = 35.9 µF/cm^2, after EP: *R2* = 3.04 Ω·cm^2, after CPS *R2* = 5.69 Ω·cm^2.	[73]
—	—	Corrosion resistance/316L, stents: Corrosion tests were conducted in tyrode solution at 37 °C constant temperature. Calomel electrode (SCE) as a reference electrode and platinum plate served as an auxiliary electrode.	Corrosion resistance: After mechanical processing: *Ecor* = −0.533 V, after EP: *Ecor* = −0.324 V, After mechanical processing: *Vcor* = 0.04 mm/year, after EP: *Vcor* = 0.119 mm/year.	[74,75]
U = 3 V, j = 1.25–25.5 A/dm^2, t = 5–25 min.	H_3PO_4 (55 wt.%), H_2SO_4 (14 wt.%), H_2O (31 wt.%).	Gloss, corrosion resistance, chemical composition, adhesion/304.	SS: Gloss G = 400, after EP: G = 1700–2500. Chemical composition: wt.% Cr increase after EP: from 18.83% to 19.33%.	[76]
j = 15.5; 31.0; 46.5 A/dm^2, T = 333, 343, 353 K, t = 5–12 min.	1—H_3PO_4 (500 mL/L), H_2SO_4 (360 mL/L), monoethanolamine 20 (mL/L), 2—H_3PO_4 (500 mL/L), H_2SO_4 (360 mL/L), diethanolamine 20 mL/L, 3—H_3PO_4 (500 mL/L), H_2SO_4 (360 mL/L), triethanolamine 20 mL/L.	Gloss, bath composition/304, Sample dimensions: 25 × 25 × 1 mm.	Bath 1: (9 min, 31 A/dm^2 at 333, 353K)-reflection coefficient = 98%, bath 2: (46.5 A/dm^2)-max reflection coefficient 98%, bath 3: (9 and 12 min, 15.5 A/dm^2, 333K or 343K and 353K)-reflection coefficient = 99%.	[5]

Table 3. *Cont.*

EP Process Parameters	EP Bath	Type of Analyses/Material	Test Results	Source
$j = 50 \pm 2$, 1000 ± 10 A/dm^2, $T = 65 \pm 5$, 55 ± 5 °C.	1—H_3PO_4 (20% *v*/*v*), H_2SO_4 (80% *v*/*v*). 2—H_3PO_4 (80% *v*/*v*), H_2SO_4 (20% *v*/*v*).	Chemical composition of surface, Young modulus/304L, 316L, sample dimensions: 30 × 5 × 1 mm.	XPS analysis: Cr/Fe ratio after EP (1000 A/dm^2) = 1.5 in bath 1, Cr/Fe ratio after EP (1000 A/dm^2) = 2.7 in bath 2, Bath 1: Young modulus higher for EP (50A /dm^2), than for EP (1000 A/dm^2). Thickness of *h* layer for 316L steel, after EP (50 A/dm^2): $PO_4^{3-} >> SO_4^{2-}$ $h = 10$ nm, $PO_4^{3-} > SO_4^{2-}$ $h = 3$ nm, PO_4^{3-} $h = 12$ nm, after EP (1000 A/dm^2): $PO_4^{3-} \approx SO_4^{2-}$ $h = 7$ nm, $PO_4^{3-} > SO_4^{2-}$ $h = 3$ nm, PO_4^{3-} $h = 3$ nm, Bath 2: Young modulus higher after EP (1000 A/dm^2), than after EP (50 A/dm^2). Thickness of *h* layer for 316L steel, after EP (50 A/dm^2): $PO_4^{3-} >> SO_4^{2-}$ $h = 10$ nm, $PO_4^{3-} > SO_4^{2-}$ $h = 5$ nm, PO_4^{3-} $h = 10$ nm, after EP (1000 A/dm^2): $PO_4^{3-} = 3$ nm, $PO_4^{3-} >> SO_4^{2-} = 17$ nm, $PO_4^{3-} = 15$ nm.	[77,78]
$I = 1$ A, $j = 12.9$ A/dm^2, $T = 55 \pm 5$ °C, $S = 7.78$ cm^2.	H_3PO_4, H_2SO_4, H_2O.	Chemical composition of surface/316L, sample dimensions: 19 × 19 × 0.737 mm.	Increase in at.% Cr content in the top layer from 16% after mechanical processing to 20% after EP. Decrease in at.% Fe content in the top layer from 18% after mechanical processing to 10% after EP.	[79]
$j = 0.29$ A/dm^2, $t = 1$ h.	H_3PO_4 (430 mL), H_2SO_4 (25 mL), CrO_3 (30 g).	Chemical composition of the surface/316.	AES SS: at.% Fe = 27%, pickled: at.% Fe = 25%, sanded: at.% Fe = 20%, pickled + EP: at.% Fe = 16%, sanded + EP: at.% Fe = 30%, SS: at.% Cr = 7%, pickled: at.% Cr = 8%, sanded: at.% Cr = 5%, pickled + EP: at.% Cr = 3%, sanded + EP: at.% Cr = 16%, SS: at.% Ni = 4%, pickled: at.% Ni = 4%, sanded: at.% Ni = 3%, pickled + EP: at.% Ni ≤ 2%, sanded + EP: at.% Ni = 3%, sanded: at.% Si = 22%.SS: at.% C = 4%, pickled: at.% C = 15%, sanded: at.% C = 25%, SS: at.% O = 50%, pickled: at.% O = 45%, sanded: at.% O = 16%, pickled + EP: at.% O = 56%, sanded + EP: at.% O = 40%.	[80]
$U = 40$ V, $t = 75$ s.	$HClO_4$ (20% *v*/*v*), CH_3COOH (80% *v*/*v*).	Chemical composition of the surface/316L Sample dimensions: 12 × 5 × 3 mm.	EDS SS: at.% Fe = 33.21 ± 0.89%, after MP (mechanical processing): at.% Fe = 34.72 ± 3.22%, after EP: at.% Fe = 31.95 ± 2.72%, SS: at.% Cr = 1.18 ± 0.03%, after MP: at.% Cr = 1.21 ± 0.02%, after EP: at.% Cr = 1.31 ± 0.19%, SS: at.% Ni = 2.32 ± 0.35%, after MP: at.% Ni = 2.72 ± 0.41%, after EP: at.% Ni = 4.63 ± 1.00% SS: at.% O = 63.12 ± 0.9%, after MP: at.% O = 61.35 ± 2.83%, after EP: at.% O = 62.10 ± 3.89%.	[81]
$j = 500$ A/dm^2, 1000 A/dm^2, $T = 60 \pm 1$ °C.	H_3PO_4, H_2SO_4.	Chemical composition of the surface/316L Sample dimensions: 25 × 5 × 1 mm.	XPS: Cr/ΣE4 ratio (Fe, Cr, Ni, O) after EP (500 A/dm^2): Cr/ΣE4 = 0.0807, Cr/ΣE4 ratio (Fe, Cr, Ni, O) after MEP (225/50 A/dm^2): Cr/ΣE4 = 0.0397, Cr/ΣE4 ratio (Fe, Cr, Ni, O) after MEP (225/1000 A/dm^2): Cr/ΣE4 = 0.0673.	[82,83]
$j = 50 \pm 2$ A/dm^2, 1000 ± 10 A/dm^2, $T = 65 \pm 5$, 75 ± 5 °C	1—H_3PO_4 (60% *v*/*v*), H_2SO_4 (40% *v*/*v*) 2—H_3PO_4 (40% *v*/*v*), H_2SO_4 (60% *v*/*v*)	Chemical composition of the surface/316L Sample dimensions: 30 × 5 × 1 mm.	XPS: Fe2p signal lower after EP (1000 A/dm^2) in bath 1 than after EP (50 A/dm^2), Min. Fe content (at.% Fe = 2.5%) after EP (1000 A/dm^2) in bath 1, at.% Cr2p (Cr6+) after EP: (50 A/dm^2) = 1.6%, at.% Cr2p (Cr6+) after EP: (1000 A/dm^2) = 23.2%.	[84]
$U = 10$ V, $T = 60$ °C.	H_2SO_4, H_3PO_4—vol. 1:3.	Chemical composition of the surface/316LVM.	XPS: after EP: at.% Cr = 1.8%, after MEP (magnetic electrochemical polishing): at.% Cr = 10.1%, after EP: at.% Mo = 0.4%, after MEP: at.% Mo = 1.9%, after EP: at.% Fe = 15.4%, after MEP: at.% Fe = 9.7%, after EP: at.% S = 1.5%, after MEP: at.% S = 3.9%, after EP: at.% P = 17.2%, after MEP: at.% P = 16.4%.	[85,86]

Table 3. *Cont.*

EP Process Parameters	EP Bath	Type of Analyses/Material	Test Results	Source
I = 2.5 A, j = 65 ± 5 A/dm^2, T = 65 ± 5 °C, t = 3 min, S = 3.854 cm^2, q = 32 mAh/cm^2.	H_3PO_4, H_2SO_4—vol. 4:6.	Chemical composition of the surface/430 SS, Sample dimensions: 30 × 5 × 1.22 mm.	XPS: after MP (mechanical processing) Cr/Fe ratio = 0.4, after EP: Cr/Fe ratio = 2.25, after EP (electrochemical polishing with mixing) Cr/Fe ratio = 0.96, after MEP (magnetic electrochemical polishing) Cr/Fe ratio = 0.7.	[87]
j = 50 ± 1 A/dm^2, 1000 ± 10 A/dm^2, T = 65 ± 5 °C, 55 ± 5 °C.	H_3PO_4 (20% v/v), H_2SO_4 (80% v/v).	Chemical composition of the surface/2205 duplex steel. Sample dimensions: 30 × 5 × 1 mm.	Fe2p signal lower after EP (1000 A/dm^2) than after EP (1000 A/dm^2), after EP (50 A/dm^2) Cr/Fe ratio = 1.9, after EP (1000 A/dm^2) Cr/Fe ratio = 1.7, after EP (50 A/dm^2) P/S ratio = 0.5, after EP (1000 A/dm^2) P/S ratio = 0.3.	[88]
U = 8 V, 9.5 V; t = 15 min, 20 min.	H_2SO_4 96%, H_3PO_4 85% —mas. 1:1.	Hardness/316L.	Vickers microhardness SS: VHN = 168.27 kg/cm^2, after sanding VHN = 283 kg/cm^2, after sanding and EP VHN < 205 kg/cm^2.	[89]
j = 20 A/dm^2, T = 55 ± 2 °C, t = 2–4 min S = 20 cm^2, q = 7–13 mAh/cm^2.	H_3PO_4 (35% v/v), H_2SO_4 (60.5% v/v), triethanolamine (4.5% v/v.).	Hardness/316L Sample dimensions: 80 × 20 × 2 mm.	Vickers microhardness *SS*: 188 HV, after MP: 318 HV, after EP samples were characterized by lower microhardness 230 HV.	[90]
j = 4–8 A/dm^2, T = 35–55 °C, t = 15–45 min.	H_3PO_4, H_2SO_4, triethanolamine (3 wt.%).	Surface defects, baths containing/304 surface area of approx. in industrial conditions 33.3 dm^2, in laboratory conditions 0.4 dm^2.	Not optimal variants of process parameters (current density, temperature) led to the emergence of defects on the surface of the electrochemical polished samples. Poor quality of surface was reflected high roughness results, exceeding 0.24 µm and even reaching 0.55 µm, and low gloss values, below 500 GU. Recommended parameters: T = 35 °C, j = 8 A/dm^2.	[46]

Current—*I* (A), voltage—*U* (V), current density—*j* (A/cm^2, A/dm^2), electric load—*q* (Ah/cm^2), temperature—*T* (°C, K), time—*t* (s, min, h), equals the composition of the electrochemical polishing baths and main results of the tests. The specific publications mainly analysed the following: surface roughness (with use of AFM or profile meter), chemical composition of the surface (EDX, AES, and XPS), surface gloss, corrosion resistance, chemical composition of the electropolishing bath, adhesion of cells/bacteria to the processed surface, and the hardness of sample surfaces. Experiments were conducted for stainless steels: 304, 316L, 316LVM, 430, and duplex 2205. Measured 2D and 3D roughness indicators:

- *Ra* (μm, nm)—mean arithmetic deviation of surface profile from the average line measured along the measurement or elementary section;
- *Rq* (μm, nm)—root mean square deviation of surface profile from the average line measured along the measurement or elementary section;
- *Rz* (μm, nm)—maximum height of roughness from the average line measured along the measurement or elementary section;
- *Rp* (μm, nm)—maximum profile peak height;
- *Rv* (μm, nm) —depth of the deepest profile indentation;
- *Sa* (μm, nm) —mean arithmetic deviation of surface roughness from the reference plane;
- *Srms* (μm, nm) —root mean square surface roughness.

Corrosion tests indicators:

- *OCP* (V)—open circuit potential;
- *Ecor* (mV, V)—corrosion potential;
- *Epit* (V)—pitting potential;
- *Vp* (mm/y)—corrosion rate.

Raw/ground samples were designated as SS, samples after mechanical polishing as after MP, after electrochemical polishing after EP, after magnetic electrochemical polishing—after MEP.

In the studies presented above, the surface quality and the surface smoothening effect were usually controlled by analysing the roughness profile *Ra* (nm, μm). After electrochemical polishing, it was possible to obtain lower *Ra* values for raw samples and samples previously subjected to mechanical or chemical processing. Electropolishing lowered the surface roughness *Ra* of samples of implants made from 316L steel from 18.4 ± 4.5 nm (AFM) and 183.1 ± 80.6 nm (profile meter) to 2.1 ± 0.8 nm and 76.2 ± 67.9 nm [58]. Chemical pickling and mechanical processing also reduced the roughness parameter *Ra*, compared to the initial sample—SS: *Ra* = 120.52 ± 26.65 nm, after pickling: *Ra* = 126.07 ± 37.13 nm, after relaxing: *Ra* = 142.71 ± 26.20 nm, after EP: *Ra* = 13.13 ± 1.56 nm [61]. Lower *Ra* parameter of *EP* samples (*Ra* < 0.11 μm), after previous pickling (*Ra* = 0.37–0.43 μm) compared to the initial samples (*Ra* = 0.17–0.25 μm) was obtained in publication [17].

The chemical composition of samples was mainly analysed using X-ray photoelectron spectroscopy (XPS). The sample surface was analysed both in the initial state, directly after processing or after additional cleaning with Ar^+ ions. Testing revealed that the composition of the passive layer after EP was enriched with chromium (chromium oxides and hydroxides) whose content, for the SS sample, was: at.% Cr = 5.7% at. and after EP: at.% Cr = 11.5% at. [60]. A similar correlation was obtained in other studies, where an increase in at.% Cr was observed from 16% (mechanically processed sample) to 20% after EP, along with a decrease in the Fe (iron oxides) content in the top layer: after mechanical processing: at.% Fe = 18%, after EP: at.% Fe = 10% [79]. The low ratio of iron oxides content to the total chromium oxide and hydroxide content may contribute to improved corrosion resistance of the passive layers obtained as a result of EP.

Qualitative analysis of the chemical composition of the surface was also performed using energy-dispersive X-ray spectroscopy (EDS) [76]. The chromium content in the passive layer increased from 18.83 wt.% to 19.33 wt.%, after electrochemical polishing. The tests also analysed surface gloss before and after EP. After initial preparation of samples (grinding) gloss $G = 400$, while after EP the gloss increased to $G = 1700–2500$.

Corrosion resistance was determined in potentiodynamic studies [59,71,72,74,75]. Based on results, it was determined that the corrosion resistance of 304 and 316L stainless steel improved significantly. High chromium content in both these types of steel enables the formation of the passive layer that protects the material against a corrosive environment on the surface of elements. Polarisation measurements revealed a noticeable improvement in pitting corrosion resistance and a shift in the pitting potential towards higher potential values. For samples after EP, *Epit* was 0.57 V and it was noticeably higher than for pickled samples *Epit* = 0.38 V and raw samples *Epit* = 0.30 V (versus satd. calomel electrode) [71,72]. In other studies, *Ecor* for mechanically processed surface was *Ecor* = −0.533 V, while after EP: *Ecor* = −0.324 V [74,75]. A similar correlation was found in study [59]. After pickling *Ecor* was −0.326 V, and after electrochemical polishing *Ecor* = −0.173 V. The presented research results allow us to state that electropolishing improves corrosion resistance compared to samples that were raw, pickled or mechanically processed.

Another interesting research project analysed the adhesion of blood cells to the surface of samples from 316L-SS steel, after electrochemical polishing, at direct current voltage $U = 2.5$, 4 and 10 V ($T = 65–70$ °C, $t = 3$ min) [64]. Along with the decrease in *SS* surface roughness: $Ra = 188 \pm 9$ nm, EP (2.5 V) $Ra = 107 \pm 6$ nm, EP (4 V) $Ra = 77 \pm 4$ nm, EP (10 V) $Ra = 97 \pm 11$ nm and increase in corrosion resistance: SS: *OCP* = 0.34 V, EP (2.5 V) *OCP* = 0.29 V, EP (4 V) *OCP* = 0.18 V, reduced adhesion of blood cells to the surface of 316L steel was observed for samples after EP: at 2.5 V by 71%, at 4 V by 89% and at 10 V by 93%. This study permits us to state that electropolishing significantly contributes to the reduced adhesion of blood cells to the surface of the processed elements (stents), which has a significant influence on thrombogenicity.

Electrochemical polishing also reduces the adhesion of bacteria to the surface of stainless-steel elements used in heavy industry, food industry, pharmaceutical and medical industry [91–94]. The adhesion of bacteria to the surface of devices (the formation of biofilm) may result in the contamination of the product by penetration of pathogens from the biofilm [95–97] and in damage or destruction of the stainless-steel surface [98,99]. After EP, the number of bacteria on the surface of stainless steel was reduced by 79% and 94% [91]. Electrochemical processing may have a significant influence on enhancing resistance to the formation of bacterial biofilm on the surface of devices [100].

The electropolishing process also influences the microhardness of the processed elements [89,90]. The microhardness of raw samples was 168.27 kg/cm^2, after sanding −283 kg/cm^2, while after sanding and EP < 205 kg/cm^2. The above results indicate that the microhardness of material subjected to mechanical processing is higher than after electrochemical polishing. Sanding causes cold-work hardening, which has a significant influence on the microhardness results. Compression after mechanical processing may be partly removed by electropolishing. Samples after mechanical processing were characterised by higher microhardness than samples after mechanical processing and electrochemical polishing.

Surface defects were presented after the electropolishing process in industrial conditions [46]. A bath containing H_3PO_4 and H_2SO_4 and triethanolamine contaminated in industrial conditions was used for the tests, and the iron ions contamination was about 3% by wt. Samples with an area of approximately 33.3 dm^2 were used in the industrial tests. For comparison, laboratory-scale tests were carried out with the area of the samples 0.4 dm^2. The results of laboratory tests make it possible to estimate the weight loss that will be obtained in industrial conditions, taking into account the use of the same electrolyte. Due to the limitations in industrial conditions and repeated electropolishing of large-size work pieces, the current density range used was 4–8 A/dm^2, the temperature range 35–55 °C.

The best effects were obtained for the low temperature and current density 8 A/dm^2. The combination of low current density and high temperature resulted in an uneven polishing of the sample and "orange peel effect". On one hand the surface is polished, on the other this kind of textural irregularities are similar to the skin of an orange (Figure 4). Besides, the "orange peel effect" occurs in metals and alloys with a thick crystalline structure, because the oxide film, adsorption film and diffusion film, which shift the anode's potential far from the thermodynamic state characteristic of the anodic etching process, are not able to overcome the effect of the privileged influence for the digestion anode places in the case of metal digestion under thermodynamic conditions, called active centres. This is the case when the surface is shiny but not smooth.

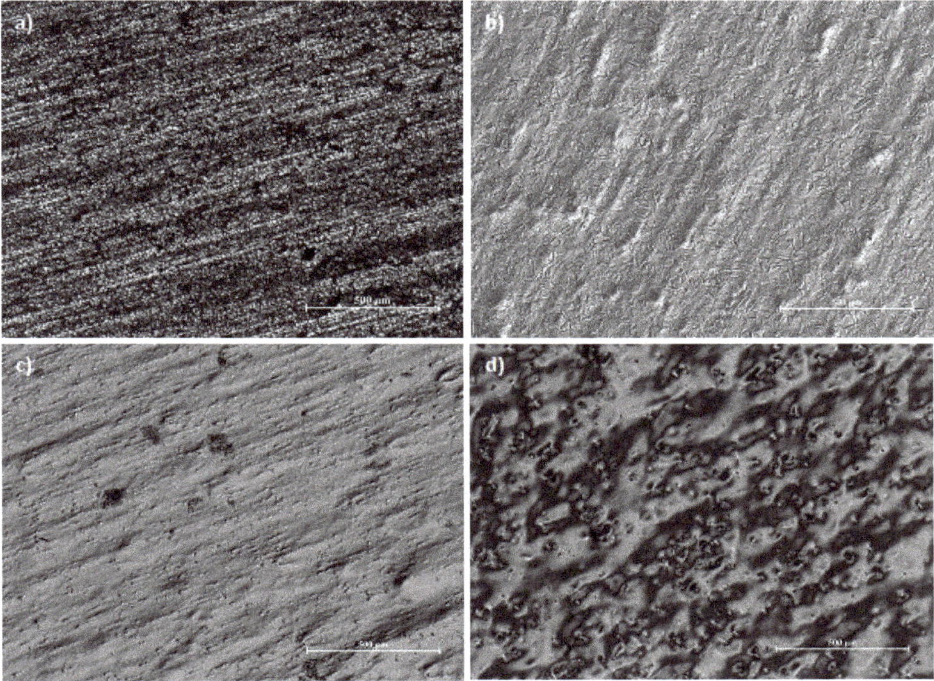

Figure 4. Metallographic photos of austenitic chromium–nickel stainless steel: (**a**) as-received, 2B surface finish, (**b**) high gloss surface after EP (T = 35 °C, 8 A/dm^2), (**c**) edge of the sample with high-gloss surface after EP (T = 55 °C, j = 4 A/dm^2), (**d**) internal area of the sample with orange peel effect after EP (T = 55 °C, j = 4 A/dm^2).

In Reference [101], defects were presented such electropolishing streaks (oxygen formation and movement). After the electropolishing process, the surface may contain certain defects, including those resulting from long-term use of electrolyte, improper current density, or local overheating of the electrolyte. Defects may be a result of non-optimal parameters and bath composition and contamination. The most common defects include orange peel effect, shadows, smudges, streaks, and uneven polishing of the material [102].

The electropolishing process parameters, bath composition and contaminations have a major role in the possibility of the occurrence of defects. Also, the selection of optimal process parameters mainly depends on the type of electrolyte used and the degree of its contamination. Undoubtedly, mixtures of H_3PO_4 and H_2SO_4 are most often used, sometimes with the addition of organic additives, e.g., glycerine. When choosing process parameters, the shape and size of electropolished samples should also be considered. Some researchers even used high temperatures 75–95 °C for electrochemical polishing of small samples. When electropolishing large-size workpieces and the need to use the lowest possible current densities, often in the range of 4–5 A/dm^2, depending on the rate of electrolyte contamination, temperatures below 50 °C could be also used.

About the right course for the proper conduct of the electropolishing process and other electrochemical processes, decisive is the value of the electrode potential, and the process parameters must be selected so that the anode potential has the desired value. Therefore, for large-size workpieces, high values of the current supplied to the anode must be used, which requires an appropriate power supply and may cause "burns" on the workpiece being processed. The places occupied by the current contacts, of course, remain raw because there is no processing. Designing such a bath so that low current densities can be used obviously requires lower currents at the same surface size, which limits these adverse effects. Baths that require lower current densities are always better.

The paper focused on the results of electrochemical polishing of stainless-steel parameters and the effect of the process on surface roughness, chemical composition, corrosion resistance, adhesion of blood cells, adhesion of bacteria and influences the microhardness of the processed elements. The paper could help in selecting the best EP process parameter—depending on what the final part of stainless steel is being used for and which surface quality meets individual expectations.

6. Conclusions

The large number of hypotheses concerning the electrochemical polishing mechanism results from the complexity of the process. Apart from the anodic diffusion layer of high density and electric resistance, also called the viscous layer, thatt is formed in the proximity of the anode, an oxide layer is also formed on the surface of the anode itself. The presence of these layers limits the processes of diffusion of water particles and migration of hydrated anions and cations in the electric field. A common phenomenon that results from the simultaneous formation of both these layers is the process of diffusion of water particles and migration of anions towards the anode as well as the migration of anodic dissolution products away from the anode.

Water particles and anions, which are referred to as the acceptors of the anodic dissolution process, are necessary in the process of electrochemical formation and chemical dissolution of the oxides that constitute the passive layer. Due to the competitiveness of both these processes, even small differences in the rate of migration and diffusion of the acceptors of the dissolution process may lead to significant differences in the rate of dissolution of individual elements of the anode surface structure.

For designing the composition of the electrochemical polishing bath solution and selecting the parameters of the electrolytic polishing process, it is beneficial to assume the occurrence of three, at the same time, independent physicochemical processes:

- anodic passivation (formation of an oxide film);
- adsorption of surface-active substances;
- formation of a diffusion layer with increased viscosity and density and reduced water content concerning the rest of the solution.

The most popular additions to phosphate–sulphate solutions used as electropolishing baths are glycerol (glycerine) or chromic acid anhydride, CrO_3. Glycerol increases the viscosity and density of the solution, which increases the thickness of the diffusion film and the greater distance of the anode potential from the thermodynamic value. Glycerol also reduces the relative concentration of water in the near-anode area and as a substance with active oxygen atoms forms an adsorption film on the surface of the anode. Chromic anhydride also increases the viscosity and density of the solution but also has oxidizing properties, which facilitate the formation of the oxide film. Of course, chromium anhydride and glycerine should not be added at the same time, as there would be a reaction between the oxidant (chromic acid) and the reducing agent (glycerine).

The main method of modifying the surface (preceding mechanical polishing and electrochemical polishing) or the final processing of metal elements is abrasive processing. Mechanical processing (machining, blast cleaning, or tumbling) cannot ensure the desired smoothness and visual properties of elements of a complex internal structure. After electropolishing, samples are characterised by lower roughness *Ra* and higher corrosion resistance. Electrochemical polishing reduces the adhesion of cells and bacteria to the processed surfaces, positively influencing the usability and durability of the devices. Errors in designing and assembling devices made from 304 and 304L stainless steel used in constructing technological equipment installed in wastewater treatment plants may result in a fast rate of corrosion. The advantage of electropolishing is that it may be applied both as a finishing process and after previous mechanical or chemical processing.

Author Contributions: Conceptualization, E.Ł.-W.; formal analysis, E.Ł.-W.; investigation, E.Ł.-W., P.L. and G.N.; data curation, E.Ł.-W.; writing—original draft preparation, E.Ł.-W.; writing—review and editing, P.L. and G.N.; visualization, E.Ł.-W.; supervision, G.N. and P.L.; project administration, P.L.; funding acquisition, P.L. All authors have read and agreed to the published version of the manuscript.

Funding: This research was funded by the National Centre for Research and Development (NCBR) within the project IonsMonit as part of the Lider programme (LIDER/22/0187/L-7/15/NCBR/2016). The publication stage was supported by the Leading Reasarch Groups support project from the subsidy increased for the period 2020–2025 in the amount of 2% of the subsidy referred to Art. 387 (3) of the Law of 20 July 2018 on Higher Education and Science, obtained in 2019.

Conflicts of Interest: The authors declare no conflict of interest.

References

1. Clerc, C.; Datta, M.; Landolt, D. On the theory of anodic levelling: Model experiments with triangular nickel profiles in chloride solution. *Electrochim. Acta* **1984**, *29*, 1477–1486. [CrossRef]
2. Taylor, E.J.; Inman, M. Electrochemical Surface Finishing. *Electrochem. Soc. Interface* **2014**, *23*, 57–61. [CrossRef]
3. Bagdach, S. *Poradnik Galwanotechnika. Praca Zbiorowa*; Wydawnictwo Naukowo-Techniczne: Warszawa, Poland, 2002.
4. Kao, P.; Hocheng, H. Optimization of electrochemical polishing of stainless steel by grey relational analysis. *J. Mater. Process. Technol.* **2003**, *140*, 255–259. [CrossRef]
5. Jeyashree, G.; Subramanian, A.; Vasudevan, T.; Mohan, S.; Venkatachalam, R. Electropolishing of stainless steel. *Bull. Electrochem.* **2000**, *16*, 388–391.
6. Dobrev, T.; Pham, D.T.; Dimov, S. *Electrochemical Polishing: A Technique for Surface Improvements after Laser Milling*; Manufacturing Engineering Centre, Cardiff University: Cardiff, UK, 2006.
7. Niveen, J.A.; Hussain, M. Study of Electrochemical Polishing Applications in some alloys to obtain high surface finish. In Proceedings of the 2012 International Conference on Industrial Engineering and Operations Management, Istanbul, Turkey, 3–6 July 2012.
8. Kaladhar, M.; Subbaiah, K.V.; Rao, C.H.S. Machining of austenitic stainless steels—A review. *Int. J. Mach. Mach. Mater.* **2012**, *12*, 178–192. [CrossRef]
9. Basmaci, G.; Ay, M. Optimization of Cutting Parameters, Condition and Geometry in Turning AISI 316L Stainless Steel Using the Grey-Based Taguchi Method. *Acta Phys. Pol. A* **2017**, *131*, 354–359. [CrossRef]
10. Zhao, O.; Afshan, S.; Gardner, L. Structural response and continuous strength method design of slender stainless steel cross-sections. *Eng. Struct.* **2017**, *140*, 14–25. [CrossRef]
11. Lee, S.-J.; Lai, J.-J. The effects of electropolishing (EP) process parameters on corrosion resistance of 316L stainless steel. *J. Mater. Process. Technol.* **2003**, *140*, 206–210. [CrossRef]
12. Ziemniak, S.; Hanson, M. Corrosion behavior of 304 stainless steel in high temperature, hydrogenated water. *Corros. Sci.* **2002**, *44*, 2209–2230. [CrossRef]
13. Ziemniak, S.E.; Hanson, M.; Sander, P.C. Electropolishing effects on corrosion behavior of 304 stainless steel in high temperature, hydrogenated water. *Corros. Sci.* **2008**, *50*, 2465–2477. [CrossRef]
14. Yang, G.; Wang, B.; Tawfiq, K.; Wei, H.; Zhou, S.; Chen, G. Electropolishing of surfaces: Theory and applications. *Surf. Eng.* **2016**, *33*, 149–166. [CrossRef]
15. Tam, S.; Loh, N.; Mah, C.; Loh, N. Electrochemical polishing of biomedical titanium orifice rings. *J. Mater. Process. Technol.* **1992**, *35*, 83–91. [CrossRef]
16. Simka, W.; Nawrat, G.; Chłodek, J.; Maciej, A.; Winiarski, A.; Szade, J.; Radwański, K.; Gazdowicz, J. Electropolishing and anodic passivation of Ti6Al7Nb alloy. *Przemysł Chem.* **2011**, *90*, 84–90.
17. Lochyński, P.; Łyczkowska, E.; Pawełczyk, A.; Szczygieł, B. Effect of bath exploitation on steel electropolishing process efficiency. *Przemysł Chem.* **2012**, *91*, 846–848.

18. Nawrat, G.; Bołd, T.; Simka, W.; Waś, J.; Gonet, M.; Gardela, A.; Nieużyła, Ł. The influence of surface treatment of coronary stents on their corrosion resistance. *Ochr. Przed Korozją* **2012**, *5*, 256–261.
19. Raman, S.G.S.; Padmanabhan, K.A. Effect of electropolishing on the room temperature low-cycle fatigue bahaviour AISI 304LN stainless steel. *Int. J. Fatigue* **1995**, *17*, 179–182. [CrossRef]
20. Nawrat, G. *Elektrochemiczne Metody Inżynierii Powierzchni, Monograph*; Wydawnictwo Politechniki Śląskiej: Gliwice, Poland, 2010.
21. *PN-EN 1672:2009 Food Processing Machinery—Basic Concepts-Part 2: Hygiene Requirements*; Polish Committee for Standardization: Warsaw, Poland, 2009.
22. *PN-EN 14630:2013 Non-Active Surgical Implants—General Requirements*; Polish Committee for Standardization: Warsaw, Poland, 2013.
23. Metz, F.I. Electropolishing of metals. Ph.D. Thesis, Iowa State University, Ames, IA, USA, 1960.
24. Smith, E.L.; Abbott, A.P.; Ryder, K.S. Deep Eutectic Solvents (DESs) and Their Applications. *Chem. Rev.* **2014**, *114*, 11060–11082. [CrossRef]
25. Jacquet, P.A. Electrolytic Method for obtaining Bright Copper Surfaces. *Nature* **1935**, *135*, 1076. [CrossRef]
26. Lee, S.J.; Chen, Y.H.; Hung, J.H. The Investigation of Surface Morphology Forming Mechanisms in Electropolishing Process. *Int. J. Electrochem. Sci.* **2012**, *7*, 12495–12506.
27. Buhlert, M. *Elektropolieren*; Eugen, G., Ed.; Leuze Verlag: Bad Saulgau, Germany, 2009.
28. Datta, M.; Landolt, D. Fundamental aspects and applications of electrochemical microfabrication. *Electrochim. Acta* **2000**, *45*, 2535–2558. [CrossRef]
29. Landolt, D.; Chauvy, P.-F.; Zinger, O. Electrochemical micromachining, polishing and surface structuring of metals: Fundamental aspects and new developments. *Electrochim. Acta* **2003**, *48*, 3185–3201. [CrossRef]
30. Maltosz, M. Modeling of impedance mechanisms in electropolishing. *Electrochim. Acta* **1995**, *40*, 393–401.
31. Hryniewicz, T. *Physico-Chemical and Technological Fundamentals of Electropolishing Steels (Fizykochemiczne i Technologiczne Podstawy Procesu Elektropolerowania Stali)*; Wyższa Szkoła Inżynierska w Koszalinie: Koszalin, Poland, 1989.
32. Ciszewski, A. *Technologia Chemiczna, Procesy Elektrochemiczne*; Wydawnictwo Politechniki Poznańskiej: Poznań, Poland, 2008.
33. Lin, C.-C.; Hu, C.-C. Electropolishing of 304 stainless steel: Surface roughness control using experimental design strategies and a summarized electropolishing model. *Electrochim. Acta* **2008**, *53*, 3356–3363. [CrossRef]
34. Lin, C.-C.; Hu, C.-C.; Lee, T.-C. Electropolishing of 304 stainless steel: Interactive effects of glycerol content, bath temperature, and current density on surface roughness and morphology. *Surf. Coat. Technol.* **2009**, *204*, 448–454. [CrossRef]
35. Hryniewicz, T. *Wstęp do Obróbki Powierzchniowej Biomateriałów Metalowych*; Wydawnictwo Uczelniane Politechniki Koszalińskiej: Koszalin, Poland, 2007.
36. Available online: https://www.hanser-elibrary.com/doi/pdf/10.12850/9783874803052.fm (accessed on 2 June 2020).
37. Faust, C.L.; Makio, S. Surface Preparation by Electropolishing. *J. Electrochem. Soc.* **1949**, *95*, 62C–72C. [CrossRef]
38. Lee, E.S. Machining Characteristics of the Electropolishing of Stainless Steel (AISI 316L). *Int. J. Adv. Manuf. Technol.* **2000**, *16*, 591–599. [CrossRef]
39. Nazneen, F.; Galvin, P.; Arrigan, D.; Thompson, M.; Benvenuto, P.; Herzog, G. Electropolishing of medical-grade stainless steel in preparation for surface nano-texturing. *J. Solid State Electrochem.* **2011**, *16*, 1389–1397. [CrossRef]
40. Jullien, C.; Benezech, T.; Carpentier, B.; Lebret, V.; Faille, C. Identification of surface characteristics relevant to the hygienic status of stainless steel for the food industry. *J. Food Eng.* **2003**, *56*, 77–87. [CrossRef]
41. Zaborski, S. *Obróbka Elektrochemiczno-Ścierna: Podstawy i Zastosowania*; Oficyna Wydawnicza Politechniki Wrocławskiej: Wrocław, Poland, 2007.
42. Chen, S.; Tu, G.; Huang, C.A. The electrochemical polishing behavior of porous austenitic stainless steel (AISI 316L) in phosphoric-sulfuric mixed acids. *Surf. Coat. Technol.* **2005**, *200*, 2065–2071. [CrossRef]
43. Bhuyan, A.; Gregory, B.; Lei, H.; Yee, S.Y.; Gianchandani, Y.B. Pulse and DC Electropolishing of Stainless Steel for Stents and Other Devices. In Proceedings of the IEEE Sensors 2005, Irvine, CA, USA, 30 October–3 November 2005.
44. Núñez, P.J.; Plaza, E.G.; Prada, M.H.; Coronel, R.T.; López, P.J.N. Electrolyte Effect on the Surface Roughness Obtained by Electropolishing of AISI 316L Stainless Steel. *Mater. Sci. Forum* **2014**, *797*, 133–138. [CrossRef]

45. Maitak, G.P.; Yudenkova, I.N.; Pasechnik, M.G.; Drozd, N.A. Electrochemical Polishing Solution. USSR Patent No. 510537A1, 15 April 1976.
46. Lochyński, P.; Charazińska, S.; Łyczkowska-Widłak, E.; Sikora, A. Electropolishing of Stainless Steel in Laboratory and Industrial Scale. *Metals* **2019**, *9*, 854. [CrossRef]
47. Alekseev, G.I.; Zot'eva, G.A.; Golovanov, V.N.; Boitsova, T.V.; Kuznetsov, E.A. Electrolyte for Polishing Stainless Steels. USSR Patent No. 396428A1, 29 August 1973.
48. Hensel, K.B. Electropolishing. *Met. Finish.* **2000**, *98*, 440–446. [CrossRef]
49. Hryniewicz, T. Surface Electrochemistry for Materials and Mechanical Engineering. In Proceedings of the International Science Conference "Challenges to Civil and Mechanical Engineering in 2000 and Beyond", Wrocław, Poland, 2–5 June 1997.
50. Taguchi, C. Electrolytic Solution for Use in Electropolishing Process for Stainless Steel. Japan Patent No. 2007332416A, 27 December 2007.
51. Taguchi, C. Electrolytic Solution to Be Used for Electrolytic Polishing Method for Stainless Steel. Japan Patent No. 2007231413A, 13 September 2007.
52. Gellér, Z.E.; Albrecht, K.; Dobránszky, J. Electropolishing of Coronary Stents. *Mater. Sci. Forum* **2008**, *589*, 367–372. [CrossRef]
53. Faust, C.L. Electropolishing-Stainless Steel. *Met. Finish.* **1982**, *80*, 89–93.
54. Faust, C.L. Electropolishing-Stainless Steel, Part I. *Met. Finish.* **1982**, *80*, 21–25.
55. Faust, C.L. Electropolishing-Stainless Steel, Part II. *Met. Finish.* **1982**, *81*, 53–56.
56. Faust, C.L. Electropolishing, Carbon and Low Alloy Steel, Part I. *Met. Finish.* **1983**, *81*, 47–51.
57. Abbott, A.P.; Capper, G.; McKenzie, K.J.; Ryder, K.S. Voltammetric and impedance studies of the electropolishing of type 316 stainless steel in a choline chloride based ionic liquid. *Electrochim. Acta* **2006**, *51*, 4420–4425. [CrossRef]
58. Eliaz, N.; Nissan, O. Innovative processes for electropolishing of medical devices made of stainless steels. *J. Biomed. Mater. Res. Part A* **2007**, *83*, 546–557. [CrossRef]
59. Sojitra, P.; Engineer, C.; Kothwala, D.; Raval, A.; Kotadia, H.; Mehta, G. Electropolishing of 316LVM Stainless Steel Cardiovascular Stents: An Investigation of Material Removal. Surface Roughness and Corrosion Behaviour. *Trends Biomater.* **2010**, *23*, 115–121.
60. Haïdopoulos, M.; Turgeon, S.; Sarra-Bournet, C.; Laroche, G.; Mantovani, D. Development of an optimized electrochemical process for subsequent coating of 316 stainless steel for stent applications. *J. Mater. Sci. Mater. Electron.* **2006**, *17*, 647–657. [CrossRef] [PubMed]
61. Zhao, H.; van Humbeeck, J.; Sohier, J.; de Scheerder, I. Electrochemical Polishing of 316L Stainless Steel Slotted Tube Coronary Stents: An Investigation of Material Removal and Surface Roughness. *Prog. Biomed. Res.* **2003**, *8*, 70–81.
62. Löber, L.; Flache, C.; Petters, R.; Kühn, U.; Eckert, J. Comparison of different post processing technologies for SLM generated 316l steel parts. *Rapid Prototyp. J.* **2013**, *19*, 173–179. [CrossRef]
63. Lochyński, P.; Kowalski, M.; Szczygieł, B.; Kuczewski, K. Improvement of the stainless steel electropolishing process by organic additives. *Pol. J. Chem. Technol.* **2016**, *18*, 76–81. [CrossRef]
64. Habibzadeh, S.; Li, L.; Shum-Tim, D.; Davis, E.C.; Omanovic, S. Electrochemical polishing as a 316L stainless steel surface treatment method: Towards the improvement of biocompatibility. *Corros. Sci.* **2014**, *87*, 89–100. [CrossRef]
65. Rahman, Z.U.; Deen, K.; Cano, L.; Haider, W. The effects of parametric changes in electropolishing process on surface properties of 316L stainless steel. *Appl. Surf. Sci.* **2017**, *410*, 432–444. [CrossRef]
66. Latifi, A.; Imani, M.; Khorasani, M.T.; Joupari, M.D. Electrochemical and chemical methods for improving surface characteristics of 316L stainless steel for biomedical applications. *Surf. Coat. Technol.* **2013**, *221*, 1–12. [CrossRef]
67. Łyczkowska, E.; Lochyński, P.; Chlebus, E. Electropolishing of a stainless steel. *Przemysł Chem.* **2013**, *92*, 1364–1366.
68. Hocheng, H.; Kao, P.; Chen, Y. Electropolishing of 316L Stainless Steel for Anticorrosion Passivation. *J. Mater. Eng. Perform.* **2001**, *10*, 414–418. [CrossRef]
69. Oravcová, M.; Palček, P.; Zatkalíková, V.; Tański, T.; Król, M. Surface treatment and corrosion behaviour of austenitic stainless steel biomaterial. *IOP Conf. Ser. Mater. Sci. Eng.* **2017**, *175*, 12009. [CrossRef]

70. Núñez, P.; García-Plaza, E.; Hernando, M.; Trujillo, R. Characterization of Surface Finish of Electropolished Stainless Steel AISI 316L with Varying Electrolyte Concentrations. *Procedia Eng.* **2013**, *63*, 771–778. [CrossRef]
71. Lochyński, P.; Łyczkowska, E.; Kuczewski, K.; Szczygieł, B. Pitting corrosion of pickled and electropolished Cr-Ni stainless steel. *Przemysł Chem.* **2014**, *93*, 762–765.
72. Lochyński, P.; Sikora, A.; Szczygieł, B. Surface morphology and passive film composition after pickling and electropolishing. *Surf. Eng.* **2016**, *33*, 395–403. [CrossRef]
73. Han, Y.; Mei, J.; Peng, Q.; Han, E.H.; Ke, W. Effect of electropolishing on corrosion of nuclear grade 316L stainless steel in deacerated high temperature water. *Corros. Sci.* **2016**, *112*, 625–634. [CrossRef]
74. Baron, A.; Simka, W.; Nawrat, G.; Szewieczek, D.; Krzyżak, A. Influence of electrolytic polishing on electrochemical behaviour of austenitic steel. *J. Achiev. Mater. Manuf. Eng.* **2006**, *18*, 55–58.
75. Baron, A.; Simka, W.; Nawrat, G.; Szewieczek, D. Electropolishing and chemical passivation of austenitic steel. *J. Achiev. Mater.* **2008**, *31*, 197–202.
76. Awad, A.; Ghazy, E.; El-Enin, S.A.; Mahmoud, M. Electropolishing of AISI-304 stainless steel for protection against SRB biofilm. *Surf. Coat. Technol.* **2012**, *206*, 3165–3172. [CrossRef]
77. Rokosz, K.; Hryniewicz, T.; Lukeš, J.; Sepitka, J. Nanoindentation studies and modeling of surface layers on austenitic stainless steels by extreme electrochemical treatments. *Surf. Interface Anal.* **2015**, *47*, 643–647. [CrossRef]
78. Rokosz, K.; Lahtinen, J.; Hryniewicz, T.; Rzadkiewicz, S. XPS depth profiling analysis of passive surface layers formed on austenitic AISI 304L and AISI 316L SS afterhigh-current-density electropolishing. *Surf. Coat. Technol.* **2015**, *276*, 516. [CrossRef]
79. Selvaduray, G.; Trigwell, S. Effect of Surface Treatment on the Surface Characteristics of AISI 316L Stainless Steel. In Proceedings of the Conference of Materials and Processes for Medical Devices, Boston, MA, USA, 14–18 November 2005.
80. Rao, T.V. Effect of surface treatments on near surface composition of 316 nuclear grade stainless steel. *J. Vac. Sci. Technol. A* **1986**, *4*, 1604–1607. [CrossRef]
81. Han, G.; Lu, Z.; Ru, X.; Chen, J.; Xiao, Q.; Tian, Y. Improving the oxidation resistance of 316L stainless steel in simulated pressurized water reactor primary water by electropolishing treatment. *J. Nucl. Mater.* **2015**, *467*, 194–204. [CrossRef]
82. Hryniewicz, T.; Rokosz, K. Analysis of XPS results of AISI 316L SS electropolished and magnetoelectropolished at varying conditions. *Surf. Coat. Technol.* **2010**, *204*, 2583–2592. [CrossRef]
83. Hryniewicz, T.; Rokosz, K.; Micheli, V. Auger/AES surface film measurements on AISI 316L biomaterial after magnetoelectropolishing. *Pomiary Autom. Kontrola* **2011**, *57*, 609–614.
84. Rokosz, K.; Hryniewicz, T.; Rzadkiewicz, S.; Raaen, S. High-current-density electropolishing (HDEP) of AISI 316L (EN 1.4404) stainless steel. *Tehnički Vjesnik* **2015**, *22*, 415–424. [CrossRef]
85. Rokosz, K.; Hryniewicz, T.; Rokicki, R. XPS Measurements of AISI 316LVM SS Biomaterial Tubes after Magnetoelectropolishing. *Tehnički Vjesnik* **2014**, *21*, 799–805.
86. Rokosz, K.; Hryniewicz, T.; Raaen, S. Cr/Fe Ratio by XPS Spectra of Magnetoelectropolished AISI 316L SS Fitted by Gaussian-Lorentzian Shape Lines. *Tehnički Vjesnik* **2014**, *21*, 533–538.
87. Hryniewicz, T.; Rokosz, K.; Raaen, S. XPS Measurements of AISI 430 SS Surface after Electropolishing Operations, in a Transpassive Region of Polarisation Characteristics. *Pomiary Autom. Kontrola* **2012**, *58*, 126–129.
88. Rokosz, K.; Hryniewicz, T.; Simon, F.; Rzadkiewicz, S. Comparative XPS analyses of passive layers composition formed on Duplex 2205 SS after standard and high-current-density electropolishing. *Tehnički Vjesnik* **2016**, *23*, 731–735. [CrossRef]
89. Suyitno, S. The Influence of Sandblasting and Electropolishing on the Surface Hardness of AISI 316L Stainless Steel. *Adv. Mater. Res.* **2014**, *896*, 517–520. [CrossRef]
90. Lyczkowska-Widlak, E.; Lochyński, P.; Nawrat, G.; Chlebus, E. Comparison of electropolished 316L steel samples manufactured by SLM and traditional technology. *Rapid Prototyp. J.* **2019**, *25*, 566–580. [CrossRef]
91. Arnold, J.W.; Boothe, D.H.; Suzuki, O.; Bailey, G.W. Multiple imaging techniques demonstrate the manipulation of surfaces to reduce bacterial contamination and corrosion. *J. Microsc.* **2004**, *216*, 215–221. [CrossRef] [PubMed]

92. Shih, C.-C.; Shih, C.-M.; Su, Y.-Y.; Su, L.H.J.; Chang, M.-S.; Lin, S.-J. Effect of surface oxide properties on corrosion resistance of 316L stainless steel for biomedical applications. *Corros. Sci.* **2004**, *46*, 427–441. [CrossRef]
93. Harris, L.; Meredith, D.O.; Eschbach, L.; Richards, R.G. Staphylococcus aureus adhesion to standard micro-rough and electropolished implant materials. *J. Mater. Sci. Mater. Electron.* **2007**, *18*, 1151–1156. [CrossRef] [PubMed]
94. Verran, J.; Rowe, D.; Cole, D.; Boyd, R. The use of the atomic force microscope to visualise and measure wear of food contact surfaces. *Int. Biodeterior. Biodegrad.* **2000**, *46*, 99–105. [CrossRef]
95. Gündüz, G.T.; Tuncel, G. Biofilm formation in an ice cream plant. *Antonie Leeuwenhoek* **2006**, *89*, 329–336. [CrossRef]
96. Poulsen, L.V. Microbial Biofilm in Food Processing. *LWT-Food Sci. Technol.* **1999**, *32*, 321–326. [CrossRef]
97. Bryers, J.D. Biofilms and the technological implications of microbial cell adhesion. *Colloids Surf. B Biointerfaces* **1994**, *2*, 9–23. [CrossRef]
98. Zottola, E.A. Microbial attachment and biofilm formation: A new problem for the food industry. *Food Technol.* **1994**, *48*, 107–114.
99. Myszka, K.; Czaczyk, K. Bacterial Biofilms on Food Contact Surfaces—A Review. *Pol. J. Food Nutr. Sci.* **2011**, *61*, 173–180. [CrossRef]
100. Lechevallier, M.W.; Babcock, T.M.; Lee, R.G. Examination and characterization of distribution system biofilms. *Appl. Environ. Microbiol.* **1987**, *53*, 2714–2724. [CrossRef]
101. Lochynski, P.; Charazińska, S.; Łyczkowska-Widłak, E.; Sikora, A.; Karczewski, M. Electrochemical Reduction of Industrial Baths Used for Electropolishing of Stainless Steel. *Adv. Mater. Sci. Eng.* **2018**, *2018*, 1–11. [CrossRef]
102. Gadali' nska, E.; Wronicz, W. Electropolishing procedure dedicated to in-depth stress measurements withx-ray diractometry. *Fatigue Aircr. Struct.* **2016**, *2016*, 65–72. [CrossRef]

© 2020 by the authors. Licensee MDPI, Basel, Switzerland. This article is an open access article distributed under the terms and conditions of the Creative Commons Attribution (CC BY) license (http://creativecommons.org/licenses/by/4.0/).

Article

Effect of Cooling Rate at the Eutectoid Transformation Temperature on the Corrosion Resistance of Zn-4Al Alloy

Marzena M. Lachowicz [1,*] and Robert Jasionowski [2]

[1] Faculty of Mechanical Engineering, Wroclaw University of Technology and Science, 50-370 Wrocław, Poland
[2] Faculty of Marine Engineering, Maritime University of Szczecin, 70-500 Szczecin, Poland; r.jasionowski@am.szczecin.pl
* Correspondence: marzena.lachowicz@pwr.edu.pl

Received: 18 March 2020; Accepted: 3 April 2020; Published: 5 April 2020

Abstract: The main purpose of this work was to experimentally determine the effect of the cooling rate during the eutectoid transformation on the corrosion resistance of a hypoeutectic Zn-4Al cast alloy in 5% NaCl solution. This was considered in relation to the alloy microstructure. For this purpose, metallographic and electrochemical studies were performed. It was found that the faster cooling promoted the formation of finer ($\alpha + \eta$) eutectoid structures, which translated into a higher hardness and lower corrosion current density. In the initial stage of corrosion processes the eutectoid structure in the eutectic areas were attacked. At the further stages of corrosion development, the phase η was dissolved, and the α phase appears to be protected by the formation of corrosion products.

Keywords: zinc alloys; Zamak; Zn-Al alloys; heat treatment; corrosion resistance; microstructure

1. Introduction

Zinc-based alloys have good tribological properties, relatively high mechanical strength and hardness values, and show good castability due to their low melting points. These features make them good candidates for use in automotive and electronics applications, and they have also been used in the production of small components and plain bearings. Studies have shown that these alloys have superior wear resistance to common copper-bearing alloys. Adhesion and smearing are the main wear mechanisms of zinc-based alloys, while abrasive wear is the predominant wear mode in bronzes [1–5]. In the last decade, zinc has been extensively studied as a potential biocompatible and biodegradable metal for medical applications [4,6–10].

Zinc has one of the lowest electrode potentials, and machines made from it are highly susceptible to electrochemical corrosion due to the formation of a galvanic cell. The presence of extensive corrosion may also affect other co-working components made of different materials. The resulting corrosion products affect the pH of the surrounding environment, which, in turn, may accelerate the degradation of lubricants. On the other hand, a low electrochemical potential gives zinc and its alloys broad application prospects as cathodic protection coatings. Thus, Zn-Al alloys may be used to replace traditional zinc galvanic coatings [11–18] The commercial Galfan alloy has found broad applications in this area [18] but it exhibits several serious drawbacks, including a low creep resistance, low shape stability associated with aging, insufficient corrosion resistance in acidic and alkaline environments, and a low cavitation erosion resistance [5,19–21]

The main alloying elements in Zn-based alloys are aluminum, magnesium, and copper. Cast Zn-Al alloys are commercially available under the Zamak tradename, the most popular of which is the Zamak 3 alloy which has a nominal composition of 4% Al. This aluminum content classifies this alloy as hypoeutectic (Figure 1) whose microstructure is composed of a η-Zn(Al) dendrite solid solution and (α+η) eutectic phases, in which the α solid solution is Al(Zn).

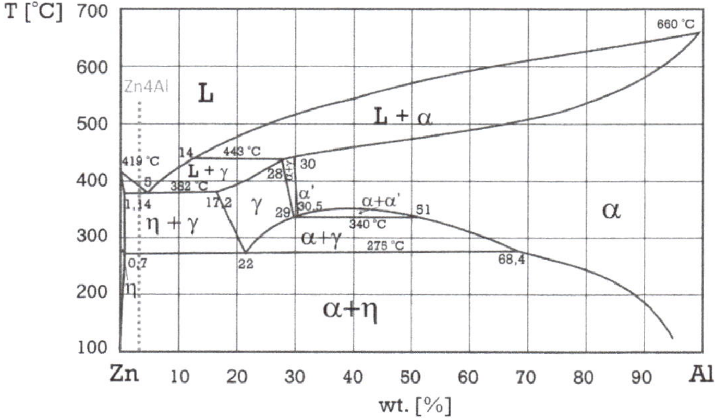

Figure 1. Zn-Al phase diagram adapted from [5].

Chloride ions are one of the primary catalysts responsible for the corrosion of zinc and its alloys. Components made from these alloys are exposed to chloride ions in both seawater and also in seaside environments. Micron-sized salt aerosol particles can be deposited on elements located as far as 10 km from the shoreline. The threat in engineering practice may be intergranular corrosion by chloride ions [10,22,23], which was the main reason that a solution rich in these ions was used as a corrosive environment.

Analyzing the corrosion mechanisms of Zn-based alloys is challenging due to their complex microstructures. However, previous studies have only focused on the influence of factors affected by the crystallization conditions (from the temperature of the liquid phase). This translates into an effect on the size and branching of dendrites, as well as eutectic dispersion. Thus, practical considerations are important during casting. The grain size and microstructure morphology (affected by the crystallization conditions) [5,16,17,19,24] and cooling rate have been shown to affect the corrosion resistance of zinc alloys [12,24–27]. Finer dendrites were shown to improve the corrosion resistance of hypoeutectic alloys, whereas a coarse microstructure was more preferable for hypereutectic alloys [5,17]. The aim of this work was to determine the effect of the cooling rate during the eutectoid transformation on the corrosion resistance of a hypoeutectic Zn-4Al alloy. These changes apply to the crystallized alloy and are relevant to determining a heat treatment process. As a part of this research, samples were heat-treated at temperatures higher and lower than the eutectoid transformation, and electrochemical studies were combined with metallographic studies to confirm the effect of heat treatment on the alloy's microstructure. The surface condition of the alloy was assessed after electrochemical tests to determine the role of microstructure during corrosion.

2. Materials and Methods

The investigated material was a Zn-4Al alloy that was fabricated by melting and casting pure elements (99.995% Zn and 99.7% Al) in a PIT10 induction furnace. The obtained material was subjected to heat treatment by annealing for 1 h at 250 °C and 300 °C, followed by cooling. The samples were quenched in water and cooled in air or in a furnace. The scheme of the research design is shown in Table 1.

Hardness measurements were performed using the Vickers method. Microscopic examinations were carried out using a stereoscopic microscope (Leica M205 C, Leica Microsystems, Wetzlar, Germany), a light microscope (Nikon Eclipse MA 200, Nikon Instruments Inc., Tokyo, Japan), and a scanning electron microscope (SEM) (Phenom World ProX, Thermo Fisher Scientific, Waltham, Massachusetts, USA). Light microscopy was used to examine metallographic sections to identify microstructural features after Nital etching (3% nitric acid in ethanol) and 5% NaCl solution. Stereoscopic and SEM microscopes were used after electrochemical measurements to illustrate the corrosion progress.

Table 1. Scheme of the research design.

Material		Methods
As delivered		Hardness measurements, Microstructural examination, Electrochemical examination, SEM surface evaluation
Heat treatment at 250 °C	Furnace cooled Air cooled Water quenched	Hardness measurements
Heat treatment at 300 °C	Furnace cooled Air cooled Water quenched	Hardness measurements, Microstructural examination, Electrochemical examination, SEM surface evaluation

Polarization tests were performed using a three-electrode cell with a potentiostat (ATLAS 0531 ELEKTROCHEMICAL UNIT & IMPEDANCE ANALYSER, Atlas-Sollich, Gdansk, Poland). The auxiliary electrode was made of austenitic stainless steel, while a saturated Ag/AgCl electrode was used as the reference electrode. Just before the experiments, samples were subjected to mechanical grinding with 800 SiC emery papers. The surface area of the working electrode (the sample) was 0.785 cm^2. Before experiments, each sample was immersed for 20 min in 250 mL of a 5% NaCl solution at room temperature. After that, the open circuit potential (E_{OCP}) was measured. Polarization tests were conducted in the same solution by stepping the potential in the anodic direction using a scanning rate of 1 mV/s from −250 mV relative to the open-circuit potential. The pH of the applied solution was 7.5. Four anodic and cathodic polarization curves were recorded for the as-delivered material. The initial potential value was 200 mV lower than the E_{OCP} value. The polarization of each tested sample was terminated at different potential values. Thus, the potentiodynamic curves were stopped at potentials of +150, +225, +300, and +450 mV vs. E_{corr}. Polarization curves were also obtained for samples heat-treated at 300 °C. Three curves per series were determined for the heat-treated alloy. The polarization curves were plotted using an automatic data acquisition system, and the corrosion potential (E_{corr}) and corrosion current density (I_{corr}) were estimated by Tafel plot extrapolation.

3. Results

3.1. Hardness Measurements

The hardness measurement results and their standard deviations in Figure 2 show that the hardness of the as-delivered material was 58 ± 2 HV1. Changing the cooling rate affected the hardness of samples heat-treated at 300 °C. The hardness increased by more than 20 HV1 for the water-quenched sample compared with the material that was furnace-cooled from the same temperature (300 °C). This is due to the eutectoid transformation which occurred at 275 °C. Faster cooling promoted the formation of finer ($\alpha + \eta$) eutectoid structures from the γ phase, while slower cooling allowed the alloy to form a coarser eutectoid structure, which translated into a lower hardness. The increased hardness due to the increased cooling rate realized from the beginning crystallization temperature and microstructure refinement has been observed by other Authors [27].

To provide a comparison, heat treatment was also carried out at a temperature lower than the eutectoid transformation, i.e., 250 °C, and various cooling rates were also used. The cooling rate had no effect on the material hardness at this temperature, which indicates that the formed microstructure was stable. The slight differences in the hardness values were within the standard deviation.

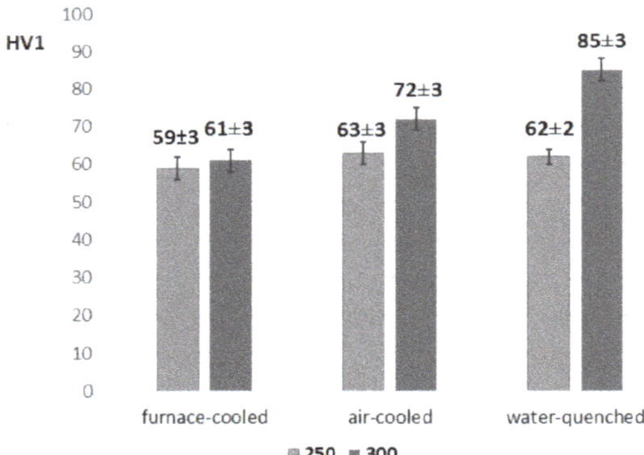

Figure 2. Hardness values obtained for heat-treated Zn-4Al alloy at 250 and 300 °C using various cooling rates.

3.2. Microstructural Examination

The microstructure of the material in delivered state was typical of hypoeutectic Zn-Al alloys (Figure 3). Dendrites of the Zn-base solid solution (η) and an (α + η) eutectic lamellar structure were visible. The microstructure contained the product of eutectoid decomposition because the γ phase was transformed into (α + η) phase at 275 °C, as shown in Figure 1. On the other hand, rod-like eutectic features that may have been formed due to rapid quenching were not observed [11].

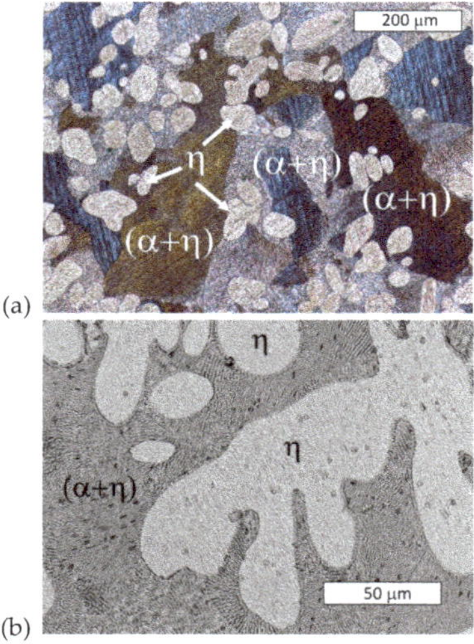

Figure 3. Microstructure of examined Zn-4Al alloy (as-delivered). Visible dendrites of η phase and a eutectic lamellar morphology (α + η). Etched with 10% NaCl solution. (**a**) Light Microscopy, (**b**) SEM.

After heat treatment at 300 °C, the effect of the cooling rate on the phase distribution in the eutectoid structure was examined (Figures 4 and 5). During the applied heat treatment, only the morphology of the microconstituents inside the lamellar structure was affected by eutectoid decomposition. Some divorced eutectic structure was also observed along the grain boundaries. Despite the eutectoid decomposition, the morphology of the interdendritic lamellar eutectic structure was not affected because it was not subjected to any solid-state transformation (Figure 4).

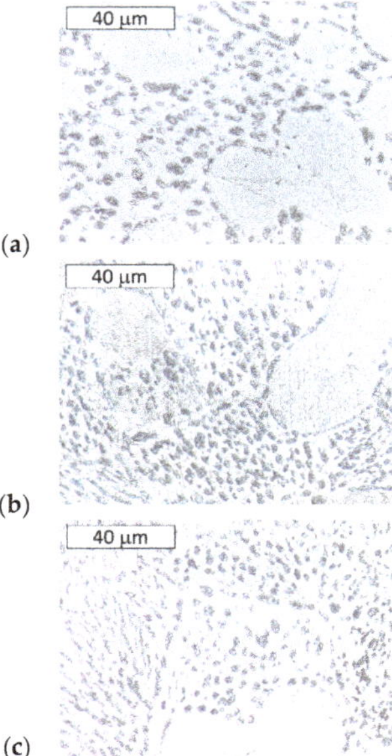

Figure 4. Microstructure of Zn-4Al alloy after heat treatment at 300 °C: (**a**) furnace-cooled, (**b**) air-cooled, (**c**) water-quenched. The visible morphology of the eutectic structure was not affected by the eutectoid transformation. Etched with Nital. Light Microscopy.

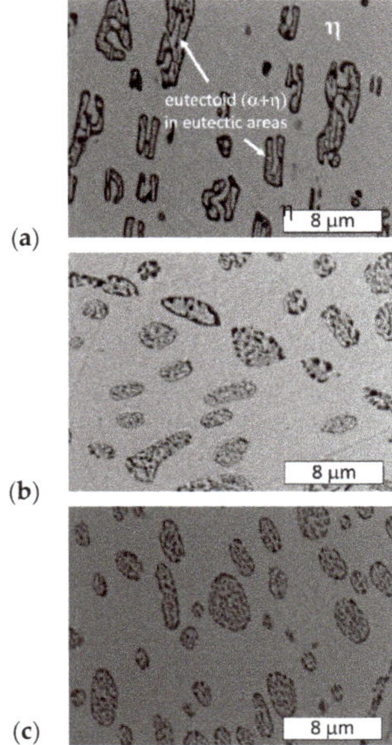

Figure 5. Microstructure of the Zn-4Al alloy after heat treatment at 300 °C: (**a**) furnace-cooled, (**b**) air-cooled, (**c**) water-quenched. A finer ($\alpha + \eta$) eutectoid phase was formed from the γ phase. Etched with Nital. SEM.

3.3. Electrochemical Examinations

Figure 6 shows a comparison between the polarization curves of investigated samples. The corrosion current density and corrosion potential were estimated from the polarization curves using the Tafel extrapolation method (Table 2). The corrosion test results of the as-delivered material are presented as the average of four measurements. For the heat-treated samples measurements are shown as the average of three measurements. As expected, the Zn-4Al alloy had a negative corrosion potential, and the four curves obtained for the as delivered material were similar. This value was consistent with the results of other Authors [6,16]. A strong increase in the current density during the initial stage of the anodic curve was found, which indicates highly intense electrochemical processes.

The microstructure of the investigated material was composed of two phases—an α aluminum-rich solid solution and a η zinc-rich solid solution. The resultant electrochemical potential was closely related to the phase heterogeneity of the zinc alloy, i.e., to the corrosion potential of each phase. The phases with various electrode potentials became anodic and cathodic during contact between the alloy and the electrolyte [28]. Al has a nobler electrochemical behavior than Zn [17,29], and similar behavior should be attributed to Al-base and Zn-base solid solutions. It was previously shown that the anodic nature of the η phase depends on the pH of the corrosive agent [16,24,30]. In slightly acidic or neutral environments, the α phase is nobler than the η phase, so it may act as a cathode. Conversely, in alkaline environments, the α phase may play the role of the anode [24,30]. Shihirova at al. [31] indicated that the electrochemical behavior of phases may be associated with local pH changing and their thermodynamic stability in this corrosive environment. In this study, experiments were carried out at a slightly alkaline pH of 7.5. However, the anodic processes lead to a local reduction pH due to the H^+ produced from the hydrolysis of Al^{3+} [32].

The test results show that the corrosion current density and corrosion potential change as the microstructural morphology changes. The other morphologies were obtained due to different cooling rates during the eutectoid reaction. A very important factor in galvanic corrosion is the ratio of the anodic to cathodic area. If the surface of the cathode is larger than the anode, then more oxygen reduction or another cathodic reaction can occur, which increases the galvanic current. However, in this case, it remained at the same level, but the distance between the anode and cathode changed.

In this case, we had a corrosion microcell, in which the anodes and cathodes were separated by just a few microns. Previous electrochemical research determined that finer structures show a lower I_{corr} compared with a coarse structure. E_{corr} remained rather constant, although it showed a slight decrease. It can be observed that the furnace-cooled structure was related to a corrosion current density and a corrosion potential of 7.01 μA/cm^2 and −1.06 V (vs. Ag/AgCl), respectively, compared with 4.74 μA/cm^2 and −1.07 V (vs. Ag/AgCl), respectively, for the water-cooled structure. Increasing the dispersion of cathode inclusions usually increases the cathode activity. However, if anode passivation occurs or a surface film of corrosion products forms, its activity can be decreased, and the anode process will be inhibited. On the other hand, the short phase distances typical of eutectoid structures may have protected the anode phase. This effect may be clearer due to the finer eutectoid structure.

The E_{corr} value was more electronegative than E_{OCP}. The differences between the E_{OCP} and E_{corr} values were due to the diffusive nature of the cathode potential curve, which has been previously observed during anodic polarization [29].

Table 2. The electrochemical parameters obtained for as delivered samples, as well as samples heat-treated at 300 °C and subjected to different cooling rates.

Sample	I_{corr} (μA/cm^2)	E_{corr} (V) vs. Ag/AgCl	E_{OCP} (V)
As delivered	9.45 ± 0.36	−1.05 ± 0.01	−1.02 ± 0.01
Furnace-cooled	7.01 ± 0.23	−1.06 ± 0.01	−1.02 ± 0.01
Air-cooled	5.47 ± 0.9	−1.06 ± 0.01	−1.03 ± 0.01
Water-quenched	4.74 ± 0.20	−1.07 ± 0.01	−1.05 ± 0.02

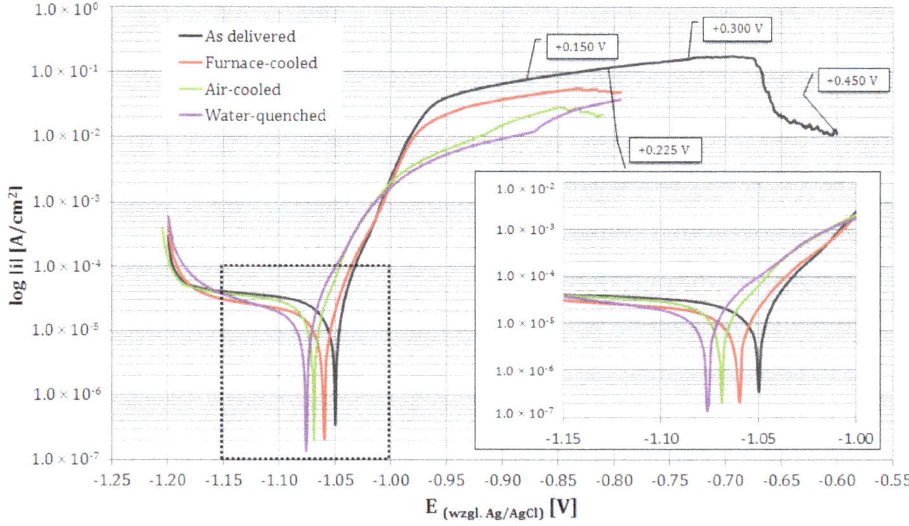

Figure 6. Example potentiodynamic polarization curves of the as-delivered Zn-4Al alloy and heat-treated at 300 °C in 5% NaCl solution. In the curve of the sample polarized to the highest potential value, the potential values (relative to E_{corr}) were marked where polarization was stopped.

3.4. SEM Surface Evaluation after Corrosion Tests

The surfaces of the as-delivered material after electrochemical study were examined using SEM. Samples whose anodic polarization was terminated at different potential values were examined in order to illustrate the corrosion progress in chloride-containing media. The results were discussed in relation to the structural features of the alloys.

Corrosion began locally with the formation of aluminum-rich corrosion products (Figure 7, Table 3). The microscopic observations of the sample tested after reaching a potential of +150 mV versus E_{corr}, did not permit the determination of which structural features underwent corrosion at this stage of development. However, the high aluminum content in the corrosion products on the surface suggested that degradation mainly involved eutectic areas. The formation of aluminum-rich corrosion products first may be unfavorable from the point of view using the alloy as a biomaterial.

Previous works have reported the preferential oxidation of Al-rich areas [29,33]. Other authors have shown that the α phase was protected at the initial stages of corrosion due to the formation of a corrosion product surface film that contained various aluminum-rich phases [24,34–36]. The presence of chlorine indicates that chlorides play an active role in the formation of corrosion products (Table 3). The simultaneous presence of Zn, Al, and Cl in the EDX spectra may be attributed to the formation of $Zn_2Al(OH)_6Cl{\cdot}2H_2O$, which has been reported to form during the early stage of corrosion [18]. Other Authors have observed an $Al_2(OH)_5Cl{\cdot}2H_2O$ phase [35,36]. In this case, zinc may be associated with the base material. It is believed that, regardless of the chemical composition, these phases provide excellent protection against further corrosion.

Based on the electrochemical tests and the above literature data, it can be hypothesized that the finer Al-base phase in the eutectoid structure may result in the formation of a more compact corrosion product film that increases the temporary corrosion protection. The formation of a corrosion product film on the α phase can help reduce the corrosion current density as the distance between eutectoid components decreases. Consequently, the finer distribution of the two phases that formed during eutectoid decomposition in the eutectic mixture tended to decrease their corrosion rate.

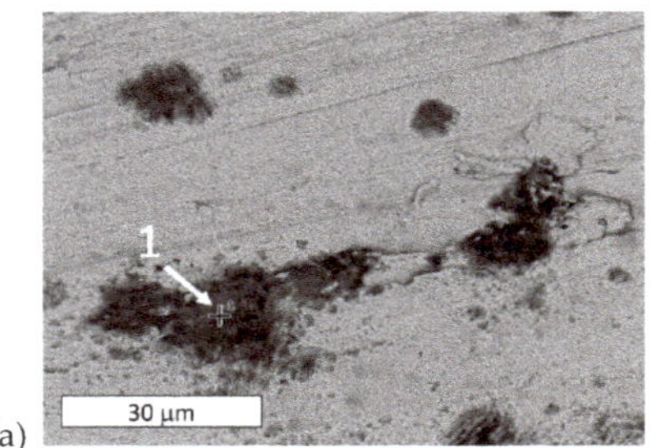

(a)

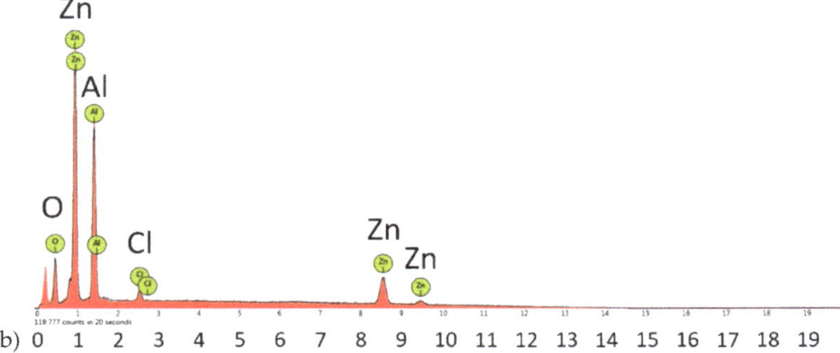

Figure 7. (**a**) SEM image of the surface of a sample after polarization up to a potential of +150 mV vs. E_{corr}. Corrosion initiation areas are visible. The corrosion products are rich in aluminum and chlorine (marked with point 1 and summarized in Table 3; (**b**) characteristic X-ray emission spectrum obtained from point 1 in Figure 7a.

Table 3. Chemical composition obtained from EDX analysis of point 1 in Figure 7a.

Element	Atomic %	Weight %
Zn	22.15	45.88
Al	39.14	33.47
O	37.04	18.78
Cl	1.67	1.87

As corrosion progressed and the potential increased to +225 mV vs. $E_{corr,}$ the alloy selectively dissolved. At this stage, due to the formation of an electrochemical cell between the α and η phases, the eutectoid (α + η) became susceptible to corrosion (Figure 8). Thus, the anode phase was present only in eutectoid areas, which suggested the α phase. When immersed in the corrosive solution, the hypoeutectic Zn-4Al alloy displayed Al-rich regions (the phase of the eutectic structure) which acted as anodic barriers that protected the η phase. Corrosion gradually occurred throughout the entire eutectic area (Figure 9), which was also reflected by a macroscopically visible color change over the sample surface where eutectics formed. The selective dissolution of eutectic areas has also been documented in other works [18,24,37]. Despite this, the local dissolution of η phase dendrites was also observed at higher magnifications (Figure 9a).

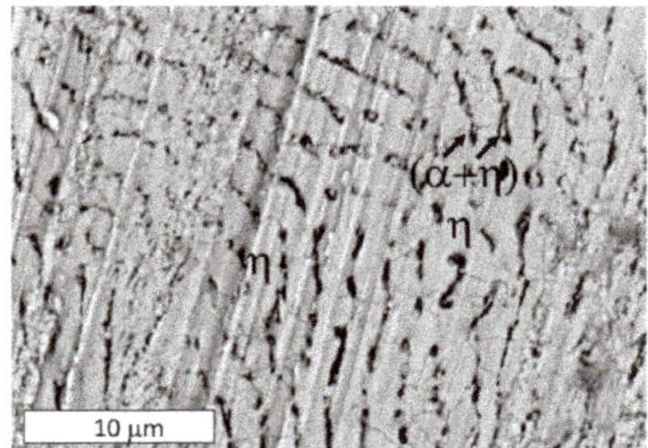

Figure 8. SEM image of the surface of a sample after polarization up to a potential of +225 mV vs. E_{corr}. Selective dissolution of the eutectoid (α + η) in eutectic areas of the Zn-4Al alloy is visible.

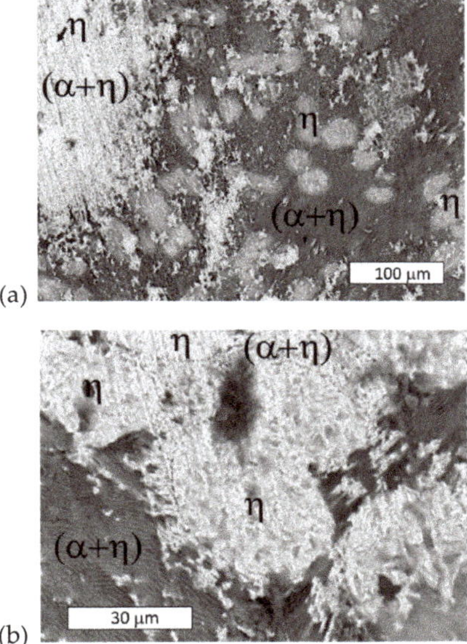

Figure 9. (**a**) SEM image of the surface of a sample after polarization up to a potential of +225 mV vs. E_{corr}. Selective dissolution of the Zn-4Al alloy is visible. The dark areas represent areas in which the eutectic structure has dissolved. (**b**) Magnified image.

The sample polarized up to a potential of +300 mV vs. E_{corr} experienced more extensive corrosion of the eutectic areas over its entire surface (Figure 10). At this stage, the dissolution of the η phase and the revealed α crystals (or products of its corrosion), was observed at the macroscopic scale (Figure 11). This is consistent with the observations that the lamellar structure enables the storage of corrosion products in areas of the corroded α phase, thereby delaying the corrosion process in the eutectic η phase [36]. A higher oxygen content was observed in the dendritic regions (Figure 12). The dendritic η phase dissolved and underwent anodic dissolution reactions [29] which resulted in a constant increase of the current density with the increased polarization potential (Figure 6).

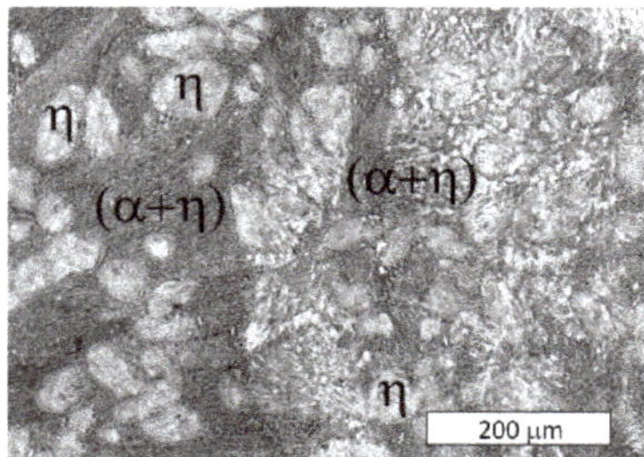

Figure 10. SEM image of the surface of a sample after polarization up to a potential of +300 mV vs. E_{corr}. Corrosion initiation locations of the Zn-4Al alloy are visible. The dark areas represent areas of the eutectic alloy affected by corrosion.

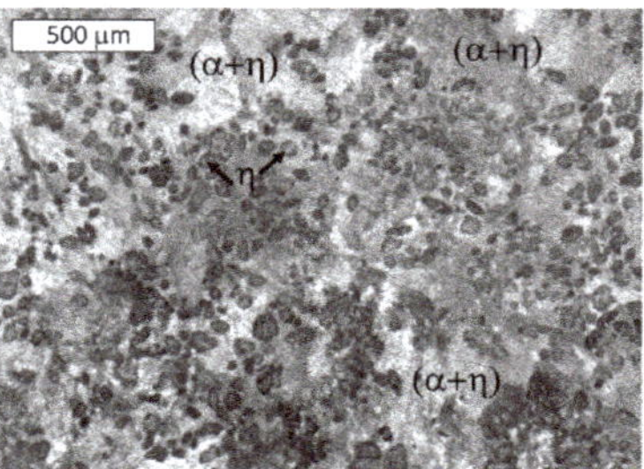

Figure 11. Stereoscopic image of a sample after polarization up to a potential of +300 mV vs. E_{corr}. Selective dissolution of the dendritic η phase at the macroscopic scale.

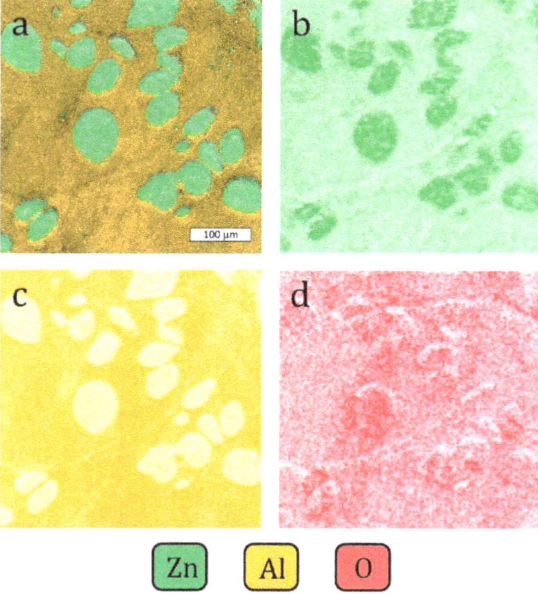

Figure 12. SEM and EDX images showing the distribution of elements on the sample surface after polarization up to the potential +300 mV vs. E_{corr}: (**a**) SEM image merged with zinc and aluminum, (**b**) zinc, (**c**) aluminum, (**d**) oxygen. Element-rich areas are darker in the image.

After reaching a potential of −675 mV vs. Ag/AgCl (+450 mV vs. E_{corr}) the current density decreased on the potentiodynamic curve (Figure 6). The SEM observations of samples tested at a higher potential revealed that at this stage, corrosion extended to all structural constituents (Figure 13). Due to the selective dissolution of the η phase and the macroscopic exposure of (α + η) eutectoid areas, surface topography was observed (Figure 14). These microscopic observations suggest the anodic character of the η phase relative to the α phase in the corrosive solution at this stage of corrosion. Zn^+ ions are formed in the Zn-rich phase (η) due to anodic reactions: $Zn \rightarrow Zn^{2+} + 2e^-$, while the Al-rich phase (α) is expected to be responsible for the cathodic reactions: $O_2 + 2H_20 + 4e^- \rightarrow 4OH^-$. This indicates that despite the initiation of corrosion in areas of the α phase, there is a change in the η phase polarity and its corrosion. This is most likely due to the formation of a corrosion product film on the surface of the α phase that protects it from further corrosion, in accordance with other works [18,24,34]. Thus, changing the anodic zone polarity due to the formation of a protective film can be used in corrosion protection [38].

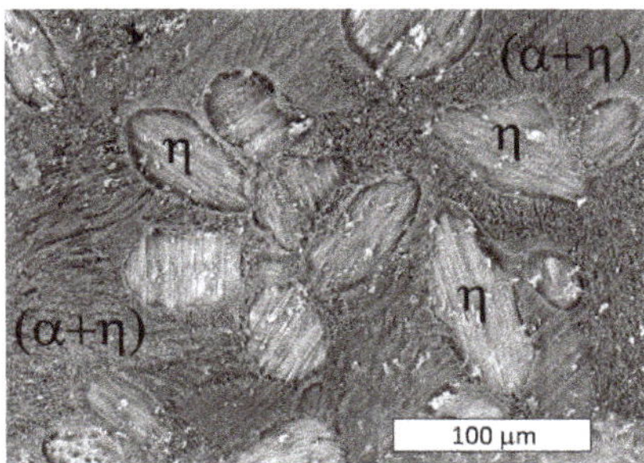

Figure 13. The surface of a sample after polarization up to a potential of +450 mV vs. E_{corr}. Changes on the sample surface due to corrosion are visible over the entire alloy surface. SEM.

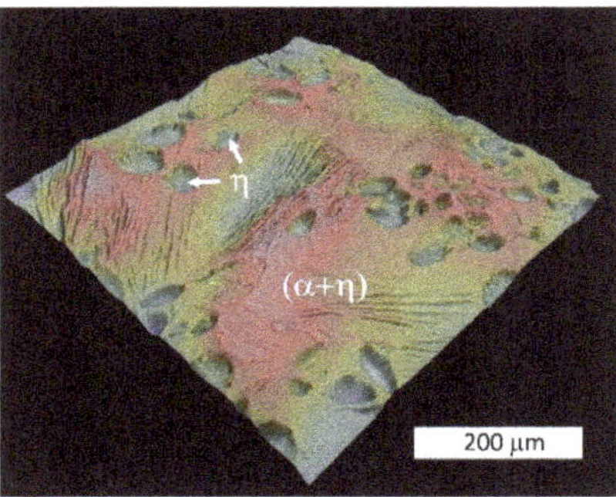

Figure 14. 3D SEM topography of the sample surface after polarization up to a potential of +450 mV vs. E_{corr}. The image based on "shape from shading" technology shows the selective dissolution of the η-phase and a macroscopic exposure of the eutectic structure. Blue indicates the lowest areas, while red represents the highest features.

4. Discussion

The results were intended to discuss the effect of the cooling rate of a Zn-4Al alloy on the corrosion processes at the eutectoid transformation temperature. The main comments are as follows:

1) The microstructure of the material was typical for hypoeutectic cast Zn-Al alloys and was composed of a dendritic η-phase: Zn(Al) solid solution and lamellar (α + η) eutectoid. It contained the products of the eutectoid reaction which transformed the γ phase to (α + η) at

275 °C. Increasing the susceptibility to corrosion by increasing the aluminum content in Zn-Al alloys has been reported in the previous literature [5,19] which may be associated with an increased volumetric fraction of the (α + η) eutectic. Upon progression of the corrosion process, the (α + η) eutectoid structure in eutectic areas was attacked first and subjected to intense corrosion. Therefore, increasing the eutectic volumetric fraction should deteriorate the corrosion resistance of Zn-Al alloys. The high corrosion tendency of eutectic areas may induce intergranular corrosion [10,22,23]. In the case of two cooperating details, the accompanying pulverisation promotes the penetration of material fragments and the corrosion products into the friction area [22,23].

2) Different cooling rates affected the hardness of samples annealed at 300 °C. Water quenching promoted the creation of a finer (α + η) eutectoid structure from the γ phase in eutectic areas of the Zn-Al alloy and obtained higher hardness values. Slower cooling formed a coarser eutectic structure in the alloy, which translated into a lower hardness. After furnace cooling, a hardness similar to the as-cast material was obtained. Heat treatment at 250 °C showed no effect on the hardness of the Zn-4Al alloy.
3) A finer eutectoid structure decreased the corrosion current density I_{corr} compared with a coarse structure, which indicates that the short phase distances of eutectoid structures may contribute to the protection of the anode phase and reduce the corrosion rate. The corrosion potential E_{corr} remained rather constant, although a slight decrease was observed.
4) In the initial corrosion stage, the α-phase Al-base solid solution served as the anode in a formed corrosion microcell in the examined corrosive environment. As corrosion further developed, it extended over the entire alloy surface. Thus, it can be stated that the dissolution of the η phase was the preferred corrosion mode due to anodic dissolution reactions. This phenomenon may have been related to the formation of an α-phase corrosion product film. The formation of this film can also explain the lower corrosion current density due to a decrease in the cathode activity due to a smaller distance between eutectoid components.
5) If there is an anode phase whose fragments are fine and homogeneously distributed within the grain, corrosion will lead to their dissolution and the material eventually becomes quasi-homogeneous. A very different situation takes place for large η phase dendrites which occurs in the microstructure of Zn-Al alloy. In this case, corrosion develops involving these structural elements, which decrease the cross-sections of components made of this material.

5. Conclusions

A finer eutectoid structure was shown to improve the corrosion resistance of the Zn-4Al alloy, which indicates that the small phase distances between the eutectoid structures may help protect the anode phase. Importantly, this was also accompanied by an increase in the alloy hardness. This issue is important from the point of view of the heat treatment design of machine components exposed to chloride ions.

Although corrosion was initiated in the α-phase, the polarity of the η-phase changed, and its corrosion was observed. This was most likely due to the formation of a corrosion product film on the surface of the α phase that protected it from further corrosion. A scheme of the corrosion mechanism is presented in Figure 15.

Figure 15. Schematic illustration of the corrosion mechanism based on as-delivered Zn-4Al alloy.

Author Contributions: Conceptualization, M.M.L.; methodology, M.M.L.; validation, R.J.; formal analysis, M.M.L.; investigation, M.M.L.; resources, M.M.L.; data curation, M.M.L.; writing—original draft preparation, M.M.L.; writing—review and editing, M.M.L. and R.J.; visualization, M.M.L.; supervision, M.M.L. and R.J.; project administration, R.J.; funding acquisition, R.J. All authors have read and agreed to the published version of the manuscript.

Funding: The APC was funded by the Polish Ministry of Science and Higher Education for statutory activities No. S/1/S/KPBMiM/20.

Conflicts of Interest: The authors declare no conflict of interest.

References

1. Savaşkan, T.; Pürçek, G.; Murphy, S. Sliding wear of cast zinc-based alloy bearings under static and dynamic loading conditions. *Wear* **2002**, *252*, 693–703. [CrossRef]
2. Murphy, S.; Savaskan, T. Comparative wear behaviour of Zn-Al-based alloys in an automotive engine application. *Wear* **1984**, *98*, 151–161. [CrossRef]
3. Calayag, T.S. Zinc alloys replace bronze in mining equipment bushings and bearings. *Min. Eng.* **1983**, *35*, 727–728.
4. Babić, M.; Ninković, R. Zn-Al alloys as tribomaterials. *Tribol. Ind.* **2004**, *26*, 3–7.
5. Ares, A.E.; Gassa, L.M. Corrosion susceptibility of Zn–Al alloys with different grains and dendritic microstructures in Nacl solutions. *Corros. Sci.* **2012**, *59*, 290–306. [CrossRef]
6. Katarivas Levy, G.; Goldman, J.; Aghion, E. The Prospects of Zinc as a Structural Material for Biodegradable Implants—A Review Paper. *Metals* **2017**, *7*, 402. [CrossRef]

7. Zhao, L.; Zhang, Z.; Song, Y.; Liu, S.; Qi, Y.; Wang, X.; Wang, Q.; Cui, C. Mechanical properties and in vitro biodegradation of newly developed porous Zn scaffolds for biomedical applications. *Mater. Des.* **2016**, *108*, 136–144. [CrossRef]
8. Kannan, M.B.; Moore, C.; Saptarshi, S.; Somasundaram, S.; Rahuma, M.; Lopata, A.L. Biocompatibility and biodegradation studies of a commercial zinc alloy for temporary mini-implant applications. *Sci. Rep.* **2017**, *7*, 15605. [CrossRef]
9. Vojtěch, D.; Kubásek, J.; Šerák, J.; Novák, P. Mechanical and corrosion properties of newly developed biodegradable Zn-based alloys for bone fixation. *Acta Biomater.* **2011**, *7*, 3515–3522. [CrossRef]
10. Bowen, P.K.; Seitz, J.-M.; Guillory, R.J.; Braykovich, J.P.; Zhao, S.; Goldman, J.; Drelich, J.W. Evaluation of wrought Zn-Al alloys (1, 3, and 5 wt % Al) through mechanical and in vivo testing for stent applications. *J. Biomed. Mater. Res. Part B Appl. Biomater.* **2018**, *106*, 245–258. [CrossRef]
11. Sullivan, J.; Penney, D.; Elvins, J.; Khan, K. The effect of ultrasonic irradiation on the microstructure and corrosion rate of a Zn–4.8wt.% Al galvanising alloy used in high performance construction coatings. *Surf. Coat. Technol.* **2016**, *306*, 480–489. [CrossRef]
12. Yasakau, K.A.; Kallip, S.; Lisenkov, A.; Ferreira, M.G.S.; Zheludkevich, M.L. Initial stages of localized corrosion at cut-edges of adhesively bonded Zn and Zn-Al-Mg galvanized steel. *Electrochim. Acta* **2016**, *211*, 126–141. [CrossRef]
13. Salgueiro Azevedo, M.; Allély, C.; Ogle, K.; Volovitch, P. Corrosion mechanisms of Zn(Mg,Al) coated steel: 2. The effect of Mg and Al alloying on the formation and properties of corrosion products in different electrolytes. *Corros. Sci.* **2015**, *90*, 482–490. [CrossRef]
14. Prosek, T.; Persson, D.; Stoulil, J.; Thierry, D. Composition of corrosion products formed on Zn–Mg, Zn–Al and Zn–Al–Mg coatings in model atmospheric conditions. *Corros. Sci.* **2014**, *86*, 231–238. [CrossRef]
15. Vu, T.N.; Volovitch, P.; Ogle, K. The effect of pH on the selective dissolution of Zn and Al from Zn–Al coatings on steel. *Corros. Sci.* **2013**, *67*, 42–49. [CrossRef]
16. Osório, W.R.; Freire, C.M.; Garcia, A. The role of macrostructural morphology and grain size on the corrosion resistance of Zn and Al castings. *Mater. Sci. Eng. A* **2005**, *402*, 22–32. [CrossRef]
17. Osório, W.R.; Freire, C.M.; Garcia, A. The effect of the dendritic microstructure on the corrosion resistance of Zn–Al alloys. *J. Alloys Compd.* **2005**, *397*, 179–191. [CrossRef]
18. Zhang, X.; Leygraf, C.; Odnevall Wallinder, I. Atmospheric corrosion of Galfan coatings on steel in chloride-rich environments. *Corros. Sci.* **2013**, *73*, 62–71. [CrossRef]
19. Ares, A.E.E.; Gassa, L.M.M.; Schvezov, C.E.E.; Rosenberger, M.R.R. Corrosion and wear resistance of hypoeutectic Zn–Al alloys as a function of structural features. *Mater. Chem. Phys.* **2012**, *136*, 394–414. [CrossRef]
20. Jasionowski, R.; Polkowski, W.; Zasada, D. Destruction Mechanism of ZnAl4 as Cast Alloy Subjected to Cavitational Erosion Using Different Laboratory Stands. *Arch. Foundry Eng.* **2016**, *16*, 19–24. [CrossRef]
21. Jasionowski, R.; Zasada, D.; Polkowski, W. Destruction Mechanism of Z10400 Zn-based Alloy Subjected to Cavitational Erosion. *Arch. Foundry Eng.* **2013**, *13*, 48–52.
22. Lachowicz, M.M.; Lachowicz, M.B. Intergranular Corrosion of the as Cast Hypoeutectic Zinc-Aluminium Alloy. *Arch. Foundry Eng.* **2017**, *17*, 79–84. [CrossRef]
23. Lachowicz, M.M. A study on the intergranular corrosion-fatigue failure of the Zn-Al alloy solenoid valve. *Eng. Fail. Anal.* **2019**, *103*, 184–194. [CrossRef]
24. Prosek, T.; Thierry, D.; Hagström, J.; Persson, D.; Fuertes, N.; Lindberg, F.; Taxén, C.; Šerák, J. Relationship between corrosion performance and microstructure of Zn-AI and Zn-AI-Mg model alloys. In Proceedings of the European Corrosion Congress, EUROCORR 2015, Graz, Austria, 6–10 September 2015; Volume 358.
25. Elvins, J.; Spittle, J.A.; Worsley, D.A. Relationship between microstructure and corrosion resistance in Zn–Al alloy coated galvanised steels. *Corros. Eng. Sci. Technol.* **2003**, *38*, 197–204. [CrossRef]
26. Elvins, J.; Spittle, J.A.; Worsley, D.A. Microstructural changes in zinc aluminium alloy galvanising as a function of processing parameters and their influence on corrosion. *Corros. Sci.* **2005**, *47*, 2740–2759. [CrossRef]
27. Krupińska, B.; Dobrzański, L.A.; Rdzawski, Z.M.; Labisz, K. Cooling rate influence on microstructure of the Zn-Al cast alloy. *Arch. Mater. Sci. Eng.* **2010**, *43*, 13–20.
28. Gnedenkov, A.S.; Sinebryukhov, S.L.; Mashtalyar, D.V.; Imshinetskiy, I.M.; Vyaliy, I.E.; Gnedenkov, S.V. Effect of Microstructure on the Corrosion Resistance of TIG Welded 1579 Alloy. *Materials* **2019**, *12*, 2615. [CrossRef]

29. Gancarz, T.; Mech, K.; Guśpiel, J.; Berent, K. Corrosion studies of Li, Na and Si doped Zn-Al alloy immersed in NaCl solutions. *J. Alloys Compd.* **2018**, *767*, 1225–1237. [CrossRef]
30. Prosek, T.; Hagström, J.; Persson, D.; Fuertes, N.; Lindberg, F.; Chocholatý, O.; Taxén, C.; Šerák, J.; Thierry, D. Effect of the microstructure of Zn-Al and Zn-Al-Mg model alloys on corrosion stability. *Corros. Sci.* **2016**, *110*, 71–81. [CrossRef]
31. Snihirova, D.; Taryba, M.; Lamaka, S.V.; Montemor, M.F. Corrosioninhibition synergies on a model Al-Cu-Mg sample studied by localized scanning electrochemical techniques. *Corros. Sci.* **2016**, *112*, 408–417.
32. Queiroz, F.M.; Donatus, U.; Prada Ramirez, O.M.; de Sousa Araujo, J.V.; Goncalves de Viveiros, B.V.; Lamaka, S.; Zheludkvich, M.; Masoumi, M.; Vivier, V.; Costa, I.; et al. Effect of unequal levels of deformation and fragmentation on the electrochemical response of friction stir welded AA2024-T3 alloy. *Electrochim. Acta* **2019**, *313*, 271–281. [CrossRef]
33. Qiu, P.; Leygraf, C.; Odnevall Wallinder, I. Evolution of corrosion products and metal release from Galvalume coatings on steel during short and long-term atmospheric exposures. *Mater. Chem. Phys.* **2012**, *133*, 419–428. [CrossRef]
34. Cao, Z.; Kong, G.; Che, C.; Wang, Y. Influence of Nd addition on the corrosion behavior of Zn-5%Al alloy in 3.5wt.% NaCl solution. *Appl. Surf. Sci.* **2017**, *426*, 67–76. [CrossRef]
35. Neto, P.D.L.; Correia, A.N.; Colares, R.P.; Araujo, W.S. Corrosion study of electrodeposited Zn and Zn-Co coatings in chloride medium. *J. Braz. Chem. Soc.* **2007**, *18*, 1164–1175. [CrossRef]
36. Liu, Y.; Li, H.; Li, Z. EIS investigation and structural characterization of different hot-dipped zinc-based coatings in 3.5% NaCl solution. *Int. J. Electrochem. Sci.* **2013**, *8*, 7753–7767.
37. Tang, N.Y.; Liu, Y. Corrosion performance of aluminum-containing zinc coatings. *ISIJ Int.* **2010**, *50*, 455–462. [CrossRef]
38. Gnedenkov, A.S.; Sinebryukhov, S.L.; Mashtalyar, D.V.; Vyaliy, I.E.; Egorkin, V.S.; Gnedenkov, S.V. Corrosion of the Welded Aluminium Alloy in 0.5 M NaCl Solution. Part 2: Coating Protection. *Materials* **2018**, *11*, 2177. [CrossRef]

© 2020 by the authors. Licensee MDPI, Basel, Switzerland. This article is an open access article distributed under the terms and conditions of the Creative Commons Attribution (CC BY) license (http://creativecommons.org/licenses/by/4.0/).

Article

Biofilm Formation Plays a Crucial Rule in the Initial Step of Carbon Steel Corrosion in Air and Water Environments

Akiko Ogawa [1,*], Keito Takakura [1], Nobumitsu Hirai [1], Hideyuki Kanematsu [2], Daisuke Kuroda [2], Takeshi Kougo [2], Katsuhiko Sano [3] and Satoshi Terada [4]

1 Department of Chemistry and Biochemistry, National Institute of Technology (KOSEN), Suzuka College, Suzuka 510-0294, Japan; takakura.0820@gmail.com (K.T.); hirai@chem.suzuka-ct.ac.jp (N.H.)

2 Department of Material Science and Engineering, National Institute of Technology (KOSEN), Suzuka College, Suzuka 510-0294, Japan; kanemats@mse.suzuka-ct.ac.jp (H.K.); daisuke@mse.suzuka-ct.ac.jp (D.K.); kougo@mse.suzuka-ct.ac.jp (T.K.)

3 D&D Company, Yokkaichi 512-1211, Japan; sano@ddcorp.co.jp

4 Department of Materials Science and Biotechnology, University of Fukui, Fukui 910-8507, Japan; terada@u-fukui.ac.jp

* Correspondence: ogawa@chem.suzuka-ct.ac.jp

Received: 28 December 2019; Accepted: 14 February 2020; Published: 19 February 2020

Abstract: In this study, we examined the relationship between the effect of a zinc coating on protecting carbon steel against biofilm formation in both air and water environments. SS400 carbon steel coupons were covered with a zinc thermal spray coating or copper thermal spray coating. Coated coupons were exposed to either air or water conditions. Following exposure, the surface conditions of each coupon were observed using optical microscopy, and quantitatively analyzed using an x-ray fluorescence analyzer. Debris on the surface of the coupons was used for biofilm analysis including crystal violet staining for quantification, Raman spectroscopic analysis for qualification, and microbiome analysis. The results showed that the zinc thermal spray coating significantly inhibited iron corrosion as well as biofilm formation in both air and water environments. The copper thermal spray coating, however, accelerated iron corrosion in both air and water environments, but accelerated biofilm formation only in a water environment. microbially-influenced-corrosion-related bacteria were barely detected on any coupons, whereas biofilms were detected on all coupons. To summarize these results, electrochemical corrosion is dominant in an air environment and microbially influenced corrosion is strongly involved in water corrosion. Additionally, biofilm formation plays a crucial rule in carbon steel corrosion in both air and water, even though microbially-influenced-corrosion-related bacteria are barely involved in this corrosion.

Keywords: thermal spray; corrosion; zinc; copper; steel

---

1. Introduction

Zinc coatings are widely used in building products for the anti-corrosive treatment of steel. Zinc coatings mainly protect steel against corrosion through barrier and galvanic protection effects [1]. In general, a metallized zinc coating can produce a thicker layer than a galvanized one [2], which guarantees beneficial effects for resistance to corrosion. The thermal coating of zinc is one type of metallic zinc coating, which is commonly applied to bridges and marine structures where atmospheric corrosion is likely to be accelerated by soluble salts such as sodium chloride [3].

Biofilm formation also enhances the metallic corrosion of steels [4]. Biofilms refer to microbially produced three-dimensional structures that consist of water (over 80%), microbes, and extracellular polymeric substrates (EPSs) produced by the microbes [5]. Biofilm-related corrosion is also referred to as biocorrosion or microbially influenced corrosion (MIC). Several mechanisms for MIC have been proposed, as follows: galvanic cells, differential aeration or chemical concentration cells, acidity of organic acids derived from bacterial metabolites, secretion of H_2S, NH_3, or PH_3 from bacteria, enzymatic oxygen reduction reactions, and direct extraction and consumption of electrons from iron [6].

Many studies have investigated either electrical corrosion or MIC [7–15], however, none have dealt with a combination of them. In this study, we investigated a corporate-steel corrosion mechanism that connects the gap between electrical corrosion and MIC under both air atmosphere and water-immersion conditions. At the same time, we examined the inhibitory effect of the zinc thermal spray coating on steel corrosion in comparison with a copper thermal spray coating. The reason why we tested the copper thermal spray coating was that we expected that copper would postpone and (or) regulate MIC. Some researchers have reported that copper can inhibit microbial activities of some bacteria [16–18]. Additionally, we found that copper downregulated the biofilm formation of marine living bacteria [19]. In order to elucidate the steel corrosion mechanism, we analyzed the corroded test sites by microscopic observation and energy-dispersive x-ray (EDX) analysis; we also analyzed any biofilms formed using a combination of quantitative and qualitative methods to explore the microbiome.

2. Materials and Methods

2.1. Thermal Sprayed Coupons

JIS SS400 carbon steel plate coupons (10 cm × 10 cm, thickness 1 mm, Sakai Netsu-Giken, Tsu, Japan) were subjected to a zinc or copper coating using a wire flame spraying process at TOCALO Co. Ltd. (Kobe, Japan). The thickness of the sprayed layer was about 1 mm. For the remainder of this manuscript, we refer to SS400 carbon steel plate, zinc thermal spray-coated SS400 carbon steel plate, and copper thermal spray-coated SS400 carbon steel plate, as SS400, Zn-coated, and Cu-coated, respectively.

2.2. Outdoor Exposure Test

Each coupon was put on an acrylic plate (Sakai Netsu-Giken) with two corners held in place by stainless steel clamps (Sakai Netsu-Giken). The acrylic plates were fixed on a wooden board at a distance of 5 cm from the board (Figure 1). This board was then left on the rooftop deck of a building (10 m tall) at the National Institute of Technology, Suzuka College (34.8502 N, 136.58132 E), located 2 km from the Shiroko coast, for two months (from 1 December 2016 to 31 January 2017).

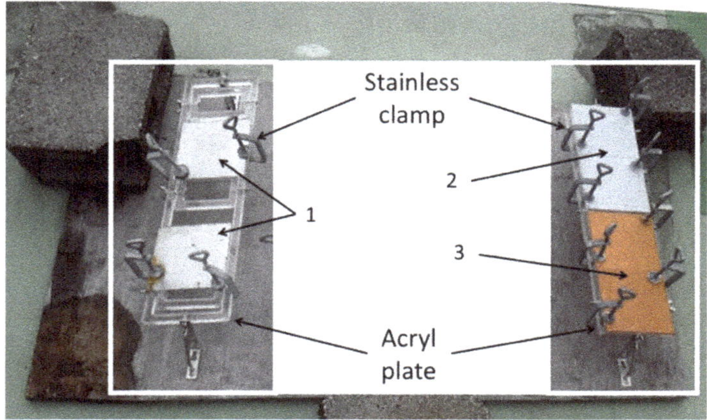

Figure 1. The apparatus for the outdoor exposure test. 1: SS400; 2: Zn-coated; 3: Cu-coated.

2.3. *Aquatic Immersion Test*

SS400 and each coated coupon were cut into 2 cm × 1 cm rectangles using a shearing machine (Komatsu, Kanazawa, Japan). Non-coated sides were masked using silicone sealant (Cemedine Co. Ltd., Tokyo, Japan). We used an open laboratory biofilm reactor (LBR) [20] for aquatic biocorrosion accelerated testing (Figure 2). This open LBR was made by Sakai Netsu-Giken, and consisted of a cylindrical column, a water tank, an air fan, and a pump. Each coupon segment was secured to an acrylic holder with an acrylic screw pin. The holder was inserted into an acrylic column that was connected to polyvinylchloride (PVC) pipes at either end. The water tank held approximately 50 L of tap water, and this water was circulated through the open LBR at 6 L/min, at 37 °C for seven days. One patterned holes plate was placed between the water tank and one end of the PVC pipe, which could trap microbes in the atmosphere drawn in by the air fan.

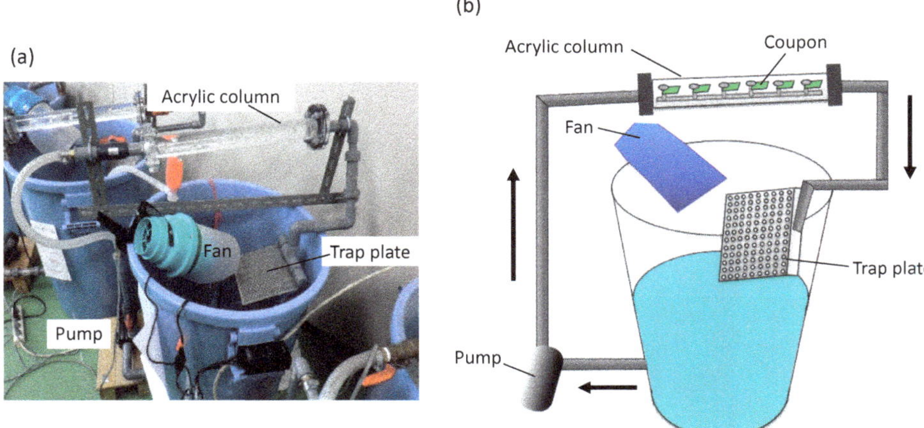

Figure 2. The laboratory biofilm reactor (LBR). (**a**) Photo image of the LBR. (**b**) The outline of the LBR. Arrows indicate water flow.

2.4. *Observation of Morphology and Element Visualization on the Surface of Coupons*

The surface morphology of coupons was investigated using field emission scanning electron microscopy (FE-SEM, Hitachi, Tokyo, Japan). In preparation for SEM observation, each coupon was ultrasonically rinsed in acetone to remove contaminants such as corrosion products. To detect specific elements such as calcium and silicon that originate from biofilms [21], iron, silicon, calcium, and zinc or copper were visualized using the mapping function of EDX (Hitachi, Tokyo, Japan) attached to the FE-SEM.

2.5. *Quantification of Elements on the Surface of Coupons*

The amount of each element on the surface of the coupons was evaluated three times using an x-ray fluorescence analyzer (Hitachi, Tokyo, Japan).

2.6. *DNA Extraction*

A PowerSoil® DNA isolation kit (MO Bio Laboratories, Carlsbad, CA, USA) was used for DNA extraction from biofilms formed on the surface of each specimen, whether from the outdoor exposure test or the aquatic biofilm formation test. For the outdoor exposure test, DNA was extracted from deposits on the surface of each outdoor-exposed coupon from two individual areas of each coupon. Biofilm samples from each specimen were scratched off using a sterile spatula, and collected in a

PowerSoil® Bead Tube. Sixty μL of C1 solution was added to each tube, tubes were then inverted several times, secured to a bead crusher (TITEC, Koshigaya, Japan), vortexed at 4600 rpm for 1 min, and placed on ice. The vortexing and cooling step was repeated nine more times. Any DNA present in the tubes was then purified according to a previously reported protocol [19]. The concentration of the purified DNA solution was measured using a Qubit fluorometer (Thermo Fisher Scientific Inc., Waltham, MA, USA) and a dsDNA high sensitivity (HS) assay kit (Thermo Fisher Scientific, Waltham, MA, USA).

2.7. 16S rRNA Gene-Based Bacterial Community Analysis

The experimental procedure performed according to our previous study in Antibiotics [22].

2.8. Raman Spectroscopy Analysis

Before and after exposure testing, the surface of each coupon was observed at 5-fold or 100-fold magnification using the attached microscope of a laser Raman spectroscope (NRS-3100, JASCO Co., Tokyo, Japan). The surfaces were then irradiated with laser light at approximately 649 cm^{-1} (500–2500 cm^{-1}) for 0.3 s, and the Raman reflection was measured. About non-exposure tested coupons, five points (in the vicinities of the center and the four corners) were randomly selected for each coupon. Regarding the exposure tested coupons, five spots where deposits were observed were selected for each coupon.

2.9. Quantitative Biofilm Formation

After Raman spectroscopic analysis, one section (1 cm × 5 cm) was cut out of each post- exposed coupon. After taking photos, each coupon section was soaked in 0.1% crystal violet solution for 30 min at 25 °C. Treated samples were removed from the solution and rinsed with tap water to remove non-specifically absorbed stain from the surface. Samples were dried for 10 min on a paper towel, then Scotch® mending tape (3M Japan, Tokyo, Japan) was affixed to the polymer-coated side. Thirty minutes later, the tape was removed and affixed to a glass slide. The stained area was measured at five positions using a color reader CR-13 (KONICA MINOLTA, Inc., Tokyo, Japan) from the opposite side to the tape affixed on the glass slide. White paper was used for calibration. Measured data were described using the L*a*b* color system: L* represents lightness (calibration value was 100), a* represents the red/green coordinate (calibration value was zero), and b* represents the yellow/blue coordinate (calibration value was zero). If the color is violet, a* assumes a positive value and b* a negative value. We calculated the three-dimensional vector values (i.e., $\sqrt{(a^*)^2 + (b^*)^2 + (100 - L^*)^2}$) to infer the extent to which the sample formed a biofilm, indicating how a sample formed a biofilm [23].

3. Results and Discussion

3.1. Outdoor Exposure Tests

All coupons were placed on a roof terrace for two months. The surface color of the SS400 coupons changed from a shiny gray to a drab gray, yellowish gray, and partly rubric brown. The surface color of Cu-coated coupons also changed from shiny bronze to a drab dark brown and rubric brown. The surface color of the Zn-coated coupons changed from a shiny blueish gray to a drab light gray and were partly white (Figure 3). Generally, the color of iron oxides varies according to their oxidative condition. When water and oxygen are present, red-brown-colored rust occurs [24], while zinc oxide is white in color and insoluble in water [25]. Considering this information, the brown-colored area on the SS400 coupons and Cu-coated coupons was probably deposited iron rust such as $Fe(OH)_3$. In addition, Cu-coated coupons were covered with larger brown-colored areas than SS400, while there were no brown-colored areas on the surface of Zn-coated coupons. These observations indicated that Cu-coated coupons were more corroded than the SS400 coupons, whereas Zn-coated coupons were not corroded.

Furthermore, the surfaces of the Zn-coated coupons were covered with zinc oxide and were therefore in a passive state.

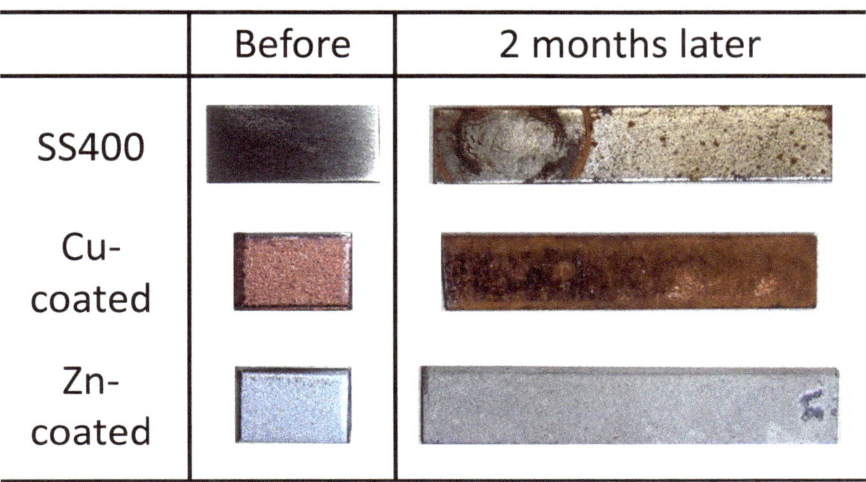

Figure 3. Digital photo images of the outdoor-exposed specimens.

Next, we observed the surface conditions of the coupons both before and after outdoor exposure testing using an optical microscope (Figure 4). For the before outdoor-testing samples, SS400 showed a flat surface; Zn-coated coupons were almost completely covered with the zinc thermal sprayed coating, although some silver color parts of the SS400 basal plate were observed; while the Cu-coated coupons were covered with the copper thermal sprayed coating, but also showed some silver parts as well as red-brown-colored parts, which were probably the SS400 basal plate and iron rust, respectively. The iron rust would be made on the surface of the Cu-coated coupons. As the Cu-coated coupons were very iron corrosive, it would suffer iron corrosion before the test. The surface of a thermal spray coated specimen is irregularly shaped and has gaps because the sprayed metals are aggregated and attached on the surface of a substrate. When moisture exists on the Cu-coated coupon, the coating element (Cu) and the substrate (Fe) are simultaneously immerged in a water solution. Then, Fe becomes an ionized state according to ionization tendency. Ionized irons move to the surface of the Cu-coated coupon through the gaps of spray-coating, and combine with oxygen to form iron oxides there. For the after outdoor-testing samples, the surface of the SS400 coupons showed many small pits (<0.1 mm), cracks, and large holes (3–5 mm). Some parts of the large holes were covered with semi-opaque brown sediments. Zn-coated coupons showed a very rough surface that was white-blue in color, with no silver or brown areas. The Cu-coated coupons also showed a very rough surface, with some parts that were dark brown in color and other parts red-brown. Since pits and holes signify corrosion, and brownish sediments suggest iron rust, the SS400 coupons and Cu-coated coupons incurred iron corrosion, while the Zn-coated coupons incurred zinc corrosion (i.e., the surfaces of Zn-coated coupons were covered with zinc oxide). It appeared that the Cu-coated coupons were more aggressively corroded than the SS400 coupons. Figure 5 shows the SEM images of the surfaces of the SS400, Cu-, and Zn-coated coupons, before and after outdoor exposure testing. Polishing scratches were observed on the SS400 coupons before exposure testing (Figure 5a). Weld splashes of thermally sprayed materials such as copper and zinc were observed in the Cu- and Zn-coated coupons before exposure testing (Figure 5b,c). In addition, many pores were observed on the Cu-coated coupons, which indicated that the density and adhesion of the thermally sprayed layer was low. All coupons were covered with corrosion products following outdoor exposure testing (Figure 5d–f). Iron oxides were detected on the surface of the SS400 and Cu-coated coupons following outdoor exposure testing, while zinc oxides were detected on the surface of the Zn-coated coupons. According to the microscopic images, it was confirmed that the SS400 and Cu-coated coupons underwent iron corrosion, while the Zn-coated coupons underwent zinc corrosion.

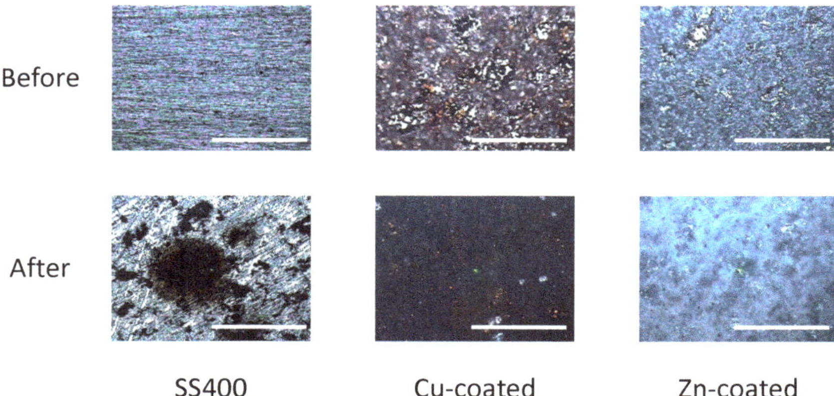

Figure 4. Optical microscopy images of the surface of the specimens before and after outdoor exposure tests. Each white bar represents 5 mm. A green center point is a laser-irradiated spot.

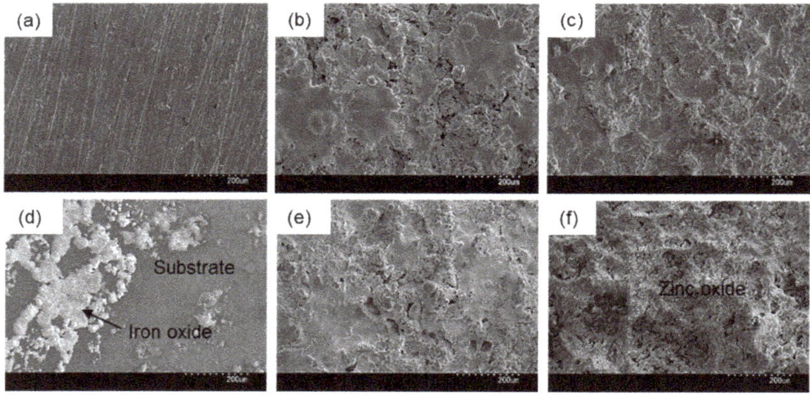

Figure 5. SEM images of the surface of coupons before (**a–c**) and after (**d–f**) outdoor exposure. (**a**) and (**d**) SS400; (**b**) and (**e**) Cu-coated; (**c**) and (**f**) Zn-coated.

Next, sediments on the surfaces of the outdoor-exposed coupons were identified as biofilms by Raman spectroscopic analysis, with the amount of biofilm then quantified using crystal violet staining. In addition, DNA extracted from the sediments was used for bacterial microbiota analysis.

A biofilm consists of water (~80%), microbes, and exstracellular polymeric substances (EPSs). EPSs are a mixture of polysaccharides, extracellular DNA (eDNA), lipids, and proteins [26], and they persist even if biofilm-forming bacteria disappear from a mature biofilm. Since many EPS components can be detected by Raman spectroscopy, several Raman peaks derived from lipids, nucleic acids, proteins, and polysaccharides have been reported [27–35]. Since most reported Raman peaks involving biological components have been detected at 500–1800 cm^{-1}, there is unfortunately relatively little information about biological components in the range of 1900–2300 cm^{-1}. Figure 6 shows the results of Raman spectroscopic analysis of outdoor-exposed coupons. All coupons were found to have several Raman peaks derived from EPSs. For SS400 coupons, amide III-derived peaks (1199–1347 cm^{-1}) [34] had the strongest relative intensity. For Zn-coated coupons, peaks in the >1700 cm^{-1} region were stronger than those in the <1700 cm^{-1} region, with the >1700 cm^{-1} region including carbonyl compound-related peaks (1706–1918 cm^{-1}) [34] and OH–NH–CH stretching vibration peaks associated with nucleic acids (2313–2500 cm^{-1} in SS400 coupon, 2290–2498 cm^{-1} in Cu-coated coupon, 2292–2500 cm^{-1} in Zn-coated coupon) [32]. For the Cu-coated coupons, tyrosine-derived peaks (643–696 cm^{-1}) [29] were the strongest. According to these results, biofilms existed on the surfaces of the SS400, Zn-, and Cu-coated coupons following the outdoor exposure testing, however, the EPS contents probably differed among them.

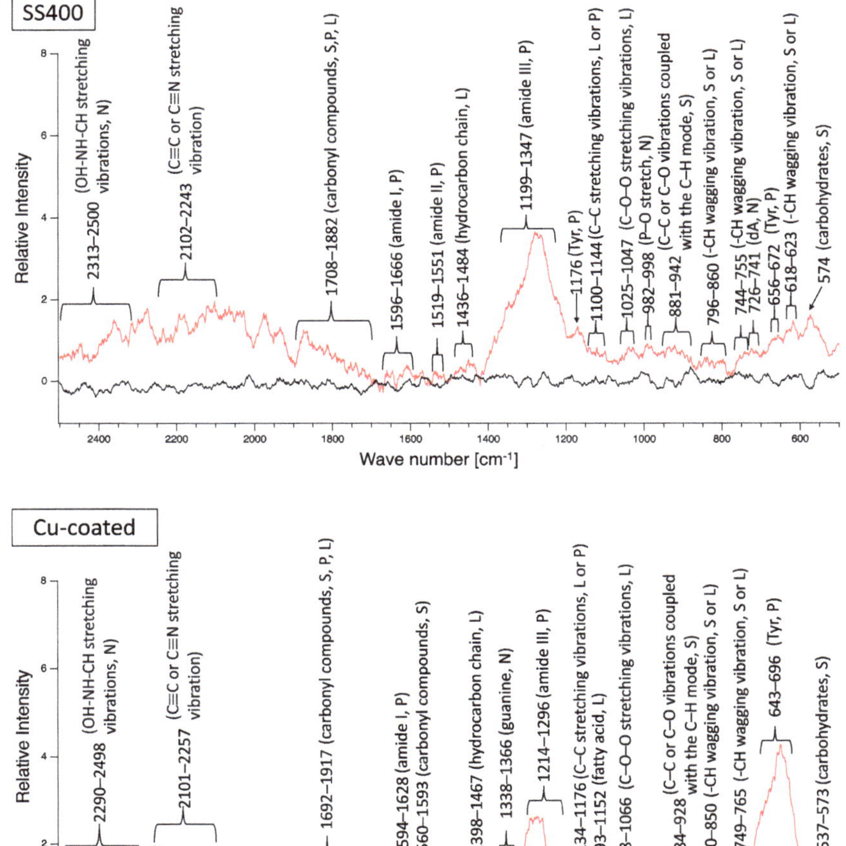

Figure 6. *Cont.*

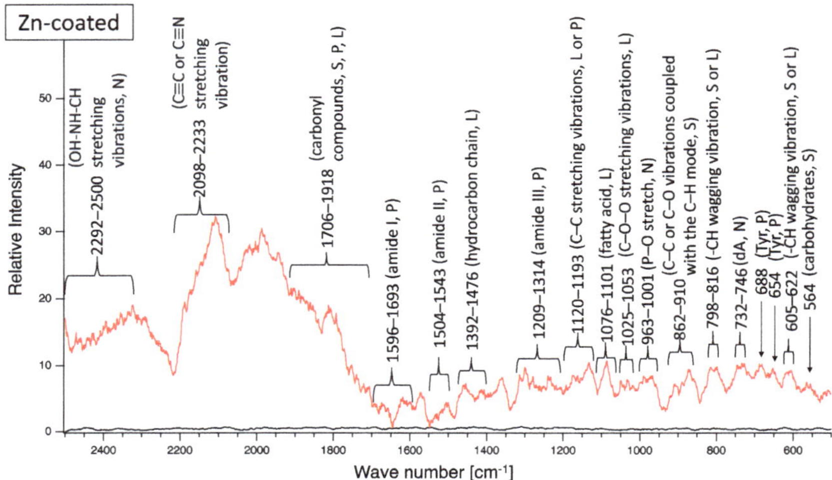

Figure 6. Raman peaks of sediments on the surface of the coupons following outdoor exposure testing (red line). Black lines show the Raman peaks of the surface of the specimens before the test. Detected Raman peaks after the test were assigned to related chemical bonds of compounds according to information in references [27–35]. N: nucleic acids; L: lipids; P: proteins; S: polysaccharides.

Crystal violet staining is one of the methods used for biofilm biomass quantification [36]. The density of crystal violet staining was measured using a colorimeter, then the modulus of the combination of chromaticity value and brightness ($\sqrt{(a^*)^2 + (b^*)^2 + (100 - L^*)^2}$) was calculated. When a biofilm is stained using crystal violet, the color changes to violet and the transparence is lower (i.e., chromaticity of a* is positive value, that of b* is negative value, and brightness, L*, is lower than 100). If a biofilm becomes thicker (mature), $\sqrt{(a^*)^2 + (b^*)^2 + (100 - L^*)^2}$ is larger. Therefore, the amounts of biofilms can be quantified using the value of $\sqrt{(a^*)^2 + (b^*)^2 + (100 - L^*)^2}$. The value for the Zn-coated coupons was significantly smaller than that of the SS400 coupons, the value for the Cu-coated coupons, however, was similar to that of the SS400 coupons (Figure 7). This result indicated that the Zn-coated coupons had a lower biofilm formation than the SS400 and Cu-coated coupons. Kanematsu et al. reported that *Pseudomonas aeruginosa* and *Pseudoalteromonas carageenavara* formed biofilms on the surface of SS400 much more than that of the other metal plated steels such as tin-, copper-, and zinc-plated ones, and iron (ion) could pull these bacteria better than other metals including tin, copper, and zinc [37]. In this study, iron oxide was detected on the surface of the coupons and that of the Cu-coated coupons after the outdoor exposure test (Figure 5), on the other hand, zinc oxide was detected on the surface of Zn-coated coupons. Therefore, it was probably that the surface of SS400 coupons and that of Cu-coated coupons were rich in iron (ions), but that of Zn-coated coupons was poor in iron, resulting in the difference of biofilm formation among these coupons.

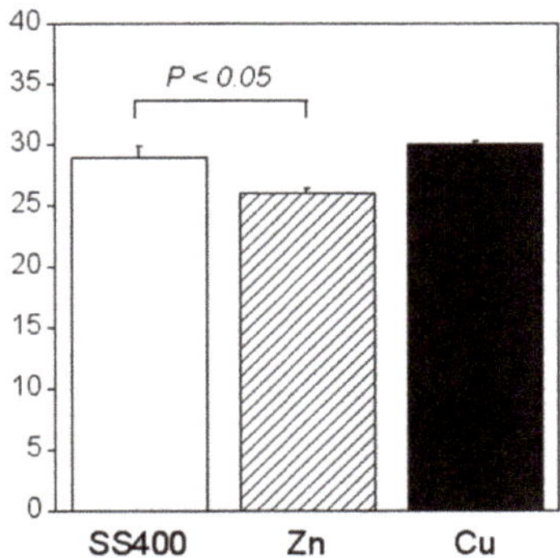

Figure 7. Biofilm quantification of outdoor-exposed coupons. Each column shows the mean of the modulus of $\sqrt{(a^*)^2 + (b^*)^2 + (100 - L^*)^2}$ (n = 5). Zn and Cu mean Zn-coated coupons and Cu-coated coupons, respectively. Error bars show the standard deviation. Student's t-test was performed between the SS400 coupons and Zn-coated coupons or SS400 coupons and Cu-coated coupons. P means the p-value.

16S rRNA gene analysis was performed to reveal which bacterial groups were related to biofilm formation and microbially influenced corrosion (MIC). Some species are known to be MIC-related bacteria. For example, *Pseudomonas aeruginosa* can induce MIC on steel and stainless steel under aerobic conditions [10,38,39]; sulfate-reducing bacteria (SRB) such as *Desulphovibrio vulgarius* produce H_2S, which triggers iron ionization under anaerobic conditions [40,41]; sulfur-oxidizing bacteria such as *Acidithiobacillus thiooxidans* (old name: *Thiobacillus thiooxidans*) produce sulfuric acid which increases acidity and induces iron oxidization [6,42,43]; and iron-oxidizing bacteria such as *Callionella* and *Leptothrix* oxidize iron [6,44]. *Desulphovibrio*, *Acidithiobacillus*, *Pseudomonas*, *Callionella*, and *Leptothrix* belong to the orders Desulfovibrionales, Acidithiobacillales, Pseudomonadales, Nitrosomonadales, and Burkholderiales, respectively. Figure 8 shows the results of bacterial microbiome analysis related to outdoor-exposed coupons. The most abundant bacterial orders were Actinomycetales (25%, #2 in Figure 8) and Burkholderiales (25%, #18 in Figure 8) in SS400–1; Actinomycetales (26%) in SS400–2; Bacillales (32%, #11 in Figure 8) and Burkholderiales (32%) in Cu-coated–1; Pseudomonadales (49%) in Cu-coated–2; Burkholderiales (29%) in Zn-coated–1; and Flavobacteriales (18%, #4 in Figure 8) and Pseudomonadales (17%, #21 in Figure 8) in Zn-coated–2. In these three kinds of coupons, bacterial order occupancies were very different between one area and the other, even on the same coupon. However, some common bacterial orders, namely Actinomycetales, Bacillales, and Pseudomonadales were detected on all coupons, although the degrees of occupancy were different among them. Actinomycetales and Pseudomonadales are dominant bacterial orders in soil communities [45,46]. From this, it can be inferred that major bacterial groups detected from the biofilms came from soil carried by the wind. In addition, Bacillales can survive starvation by forming dormant and resistant spores [47], therefore, Bacillales probably survived as spores in the biofilms of all samples. The top three most abundant genera in each sample are summarized in Table 1. The second-most abundant genus, *Pseudomonas*, seen on Zn-coated-1, was the only MIC-related bacteria, however, most bacteria are reported as being able to form biofilms [48–52]. These findings imply that MIC-related bacteria existed in very low quantities in the biofilms formed on these coupons and were rarely involved in the corrosion. Additionally, the abundant genera of biofilm-forming bacteria were very different in samples with the same coated coupons, indicating that biofilms were partly removed from the surface of the coupons when corrosive products (rusts) were eroded by wind or rain during the exposure period.

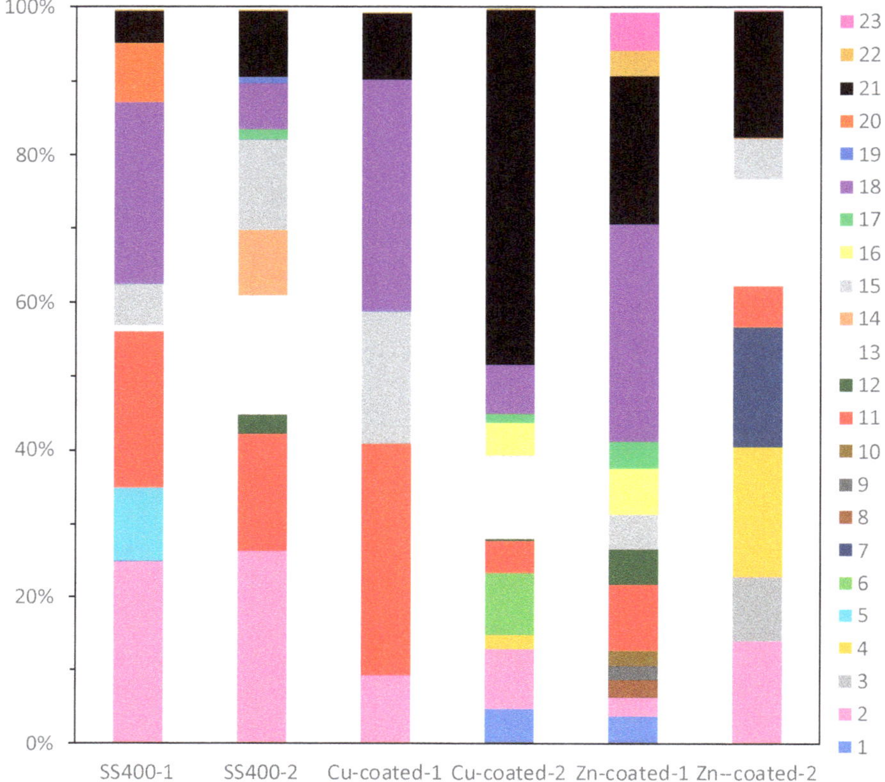

Figure 8. OTU abundance percentages of outdoor-exposed coupons. Bacterial orders present at <1.0% and unassigned OTUs are excluded from each column. (1) Unknown member of Acidobacteria-5; (2) Actinomycetales; (3) Gaiellales; (4) Flavobacteriales; (5) Sphingobacteriales; (6) Saprospirales; (7) G30-KF-AS9; (8) Streptophyta; (9) Stigonematales; (10) Chroococcales; (11) Bacillales; (12) Lactobacillales; (13) Clostridiales; (14) Pirellulales; (15) Rhizobiales; (16) Rhodospirillales; (17) Sphingomonadales; (18) Burkholderiales; (19) Campylobacterales; (20) Legionellales; (21) Pseudomonadales; (22) Xanthomonadales; (23) Chthoniobacterales.

Table 1. Abundant genera of the outdoor-exposed coupons. Each column indicates the assigned genus name and percentage occupancy in parentheses. The numbers in brackets indicate the occupancy of each genus (%).

Rank	SS400-1	SS400-2	Cu-Coated-1	Cu-Coated-2	Zn-Coated-1	Zn-Coated-2
1st	*Bacillus* (19.8)	*Staphylococcus* (15.9)	New someone of Oxalobacteraceae (18.7)	*Acinetobacter* (47.9)	New someone of Oxalobacteraceae (19.5)	*Acinetobacter* (17.0)
2nd	New someone of Oxalobacteraceae (14.8)	*Peptoniphilus* (9.1)	*Rhodoplanes* (17.8)	New someone of Chitinophagaceae (8.4)	*Pseudomonas* (19.0)	New someone of JG30-KF-AS9 (order) (16.1)
3rd	Someone of Intrasporangiaceae (12.9)	*Corynebacterium* (8.9)	*Bacillus* (16.7)	New someone of Ruminococcaceae (7.8)	*Janthinobacterium* (6.7)	*Coprococcus* (14.3)

In the outdoor exposure test, Cu-coated coupon was the most iron corroded followed by SS400 coupon, while Zn-coated coupon was barely iron corroded at all because of the sacrificial corrosion of zinc. The amount of biofilm on Zn-coated coupons was significantly lower than that on the SS400 coupons. These results implied that the tendency for iron corrosion was negatively correlated with that of biofilm formation. Meanwhile, the abundance of MIC-related bacteria was unrelated to the tendency for corrosion. Considering the little relationship between biofilm formation and corrosion, chemical corrosion (i.e., non-MIC) would dominantly progress than MIC on the surface of specimens under outdoor-exposed condition, and biofilms would play an important role in triggering corrosion because biofilms can store water, a crucial factor for corrosion in steel [53]. In addition, the outdoor exposure testing position was located 2 km from Ise Bay, so the air would be rich in sodium chloride, an accelerating factor for corrosion in atmospheric conditions (ionic conditions) [54]. Generally, the initiation of corrosion needs ionization, which tends to occur in aquatic conditions. Biofilms can retain moisture from the air on the surface of the coupons, which makes iron ionization easier. Based on these factors, we inferred a corrosion process under outdoor exposure conditions as follows (Figure 9). First, environmental bacteria were carried to the surface of SS400, Zn-, and Cu-coated coupons by the wind (Step 1). Next, these organisms attached to the surfaces, proliferated, and produced EPSs, resulting in biofilms (Step 2). The process of corrosion is the result of electrochemical reactions i.e., an iron atom changes its oxidation state, then ionized iron is released from the solid surface and combines with H_2O to form iron oxide (iron rust) [55,56]. The dissolution of iron atoms into iron ions would depend on the porosity of the surface films, and the potential difference between the substrate metal (iron) and the film constituent (metal used for thermal spray coating). Since thermal spray coated films are generally porous, the results differ from the kinds of coated metal to metal. The initiation of corrosion needs water and oxygen. Biofilms act as water storage locations where metals are dissolved and transformed, resulting in mineral formation [6], according to the ionization tendency of zinc, iron, and copper. Zinc is more easily ionized than iron, therefore it occurs as insoluble zinc oxide or zinc hydroxide, which will cover the surface of the Zn-coated coupons, terminating any further iron ionization. Meanwhile, iron is more easily ionized than copper, therefore the surface of the Cu-coated coupon was more corroded than that of SS400 coupon (Step 3). Indeed, the surfaces of the Cu-coated coupon showed more rust than those of SS400 coupon. At the same time, sodium chloride in the atmosphere dissolved in the biofilms and accelerated electrochemical corrosion (Step 4).

3.2. Aquatic Immersion Tests

All specimens were set in the open LBR, where tap water was circulated for seven days. The surface of the SS400 coupons had some brown-colored spots (iron rust), the surface of the Cu-coated coupons had more brown-colored spots than that of SS400 coupons, while the surface of the Zn-coated coupons was mostly covered in green or brown-colored sediments (Figure 10). In the optical microscopy images obtained following the LBR immersion test, black-colored concentric holes (about 2.5 mm diameter) were observed on the surface of the SS400 coupons, dark brown-colored pockets (>3 mm-long axes) were observed on the surface of the Cu-coated coupons, whereas most parts of the surfaces of the Zn-coated coupons were covered with brown or green-brown-colored sediments, while the remainder on the Zn-coated coupons were white-blue in color (Figure 11). Brown-colored holes/pockets seemed to be corroded, whereas green-brown-colored sediments seemed to be cyanobacteria-rich biofilms.

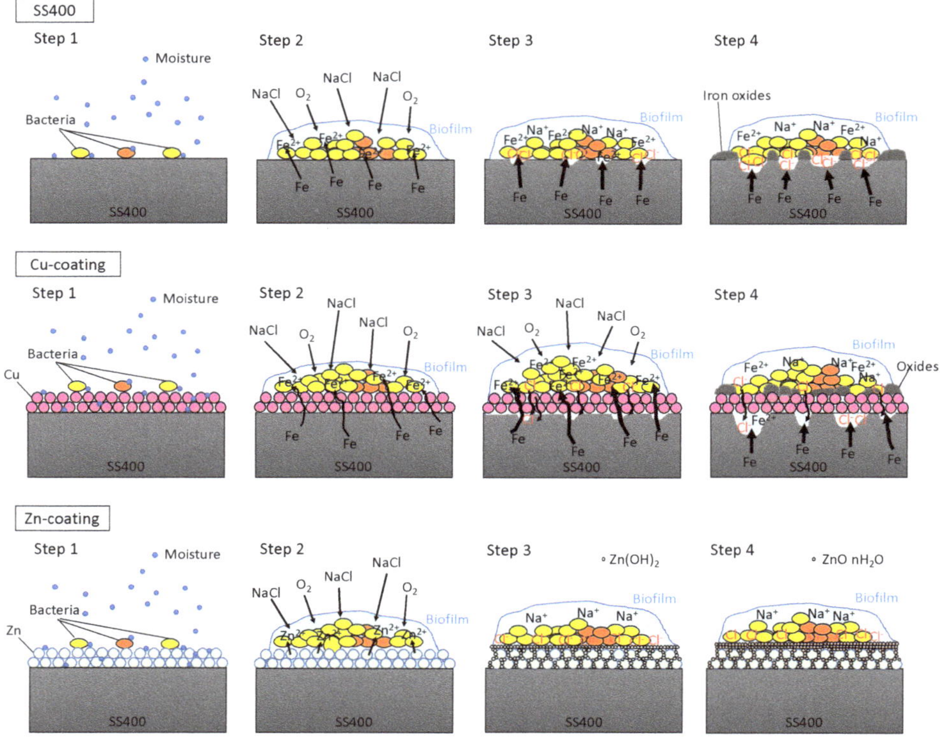

Figure 9. Predictive corrosion progression of the SS400 coupon, Zn-coated coupon, and Cu-coated coupon under outdoor conditions.

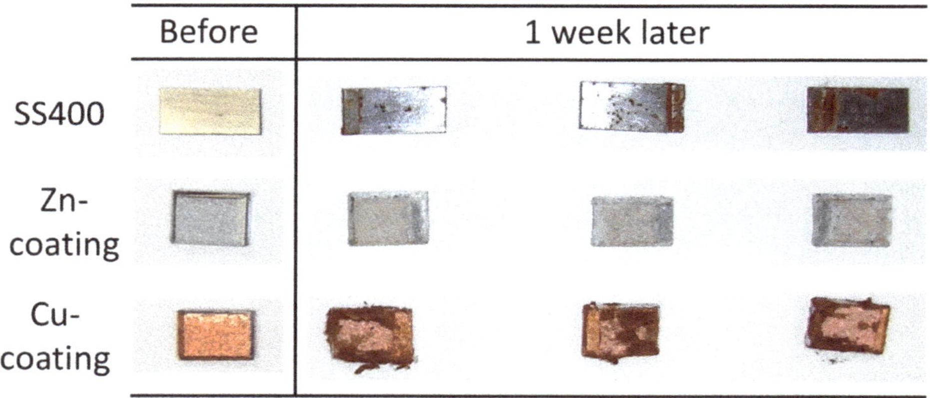

Figure 10. Digital photometric images of the surface of LBR-immerged coupons.

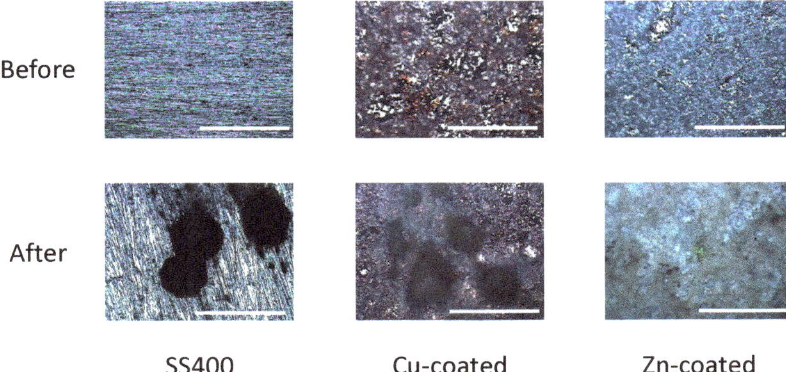

Figure 11. Optical microscopy images of the surface of the LBR-immerged coupons. Each white bar represents 5 mm.

SEM analysis showed that areas that bulged outward were observed on the surfaces of SS400, Cu- and Zn-coated coupons post-water-immersion testing (Figure 12). Since SS400, a carbon steel, consists of carbon and steel, these bulging sediments were obviously iron oxides (i.e., corroded iron products). In order to confirm whether or not the bulging areas arose from iron oxides or other compounds, an element mapping was conducted. On the surface of the Cu-coated coupons, areas with iron corresponded to areas with oxygen and silicon (Figure 13). Meanwhile, on the surface of the Zn-coated coupons, zinc areas corresponded to areas with oxygen and silicon (Figure 14). These results indicate that bulging areas of the surface of the Cu-coated and Zn-coated coupons were iron oxides and zinc oxides, respectively. Interestingly, these oxides were corrosion products. According to ionization tendency, zinc is more easily ionized than iron, while copper is barely more ionized than iron. Therefore, Zn-coated coupons and Cu-coated coupons will cause mainly zinc corrosion and iron corrosion, respectively. Indeed, detected corrosion products reflected the differences in ionization tendency between zinc and iron, or copper and iron. In addition, Kuroda et al. reported that silicon was detected with calcium in biofilms in cooling water systems [21]. Considering that silicon and calcium are derived from tap water, biofilms would be formed on the corroded sites of the Cu-coated coupons, Zn- coated coupons, and SS400 coupons by tap water and air-trapped bacteria.

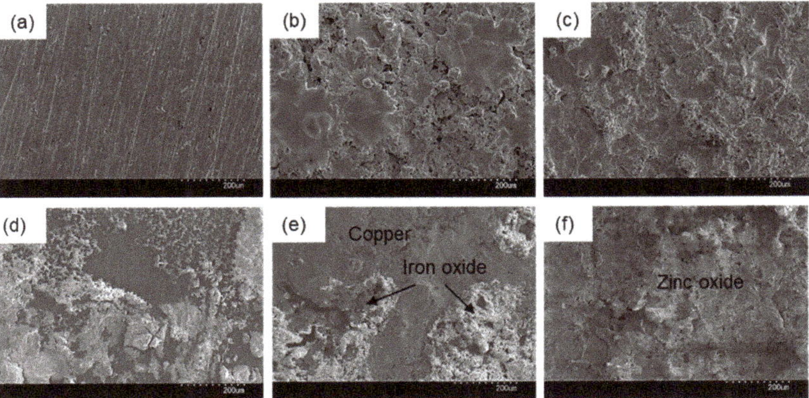

Figure 12. SEM images of the surface of coupons before (**a–c**) and after (**d–f**) LBR immersion testing. **a** and **d**: SS400; **b** and **e**: Cu-coated; **c** and **f**: Zn-coated.

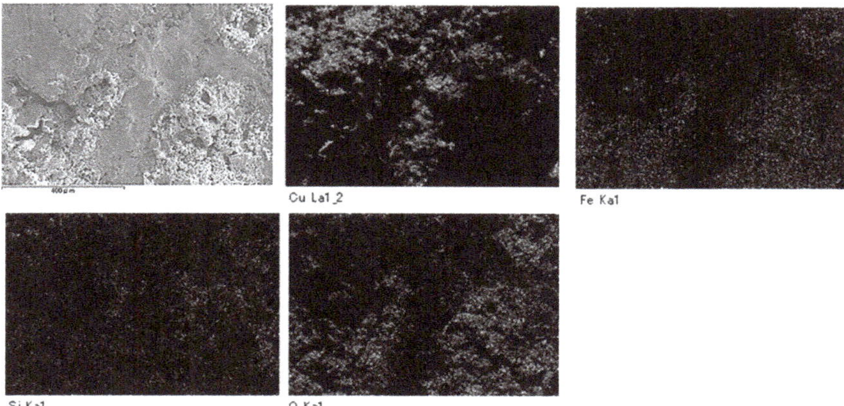

Figure 13. Element mapping images of Cu-coated coupons after LBR-immersion testing.

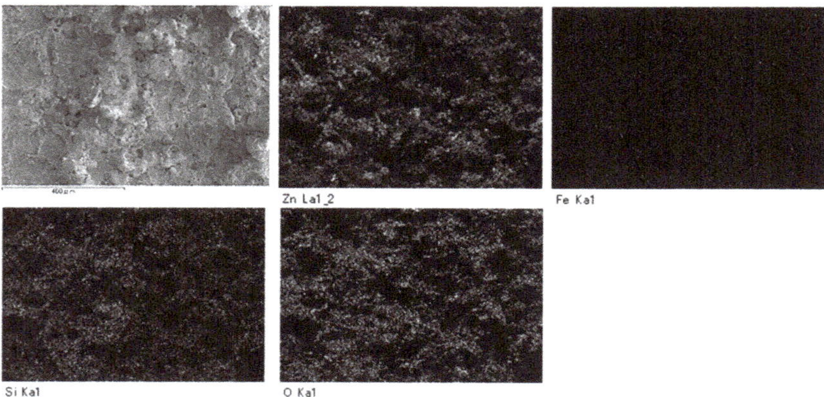

Figure 14. Element mapping images of the Zn-coated coupons after LBR-immersion testing exposure testing in water.

Next, biofilm formation was measured both quantitatively and qualitatively. Raman analysis showed that many peaks assigned to EPSs were detected on the surface of all coupons after water-immersion testing (Figure 15). These results indicated that biofilms formed on the surfaces of all coupons. Additionally, all coupons showed that relative intensities more than 2000 cm^{-1} were higher than those for less than 2000 cm^{-1}, although the pattern of Raman peaks varied among samples. Considering the results of the Raman peaks, it remains unclear as to how the biofilms formed on the SS400, Cu-, or Zn-coated coupons were similar (or different), because there is very little information about Raman peaks in the more than 2000 cm^{-1} region that have been assigned in EPSs.

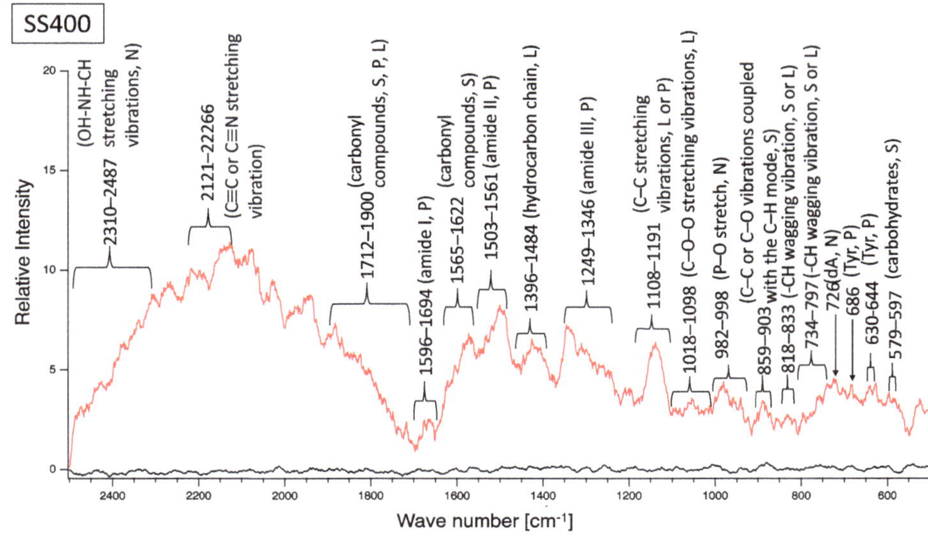

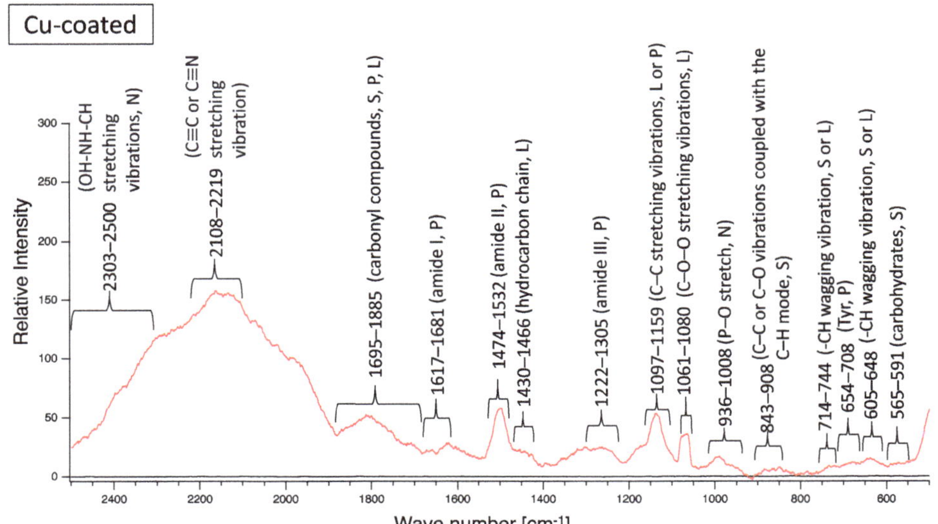

Figure 15. *Cont.*

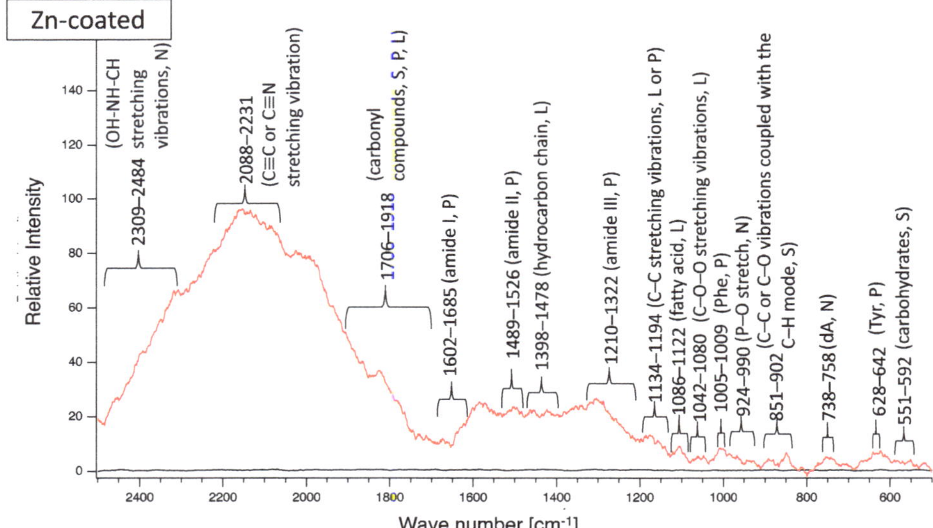

Figure 15. Raman peaks of sediments on the surface of the coupons after LBR immersion testing (red lines). Black lines show the Raman peaks of the surface of the specimens before the test. Detected Raman peaks after the test were assigned to related chemical bonds or compounds according to information in references [27–35]. N: nucleic acids; L: lipids; P: proteins; S: polysaccharides.

Compared with the amount of biofilm that formed on the SS400 coupons, significantly more biofilms formed on the Cu-coated coupons, while the amount of biofilm that formed on the Zn-coated coupons was significantly lower (Figure 16). Ranking the three specimens in descending order of biofilm quantity, the order (Cu-coated coupons > SS400 coupons > Zn-coated coupons) corresponds exactly to their corrosion tendency. This indicates that the larger the biofilms that form on the surface of coupons in an aquatic environment, the more corrosion proceeds.

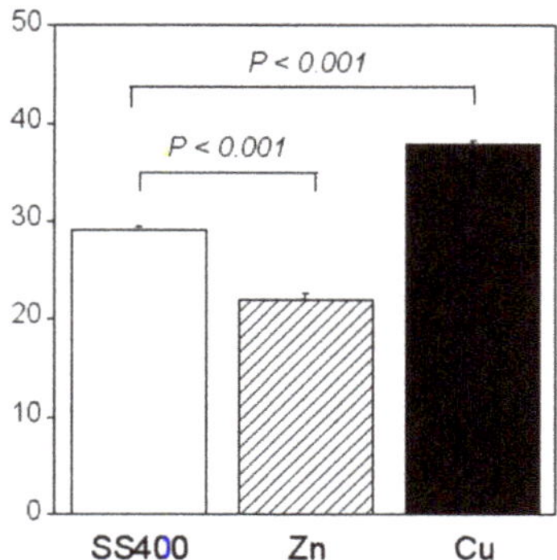

Figure 16. Biofilm quantification of the LBR-immersed coupons. Each column shows the mean of modulus of $\sqrt{(a^*)^2 + (b^*)^2 + (100 - L^*)^2}$ (n = 5). Zn and Cu mean the Zn-coated coupons and Cu-coated coupons, respectively. Error bars show the standard deviation. Student's t-test was performed between the SS400 coupons and Zn-coated coupons or SS400 coupons and Cu-coated coupons. P means the p-value.

Bacterial microbiome analysis showed that the occupancy pattern of taxonomic orders was similar in duplicate microbiomes of Zn-coated coupons and Cu-coated coupons, but not for the SS400 coupons (Figure 17). Rhodocyclales (#19 in Figure 17) was the most dominant in all samples; its occupancy on the Cu-coated coupons was just under 40%, while for the SS400 and Zn-coated coupons, it was more than 50%. However, the main dominant genera were different among them (Table 2): for the SS400 coupons, they were *Methyloversatilis* and an untitled genus, for Zn-coated coupons it was *Methyloversatilis*, but for Cu-coated coupons it was an untitled genus. Additionally, Sphingomonadales and Stramenopiles were characteristic orders found on the Cu- and Zn-coated coupons, respectively. The main genus of Sphingomonadales on Cu-coated coupons was *Sphingomonas* (Table 3), while for Stramenopiles on Zn-coated coupons, it was an unclassified new genus. Stramenopiles include ecologically important algae such as diatoms and kelp as well as heterotrophic and parasitic bacteria [57]. Algae are usually green or brown because of their chlorophylls, which will influence the color of biofilms. Stramenopiles was an abundant order (about 10%) on Zn-coated coupons, which is probably the reason why the biofilm on the Zn-coated coupons was green-brown in color (Figures 10 and 11). Many members of the Rhodocyclales have the capability to remove anthropogenic compounds from the environment and biological systems (i.e., to utilize various carbon compounds) [58]. *Methyloversatilis* comprises three species that can utilize single carbon compounds such as methanol and methylamine, as a sole source of energy [59–61]. This implies that biofilm-forming bacteria actively participate in the biodegradation of chemical compounds derived from metal coupons and/or the circulating LBR water. Moreover, *Sphingomonas* is known to be a nuisance in copper pipes used in drinking water distribution systems because it can accumulate copper ions in its cell wall and use these copper ions as binders for facilitating anodic reactions in the MIC of copper [62]. *Sphingomonas paucimobilis* is also reported to be a major biofilm producer [63], but copper can generally kill these bacteria by contact killing [64]. According to these bacterial features, the following story was inferred: a small amount of copper ions was eluted from the surface of Cu-coated coupons where *Sphingomonas* was preferentially drawn, forming a biofilm. Subsequently, a member of Rhodocyclales was recruited and became dominant due to iron ions eluted from the surface of the Cu-coated coupons, and attached to the biofilm, resulting in the progression of biofilm formation. On the surface of the SS400 coupons, iron ions were eluted and attracted many kinds of bacteria including the members of Rhodocyclales. Rhodocyclales probably proliferated dominantly because they could use various carbon compounds, then formed biofilms. On the surface of the Zn-coated coupons, zinc ions were eluted at the initial stage of this experience and also recruited some bacteria because zinc is known to be an essential element in many bacteria [65]. Next, the zinc ions bound to oxygen to make zinc oxides that inhibited the elution of zinc ions, which could limit the biofilm formation. Some bacteria are known as MIC-related bacteria, as listed in the previous section, for example, *Pseudomonas*, SRB such as *Desulphovibrio*, and iron-oxidizing bacteria such as *Acidthiobacillus* (*Thiobacillus*) and *Gallionella*. As shown in Table 4, few MIC-related bacteria were found on any of the LBR-immersed coupons. This result indicates that MIC-related bacteria were not involved in the biofilm formation or corrosion of the coupons in the LBR environment.

Table 2. The OTU percentages of order Rhodocyclales on post LBR-immersed coupons. All assigned genera are from the family Rhodocyclaceae.

Genus	SS00-1	SS400-2	Cu-Coated-1	Cu-Coated-2	Zn-Coated-1	Zn-Coated-2
Azoarcus	0.15	0.09	0	0	0.15	0.19
Azospira	0.02	0.05	0.02	0.02	0.01	0.01
Dechloromonas	1.24	2.17	1.13	1.06	0.67	0.74
Methyloversatilis	39.01	22.25	0.23	0.27	55.31	48.43
Rhodocyclus	0.03	0.01	0	0	0.03	0.03
untitled	16.92	48.76	36.27	34.25	3.74	3.40
new genus	1.57	1.22	0.18	0.12	1.68	1.61

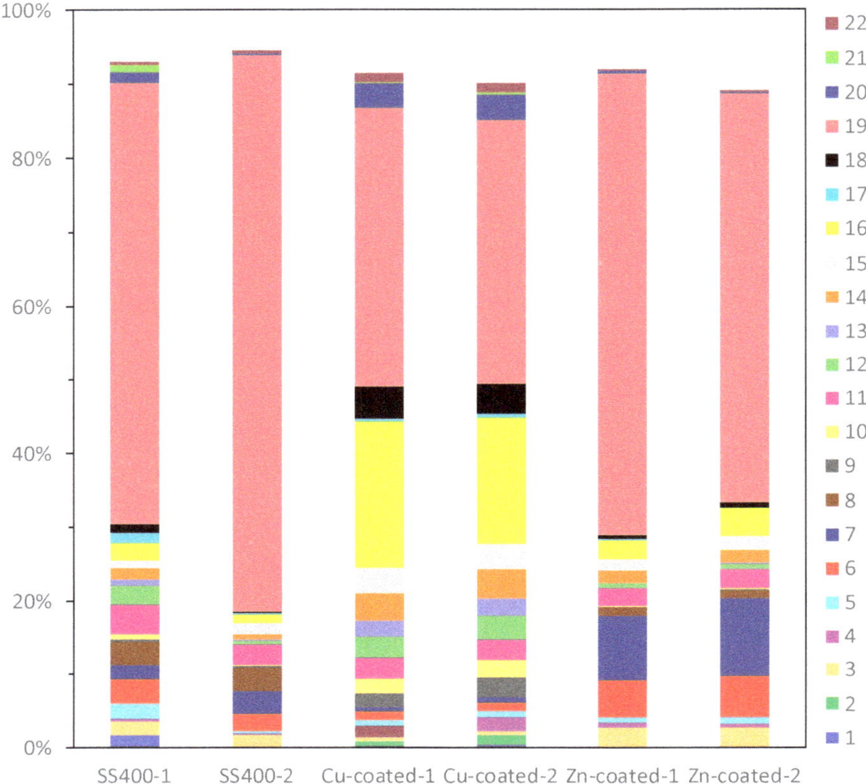

Figure 17. The OTU abundance percentages on water-immersed coupons. Bacterial orders present at <1.0% and unassigned OTUs are excluded from each column. (1) Solibacterales; (2) PK29; (3) Actinomycetales; (4) Cytophagales; (5) Saprospirales; (6) MLE1-12; (7) Stramenopiles; (8) Clostridiales; (9) Gemmatales; (10) Pirellulales; (11) Caulobacterales; (12) Rhizobiales; (13) Rhodobacterales; (14) Rhodospirillales; (15) Rickettsiales; (16) Sphingomonadales; (17) Unknown member of Betaproteobacteria; (18) Burkholderiales; (19) Rhodocyclales; (20) Myxococcales; (21) Unknown member of VHS-B5-50; (22) Unknown member of SJA-4.

Table 3. The OTU percentages of order Sphingomonadales on post LBR-immersed coupons.

Family	Genus	SS400-1	SS400-2	Cu-coated-1	Cu-coated-2	Zn-coated-1	Zn-coated-2
Erythrobacteraceae	*Azoarcus*	0	0	0.04	0.03	0	0
	Novosphingobium	0.07	0.02	0.13	0.15	0.13	0.22
	Sphingobium	0.18	0.10	0.11	0.12	1.17	2.31
Sphingomonadaceae	*Sphingomonas*	2.08	1.08	19.32	16.58	1.07	1.22
	untitled	0.01	0	0.02	0.03	0.01	0.01
	new genus	0.05	0.02	0.17	0.14	0.02	0.02

Table 4. The OTU percentages of MIC-related bacteria on post water-immersed coupons.

Family	Genus	SS400-1	SS400-2	Cu-coated-1	Cu-coated-2	Zn-coated-1	Zn-coated-2
Pseudomonadaceae	*Pseudomonas*	0.01	0	0.01	0	0.01	0
Hydrogenophilaceae	*Thiobacillus*	0	0	0	0	0	0
Gallionellaceae	*Gallionella*	0	0	0	0	0	0
Desulfarculaceae	new genus	0	0	0	0	0	0
Desulfobulbaceae	new genus	0	0	0	0	0	0
Desulfuromonadaceae	untitled	0	0	0	0	0	0
Geobacteraceae	new genus	0	0	0	0	0	0

3.3. Comparison of Corrosion in the Outdoor Exposure Test with that of the Water-Immersion Test

In both the outdoor exposure test and the water-immersion test, the Cu-coated coupons were the most intensively corroded, while the Zn-coated coupons were almost entirely covered with zinc oxide, and not iron oxide. This iron corrosion tendency of the Cu-coated coupon, Zn-coated coupon, and SS400 (Cu-coated coupon > SS400 coupon > Zn-coated coupon) was negatively correlated with the ionization tendencies of zinc, iron, and copper (zinc > iron > copper). Biofilms were detected on all coupons following exposure outdoors or in water, with the Zn-coated coupon found to have the smallest amount of biofilm in both trials. Based on these results, it seems likely that biofilm formation was initially involved in corrosion in both the outdoor and the water exposure tests. However, electrochemical corrosion was the dominant process in the outdoor exposure tests, while MIC was the dominant process in the water exposure tests. Indeed, a mixture of silicon and calcium was detected on the surface of the water-exposed SS400, Cu-, and Zn-coated coupons (Table 5). Based on a previous report that calcium and silicon are constituent elements of biofilms [21], biofilms formed well on the surface of the coupons immersed in water. After the exposure tests, the presence of coating elements (copper and zinc) decreased, while that of iron simultaneously increased. The range of decreases and increases was higher for the water exposure tests compared with the outdoor exposure tests. These results showed that an aquatic environment accelerated MIC more than an atmospheric environment.

Table 5. Main components of the surface of each coupon before and after the exposure tests (mass %).

Coupon	Test Type	Element				
		Fe	Cu	Zn	Ca	Si
SS400	before	100	0	0	0	0
	outdoor exposure	99.98	0	0	0.02	0
	water exposure	98.41	0	0	0.36	1.23
Cu-coated	before	0.68	99.01	0	0	0.31
	outdoor exposure	6.27	93.73	0	0	0
	water exposure	11.34	77.95	0	8.46	2.25
Zn-coated	before	0.09	0	99.52	0	0.39
	outdoor exposure	0.13	0	99.17	0	0.70
	water exposure	0.33	0	90.23	2.84	6.60

4. Conclusions

In this study, we investigated the effect of zinc thermal spray coated carbon steel (Zn-coated) and copper thermal spray coated carbon steel (Cu-coated) on iron corrosion in air or an aquatic environmental condition. We also explored which factor(s) such as biofilm formation (MIC) and ionization tendency (electrochemical corrosion) worked dominantly in iron corrosion. We expected that the Zn-coated and Cu-coated could inhibit iron corrosion due to the sacrificial corrosion of zinc and contact killing of biofilm-forming bacteria, respectively. In both air and water environments, the Zn-coated inhibited iron corrosion, but the Cu-coated accelerated iron corrosion, which was negatively correlated to ionization tendency (i.e., Zn > Fe > Cu). The dominant factor of iron corrosion will be electrochemical corrosion in the air environment and MIC in the water environment, however, it is probable that biofilm formation plays an important role in iron corrosion. In an air environment, biofilms can store water (moisture) that makes galvanic cells elute metallic ions according to ionization tendency. In a water environment, biofilms can accelerate iron corrosion caused by bacterial metabolites [66]. MIC-related bacteria have been found in specific environments such as oil tanks and water systems of nuclear power plants. In this study, MIC-related bacteria were barely detected in the biofilms formed on the surface of the Zn-coated, Cu-coated, and carbon steel. This implies that biofilm formation is an essential factor for iron corrosion, but MIC-related bacteria are not always necessary for it. Additionally, iron ions can predominantly attract bacteria more than zinc ions and copper ions, therefore, inhibiting the elution of iron (ions) will be an effective approach to regulate biofilm formation as well as iron corrosion.

Author Contributions: Conceptualization, A.O. and H.K.; methodology, A.O.; formal analysis, A.O.; investigation, A.O. and H.K.; resources, K.S.; data curation, A.O., K.T., H.K., D.K., T.K. and K.S.; writing—original draft preparation, A.O. and D.K.; writing—review and editing, S.T.; visualization, A.O., D.K. and H.K.; supervision, N.H.; project administration, A.O.; funding acquisition, H.K. All authors have read and agreed to the published version of the manuscript.

Funding: This research received no external funding

Acknowledgments: We had financial support from the Suga Weathering Technology Foundation (SWTF: 5-4-14, Shinjyuku, shinjyuku-ku, Tokyo, 160-0022, Japan) through a recommendation by the Japan Thermal Spray Society (JTSS: 2-2-29, Eiwa, Higashi Osaka, Osaka Pref. 577-0809, Japan) to carry out the experiments. We would like to take this opportunity to extend our special thanks to them all.

Conflicts of Interest: The authors declare no conflict of interest.

References

1. Marder, A.R. The metallurgy of zinc-coated steel. *Prog. Mater. Sci.* **2000**, *45*, 191–271. [CrossRef]
2. American Galvanizers Association. Zinc Coating 2011. Available online: https://galvanizeit.org/education-and-resources/publications/zinc-coatings (accessed on 19 February 2020).
3. Wang, P.; Zhang, D.; Qiu, R.; Wu, J.; Wan, Y. Super-hydrophobic film prepared on zinc and its effect on corrosion in simulated marine atmosphere. *Corros. Sci.* **2013**, *69*, 23–30. [CrossRef]
4. Little, B.J.; Lee, J.S.; Ray, R.I. The influence of marine biofilms on corrosion: A concise review. *Electrochim. Acta* **2008**, *54*, 2–7. [CrossRef]
5. Wilking, J.N.; Angelini, T.E.; Seminara, A.; Brenner, M.P.; Weitz, D.A. Biofilms as complex fluids. *MRS Bull.* **2011**, *36*, 385–391. [CrossRef]
6. Usher, K.M.; Kaksonen, A.H.; Cole, I.; Marney, D. Critical review: Microbially influenced corrosion of buried carbon steel pipes. *Int. Biodeterior. Biodegrad.* **2014**, *93*, 84–106. [CrossRef]
7. Xing, X.; Wang, H.; Hu, C.; Liu, L. Characterization of bacterial community and iron corrosion in drinking water distribution systems with O3-biological activated carbon treatment. *J. Environ. Sci.* **2008**, *69*, 192–204. [CrossRef]
8. Khoshnaw, F.; Gubner, F. *Corrosion Atlas Case Studies*; 2019 Edition; Elsevier Amsterdam: Oxford, UK, 2019.
9. Huttunen-Saarivirta, E.; Rajala, P.; Marja-aho, M.; Maukonen, J.; Sohlberg, E.; Carpén, L. Ennoblement, corrosion, and biofouling in brackish seawater: Comparison between six stainless steel grades. *Bioelectrochemistry* **2018**, *120*, 27–42. [CrossRef]
10. Jia, R.; Yang, D.; Xu, D.; Gu, T. Anaerobic Corrosion of 304 Stainless Steel Caused by the *Pseudomonas aeruginosa* Biofilm. *Front. Microbiol.* **2017**, *8*, 2335. [CrossRef]
11. Sano, K.; Kanematsu, H.; Kogo, T.; Hirai, N.; Tanaka, T. Corrosion and biofilm for a composite coated iron observed by FTIR-ATR and Raman spectroscopy. *Trans. IMF* **2016**, *94*, 139–145. [CrossRef]
12. Dwivedi, D.; Lepková, K.; Becker, T. Carbon steel corrosion: A review of key surface properties and characterization methods. *RSC Adv.* **2017**, *7*, 4580–4610. [CrossRef]
13. Xu, D.; Li, Y.; Gu, T. Mechanistic modeling of biocorrosion caused by biofilms of sulfate reducing bacteria and acid producing bacteria. *Bioelectrochemistry* **2016**, *110*, 52–58. [CrossRef] [PubMed]
14. Kanematsu, H.; Barry, D.M. (Eds.) *Corrosion control and surface finishing: Environmentally Friendly Approaches*; Springer Nature: Tokyo, Japan, 2016.
15. Melchers, R.E. *9—Modelling long term corrosion of steel infrastructure in natural marine environments*; Liengen, T., Féron, D., Basséguy, R., Beech, I.B., Eds.; Understanding Biocorrosion. Woodhead Publishing: Oxford, UK, 2014; pp. 213–241.
16. Faúndez, G.; Troncoso, M.; Navarrete, P.; Figueroa, G. Antimicrobial activity of copper surfaces against suspensions of Salmonella enterica and Campylobacter jejuni. *BMC Microbiol.* **2004**, *4*, 19. [CrossRef] [PubMed]
17. Mehtar, S.; Wiid, I.; Todorov, S.D. The antimicrobial activity of copper and copper alloys against nosocomial pathogens and Mycobacterium tuberculosis isolated from healthcare facilities in the Western Cape: An in vitro study. *J. Hosp. Infect.* **2008**, *68*, 45–51. [CrossRef] [PubMed]

18. Grass, G.; Rensing, C.; Solioz, M. Metallic copper as an antimicrobial surface. *Appl. Environ. Microbiol.* **2011**, *77*, 1541–1547. [CrossRef] [PubMed]
19. Ogawa, A.; Kanematsu, H.; Sano, K.; Sakai, Y.; Ishida, K.; Beech, I.; Suzuki, O.; Tanaka, T. Effect of silver or copper nanoparticles-dispersed silane coatings on biofilm formation in cooling water systems. *Materials* **2016**, *9*, 632. [CrossRef] [PubMed]
20. Kanematsu, H.; Barry, D.M.; Ikigai, H.; Yoshitake, M.; Mizunoe, Y. Biofilm evaluation methods outside body to inside -Problem presentations for the future. *Med. Res. Arch.* **2017**, *5*, 1–17.
21. Kuroda, D.; Kamakura, N.; Ito, H.; Ikegai, H.; Kanematsu, H. Adhesion of microorganisms on the surfaces of various metallic immersed in a cooling water tank of the package type cooling tower. *Tetsu Hagané* **2012**, *98*, 1–8. [CrossRef]
22. Ogawa, A.; Takakura, K.; Sano, K.; Kanematsu, H.; Yamano, T.; Saishin, T.; Terada, S. Microbiome analysis of biofilms of silver nanoparticle-dispersed silane-based coated carbon steel using a next-generation sequencing technique. *Antibiotics* **2018**, *7*, 91. [CrossRef]
23. Ogawa, A.; Kiyohara, T.; Kobayashi, Y.; Sano, K.; Kanematsu, H. Nickel, molybdenum, and tungsten nanoparticle-dispersed alkylalkoxysilane polymer for biomaterial coating: Evaluation of effects on bacterial biofilm formation and biosafety. *Biomed. Res. Clin. Pract.* **2017**, *2*, 1–7. [CrossRef]
24. Iron Corrosion Products. Available online: https://corrosion-doctors.org/Experiments/iron-products.htm (accessed on 29 December 2019).
25. PubChem Compound Summary for CID 14806. Available online: https://pubchem.ncbi.nlm.nih.gov/compound/zinc_oxide#section=Top (accessed on 29 December 2019).
26. Lewandowski, B. *Chapter 1, Fundamentals of Biofilm Research*, 2nd ed.; CRC Press: Boca Raton, FL, USA, 2014; pp. 1–65.
27. George, J.; Thomas, Jr. Raman spectroscopy of protein and nucleic acid assembles. *Annu. Rev. Biophys. Biomol. Struct.* **1999**, *28*, 1–27.
28. Yuen, S.N.; Choi, S.M.; Phillips, D.L.; Ma, C.Y. Raman and FTIR spectroscopic study of carboxymethylated non-starch polysaccharides. *Food Chem.* **2009**, *114*, 1091–1098. [CrossRef]
29. Chao, Y.; Zhang, T. Surface-enhanced Raman scattering (SERS) revealing chemical variation during biofilm formation: From initial attachment to mature biofilm. *Anal. Bioanal. Chem.* **2012**, *404*, 1465–1475. [CrossRef] [PubMed]
30. Czamara, K.; Majzner, K.; Pacia, M.Z.; Kochan, K.; Kaczor, A.; Baranska, M. Raman spectroscopy of lipids: A review. *J. Raman Spectrosc.* **2015**, *46*, 4–20. [CrossRef]
31. Benevides, J.M.; Overman, S.A.; Thomas, G.J. Raman, polarized Raman and ultraviolet resonance Raman spectroscopy of nucleic acids and their complexes. *J. Raman Spectrosc.* **2005**, *36*, 279–299. [CrossRef]
32. Movasaghi, Z.; Rehman, S.; Rehman, I.U. Raman spectroscopy of biological tissues. *Appl. Spectrosc. Rev.* **2007**, *42*, 493–541. [CrossRef]
33. Rygula, A.; Majzner, K.; Marzec, K.M.; Kaczor, A.; Pilarczyk, M.; Baranska, M. Raman spectroscopy of proteins: A review. *J. Raman Spectrosc.* **2013**, *44*, 1061–1076. [CrossRef]
34. Larkin, P. *Chapter 7—General Outline and Strategies for IR and Raman Spectral Interpretation*; Elsevier: Oxford, UK, 2011.
35. Larkin, P. *Chapter 6—IR and Raman Spectra-Structure Correlations: Characteristic Group Frequencies*; Elsevier: Oxford, UK, 2011.
36. Pantanella, F.; Valenti, P.; Natalizi, T.; Passeri, D.; Berlutti, F. Analytical techniques to study microbial biofilm on abiotic surfaces: Pros and cons of the main techniques currently in use. *Ann Ig* **2013**, *25*, 31–42.
37. Kanematsu, H.; Ikigai, H.; Yoshitake, M. Evaluation of Various Metallic Coatings on Steel to Mitigate Biofilm Formation. *Int. J. Mol. Sci.* **2009**, *10*, 559–571. [CrossRef]
38. Hamzah, E.; Hussain, M.Z.; Ibrahim, Z.; Abdolahi, A. Influence of *Pseudomonas aeruginosa* bacteria on corrosion resistance of 304 stainless steel. *Corros. Eng. Sci. Technol.* **2013**, *48*, 116–120. [CrossRef]
39. Abdolahi, A.; Hamzah, E.; Ibrahim, Z.; Hashim, S. Microbially influenced corrosion of steels by *Pseudomonas aeruginosa*. *Corros. Rev.* **2014**, *32*, 129–141. [CrossRef]
40. Beech, I.B.; Sztyler, M.; Gaylarde, C.C.; Smith, W.L.; Sunner, J. 2—Biofilms and Biocorrosion. In *Understanding Biocorrosion*; Liengen, T., Féron, D., Basséguy, R., Beech, I.B., Eds.; Woodhead Publishing: Oxford, UK, 2014; pp. 33–56.

41. Chen, Y.; Tang, Q.; Senko, J.M.; Cheng, G.; Newby, B.M.Z.; Castaneda, H.; Ju, L.K. Long-term survival of *Desulfovibrio vulgaris* on carbon steel and associated pitting corrosion. *Corros. Sci.* **2015**, *90*, 89–100. [CrossRef]
42. Donati, E.; Curutchet, G.; Pogliani, C.; Tedesco, P. Bioleaching of covellite using pure and mixed cultures of *Thiobacillus ferrooxidans* and *Thiobacillus thiooxidans*. *Process Biochem.* **1996**, *31*, 129–134. [CrossRef]
43. Jones, D.S.; Albrecht, H.L.; Dawson, K.S.; Schaperdoth, I.; Freeman, K.H.; Pi, Y.; Pearson, A.; Macalady, J.L. Community genomic analysis of an extremely acidophilic sulfur-oxidizing biofilm. *ISME J.* **2012**, *6*, 158–170. [CrossRef] [PubMed]
44. Ray, R.I.; Lee, J.S.; Little, B.J. Iron-Oxidizing Bacteria: A Review of corrosion mechanisms in fresh water and marine environments. In Proceedings of the NACE Corrosion 2010 Conference, Atlanta, GA, USA, 14–18 March 2010.
45. Yeager, C.M.; Gallegos-Graves, V.; Dunbar, J.; Hesse, C.N.; Daligault, H.; Kuske, C.R. Polysaccharide degradation capability of Actinomycetales soil isolates from a Semiarid Grassland of the Colorado Plateau. *Appl. Environ. Microbiol.* **2017**, *83*, e03020-16. [CrossRef]
46. Brahmaprakash, G.P.; Pramod Kumar Sahu, G.; Pramod Kumar Sahu, G.; Lavanya, G.; Nair, S.S.; Gangaraddi, V.K.; Gupta, A. Microbial functions of the Rhizosphere. Plant-microbe interactions in agro-ecological perspectives volume 1: Fundamental mechanisms. *Methods Funct.* **2017**, *1*, 177–210.
47. Paredes-Sabja, D.; Setlow, P.; Sarker, M.R. Germination of spores of Bacillales and Clostridiales species: Mechanisms and proteins involved. *Trends Microbiol.* **2011**, *19*, 85–94. [CrossRef]
48. Baldani, J.I.; Rouws, L.; Cruz, L.M.; Olivares, F.L.; Schmid, M.; Hartmann, A. The family Oxalobacteraceae. In *the Prokaryotes*; Rosenberg, E., DeLong, E.F., Lory, S., Stackebrandt, E., Thompson, F., Eds.; Springer: Berlin/Heidelberg, Germany, 2014.
49. Percival, S.; Knapp, J.; Wales, D.; Edyvean, R.G.J. The effect of turbulent flow and surface roughness on biofilm formation in drinking water. *J. Ind. Microbiol. Biotech.* **1999**, *22*, 152. [CrossRef]
50. Yang, L.; Liu, Y.; Wu, H.; Høiby, N.; Molin, S.; Song, Z.J. Current understanding of multi–species biofilms. *Int. J. Oral Sci.* **2011**, *3*, 74–81. [CrossRef]
51. Zhang, S.; Pang, S.; Wang, P.; Wang, C.; Guo, C.; Addo, F.G.; Li, Y. Responses of bacterial community structure and denitrifying bacteria in biofilm to submerged macrophytes and nitrate. *Sci. Rep.* **2016**, *6*, 36178. [CrossRef]
52. Pantanella, F.; Berlutti, F.; Passariello, C.; Sarli, S.; Morea, C.; Schippa, S. Violacein and biofilm production in *Janthinobacterium lividum*. *J. Appl. Microbiol.* **2007**, *102*, 992–999. [CrossRef]
53. Alec Groysman. Chapter 1 Corrosion mechanism and corrosion factors. In *Corrosion for Everybody*; Springer: Berlin/Heidelberg, Germany, 2010; pp. 1–52.
54. Ericsson, R. The influence of sodium chloride on the atmospheric corrosion of steel. *Mater. Corros.* **1978**, *29*, 400–403. [CrossRef]
55. Burns, R.M. Chemical reactions in the corrosion of metals. *J. Chem. Educ.* **1953**, *30*, 318–321. [CrossRef]
56. Tamura, H. The role of rusts in corrosion and corrosion protection of iron and steel. *Corros. Sci.* **2008**, *50*, 1872–1883. [CrossRef]
57. Derelle, R.; Lopez-Garcia, P.; Timpano, H.; Moreira, D.; Loy, A.; Schulz, C.A. Phylogenomic framework to study the diversity and evolution of Stramenopiles (=Heterokonts). *Mol. Biol. Evol.* **2016**, *33*, 2890–2898. [CrossRef]
58. Loy, A.; Schulz, C.; Lucker, S.; Schopfer-Wendels, A.; Stoecker, K.; Baranyi, C.; Lehner, A.; Wagner, M. 16S rRNA gene-based oligonucleotide microarray for environmental monitoring of the betaproteobacterial order "Rhodocyclales". *Appl. Environ. Microbiol.* **2005**, *71*, 1373–1386. [CrossRef]
59. Kalyuzhnaya, M.G.; De Marco, P.; Bowerman, S.; Pacheco, C.C.; Lara, J.C.; Lidstrom, M.E.; Chistoserdova, L. Methyloversatilis universalis gen. nov., sp. nov., a novel taxon within the Betaproteobacteria represented by three methylotrophic isolates. *Int. J. Syst. Evol. Microbiol.* **2006**, *56*, 2517–2522. [CrossRef]
60. Doronina, N.V.; Kaparullina, E.N.; Trotsenko, Y.A. Methyloversatilis thermotolerans sp. nov., a novel thermotolerant facultative methylotroph isolated from a hot spring. *Int. J. Syst. Evol. Microbiol.* **2014**, *64*, 158–164. [CrossRef]
61. Smalley, N.E.; Taipale, S.; De Marco, P.; Doronina, N.V.; Kyrpides, N.; Shapiro, N.; Woyke, T.; Kalyuzhnaya, M.G. Functional and genomic diversity of methylotrophic Rhodocyclaceae: Description of *Methyloversatilis discipulorum* sp. nov. *Int. J. Syst. Evol. Microbiol.* **2015**, *65*, 2227–2233. [CrossRef]

62. White, D.C.; Sutton, S.D.; Ringelberg, D.B. The genus Sphingomonas: Physiology and ecology. *Curr. Opin. Biotechnol.* **1996**, *7*, 301–306. [CrossRef]
63. Gusman, V.; Medić, D.; Jelesić, Z.; Mihajlović-Ukropina, M. *Sphingomonas paucimobilis* as a biofilm producer. 64. *Arch. Biol. Sci.* **2012**, *64*, 1327–1331. [CrossRef]
64. Leys, N.M.E.J.; Ryngaert, A.; Bastiaens, L.; Verstraete, W.; Top, E.M.; Springael, D. Occurrence and phylogenetic diversity of Sphingomonas strains in soils contaminated with polycyclic aromatic hydrocarbons. *Appl. Environ. Microbiol.* **2004**, *70*, 1944–1955. [CrossRef] [PubMed]
65. Blencowe, D.K.; Morby, A.P. Zn(II) metabolism in prokaryotes. *FEMS Microbiol. Rev.* **2003**, *27*, 291–311. [CrossRef]
66. Kakooei, S.; Che Ismail, M.; Ariwahjoedi, B. Mechanisms of microbiologically influenced corrosion: A review. *World Appl. Sci. J.* **2012**, *17*, 524–531.

© 2020 by the authors. Licensee MDPI, Basel, Switzerland. This article is an open access article distributed under the terms and conditions of the Creative Commons Attribution (CC BY) license (http://creativecommons.org/licenses/by/4.0/).

Article

Investigation of Biogenic Passivating Layers on Corroded Iron

Lucrezia Comensoli [1,2,†], Monica Albini [1,2,‡,§], Wafa Kooli [1,2,§], Julien Maillard [3], Tiziana Lombardo [4], Pilar Junier [1] and Edith Joseph [2,5,*]

1 Laboratory of Microbiology, Institute of Biology, University of Neuchâtel, 2000 Neuchâtel, Switzerland; lucrezia.comensoli@empa.ch (L.C.); monica.albini@cnr.it (M.A.); wafam.kooli@gmail.com (W.K.); pilar.junier@unine.ch (P.J.)

2 Laboratory of Technologies for Heritage Materials, Institute of Chemistry, University of Neuchâtel, 2000 Neuchâtel, Switzerland

3 Laboratory for Environmental Biotechnology, ENAC-IIE-LBE, Ecole Polytechnique Fédérale de Lausanne, 1015 Lausanne, Switzerland; julien.maillard@epfl.ch

4 Laboratory of conservation research, Sammlungszentrum, Swiss national museum, Lindenmoosstrasse 1, 8910 Affoltern am Albis, Switzerland; Tiziana.Lombardo@nationalmuseum.ch

5 Haute Ecole Arc Conservation-Restauration, HES-SO, 2000 Neuchâtel, Switzerland

* Correspondence: edith.joseph@unine.ch

† Present address: Laboratory for Mechanical Systems Engineering, Swiss Federal Laboratories for Materials Science and Technology, 8600 Dübendorf, Switzerland.

‡ Present address: Institute for the study of Nanostructured Materials, Italian National council of Research (CNR-ISMN), 7-00185 Rome, Italy.

§ These authors contributed equally to this work.

Received: 12 February 2020; Accepted: 3 March 2020; Published: 6 March 2020

Abstract: This study evaluates mechanisms of biogenic mineral formation induced by bacterial iron reduction for the stabilization of corroded iron. As an example, the *Desulfitobacterium hafniense* strain TCE1 was employed to treat corroded coupons presenting urban natural atmospheric corrosion, and spectroscopic investigations were performed on the samples' cross-sections to evaluate the corrosion stratigraphy. The treated samples presented a protective continuous layer of iron phosphates (vivianite $Fe^{2+}{}_3(PO_4)_2{\cdot}8H_2O$ and barbosalite $Fe^{2+}Fe^{3+}{}_2(PO_4)_2(OH)_2$), which covered 92% of the surface and was associated with a decrease in the thickness of the original corrosion layer. The results allow us to better understand the conversion of reactive corrosion products into stable biogenic minerals, as well as to identify important criteria for the design of a green alternative treatment for the stabilization of corroded iron.

Keywords: iron corrosion; SEM; Raman spectroscopy; biogenic minerals; bacterial iron reduction; cultural heritage; conservation-restoration; corrosion stabilization

1. Introduction

Cast iron and steel objects corrode rapidly when they are exposed outdoors, and complex layers composed of iron oxides and iron hydroxides are formed over time [1–3]. In the presence of moisture, these corrosion layers adsorb water and incorporate particulate matter and pollution agents from the atmosphere, which instigate further corrosion processes. Generally, in addition to iron oxides and iron hydroxides, iron sulfates are also frequently reported compounds within an atmospheric corrosion layer, especially in polluted and urban areas [4,5]. Near coastal areas, chlorine is a problematic element. Iron items exposed to such surroundings are susceptible to chloride-promoted corrosion, leading to chemical as well as mechanical damage to these objects [2,3,5]. Chloride ions diffused into the objects as counter-ions to iron(II) ions are produced by the oxidation of the metal [6]. Iron oxyhydroxides are then produced, including chloride-bearing akaganéite $FeO_{0.833}(OH)_{1.167}Cl_{0.167,}$ and another ferrous hydroxychloride β–$Fe_2(OH)_3Cl$ [6,7]. In the presence of such hygroscopic compounds, iron corrosion starts at low relative humidity (RH) values, down to 15% RH, and dramatically increases above 35% RH, leading to the fragmentation and disintegration of the objects [8].

Hence, outdoor iron objects that are exposed to these environments are subjected to severe degradation and would be damaged or lost without maintenance and appropriate conservation strategies. Therefore, several conservation and restoration methods are currently performed to stabilize the corrosion layers on outdoor iron-based structures. Organic coatings, such as waxes, resins, and oils, are applied to protect iron surfaces, acting as a physical barrier against atmospheric agents [3,9,10]. Nevertheless, these protective systems often need frequent maintenance to avoid failures of the coating film and exposure of the metal substrate to further oxidation [11]. Another approach employed is the use of corrosion inhibitors. However, most of these are hazardous compounds such as chromates, benzoates or nitrites, which have to be carefully manipulated [3]. Concerning archaeological iron artifacts, three methods are used to stabilize their corrosion layer. First, to remove the chloride ions from the corrosion layer, the object is immersed in anoxic and alkaline aqueous solutions [3,12]. Nevertheless, this method is based on osmotic diffusion, which is a slow process, and thus the solution needs to be replaced regularly when the chloride ions' concentration stops increasing or stays low (below 20 ppm). Also, a large quantity of generated waste needs to be processed afterward for safe disposal. Second, electrolytic reduction allows an increase in the porosity of the corrosion layer and thus enhances the diffusion of harmful salts from the objects [13]. However, this method is restricted to large marine finds, as there is a significant loss of the surface and a lack of control over the amount of salts extracted and the corrosion products reduced during hydrogen bubbling [3]. Finally, plasma treatment is usually applied as a pre-treatment, as it creates cracks and fissures that will facilitate the diffusion of chloride ions during a successive alkaline bath [14].

For all these reasons, there is a general consensus that, to date, an efficient and durable protective system to control iron corrosion on archaeological artifacts does not exist [15].

Microbes are frequently considered detrimental to metallic surfaces, as they are associated with the process of microbial induced corrosion (MIC) [16–18]. However, given the large metabolic diversity found within the microbial world, an increasing number of studies focus on the exploitation of microbial processes to inhibit corrosion. Experimental evidence suggests that a biofilm of aerobic bacteria attached to copper surfaces in freshwater and seawater reduces the copper corrosion rate by decreasing the oxygen content nearby [15]. In addition, several studies reported the protective behavior of specific biofilms on submerged carbon steel pipelines [19–22]. In particular, the formation of iron phosphates as a protective layer was accomplished through the exploitation of bacterial biomineralization processes. For instance, electrochemical measurements demonstrated that the presence of a biogenic layer of vivianite ($Fe^{2+}{}_3(PO_4)_2{\cdot}8H_2O$) produced by *Geobacter sulfurreducens* on the surface of carbon steel coupons had a protective effect against corrosion [23].

In the examples cited above, most of the bio-based approaches were developed for the protection of bare iron surfaces before their exposure to outdoor environments. As part of our research topic, we investigated the potential of microbes for the stabilization of already corroded iron (archaeological objects and outdoor surfaces) by converting part of the reactive corrosion layer into more stable biogenic minerals (Figure 1). In particular, we studied different bacterial species, *Shewanella loihica*, *Desulfitobacterium hafniense* and *Aeromonas* spp., for their ability to produce biogenic iron minerals on corroded steel coupons [24–28]. Hence, vivianite and siderite were produced by *S. loihica* on costal-exposed coupons, while *D. hafniense* and *Aeromonas* spp. induced the formation of vivianite and siderite on urban-exposed coupons [26–28]. In order to better understand the biomineralization process involved, corroded coupons exposed in an urban environment and treated with *D. hafniense* have been investigated through the present stratigraphic study.

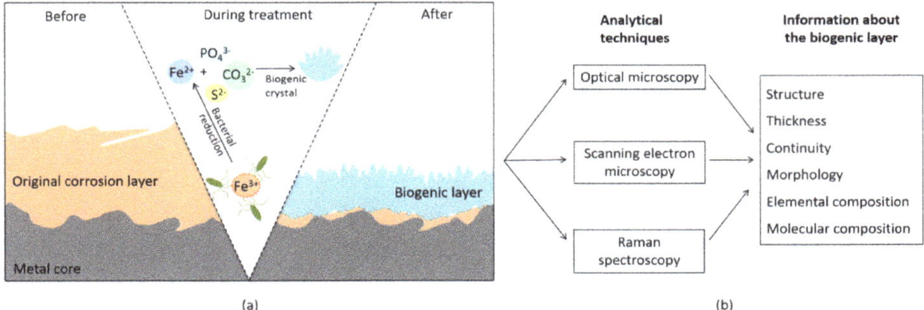

Figure 1. (**a**) Schematic cross-section of a corroded iron coupon submitted to bacterial treatment, showing microbial-induced modifications occurring; (**b**) the analytical methodology performed.

2. Materials and Methods

2.1. Description of Samples

Samples of 12.5 × 25 × 2–3 mm were obtained from a steel plate with a natural urban corrosion layer mainly composed of lepidocrocite and goethite [25,26]. The plate was exposed for about 10 years in an urban environment (Zürich, Switzerland).

2.2. Bacterial Treatment

The *D. hafniense* strain TCE1 (DSMZ-German Collection of Microorganisms and Cell Culture GmbH 12704) was used for this study. This bacterium was selected as it can use a variety of electron acceptors, especially halogenated organic compounds and metals [29,30]. In addition, a previous study revealed that this strain was able to reduce Fe^{3+}-citrate in the presence of 0.2% and 0.3% NaCl, and that it was more efficient in terms of production of iron phosphates on the surface of corroded iron coupons [25,26]. Bacterial pre-incubation was performed in the dark at 30 °C under agitation in a standard mineral medium under anoxic conditions in 500 mL serum bottles, until reaching an optical density (OD600) of 0.1–0.15, as previously described [31]. Quantities of 45 mM of lactate and 20 mM of fumarate were added as an electron donor and acceptor, respectively, as well as a buffer solution containing phosphates and carbonates (4 mM K_2HPO_4 and 1 mM NaH_2PO_4; 54 mM $NaHCO_3$ and 6 mM NH_4HCO_3) to maintain the pH at 7.3. To avoid the corrosion of iron coupons during treatment that would lead to a misinterpretation of the results obtained, passivating conditions were achieved by replacing O_2 with a mix of N_2/CO_2 (80%/20%) and by adding Na_2S as a reducing agent. The treatment of the coupons was then performed as previously described [25,26]. Before treatment, the coupons were sterilized by spraying them with ethanol 70% (wt/wt in deionized water) and exposure to UV radiation (20 minutes on each face). The samples were then placed into 50 mL serum bottles, and autoclaving was performed under anoxic conditions (as defined in pre-incubation). Next, 20 mL of bacterial solution or culture medium (abiotic control) was added. After 7 days of incubation, the coupons were taken out of the treatment solution and sterilized as above (no more bacteria or culture media were present on the treated surfaces). All of the experiments were performed in triplicates, and the results presented here were identical for each set of samples.

2.3. Analytical Techniques

After treatment, the coupons were sampled and cold-embedded in methacrylate resin using the EpoFix Kit (Struers GmbH—Zweigniederlassung Schweiz, Birmensdorf, Switzerland). Cross-polishing was performed using successive silicon carbide abrasive papers (250, 500, and 1000 grit) and Micro-Mesh abrasive cloths (1800, 2400, 3200, 3600, 4000, 6000, 8000, and 12,000 grades). The cross-sectioned samples were then analyzed with optical and scanning electron microscopy, as well as with Raman spectroscopy.

2.3.1. Optical Microscopy

Microscopic observations were performed under a Polyvar MET optical microscope (Leica Microsystems (Schweiz) AG,
Verkaufsgesellschaft, Heerbrugg, Switzerland) to characterize the corrosion layer and the biogenic crystals formed. An estimation of the conversion percentage of the original corrosion layer into biogenic crystals was extrapolated with Axio Vision LE® software (version 4.8.2.0, Carl Zeiss MicroImaging GmbH, Iéna, Germany).

2.3.2. Scanning Electron Microscopy

Scanning Electron Microscopy coupled with Energy Dispersive Spectroscopy (SEM–EDS) mapping was carried out to evaluate the elemental composition as well as the distribution of the corrosion products and biogenic minerals. Coupons were mounted onto stubs using carbon conductive tape to ensure electrical conductivity and were directly analyzed using a environmental scanning electron microscope Philips XL30 ESEM FEG (Thermo Fisher Scientific, Hillsboro, Oregon, USA) equipped with an energy-dispersive X-ray analyzer. Backscattered electron images were acquired at an acceleration potential of 10 to 25 keV and a working distance of 10 mm. For elemental mapping, a resolution of 64×50 pixels and a dwell of 1000 were employed.

2.3.3. Raman Spectroscopy

Raman spectroscopy was also performed to study the molecular composition of the corrosion layer before and after bacterial treatment. The analyses were carried out with a Horiba-Jobin Yvon Labram Aramis microscope equipped with an Nd:YAG (neodymium-doped yttrium aluminum garnet; Nd:$Y_3Al_5O_{12}$) laser of 532 nm at a power lower than 1 mW. Single-point measurements were carried out with the following conditions: spectral range 100–1600 cm^{-1}, 400 μm hole, 200 μm slit, and 10 accumulations of 10 s. Raman mapping was performed on selected areas with the same conditions and a step size of 2.5 μm in the x and y directions. Spectrum correction (automatic baseline correction) and chemical maps were elaborated using LabSpec Raman spectroscopy software suite (version 6, HORIBA France SAS, Villeneuve d'Ascq, France). The identification of the compounds present was based on literature records and a reference spectra library compiled by the authors.

3. Results and Discussion

3.1. Structure, Thickness and Continuity of the Corrosion Layer

Microscopic observations of the untreated samples revealed a corrosion layer with brown, red and orange-colored compounds (Figure 2a). The corrosion layer of abiotic control coupons had a comparable thickness and color to the untreated coupons (Figure 2b,c). On the outer part of this layer, some blue spots were also detected. However, these did not form a continuous layer (Figure 2c). The formation of these blue compounds on the abiotic control coupons was probably due to an interaction between the iron corrosion products and the buffer solution containing phosphates and carbonates present in the culture medium. On the contrary, after bacterial treatment, the surface color of the iron coupons drastically changed. Indeed, the original corrosion layer disappeared almost completely, and a continuous layer of blue biogenic crystals was observed instead (Figure 2d).

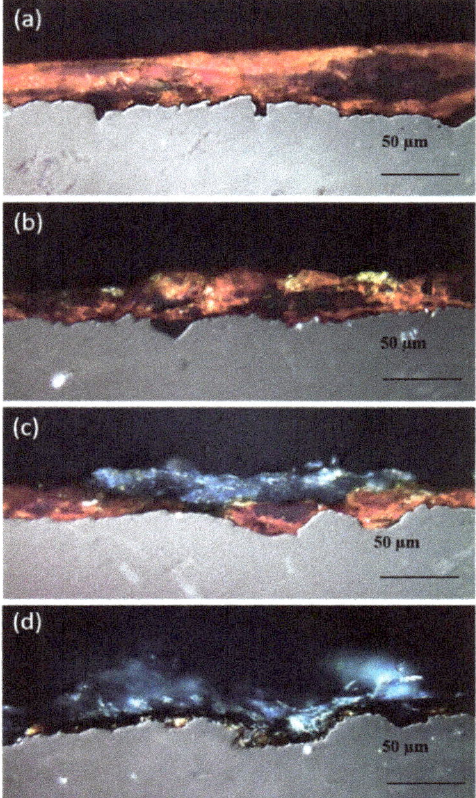

Figure 2. Optical microscope images of untreated (**a**), two different zones of the abiotic control (**b**,**c**), and bacterially treated (**d**) iron coupons.

Even if the thickness of a layer of naturally formed corrosion products is uneven, an overall decrease in the corrosion thickness was observed in the coupons treated with bacteria (Figure 3a). In fact, the mean value of the thickness of the corrosion layer decreased from nearly 28 µm (untreated coupons) to about 7 µm on the treated coupons (Figure 3a). The corrosion thickness decrease is due to the dissolutive microbial reduction of the iron phases composing the original corrosion layer. As a result, part of the iron oxyhydroxides is converted into iron biogenic minerals. The results demonstrated that this specific microbial process did convert a part of the original corrosion layer into reduced iron compounds, as least within the 7-day treatment duration. In fact, it is worth mentioning that treatment duration is a key element to assess and that the metal substrate could eventually become corroded if the growing conditions are not carefully set.

Another important feature observed was the continuity of the newly formed biogenic layer. In fact, in order to inhibit corrosion, this layer has to completely cover the remaining original corrosion layer, avoiding further contact of the metal core with atmospheric oxygen and moisture [3]. Microscopic observations of the cross-sections confirmed that after 7 days of incubation, about 92% of the analyzed surface was covered by biogenic crystals (iron phosphates), while only 55% of the original corrosion layer was covered by blue spots on the abiotic control coupons (Figure 3b). Thus, it can be concluded that bacteria are needed to produce a uniform and continuous layer of biogenic minerals and that an abiotic reduction is not enough to achieve comparable results.

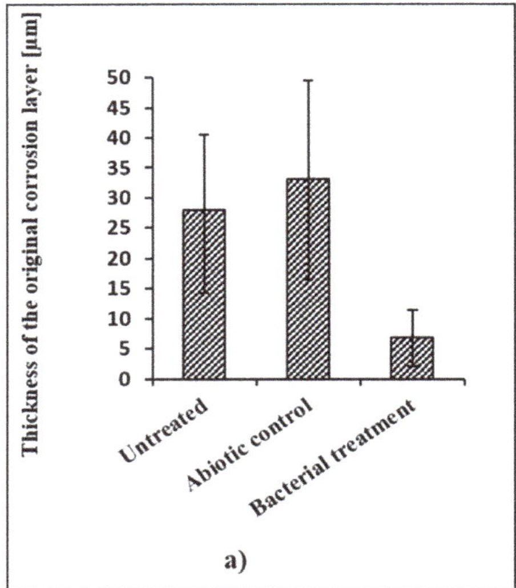

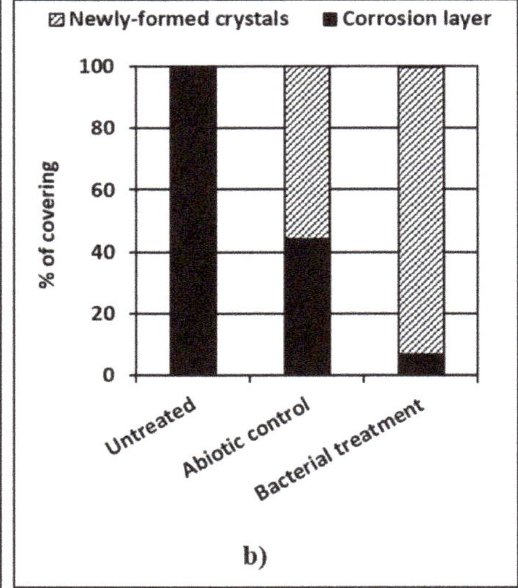

Figure 3. Efficiency of the treatment in terms of thickness and surface covering: (**a**) a graphic representation of the thickness of the corrosion layer of untreated, abiotic control, and bacterially treated iron coupons; (**b**) an estimation of the surface covered by iron phosphates by microscopic observation of the cross-sections.

3.2. Elemental Composition of the Corrosion Layer

Elemental mapping revealed that the corrosion layer of untreated coupons was mainly composed of Fe and O (Figure 4a). The same layer (numbered 1) was observed on the abiotic control coupons (Figure 4b). However, an upper layer (numbered 2) mainly composed of S was also detected. In addition, an outermost layer (numbered 3) composed of Fe, O, and P, was detected and corresponded to the areas with blue spots (Figure 4b). Regarding the treated coupons, the same stratigraphy as for the abiotic control was observed, with layer 2, composed of S, superimposed by layer 3, which is rich in Fe, O and P, corresponding to the area covered by the biogenic crystals (Figure 4c). In this case, no discontinuity within layer 3 was observed (Figure 4c). The presence of a sulfur-rich layer under the biogenic crystals is an interesting observation. In fact, a previous study showed the formation of elemental sulfur (S_8) and partially oxidized mackinawite ($Fe^{2+}/Fe^{3+}S$) on the surface of coupons used as an abiotic control, but not when incubated with bacteria [26]. Analyzing the cross-sectioned samples, the current study demonstrates that a layer mainly composed of S is also present on the bacterially treated coupons, and is localized between the remaining original corrosion layer and the biogenic minerals. This sulfur-rich layer is probably the result of an abiotic reaction between Na_2S added to ensure anoxic conditions and the reactive corroded surface of the iron coupons. Since this layer is located underneath the biogenic crystals, it can be assumed that it was produced first. It is worth mentioning that the formation of such a layer is already reported during iron corrosion in anoxic environments [32]. Even if the effect of sulfur on the corrosion process of iron is still under evaluation, experimental evidence suggests that elemental sulfur could speed up the corrosion rate of iron objects [33,34]. Indeed, this compound is known to be highly reactive, oxidizing organic and inorganic material regardless of the oxygen content [34]. In the presence of humidity, iron and steel surfaces exposed to elemental sulfur are corroded by an electrochemical reaction involving the reduction of elemental sulfur coupled with the oxidation of iron [33,34]. Hence, the sulfur-rich layer detected here and localized underneath the biogenic layer, if elemental sulfur, could be detrimental for the objects, and has to be avoided. In conclusion, for future improvements, Na_2S should be replaced by other reducing agents containing less sulfur, such as cysteine [35] or titanium(III) citrate [36].

As previously reported, the formation of biogenic iron minerals containing phosphorus can be attributed to the microbial reduction of iron phases, and the reaction of Fe^{2+} ions with phosphates (PO_4^{3-}) added in the buffered bacterial medium [26].

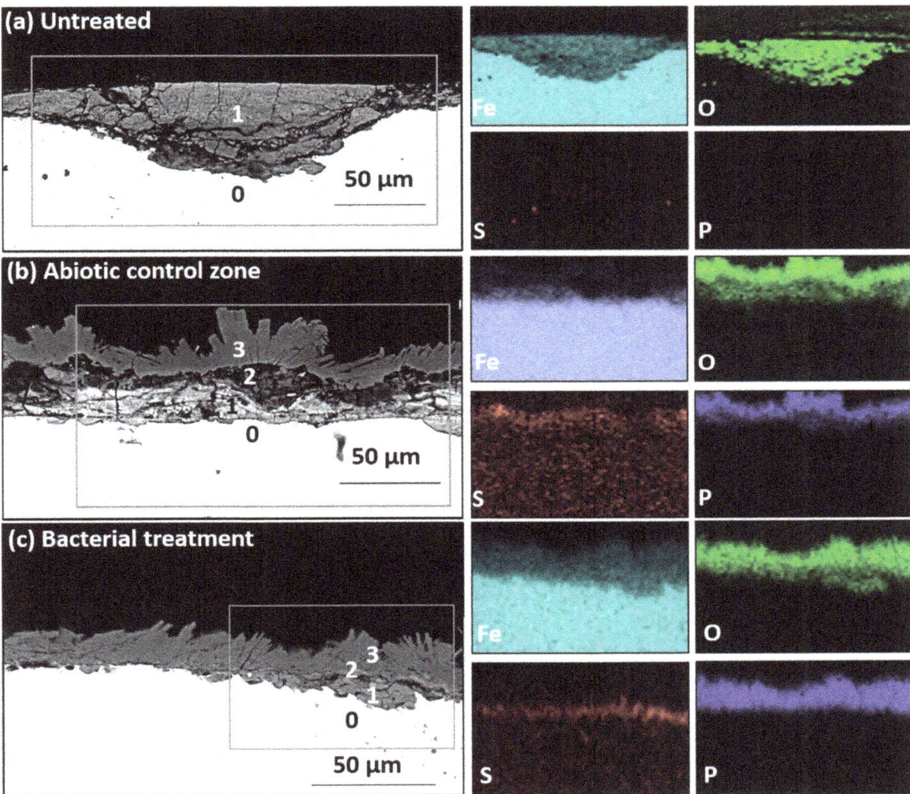

Figure 4. SEM–EDS characterization of (**a**) untreated, (**b**) abiotic control and (**c**) bacterially treated iron coupons. On the left, images of backscattered electrons with a square box indicating the analyzed area and numbers indicating the different layers present: 0 = bulk metal; 1 = original corrosion layer; 2 = newly formed S-rich layer; 3 = biogenic layer. On the right, elemental mapping showing the presence of iron, oxygen, sulfur, and phosphorus in the different layers.

3.3. *Molecular Composition of the Corrosion Layer*

Single points, as well as areas, were analyzed with Raman spectroscopy, allowing for the identification of lepidocrocite and goethite as the main compounds in the corrosion layer of untreated coupons (Figure 5).

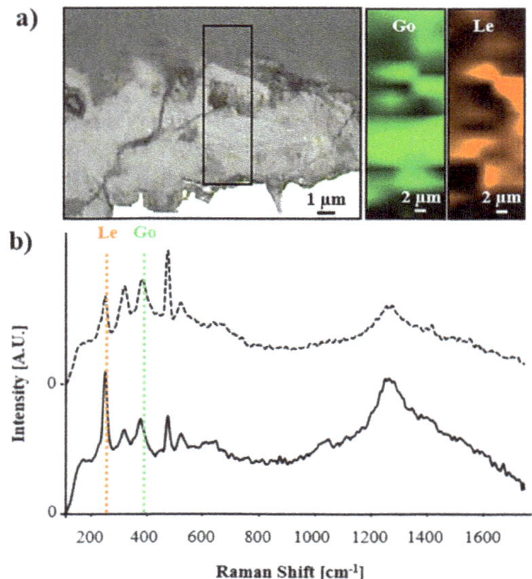

Figure 5. Molecular mapping performed by Raman spectroscopy on untreated coupons. (**a**) From left to right, a microscopic image with a square box indicating the analyzed area and chemical maps of goethite (Go) and lepidocrocite (Le). (**b**) Representative Raman spectrum of a mixture of lepidocrocite and goethite with the corresponding spectral regions selected for the elaboration of the respective chemical maps of lepidocrocite (region labeled "Le" with orange dashed lines) and goethite (region labeled "Go" with green dashed lines).

On the abiotic control, lepidocrocite and goethite were detected, but also siderite and vivianite (Figure 6). Siderite was probably produced as a consequence of the interaction between iron and the carbonaceous sources present in the culture medium (lactate, fumarate, or carbonated buffer) [35]. The formation of vivianite (an Fe^{2+} phosphate) on the abiotic control coupons can be the result of the interaction of the phosphorus buffer with Fe^{2+} ions. The latter can either be present in the original corrosion layer or be produced when the Fe^{3+} phases of the original layer are abiotically reduced by Na_2S. In fact, the reduction of iron in the abiotic control has already been documented in similar conditions [14].

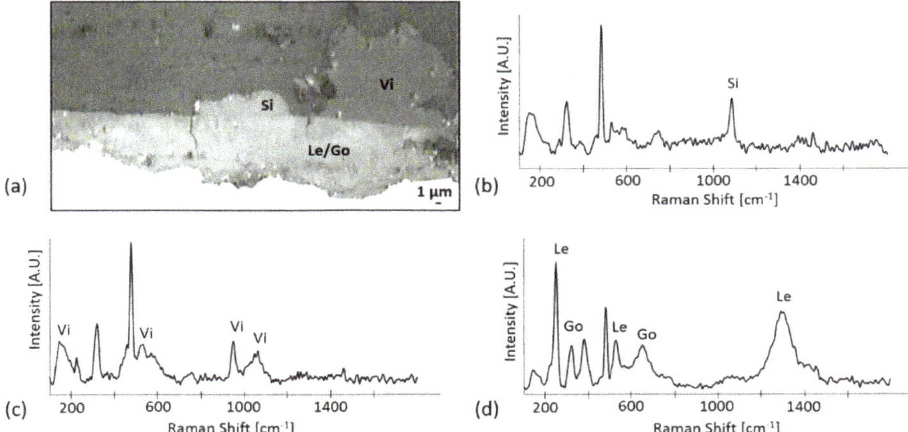

Figure 6. Raman characterization of abiotic control coupons: (**a**) Microscopic image of the analyzed area with identified compounds indicated as siderite (Si), vivianite (Vi), and a mixture of goethite (Go) and lepidocrocite (Le). Representative Raman spectra of the different analyzed regions where (**b**) siderite (Si), (**c**) vivianite (Vi) and (**d**) a mix of goethite and lepidocrocite were identified (Le/Go), respectively.

In the samples treated with bacteria, lepidocrocite and goethite were not detected. In fact, after treatment, the thickness of the original corrosion layer drastically decreased, making its detection difficult by Raman spectroscopy (spatial resolution of about 1 µm). As observed above during SEM–EDS analyses, these iron corrosion compounds have been converted into biogenic crystals through microbial reduction. These newly formed minerals were identified as a mixture of two different iron phosphates. The most abundant was vivianite (Figure 7). The vivianite spectrum displayed an intense vibrational band at 949 cm^{-1} and two less intense vibrational bands at 1014 and 1052 cm^{-1}, typical of the P-O stretching mode [37]. In order to localize this compound, its characteristic Raman shift at 949 cm^{-1} was employed for chemical mapping (Figure 7). Interestingly, another iron phosphate compound identified as barbosalite $Fe^{2+}Fe^{3+}{}_2(PO_4)_2(OH)_2$ was also detected. The same bands related to the P-O stretching mode were present but with different relative intensities. For barbosalite, the main Raman shift was at 1015 cm^{-1} (Figure 7). Since barbosalite contains both Fe^{2+} and Fe^{3+} ions, its formation could be the consequence of the interaction between Fe^{3+} ions present in the original corrosion layer, Fe^{2+} ions already present or produced from microbial iron reduction, and the $PO_4{}^{3-}$ ions contained in the buffer solution. Finally, Raman measurements did not allow the detection of the sulfur-rich inner layer revealed by SEM–EDS. This could be explained by the low thickness of this layer, probably below the spatial resolution limit of Raman spectroscopy (about 1 µm).

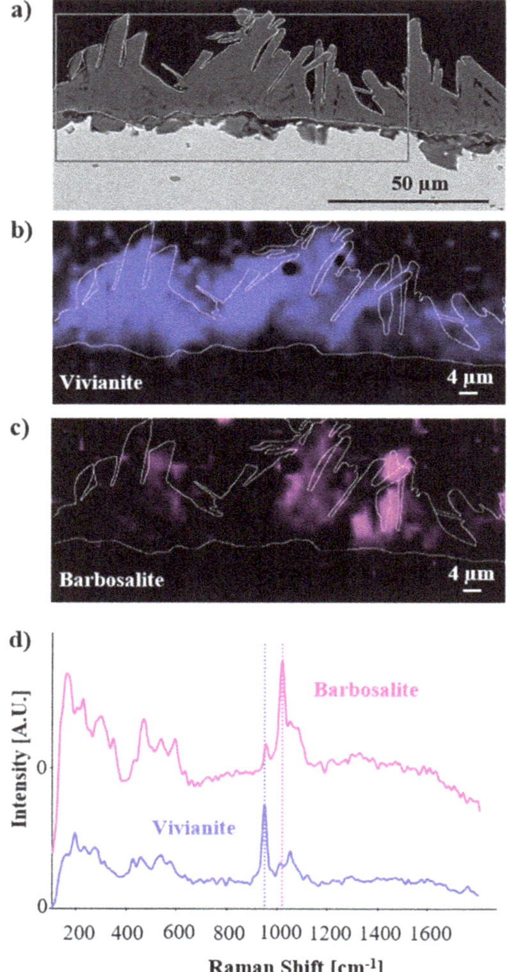

Figure 7. Molecular mapping performed by Raman spectroscopy on bacterially treated iron coupons: (**a**) image of secondary electrons with a square box indicating the area analyzed by Raman spectoscopy, chemical maps (**b**) of vivianite (Vi, blue) and (**c**) of barbosalite (Ba, pink), and (**d**) Raman spectra of vivianite and barbosalite with the corresponding spectral regions selected for the elaboration of the respective chemical maps of vivianite (indicated with blue dashed lines) and barbosalite (indicated with magenta dashed lines).

4. Conclusions

The development of new and ecologically friendly strategies to protect outdoor iron surfaces has a clear economical, as well as ecological, interest. We demonstrated the potential of bacteria as an alternative for the development of innovative and green methods to protect iron artifacts from detrimental corrosion [24–28]. Here, the obtained results allow us to better understand how reactive corrosion layers are converted into biogenic iron phosphates with *Desulfitobacterium hafniense*. These biogenic minerals covered almost all of the remaining original corrosion layer. This layer could act as a barrier isolating the unstable iron corrosion products that would eventually still be present underneath from the exposure to atmospheric oxygen and moisture that could lead to further corrosion. Vivianite is in fact a stable, poorly soluble, and non-oxidizing Fe^{2+} mineral [22]. However, great attention has to be drawn to the composition of the bacterial solution applied, which not only drives the type of biogenic minerals produced, but also potentially contaminates the corrosion layer with undesired compounds that are able to instigate further corrosion, such as sulfur-containing compounds. In fact, through the stratigraphy study carried out here, a sulfur-rich layer was detected below the biogenic iron phosphates. This layer was not detected with surface analyses of the coupons, and only stratigraphic investigations allowed us to conclude that careful attention has to be paid to the culture medium composition in order to produce a stable vivianite layer that would passivate the iron surface. Our study on cross-sectioned samples further improved the evaluation of the depth efficiency of the proposed bacterial treatment, as well as demonstrated the formation of biogenic vivianite as an adherent, even, and uniform layer. These features are important criteria when designing new protective systems to provide long-term inhibition of corrosion on iron surfaces.

Author Contributions: Conceptualization, E.J.; methodology, L.C., M.A. and W.K.; validation, L.C., J.M., T.L., P.J. and E.J.; investigation, L.C., M.A., W.K., and T.L.; resources, J.M.; data curation, L.C.; writing—original draft preparation, L.C. and E.J.; writing—review and editing, L.C., M.A., W.K., J.M., T.L., P.J. and E.J.; visualization, L.C.; supervision, P.J. and E.J.; project administration, E.J.; funding acquisition, E.J. All authors have read and agreed to the published version of the manuscript.

Funding: This research was funded by the Swiss National Science Foundation, Ambizione, grant number PZ00P2_142514, 2013–2016 and professorship grant number PP00P2_163653, 2016–2020.

Acknowledgments: The authors are also grateful to the research conservation laboratory of the Swiss National Museum for providing the steel plate used in the experiments.

Conflicts of Interest: The authors declare no conflict of interest. The funders had no role in the design of the study; in the collection, analyses, or interpretation of data; in the writing of the manuscript, or in the decision to publish the results.

References

1. Yamashita, M.; Miyuki, H.; Matsuda, Y.; Nagano, H.; Misawa, T. The long term growth of the protective rust layer formed on weathering steel by atmospheric corrosion during a quarter of a century. *Corros. Sci.* **1994**, *36*, 283–299. [CrossRef]
2. Tamura, H. The role of rusts in corrosion and corrosion protection of iron and steel. *Corros. Sci.* **2008**, *50*, 1872–1883. [CrossRef]
3. Scott, D.A.; Eggert, G. *Iron and Steel in Art*; Archetype Publications Ltd.: London, UK, 2009.
4. Narain, S.; Jain, K.K. *Iron Artifacts: History, Metallurgy, Corrosion and Conservation*; Agam Kala Prakashan: Delhi, India, 2009.
5. Moore, J.D. Long-term corrosion processes of iron and steel shipwrecks in the marine environment: A review of current knowledge. *J. Marit. Archaeol.* **2015**, *10*, 191–204. [CrossRef]
6. Reguer, S.; Neff, D.; Bellot-Gurlet, L.; Dillmann, P. Deterioration of iron archaeological artefacts: Micro-Raman investigation on Cl-containing corrosion products. *J. Raman Spectrosc.* **2007**, *38*, 389–397. [CrossRef]
7. Ståhl, K.; Nielsen, K.; Jiang, J.; Lebech, B.; Hanson, J.C.; Norby, P.; van Lanschot, J. On the akaganéite crystal structure, phase transformations and possible role in post-excavational corrosion of iron artifacts. *Corros. Sci.* **2003**, *45*, 2563–2575. [CrossRef]
8. Rimmer, M.; Watkinson, D.; Wang, Q. The impact of chloride desalination on the corrosion rate of archaeological iron. *Stud. Conserv.* **2013**, *58*, 326–337. [CrossRef]
9. Dillmann, P.; Watkinson, D.; Angelini, E.; Adriaens, A. Introduction: Conservation versus laboratory investigation in the preservation of metallic heritage artefacts. In *Corrosion and Conservation of Cultural Heritage Metallic Artefacts*, 1st ed.; Woodhead Publishing Limited on Behalf of the European Federation of Corrosion: Cambridge, UK, 2013.
10. Jegdić, B.; Bajat, J.B.; Popić, J.P.; Mišković-Stanković, V. Corrosion stability of polyester coatings on steel pretreated with different iron–phosphate coatings. *Prog. Org. Coat.* **2011**, *70*, 127–133. [CrossRef]
11. Zheng, S.; Li, J. Inorganic–organic sol gel hybrid coatings for corrosion protection of metals. *J. Sol-Gel Sci. Technol.* **2010**, *54*, 174–187. [CrossRef]

12. Selwyn, L. Overview of archaeological iron: The corrosion problem, key factors affecting treatment, and gaps in current knowledge. In *Metal 04: Proceedings of the International Conference on Metals Conservation*; Ashton, J., Hallan, D., Eds.; National Museum of Australia: Canberra, Australia, 2004; pp. 294–306.
13. Carlin, W.; Keith, D.; Rodriguez, J. Less is more: Measure of chloride removal rate from wrought iron artifacts during electrolysis. *Stud. Conserv.* **2001**, *46*, 68–76.
14. Schmidt-Ott, K.; Boissonnas, V. Low-pressure hydrogen plasma: An assessment of its application on archaeological iron. *Stud. Conserv.* **2002**, *47*, 81–87.
15. Zarasvand, K.A.; Rai, V.R. Microorganisms: Induction and inhibition of corrosion in metals. *Int. Biodeterior. Biodegrad.* **2014**, *87*, 66–74. [CrossRef]
16. Iverson, W.P. Microbial corrosion of metals. *Adv. Appl. Microbiol.* **1987**, *32*, 1–36. [CrossRef]
17. Little, B.; Wagner, P.; Mansfeld, F. Microbiologically influenced corrosion of metals and alloys. *Int. Mater. Rev.* **1991**, *36*, 253–272. [CrossRef]
18. Videla, H.A.; Herrera, L.K. Microbiologically influenced corrosion: Looking to the future. *Int. Microbiol.* **2005**, *8*, 169.
19. Volkland, H.-P.; Harms, H.; Müller, B.; Repphun, G.; Wanner, O.; Zehnder, A.J. Bacterial phosphating of mild (unalloyed) steel. *Appl. Environ. Microbiol.* **2000**, *66*, 4389–4395. [CrossRef]
20. Mansfeld, F.; Hsu, H.; Örnek, D.; Wood, T.K.; Syrett, B. Corrosion control using regenerative biofilms on aluminum 2024 and brass in different media. *J. Electrochem. Soc.* **2002**, *149*, B130–B138. [CrossRef]
21. Zuo, R.; Örnek, D.; Syrett, B.; Green, R.; Hsu, C.-H.; Mansfeld, F.; Wood, T.K. Inhibiting mild steel corrosion from sulfate-reducing bacteria using antimicrobial-producing biofilms in Three-Mile-Island process water. *Appl. Microbiol. Biotechnol.* **2004**, *64*, 275–283. [CrossRef]
22. Chongdar, S.; Gunasekaran, G.; Kumar, P. Corrosion inhibition of mild steel by aerobic biofilm. *Electrochim. Acta* **2005**, *50*, 4655–4665. [CrossRef]
23. Cote, C.; Rosas, O.; Basséguy, R. *Geobacter sulfurreducens*: An iron reducing bacterium that can protect carbon steel against corrosion? *Corros. Sci.* **2015**, *94*, 104–113. [CrossRef]
24. Comensoli, L.; Bindschedler, S.; Junier, P.; Joseph, E. Iron and fungal physiology: A review of biotechnological opportunities. *Adv. Appl. Microbiol.* **2017**, *98*, 31–60. [CrossRef]
25. Joseph, E.; Bindschedler, S.; Albini, M.; Comensoli, L.; Kooli, W.; Mathys, L. Microorganisms for Safeguarding Cultural Heritage. In *The Fungal Community: Its Organization and Role in the Ecosystem*, 4th ed.; Dighton, J., White, J.F., Eds.; CRC Press: Boca Raton, FL, USA, 2017; pp. 509–518.
26. Comensoli, L.; Maillard, J.; Albini, M.; Sandoz, F.; Junier, P.; Joseph, E. Use of bacteria to stabilize archaeological iron. *Appl. Environ. Microbiol.* **2017**, *83*, e03478-16. [CrossRef] [PubMed]
27. Kooli, W.M.; Comensoli, L.; Maillard, J.; Albini, M.; Gelb, A.; Junier, P.; Joseph, E. Bacterial iron reduction and biogenic mineral formation for the stabilisation of corroded iron objects. *Sci. Rep.* **2018**, *8*, 764. [CrossRef] [PubMed]
28. Kooli, W.M.; Junier, T.; Shakya, M.; Monachon, M.; Davenport, K.W.; Vaideeswaran, K.; Vernudachi, A.; Marozau, I.; Monrouzeau, T.; Gleasner, C.D. Remedial treatment of corroded iron objects by environmental *Aeromonas* isolates. *Appl. Environ. Microbiol.* **2019**, *85*, e02042-18. [CrossRef] [PubMed]
29. Villemur, R.; Lanthier, M.; Beaudet, R.; Lépine, F. The Desulfitobacterium genus. *FEMS Microbiol. Rev.* **2006**, *30*, 706–733. [CrossRef]
30. Futagami, T.; Furukawa, K. The Genus *Desulfitobacterium*. In *Organohalide-Respiring Bacteria*; Adrian, L., Löffler, F.E., Eds.; Springer: Berlin/Heidelberg, Germany, 2016; pp. 173–207. [CrossRef]
31. Prat, L.; Maillard, J.; Grimaud, R.; Holliger, C. Physiological Adaptation of *Desulfitobacterium hafniense* Strain TCE1 to Tetrachloroethene Respiration. *Appl. Environ. Microbiol.* **2011**, *77*, 3853–3859. [CrossRef]
32. Bellot-Gurlet, L.; Neff, D.; Reguer, S.; Monnier, J.; Saheb, M.; Dillmann, P. Raman studies of corrosion layers formed on archaeological irons in various media. *J. Nano Res.* **2009**, *8*, 147–156. [CrossRef]
33. Schmitt, G. Effect of elemental sulfur on corrosion in sour gas systems. *Corrosion* **1991**, *47*, 285–308. [CrossRef]
34. Leonard, S.E.; Carroll, K.S. Chemical 'omics' approaches for understanding protein cysteine oxidation in biology. *Curr. Opin. Chem. Biol.* **2011**, *15*, 88–102. [CrossRef]
35. Legrand, L.; Savoye, S.; Chausse, A.; Messina, R. Study of oxidation products formed on iron in solutions containing bicarbonate/carbonate. *Electrochim. Acta* **2000**, *46*, 111–117. [CrossRef]

36. Zehnder, A.; Wuhrmann, K. Titanium (III) citrate as a nontoxic oxidation-reduction buffering system for the culture of obligate anaerobes. *Science* **1976**, *194*, 1165–1166. [CrossRef]
37. Frost, R.L.; Martens, W.; Williams, P.A.; Kloprogge, J.T. Raman and infrared spectroscopic study of the vivianite-group phosphates vivianite, baricite and bobierrite. *Mineral. Mag.* **2002**, *66*, 1063. [CrossRef]

© 2020 by the authors. Licensee MDPI, Basel, Switzerland. This article is an open access article distributed under the terms and conditions of the Creative Commons Attribution (CC BY) license (http://creativecommons.org/licenses/by/4.0/).

Article

Chitosan/Lignosulfonate Nanospheres as "Green" Biocide for Controlling the Microbiologically Influenced Corrosion of Carbon Steel

Pathath Abdul Rasheed [1], Ravi P. Pandey [1], Khadeeja A. Jabbar [1], Ayman Samara [1], Aboubakr M. Abdullah [2] and Khaled A. Mahmoud [1,3,*]

1 Qatar Environment and Energy Research Institute (QEERI), Hamad Bin Khalifa University (HBKU), Qatar Foundation, Doha P.O. Box 34110, Qatar; pabdul.rasheed@gmail.com (P.A.R.); rpandey@hbku.edu.qa (R.P.P.); KAbdulJabbar@hbku.edu.qa (K.A.J.); asamara@hbku.edu.qa (A.S.)

2 Center for Advanced Materials, Qatar University, Doha P.O. Box 2713, Qatar; bakr@qu.edu.qa

3 Department of Physics & Mathematical Engineering, Faculty of Engineering, Port Said University, Port Said 42523, Egypt

* Correspondence: kmahmoud@hbku.edu.qa; Tel.: +974-4454-1694

Received: 10 May 2020; Accepted: 22 May 2020; Published: 29 May 2020

Abstract: In this work, uniform cross-linked chitosan/lignosulfonate (CS/LS) nanospheres with an average diameter of 150–200 nm have been successfully used as a novel, environmentally friendly biocide for the inhibition of mixed sulfate-reducing bacteria (SRB) culture, thereby controlling microbiologically influenced corrosion (MIC) on carbon steel. It was found that 500 $\mu g \cdot mL^{-1}$ of the CS/LS nanospheres can be used efficiently for the inhibition of SRB-induced corrosion up to a maximum of 85% indicated by a two fold increase of charge transfer resistance (R_{ct}) on the carbon steel coupons. The hydrophilic surface of CS/LS can readily bind to the negatively charged bacterial surfaces and thereby leads to the inactivation or damage of bacterial cells. In addition, the film formation ability of chitosan on the coupon surface may have formed a protective layer to prevent the biofilm formation by hindering the initial bacterial attachment, thus leading to the reduction of corrosion.

Keywords: chitosan; lignosulfonate; sulfate reducing bacteria; biofilm; microbial corrosion; impedance spectroscopy

---

1. Introduction

Microbiologically influenced corrosion (MIC) of carbon steel is a major cause of metal corrosion and pipeline failure [1–3]. It is estimated that MIC accounts for about 20% of the corrosion damage in the oil and gas sector [4]. Despite the tremendous efforts made thus far to improve corrosion management, MIC has remained a pressing issue for the oil/gas sector, where there is exposure of metals to bacteria found in water. Several types of microorganism are responsible for MIC, including sulfate-reducing bacteria (SRB), iron-oxidizing bacteria (IOB), slime-forming bacteria, and iron-reducing bacteria (IRB). Amongst those, SRB are the main microorganisms responsible for MIC by generating sulfide species anaerobically, which causes progressive biocorrosion in the water transport systems [5,6]. The SRB strains produce hydrogen sulfides (H_2S), metal sulfides, and sulfates as a result of biogenic oxidation/reduction reactions [7,8]. In particular, the production of H_2S at elevated concentrations creates intrinsic heterogeneity, which accelerates the corrosion process by favoring electrochemical reactions [9–11].

Biocorrosion control methods are mainly based on either inhibiting the metabolic/growth activities or altering the corrosive conditions to reduce the adaptation of microorganisms. Different types of approach, such as cathodic protection [12], protective coatings [13], corrosion inhibitors [14], and biocides [15], have been used to control/minimize biocorrosion. Oil/gas industries usually need high concentrations of biocides for water disinfection and controlling SRB biofilm formation thereby reducing biocorrosion [16]. However, the use of conventional biocides may cause harmful impact to the environment since it produces disinfection byproducts in addition to the low efficiency against biofilms, and high operational cost [17]. Different nanomaterials demonstrate strong antimicrobial activities, rendering them potential alternatives to conventional biocides [18–22]. Nanomaterials, such as AgNPs [23], ZnONPs [24], TiO_2NPs [4], FeNPs [25], graphene [26], CuONPs [27], and metal-nanocomposites [22], have been used for the inhibition of SRB-induced biofilm and subsequent MIC. However, the environmental impact of nanomaterials due to their biological toxicity has restricted their use in practical applications [28,29]. The use of green biocides with lower toxicity, environmentally benign, and ease of use can overcome these issues [30].

Chitosan (CS) is a biodegradable polymer abundant in nature with high hydrophilicity, nontoxicity, antimicrobial properties, and low cost [31]. The antimicrobial activity of CS has been widely established against many microorganisms and it shows a high inhibition rate against both Gram-positive and Gram-negative bacteria [32–35]. CS also displays anti-biofilm activities with a high ability to damage biofilms formed by microbes [36–38]. Due to its cationic nature, CS has been able to penetrate biofilms by disrupting negatively charged cell membranes through electrostatic interaction when microbes settle on the surface [36]. Recently, our research group used ZnO-interlinked chitosan nanoparticles (CZNC-10) as stable biocide formulations against SRBs from industrial waste sludge [24]. The nanoparticles achieved a concentration-dependent SRBs inhibition with over 73% efficiency at 250 $\mu g \cdot mL^{-1}$. Also, CZNC-10 demonstrated 74% MIC inhibition on carbon steel [39]. The dose of CZNC was limited to 250 $\mu g \cdot mL^{-1}$ due to the ZnO content in the CZNC biocide considering the environmental toxicity issues. In order to develop a more "green" and efficient chitosan-based biocide, the metal or metal oxide nanoparticles need to be replaced with more environmentally benign alternatives.

Lignin is an abundant natural resource that has been widely used as a potential source for fuel and chemical production [40]. Lignin can be incorporated into different polymeric systems such as dispersants, bioadhesives, biosurfactants, polyurethane foams, and epoxy resins, etc., depending on its solubility and reactivity characteristics [41]. Lignosulfonate (LS) exhibits good water-solubility and anionic characteristic [42]. LS also exhibits antioxidant and antimicrobial properties that extend its potential applications to different fields [43,44]. Both CS and LS are good bactericidal agents, and therefore it is expected that CS-LS complexes can be used as highly efficient and environmentally friendly chitosan-based biocides against SRB induced biocorrosion.

The synthesis of CS-LS polyelectrolyte complexes is mainly based on ionic interaction or ultrasonic homogenization [45,46]. CS cross-linked graphene oxide (GO)/LS composite aerogels have been synthesized by the simple mixing of GO, LS, and CS solutions [47]. The aerogels demonstrated 3D porous structure. These CS-LS hybrids showed non-uniform sizes/shapes and instability above pH 4.5, which restricts their practical applications. To solve these issues, we have introduced a new crosslinking strategy towards the synthesis of stable cross-linked chitosan-lignosulfonate (CS/LS) nanospheres [31]. The optimum composite structure was formed at 1:1 ratio of CS:LS. These 150–200 nm nanospheres demonstrated the highest thermal, mechanical, and bactericidal effect against aerobic *Escherichia coli* (*E. coli*) and *Bacillus subtilis* (*B. subtilis*) bacteria as well as anaerobic SRB. As a proof of concept, 100 mg/L CS/LS-1:1 was able to inhibit SRB growth, as demonstrated by 48.8% lower sulfate reduction [31].

Here, we have investigated the ability of the new "green" CS/LS nanospheres with an optimal 1:1 (CS:LS) ratio to treat SRB induced MIC on SS400 carbon steel. The nature and kinetics of SRB inhibition induced by the CS/LS are thoroughly studied with electrochemical impedance spectroscopy (EIS) and surface characterization methods.

2. Materials and Methods

2.1. Materials

Chitosan (low molecular weight) with 85% deacetylation (CS), Lignosulfonic acid sodium salt (LS), $MgSO_4$, sodium citrate, Absolute ethanol, CH_3CH_2OH (≥99.8%), $CaSO_4$, NH_4Cl, NaCl, Na_2SO_4, KCl, $SrCl_2$, KBr, K_2HPO_4, HCl, NaOH, hexamethylenetetramine, sodium lactate, and yeast extract were obtained from Sigma-Aldrich (St. Louis, MO, USA). All analytical grade chemicals were used as received. Carbon steel (SS400) rods of 10 mm diameter were obtained locally. The chemical composition of SS400 is 99.25–100% Fe, 0–0.4% Si, 0–0.26% C, 0–0.05% S, and 0–0.04% P. PhenoCure^{TM} (phenolic resin) was procured from Buehler, Lake Bluff, IL, USA.

2.2. Synthesis and Characterization of Cross-Linked CS/LS Nanospheres

CS/LS nanospheres (1:1 CS:LS) were prepared according to our previous work [31]. Briefly, 30 mL of both CS and LS solutions were stirred together at room temperature for 30 min. Then, 450 μL of the cross-linking solution were added gradually, and continued stirring for additional 30 min. The cross-linking agent is a mixture of formaldehyde and sulfuric acid ($HCHO/H_2SO_4$, 40/60 *w*/*w*). The resulting solution was purified by centrifuging at 10,000 rpm followed by washing five times with DI water to obtain CS/LS. The size and morphology of the CS/LS were characterized by FEI Quanta 650 FEG SEM (Hillsboro, OR, USA) and FEI Talos F200X TEM (Hillsboro, OR, USA). X-ray diffractogram (XRD) of CS/LS were measured by D8 Advance (Bruker AXS, Bremen, Germany). The X-ray diffractometer is equipped with Cu-Kα radiation (λ = 1.54056 Å) at 40 kV, 40 mA with a step scan of 0.02° per step and scanning speed of 1° min^{-1}. The hydrodynamic radius and Zeta potential were measured by Malvern Zetasizer Ultra (Malvern Panalytical Ltd., Malvern, UK).

2.3. Fabrication of Coupons and SRB Culture

The coupons of 8 mm diameter steel bar were fabricated by using SimpliMet 3000 mounting press (Buehler, Lake Bluff, IL, USA) and PhenoCureTM as an outer shell. Then the coupons were polished with EcoMet 2500 Polisher (Buehler, Lake Bluff, IL, USA) in a sequence from 240 to 1200 silicon carbide paper and 6, 3 and 1 μm diamond slurry to get a mirror-like finish. The coupons were cleaned with acetone and sterilized with absolute ethanol followed by storing the dried coupons in a desiccated environment until further use. The surface roughness was analyzed by KLA P17 stylus profiler (Milpitas, CA, USA) and SEM was used the image the surface morphology.

Mixed SRB culture was enriched from biofilm samples from a local oil field in Qatar as described earlier [24]. The SRB was further cultured in Postage's C medium in a saline media [39,48]. The composition of the inject seawater is given in the Supplementary Table S1. The concentration of SRB biomass was represented by volatile suspended solids (VSS) available in the culture media [24]. The known volume of the SRB biomass were taken and oven dried at inert atmosphere to calculate the VSS. The carbon steel coupons were incubated in 200 mL bottles containing 250 mg VSS/L SRB biomass and were kept shaken at 120 rpm, 37 °C. Control experiment was considered as the coupons incubated in a media containing 250 mg VSS/L SRB in the absence of CS/LS. The optimum concentration of CS/LS was directly added to the media in the range of 0–1000 μg·mL^{-1} CS/LS and EIS analysis was performed after 10 days of incubation. The R_{ct} values from the Nyquist plots were used to evaluate the corrosion inhibition. Next, corrosion inhibition of optimum CS/LS concentration was evaluated at different time intervals (0, 7, 10, 15, 21, 28, and 35 days) [39,49,50]. The abiotic conditions were made by incubating coupons in the SRB-free media at sterile conditions and in absence of CS/LS to differentiate the chemical corrosion [51]. For comparison, 5% glutaraldehyde (GA) was used as a conventional biocide [52,53]. All experiments were performed in an anaerobic chamber at anaerobic conditions. After this, the sealed reaction bottles were kept shaking at different times at 37 °C. After particular time intervals, coupons were drawn from the incubation mixture and gently washed with DI water before analysis.

2.4. Electrochemical Measurements

Electrochemical measurements were recorded by Gamry potentiostat (Ref 600$^+$, (Gamry Instruments, Warminster, PA, USA). Treated SS400 carbon-steel coupons were the working electrodes, saturated calomel electrode (SCE) as the reference and graphite disk as the counter electrode. The treated coupons were mounted into a Gamry ParaCell™ (Gamry Instruments, Warminster, PA, USA) Electrochemical Cell Kit (Part No.992-80) for electrochemical analysis. Simulated inject seawater was used as the electrolyte for all electrochemical experiments. The EIS measurements were recorded over 0.01–10^5 Hz with 10 mV sinusoidal signal. The EIS measurements were performed after achieving the steady-state condition by keeping the setup for 30 min at open circuit potential (OCP). Gamry Echem Analyst software (Version 7.05) (Gamry Instruments, Warminster, PA, USA) was used to analyze the experimental data.

2.5. Biofilm, Corrosion Products, and Coupon Surface Characterization

For SEM and XPS analysis of SRB biofilm, the incubated coupons were fixed with 2% GA solution for 2 h. After washing the coupons with DI water, dehydration was performed with ethanol, followed by washing with water [39]. The coupons were stored under dry nitrogen before each analysis. The SEM and EDS analyses were performed with FEI Quanta 650 FEG (Hillsboro, OR, USA) SEM after gold (3 nm) coating. XPS analysis was carried out with ESCALAB 250X (Thermo Fisher Scientific, Hillsboro, OR, USA) with AlKα excitation (25 W, hυ = 1486.5 eV) and 1 eV resolution. The X-ray fluorescence (XRF) analysis was carried out with an XGT-7200V X-ray Analytical Microscope (Horiba Scientific, Piscataway, NJ, USA). The X-ray source was operated at 50 kV and 0.8 mA, and generates an X-ray beam from Rh anode that is focused to a spot size of 1.2 mm. To study the post corrosion morphology of carbon steel coupons, biofilms were removed by sonication in ethanol (10 s intervals), 5 mL·L^{-1} HCl, and 3.5 g·L^{-1} hexamethylenetetramine for 5 min followed by drying with nitrogen flow [54]. The post-corrosion morphology of the coupons after 35 days of incubation with and without CS/LS nanospheres were analyzed by SEM and profilometry. In addition, the bare coupon was analyzed for comparison. KLA-Tencor P17 stylus profilometer (Milpitas, CA, USA) at 2 μm resolution, with a loading force of 2 mg was used to capture surface profile images of the coupons. Seven measurements were made at each position on each coupon with a scan size of 400 × 400 μm^2 for each frame. Apex 3d-7 software (Milpitas, CA, USA) was used to calculate the average surface roughness (S_a).

3. Results and Discussion

3.1. Characterization of Cross-Linked CS/LS Nanospheres

The uniform CS/LS nanospheres have been prepared by one-step covalent cross-linking between chitosan (CS) and lignosulfonate (LS) according to our reported method [31]. As shown in the SEM image (Figure 1A), well-dispersed spherical nanoparticles were obtained with an average diameter of 150–200 nm. TEM image further confirmed the well-defined shape of a single CS/LS nanosphere (Figure 1B). CS/LS at 1:1 CS:LS ratio has been considered as optimum ratio for preparing the nanospheres. According to the XRD pattern in Figure 1C, pure CS has two characteristic peaks at 10° and 20° while LS having a broad peak at 22.6° [55–57]. The formation of CS/LS is confirmed by the diminishing of the peak at 10°, as well as a reduction in intensity, and broadening of the peak at ~20°. This weaker and broader peak can be attributed to the formation of a new binary framework in which the original structure of both CS and LS can be disrupted. The detailed characterization of this material has been described elsewhere [31]. The average size distribution of CS/LS in the aqueous suspension was stable between 20 and 60 °C at pH 3–8 [31]. The stability of CS/LS in simulated seawater was evaluated by measuring the hydrodynamic radius and Zeta potential at varying pH between 3 and 9. The average size of the particles remained within the range of 230–250 nm at acidic media until pH 6. After this pH, the nanospheres have shrunk slightly and remained almost unchanged at pH above 7. This can be attributed to the collapse and deformation of CS/LS at alkaline pH values [31]. As shown in Figure 1D, the Zeta potential changed to a negative value with point of zero charge at pH 5.8, indicating that the CS/LS nanospheres are negatively charged at neutral pH conditions.

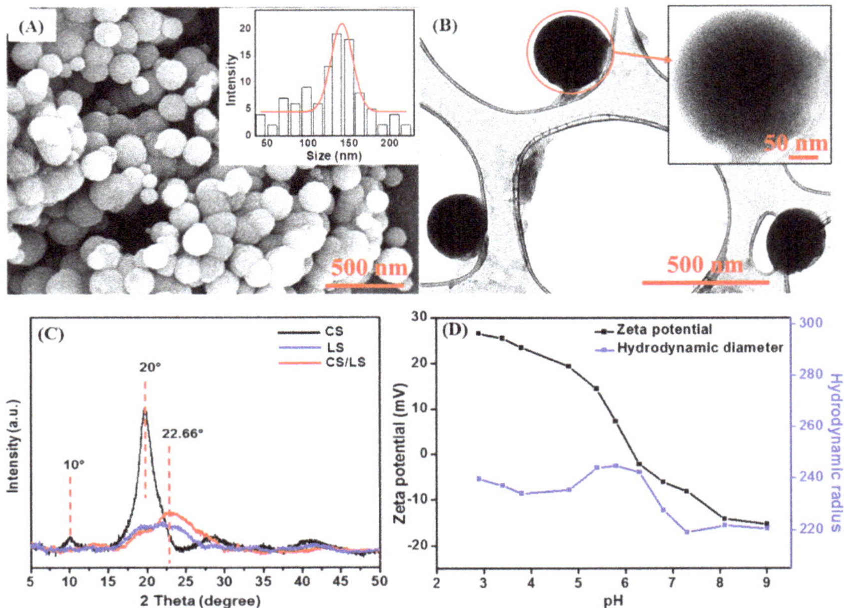

Figure 1. (**A**) SEM images of CS/LS nanospheres (Inset shows the size distribution); (**B**) TEM of CS/LS nanospheres (Inset: TEM image of a single nanosphere); (**C**) XRD for CS, LS, CS/LS and (**D**) The size and zeta potential changes with pH of the CS/LS in simulated seawater.

3.2. Investigation of SRB-Induced Corrosion on Carbon Steel

It is important to investigate the SRB induced MIC in our experimental conditions to make a valid comparison. The SS400 carbon steel coupons were incubated in a solution containing SRB in Postage's C containing simulated seawater. SRB have been enriched from a mixed-culture bacterial sludge obtained from in a real oil filed sample (see experimental section). The coupons were analyzed by EIS after 7, 10, 15, 21, 28, and 35 days of incubation times.

As observed in Figure 2A, the R_{ct} values of Nyquist semicircle is higher after 7 days compared to the ones from longer incubation times. A complete biofilm formation is expected to reach the highest protection capacity at 7 days [50,58]. After which, the semicircles diameters gradually decrease with time, indicating a gradual breakdown in the corrosion protection by the biofilm, i.e., faster corrosion rates. A small capacitive semicircle loop appeared at high frequencies, only for 7 days of incubation, mainly due to the precipitation of the corrosion product along with biofilm results in a porous and more adherent outer layer [59]. This high frequency semicircle started to diminish at longer incubation times, and the Nyquist plot behavior is also different compared to the longer incubation times. However, when steady state is reached at around 7 days, mass transfer limitations was dominant over the interfacial activation, which changes the shape of Nyquist from semicircle to a straight line in the low frequency regime.

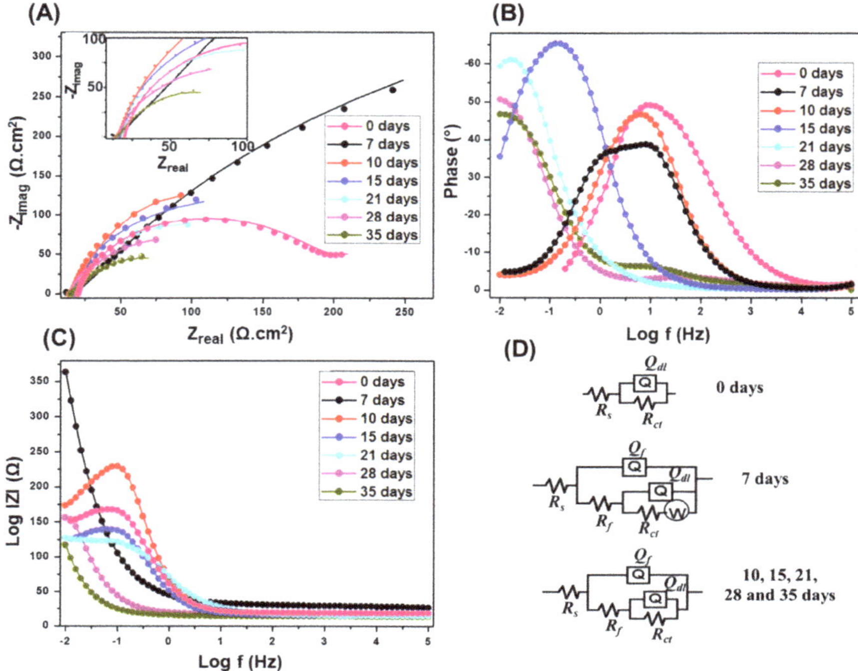

Figure 2. Nyquist (**A**); Bode (**B**,**C**) plots at SRB enriched media. The inset of (**A**) is a magnification of the low impedance region. The EIS were obtained at OCP of 10 mV and sinusoidal signal of 0.01–10^5 Hz. (**D**) Equivalent circuits used to fit the experimental results.

The phase angle, θ, vs. frequency, in logarithmic scale plot (Figure 2B) presents the phase angle peak shifts to lower frequency with the increasing incubation times. The frequency shift confirms the increase in mass or thickness of the corrosion product layer with high electrical capacitance as a result of SRB activity onto carbon steel coupon [58]. Meanwhile, an increase in the capacitance is a result of increases in the mass and thickness of the porous layer, and consequently the surface area. From the impedance modulus |Z| vs. frequency, in logarithmic scale plot (Figure 2C), the |Z| values at low frequencies are maximum after 7 days, after which it decreases with longer incubation. The higher values of |Z| at low frequencies indicating lower corrosion rates [60].

Figure 2D represents the equivalent circuits of the fitted EIS data. R_s represents the resistance of electrolyte, Q_f is the constant phase element (CPE) at the film/solution interface, R_f exemplify the pore film resistance, Q_{dl} is the CPE of coupon/solution interface, R_{ct} is the charge transfer resistance at the coupon/solution interface, and W represent Warburg impedance elements. Both circuits have constant phase elements (Q_f) instead of the ideal electrical double layer capacitors. This was attributed to the surface roughness, inhomogeneous reaction rates distribution, non-uniform thickness, non-uniform composition of the double layer, and/or non-uniform current distribution [61–65]. Since the SRB biofilm provides a more prominent effect during the initial days of incubation, an additional element W is used in the equivalent circuit, corresponding to the diffusion controlled electrochemical process as a result of the complete biofilm formation [59]. As the incubation time increases, the biofilm starts to degrade so the mass transfer is not the controlling factor anymore, so W is removed from the equivalent circuit.

The EIS analysis of the coupons in the abiotic media in the absence CS/LS is shown in Figure S1. The semicircle's behavior is different at 7 and 10 days in the absence of SRB compared to the longer incubation intervals. The precipitation of iron phosphide on the carbon steel surface can be detected from the change of medium frequency capacitive loops at 15 and 21 days of incubations [66]. Iron phosphide can be homogeneously distributed after precipitating with the ferrous ion produced by the steel dissolution under abiotic conditions [66]. A gradual decrease of the Nyquist plot diameter and the phase peak shift to a lower frequency was observed at a longer incubation time, and confirmed the low corrosion rate in the absence of SRB.

Table 1 gives the R_{ct}, as well as R_f values of the carbon steel coupons incubated in the presence of SRB after EIS fitting, and Supplementary Table S2 shows the complete EIS fitting data. The R_{ct} value is highest at 7 days compared to other incubation times due to maximum protection of the complete biofilm. Afterward, the R_{ct} values keep decreasing as the incubation time increases. The decrease in R_{ct} value results in an increase in the dissolution kinetics of the metallic surface due to the fast corrosion rates induced by the breakdown of the biofilm that accelerates the corrosion process. Similarly, the R_f values are highest at 7 days. The decrease in R_f value could be a result of higher porosity of the biofilm on the coupon surface, resulting in the observed accelerated corrosion.

Table 1. R_{ct}, R_f and IE values after EIS fitting.

Incubation Media	Incubation Time (Days)	R_f ($\Omega{\cdot}cm^2$)	R_{ct} ($\Omega{\cdot}cm^2$)	IE (%)
SRB alone	7	79.7	363.6	-
	10	75.6	256.1	-
	15	69.4	173.8	-
	21	45.4	137.2	-
	28	38.6	107.9	-
	35	23.6	88.4	-
CS/LS alone	7	136.1	74.4	
SRB with CS/LS	7	292.7	609	68
	10	249.2	468	82
	15	209	312.8	80
	21	129.7	254.6	85
	28	90.6	193.2	80
	35	71.2	157.3	78

3.3. *Investigation of CS/LS Nanospheres Inhibitory Effect on SRB Induced Corrosion*

The previous aqueous media analysis indicated that CS/LS-1:1 demonstrates strong inhibition of SRB activities at 100 $\mu g{\cdot}mL^{-1}$. The CS/LS-1:1 demonstrated 48.8% inhibition of sulfate reduction and 54.26% reduction of total organic carbon (TOC) removal [31]. Here, we investigate the ability of the new CS/LS nanospheres to inhibit the biofilm formation and control MIC on the coupon surface. The first step was to identify the optimum concentration of CS/LS that gives maximum corrosion inhibition in a concentration range from 100 to 1000 $\mu g{\cdot}mL^{-1}$. SRB induced corrosion starts progressing after 10 days of incubation. Hence, the impedance analysis was performed after 10 days of incubation [39]. The Nyquist plots are shown in Supplementary Figure S2A and the equivalent circuit used for fitting the impedance plots is shown in Figure 2D. The relation between R_{ct} values and CS/LS concentration are given in Supplementary Figure S2B. The R_{ct} values of CS/LS are 256, 287, 337, 468 and 307 $\Omega{\cdot}cm^2$ for 0, 100, 200, 500 and 1000 $\mu g{\cdot}mL^{-1}$ respectively. From the R_{ct} values, it is found that 500 $\mu g{\cdot}mL^{-1}$ resulted in the maximum corrosion inhibition. At 1000 $\mu g{\cdot}mL^{-1}$, precipitation started to be observed in the reaction medium. Hence, 500 $\mu g{\cdot}mL^{-1}$ has been selected as the optimum concentrations of CS/LS for further experiments.

The effect of incubation time on the carbon steel coupons was investigated by EIS after 7, 10, 15, 21, 28, and 35 days in the media containing SRB at 500 $\mu g{\cdot}mL^{-1}$ CS/LS. The Nyquist plot (Figure 3A) showed the same trend as in the presence of SRB but the diameter is higher at the corresponding incubation times. The equivalent circuit used is shown in Figure 3D. The increase in the diameter of the semicircle in the impedance spectrum implies corrosion inhibition in the presence of CS/LS. The R_{ct} values are 609 and 363.6 $\Omega{\cdot}cm^2$ respectively (Table 1) for the coupon incubated with and without CS/LS respectively after 7 days. However, the Nyquist plot behavior is different for 7 days of incubation compared to the longer incubation periods, and there is no high-frequency capacitive loop after 7 days or even at higher incubation times in presence of CS/LS [59,67]. This can be attributed to the formation of CS/LS layer on the metal surface as also confirmed by the increase in R_f values.

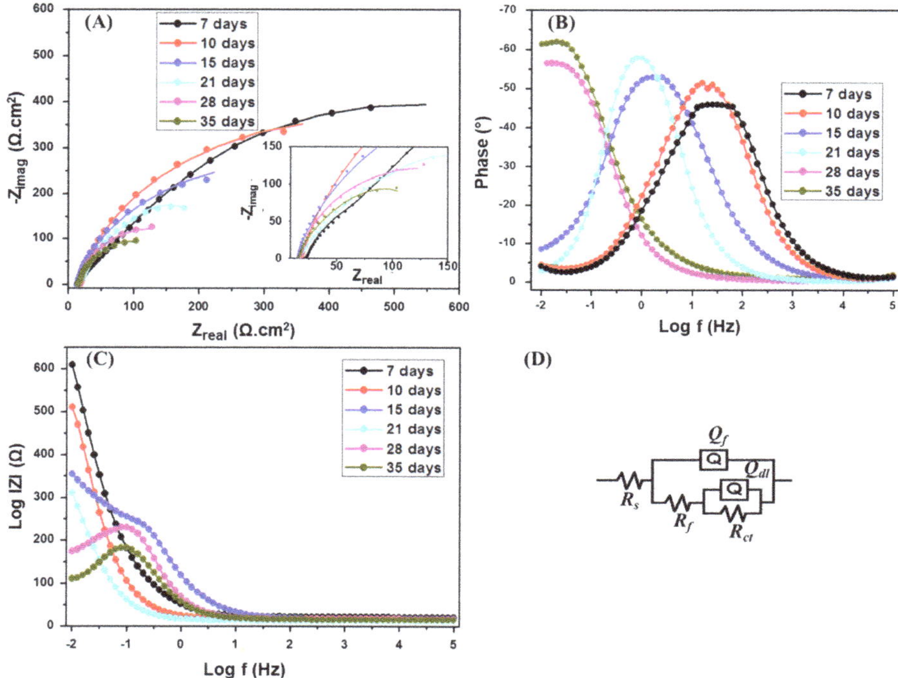

Figure 3. Nyquist (**A**); Bode (**B**,**C**) plots of SRB with CS/LS. The inset is a zoom of the low impedance region. The EIS were recorded at a range of 0.01–10^5 Hz. (**D**) Equivalent circuit used for fitting.

From the phase angle θ vs. frequency, in the logarithmic scale plot (Figure 3B), the phase peak has shifted to a lower frequency with increasing incubation time similar to the SRB corrosion experiments. However, the intensity of the lower frequency shift is less compared to the presence of SRB. Similarly, from the impedance modulus $|Z|$ vs. frequency, in logarithmic scale plot (Figure 3C), the $|Z|$ value at low frequencies is high in the case of 7 days incubation.

The R_{ct} value is maximum at 7 days and it has decreasing with the higher incubation time (Table 1). However, the R_{ct} values are about two times more than the SRB corrosion rate at the corresponding incubation intervals and this enhancement in the R_{ct} value is due to the corrosion inhibition effect of CS/LS. The inhibitory effect of CS/LS in absence of the SRB is evaluated after 7 days of incubation (Supplementary Figure S3). The CS/LS may compete with the biofilm and form a spatial layer on the coupon surface, and this can be verified by comparing the R_f values during the initial incubation times [67]. The R_f values are 3–3.7 times higher in the presence of CS/LS compared with SRB alone during 7–15 days of incubation. The biofilm breakdown takes place as time lapses, which is confirmed by the decrease in the R_f values with longer incubation times.

Corrosion inhibition efficiency (IE) is obtained from:

$$IE = (R_{ct'} - R_{ct})/R_{ct'} \quad (1)$$

where $R_{ct'}$ is obtained in the presence of SRB with CS/LS and R_{ct} is in the presence of SRB alone. The IE at different incubation time intervals is given in Table 1. The maximum corrosion inhibition was found to be 85% in presence of CS/LS. In our previous study, the CZNC inhibitor was able to provide 74% maximum corrosion inhibition with lesser dose of 250 $\mu g \cdot mL^{-1}$ [24]. There was no significant enhancement in the corrosion inhibition, even at the higher dose of 500 $\mu g \cdot mL^{-1}$. However, the inhibitor dose of CZNC was limited to 250 $\mu g \cdot mL^{-1}$ due to the ZnO content in the CZNC biocide. Here we were able to use a higher dose of 500 $\mu g \cdot mL^{-1}$ since CS/LS are metal-free and made of renewable components with expected low toxicity. Nevertheless, the toxicity range and environmental impact of the new nanospheres need to be investigated in future studies.

The corrosion inhibition capability of CS/LS has been compared with the commercial GA biocide by EIS analysis. Supplementary Figure S4 shows the Nyquist plots of the coupons incubated in 5% GA after 15 days of incubation showing a lower diameter of the semicircle in the impedance spectrum as compared with CS/LS, which can be attributed to the better corrosion inhibition induced by the CS/LS. The R_f and R_{ct} values are low (136.4 and 158.8 $\Omega \cdot cm^2$, respectively) when SRB is incubated with GA as compared with CS/LS (209 and 312.8 $\Omega \cdot cm^2$, respectively). In the case of SRB with 5% GA, complete bacterial growth can be inhibited; however, a side reaction can be observed between SRB media and GA which is evidenced by the change of media color to pink instead of the expected black (Supplementary Figure S4). Despite the highly efficiency of GA as a biocide, its toxicity effect to the aquatic ecosystems limited their use [6]. Moreover, GA did not entirely suppress the corrosion in the studied medium during longer incubation times, most likely due to the accumulation of some corrosion products in the cracks of the carbon steel surface. Therefore, CS/LS can provide a more benign alternative to minimize or replace the utilization of GA as a biocide.

3.4. Biofilm and Corrosion Products Characterization

The inhibition of SRB by CS/LS (500 $\mu g \cdot mL^{-1}$) and the consequent formation of biofilm and corrosion products on the carbon steel were examined by SEM, EDS, and XPS analysis. Generally, the presence of exopolysaccharides (EPS), which are excreted by the bacteria to adhere to the metal surface, is visible after four days of incubation along with SRB cells (Supplementary Figure S5). In the presence of CS/LS, different morphology of EPS is visible by SEM due to the possible complex formation with the exopolysaccharide component of CS/LS. Few SRB cells are present on the surface but with deformed cell morphology (Figure S5). After 7 days, the adhesion of numerous active SRBs can be seen on the coupon surface (Figure 4A). Meanwhile, the number of attached SRB cells is significantly reduced in presence of CS/LS, even with noticeable damage at the bacterium cell surface (Figure 4B). The CS/LS particles on the metal surface may have hindered the bacterial attachment as indicated by the smaller number of bacteria on the surface. In addition, an obvious damage to the bacterial cell walls can be seen [31]. Similar observations of cell wall damage were observed as the effect of different nanomaterials on other bacteria [28,68]. For example, predominant damage can be induced on most of the SRB cells in presence of 100 $\mu g \cdot mL^{-1}$ of ZnONPs [28]. Another work reported a progressive damage to the cell wall of *Staphylococcus aureus* and *Pseudomonas aeruginosa*, causing a total lysis of cells in contact with the nanocomposites [68]. Chitosan-based CZNC-10 showed a similar behavior on biofilm formation which was attributed to the synergic influence of CZNC-10 bactericidal impact on the SRB together with the formation of a protective coating on the coupon surface [6,24,67].

Figure 4. SEM images of the biofilm after 7 days in SRB media (**A**), 21 days (**C**) and 35 days (**E**) of incubations. SEM of the biofilm after 7 (**B**), 21 (**D**) and 35 days (**F**) in SRB media containing 500 $\mu g \cdot mL^{-1}$ CS/LS.

After 21 days, uneven deposits of corrosion products were visible on the coupon surface along with complex porous structure of the biofilm when exposed to SRB alone (Figure 4C) [69]. In the presence of CS/LS, SEM showed different corrosion products and biofilm morphology with few deformed bacteria on the surface (Figure 4D). After 35 days, corrosion products were dominant on the surface along with limited biofilm structures in the case of SRB alone (Figure 4E,F). In the presence of CS/LS, similar observation is present but with different morphology of the corrosion products. No bacteria were visible on the coupon surface in both cases, which could be covered by the corrosion products layer. The different morphology is an implication of heterogeneous corrosion products in presence of the CS/LS nanoparticles. EDS and XRF analysis have quantified the sulfur and iron content in both biofilm and corrosion products. According to the EDS analysis after 35 days, a reduction in the concentration of Fe and S content by 43% and 31% respectively was observed in the presence of CS/LS (Figure S6). The XRF analysis also showed the reduction in Fe and S contents by 56% and 29%, respectively, in the presence of CS/LS, confirming the inhibition of SRB (Supplementary Table S3).

Figure 5 shows the XPS survey of corrosion products after 35 days in SRB media with and without 500 $\mu g \cdot mL^{-1}$ of CS/LS. SRB induced biocorrosion follows the sulfate reduction pathway at the metal surface with the help of a hydrogen intermediate [70]. This sulfide can appear as H_2S, HS^- ions, S^{2-} ions or metal sulfides, according to the different conditions which build up at the metal surface and catalyze the corrosion process [71]. As a result, the Fe is oxidized to Fe^{2+} and sulfate is reduced to sulfide followed by the formation of FeS. The overall reaction can be written as:

$$4\,Fe^0 + SO_4{}^{2-} + 3\,HCO_3{}^- + 5\,H^+ \rightarrow FeS + 3\,FeCO_3 + 4\,H_2O$$

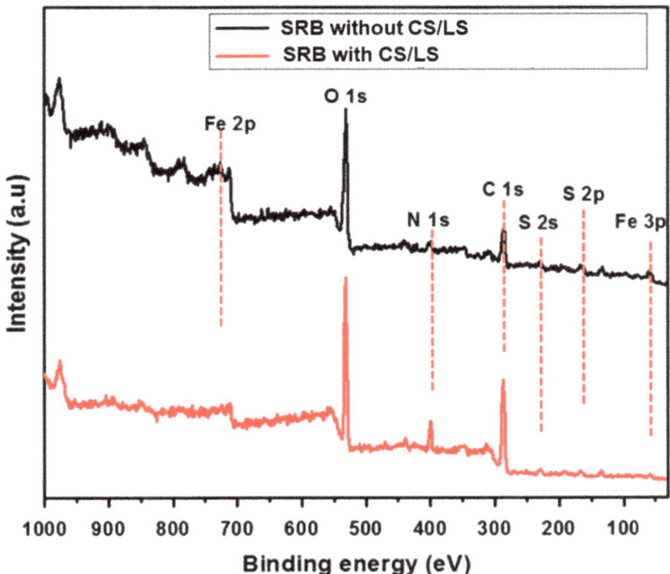

Figure 5. XPS survey spectra of the coupon surface incubated in SRB enriched media with and without CS/LS after 35 days incubation.

The peaks of Fe 3p, Fe 2p, C 1s, O 1s, S 2p, and S 2s are observed in both spectra, which can be attributed to corrosion products and biofilm on the coupon. The S 2p and S 2s peaks confirmed the presence of sulfide and organic sulfur which are mainly formed by the SRB activity. However, the smaller peaks of Fe 2p, S 2s and S 2p were observed from coupon incubated with CS/LS. In addition, the high-resolution spectra for Fe 2p and S 2p are examined to confirm the reduction in the intensity of XPS peak as well as to quantify the corrosion products.

The fitted Fe 2p peaks after 35 days with and without CS/LS are shown in Figure 6A,B respectively. Two sharp peaks at 709.6 eV and 707.6 eV (Fe $2p_{3/2}$) corresponding to FeO (pink curve) and mackinawite ($Fe_{1+x}S$) (green curve), respectively. Pyrite (FeS_2) is present in both spectra (green curve) [50]. In addition, a peek at around 712.4 eV corresponds to Fe^{3+} (originated from Fe_2O_3) is present in both the coupons (black curve) [72]. A sharp peak at around 710.4 eV (Fe $2p_{3/2}$) corresponding to FeS (cyan curve) is present in the coupon exposed to SRB alone [73]. The peak at 713.7 eV of (Fe $2p_{3/2}$) corresponds to Fe(III)O from Fe_2O_3 and is found in presence of CS/LS [74]. From the XPS analysis, the corrosion products are mostly FeO, FeS, FeS_2, and Fe_2O_3. However, FeS peak is prominent only in the coupon exposed to SRB alone which confirmed the reduction in SRB activity in the presence of CS/LS.

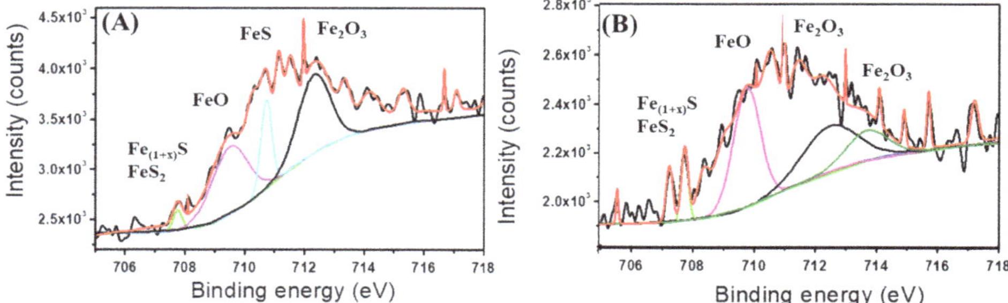

Figure 6. Fitted Fe 2p spectra from coupons incubated in SRB enriched media without CS/LS (**A**) and with 500 $\mu g \cdot mL^{-1}$ CS/LS (**B**) after 35 days of incubation. Fitted curves with different colour correspond to different chemical states; light green corresponds to FeS_2 and mackinawite ($Fe_1 + _xS$), pink corresponds to FeO, Cyan corresponds to FeS, black and green correspond to Fe_2O_3.

SEM and surface profilometry were used to analyze the coupon surface after removing biofilms and corrosion products [75–77]. Figure 7 shows the formed pits on the coupon surface. Pits diameter is greater when coupon is incubated in SRB alone as compared with SRB/CS/LS mixture. The widest pit is observed in the coupon incubated with SRB alone is around 8.2 μm diameter while it is only 4 μm in the presence of CS/LS. These results are matching with EIS and XPS data. In addition, the surface roughness of both coupons was also calculated form profilometry. For comparison, the 2D and 3D profilometry images of the bare coupon is shown in Supplementary Figure S7A,B. The average roughness of the bare coupon is 17 ± 2 nm, which is appropriate for bacterial attachment [78–80]. Supplementary Figure S8 displays high-resolution spectra of the coupon surface after removing the corrosion products. After 35 days, the average roughness of the coupons surface was 780 ± 19 nm in the absence of CS/LS inhibitor. While in the presence of CS/LS inhibitor, the average surface roughness is reduced to 458 ± 16 nm, i.e., the surface roughness of the coupon is reduced to approx. 40% in presence of the CS/LS.

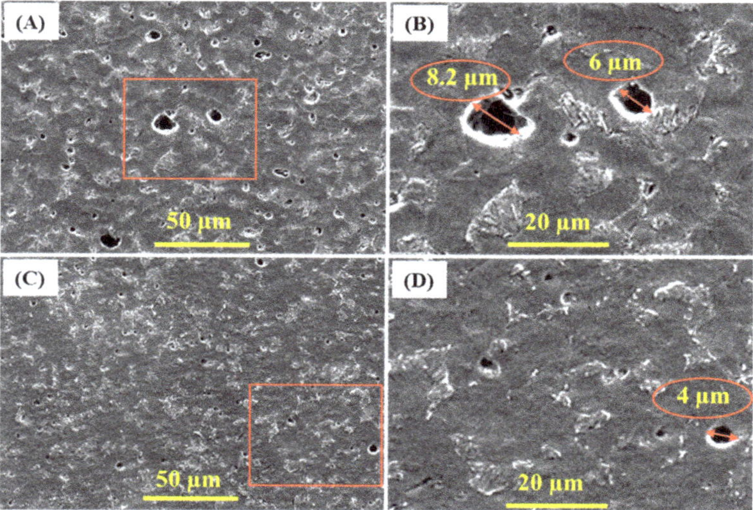

Figure 7. SEM micrograph of corrosion pits after removing the corrosion products from SS400 coupons incubated in SRB alone (**A**,**B**) and in presence of CS/LS nanospheres (**C**,**D**) after 35 days of incubation.

In general, the corrosion inhibition mechanism of nanomaterials can be originated mainly from their antibacterial effect [6]. The CS/LS antibacterial activity can be explained by their surface charge and active surface of CS and LS [31]. Due to the nanostructure and hydrophilic nature of CS/LS (originated from the presence of large number of amino groups of CS component), CS/LS can readily bind to the negatively charged bacteria, leading to cell membrane disruption [24]. Lignosulfonate component can cause oxidative stress to proteins and DNA in bacteria generated by reactive oxygen species. These two processes can cause a reduction in the cell viability by severe cytoplasmic leakage and loss of cell integrity and EPS contents. Sulfate reduction and co-substrate oxidation assays confirmed the CS/LS inhibitory effect on SRB [31]. The SEM analysis of SRB cells showed cell aggregation and prevalent surface damage after CS/LS exposure [31]. In addition, formation of CS/LS layer on the coupon surface protects the surface from the initial bacterial attachment [81]. Morphological results and EIS experiments suggest that CS/LS have hindered the formation of biofilm and unstable corrosion products on the carbon steel surface.

4. Conclusions

CS/LS nanospheres have been successfully evaluated as a novel biocide for the inhibition of SRB induced biocorrosion. The R_{ct} values are approximately doubled in the presence of CS/LS compared with the CS/LS-free media, irrespective of incubation intervals from the electrochemical analysis, with a corrosion inhibition efficiency of 85% at 500 $\mu g \cdot mL^{-1}$ CS/LS. Post-corrosion analysis with SEM and profilometry showed fewer surface defects on the coupon incubated with CS/LS, indicating less corrosion. Two synergic effects can explain the biocidal effect of CS/LS. The hydrophilic CS/LS can readily bind to the bacterial surfaces and thereby damage the bacterial cell wall. Meanwhile, the film forming capability of CS prevents the initial bacterial attachment on the metal surface and thereby leads to a reduction of biofilm formation. In short, due to the biodegradable nature, promising antimicrobial properties of the building blocks, and high biocorrosion inhibition efficiency, the CS/LS nanospheres can present a renewable, cost efficient, and environmentally benign biocide for the inhibition of SRB induced MIC on carbon steel systems.

Supplementary Materials: The following are available online at http://www.mdpi.com/1996-1944/13/11/2484/s1, Figure S1: The Nyquist plot of the coupon incubated in the abiotic media (A). The corresponding Bode plots are given in (B) and (C). Figure S2: (A) The Nyquist plot of the coupon incubated with SRB with CS/LS with different concentrations from 0 to 1000 $\mu g \cdot mL^{-1}$. The impedance analysis was performed after 10 days of incubation. (B) The R_{ct} vs concentration of CS/LS after 10 days of incubation. The error bar indicates the standard deviation from the three independent measurements. Figure S3: The Nyquist plot of the coupon after 7 days of incubation in presence of CS/LS. Figure S4: Nyquist plot of the incubated coupon after 15 days in SRB enriched media with 5% GA. Inset shows the fitting equivalent circuit and the photographs of the incubation mixtures after 15 days of incubation. Figure S5: SEM images of the biofilm grown on the coupon surface after being incubated with SRB enriched media for 4 days without inhibitor (A) and in presence of 500 $\mu g \cdot mL^{-1}$ CS/LS (B). Figure S6: EDS analysis of biofilm incubated in SRB enriched media without CS/LS (A) and with 500 $\mu g \cdot mL^{-1}$ CS/LS (B) after 35 days of incubation. Figure S7: Profilometry of the bare carbon steel coupon surface. (A) 2D and (B) 3D images. Figure S8: Profilometry images (2D and 3D) of the cleaned carbon steel coupon surface after 35 days of incubation in SRB without CS/LS (A and B) and SRB with 500 $\mu g \cdot mL^{-1}$ CS/LS (C and D), respectively. Table S1: Composition of simulated seawater. Table S2. EIS fitting data. Table S3: XRF data of the biofilm incubated in SRB enriched media without CS/LS (A) and with 500 $\mu g \cdot mL^{-1}$ CS/LS (B) after 35 days of incubation.

Author Contributions: Conceptualization, P.A.R. and K.A.M.; methodology, P.A.R. and K.A.J.; validation, R.P.P. and K.A.M.; formal analysis, P.A.R. and K.A.J.; investigation, P.A.R.; data curation, P.A.R. and R.P.P.; writing—Original draft preparation, P.A.R. and R.P.P.; writing—Review and editing, A.M.A. and A.S.; visualization, P.A.R.; supervision, K.A.M. All authors have read and agreed to the published version of the manuscript.

Funding: This research was funded by Qatar National Research Fund QNRF, Qatar Foundation through the NPRP grant # NPRP8-286-2-118. The publication of this article was funded by the Qatar National Library.

Acknowledgments: The authors are grateful to V. Madhavan, J. Ponraj, M. Pasha and M. Helal at the Core lab of QEERI/HBKU, Qatar for XPS, TEM and SEM analysis.

Conflicts of Interest: The authors declare no conflict of interest.

References

1. Vanaei, H.R.; Eslami, A.; Egbewande, A. A review on pipeline corrosion, in-line inspection (ILI), and corrosion growth rate models. *Int. J. Press. Vessel. Pip.* **2017**, *149*, 43–54. [CrossRef]
2. Xu, D.; Huang, W.; Ruschau, G.; Hornemann, J.; Wen, J.; Gu, T. Laboratory investigation of MIC threat due to hydrotest using untreated seawater and subsequent exposure to pipeline fluids with and without SRB spiking. *Eng. Fail. Anal.* **2013**, *28*, 149–159. [CrossRef]
3. Skovhus, T.L.; Enning, D.; Lee, J.S. *Microbiologically Influenced Corrosion in the Upstream Oil and Gas Industry*; CRC Press: Boca Raton, FL, USA, 2017.
4. Wang, H.; Wang, Z.; Hong, H.; Yin, Y. Preparation of cerium-doped TiO2 film on 304 stainless steel and its bactericidal effect in the presence of sulfate-reducing bacteria (SRB). *Mater. Chem. Phys.* **2010**, *124*, 791–794. [CrossRef]
5. Enning, D.; Venzlaff, H.; Garrelfs, J.; Dinh, H.T.; Meyer, V.; Mayrhofer, K.; Hassel, A.W.; Stratmann, M.; Widdel, F. Marine sulfate-reducing bacteria cause serious corrosion of iron under electroconductive biogenic mineral crust. *Environ. Microbiol.* **2012**, *14*, 1772–1787. [CrossRef]
6. Rasheed, P.A.; Jabbar, K.A.; Mackey, H.R.; Mahmoud, K.A. Recent advancements of nanomaterials as coatings and biocides for the inhibition of sulfate reducing bacteria induced corrosion. *Curr. Opin. Chem. Eng.* **2019**, *25*, 35–42. [CrossRef]
7. Rajasekar, A.; Anandkumar, B.; Maruthamuthu, S.; Ting, Y.P.; Rahman, P. Characterization of corrosive bacterial consortia isolated from petroleum-product-transporting pipelines. *Appl. Microbiol. Biotechnol.* **2010**, *85*, 1175–1188. [CrossRef]
8. Vance, I.; Thrasher, D.R. *Reservoir souring: Mechanisms and prevention*; American Society of Microbiology: Washington, DC, USA, 2005.
9. Enning, D.; Garrelfs, J. Corrosion of Iron by Sulfate-Reducing Bacteria: New Views of an Old Problem. *Appl. Environ. Microbiol.* **2014**, *80*, 1226–1236. [CrossRef]
10. Sun, C.; Xu, J.; Wang, F. Interaction of Sulfate-Reducing Bacteria and Carbon Steel Q 235 in Biofilm. *Ind. Eng. Chem. Res.* **2011**, *50*, 12797–12806. [CrossRef]
11. Antony, P.J.; Raman, R.K.S.; Raman, R.; Kumar, P. Role of microstructure on corrosion of duplex stainless steel in presence of bacterial activity. *Corros. Sci.* **2010**, *52*, 1404–1412. [CrossRef]
12. Esquivel, R.G.; Olivares, G.Z.; Gayosso, M.J.H.; Trejo, A.G. Cathodic protection of XL 52 steel under the influence of sulfate reducing bacteria. *Mater. Corros.* **2011**, *62*, 61–67. [CrossRef]
13. Krishnamurthy, A.; Gadhamshetty, V.; Mukherjee, R.; Chen, Z.; Ren, W.; Cheng, H.M.; Koratkar, N. Passivation of microbial corrosion using a graphene coating. *Carbon* **2013**, *56*, 45–49. [CrossRef]
14. Duncan, K.E.; Perez-Ibarra, B.M.; Jenneman, G.; Harris, J.B.; Webb, R.; Sublette, K. The effect of corrosion inhibitors on microbial communities associated with corrosion in a model flow cell system. *Appl. Microbiol. Biotechnol.* **2014**, *98*, 907–918. [CrossRef] [PubMed]
15. Narenkumar, J.; Parthipan, P.; Usha Raja Nanthini, A.; Benelli, G.; Murugan, K.; Rajasekar, A. Ginger extract as green biocide to control microbial corrosion of mild steel. *3 Biotech.* **2017**, *7*, 133. [CrossRef] [PubMed]
16. Xue, Y.; Voordouw, G. Control of Microbial Sulfide Production with Biocides and Nitrate in Oil Reservoir Simulating Bioreactors. *Front. Microbiol.* **2015**, *6*, 1387. [CrossRef]
17. Nguyen, T.; Roddick, F.A.; Fan, L. Biofouling of water treatment membranes: A review of the underlying causes, monitoring techniques and control measures. *Membranes* **2012**, *2*, 804–840. [CrossRef]
18. Rasool, K.; Helal, M.; Ali, A.; Ren, C.E.; Gogotsi, Y.; Mahmoud, K.A. Antibacterial Activity of Ti3C2Tx MXene. *ACS Nano* **2016**, *10*, 3674–3684. [CrossRef]
19. Hajipour, M.J.; Fromm, K.M.; Akbar Ashkarran, A.; Jimenez de Aberasturi, D.; Larramendi, I.R.d.; Rojo, T.; Serpooshan, V.; Parak, W.J.; Mahmoudi, M. Antibacterial properties of nanoparticles. *Trends Biotechnol.* **2012**, *30*, 499–511. [CrossRef]
20. Mathur, A.; Bhuvaneshwari, M.; Babu, S.; Chandrasekaran, N.; Mukherjee, A. The effect of TiO2 nanoparticles on sulfate-reducing bacteria and their consortium under anaerobic conditions. *J. Environ. Chem. Eng.* **2017**, *5*, 3741–3748. [CrossRef]
21. Khowdiary, M.M.; El-Henawy, A.A.; Shawky, A.M.; Sameeh, M.Y.; Negm, N.A. Synthesis, characterization and biocidal efficiency of quaternary ammonium polymers silver nanohybrids against sulfate reducing bacteria. *J. Mol. Liq.* **2017**, *230*, 163–168. [CrossRef]

22. Fathy, M.; Badawi, A.; Mazrouaa, A.M.; Mansour, N.A.; Ghazy, E.A.; Elsabee, M.Z. Styrene N-vinylpyrrolidone metal-nanocomposites as antibacterial coatings against Sulfate Reducing Bacteria. *Mater. Sci. Eng. C* **2013**, *33*, 4063–4070. [CrossRef]
23. Wan, D.; Yuan, S.; Neoh, K.G.; Kang, E.T. Surface Functionalization of Copper via Oxidative Graft Polymerization of 2,2′-Bithiophene and Immobilization of Silver Nanoparticles for Combating Biocorrosion. *Acs Appl. Mater. Interfaces* **2010**, *2*, 1653–1662. [CrossRef] [PubMed]
24. Rasool, K.; Nasrallah, G.K.; Younes, N.; Pandey, R.P.; Abdul Rasheed, P.; Mahmoud, K.A. "Green" ZnO-Interlinked Chitosan Nanoparticles for the Efficient Inhibition of Sulfate-Reducing Bacteria in Inject Seawater. *Acs Sustain. Chem. Eng.* **2018**, *6*, 3896–3906. [CrossRef]
25. Kumar, N.; Omoregie, E.O.; Rose, J.; Masion, A.; Lloyd, J.R.; Diels, L.; Bastiaens, L. Inhibition of sulfate reducing bacteria in aquifer sediment by iron nanoparticles. *Water Res.* **2014**, *51*, 64–72. [CrossRef] [PubMed]
26. Krishnamurthy, A.; Gadhamshetty, V.; Mukherjee, R.; Natarajan, B.; Eksik, O.; Ali Shojaee, S.; Lucca, D.A.; Ren, W.; Cheng, H.-M.; Koratkar, N. Superiority of Graphene over Polymer Coatings for Prevention of Microbially Induced Corrosion. *Sci. Rep.* **2015**, *5*, 13858. [CrossRef] [PubMed]
27. Alasvand Zarasvand, K.; Rai, V.R. Inhibition of a sulfate reducing bacterium, Desulfovibrio marinisediminis GSR3, by biosynthesized copper oxide nanoparticles. *3 Biotech.* **2016**, *6*, 84. [CrossRef]
28. Rasool, K.; Lee, D.S. Effect of ZnO nanoparticles on biodegradation and biotransformation of co-substrate and sulphonated azo dye in anaerobic biological sulfate reduction processes. *Int. Biodeterior. Biodegrad.* **2016**, *109*, 150–156. [CrossRef]
29. Yan, X.; Rong, R.; Zhu, S.; Guo, M.; Gao, S.; Wang, S.; Xu, X. Effects of ZnO Nanoparticles on Dimethoate-Induced Toxicity in Mice. *J. Agric. Food Chem.* **2015**, *63*, 8292–8298. [CrossRef]
30. Ashraf, M.A.; Ullah, S.; Ahmad, I.; Qureshi, A.K.; Balkhair, K.S.; Abdur Rehman, M. Green biocides, a promising technology: Current and future applications to industry and industrial processes. *J. Sci. Food Agric.* **2014**, *94*, 388–403. [CrossRef]
31. Pandey, R.P.; Rasool, K.; Rasheed, P.A.; Gomez, T.; Pasha, M.; Mansour, S.A.; Lee, O.-S.; Mahmoud, K.A. One-step synthesis of an antimicrobial framework based on covalently cross-linked chitosan/lignosulfonate (CS@LS) nanospheres. *Green Chem.* **2020**, *22*, 678–687. [CrossRef]
32. Hosseinnejad, M.; Jafari, S.M. Evaluation of different factors affecting antimicrobial properties of chitosan. *Int. J. Biol. Macromol.* **2016**, *85*, 467–475. [CrossRef]
33. Yuan, G.; Lv, H.; Tang, W.; Zhang, X.; Sun, H. Effect of chitosan coating combined with pomegranate peel extract on the quality of Pacific white shrimp during iced storage. *Food Control.* **2016**, *59*, 818–823. [CrossRef]
34. Piras, A.M.; Maisetta, G.; Sandreschi, S.; Gazzarri, M.; Bartoli, C.; Grassi, L.; Esin, S.; Chiellini, F.; Batoni, G. Chitosan nanoparticles loaded with the antimicrobial peptide temporin B exert a long-term antibacterial activity in vitro against clinical isolates of Staphylococcus epidermidis. *Front. Microbiol.* **2015**, *6*, 372. [CrossRef] [PubMed]
35. Ramezani, Z.; Zarei, M.; Raminnejad, N. Comparing the effectiveness of chitosan and nanochitosan coatings on the quality of refrigerated silver carp fillets. *Food Control.* **2015**, *51*, 43–48. [CrossRef]
36. Zhang, A.; Mu, H.; Zhang, W.; Cui, G.; Zhu, J.; Duan, J. Chitosan Coupling Makes Microbial Biofilms Susceptible to Antibiotics. *Sci. Rep.* **2013**, *3*, 3364. [CrossRef]
37. Martinez, L.R.; Mihu, M.R.; Han, G.; Frases, S.; Cordero, R.J.B.; Casadevall, A.; Friedman, A.J.; Friedman, J.M.; Nosanchuk, J.D. The use of chitosan to damage Cryptococcus neoformans biofilms. *Biomaterials* **2010**, *31*, 669–679. [CrossRef]
38. Andersen, T.; Mishchenko, E.; Flaten, G.; Sollid, J.; Mattsson, S.; Tho, I.; Škalko-Basnet, N. Chitosan-Based Nanomedicine to Fight Genital Candida Infections: Chitosomes. *Mar. Drugs* **2017**, *15*, 64. [CrossRef]
39. Rasheed, P.A.; Jabbar, K.A.; Rasool, K.; Pandey, R.P.; Sliem, M.H.; Helal, M.; Samara, A.; Abdullah, A.M.; Mahmoud, K.A. Controlling the biocorrosion of sulfate-reducing bacteria (SRB) on carbon steel using ZnO/chitosan nanocomposite as an eco-friendly biocide. *Corros. Sci.* **2019**, *148*, 397–406. [CrossRef]
40. Azadi, P.; Inderwildi, O.R.; Farnood, R.; King, D.A. Liquid fuels, hydrogen and chemicals from lignin: A critical review. *Renew. Sustain. Energy Rev.* **2013**, *21*, 506–523. [CrossRef]
41. Berlin, A.; Balakshin, M. Chapter 18-Industrial Lignins: Analysis, Properties, and Applications. In *Bioenergy Research: Advances and Applications*; Gupta, V.K., Tuohy, M.G., Kubicek, C.P., Saddler, J., Xu, F., Eds.; Elsevier: Amsterdam, The Netherland, 2014; pp. 315–336. [CrossRef]

42. Kim, S.; Fernandes, M.M.; Matamá, T.; Loureiro, A.; Gomes, A.C.; Cavaco-Paulo, A. Chitosan–lignosulfonates sono-chemically prepared nanoparticles: Characterisation and potential applications. *Colloids Surf. B Biointerfaces* **2013**, *8*, 1–8. [CrossRef]
43. Lora, J.H.; Glasser, W.G. Recent Industrial Applications of Lignin: A Sustainable Alternative to Nonrenewable Materials. *J. Polym. Environ.* **2002**, *10*, 39–48. [CrossRef]
44. Dong, X.; Dong, M.; Lu, Y.; Turley, A.; Jin, T.; Wu, C. Antimicrobial and antioxidant activities of lignin from residue of corn stover to ethanol production. *Ind. Crops Prod.* **2011**, *34*, 1629–1634. [CrossRef]
45. Fredheim, G.E.; Christensen, B.E. Polyelectrolyte Complexes: Interactions between Lignosulfonate and Chitosan. *Biomacromolecules* **2003**, *4*, 232–239. [CrossRef] [PubMed]
46. Al-Rashed, M.M.; Niknezhad, S.; Jana, S.C. Mechanism and Factors Influencing Formation and Stability of Chitosan/Lignosulfonate Nanoparticles. *Macromol. Chem. Phys.* **2019**, *220*, 1800338. [CrossRef]
47. Yan, M.; Huang, W.; Li, Z. Chitosan cross-linked graphene oxide/lignosulfonate composite aerogel for enhanced adsorption of methylene blue in water. *Int. J. Biol. Macromol.* **2019**, *136*, 927–935. [CrossRef] [PubMed]
48. Javed, M.A.; Neil, W.C.; McAdam, G.; Wade, S.A. Effect of sulphate-reducing bacteria on the microbiologically influenced corrosion of ten different metals using constant test conditions. *Int. Biodeterior. Biodegrad.* **2017**, *125*, 73–85. [CrossRef]
49. Javed, M.A.; Stoddart, P.R.; Wade, S.A. Corrosion of carbon steel by sulphate reducing bacteria: Initial attachment and the role of ferrous ions. *Corros. Sci.* **2015**, *93*, 48–57. [CrossRef]
50. Yuan, S.; Liang, B.; Zhao, Y.; Pehkonen, S.O. Surface chemistry and corrosion behaviour of 304 stainless steel in simulated seawater containing inorganic sulphide and sulphate-reducing bacteria. *Corros. Sci.* **2013**, *74*, 353–366. [CrossRef]
51. Huttunen-Saarivirta, E.; Rajala, P.; Carpén, L. Corrosion behaviour of copper under biotic and abiotic conditions in anoxic ground water: Electrochemical study. *Electrochim. Acta* **2016**, *203*, 350–365. [CrossRef]
52. Yin, B.; Williams, T.; Koehler, T.; Morris, B.; Manna, K. Targeted microbial control for hydrocarbon reservoir: Identify new biocide offerings for souring control using thermophile testing capabilities. *Int. Biodeterior. Biodegrad.* **2018**, *126*, 204–207. [CrossRef]
53. Kahrilas, G.A.; Blotevogel, J.; Stewart, P.S.; Borch, T. Biocides in Hydraulic Fracturing Fluids: A Critical Review of Their Usage, Mobility, Degradation, and Toxicity. *Environ. Sci. Technol.* **2015**, *49*, 16–32. [CrossRef]
54. Chen, S.Q.; Wang, P.; Zhang, D. The influence of sulphate-reducing bacteria on heterogeneous electrochemical corrosion behavior of Q235 carbon steel in seawater. *Mater. Corros.* **2016**, *67*, 340–351. [CrossRef]
55. Wang, Y.; Pitto-Barry, A.; Habtemariam, A.; Romero-Canelon, I.; Sadler, P.J.; Barry, N.P.E. Nanoparticles of chitosan conjugated to organo-ruthenium complexes. *Inorg. Chem. Front.* **2016**, *3*, 1058–1064. [CrossRef]
56. Li, P.-C.; Liao, G.M.; Kumar, S.R.; Shih, C.-M.; Yang, C.-C.; Wang, D.-M.; Lue, S.J. Fabrication and Characterization of Chitosan Nanoparticle-Incorporated Quaternized Poly(Vinyl Alcohol) Composite Membranes as Solid Electrolytes for Direct Methanol Alkaline Fuel Cells. *Electrochim. Acta* **2016**, *187*, 616–628. [CrossRef]
57. Wang, C.H.; Fang, G.Z.; Ai, Q.; Zhao, Y.F. Preparation of Lignosulfonate-Chitosan Polyelectrolyte Complex. *Adv. Mater. Res.* **2011**, *197–198*, 1249–1252.
58. AlAbbas, F.M.; Williamson, C.; Bhola, S.M.; Spear, J.R.; Olson, D.L.; Mishra, B.; Kakpovbia, A.E. Influence of sulfate reducing bacterial biofilm on corrosion behavior of low-alloy, high-strength steel (API-5L X80). *Int. Biodeterior. Biodegrad.* **2013**, *78*, 34–42. [CrossRef]
59. Castaneda, H.; Benetton, X.D. SRB-biofilm influence in active corrosion sites formed at the steel-electrolyte interface when exposed to artificial seawater conditions. *Corros. Sci.* **2008**, *50*, 1169–1183. [CrossRef]
60. Su, C.; Wu, W.; Li, Z.; Guo, Y. Prediction of film performance by electrochemical impedance spectroscopy. *Corros. Sci.* **2015**, *99*, 42–52. [CrossRef]
61. Kim, C.-H.; Pyun, S.-I.; Kim, J.-H. An investigation of the capacitance dispersion on the fractal carbon electrode with edge and basal orientations. *Electrochim. Acta* **2003**, *48*, 3455–3463. [CrossRef]
62. Mulder, W.H.; Sluyters, J.H.; Pajkossy, T.; Nyikos, L. Tafel current at fractal electrodes: Connection with admittance spectra. *J. Electroanal. Chem. Interfacial Electrochem.* **1990**, *285*, 103–115. [CrossRef]
63. Schiller, C.A.; Strunz, W. The evaluation of experimental dielectric data of barrier coatings by means of different models. *Electrochim. Acta* **2001**, *46*, 3619–3625. [CrossRef]

64. Jorcin, J.-B.; Orazem, M.E.; Pébère, N.; Tribollet, B. CPE analysis by local electrochemical impedance spectroscopy. *Electrochim. Acta* **2006**, *51*, 1473–1479. [CrossRef]
65. Oldham, K.B. The RC time "constant" at a disk electrode. *Electrochem. Commun.* **2004**, *6*, 210–214. [CrossRef]
66. Liu, H.; Fu, C.; Gu, T.; Zhang, G.; Lv, Y.; Wang, H.; Liu, H. Corrosion behavior of carbon steel in the presence of sulfate reducing bacteria and iron oxidizing bacteria cultured in oilfield produced water. *Corros. Sci.* **2015**, *100*, 484–495. [CrossRef]
67. Gupta, N.K.; Joshi, P.G.; Srivastava, V.; Quraishi, M.A. Chitosan: A macromolecule as green corrosion inhibitor for mild steel in sulfamic acid useful for sugar industry. *Int. J. Biol. Macromol.* **2018**, *106*, 704–711. [CrossRef] [PubMed]
68. Regiel-Futyra, A.; Kus-Liśkiewicz, M.; Sebastian, V.; Irusta, S.; Arruebo, M.; Stochel, G.; Kyzioł, A. Development of Noncytotoxic Chitosan–Gold Nanocomposites as Efficient Antibacterial Materials. *Acs Appl. Mater. Interfaces* **2015**, *7*, 1087–1099. [CrossRef] [PubMed]
69. Chen, S.; Wang, P.; Zhang, D. Corrosion behavior of copper under biofilm of sulfate-reducing bacteria. *Corros. Sci.* **2014**, *87*, 407–415. [CrossRef]
70. Lin, J.; Ballim, R. Biocorrosion control: Current strategies and promising alternatives. *Afr. J. Biotechnol.* **2012**, *11*, 15736–15747. [CrossRef]
71. Kan, J.; Chellamuthu, P.; Obraztsova, A.; Moore, J.E.; Nealson, K.H. Diverse bacterial groups are associated with corrosive lesions at a Granite Mountain Record Vault (GMRV). *J. Appl. Microbiol.* **2011**, *111*, 329–337. [CrossRef]
72. Grosvenor, A.P.; Kobe, B.A.; Biesinger, M.C.; McIntyre, N.S. Investigation of multiplet splitting of Fe 2p XPS spectra and bonding in iron compounds. *Surf. Interface Anal.* **2004**, *36*, 1564–1574. [CrossRef]
73. Zheng, B.; Li, K.; Liu, H.; Gu, T. Effects of Magnetic Fields on Microbiologically Influenced Corrosion of 304 Stainless Steel. *Ind. Eng. Chem. Res.* **2014**, *53*, 48–54. [CrossRef]
74. Wang, W.P.; Yang, H.; Xian, T.; Jiang, J.L. XPS and magnetic properties of $CoFe_2O_4$ nanoparticles synthesized by a polyacrylamide gel route. *Mater. Trans.* **2012**, *53*, 1586–1589. [CrossRef]
75. Finšgar, M. 2-Mercaptobenzimidazole as a copper corrosion inhibitor: Part, I. Long-term immersion, 3D-profilometry, and electrochemistry. *Corros. Sci.* **2013**, *72*, 82–89. [CrossRef]
76. Finšgar, M.; Merl, D.K. 2-Mercaptobenzoxazole as a copper corrosion inhibitor in chloride solution: Electrochemistry, 3D-profilometry, and XPS surface analysis. *Corros. Sci.* **2014**, *80*, 82–95. [CrossRef]
77. Xu, D.K.; Birbilis, N.; Lashansky, D.; Rometsch, P.A.; Muddle, B.C. Effect of solution treatment on the corrosion behaviour of aluminium alloy AA7150: Optimisation for corrosion resistance. *Corros. Sci.* **2011**, *53*, 217–225. [CrossRef]
78. Crawford, R.J.; Webb, H.K.; Truong, V.K.; Hasan, J.; Ivanova, E.P. Surface topographical factors influencing bacterial attachment. *Adv. Colloid Interface Sci.* **2012**, *179–182*, 142–149. [CrossRef] [PubMed]
79. Ploux, L.; Ponche, A.; Anselme, K. Bacteria/Material Interfaces: Role of the Material and Cell Wall Properties. *J. Adhes. Sci. Technol.* **2010**, *24*, 2165–2201. [CrossRef]
80. Anselme, K.; Davidson, P.; Popa, A.M.; Giazzon, M.; Liley, M.; Ploux, L. The interaction of cells and bacteria with surfaces structured at the nanometre scale. *Acta Biomater.* **2010**, *6*, 3824–3846. [CrossRef]
81. Dutta, P.K.; Tripathi, S.; Mehrotra, G.K.; Dutta, J. Perspectives for chitosan based antimicrobial films in food applications. *Food Chem.* **2009**, *114*, 1173–1182. [CrossRef]

© 2020 by the authors. Licensee MDPI, Basel, Switzerland. This article is an open access article distributed under the terms and conditions of the Creative Commons Attribution (CC BY) license (http://creativecommons.org/licenses/by/4.0/).

Article

Inhibitive Properties of Benzyldimethyldodecylammonium Chloride on Microbial Corrosion of 304 Stainless Steel in a *Desulfovibrio desulfuricans*-Inoculated Medium

Chung-Wen Hsu [1], Tzu-En Chen [1], Kai-Yin Lo [2] and Yueh-Lien Lee [1,*]

1 Department of Engineering Science and Ocean Engineering, National Taiwan University, Taipei 106, Taiwan; r05525021@ntu.edu.tw (C.-W.H.); r07525016@ntu.edu.tw (T.-E.C.)
2 Department of Agricultural Chemistry, National Taiwan University, Taipei 106, Taiwan; kaiyin@ntu.edu.tw
* Correspondence: yuehlien@ntu.edu.tw; Tel.: +886-2-33665740; Fax: +886-2-23929885

Received: 13 December 2018; Accepted: 17 January 2019; Published: 18 January 2019

Abstract: Biocides are frequently used to control sulfate-reducing bacteria (SRB) in biofouling. The increasing restrictions of environmental regulations and growing safety concerns on the use of biocides result in efforts to minimize the amount of biocide use and develop environmentally friendly biocides. In this study, the antimicrobial activity and corrosion inhibition effect of a low-toxic alternative biocide, benzyldimethyldodecylammonium chloride (BDMDAC), on a 304 stainless steel substrate immersed in a *Desulfovibrio desulfuricans* (*D. desulfuricans*)-inoculated medium was examined. Potentiodynamic polarization curves were used to analyze corrosion behavior. Biofilm formation and corrosion products on the surfaces of 304 stainless steel coupons were examined using scanning electron microscopy (SEM), energy-dispersive X-ray spectrum, and confocal laser scanning microscopy (CLSM). Results demonstrated that this compound exhibited satisfactory results against microbial corrosion by *D. desulfuricans*. The corrosion current density and current densities in the anodic region were lower in the presence of BDMDAC in the *D. desulfuricans*-inoculated medium. SEM and CLSM analyses revealed that the presence of BDMDAC mitigated formation of biofilm by *D. desulfuricans*.

Keywords: microbial corrosion; sulfate-reducing bacteria; biocides; electrochemical test; confocal laser scanning microscopy

1. Introduction

Microbiologically influenced corrosion (MIC), which is caused by interactions between various microorganisms, is a long-term concern in many industries, including those involved with underground pipelines, storage vessels, shipping, and marine equipment. Reports indicate that MIC was responsible for at least 20% of all damaging corrosion, with a direct cost of $30–$50 billion annually worldwide [1,2]. Among the various types of micro-organisms, SRB are considered the major bacterial group responsible for microbial corrosion under anaerobic conditions [3,4]. Several studies have been conducted to investigate SRB-induced corrosion of iron substrate, and SRB are widely accepted to play a crucial role in the anaerobic MIC of iron, low-alloy steel, and stainless steel [5–17]. Various mechanisms explain the enhanced corrosion caused by SRB. Among these, the primary mechanism sees SRB reduce inorganic sulfate to hydrogen sulfide, resulting in the bulk equation ($Fe + H_2S \rightarrow FeS + H_2$) [18–21]. Moreover, extracellular polymer substances and corrosive metabolites secreted by SRB [22,23], consumption of cathodic hydrogen and iron-derived electron transfers [24], and anodic depolarization resulting from the local acidification at the anode [25], all affect the corrosion behavior of iron.

Because of economic losses and safety hazards, it is important to control microbial corrosion by SRB and aggressive sulfide anions when they contact metal substrates. Several methods such as biocide treatment, cathodic protection, and addition of nitrate (or nitrite) have been developed to minimize the risks resulting from SRB activity [23,26]. Among these, biocide treatment is the most common method of controlling microbial corrosion. Organotins, such as dibutyltin and tributyltin (TBT), have been extensively used as antifouling agents in paints since the early 1970s. Despite their high resistance to microbial corrosion in marine environments, these compounds were banned in numerous parts of the world in the early 1990s due to the severe environmental and human health risks that they pose [27–33]. Therefore, nontoxic or less-toxic alternative biocides must be developed to substitute for TBT.

Quaternary ammonium compounds (QACs) are considered candidates for biocides because of their excellent stability and antimicrobial properties [34–36]. QACs are surfactants with a hydrocarbon water-repellent (hydrophobic) group and a water-attracting group (hydrophilic). The antimicrobial properties of QACs mainly depend on the length of the long-chain alkyl group [37]. QACs have wide applications ranging from clinical to industrial purposes. For example, they are used for the disinfection of surfaces, equipment, and medical devices. Benzyldimethyldodecylammonium chloride (BDMDAC) is a quaternary ammonium compound with a C12-alkyl chain. The negative charge of a cell membrane easily attracts the positive charge of the ammonium group in BDMDAC. The hydrophobic C12-alkyl chain inserts into the membrane, causing disruption of bacterial cells [38]. Thus, BDMDAC is regarded as highly bactericidal. Comparing the median lethal dose (LD_{50}) values of BDMDAC with the TBT obtained from their respective material safety datasheets (Alfa Aesar, Ward Hill, MA, USA, 2015), the LD_{50} value for BDMDAC is two to three times higher than that of TBT (LD_{50} of BDMDAC: 400 mg/kg in rats; LD_{50} of TBT: 132 mg/kg in rats), suggesting BDMDAC is less toxic than TBT.

The objective of this study was to evaluate the antimicrobial properties of BDMDAC in anaerobic conditions by examining the effect of BDMDAC as a biocide on *D. desulfuricans*. The influence of BDMDAC and *D. desulfuricans* on the corrosion behavior of 304 stainless steel (304SS) was studied using polarization curves.

2. Materials and Methods

2.1. Sample Preparation

The 304SS coupons ($2 \times 1 \times 0.1$ cm^3) with a testing area of 2 cm^2 were prepared for the subsequent corrosion studies and biofilm observation. All the coupons were mechanically ground using emery papers of 200–1200 grit, rinsed with deionized water, and then washed with alcohol in an ultrasonic bath.

2.2. Bacteria and Culture Medium

Desulfovibrio desulfuricans subsp. *desulfuricans*, the SRB used in this study, was obtained from the Bioresource Collection and Research Center, Hsinchu, Taiwan. The medium used for *D. desulfuricans*, Postgate's medium (DSMZ, *Desulfovibrio* medium, Medium 63), was prepared as follows [39–41]: A 980 mL solution A (0.5 g of K_2HPO_4, 1 g of NH_4Cl, 2 g of $MgSO_4 \cdot 7H_2O$, 1 g of Na_2SO_4, 0.1 g of $CaCl_2 \cdot 2H_2O$, 1 g of yeast extract, and 2 g of sodium lactate) was boiled, and the dissolved oxygen in the solution was removed using a mechanical deaerator with nitrogen. A 10 mL solution B (0.1 g of ascorbic acid and 0.1 g of sodium thioglycolate) and a 10 mL solution C (0.5 g of $FeSO_4 \cdot 7H_2O$) were added to solution A. The medium was adjusted to pH 7.8 with NaOH and autoclaved at 120 °C for 15 min. *D. desulfuricans* was inoculated in the medium and cultured at 37 °C.

2.3. Bactericidal Assay of BDMDAC

Optical density at 600 nm (OD_{600}) was measured using UV/VIS spectrophotometer (Optizen Pop, Mecasys, Daejeon, Korea). The *D. desulfuricans* culture was left overnight and diluted to an OD_{600}

of 0.1 with a fresh medium. Various concentrations of BDMDAC were added to the culture and the absorbance at 600 nm was recorded as bacterial growth curves.

2.4. Electrochemical Measurements

Potentiodynamic polarization curves were deduced using a Gamry Reference 600 potentiostat (Warminster, PA, USA) to evaluate corrosion performance [5,6,8,42–44]. A standard three-electrode system comprising a graphite counter electrode and saturated calomel electrode (SCE) as a reference electrode was used in all electrochemical tests. Three polarization curve measurements were performed under different working solutions as follows: (1) 304SS coupons in a culture medium without *D. desulfuricans* inoculation and BDMDAC (labeled as a blank solution); (2) 304SS coupons in a *D. desulfuricans*-inoculated medium without BDMDAC (labeled as an SRB solution); and (3) 304SS coupons in a *D. desulfuricans*-inoculated medium with BDMDAC (labeled as a BDMDAC solution). Potentiodynamic polarization curve measurements were obtained by sweeping the potential from −0.5 to 1.5 V versus open circuit potential at a scan rate of 1 mV/s [42]. Corrosion potential (E_{corr}) and current density (I_{corr}) were determined through Tafel extrapolation. The test area on the 304SS coupons for all electrochemical tests was 2 cm^2.

2.5. Surface Characterization

Biofilm formation and corrosion products formed on the 304SS coupons after different immersion times in various working solutions were analyzed through scanning electron microscopy (SEM) (JSM-6510, JEOL, Tokyo, Japan) and energy-dispersive X-ray spectrum (EDS) (Inca x-act, Oxford Analytical Instruments, Abington, UK). After the immersion tests, the 304SS coupons were extracted, rinsed with distilled water, and then fixed with 2.5 wt % glutaraldehyde for 15 min, followed by dehydration in a graded series of ethanol solutions (30%, 50%, 70%, 90%, and 100% for 15 min each) and air drying [7,45–47]. The dried coupons were coated with plate platinum on the surface and then studied using SEM. EDS was performed to analyze the chemical compositions of the biofilms and corrosion products. The surface morphology of 304SS coupons after polarization curve measurements was examined under an optical microscope (OM) (SG-3006HM, SAGE Vision, New Taipei City, Taiwan).

2.6. CLSM

Confocal laser scanning microscopy (CLSM) (LSM780, Carl Zeiss, Jena, Germany) was used in this study to detect the live and dead cells in the biofilms [46–49]. The 304SS coupons were immersed in a culture medium without inoculation, or with *D. desulfuricans* inoculum in the absence or presence of 25 ppm BDMDAC at 37 °C at different time intervals up to 28 days. After cleaning the surface once with phosphate-buffered saline, the biofilms were stained with LIVE/DEAD™ *Bac*Light™ bacterial viability kit (L7012, Thermo Fisher Scientific, Waltham, MA, USA). After staining, the coupons were observed under a fluorescence microscope. Live cells and dead cells appeared green and red, respectively, in the biofilm. The three-dimensional (3D) scanning images obtained by CLSM were used to measure biofilm thicknesses [46,47].

3. Results and Discussion

3.1. Bactericidal Assay of BDMDAC

To test the antimicrobial activity of BDMDAC, the growth of *D. desulfuricans* was monitored in different concentrations of BDMDAC. The cell growth in 10 ppm BDMDAC did not differ from control (0 ppm) (Figure 1). When the BDMDAC concentration was elevated to 15 and 20 ppm, cell growth was slowed, with doubling time increasing from 12 to 18 h. Furthermore, when 25 and 50 ppm BDMDAC was added to the SRB solution, the cells did not grow. Therefore, the minimum inhibition concentration of BDMDAC for *D. desulfuricans* is 25 ppm.

To validate the long-term antimicrobial effect, *D. desulfuricans* was inoculated in a medium containing 25 ppm BDMDAC, which was then incubated for 4 weeks at 37 °C. As shown in Figure 2, no growth was observed during this period. Thus, BDMDAC has a favorable inhibitory effect on the growth of *D. desulfuricans*.

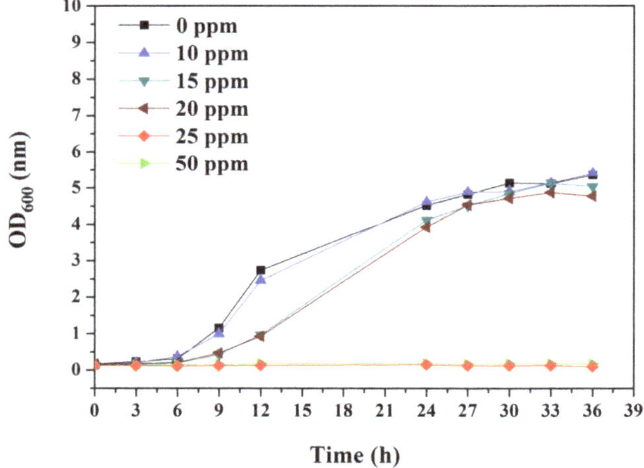

Figure 1. Effects of BDMDAC concentration on the growth curve of *D. desulfuricans* as a function of time in hours.

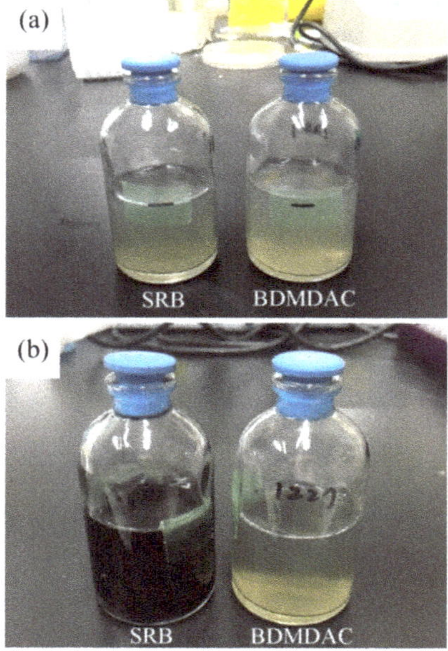

Figure 2. Visual inspection of the *D. desulfuricans*-inoculated medium with and without BDMDAC after (**a**) 7 days and (**b**) 28 days of incubation.

3.2. SEM

Figure 3 displays the morphological characteristics of the biofilms grown on 304SS coupons after different immersion times in the *D. desulfuricans*-inoculated medium with and without BDMDAC addition. The surface morphologies of the 304SS coupons immersed in the culture medium without inoculation are shown for comparison. The SEM images reveal that the surfaces of the 304SS coupons were strongly influenced in the absence of *D. desulfuricans* and BDMDAC. First, Figure 3a,c show no visible corrosion products or biofilms formed on the 304SS coupons in the blank and BDMDAC solutions after 14 days of immersion, whereas Figure 3b displays corrosion products and biofilms on the 304SS coupon in the SRB solution. Some pitting is observable on the surface of the 304SS coupons in the blank and BDMDAC solutions, but this is only in a few locations. This observation is attributable to the presence of Cl^- ions in the culture medium [50–52]. The EDS elemental analysis results (Table 1) reveal that, in addition to characteristic corrosion products (Fe and O), sulfur elements were detected after 14 days of immersion in the SRB solution. Sulfur elements mainly result from the metabolic activity of SRB (formation of FeS) [53]. Thus, the EDS results suggest that the addition of BDMDAC in the SRB solution could decrease the activity of SRB and reduce the formation of biofilm. After extending the immersion time to 28 days, large areas of corrosion products and biofilms were observed on the coupons immersed in the SRB solution (Figure 3e), whereas the 304SS surface was largely intact in the blank and BDMDAC solutions (Figure 3d,f). Moreover, the EDS elemental analysis results revealed increased sulfur elements on the coupons immersed in the SRB solution. By contrast, sulfur elements were not observed on the 304SS coupons immersed in the blank and BDMDAC solutions. These results confirm the inhibition efficiency of 25 ppm BDMDAC.

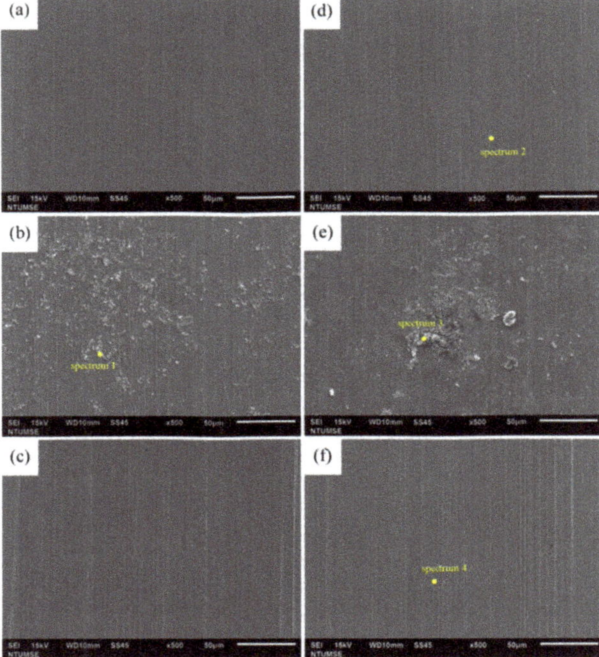

Figure 3. SEM images of the 304SS coupons after different immersion times in various solutions: (**a**) blank for 14 days; (**b**) SRB for 14 days; (**c**) BDMDAC for 14 days; (**d**) blank for 28 days; (**e**) SRB for 28 days; and (**f**) BDMDAC for 28 days.

Table 1. EDS analysis results after 14 and 28 days of immersion.

Immersion Times	Elements (wt %)	O	S	Cr	Fe	Ni	Si
14 days	Spectrum 1	8.39	4.67	14.66	64.57	5.42	2.29
28 days	Spectrum 2	-	-	19.59	71.51	8.23	0.67
	Spectrum 3	32.26	14.34	-	53.40	-	-
	Spectrum 4	-	-	19.36	72.53	8.11	-

3.3. CLSM

To evaluate the inhibitory effect of BDMDAC to biofilm formation, CLSM was used for observation. Figure 4 shows the CLSM images of the 304SS coupons after 14 and 28 days of immersion in various solutions. In the culture medium without *D. desulfuricans* inoculation and BDMDAC, no growth was observed (Figure 4a,d). After incubation with *D. desulfuricans*, numerous green dots were observed, as shown in Figure 4b,e. The cells in the biofilm were stained green after staining, suggesting that most of the cells were alive. Furthermore, Figure 4b,e show that the thickness of biofilm increased to 18 and 40 μm for 14 and 28 days of immersion, respectively. In the prescence of BDMDAC, the thickness of biofilm was less than 20 μm, even after 4 weeks. Furthermore, most cells in the biofilm were now dead, as indicated by the red stains (Figure 4c,f). Notably, the biofilm grown on the 304SS coupons in the BDMDAC solution was not visible in the previous SEM images (Figure 3c,f). This can be attributed to the fact that killed bacteria or disrupted biofilm due to BDMDAC addition are easily removed during the cleaning process of SEM sample preparation.

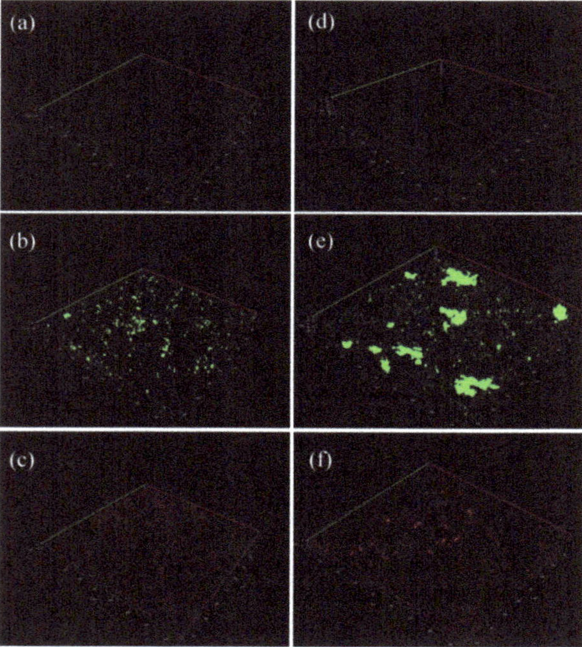

Figure 4. CLSM 3D images of the 304SS coupons after different immersion times in various solutions: (**a**) blank for 14 days; (**b**) SRB for 14 days; (**c**) BDMDAC for 14 days; (**d**) blank for 28 days; (**e**) SRB for 28 days; and (**f**) BDMDAC for 28 days.

3.4. Potentiodynamic Polarization

Figure 5 illustrates the potentiodynamic polarization curves of coupons under different immersion environments after 14- and 28-day immersion periods. The values of E_{corr} and I_{corr} according to the potentiodynamic polarization curve are listed in Table 2. Table 2 shows that the corrosion current densities I_{corr} in the presence of bacteria were higher than that in the control medium for each corresponding exposure time, which suggests the occurrence of MIC. The I_{corr} value was highest in the presence of *D. desulfuricans*. Furthermore, the current density of the anodic branch in the 304SS coupons in the SRB solution was higher than that of those immersed in blank solutions for each corresponding immersion time. These observations can be attributed to the aggressive role of bacteria in enhancing the corrosion of stainless steel by destroying the oxide layer on stainless steel [54]. The I_{corr} value and current density of the anodic branch for the 304SS coupons immersed in the BDMDAC solution significantly decreased compared with coupons after 14 and 28 days of immersion in the SRB solution. These results indicate the ability of BDMDAC to protect stainless steel substrate against *D. desulfuricans*. Notably, the potentiodynamic polarization curves of coupons exposed to the BDMDAC solution still exhibited a little positive shift compared with those of the blank solution curves. This observation can be attributed to the fact that the bacterial adhesion and biofilm formation at very beginning of immersion test could affect the physical and chemical conditions of 304SS coupons, and result in adverse effects on their corrosion performance. Figure 6 shows the OM images of 14-day immersed coupons after potentiodynamic polarization curve measurements. As can be seen, the obvious localized corrosion was found on the 304SS coupon in the SRB solution (as indicated by red arrows in the Figure 6b). This can be attributed to the MIC effect, leading to the deterioration of passive film by sulfide produced by SRB [42] after being immersed in the SRB solution. These results provide considerable support for our potentiodynamic polarization curve results that the lower current density of the anodic branch and poor passivity were found on the 304SS coupon immersed in the SRB solution.

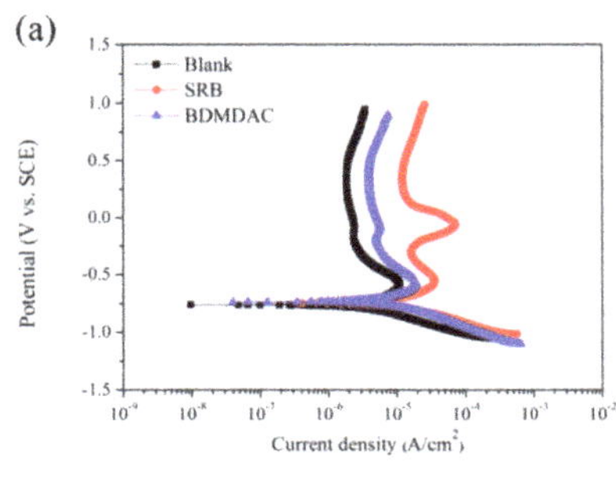

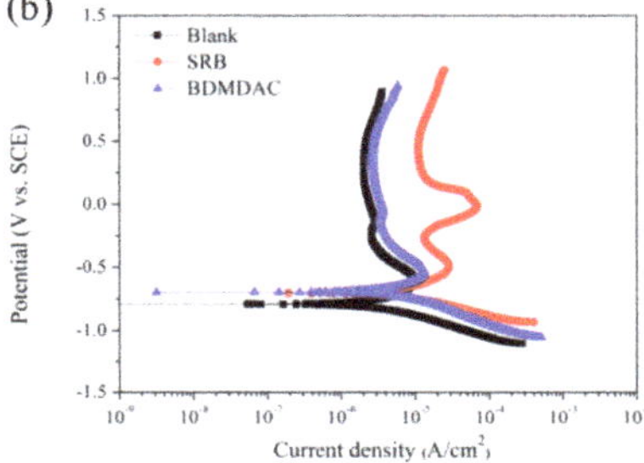

Figure 5. Potentiodynamic polarization curves for the 304SS coupons after 14 days (**a**) and 28 days (**b**) of immersion in various solutions.

Table 2. Corrosion potential and corrosion current density measured according to potentiodynamic polarization curves after 14 and 28 days of immersion.

Samples	Blank 14 days	SRB 14 days	BDMDAC 14 days	Blank 28 days	SRB 28 days	BDMDAC 28 days
E_{corr} (mV vs. SCE)	−755 ± 36	−733 ± 38	−758 ± 58	−751 ± 24	−765 ± 23	−764 ± 32
I_{corr} (μA/cm^2)	1.06 ± 0.06	2.39 ± 0.25	1.57 ± 0.02	0.95 ± 0.05	3.06 ± 0.25	1.83 ± 0.08

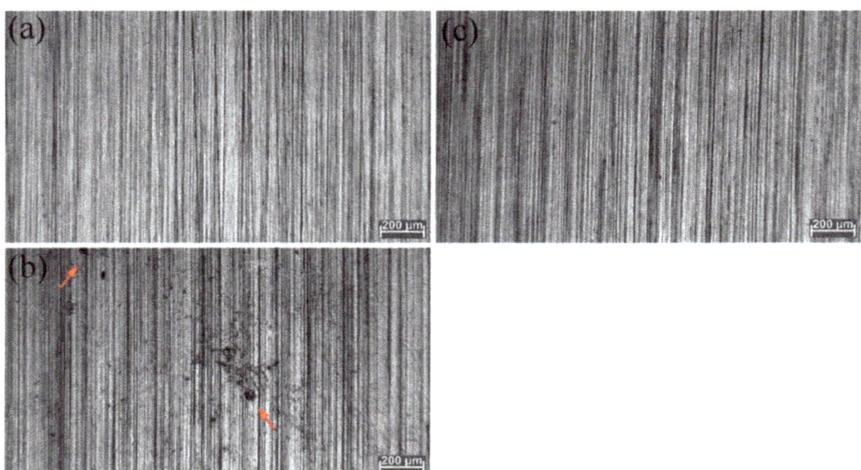

Figure 6. OM images of 304SS coupons after potentiodynamic polarization curve measurements. These surfaces were initially exposed to (**a**) blank for 14 days; (**b**) SRB for 14 days; and (**c**) BDMDAC for 14 days.

3.5. Limitations of This Study

Electrochemical techniques include potentiodynamic polarization are useful for examining the corrosion performance of metallic substrates. However, some limitations of electrochemical techniques for investigating microbial corrosion have been proposed. Javaherdashti has mentioned that large polarizations could be deleterious to micro-organisms in the biofilm [55]. However, it should be noticed that the two examples provided in this reference focused on cathodic protection and its effect(s) on MIC. A possible mechanism is that the application of cathodic potential will increase local pH (due to ORR reaction: $O_2 + 2H_2O + 4e^- \rightarrow 4OH^-$) and result in inhibiting the bacterial reproduction of microbes in such a high alkaline environment [56,57]. In addition, the time staying in the cathodic polarization potential is another important issue to study the effects of cathodic polarization on MIC. Little et al. [58] and Romero et al. [59] had polarized their samples at –400 mV (vs. SCE) for 72 h and –1000 mV (vs. SCE) for 3–72 h, respectively, to investigate the cathodic polarization effects on MIC. In contrast, the length of sample staying in the cathodic polarization potential is only 500 s (from –500 mV to 0 mV vs. open circuit potential (OCP), scan rate: 1 mV/s) in the present study. As a result, it is believed that the electrochemical experimental setup used in this study is likely to have less influence on the microbial environment.

On the other hand, potentiodynamic polarization curves were not periodically made during the immersion test in the present study. These measurements were applied and started applying potential on the 304SS coupons only after 14 or 28 days of immersion. Although the bacterial activity may be influenced during the polarization test, MIC-induced changes in the electrochemical conditions of 304SS coupons during 14 or 28 days of immersion had resulted in adverse effects on their corrosion performance. The potentiodynamic polarization curve measurement of coupons under different corrosion conditions (with and without bacteria) after different exposure times had also been used in numerous relevant studies [5,42–44]. In addition, the coupons for 14- and 28-day immersion tests were prepared in the different bottles, indicating the polarization potential applied on the 14-day immersed coupons would not have influence on the bacteria and culture medium in 28-day immersed ones. For all the above reasons, the electrochemical analysis used in this study gives useful information for research purposes.

4. Conclusions

In this study, potentiodynamic polarization curves were employed together with SEM and CLSM techniques to investigate corrosion and microbial corrosion inhibition of BDMDAC for 304SS substrates. The results reveal that BDMDAC achieved satisfactory results against microbial corrosion by *D. desulfuricans* in anaerobic conditions. Addition of 25 ppm BDMDAC could effectively inhibit the growth of *D. desulfuricans*. Surface characterization results revealed no biofilm on the 304SS coupons surfaces after BDMDAC was added to the *D. desulfuricans*-inoculated medium. The electrochemical results revealed deceleration of corrosion caused by the addition of BDMDAC in the *D. desulfuricans*-inoculated medium, which was demonstrated by the decrease of current densities in the anodic region and I_{corr}.

Author Contributions: Conceptualization, Y.-L.L.; Methodology, K.-Y.L. and Y.-L.L.; Validation, C.-W.H. and T.-E.C.; Formal Analysis, C.-W.H. and Y.-L.L.; Investigation, C.-W.H. and T.-E.C.; Writing—Original Draft Preparation, C.-W.H. and Y.-L.L.; Writing—Review & Editing, K.-Y.L. and Y.-L.L.; Supervision, Y.-L.L.

Funding: This work was funded by the Ministry of Science and Technology, Taiwan, under grant No. MOST 106-2218-E-002-033.

Conflicts of Interest: The authors declare no conflicts of interest.

References

1. Sorkhabi, H.A.; Haghighi, M.M.; Zarrini, G.; Javaherdashti, R. Corrosion behavior of carbon steel in the presence of two novel iron-oxidizing bacteria isolated from sewage treatment plants. *Biodegradation* **2012**, *23*, 69–79. [CrossRef] [PubMed]
2. Lv, L.; Yuan, S.; Zheng, Y.; Liang, B.; Pehkonen, S.O. Surface modification of mild steel with thermally cured antibacterial poly(vinylbenzyl chloride)–polyaniline bilayers for effective protection against sulfate reducing bacteria induced corrosion. *Ind. Eng. Chem. Res.* **2014**, *53*, 12363–12378. [CrossRef]
3. Venzlaff, H.; Enning, D.D.; Srinivasan, J.; Mayrhofer, K.J.J.; Hassel, A.W.; Widdel, F.; Stratmanna, M. Accelerated cathodic reaction in microbial corrosion of iron due to direct electron uptake by sulfate-reducing bacteria. *Corros. Sci.* **2013**, *66*, 88–96. [CrossRef]
4. Enning, D.; Venzlaff, H.; Garrelfs, J.; Dinh, H.; Meyer, V.; Mayrhofer, K.; Hasse, A.W.; Stratmann, M.; Widdel, F. Marine sulfate-reducing bacteria cause serious corrosion of iron under electroconductive biogenic mineral crust. *Environ. Microbiol.* **2012**, *14*, 1772–1787. [CrossRef] [PubMed]
5. Xu, J.; Wang, K.; Sun, C.; Wang, F.; Li, X.; Yang, J.; Yu, C. The effects of sulfate reducing bacteria on corrosion of carbon steel Q235 under simulated disbonded coating by using electrochemical impedance spectroscopy. *Corros. Sci.* **2011**, *53*, 1554–1562. [CrossRef]
6. Bao, Q.; Zhang, D.; Lv, D.; Wang, P. Effects of two main metabolites of sulphate-reducing bacteria on the corrosion of Q235 steels in 3.5 wt.% NaCl media. *Corros. Sci.* **2012**, *65*, 405–413. [CrossRef]
7. Liu, H.; Fu, C.; Gu, T.; Zhang, G.; Lv, Y.; Wang, H.; Liu, H. Corrosion behavior of carbon steel in the presence of sulfate reducing bacteria and iron oxidizing bacteria cultured in oilfield produced water. *Corros. Sci.* **2015**, *100*, 484–495. [CrossRef]
8. Sheng, X.; Ting, Y.P.; Pehkonen, S.O. The influence of sulphate-reducing bacteria biofilm on the corrosion of stainless steel AISI 316. *Corros. Sci.* **2007**, *49*, 2159–2176. [CrossRef]
9. Antony, P.J.; Raman, P.K.S.; Mohanram, R.; Kumar, P.; Raman, R. Influence of thermal aging on sulfate-reducing bacteria (SRB)-influenced corrosion behaviour of 2205 duplex stainless steel. *Corros. Sci.* **2008**, *50*, 1858–1864. [CrossRef]

10. Yuan, S.; Liang, B.; Zhao, Y.; Pehkonen, S.O. Surface chemistry and corrosion behaviour of 304 stainless steel in simulated seawater containing inorganic sulphide and sulphate-reducing bacteria. *Corros. Sci.* **2013**, *74*, 353–366. [CrossRef]
11. Starosvetsky, D.; Khaselev, O.; Starosvetsky, J.; Armon, R.; Yahalom, J. Effect of iron exposure in SRB media on pitting initiation. *Corros. Sci.* **2000**, *42*, 345–359. [CrossRef]
12. Liu, H.; Xu, D.; Yang, K.; Liu, H.; Cheng, Y.F. Corrosion of antibacterial Cu-bearing 316L stainless steels in the presence of sulfate reducing bacteria. *Corros. Sci.* **2018**, *132*, 46–55. [CrossRef]
13. Dong, Z.H.; Wei, S.; Hong, M.R.; Guo, A.Z. Heterogeneous corrosion of mild steel under SRB-biofilm characterised by electrochemical mapping technique. *Corros. Sci.* **2011**, *53*, 2978–2987. [CrossRef]
14. Liu, H.; Cheng, Y.F. Mechanism of microbiologically influenced corrosion of X52 pipeline steel in a wet soil containing sulfate-reduced bacteria. *Electrochim. Acta* **2017**, *253*, 368–378. [CrossRef]
15. Javed, M.A.; Stoddart, P.R.; Wade, S.A. Corrosion of carbon steel by sulphate reducing bacteria: Initial attachment and the role of ferrous ions. *Corros. Sci.* **2015**, *93*, 48–57. [CrossRef]
16. Chen, X.; Wang, G.; Gao, F.; Wang, Y.; He, C. Effects of sulphate-reducing bacteria on crevice corrosion in X70 pipeline steel under disbonded coatings. *Corros. Sci.* **2015**, *101*, 1–11. [CrossRef]
17. Xu, D.; Li, Y.; Gu, T. Mechanistic modeling of biocorrosion caused by biofilms of sulfate reducing bacteria and acid producing bacteria. *Bioelectrochemistry* **2016**, *110*, 52–58. [CrossRef]
18. Lee, W.; Lewandowski, Z.; Nielsen, P.H.; Hamilton, W.A. Role of sulfate-reducing bacteria in corrosion of mild steel: A review. *Biofouling* **1995**, *8*, 165–194. [CrossRef]
19. Sungur, E.I.; Cansever, N.; Cotuk, A. Microbial corrosion of galvanized steel by a freshwater strain of sulphate reducing bacteria (*Desulfovibrio* sp.). *Corros. Sci.* **2007**, *49*, 1097–1109. [CrossRef]
20. Sungur, E.I.; Cotuk, A. Microbial corrosion of galvanized steel in a simulated recirculating cooling tower system. *Corros. Sci.* **2010**, *52*, 161–171. [CrossRef]
21. Sherar, B.W.A.; Power, I.M.; Keech, P.G.; Mitlin, S.; Southam, G.; Shoesmith, D.W. Characterizing the effect of carbon steel exposure in sulfide containing solutions to microbially induced corrosion. *Corros. Sci.* **2011**, *53*, 955–960. [CrossRef]
22. Beech, I.B.; Cheung, C.W.S. Interactions of exopolymers produced by sulphate-reducing bacteria with metal ions, Int. *Biodeterior. Biodegrad.* **1995**, *35*, 59–72. [CrossRef]
23. Xu, D.; Li, Y.; Song, F.; Gu, T. Laboratory investigation of microbiologically influenced corrosion of C1018 carbon steel by nitrate reducing bacterium *Bacillus licheniformis*. *Corros. Sci.* **2013**, *77*, 385–390. [CrossRef]
24. Wan, Y.; Zhang, D.; Liu, H.; Li, Y.; Hou, B. Influence of sulphate-reducing bacteria on environmental parameters and marine corrosion behavior of Q235 steel in aerobic conditions. *Electrochim. Acta* **2010**, *55*, 1528–1534. [CrossRef]
25. Hamilton, W.A. Microbially influenced corrosion as a model system for the study of metal microbe interactions: A unifying electron transfer hypothesis. *Biofouling* **2003**, *19*, 65–76. [CrossRef] [PubMed]
26. Gieg, L.M.; Jack, T.R.; Foght, J.M. Biological souring and mitigation in oil reservoirs. *Appl. Microbiol. Biotechnol.* **2011**, *92*, 263–282. [CrossRef]
27. Mitra, S.; Gera, R.; Singh, V.; Khandelwal, S. Comparative toxicity of low dose tributyltin chloride on serum, liver, lung and kidney following subchronic exposure. *Food Chem. Toxicol.* **2014**, *64*, 335–343. [CrossRef]
28. Zhang, C.N.; Zhang, J.L.; Ren, H.T.; Zhou, B.H.; Wu, Q.J.; Sun, P. Effect of tributyltin on antioxidant ability and immune responses of zebrafish (*Danio rerio*). *Ecotoxicol. Environ. Saf.* **2017**, *138*, 1–8. [CrossRef]
29. Cooke, G.M.; Tryphonas, H.; Pulido, O.; Caldwell, D.; Bondy, G.S.; Forsyth, D. Oral (gavage), in utero and postnatal exposure of Sprague–Dawley rats to low doses of tributyltin chloride. Part 1: Toxicology, histopathology and clinical chemistry. *Food Chem. Toxicol.* **2004**, *42*, 211–220. [CrossRef]
30. Cooke, G.M. Toxicology of tributyltin in mammalian animal models. *Immunol. Endocr. Metab. Agents Med. Chem.* **2006**, *6*, 63–71. [CrossRef]
31. Ferreira, M.; Blanco, L.; Garrido, A.; Vieites, J.M.; Cabado, A.G. In vitro approaches to evaluate toxicity induced by organotin compounds Tributyltin (TBT), Dibutyltin (DBT), and Monobutyltin (MBT) in Neuroblastoma cells. *J. Agric Food. Chem.* **2013**, *61*, 4195–4203. [CrossRef] [PubMed]
32. Antizar-Ladislao, B. Environmental levels, toxicity and human exposure to tributyltin (TBT)-contaminated marine environment. *Environ. Int. Rev.* **2008**, *34*, 292–308. [CrossRef] [PubMed]

33. Araújo, J.F.P.; Podratz, P.L.; Sena, G.C.; Merlo, E.; Lima, L.C.F.; Ayub, J.G.M.; Pereira, A.F.Z.; Silva, A.P.S.; Alves, L.M.; Silva, I.V.; et al. The obesogen tributyltin induces abnormal ovarian adipogenesis in adult female rats. *Toxicol. Lett.* **2018**, *295*, 99–114. [CrossRef] [PubMed]
34. Gerike, P.; Fischer, W.K.; Jasiak, W. Surfactant quaternary ammonium salts in aerobic sewage digestion. *Water Res.* **1978**, *12*, 1117–1122. [CrossRef]
35. Ferreira, C.; Rosmaninho, R.; Simoes, M.; Pereira, M.C.; Bastos, M.M.; Nunes, O.C.; Coelho, M.; Melo, L.F. Biofouling control using microparticles carrying a biocide. *Biofouling* **2009**, *26*, 205–212. [CrossRef] [PubMed]
36. Liu, F.; Chang, X.; Yang, F.; Wang, Y.Q.; Wang, F.Y.; Dong, W.W.; Zhao, C.C. Effect of oxidizing and non-oxidizing biocides on biofilm at different substrate levels in the model recirculating cooling water system. *World J. Microbiol. Biotechnol.* **2011**, *27*, 2989–2997. [CrossRef]
37. Tomlinson, E.; Brown, M.R.W.; Davis, S.S. Effect of colloidal association on the measured activity of alkylbenzyldimethylammonium chlorides against *Pseudomonas aeruginosa*. *J. Med. Chem.* **1977**, *20*, 1277–1282. [CrossRef]
38. Ferreira, C.; Pereira, A.M.; Pereira, M.C.; Melo, L.F.; Simões, M. Physiological changes induced by the quaternary ammonium compound benzyldimethyldodecylammonium chloride on *Pseudomonas fluorescens*. *J. Antimicrob. Chemother.* **2011**, *66*, 1036–1043. [CrossRef]
39. Angelov, A.; Bratkova, S.; Loukanov, A. Microbial fuel cell based on electroactive sulfate-reducing biofilm. *Energy Convers. Manag.* **2013**, *67*, 283–286. [CrossRef]
40. Ghazy, E.A.; Mahmoud, M.G.; Asker, M.S.; Mahmoud, M.N.; Elsoud, M.M.A.; Sami, M.E.A. Cultivation and detection of sulfate reducing bacteria (SRB) in sea water. *J. Am. Sci.* **2011**, *2*, 604–608.
41. Omran, B.A.; Fatthalah, N.A.; El-Gendy, N.S.; El-Shatoury, E.H.; Abouzeid, M.A. Green biocides against sulphate reducing bacteria and macrofouling organisms. *J. Pure Appl. Microbiol.* **2013**, *7*, 2219–2232.
42. Unsal, T.; Ilhan-Sungur, E.; Arkan, S.; Cansever, N. Effects of Ag and Cu ions on the microbial corrosion of 316 L stainless steel in the presence of *Desulfovibrio* sp. *Bioelectrochemistry* **2016**, *110*, 91–99. [CrossRef] [PubMed]
43. Xu, C.; Zhang, Y.; Cheng, G.; Zhu, W. Pitting corrosion behavior of 316L stainless steel in the media of sulphate-reducing and iron-oxidizing bacteria. *Mater. Charact.* **2008**, *59*, 245–255. [CrossRef]
44. Nan, L.; Xu, D.; Gu, T.; Song, X.; Yang, K. Microbiological influenced corrosion resistance characteristics of a 304 L-Cu stainless steel against *Escherichia coli*. *Mater. Sci. Eng. C* **2015**, *48*, 228–234. [CrossRef] [PubMed]
45. Wen, J.; Zhao, K.; Gu, T.; Raad, I.I. A green biocide enhancer for the treatment of sulfate-reducing bacteria (SRB) biofilms on carbon steel surfaces using glutaraldehyde. *Int. Biodeterior. Biodegrad.* **2009**, *63*, 1102–1106. [CrossRef]
46. Li, H.; Zhou, E.; Ren, Y.; Zhang, D.; Xu, D.; Yang, C.; Feng, H.; Jiang, Z.; Li, X.; Gu, T.; et al. Investigation of microbiologically influenced corrosion of high nitrogen nickel-free stainless steel by *Pseudomonas aeruginosa*. *Corros. Sci.* **2016**, *111*, 811–821. [CrossRef]
47. Li, H.; Yang, C.; Zhou, E.; Yang, C.; Feng, H.; Jiang, Z.; Xu, D.; Gu, T.; Yang, K. Microbiologically influenced corrosion behavior of S32654 super austenitic stainless steel in the presence of marine *Pseudomonas aeruginosa* biofilm. *J. Mater. Sci. Technol.* **2017**, *33*, 1596–1603. [CrossRef]
48. Wikiel, A.J.; Datsenko, I.; Vera, M.; Sand, W. Impact of *Desulfovibrio alaskensis* biofilms on corrosion behaviour of carbon steel in marine environment. *Bioelectrochemistry* **2014**, *97*, 52–60. [CrossRef]
49. Lawrence, J.R.; Neu, T.R. Confocal laser scanning microscopy for analysis of microbial biofilms. In *Methods in Enzymology*; Doyle, R.J., Ed.; Academic Press: San Diego, CA, USA, 1999; Volume 310, pp. 131–144.
50. Freire, L.; Carmezim, M.J.; Ferreira, M.G.S.; Montemor, M.F. The electrochemical behaviour of stainless steel AISI 304 in alkaline solutions with different pH in the presence of chlorides. *Electrochim. Acta* **2011**, *56*, 5280–5289. [CrossRef]
51. Wu, K.; Jung, W.-S.; Byeon, J.-W. In-situ monitoring of pitting corrosion on vertically positioned 304 stainless steel by analyzing acoustic-emission energy parameter. *Corros. Sci.* **2016**, *105*, 8–16. [CrossRef]
52. Wang, Y.; Cheng, G.; Wu, W.; Qiao, Q.; Li, Y.; Li, X. Effect of pH and chloride on the micro-mechanism of pitting corrosion for high strength pipeline steel in aerated NaCl solutions. *Appl. Surf. Sci.* **2015**, *349*, 746–756. [CrossRef]
53. Liu, H.; Gu, T.; Zhang, G.; Liu, H.; Cheng, Y.F. Corrosion of X80 pipeline steel under sulfate-reducing bacterium biofilms in simulated CO_2-saturated oilfield produced water with carbon source starvation. *Corros. Sci.* **2018**, *136*, 47–59. [CrossRef]

54. Yuan, S.J.; Xu, F.J.; Pehkonen, S.O.; Ting, Y.P.; Neoh, K.G.; Kang, E.T. Grafting of antibacterial polymers on stainless steel via surface-initiated atom transfer radical polymerization for inhibiting biocorrosion by *Desulfovibrio desulfuricans*. *Biotechnol. Bioeng.* **2009**, *103*, 268–281. [CrossRef] [PubMed]
55. Javaherdashti, R. *Microbiologically Influenced Corrosion: An Engineering Insight*; Springer: Berlin, Germany, 2008.
56. Pedersen, K. Subterranean micro-organisms and radioactive waste disposal in Sweden. *Eng. Geol.* **1999**, *52*, 163–176. [CrossRef]
57. Stein, A. *A Pratical Manual on Microbiologically Influenced Corrosion*; Kobrin, G., Ed.; NACE International: Houston, TX, USA, 1993; Volume 1, pp. 101–112.
58. Little, B.J.; Wagner, P.A.; Hart, K.R.; Ray, R.L. Spatial relationships between bacteria and localized corrosion. *Corrosion* **1996**, *96*, 278–284.
59. De Romero, M.F.; Parra, J.; Ruiz, R.; Ocando, L.; Bracho, M.; de Rincon, O.T.; Romero, G.; Quintero, A. Cathodic polarization effect on sessile SRB growth and iron protection. *Corrosion* **2006**, *6*, 652–656.

© 2019 by the authors. Licensee MDPI, Basel, Switzerland. This article is an open access article distributed under the terms and conditions of the Creative Commons Attribution (CC BY) license (http://creativecommons.org/licenses/by/4.0/).

Article

Effect of Electron Donating Functional Groups on Corrosion Inhibition of J55 Steel in a Sweet Corrosive Environment: Experimental, Density Functional Theory, and Molecular Dynamic Simulation

Ambrish Singh [1,2,*], Kashif R. Ansari [3], Mumtaz A. Quraishi [3] and Hassane Lgaz [4]

1 School of Materials Science and Engineering, Southwest Petroleum University, Chengdu 610500, Sichuan, China

2 State Key Laboratory of Oil and Gas Reservoir Geology and Exploitation, Southwest Petroleum University, Chengdu 610500, Sichuan, China

3 Centre of Research Excellence in Corrosion, Research Institute, King Fahd University of Petroleum and Minerals, Dhahran 31261, Saudi Arabia; Ka3787@gmail.com (K.R.A.); maquraishi.apc@iitbhu.ac.in (M.A.Q.)

4 Department of Applied Bioscience, College of Life & Environment Science, Konkuk University, 120 Neungdong-ro, Gwangjin-gu, Seoul 05029, Korea; lgaz.hassane@gmail.com

* Correspondence: 201699010016@swpu.edu.cn or vishisingh4uall@gmail.com; Tel.: +8618384155035

Received: 3 December 2018; Accepted: 18 December 2018; Published: 21 December 2018

Abstract: Benzimidazole derivatives were synthesized, characterized, and tested as a corrosion inhibitor for J55 steel in a 3.5 wt % NaCl solution saturated with carbon dioxide. The experimental results revealed that inhibitors are effective for steel protection, with an inhibition efficiency of 94% in the presence of 400 mg/L of inhibitor. The adsorption of the benzimidazole derivatives on J55 steel was found to obey Langmuir's adsorption isotherm. The addition of inhibitors decreases the cathodic as well anodic current densities and significantly strengthens impedance parameters. X-ray photoelectron spectroscopy (XPS) was used for steel surface characterization. Density functional theory (DFT) and molecular dynamic simulation (MD) were applied for theoretical studies.

Keywords: corrosion; XPS; J55 steel; carbon dioxide; molecular dynamic simulation

1. Introduction

Carbon dioxide corrosion is the most commonly faced problem in the petroleum industry, and thus it has been a hot research topic for many years. Corrosion prevention consists of different methods, but one of these methods, the introduction of organic compounds as corrosion inhibitors, is both effective and cheap [1–3].

The inhibition action of organic compounds depends on the nature of the molecular structure, inhibitor planarity, electron donating functional groups, non-bonding electrons on heteroatoms, i.e., oxygen, nitrogen, and sulfur, and presence of π bonds in the aromatic ring [4]. In recent years, corrosion scientists have been interested in finding green and environment-friendly inhibitors [5]. Benzimidazole derivatives that show antitumor and antimicrobial activities have been categorized in the group of green compounds [6,7].

A survey of the literature reveals that in the past decades, various imidazole and benzimidazole derivatives have been used as anti-sweet corrosion inhibitors [8–15]. However, no literature exits about using benzimidazole as a corrosion inhibitor in a brine solution saturated with carbon dioxide. Benzimidazole has planar structure with two nitrogen atoms that provides a closer approach for interaction with the metal surface, aromatic properties, and an option for introducing different substituents.

The main purpose of the present paper was to elucidate the corrosion inhibition effect of the number of electron donating methoxy groups on the phenyl ring of the three synthesized benzimidazole derivatives, namely, 2-(3,4,5-Trimethoxyphenyl)-1*H*-benzo[*d*] imidazole (TMI), 2-(3,4-Dimethoxyphenyl)-1*H*-benzo[*d*] imidazole (DMI), and 2-(4-Methoxyphenyl)-1*H*-benzo[*d*] imidazole (MMI), for J55 steel saturated with CO_2 in a 3.5% NaCl solution. Corrosion inhibition properties of the benzimidazole derivatives were analyzed using the static weight-loss method and electrochemical methods, i.e., impedance spectroscopy (EIS) and potentiodynamic polarization. Meanwhile, the J55 steel surface was examined by scanning electron microscope (SEM) and X-ray photoelectron spectroscopy (XPS). The potential site for protonation was estimated using density functional theory (DFT). The interaction of the benzimidazole derivatives with the J55 steel surface was studied by molecular dynamic simulation (MD).

2. Experiment

2.1. *J55 Steel Sample*

J55 steel specimens were used in all experiments. All specimens used for weight-loss experiments were machined to rectangle coupons. Before the initiation of the experiment, the metal surface was mechanically grounded with emery papers graded 400–1200, washed with acetone and double distilled water, and lastly dried using a dryer. The J55 steel composition was (wt %): C (0.31), Mn (0.92), Si (0.19), P (0.01), Cr (0.2), S (0.008), and Fe in balance. The dimension of the steel coupons used for weight-loss and electrochemical experiments were 5.0 cm × 2.5 cm × 0.2 cm and 1 cm^2.

2.2. *Corrosive Medium*

The corrosive medium (3.5 wt % NaCl) was prepared using analytical grade NaCl and double distilled water. The concentration ranges of each tested inhibitor used in the course of the experiments were 50 to 400 mg/L. Before the experiments, N_2 gas was bubbled for 3 h in the corrosive solution in order to remove the oxygen. Then, the solution was deoxygenated by purging CO_2 gas for 4 h. The specimens were then immersed into the solution while the CO_2 gas-purging at a pressure of 6 MPa was maintained to ensure a full saturation throughout the test. The electrochemical setup was sealed during the experiment. The initial pH of the corrosive medium was 4.

2.3. *Synthesis of Inhibitor*

The synthesis of benzimidazole derivatives was carried out using the reported method [16]. In a round bottom flask, aromatic aldehyde (2 mmol), o-phenylenediamine (2 mmol), boric acid (0.1 g), and water (10 mL) were stirred at room temperature for 15–30 min. After the completion of the reaction, 5 mL water was added and the mixture was further stirred for 10 min. The obtained precipitate was filtered and purified by recrystallization from ethanol. The characterization of benzimidazole derivatives was done by ^{1}H NMR and ^{13}C NMR (AVH D 500 ADVANCE III HD One Bay NMR Spectrometer, Bruker Bio Spin International AG, Billerica, MA, USA). ^{1}H and ^{13}C spectra were recorded at 400 MHz and 100 MHz, respectively, using $CDCl_3$ as a solvent. The synthesis scheme and molecular structure are shown in Figure 1. The ^{1}H NMR and ^{13}C NMR spectra are given in the supplementary file (Figures S1 and S2).

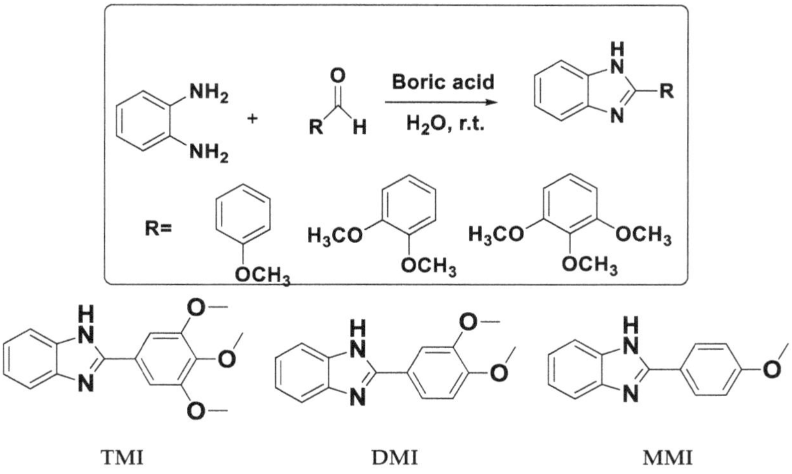

Figure 1. Synthetic scheme and molecular structure of the benzimidazole derivatives. 2-(3,4,5-Trimethoxyphenyl)-1*H*-benzo[*d*] imidazole (TMI); 2-(3,4-Dimethoxyphenyl)-1*H*-benzo[*d*] imidazole (DMI); and 2-(4-Methoxyphenyl)-1*H*-benzo[*d*] imidazole (MMI).

2.4. NMR Data for Synthesized Inhibitors

2.4.1. 2-(4-Methoxyphenyl)-1*H*-Benzo[*d*]Imidazole (MMI)

^{1}H NMR (300 MHz, DMSO-d_6) δ (ppm): 12.72 (brs, 1H, *NH*), 8.10 (d, *J* = 8.39 Hz, 2H), 7.54 (brs, 2H), 7.19–7.07 (m, 4H), 3.82 (s, 3H).

^{13}C NMR, δ (ppm): 55.38, 114.43, 122.34, 123.54, 128.58, 151.58, 161.78.

2.4.2. 2-(3,4-Dimethoxyphenyl)-1*H*-Benzo[*d*]Imidazole (DMI)

^{1}H NMR (300 MHz, DMSO-d6) δ (ppm): 12.76 (s, 1H, *NH*), 7.80–7.71 (m, 2H), 7.66–7.44 (m, 2H), 7.20–7.06 (m, 3H), 3.87 (s, 3H), 3.82 (s, 3H).

^{13}C NMR δ (ppm): 151.52, 150.31, 148.92, 143.76, 134.95, 122.74, 121.85, 119.30, 118.22, 111.81, 111.08, 109.71, 55.62, 55.59.

2.4.3. 2-(3,4,5-Trimethoxyphenyl)-1*H*-Benzo[*d*]Imidazole (TMI)

^{1}H NMR (300 MHz, DMSO-d6) δ (ppm): 12.87 (s, 1H, *NH*), 7.71–7.48 (m, 5H), 7.19 (s, 2H), 3.88 (s, 6H), 3.69 (s, 3H).

^{13}C NMR δ (ppm): 153.32, 151.33, 143.79, 138.87, 135.00, 125.59, 122.66, 121.82, 118.78, 111.36, 103.81, 60.31, 56.12.

2.5. Weight Loss

ASTM G31-2004 standard was used to determine the duration of the test for the weight-loss experiments. The test duration was 24 h. The corrosion rates were calculated using the weight-loss method data. After 24 h of exposure, the steel coupons were rinsed with distilled water and Clarke's solution for 5 min. Finally, the coupons were rinsed with distilled water and dried. All weight-loss experiments were performed in triplicate at 333 K for 24 h. Through the weight-loss method, the corrosion rates were calculated as follows:

$$C_R = \frac{8.76 \times \Delta m}{\rho a t} \tag{1}$$

where W is the average weight loss of the J55 steel specimen (mg), a is total area of the J55 steel specimen, t is the immersion time (24 h), and D is density of the J55 steel in (g cm^{-3}).

2.6. Electrochemical Analysis

An Autolab Potentiostat device (Metrohm, the Netherland) was used for electrochemical analysis. A three electrode setup was attached to the potentiostat that had a saturated calomel electrode (SCE) as a reference electrode, a graphite rod as an auxiliary electrode, and the J55 steel as the working electrode. At first, the working electrode was immersed in the test medium, i.e., 3.5% NaCl saturated with carbon dioxide for 30 min at 303 K before each experiment to maintain the steady state corrosion potential (E_{corr}).

The electrochemical impedance spectroscopy (EIS) was performed in the range of frequency 100 kHz to 10 mHz at an open circuit potential, by setting 10 mV as the AC sine wave amplitude frequency per decade. Calculations of inhibition efficiencies were done as follows:

$$\eta\% = \left(1 - \frac{R_{ct}}{R_{ct(i)}}\right) \times 100 \tag{2}$$

where R_{ct} and $R_{ct(i)}$ are representative of the resistance of charge transfer without and with the studied inhibitors, respectively.

The potentiodynamic polarization study of the J55 steel without and with the inhibitors was conducted in the range of −250 mV to +250 mV potential and the scan rate used was 1 mV/s. The following equation was used for the inhibition efficiency calculation:

$$\eta\% = \left(1 - \frac{i_{corr(i)}}{i_{corr}}\right) \times 100 \tag{3}$$

where, i_{corr} and $i_{corr(i)}$ represent the values of corrosion current densities without and with inhibitors, respectively.

2.7. X-ray Photoelectron Spectroscopy (XPS)

XPS (VG ESCALAB 220 XL spectrometer instrument, Thermo Scientific, Waltham, MA, USA) was used to analyze the chemical composition of corrosion products on the specimen after testing in the test solution. The processing of XPS data was achieved using XPS Peak-Fit 4.1 software (Hong Kong, China). The high resolution XPS spectra of C 1s, N 1s, O 1s, and Fe 2p of the TMI inhibitor were analyzed.

2.8. Quantum Chemical Calculation

The quantum chemical calculation was performed using density functional theory (DFT). The basis sets used in the present investigation were the DFT/B3LYP methods using 6-311G (d, p) and the Gaussian 09 program package (Wallingford, CT, USA) [17].

2.9. MD Simulations and Radial Distribution Function

BIOVIA Materials Studio software 7.0 (San Diego, CA, USA) were used for simulations [18]. A slab size of the 5 Å Fe (110) surface was selected due to its packed and stable configuration [19]. To allow for better metal-inhibitor interaction-analysis, a simulation box with dimensions of 24.82 × 24.82 × 35.69 $Å^3$ was used. Also in the simulation box, corrosive particles such as $9Cl^-$, $491H_2O$, $9CO_3^{2-}$, and benzimidazole derivatives in their neutral and protonated forms were added. All simulations were executed at a temperature of 303 K and an Andersen thermostat was used to maintain the constant temperature. A COMPASS force field was used for energy minimization and MD calculation processes [20,21].

The radial distribution functions (RDFs) can be defined as the probability of finding particle B around particle A at a definite range. The RDF calculations were performed using simulation trajectories [22]. The radial distribution function can be represented as [23]

$$g_{AB}(r) = \frac{1}{\langle \rho_B \rangle_{\text{local}}} \times \frac{1}{N_A} \sum_{i \in A}^{N_A} \sum_{j \in B}^{N_B} \frac{\delta(r_{ij} - r)}{4\pi r^2} \tag{4}$$

where $\rho_{B\text{local}}$ is the density of particle B averaged over all shells around particle A.

3. Results and Discussion

3.1. Weight-Loss Experiment

3.1.1. Concentration Effect

The effect of the benzimidazole derivative concentrations on the protection ability of the metal surface is represented in the form of a concentration vs. inhibition efficiency plot (Figure 2a) Table 1.

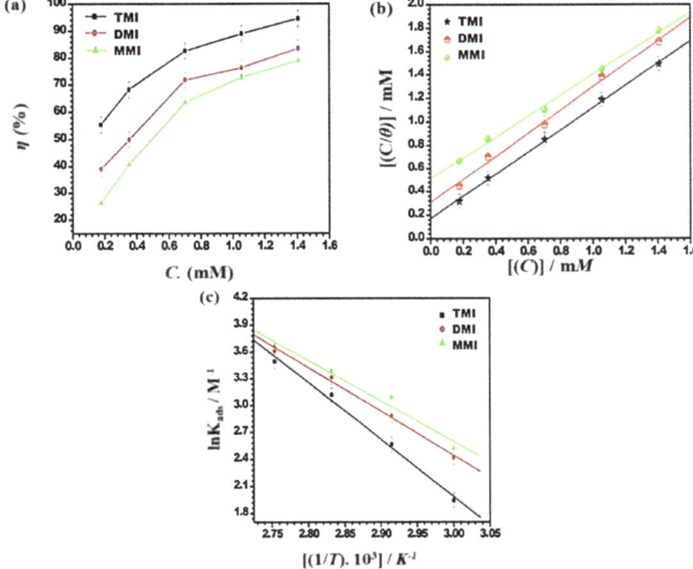

Figure 2. (**a**) Variation of inhibition efficiency (η %) with inhibitor concentration at 333 K; (**b**) Langmuir Isotherm plots for adsorption of inhibitors; (**c**) The relationship between ln K_{ads} and $1000/T$ at optimum concentration of inhibitors.

Table 1. Corrosion inhibition efficiency with the inhibitor concentrations.

Concentrations (mM)	η (%)		
	TMI	DMI	TMI
0.176	55.4	39.0	26.3
0.352	68.1	50.0	40.9
0.703	82.7	71.8	63.6
1.055	89.0	76.3	72.7
1.407	94.5	83.6	79.0

From Figure 2a–c and Table 1, it is obvious that the inhibitory performance of benzimidazole derivatives increased with an increase in their concentration and attained values of 94% (TMI), 83.7% (DMI), and 79% (MMI) at 400 mg/L. The increase in inhibitory performance is due to more coverage of the metal surface because of the adsorption of the benzimidazole derivatives molecules onto the J55 steel surface, which finally reduced the attack of acid. This finding suggests that the molecular structure of benzimidazole derivative molecules has a great influence on the inhibition efficiency values. In this study the inhibitor molecules had π electrons in the aromatic ring and non-bonding electrons on the heteroatoms, such as oxygen and nitrogen, which helped the inhibitor to adsorb onto the J55 steel surface [24].

The number of electron donating functional groups could also affect the adsorption tendency of the inhibitor, i.e., with more electron donating functional groups, the adoption would be stronger and the inhibition efficiency would be higher. Thus, in the present case, TMI had the highest protection ability due to the presence of three OCH_3 groups. Therefore, the inhibition efficiency order was TMI > DMI > MMI.

3.1.2. Adsorption Isotherm of Inhibitor on J55 Steel

Many adsorption isotherms like Langmuir, Frumkin, Flory Huggins, and Temkin were investigated to find a good fit with the experimental study. Out of these isotherms, the Langmuir isotherm, i.e., C_{inh}/θ vs. the inhibitor concentration (C_{inh}), was found to be the best fit due to the slope and regression coefficient (R^2) values approaching towards unity (Figure 2b). The Langmuir isotherm is given by the following formula [25]:

$$\frac{C_{inh}}{\theta} = \frac{1}{K_{ads}} + C_{inh} \quad (5)$$

where C_{inh} is the benzimidazole derivatives concentration (mg/L) and θ and K_{ads} represent the surface coverage and equilibrium adsorption constant, respectively. Although examination of slope values suggests a good fit, it slightly deviates from unity, which is not consistent with the Langmuir adsorption isotherm assumption of monolayer adsorption of inhibitor molecules on the metal surface. According to Eduok and Khaled [26], the discrepancies in slope values are related to the adsorption phenomena, and thus it is important to consider another physical characteristic of the adsorption isotherm. The Langmuir adsorption isotherm can be mathematically represented in terms of the dimensionless separation constant (K_L), and is given by the following equation [26]:

$$K_L = \frac{1}{1 + K_{ads}C} \quad (6)$$

where K_L is the dimensionless separation factor of inhibitor-adsorption. The mean values of the calculated K_L are given in Table 2. Ideally, when the value of K_L is less than unity, the adsorption process is considered to be favorable and the experimental data fit the Langmuir adsorption isotherm. The adsorption process is unfavorable when K_L is greater than unity, and irreversible at K_L = 1. The mean values of K_L were less than unity, suggesting that the adsorption process was favorable. The supplementary file (Figure S3) contains the Frumkin, Flory Huggins, and Temkin isotherm plots. The K_{ads} values were determined from the intercept of the Langmuir plots and are listed in Table 2. The strength of the adsorption of the benzimidazole derivatives molecules on J55 steel are represented by the values of K_{ads}. From the table it can be observed that as the value of the OCH_3 group increased, the values of K_{ads} increased, and the highest K_{ads} was for TMI, which suggests it is had the strongest adsorption onto the metal surface [27,28].

Table 2. Langmuir adsorption isotherm and thermodynamic parameters for the synthesized inhibitors.

Inhibitor	Slope	Regression Coefficient (R^2)	K_{ads} (10^3 M^{-1})	ΔG^o_{ads} (kJ/mol)	ΔH^o_{ads} (kJ/mol)	K_L (mean)
TMI	0.949	0.998	5.88	−35.15	−81.87	0.251
DMI	0.988	0.995	3.22	−33.49	−57.59	0.364
MMI	0.888	0.997	1.93	−32.07	−52.74	0.472

3.1.3. Thermodynamic Parameters of Adsorption

The adsorption behavior and the nature of the adsorption were determined by calculating the thermodynamic parameters of adsorption. The standard free energy of adsorption, i.e., ΔG^o_{ads}, was correlated to K_{ads} according to the following equation [29]:

$$\Delta G^o_{ads} = -2.303RT\log(55.5K_{ads}) \tag{7}$$

where the absolute temperature and universal gas constants are represented by T and R, respectively, and 55.5 is the magnitude of the water molecules concentration. Table 2 reveals that the ΔG^o_{ads} values are negative, suggesting a spontaneous adsorption process [30].

Thermodynamically, K_{ads} is related to the standard enthalpy and entropy of adsorption, i.e., ΔH^o_{ads} and ΔS^o_{ads} and can be calculated using the Van't Hoff equation:

$$\ln K_{ads} = \frac{-\Delta H^o_{ads}}{RT} + \frac{\Delta S^o_{ads}}{R} - \ln C_{H_2O} \tag{8}$$

where ΔH^o_{ads} and ΔS^o_{ads} are the standard enthalpy and entropy of adsorption. The graph of ln K_{ads} vs. $1/T$ is given in Figure 2c. The slopes of ($\Delta H^o_{ads}/R$) were used for the calculation of ΔH^o_{ads} values and are shown in Table 2. In general, the adsorption process is exothermic in nature, which means it is accompanied by the release of energy. All values of ΔH^o_{ads} were negative, so the process of adsorption of inhibitor molecules was exothermic in nature. It should be noted that in the present investigation, ΔH^o_{ads} values were between −52 and −81 kJ/mol, which reveals that the adsorption of the benzimidazole derivatives was both physical and chemical in nature [31]. One more important parameter is that ΔG^o_{ads} was between −40 kJ/mol and −20 kJ/ mol, which further confirms that the adsorption of inhibitor molecules on the J55 steel surface was both physical and chemical [32–35].

3.2. *Electrochemical Studies*

3.2.1. Electrochemical Impedance Spectroscopy (EIS) Studies

The capacitive and inductive behaviors of J55 steel were studied by an electrochemical impedance study. The impedance behavior of the metal is represented in the form of Nyquist plots (Figure 3a–c) at 308 K. According to Figure 3a–c, at a high frequency, a capacitive loop arises that consists of a depressed semicircle. This phenomenon is due to the charge-transfer and capacitance created by the double layer [36,37]. Also, the capacitive loop diameter in the presence of benzimidazole derivatives is larger as compared to that of the blank. Meanwhile, the capacitive loop diameter increased as the benzimidazole concentration increased, which is because of the adsorption of the benzimidazole derivative molecules onto the metal, which formed a barrier of inhibitor molecules and, in turn, enhanced the corrosion-resistance property of the metal [38–40].

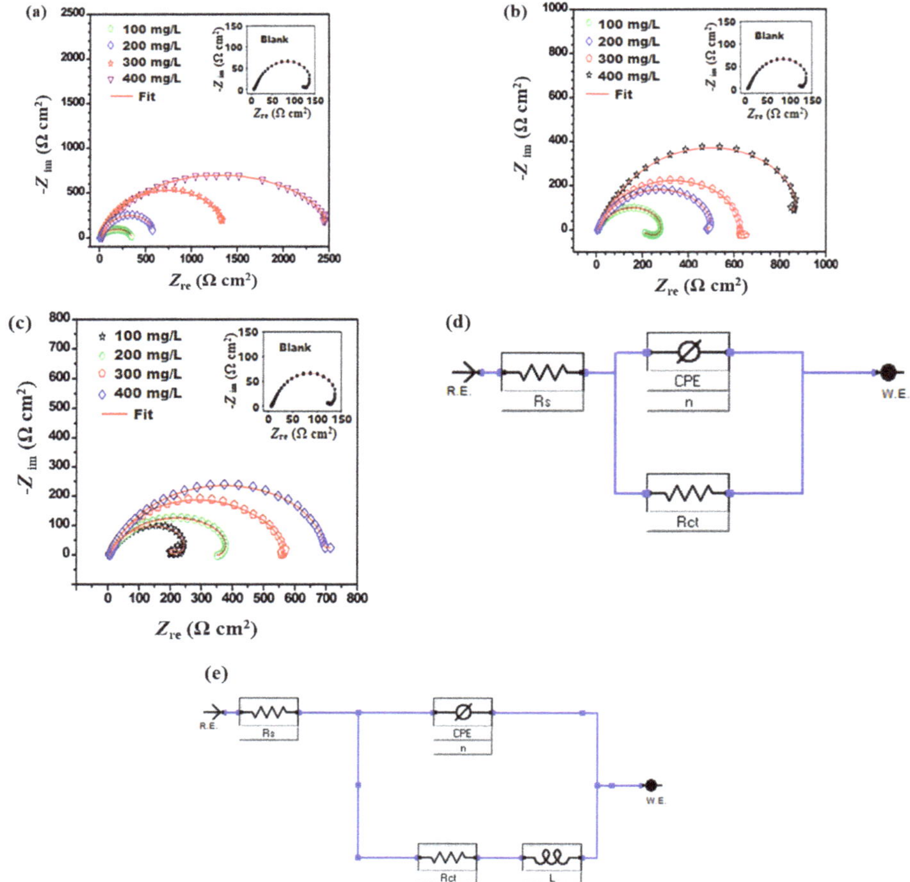

Figure 3. Nyquist plots in absence and presence of different concentration of inhibitors: (**a**) TMI; (**b**) DMI; (**c**) MMI; (**d**,**e**) Equivalent circuits used.

It is interesting to note that in the absence of inhibitors and at 100 mg/L concentration of inhibitors, an inductive loop was also observed due to the adsorption of the intermediate product, i.e., $FeOH_{ads}$, which was formed during the dissolution of the J55 steel [41]. However, as the concentration of benzimidazole derivatives increased, only capacitive loops were observed and inductive loops disappeared. This may be due to a larger area of the J55 steel being covered by benzimidazole derivative molecules with increasing concentration, which finally reduces the J55 steel corrosion.

For impedance data calculations, two circuits were used and are shown in Figure 3d,e. The elements used for construction of the circuits consisted of a resistor for charge transfer (R_{ct}), inductor (L), solution resistor (R_s), and constant phase element (CPE). The CPE was used for accurate fitting of the circuit because it can compensate for the effects that cause deviations such as surface roughness, dislocation, imperfection, impurities, inhibitor adsorption, etc. [42–46].

Table 3 represents the fitted-curve impedance results. It can be noted that the R_{ct} and Y_0 values at each concentration of all benzimidazole derivatives are showing an opposite trend, in other words, Y_0 decreases and R_{ct} increases. This phenomenon is because of the adsorption of the benzimidazole molecules onto the J55 steel, which finally enhances the J55 steel corrosion-resistance property [47]. The values of R_{ct} at 400 mg/L for TMI, DMI, and MMI were 2466 Ω cm^2, 866 Ω cm^2, and 697 Ω cm^2, respectively. Thus, the TMI derivative provided the maximum resistance towards corrosion. This is because of the three electron donating OCH_3 groups in TMI. The increase in n values with the addition of bezimidazole derivatives is due to the adsorption of the derivatives, which enhances the homogeneity of the derivatives [48]. Thus, with an increase in the number of electron donating functional groups, the corrosion inhibition property is increased. Therefore, the effectiveness of the benzimidazole derivatives as corrosion inhibitors can be given as TMI > DMI > MMI.

Table 3. Electrochemical impedance parameters in absence and presence of different concentrations of inhibitors at 308 K.

C_{inh} (mg L^{-1})	R_s (Ω)	R_{ct} (Ω cm^2)	Y_0 (μF/cm^2)	n	L (H)	η (%)
Blank	4.817	135.57	512.7	0.601	8.04	–
			TMI			
100	4.927	350.33	278.3	0.789	–	61.3
200	5.190	573.21	210.6	0.854	–	76.3
300	5.562	1400.01	140.9	0.855	–	90.3
400	5.601	2466.17	42.45	0.879	–	94.5
			DMI			
100	6.159	269.84	296.1	0.771	60.02	49.7
200	5.930	485.34	285.9	0.816	–	72.0
300	5.593	624.24	156.5	0.839	–	78.2
400	5.473	866.87	76.56	0.848	–	84.3
			MMI			
100	5.255	231.39	325.2	0.769	35.46	41.4
200	6.01	377.24	291.4	0.804	–	64.0
300	5.416	557.11	234.5	0.812	–	75.6
400	5.601	697.71	125.5	0.827	–	80.5

3.2.2. Potentiodynamic Polarization Analysis

The Tafel plots of the J55 steel in 3.5% NaCl saturated with carbon dioxide without and with different concentrations of benzimidazole derivatives are represented in Figure 4a–c. Some important electrochemical parameters such as corrosion current density (i_{corr}), corrosion potential (E_{corr}), cathodic Tafel slope (β_c), anodic Tafel slope (β_c), and inhibition efficiency (IE%) are shown in Table 4. The observation of Table 3 shows that the corrosion-current density values shift from 104.4 μA/cm^2 (Blank) to 5.5 μA/cm^2 (TMI), 13.2 μA/cm^2 (DMI), and 21.5 μA/cm^2 (MMI), and this represents that benzimidazole derivatives are effective corrosion inhibitors. The values of the maximum inhibition efficiency are 94%, 87%, and 79% for TMI, DMI, and MMI, respectively at 400 mg/L. Also, the addition of the benzimidazole derivative molecules to the aggressive medium caused reduction of both the anodic and cathodic current density. All the studied benzimidazole derivative molecules with the increment of concentration reduced more H^+ ions in the cathodic reactions as compared to anodic dissolution reactions. According to Figure 4a–c, the cathodic Tafel lines are parallel, suggesting that the activation-control evolution of H_2 gas and the mechanism of H^+ to H_2 conversion/reduction are not modified by the presence of benzimidazole derivatives. Also, with increased concentration of benzimidazole derivatives, the β_c values were changed, indicating that benzimidazole derivatives affected the hydrogen evolution kinetics. This is because of the diffusion or barrier effect [49]. Similarly, the slope values of the anodic Tafel lines, i.e., β_a, also underwent changes with increased benzimidazole derivative concentration, suggesting that the studied inhibitor molecules were initially adsorbed over the J55 steel surface, reducing the process of corrosion by blocking the reactive sites presented on the J55 steel without altering the mechanism of the anodic reaction [50].

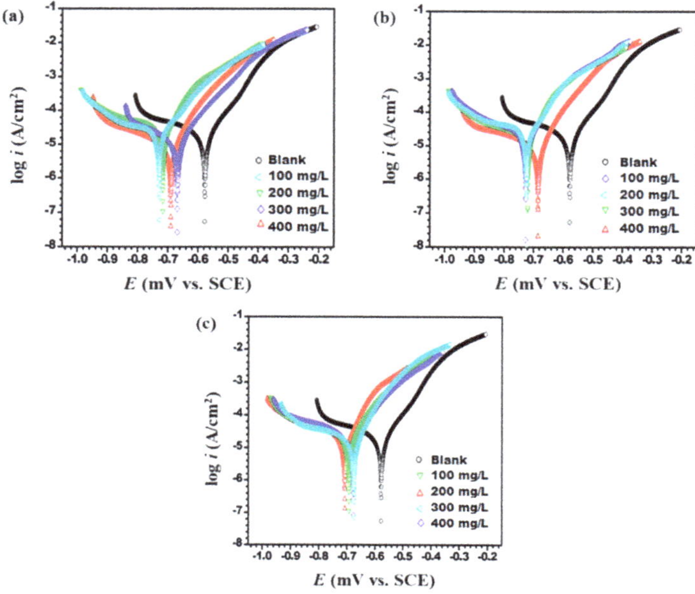

Figure 4. Potentidynamic polarization curves in absence and presence of different concentrations of inhibitors: (**a**) TMI; (**b**) DMI; (**c**) MMI.

Table 4. Electrochemical polarization parameters in the absence and presence of different concentrations of inhibitors at 308 K.

Inhibitor	E_{corr} (mV/SCE)	i_{corr} (μA/cm^2)	β_a (mV/dec)	$-\beta_c$ (mV/dec)	η (%)
Blank	−576	104.4	154	590	–
		TMI			
100	−666	45.0	192	683	56.8
200	−717	29.3	92	443	71.9
300	−669	20.0	114	499	80.7
400	−691	5.5	80	382	94.7
		DMI			
100	−727	51.4	190	693	50.5
200	−723	34.4	106	425	66.9
300	−720	25.2	82	489	75.7
400	−685	13.2	85	499	87.2
		MMI			
100	−694	65.2	183	796	37.2
200	−709	45.9	118	846	55.8
300	−678	27.7	105	542	73.2
400	−676	21.5	183	511	79.3

Tafel curves showed that the addition of benzimidazole derivatives inhibited both the cathodic and anodic reactions. Thus, the inhibitor is said to be a mixed type. However, the shift in the E_{corr} values in the inhibited solution was towards the cathodic direction, i.e., negative with respect to the uninhibited solution, revealing that while benzimidazole derivatives were predominantly cathodic, overall they were a mixed-type inhibitor. It is interesting to note here that as the number of OCH_3 groups increased, a larger reduction in the corrosion current density occurred. This is due to the donation of an electron by the OCH_3 group, which facilitates formation of stronger bonds between benzimidazole derivative molecules and the J55 steel. Thus, the inhibition efficiency order was TMI > DMI > MMI.

3.3. X-Ray Photoelectron Spectroscopy (XPS)

For the confirmation of TMI derivative adsorption on J55 steel, XPS analyses were performed and elaborated. The XPS spectra obtained for TMI adsorbed on the J55 steel at a concentration of 400 mg/L after 24 h immersion in a corrosive solution are shown in Figure 5a–d. Spectra of the XPS consist of the following peaks; C 1s, N 1s, O 1s, and Fe 2p. All spectra are complex in nature, and in order to assign the peaks of the corresponding adsorbed species, a deconvolution fitting procedure was used.

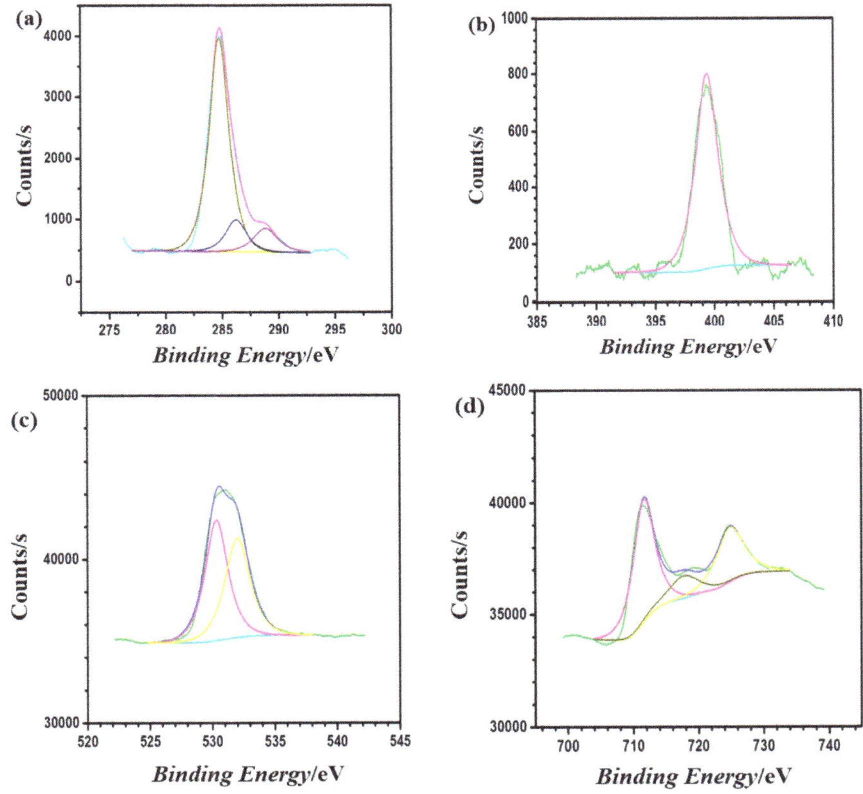

Figure 5. XPS spectra: (**a**) C 1s, (**b**) N 1s, (**c**) O 1s and (**d**) Fe 2p of the TMI inhibitor.

Three main peaks are in C 1s spectra. Appearance of first peak occurs at the binding energy of 284.79 eV and is attributed to the aromatic bonds between C–C, C=C, and C–H [51–53]. The appearance of the second peak occurs at 286.21 eV and corresponds to the carbon atoms bonded to nitrogen in

C–N and C=N bonds present in the imidazole ring [51–53]. The third peak appears at 288.87 eV and represents the nitrogen bonded carbon atom of the imidazole ring, i.e., C=N+ [54], that results because of protonation of the =N– atom in the imidazole ring and/or the imidazole ring nitrogen coordination with the J55 steel. The N 1s XPS spectrum consists of one peak at 399.27 eV. The appearance of this peak is due to the C–N and the unprotonated N atom (=N– structure) in the imidazole ring [55–57].

The spectrum of O 1s has three main peaks. The peak that appears at approximately 530.35 eV is attributed to O^{2-}, and it corresponds to the oxides of iron and oxygen, i.e., Fe_2O_3 and/or Fe_3O_4 [58]. The appearance of the second peak at a binding energy of 531.33 eV is because of OH^- and is attributed to the existence of the hydrous form of iron oxides, such as FeOOH [58]. Lastly, oxygen of adsorbed water molecules appears as a third peak at 532.05 eV [57].

The Fe 2p spectrum for the J55 steel surface covered with TMI derivatives consists of two doublets, one at 711.78 eV (Fe $2p_{3/2}$) and the second at 724.98 eV (Fe $2p_{1/2}$). The deconvolution of the high resolution Fe 2p3/2 XPS spectrum consists of two peaks. The first peak at the binding energy of 709.4 eV is attributed to metallic iron, i.e., iron in the zero oxidation state [59,60]. The appearance of the second peak at 711.78 eV is due to Fe^{3+} [61] and is attributable to the oxides of iron such as Fe_2O_3, Fe_2CO_3, and FeOOH (i.e., oxyhydroxyde) [62,63]. The comparison between Fe $2p_{3/2}$ XPS results of the TMI treated steel with that of untreated steel (as described previously) [64] shows that there is a significant decrease in the amount of Fe^0 that indicates an increment in the oxide layer thickness. The formed oxide layer of FeOOH is insoluble and stable, which reduces the diffusion of metal ions and thus enhances the corrosion-resistive property of J55 steel in an aggressive media.

3.4. Quantum Chemical Calculation

Calculation of Preferred Site for Protonation

In an aqueous medium, organic molecules can easily undergo protonation, and this protonated form of the inhibitor takes part in the adsorption process. In the present case, the number of nitrogen atoms in the imidazole ring with the most negative Mullikien charge is two (Table 5). Therefore, there are protonated N_2 and N_5 nitrogen atoms. However, the most preferential nitrogen atom that can undergo protonation was selected by calculating the proton affinity (*PA*) at both N_2 and N_5. The equation used for calculating *PA* is given below:

$$PA = E_{prot} - (E_{neutral} + E_{H^+}) \tag{10}$$

$$E_{H^+} = E_{H_3O^+} - E_{H_2O} \tag{11}$$

where the total energies of the protonated inhibitor forms and the neutral inhibitor forms are represented by E_{prot} and $E_{neutral}$, respectively. E_{H_2O} is the water molecule total energy and $E_{H_3O^+}$ is the hydronium ion total energy. The site having the most negative value of *PA* is selected as the most preferable site for protonation. Thus, in this paper, the calculated value of PA for N_5 was the most negative and so it was selected as the preferential site for protonation (Table 5).

Table 5. Atomic charges on hetroatoms and proton affinity values. *PA* = proton affinity.

Inhibitors	N_2	N_5	*PA* (kcal/mol)	
			N_2	N_5
TMI	−0.495	−0.402	5.65	−30.75
DMI	−0.492	−0.403	6.27	−30.12
MMI	−0.490	−0.403	5.65	−29.34

3.5. Molecular Dynamic Simulations

Despite extensive research having been conducted in recent years, there is uncertainty regarding the corrosion inhibition mechanisms of CO_2 corrosion, and more critical investigations should be conducted. In this regard, MD simulations could be a good way to improve scientific knowledge in the field. In the course of this study, we investigated the adsorption of neutral and protonated inhibitor molecules on an iron surface in the presence of a simulated electrolyte with the aim of mimicking the experimental conditions and assessing whether any relationship exists between the theoretical and experimental results, and, if so, how significant the MD results are in explaining the inhibition process. Simulations were run until the systems reached an equilibrium state, then, the interaction energies were estimated by calculating the single point energies of all system constituents [65]. The obtained equilibrium configurations of neutral and protonated forms of inhibitor molecules on the Fe (110) surface in solution are shown in Figures 6 and 7. It can be observed from the results in Figures 6 and 7 that both forms of the inhibitor molecules are adsorbed on the iron surface in a parallel manner and are near to the iron surface. Such situations can help to produce chemical interactions and thereby increase the adsorption rate of tested inhibitors. The benzimidazole itself is a good corrosion inhibitor, thus, with the addition of a methoxy group in the phenyl ring, the interactive forces and the affinity toward the J55 steel of our compounds strongly increased, which lead to more interactions with the steel surface [66].

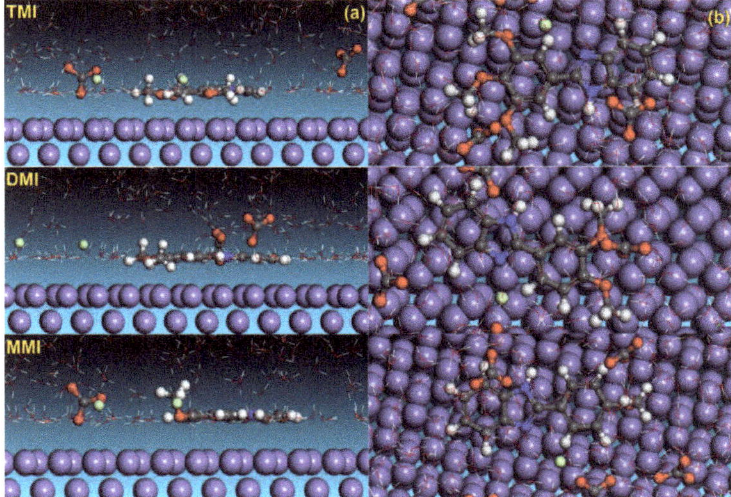

Figure 6. Side and top views of the final adsorption of neutral forms of inhibitor molecules on the Fe (110) surface in solution: (**a**) side view; (**b**) top view.

The interaction and binding energies ($E_{\text{Binding}} = -E_{\text{interaction}}$) of the obtained inhibitor molecules under equilibrium conditions for neutral and protonated forms can also be useful information to assess the extent of adsorption of the three compounds. The results in Table 6 show higher energy values that may explain the higher interaction between the inhibitor molecules and the steel surface and the stability of formed films [67,68]. The energy values are slightly decreased in a protonated form that can be explained mainly by the higher contribution of the physical interactions between protonated inhibitor molecules and the positively charged metal surface. We can also note that the energy values follow the same trend of the inhibition efficiency values, thus confirming the crucial role of the number of methoxy groups as powerful electro donating groups in increasing the adsorption abilities of the tested compounds.

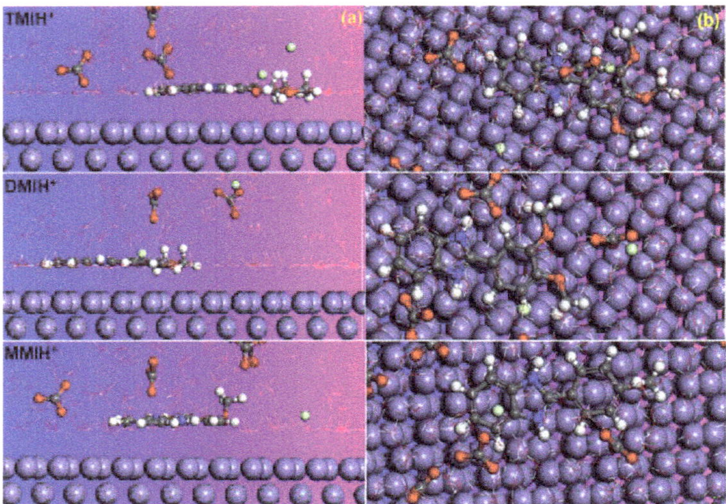

Figure 7. Side and top views of the final adsorption of the protonated forms of the inhibitor molecules on the Fe (110) surface in solution: (**a**) side view; (**b**) top view.

Table 6. Selected energy parameters obtained from molecular dynamic (MD) simulations for adsorption of inhibitors on the Fe (110) surface.

System	Neutral Form	Protonated Form
	$E_{interaction}$ (kJ/mol)	$E_{interaction}$ (kJ/mol)
Fe + TMI	−564.09	−559.77
Fe + DMI	−507.34	−497.19
Fe + MMI	−453.67	−444.31

3.6. Radial Distribution Function (RDF)

RDF analysis provides further insights into the interactive force of an inhibitor molecule and its affinity towards the iron surface [23]. Here, the total radial distribution function was calculated for both inhibitor forms using MD simulation trajectories. Whether the interactions of an inhibitor with iron atoms are meaningful can be judged by comparison of the first prominent peaks in the RDF curves. If the peak occurs at 1 Å ~ 3.5 Å, it is an indication of a small bond length, which correlates to chemisorption, while the physical interactions are associated with the peaks longer than 3.5 Å [69]. Figure 8 shows the RDF results of neutral and protonated forms. We can see that the first prominent peak for both inhibitor forms is located at a distance smaller than 3.5 Å. From Figure 8, one can easily observe that the first peak increased with the decreased inhibition efficiency of the tested compounds. A further increase was observed in the protonated state of the inhibitor molecules. All the inhibitor molecules in their neutral or protonated forms retained significant interaction with the iron surface.

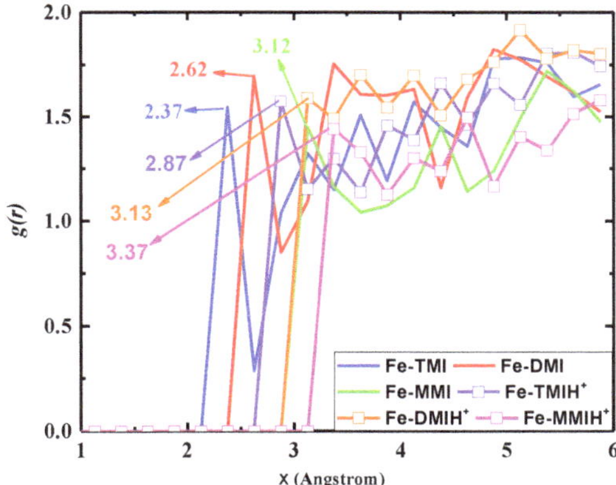

Figure 8. Radial Distribution Functions (RDFs) of neutral and protonated forms of the tested corrosion inhibitors adsorbed on the Fe (110) surface in solution.

4. Conclusions

(1) The tested benzimidazole derivatives are good inhibitors in the aggressive media of 3.5% NaCl solution saturated with carbon dioxide at 333 K.

(2) Experimental and theoretical investigations suggest that as the number of methoxy groups increase so too does the corrosion protection ability of the inhibitors, and thus TMI is the best inhibitor.

(3) The potentiodynamic polarization measurement supports the mixed mode of inhibitors with predominantly cathodic effects.

(4) Langmuir adsorption is the preferred isotherm for all inhibitors.

(5) XPS micrographs support the benzimidazole derivative adsorption.

(6) The DFT study confirms that the imine nitrogen (N_5) is the most preferred site for protonation.

(7) MD results support that TMI has a stronger adsorption ability than that of both DMI and MMI.

(8) Results of the RDF study confirmed that both the neutral and protonated form of the inhibitor show significant interaction with the steel surface.

Supplementary Materials: The following are available online at http://www.mdpi.com/1996-1944/12/1/17/s1, Figure S1: ^{1}H NMR spectra, Figure S2: ^{13}C NMR spectra, Figure S3. Adsorption isotherm plots: (a) Flory-Huggins; (b) Frumkin; and (c) Temkin.

Author Contributions: The conceptualization was done by A.S. and K.R.A.; methodology was performed by A.S. and H.L.; software for the study was used by H.L. and K.R.A.; validation of data was done by M.A.Q., A.S. and K.R.A.; formal analysis was done by A.S.; investigation was carried out by Ambrish Singh and K.R.A.; resources were provided by M.A.Q.; data curation was done by K.R.A.; writing—original draft preparation was done by A.S.; writing—review and editing was performed by A.S. and K.R.A.; supervision was of M.A.Q.

Funding: This research was funded by Sichuan 1000 Talent Fund.

Conflicts of Interest: The authors declare no conflict of interest.

References

1. Yildirim, A.; Cetin, M. Synthesis and evaluation of new long alkyl side chain acetamide, isoxazolidine and isoxazoline derivatives as corrosion inhibitors. *Corros. Sci.* **2008**, *50*, 155–165. [CrossRef]
2. Durnie, W.; De Marco, R.; Jefferson, A.; Kinsella, B. Development of astructure-activity relationship for oil field corrosion inhibitors. *J. Electrochem. Soc.* **1999**, *146*, 1751–1756. [CrossRef]
3. Ramachandran, S.; Jovancicevic, V. Molecular modeling of the inhibition ofmild steel carbon dioxide corrosion by imidazolines. *Corrosion* **1999**, *55*, 259–267. [CrossRef]

4. Yadav, D.K.; Quraishi, M.A. Application of Some Condensed Uracils as Corrosion Inhibitors for Mild Steel: Gravimetric, Electrochemical, Surface Morphological, UV-Visible, and Theoretical Investigations. *Ind. Eng. Chem. Res.* **2012**, *51*, 14966–14979. [CrossRef]
5. Sastri, V.S. *Green Corrosion Inhibitors. Theory and Practice*, 1st ed.; John Wiley & Sons: Hoboken, NJ, USA, 2011.
6. Denny, W.A.; Rewcastle, G.W.; Baguley, B.C. Potential antitumor agents. 59. Structure-activity relationships for 2-phenylbenzimidazole-4-carboxamides, a new class of minimal DNA-intercalating agents which may not act via topoisomerase II. *J. Med. Chem.* **1990**, *33*, 814–819. [CrossRef] [PubMed]
7. Fonseca, T.; Gigante, B.; Gilchrist, T.L. A short synthesis of phenanthro[2,3-d]imidazoles from dehydroabietic acid. Application of the methodology as a convenient route to benzimidazoles. *Tetrahedron* **2001**, *57*, 1793–1799. [CrossRef]
8. Jevremovic´, I.; Singer, M.; Nešic´, S.; Miškovic´-Stankovic´, V. Inhibition properties of self-assembled corrosion inhibitor talloildiethylenetriamineimidazoline for mild steel corrosion in chloride solution saturated with carbon dioxide. *Corros. Sci.* **2013**, *77*, 265–272. [CrossRef]
9. Okafor, P.C.; Liu, C.B.; Zhu, Y.J.; Zheng, Y.G. Corrosion and corrosion inhibition behavior of N80 and P110 carbon steels in CO_2-saturated simulated formation water by rosin amide Imidazoline. *Ind. Eng. Chem. Res.* **2011**, *50*, 7273–7281. [CrossRef]
10. Ortega-Sotelo, D.M.; Gonzalez-Rodriguez, J.G.; Neri-Flores, M.A.; Casales, M.; Martinez, L.; Martinez-Villafañe, A. CO_2 corrosion inhibition of X-70 pipeline steel by carboxyamidoimidazoline. *J. Solid State Electrochem.* **2010**, *15*, 1997–2004. [CrossRef]
11. Guadalupe, H.J.; García-Ochoa, E.; Maldonado-Rivas, P.J.; Cruz, J.; Pandiyan, T. A combined electrochemical and theoretical study of *N*,N_0-bis(benzimidazole-2yl-ethyl)-1,2-diaminoethane as a new corrosion inhibitor for carbon steel surface. *J. Electroanal. Chem.* **2011**, *655*, 164–172. [CrossRef]
12. Obot, I.B.; Edouk, U.M. Benzimidazole: Small planar molecule with diverse anti-corrosion potentials. *J. Mol. Liq.* **2017**, *246*, 66–90. [CrossRef]
13. He, Y.; Yang, R.; Zhou, Y.; Ma, L.; Zhang, L.; Chen, Z. Water soluble Thiosemicarbazideimidazole derivative as an efficient inhibitor protecting P110 carbon steel from CO_2 corrosion. *Anti Corros. Methods Mater.* **2016**, *63*, 437–444. [CrossRef]
14. He, Y.; Zhou, Y.; Yang, R.; Ma, L.; Chen, Z. Imidazoline derivative with four imidazole reaction centers as an efficient corrosion inhibitor for anti-CO_2 corrosion. *Russ. J. Appl. Chem.* **2015**, *88*, 1192–1200. [CrossRef]
15. Zhou, F.; Wang, H.; Dai, Q. Study on the compound of Imidazoline Corrosion Inhibitor. *IOP Conf. Ser. Earth Environ. Sci.* **2018**, *153*, 052001. [CrossRef]
16. Jaberi, Z.K.; Amiri, M. An Efficient and Inexpensive Synthesis of 2-Substituted Benzimidazoles in Water Using Boric Acid at Room Temperature. *E-J. Chem.* **2012**, *9*, 167–170. [CrossRef]
17. Frisch, M.J.; Trucks, G.W.; Schlegel, H.B.; Frisch, M.J.; Trucks, G.W.; Schlegel, H.B.; Scuseria, G.E.; Robb, M.A.; Cheeseman, J.R.; Scalmani, G.; et al. *Gaussian 09*; revision A.02; Gaussian Inc.: Wallingford, CT, USA, 2016.
18. *Materials Studio*; revision 6; Accelrys Inc.: San Diego, CA, USA, 2013.
19. Guo, L.; Obot, I.B.; Zheng, X.; Shen, X.; Qiang, Y.; Kaya, S.; Kaya, C. Theoretical insight into an empirical rule about organic corrosion inhibitors containing nitrogen, oxygen, and sulfur atoms. *Appl. Surf. Sci.* **2017**, *406*, 301–306. [CrossRef]
20. Sun, H. COMPASS: An ab Initio Force-Field Optimized for Condensed-Phase Applications Overview with Details on Alkane and Benzene Compounds. *J. Phys. Chem. B* **1998**, *102*, 7338–7364. [CrossRef]
21. Lgaz, H.; Salghi, R.; Subrahmanya Bhat, K.; Chaouiki, A.; Shubhalaxmi; Jodeh, S. Correlated experimental and theoretical study on inhibition behavior of novel quinoline derivatives for the corrosion of mild steel in hydrochloric acid solution. *J. Mol. Liq.* **2017**, *244*, 154–168. [CrossRef]
22. Wu, R.; Qiu, X.; Shi, Y.; Deng, M. Molecular dynamics simulation of the atomistic monolayer structures of N-acyl amino acid-based surfactants. *Mol. Simul.* **2017**, *43*, 491–501. [CrossRef]
23. Hansen, J.-P.; McDonald, I.R. *Theory of Simple Liquids: With Applications to Soft Matter*, 4th ed.; Academic Press: Cambridge, MA, USA, 2013.
24. Singh, A.; Ansari, K.R.; Haque, J.; Dohare, P.; Lgaz, H.; Salghi, R.; Quraishi, M.A. Effect of electron donating functional groups on corrosion inhibition of mild steel in hydrochloric acid: Experimental and quantum chemical study. *J. Taiwan Inst. Chem. Eng.* **2018**, *82*, 233–251. [CrossRef]
25. Ansari, K.R.; Quraishi, M.A. Experimental and computational studies of naphthyridine derivatives as corrosion inhibitor for N80 steel in 15% hydrochloric acid. *Physica E* **2015**, *69*, 322–331. [CrossRef]

26. Eduok, U.M.; Khaled, M. Corrosion inhibition for low-carbon steel in 1 M H_2SO_4 solution by phenytoin: Evaluation of the inhibition potency of another "anticorrosive drug". *Res. Chem. Intermed.* **2014**. [CrossRef]
27. Li, X.H.; Deng, S.D.; Fu, H.; Mu, G.N. Inhibition by tween-85 of the corrosion of cold rolled steel in 1.0 M hydrochloric acid solution. *J. Appl. Electrochem.* **2009**, *39*, 1125–1135. [CrossRef]
28. Singh, A.; Ansari, K.R.; Quraishi, M.A.; Lgaz, H.; Lin, Y. Synthesis and investigation of pyran derivatives as acidizing corrosion inhibitors for N80 steel in hydrochloric acid: Theoretical and experimental approaches. *J. Alloys Compd.* **2018**, *762*, 347–362. [CrossRef]
29. Ansari, K.R.; Quraishi, M.A.; Singh, A. Schiff's base of pyridyl substituted triazoles as new and effective corrosion inhibitors for mild steel in hydrochloric acid solution. *Corros. Sci.* **2014**, *79*, 5–15. [CrossRef]
30. Haque, J.; Ansari, K.R.; Srivastava, V.; Quraishi, M.A.; Obot, I.B. Pyrimidine derivatives as novel acidizing corrosion inhibitors for N80 steel useful for petroleum industry: A combined experimental and theoretical approach. *J. Ind. Eng. Chem.* **2017**, *49*, 176–188. [CrossRef]
31. Vonopen, B.; Kordel, W.; Klein, W. Sorption of nonpolar and polar compoundsto soils: Processes, measurement and experience with the applicability of the modified OECD-guideline. *Chemosphere* **1991**, *22*, 285–304. [CrossRef]
32. Bentiss, F.; Lebrini, M.; Lagrenée, M. Thermodynamic characterization of metal dissolution and inhibitor adsorption processes in mildsteel/2,5-bis(n-thienyl)-1,3,4-thiadiazoles/hydrochloric acid system. *Corros. Sci.* **2005**, *47*, 2915–2931. [CrossRef]
33. Soltani, N.; Salavati, H.; Rasouli, N.; Paziresh, M.; Moghadas, A. Adsorption andcorrosion inhibition effect of Schiff base ligands on low carbon steelcorrosion in hydrochloric acid solution. *Chem. Eng. Commun.* **2016**, *203*, 840–854.
34. Kowsari, E.; Payami, M.; Amini, R.; Ramezanzadeh, B.; Javanbakht, M. Task-specific ionic liquid as a new green inhibitor of mild steel corrosion. *Appl. Surf. Sci.* **2014**, *289*, 478–486. [CrossRef]
35. Yilmaz, N.; Fitoz, A.; Ergun, Ü.; Emregül, K.C. A combined electrochemical andtheoretical study into the effect of 2-((thiazole-2-ylimino)methyl) phenol as a corrosion inhibitor for mild steel in a highly acidic environment. *Corros. Sci.* **2016**, *111*, 110–120. [CrossRef]
36. Ghada, M.; Abd El-Hafez, W.; Badawy, A. The use of cysteine, N-acetyl cysteine and methionine as environmentally friendly corrosion inhibitors for Cu–10Al–5Ni alloy in neutral chloride solutions. *Electrochim. Acta* **2013**, *108*, 860–866.
37. Singh, A.; Ebenso, E.E.; Quraishi, M.A.; Lin, Y. 5,10,15,20-Tetra(4-pyridyl)-21H,23H-porphine as an effective corrosion inhibitor for N80 steel in 3.5% NaCl solution. *Int. J. Electrochem. Soc.* **2014**, *9*, 7495–7505.
38. Solmaz, R.; Altunbas, E.; Kardas, G. Adsorption and corrosion inhibition effect of 2-((5-mercapto-1,3,4-thiadiazol-2-ylimino)methyl)phenol Schiff base on mild steel. *Mater. Chem. Phys.* **2011**, *125*, 796–801. [CrossRef]
39. Doner, A.; Kardas, G. N-Aminorhodanine as an effective corrosion inhibitor for mild steel in 0.5M H_2SO_4. *Corros. Sci.* **2011**, *53*, 4223–4232. [CrossRef]
40. Yildiz, R. An electrochemical and theoretical evaluation of 4,6-diamino-2-pyrimidinethiol as a corrosion inhibitor for mild steel in HCl solutions. *Corros. Sci.* **2015**, *90*, 544–553. [CrossRef]
41. Zhang, G.A.; Cheng, Y.F. On the fundamentals of electrochemical corrosion of X65 steel in CO_2-containing formation water in the presence of acetic acid in petroleum production. *Corros. Sci.* **2009**, *51*, 87–94. [CrossRef]
42. Popova, A.; Sokolova, E.; Raicheva, S.; Christov, M. AC and DC study of the temperature effect on mild steel corrosion in acid media in the presence of benzimidazole derivatives. *Corros. Sci.* **2003**, *45*, 33–58. [CrossRef]
43. Growcock, F.B.; Jasinski, J.H. Time-Resolved Impedance Spectroscopy of Mild Steel in Concentrated Hydrochloric Acid. *J. Electrochem. Soc.* **1989**, *136*, 2310–2314. [CrossRef]
44. Macdonald, J.R.; Johanson, W.B. *Theory in Impedance Spectroscopy*; Macdonald, J.R., Ed.; John Wiley& Sons: New York, NY, USA, 1987.
45. Stoynov, Z.B.; Grafov, B.M.; Savova-Stoynova, B.; Elkin, V.V. *Electrochemical Impedance*; Nauka: Moscow, Russia, 1991.
46. Ansari, K.R.; Quraishi, M.A.; Singh, A. Pyridine derivatives as corrosion inhibitors for N80 steel in 15% HCl: Electrochemical, surface and quantum chemical studies. *Measurement* **2015**, *76*, 136–147. [CrossRef]
47. Ansari, K.R.; Quraishi, M.A.; Singh, A. Isatin derivatives as a non-toxic corrosion inhibitor for mild steel in 20% H_2SO_4. *Corros. Sci.* **2015**, *95*, 62–70. [CrossRef]

48. Singh, A.; Ansari, K.R.; Kumar, A.; Liu, W.; Songsong, C.; Lin, Y. Electrochemical, surface and quantum chemical studies of novel imidazole derivatives as corrosion inhibitors for J55 steel in sweet corrosive environment. *J. Alloys Compd.* **2017**, *712*, 121–133. [CrossRef]
49. Ahamad, I.; Prasad, R.; Quraishi, M.A. Adsorption and inhibitive properties of some new Mannich bases of Isatin derivatives on corrosion of mild steel in acidic media. *Corros. Sci.* **2010**, *52*, 1472–1481. [CrossRef]
50. Abdel-Rehim, S.S.; Ibrahim, M.A.M.; Khaled, K.F. The inhibition of 4-(2′-amino-5′-methylphenylazo) antipyrine on corrosion of mild steel in HCl solution. *Mater. Chem. Phys.* **2001**, *70*, 268–273. [CrossRef]
51. Bentiss, F.; Jama, C.; Mernari, B.; El Attari, H.; El Kadi, L.; Lebrini, M.; Traisnel, M.; Lagrenée, M. Corrosion control of mild steel using 3,5-bis(4-methoxyphenyl)-4-amino-1,2,4-triazole in normal hydrochloric acid medium. *Corros. Sci.* **2009**, *51*, 1628–1635. [CrossRef]
52. Nakayama, N.; Obuchi, A. Inhibitory effects of 5-aminouracil on cathodic reactions of steels in saturated $Ca(OH)_2$ solutions. *Corros. Sci.* **2003**, *45*, 2075–2092. [CrossRef]
53. Schick, G.A.; Sun, Z. Spectroscopic characterization of sulfonyl chloride immobilization on silica. *Langmuir* **1994**, *10*, 3105–3110. [CrossRef]
54. Watts, J.F.; Wolstenholme, J. *An Introduction to Surface Analysis by XPS and AES*; John Wiley and Sons Inc.: London, UK, 2003.
55. Bentiss, F.; Outirite, M.; Traisnel, M.; Vezin, H.; Lagrenee, M.; Hammouti, B.; Al- Deyab, S.S.; Jama, C. Improvement of corrosion resistance of carbon steel in hydrochloric acid medium by 3,6-bis(3-Pyridyl) pyridazine. *Int. J. Electrochem. Soc.* **2012**, *7*, 1699–1723.
56. Temesghen, W.; Sherwood, P.M.A. Analytical utility of valence band X-ray photoelectron spectroscopy of iron and its oxides, with spectral interpretation by cluster and band structure calculations. *Anal. Bioanal. Chem.* **2002**, *373*, 601–608. [CrossRef]
57. Babic-Samardzija, K.; Lupu, C.; Hackerman, N.; Barron, A.R.; Luttge, A. Inhibitive properties and surface morphology of a group of heterocyclic diazoles as inhibitors for acidic iron corrosion. *Langmuir* **2005**, *21*, 12187–12196. [CrossRef]
58. Swift, A.J. Surface Analysis of Corrosion Inhibitor Films by XPS and ToFSIMS. *Mikrochim. Acta* **1995**, *120*, 149–158. [CrossRef]
59. Devaux, R.; Vouagner, D.; de Becdelievre, A.M.; Duret-Thual, C. Electrochemical and surface studies of the ageing of passive layers grown on stainless steel in neutral chloride solution. *Corros. Sci.* **1994**, *36*, 171–186. [CrossRef]
60. Dicastro, V.; Ciampi, S. XPS study of the growth and reactivity of Fe/MnO thin films. *Surf. Sci.* **1995**, *331*, 294–299. [CrossRef]
61. Olivares-Xometl, O.; Likhanova, N.V.; Martı´nez-Palou, R.; Domı´nguez-Aguilar, M.A. Electrochemistry and XPS study of an imidazoline as corrosion inhibitor of mild steel in an acidic environment. *Mater. Corros.* **2009**, *60*, 14–21. [CrossRef]
62. Pech-Canul, M.A.; Bartolo-Perez, P. Inhibition effects of N-phosphono-methyl-glycine/Zn^{2+} mixtures on corrosion of steel in neutral chloride solutions. *Surf. Coat. Technol.* **2004**, *184*, 133–140. [CrossRef]
63. Bouanis, F.Z.; Bentiss, F.; Traisnel, M.; Jama, C. Enhanced corrosion resistance properties of radiofrequency cold plasma nitrided carbon steel: Gravimetric and electrochemical results. *Electrochim. Acta* **2009**, *54*, 2371–2378. [CrossRef]
64. Bouanis, F.Z.; Bentiss, F.; Bellayer, S.; Traisnel, M.; Vogt, J.B.; Jama, C. Radiofrequency cold plasma nitrided carbon steel: Microstructural and micromechanical characterizations. *Mater. Chem. Phys.* **2011**, *127*, 329–334. [CrossRef]
65. Kumar, R.; Chahal, S.; Kumar, S.; Lata, S.; Lgaz, H.; Salghi, R.; Jodeh, S. Corrosion inhibition performance of chromone-3-acrylic acid derivatives for low alloy steel with theoretical modeling and experimental aspects. *J. Mol. Liq.* **2017**, *243*, 439–450. [CrossRef]
66. Singh, A.; Ansari, K.R.; Lin, Y.; Quraishi, M.A; Lgaz, H.; Chung, I. Corrosion inhibition performance of imidazolidine derivatives for J55 pipeline steel in acidic oilfield formation water: Electrochemical, surface and theoretical studies. *J. Taiwan Inst. Chem. Eng.* **2018**. [CrossRef]
67. Obot, I.; Obi-Egbedi, N.; Ebenso, E.; Afolabi, A.; Oguzie, E. Experimental, quantum chemical calculations, and molecular dynamic simulations insight into the corrosion inhibition properties of 2-(6-methylpyridin-2-yl) oxazolo [5,4-f][1,10] phenanthroline on mild steel. *Res. Chem. Intermed.* **2013**, *39*, 1927–1948. [CrossRef]

68. Zhou, J.; Chen, S.; Zhang, L.; Feng, Y.; Zhai, H. Studies of protection of self-assembled films by 2-mercapto-5-methyl-1,3,4-thiadiazole on iron surface in 0.1 M H_2SO_4 solutions. *J. Electroanal. Chem.* **2008**, *612*, 257–268. [CrossRef]
69. Xie, S.-W.; Liu, Z.; Han, G.-C.; Li, W.; Liu, J.; Chen, Z. Molecular dynamics simulation of inhibition mechanism of 3,5-dibromo salicylaldehyde Schiff's base. *Comput. Theor. Chem.* **2015**, *1063*, 50–62. [CrossRef]

© 2018 by the authors. Licensee MDPI, Basel, Switzerland. This article is an open access article distributed under the terms and conditions of the Creative Commons Attribution (CC BY) license (http://creativecommons.org/licenses/by/4.0/).

Article

Construction of a Novel Lignin-Based Quaternary Ammonium Material with Excellent Corrosion Resistant Behavior and Its Application for Corrosion Protection

Chao Gao, Shoujuan Wang, Xinyu Dong, Keyin Liu, Xin Zhao * and Fangong Kong *

State Key Laboratory of Biobased Material and Green Papermaking, Key Laboratory of Pulp & Paper Science and Technology of Shandong Province/Ministry of Education, Shandong Academy of Sciences, Qilu University of Technology, Jinan 250353, China; 15864029637@163.com (C.G.); nancy5921@163.com (S.W.); d18954536225@163.com (X.D.); keyinliu@163.com (K.L.)

* Correspondence: zhaoxin_zixi@126.com (X.Z.); kfgwsj1566@163.com (F.K.); Tel.: +86-531-896-319-88 (F.K.)

Received: 11 May 2019; Accepted: 29 May 2019; Published: 31 May 2019

Abstract: A novel lignin-based quaternary ammonium material (lignin-DMC) with excellent corrosion resistant behavior was synthesized by grafting DMC (methacrylatoethyl trimethyl ammonium chloride) onto kraft lignin. The structure and anti-corrosion performance of lignin-DMC was investigated using many methods, for instance the scanning electron microscope (SEM), atomic force microscopy (AFM), charge density analysis, molecular weight analysis, electrochemical methods. Fourier-transform infrared spectroscopy (FT-IR) confirmed the formation of the lignin-DMC. The experiment results indicated that maximum corrosion inhibition efficiency (87.65%) occurred at a concentration of 75 mg/L via weight loss measurement. Polarization curves indicated that lignin-DMC was a mixed-type inhibitor with an efficient anti-corrosion performance in an acid medium. Electrochemical impedance spectroscopy (EIS) results indicated that lignin-DMC could create a shielding effectiveness and achieve a protective effectiveness in the HCl solution. Moreover, lignin-DMC displayed a physical and chemical adsorption process between 20 KJ/mol and 40 KJ/mol, which followed the Langmuir adsorption isotherm model.

Keywords: kraft lignin; polymer; corrosion resistant; iron; acid inhibition

1. Introduction

During daily production, metallic materials usually react with the contacted medium (air, climate, salt spray, solution, etc.), causing the metal products to suffer from various degrees of corrosion, especially the carbon steel, it is a mild and commonly used material. In an aggressive acidic solution, carbon steel will be destroyed, which leads to significant economic losses [1,2]. Many techniques, such as the cathodic protection method and electroplates, coating processes, have been applied to protect metal bases from corrosion [3–5]. Among all of these techniques, the inhibitors is one of the most useful ways [6–12]. Corrosion inhibitors are a class of substances that can effectively inhibit metal corrosion and protect metal materials by adding another small amount of material. However, compared with other methods, the use of corrosion inhibitors has the following advantages: First, it does not substantially change the corrosive environment, secondly, it does not substantially increase equipment investment. Third, the corrosion inhibition effect is not affected by the shape of the equipment and the same formulation can sometimes, at the same time, prevent corrosion of various metals in different environments. [13–15]. In general, inhibitors contain N, O, S and other heteroatoms that can be easily adsorbed onto the surface of iron and that then compete with the corrosive ions to reduce the contact between the corrosive substances and active sites to attain the effect of corrosion inhibition [16–24]. Since the inhibitor is less costly and has strong applicability, it is generally used in petroleum exploration, chemical cleaning, storage and the transportation of metal products. With improvements in the awareness of environmental protection, new requirements have also been put forward for the development and application of corrosion inhibitors. Researching and developing biological inhibitors that do not pose a destructive effect on the environment is the development direction of the future [25,26]. The corrosion inhibitors from natural plants and animals have the advantages of nontoxicity, low pollution, and low cost [27–29]. Therefore, natural substance could be seen as a promising source of corrosion inhibitor [30–32]. In the last few years, several naturally occurring polymers have been used as anti-corrosion in a corrosive environment, including starch, chitosan, cellulose, etc. [30,33–36]. Due to the existence of functional groups in natural polymers, the iron surface can be protected from corrosion.

Lignin is a byproduct of the pulping process. It is a threat to the environment if not handled properly. Lignin seems to have several current and potential applications in many fields [37,38], however, the application as an anti-corrosion material was rarely reported. As corrosion inhibitors, the main groups of lignin that are responsible in inhibiting corrosion are the hydroxyl, methyl and carboxyl groups, which can adsorb onto iron or other metal surfaces to slow down the corrosion rate [39]. However, due to the insolubility of lignin, the anti-corrosion behavior of lignin is always low. Therefore, it needs to be modified by chemicals to improve its the water-solubility and to also make it meet application requirements. In one study, acrylamide (AM) was grafted onto lignin to research its anti-corrosion performance [35]. Its inhibition efficiency was found to be 77.10% at a 100 mg/L concentration. In other study, Hussin and Rahim produced a successful inhibitor by 2-naphthol and 1,8-dihydroxyanthraquinone grafted with kraft lignin and the resulting efficiency was 60.77% when the concentration was 100 mg/L [40]. However, little work has been conducted in studying the corrosion inhibition of metal by lignin that has been chemically modified with methacrylatoethyl trimethyl ammonium chloride (DMC).

In this paper, kraft lignin was copolymerized with a cationic monomer, DMC, to prepare the lignin-DMC copolymer as a corrosion inhibitor. The inhabitation performance of the lignin-DMC inhibitor was investigated in detail by the weight-loss method, electrochemical, SEM, AFM.

2. Materials and Experimental Method

2.1. Steel Specimen Preparation

The chemical composition of the iron-based materials was C 3.05%, Al 1.45%, Mg 0.39%, Si 0.49%, and balance Fe. The specimens were first burnished using various grades of emery papers from #60 to #500 and then washed, degreased with acetone and dried at room temperature.

2.2. Surface Analysis

The morphology of the iron-based materials specimen before and after adding corrosion inhibitor into the HCl solution were measured via SEM (FEI NOVA NANO SEM 450, FEI company, Hillsboro, OH, USA) and AFM (Multimode 8, Bruker, Santa Barbara, CA, USA). The iron-based materials of 5.0 cm × 2.0 cm × 1.0 mm were polished and immersed in 1.0 mol/L HCl solutions with and without 75 mg/L of the lignin-DMC inhibitor at 25 °C. Afterwards, the specimens were taken out of the solution and rinsed with deionized water. The washed specimens were dried at 105 °C for 12 h.

2.3. Weight Loss Method

The weight loss of iron-based materials were evaluated after 2 h of immersion in 100 mL of 1 mol/L HCl solutions with and without different concentrations of the lignin-DMC at room temperature. The samples were taken out, washed, dried and accurately reweighed. The inhibition efficiency was calculated according to Equation (1) [41]

$$IE\ (\%) = \frac{(W_0 - W')}{W_0} \times 100 \quad (1)$$

The difference between W_0 and W' represents the weight loss of the specimen at various concentrations of inhibitor. The corrosion rate (CR) can be calculated using Equation (2) [42]

$$CR\ (mg/(cm^2{\cdot}h)) = \frac{\Delta W}{At} \quad (2)$$

where ΔW is the weight loss, W_0–W, mg; A is the exposed area of the specimen, cm^2; and t is the exposed time, h [43].

2.4. Electrochemical Method

For the electrochemical method, we started at a usual three-electrode cell. A platinum filament was used as a counter electrode (PE), a saturated calomel electrode (SCE) was the reference electrode (RE). The iron-based material was deemed as the working electrode (WE) with an uncovered area of 1.00 cm^2. The electrochemical impedance spectroscopy (EIS) was investigated by using the IVIUM electrochemical workstation (IviumTechnologies BV Co., Eindhoven, Noord-Brabant, Netherlands), made in the Netherlands. The process used for the polarization method is similar to other articles. At first, the work electrode was submerged in an HCl solution for 1 h to achieve a stable state to determine the OCP (Open Circuit Potential) before measurement. The scanning electric potential was ± 250 mV (compared to open circuit potential) OCP (compared to SCE), and the scanning frequency rate was 30 mV/s. All electrochemical parameters were obtained by Tafel lines. The IE (Inhibition Efficiency) results were based on Equation (3).

$$IE\ (\%) = \frac{(i\,corr - i'corr)}{icorr} \times 100 \quad (3)$$

where icorr and i′corr stand for the corrosion current densities absent and with an inhibitor, separately.

EIS was measured at OCP and a frequency range of 0.01–100 k Hz, the amplitude of the AC (Alternating Current) signal was 10 mv. The date of the impedance data were found in the Nyquist plots by calculating the difference in the values of the impedance at different frequencies [44]. Then, the inhibition efficiency can be calculated by formula (4).

$$IE\ (\%) = \frac{(Rct - R'ct)}{R'ct} \times 100 \quad (4)$$

where Rct and R′ct denote the charge transfer resistance values with and without different concentrations of the inhibitor, respectively.

3. Results and Discussion

3.1. Surface Analysis

The synthesis and characterization of lignin-DMC was presented in supporting information. The synthesis process was shown in Figures S1 and S2, briefly, 2.0 g of lignin were dissolved in water in a 250 mL three-neck glass flask. DMC (lignin and DMC chloride molar ratio of 1:1.6) was added to the solution and the pH of the medium was adjusted to 4–5. $K_2S_2O_8$ (0.03 g) was then dissolved in 5 mL of deionized water and added drop wisely to the suspension. Then, the polymerization was conducted at 80 °C for 3 h, after completion, the polymerization's solution was further purified by precipitation with 200 mL of ethanol and centrifuging. The lignin-DMC polymer contains the characteristic absorption peak of lignin and DMC, confirming that the lignin-DMC was successfully synthesized (Figure S3). In Tables S1 and S2, the molecular weight of the lignin-DMC polymer was significantly higher than that of kraft lignin, besides, the nitrogen contents of the lignin-DMC polymer were higher than that of kraft lignin, which were in accordance with the FT-IR analysis (Figure S3). Figure 1 presents the SEM images of the samples with and without corrosion inhibitor by immersion in 1.0 mol/L HCl medium for 2 h. Samples without a corrosion inhibitor showed a rough and uneven state (Figure 1a), indicating that the iron was seriously corroded. However, the surfaces of the materials immersed in the HCl medium containing lignin-DMC were relatively flat (Figure 1b), showing that the corrosion degree was greatly reduced and that the lignin-DMC displayed a significant inhibitory effect in HCl. In addition, micrographs of steel without HCl and Fe after submersion in the 100 mg/L lignin-DMC + 1.0 mol/L HCl was shown in Figure S4.

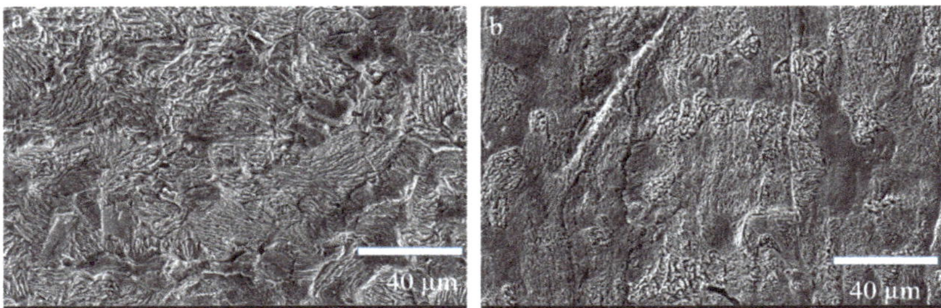

Figure 1. SEM micrographs of the Fe after submersion in 1.0 mol/L (**a**) HCl and 75 mg/L lignin-based quaternary ammonium material (lignin-DMC) + 1.0 mol/L HCl (**b**).

For research, the roughness of the material surface clearly, and the AFM tests of the samples without and with the inhibitor by immersion in 1.0 mol/L HCl medium for 0.5 h were shown in Figure 2. The sample without the inhibitor, the surface of Fe shows rough and inhomogeneity. The average roughness (Ra) for the samples without a corrosion inhibitor is 81.9 nm. However, lignin-DMC displayed that the surface was smooth. This roughness has been reduced to 44 nm with the inhibitor. This result shows that the inhibitor plays a role in anticorrosion behavior.

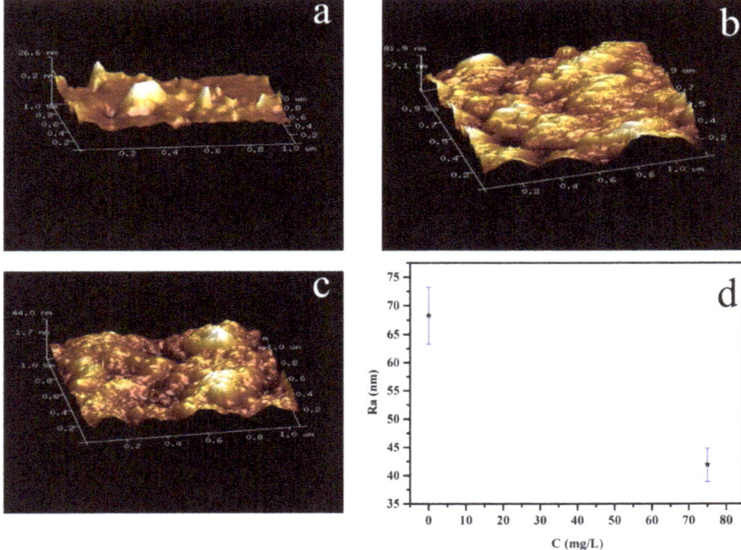

Figure 2. Images of the Fe surface (**a**), Fe after submersion in 1.0 mol/L (**b**) HCl, 75 mg/L lignin-DMC + 1.0 mol/L HCl (**c**) and (**d**) the average roughness (Ra) of the Fe surface in (**b**,**c**).

3.2. Weight Loss Experiment

The weight loss measurement, which is an easy and convenient method, was used first for initially testing the anti-corrosion performance of the inhibitor. Table 1 shows the CR of iron in a 1.0 mol/L HCl at 25 °C and the inhibition rate ($\eta_{IE)}$ of different concentrations of the lignin-DMC inhibitor (c) (weight loss test). For the blank solution, absent of the corrosion inhibitor, a large amount of H_2 was produced when the iron was exposed to the HCl, and the corrosion rate was as high as 2.65 mg/ (cm^2·h). However, the corrosion rate decreased immediately and showed a tendency of first decreasing and then increasing after adding the lignin-DMC. When the concentration was 75 mg/L, the corrosion rate was the lowest and decreased to 0.33 mg/(cm^2·h). On the other hand, the corrosion inhibition rate expressed the opposite trend. At 75 mg/L, the corrosion inhibition rate was the highest at 87.65% and at 50 mg/L was the lowest at 70.39%. These results exhibited that the lignin-DMC can be an efficient inhibitor in an HCl medium.

Table 1. Corrosion parameters and IE (Inhibition Efficiency) obtained from weight loss of iron-based materials in a 1.0 mol/L HCl at different concentrations of lignin-DMC at 25 °C.

Concentration (ppm)	CR (mg /(cm^2·h))	IE (%)
0	2.65	0.00
50	0.80	69.81%
75	0.33	87.54%
100	0.37	86.04%
125	0.71	73.21%

3.3. Electrochemical Measurements

3.3.1. Open Circuit Potential

Figure 3 shows the *OCP* immersion time (t) at 25 °C for iron in HCl with and without various conventions of lignin-DMC. It can be seen that the open circuit potential gradually increases over time and gradually stabilizes. A blank HCl solution without a corrosion inhibitor began to reach a relatively stable state at 1000 s. However, the HCl solution containing 75 mg /L lignin-DMC reached a final equilibrium state at after 2000 s. The potential polarization curves and electrochemical impedance spectroscopy (EIS) test achieved a stable state after an immersion time of 1 h.

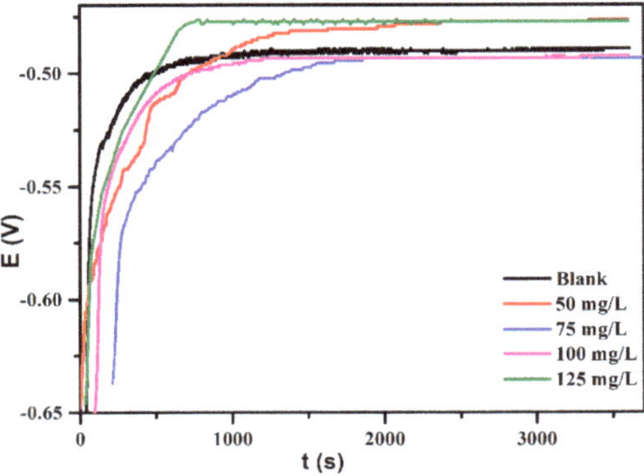

Figure 3. OCP-time curves for iron with and without lignin-DMC at 25 °C.

3.3.2. Polarization Studies

The IVIUM electrochemical workstation was used to determine the polarization curves at several different concentration of lignin-DMC. In the Tafel curves, the vertical axis is the logarithm of the corrosion current density and the horizontal axis is the corrosion potential. The results are shown in Figure 4. The polarization curves generated by the electrochemical parameters values, including the corrosion current density (Icorr), corrosion potential (Ecorr), the anodic Tafel slope (ba), the cathodic Tafel slope (bc), and the inhibit efficiency in 1.0 mol/L of HCl with and without various concentrations of lignin-DMC are presented in Table 2. The data in Table 2 indicates that at a concentration of 75 mg/L lignin-DMC the corrosion current density was at its relative lowest, which is a satisfactory agreement with the weight loss measurement. The addition of the inhibitor molecules at different concentrations did not cause a significant change in the cathode and anode curves, and the polarization curves at each concentration were substantially parallel to the Tafel curve of the blank solution. It indicates that the addition of the compound does not change the metal anode dissolution and the cathode hydrogen evolution reaction. The corrosion inhibition mechanism only inhibits the active point of the reaction by forming a protective film on the surface of the carbon steel. With various concentrations of the inhibitor, the Tafel slope of the anode of the polarization curve increased from 94 mV to 154 mV, and the Tafel slope of the cathode increased from 114 mV to 185 mV, indicating that a certain inhibitory effect of the corrosion inhibitor occurred on the iron of the anode and cathode reactions, but the cathode Tafel slope increased more rapidly than the anode Tafel slope did, further proving that the lignin-DMC belongs to an inhibition mixed-type inhibitor. The displacement in the Ecorr value is less than 85 mV, which is also evidence for the lignin-DMC acting as a mixed type inhibitor [32].

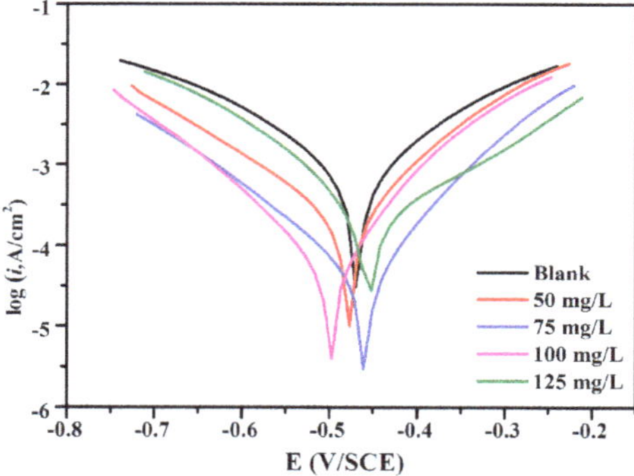

Figure 4. Polarization curves in a 1.0 mol/L HCl solution containing various inhibitor concentrations at 25 °C.

Table 2. Different electrochemical parameters of iron sheet in 1.0 mol/L HCl solutions at various lignin-DMC concentrations.

C (mg/L)	Ecorr (mV/SCE)	icorr (uA/cm^2)	ba (mV)	bc (mV)	IE (%)
0	−461.80	891	154	185	0.00
50	−427.60	246	164	148	72.39%
75	−460.30	45	94	114	94.95%
100	−508.40	78	104	125	91.25%
125	−488.30	231	122	147	74.07%

3.3.3. EIS Studies

Electrochemical impedance spectroscopic (EIS) method allows for the assessment of the performance of adsorbed inhibitor film against the corrosion kinetics [21]. Electrochemical impedance spectroscopy can quickly evaluate the corrosion resistance and obtain the electrochemical information of the metal under the film that was formed by lignin-DMC adsorption on the surface of iron. High-frequency is corresponding to the layer information and the high-frequency capacitive reactance further responses to the dielectric properties and shielding performance of the film layer. A low-frequency end can give information on the metal/solution interface reaction, and the size of the low-frequency capacitance represents the metal charge transfer resistance during corrosion. The Nyquist and Bode diagrams in the HCl and solutions contained in the lignin-DMC are shown in Figure 5. The impedance spectrum is mainly composed of a single high-frequency end-capacity arc, which is characterized by only one time constant. The single capacitive reactance arc change indicates that the electrode surface corrosion process is mainly controlled by the charge transfer step. The semicircle size increased after adding the inhibitors, which was due to the charge transfer [45]. With an increase in the lignin-DMC concentration, the size increased [46,47]. Increase in Rct values with composite concentration within the acid electrolyte is consistent with the formation of polymer film on the metal surface due to charge transfer action. At a concentration of 75 mg/L, the semicircles size showed their largest value.

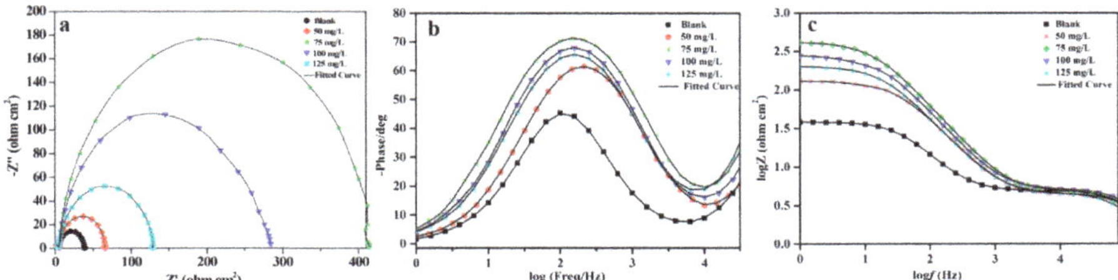

Figure 5. (**a**) Nyquist plots, (**b**) bode plots and phase angle plots (**c**) of steel in an HCl solution containing different inhibitor concentrations.

The interface layer between the electrode and the solution, also known as the electric double layer, is generally represented by an equivalent capacitance. The equivalent circuit shown in Figure 6 was used to simulate the impedance data with the values shown in Table 3, where R_s acts as the charge solution resistance. The blank solution and 50 mg/L fitted (1), and the rest fitted (2). R_c acts as the film resistance. In addition, R_{ct} represents the charge transfer resistance, C_c is the film capacitance of the double layer, and Cdl are the electric double layer capacitors. At a high frequency the membrane resistance became larger and the membrane capacitance decreased, indicating that after the addition of the lignin-DMC in the HCl solution the lignin-DMC molecules removed the water molecules, which were originally adsorbed onto the iron-based material, thereby creating a shielding effectiveness and arrived at a protective effectiveness. As the radius of the capacitive arc of the low frequency increased, the charge transfer value R_{ct} became larger, and the charge transfer capacitance decreased. The increase in R_{ct} was due to the considerable surface coverage by the inhibitor molecules on the metal surface through bonding. The decrease in Cdl can be explained by a decrease in the local dielectric constant and/or an increase in the thickness of the electrical double layer, which indicates the adsorption of the inhibitor [48]. Hence, the lignin-DMC had an excellent inhibitory effectiveness on the iron corrosion in acidic media.

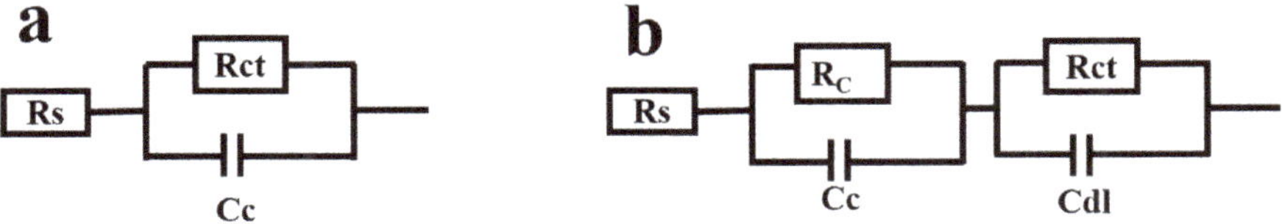

Figure 6. The electrical equivalent circuit of the capacitive loop for electrochemical impedance spectroscopy (EIS) without inhibitor (**a**) and the electrical equivalent circuit in different concentration of inhibitors at 25 °C (**b**).

Table 3. Fitted EIS results of iron corroding in 1 mol/L HCl solutions at different lignin-DMC.

Concentration(mg/L)	R_s(ohm)	C_c(uF)	R_c(ohm)	R_{ct}(ohm)	Cdl(uF)	IE(%)
0	5.729	–	–	29.60	254.60	–
50	4.701	–	–	59.74	240.60	50.45%
75	5.508	91.19	178.90	227.00	34.09	86.96%
100	4.760	180.70	96.97	101.00	45.91	70.69%
125	4.780	231.90	50.32	74.24	42.70	60.12%

3.4. Effect of Temperature

Adsorption thermodynamics and adsorption kinetics are effective research methods for determining the corrosion inhibitor adsorption behavior. To attain a better performance of the lignin-DMC of adsorption and activation processes, the effect of the temperature was studied. By studying the adsorption isotherm model and the corresponding formula, the corresponding adsorption parameters, such as the adsorption equilibrium constant and adsorption free energy, can be calculated to investigate the adsorption mechanism of this corrosion inhibitor. The values of the surface coverage (θ) are defined and calculated as IE/100 from the weight loss measurements using the following equation [41]: Several isothermal adsorption methods were matched by the experimental results, following the Langmuir adsorption isotherm model. The adsorption isotherm is as follows and the plot is shown in Figure 7.

$$\frac{C}{\theta} = \frac{1}{K} + C \tag{5}$$

θ represents the surface coverage. K is the equilibrium constant and C is the concentration of the inhibitor, mg/ L.

The inhibitor molecules gradually adsorb onto the metal surface by changing the adsorbed water molecules. The free energy of the adsorption ΔG was calculated based on Equation (6):

$$K = \frac{1}{55.5} \exp\left(\frac{-\Delta G}{RT}\right) \tag{6}$$

where ΔG is the free energy of adsorption, K is the adsorption–desorption equilibrium constant, R is the universal gas constant, and T is absolute temperature in Kelvin, K.

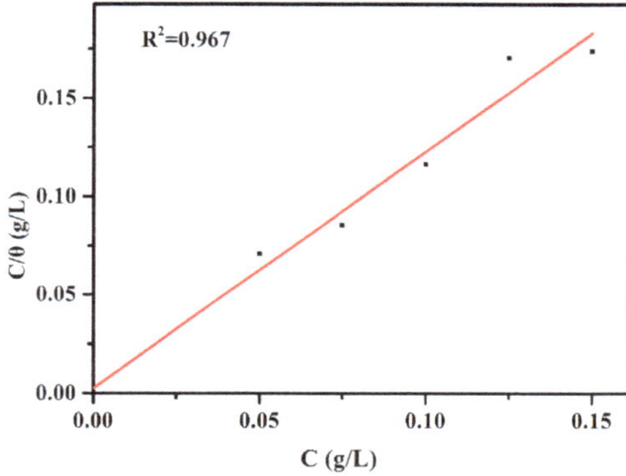

Figure 7. Langmuir adsorption model of iron in 1 mol/L HCl solution at room temperature.

The Langmuir isotherm adsorption model of lignin-DMC in 1 mol/L HCl at 25 °C is shown in Figure 7. The values of ΔG were −27.51 kJ/mol, −28.42 kJ/mol and −29.33 kJ/ mol at temperatures of 25, 35 and 45 °C, respectively, with the inhibitor at 75 mg/L. Negative values indicate that the adsorption was spontaneous. If the absolute value of G is approximately 20 kJ/mol, the adsorption mode then belongs to the physical adsorption. Additionally, the adsorption mode follows chemical adsorption if the absolute value of ΔG is approximately 40 KJ/mol. Physical adsorption and chemical adsorption (mixed adsorption) would occur if the values are between 20 to 40 kJ/mol [6]. Hence, the adsorption mode of lignin-DMC was physical adsorption and chemical adsorption.

The relationship between the temperature and corrosion rate is expressed by the following equation and its alternative formulation, called the transition state equation, was applied to determine the activation entropy (ΔS) and activation enthalpy (ΔH)

$$\ln(\mathrm{CR}) = \mathrm{A}\ \exp\ \frac{-\mathrm{E_a}}{\mathrm{RT}} \tag{7}$$

The transition state equation is as follows:

$$\mathrm{CR} = \frac{\mathrm{RT}}{\mathrm{Nh}} \exp \frac{\Delta \mathrm{S}}{\mathrm{R}} \exp \ \frac{-\Delta \mathrm{H}}{\mathrm{RT}} \tag{8}$$

where CR is the corrosion rate, A is the Arrhenius pre-exponential constant, R is the universal gas constant (8.314 J/mol/K), and h is Plank's constant (6.63 $\times 10^{34}$ J·s) [48].

The plot of CR/T vs 1/T is expressed in Figure 8, and the values of the activation enthalpy (ΔH) and the activation entropy (ΔS) were obtained by the following equation:

$$\Delta \mathrm{H} = -\mathrm{slope} \times \mathrm{R\ and\ } \Delta \mathrm{S} = (\mathrm{intercept} - \ln\left[\frac{\mathrm{R}}{\mathrm{Nh}}\right])\cdot \mathrm{R} \tag{9}$$

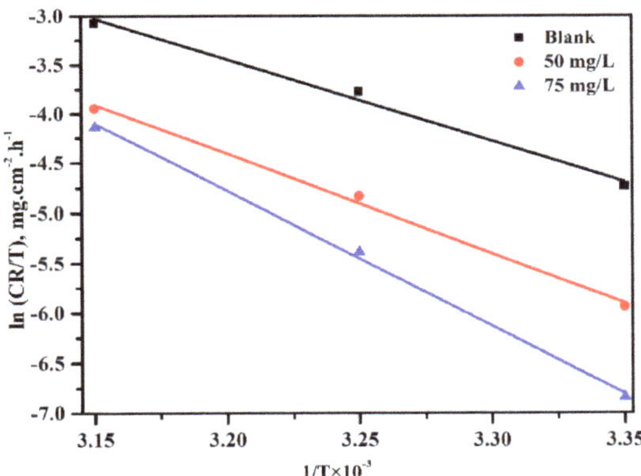

Figure 8. Arrhenius plots of the corrosion rate (CR/T) in an HCl solution in the absence and presence of Lignin-DMC.

The values of ΔH were 69.29 KJ/mol, 82.93 KJ/mol and 112.4 KJ/mol at different concentrations, which indicated it was an endothermic process. The values of ΔS were 146.5 KJ/mol 182.1 KJ/mol and 273.5 KJ/mol, respectively. The value increased (more positive) in the presence of the lignin-DMC compared to the blank solution, which illustrated that the system passes from a less orderly to a more random arrangement [49]. The results may be explained by the adsorption of organic inhibitor molecules and considered to be a quasi-substitution process between the organic inhibitor molecules in the aqueous phase and the water molecules on the surface of low carbon steel. The adsorption process of the inhibitors was the displacement reaction of the adsorption water molecules removed from the metal surface [50].

$$\mathrm{Org(sol)} + n\mathrm{H_2O\ (ads)} \rightarrow \mathrm{Org(ads)} + n\mathrm{H_2O} \tag{10}$$

Org(sol) and Org(ads) are organic molecules that are dissolved in the solution and adsorbed onto the iron surfaces. H_2O (ads) are the water molecules on the metal's surface, where n is the factor that indicates the substitution of the water molecules by the inhibitor units [39].

3.5. Mechanism of the Corrosion Inhibitor

Lignin-DMC is composed of polar groups consisting mainly of N and O atoms with higher electronegativity and nonpolar groups consisting of C and H non-polar atoms, which can be adsorbed onto the steel surface to change electric double layer structure of the metal and improve the activation energy of the metal ionization process. Nonpolar groups move away from the metal surface to form a layer of hydrophobic film, impeding the transfer of charges and substances and thus greatly reducing the metal corrosion rate. The mechanism of the corrosion inhibitor is shown in Figure 9. In HCl solutions, the unshared electron pair at the central atom of the organic corrosion inhibitor formed an onium ion with a hydrogen ion in HCl solutions. Under the effect of the electrostatic attraction, the onium iron was adsorbed onto the cathode region of the metal surface (Cl^-), making the metal surface seem positive. The onium irons then began to compete with H^+ [16], so the H^+ ions in the acid solution have difficultly moving close to the metal, which greatly slows down the corrosion rate. Another adsorption mode was chemisorption. In this adsorption, the central atoms of the polar group of the organic corrosion inhibitor molecules, N and O, have unshared electron pairs. When the metal has an empty orbit, the lone pair of electrons at the central atoms of the polar groups may combine with the empty orbitals to form coordination bonds, so the lignin-DMC are adsorbed onto the steel surface to slow down the corrosion rate.

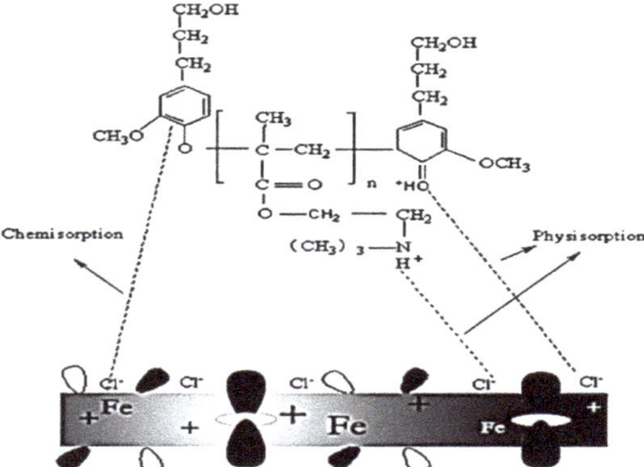

Figure 9. Schematic representation of lignin-DMC with the metal surface in 1 mol/L HCl.

4. Conclusions

Cationic lignin-based polymer (Lignin-DMC) was synthesized by copolymizing DMC with kraft lignin. The prepared lignin-DMC was an effective corrosion inhibitor in the 1 mol/L HCl solution. The corrosion inhibition effect was best at 75 mg/L at room temperature, reaching a state at 87.65%. Meanwhile, the results revealed that the proper dosage may be used as an inhibitor. However, a superfluous dosage had a negative effect on the anticorrosion performance. Additionally, the corrosion inhibitor-lignin-DMC was a mixed corrosion inhibitor. The adsorption mode of lignin-DMC was the Langmuir adsorption isotherm. The mechanism of adsorption was found to be a physical and chemical model in nature. The adsorption model indicated that the corrosion inhibitor was an adsorption type corrosion inhibitor

Supplementary Materials: The following are available online at http://www.mdpi.com/1996-1944/12/11/1776/s1, Figure S1. Schematic diagram of a lignin-DMC synthesis device, Figure S2. (a) The reaction mechanism and (b) The sketch of the lignin grafting reaction, Figure S3. FT-IR spectra of kraft lignin and lignin-DMC polymer, Figure S4. SEM micrographs of the Fe surface without 1.0 mol/L HCl (a), Fe surface after submersion in 1.0 mol/L HCl (b), 1.0 mol/L HCl with the addition of 75 mg/L lignin-DMC (c), and (d) Fe immersed in 1.0 mol/L HCl + 100 mg/L Lignin-DMC solutions, Table S1. Molecular weight and charge density of lignin, and lignin-DMC polymer, Table S2. Elemental analysis of lignin and lignin-DMC polymer.

Author Contributions: S.W., X.Z., F.K. conceived and designed the experiments; C.G., X.D. performed the experiments, C.G., K.L. wrote the main manuscript.

Funding: This research was funded by the National Key R&D Program of China (Grant No. 2017YFB0308000), the National Natural Science Foundation of China (Grant No. 31570566, 31500489, 31800499), the Key Research and Development Program of Shandong Province (Grant No. 2017GSF17130) and the Foundation of Guangxi Key Laboratory of Clean Pulp & Papermaking and Pollution Control of China (Grant No. KF201717).

Acknowledgments: We acknowledge support from the National Key R&D Program of China, the National Natural Science Foundation of China, the Key Research and Development Program of Shandong Province and the Foundation of Guangxi Key Laboratory of Clean Pulp & Papermaking and Pollution Control of China (KF201717).

Conflicts of Interest: The authors declare no conflict of interest.

References

1. Khadiri, A.; Saddik, R.; Bekkouche, K.; Aouniti, A.; Hammouti, B.; Benchat, N.; Bouachrine, M.; Solmaz, R. Gravimetric, electrochemical and quantum chemical studies of some pyridazine derivatives as corrosion inhibitors for mild steel in 1 M HCl solution. *J. Taiwan. Inst. Chem. Eng.* **2016**, *58*, 552–564. [CrossRef]
2. Kowsari, E.; Arman, S.Y.; Shahini, M.H.; Zandi, H.; Ehsani, A.; Naderi, R.; PourghasemiHanza, A.; Mehdipour, M. In situ synthesis, electrochemical and quantum chemical analysis of an amino acid-derived ionic liquid inhibitor for corrosion protection of mild steel in 1 M HCl solution. *Corros. Sci.* **2016**, *112*, 73–85. [CrossRef]
3. Baux, J.; Caussé, N.; Esvan, J.; Delaunay, S.; Tireau, J.; Roy, M.; You, D.; Pébère, N. Impedance analysis of film-forming amines for the corrosion protection of a carbon steel. *Electrochim. Acta* **2018**, *283*, 699–707. [CrossRef]
4. Fotovvati, B.; Namdari, N.; Dehghanghadikolaei, A. On coating techniques for surface protection: A review. *J. Mater. Process.* **2019**, *3*, 28. [CrossRef]
5. Ulaeto, S.B.; Pancrecious, J.K.; Rajan, T.P.D.; Pai, B.C. *Noble Metal-Metal Oxide Hybrid Nanoparticles*; Woodhead Publishing: Cambridge, UK, 2019; pp. 341–372.
6. Abu-Dalo, M.A.; Al-Rawashdeh, N.A.; Mutlaq, A.A. Green approachto corrosion inhibition of mild steel by ligninsulfonatein acidic media. *J. Iron. Steel. Res. Int.* **2016**, *23*, 722–732. [CrossRef]
7. Wang, J.; Zhang, W.; Zheng, Y.; Zhang, N.; Zhang, C. Multi-functionalization of magnetic graphene by surface-initiated icar atrp mediated by polydopamine chemistry for adsorption and speciation of arsenic. *Appl. Surf. Sci.* **2019**, *478*, 15–25. [CrossRef]
8. Maharana, H.S.; Katiyar, P.K.; Mondal, K. Structure dependent super-hydrophobic and corrosion resistant behavior of electrodeposited ni-mose2-mwcnt coating. *Appl. Surf. Sci.* **2019**, *478*, 26–37. [CrossRef]
9. Feizi Mohazzab, B.; Jaleh, B.; Kakuee, O.; Fattah-alhosseini, A. Formation of titanium carbide on the titanium surface using laser ablation in N-heptane and investigating its corrosion resistance. *Appl. Surf. Sci.* **2019**, *478*, 623–635. [CrossRef]
10. Noor, E.A. The inhibition of mild steel corrosion in phosphoric acid solutions by some N-heterocyclic compounds in the salt form. *Corros. Sci.* **2005**, *47*, 33–55. [CrossRef]
11. Popova, A.; Christov, M.; Raicheva, S.; Sokolova, E. Adsorption and inhibitive properties of benzimidazole derivatives in acid mild steel corrosion. *Corros. Sci.* **2004**, *46*, 1333–1350. [CrossRef]
12. Wang, H.-L.; Liu, R.-B.; Xin, J. Inhibiting effects of some mercapto-triazole derivatives on the corrosion of mild steel in 1.0 M HCl medium. *Corros. Sci.* **2004**, *46*, 2455–2466. [CrossRef]
13. Gan, T.; Zhang, Y.; Yang, M.; Hu, H.; Huang, Z.; Feng, Z.; Chen, D.; Chen, C.; Liang, J. Synthesis, characterization, and application of a multifunctional cellulose derivative as an environmentally friendly corrosion and scale inhibitor in simulated cooling water systems. *J. Taiwan. Inst. Chem. Eng.* **2018**, *57*, 10786–10797. [CrossRef]

14. Dong, S.; Yuan, X.; Chen, S.; Zhang, L.; Huang, T. A novel hpei-based hyperbranched scale and corrosion inhibitor: Construction, performance, and inhibition mechanism. *Ind. Eng. Chem. Res.* **2018**, *57*, 13952–13961. [CrossRef]
15. Umoren, S.A.; Banera, M.J.; Alonso-Garcia, T.; Gervasi, C.A.; Mirífico, M.V. Inhibition of mild steel corrosion in HCl solution using chitosan. *Cellulose* **2013**, *20*, 2529–2545. [CrossRef]
16. Jokar, M.; Farahani, T.S.; Ramezanzadeh, B. Electrochemical and surface characterizations of morus alba pendula leaves extract (maple) as a green corrosion inhibitor for steel in 1 M HCl. *J. Taiwan. Inst. Chem. Eng.* **2016**, *63*, 436–452. [CrossRef]
17. Odewunmi, N.A.; Umoren, S.A.; Gasem, Z.M.; Ganiyu, S.A.; Muhammad, Q. L-citrulline: An active corrosion inhibitor component of watermelon rind extract for mild steel in HCl medium. *J. Taiwan. Inst. Chem. Eng.* **2015**, *51*, 177–185. [CrossRef]
18. Li, X.; Deng, S.; Lin, T.; Xie, X.; Du, G. Cassava starch-sodium allylsulfonate-acryl amide graft copolymer as an effective inhibitor of aluminum corrosion in HCl solution. *J. Taiwan. Inst. Chem. Eng.* **2018**, *86*, 252–269. [CrossRef]
19. Yan, R.; He, W.; Zhai, T.; Ma, H. Anticorrosion organic–inorganic hybrid films constructed on iron substrates using self-assembled polyacrylic acid as a functional bottom layer. *Electrochim. Acta* **2019**, *295*, 942–955. [CrossRef]
20. Elayyachy, M.; El Idrissi, A.; Hammouti, B. New thio-compounds as corrosion inhibitor for steel in 1 M HCl. *Corros. Sci.* **2006**, *48*, 2470–2479. [CrossRef]
21. Eduok, U.; Ohaeri, E.; Szpunar, J. Electrochemical and surface analyses of x70 steel corrosion in simulated acid pickling medium: Effect of poly (N-vinyl imidazole) grafted carboxymethyl chitosan additive. *Electrochim. Acta* **2018**, *278*, 302–312. [CrossRef]
22. Emregül, K.C.; Hayvalí, M. Studies on the effect of a newly synthesized schiff base compound from phenazone and vanillin on the corrosion of steel in 2 M HCl. *Corros. Sci.* **2006**, *48*, 797–812. [CrossRef]
23. Moretti, G.; Guidi, F.; Fabris, F. Corrosion inhibition of the mild steel in 0.5 M HCl by 2-butyl-hexahydropyrrolo [1,2-b][1,2]oxazole. *Corros. Sci.* **2013**, *76*, 206–218. [CrossRef]
24. Asadi, N.; Ramezanzadeh, M.; Bahlakeh, G.; Ramezanzadeh, B. Utilizing lemon balm extract as an effective green corrosion inhibitor for mild steel in 1 M HCl solution: A detailed experimental, molecular dynamics, monte carlo and quantum mechanics study. *J. Taiwan. Inst. Chem. Eng.* **2019**, *95*, 252–272. [CrossRef]
25. Babic-Samardzija, K.; Lupu, C.; Hackerman, N.; Barron, A.R. Inhibitive properties, adsorption and surface study of butyn-1-ol and pentyn-1-ol alcohols as corrosion inhibitors for iron in HCl. *J. Mater. Chem.* **2005**, *15*, 1908–1916. [CrossRef]
26. Srivastava, V.; Chauhan, D.S.; Joshi, P.G.; Maruthapandian, V.; Sorour, A.A.; Quraishi, M.A. Peg-functionalized chitosan: A biological macromolecule as a novel corrosion inhibitor. *ChemistrySelect.* **2018**, *3*, 1990–1998. [CrossRef]
27. Haque, J.; Srivastava, V.; Verma, C.; Lgaz, H.; Salghi, R.; Quraishi, M.A. N-methyl-N,N,N-trioctylammonium chloride as a novel and green corrosion inhibitor for mild steel in an acid chloride medium: Electrochemical, dft and md studies. *New. J. Chem.* **2017**, *41*, 13647–13662. [CrossRef]
28. Rahim, A.A.; Rocca, E.; Steinmetz, J.; Jain Kassim, M. Inhibitive action of mangrove tannins and phosphoric acid on pre-rusted steel via electrochemical methods. *Corros. Sci.* **2008**, *50*, 1546–1550. [CrossRef]
29. Ochoa, N.; Bello, M.; Sancristóbal, J.; Balsamo, V.; Albornoz, A.; Brito, J.L. Modified cassava starches as potential corrosion inhibitors for sustainable development. *Mater. Res.* **2013**, *16*, 1209–1219. [CrossRef]
30. Menaka, R.; Subhashini, S. Chitosan schiff base as eco-friendly inhibitor for mild steel corrosion in 1 M HCl. *J. Adhes. Sci. Techonl.* **2016**, *30*, 1622–1640. [CrossRef]
31. Abiola, O.K.; James, A.O. The effects of aloe vera extract on corrosion and kinetics of corrosion process of zinc in HCl solution. *Corros. Sci.* **2010**, *52*, 661–664. [CrossRef]
32. Garai, S.; Garai, S.; Jaisankar, P.; Singh, J.K.; Elango, A. A comprehensive study on crude methanolic extract of artemisia pallens (asteraceae) and its active component as effective corrosion inhibitors of mild steel in acid solution. *Corros. Sci.* **2012**, *60*, 193–204. [CrossRef]
33. Li, X.; Deng, S. Cassava starch graft copolymer as an eco-friendly corrosion inhibitor for steel in H_2SO_4 solution. *Korean. J. Chem. Eng.* **2015**, *32*, 2347–2354. [CrossRef]
34. Kalaiselvi, P.; Chellammal, S.; Palanichamy, S.; Subramanian, G. Artemisia pallens as corrosion inhibitor for mild steel in HCl medium. *Mater. Chem. Phys.* **2010**, *120*, 643–648. [CrossRef]

35. Eddy, N.O.; Ebenso, E.E.; Ibok, U.J. Adsorption, synergistic inhibitive effect and quantum chemical studies of ampicillin (amp) and halides for the corrosion of mild steel in H_2SO_4. *J. Appl. Electrochem.* **2009**, *40*, 445–456. [CrossRef]
36. Sangeetha, Y.; Meenakshi, S.; Sairam Sundaram, C. Corrosion inhibition of aminated hydroxyl ethyl cellulose on mild steel in acidic condition. *Carbohyd. Polym.* **2016**, *150*, 13–20. [CrossRef]
37. Ren, Y.; Luo, Y.; Zhang, K.; Zhu, G.; Tan, X. Lignin terpolymer for corrosion inhibition of mild steel in 10% hydrochloric acid medium. *Corros. Sci.* **2008**, *50*, 3147–3153. [CrossRef]
38. Stewart, D. Lignin as a base material for materials applications: Chemistry, application and economics. *Carbohyd. Polym.* **2008**, *27*, 202–207. [CrossRef]
39. Fox, S.C.; McDonald, A.G. Chenical and thermal characterization of three inndustrial lignins and their corresponding lignin esters. *Bioresources* **2010**, *5*, 990–1009.
40. Hussin, M.H.; Rahim, A.A.; Mohamad Ibrahim, M.N.; Brosse, N. Improved corrosion inhibition of mild steel by chemically modified lignin polymers from elaeis guineensis agricultural waste. *Mater. Chem. Phys.* **2015**, *163*, 201–212. [CrossRef]
41. Akbarzadeh, E.; Ibrahim, M.N.M.; Rahim, A.A. Monomers of lignin as corrosion inhibitors for mild steel: Study of their behaviour by factorial experimental design. *Corros. Eng. Sci. Technol.* **2013**, *47*, 302–311. [CrossRef]
42. Li, L.; Dong, C.; Liu, L.; Li, J.; Xiao, K.; Zhang, D.; Li, X. Preparation and characterization of pH-controlled-release intelligent corrosion inhibitor. *Mater. Lett.* **2014**, *116*, 318–321. [CrossRef]
43. Umoren, S.A.; Ogbobe, O.; Igwe, I.O.; Ebenso, E.E. Inhibition of mild steel corrosion in acidic medium using synthetic and naturally occurring polymers and synergistic halide additives. *Corros. Sci.* **2008**, *50*, 1998–2006. [CrossRef]
44. Benali, O.; Larabi, L.; Traisnel, M.; Gengembre, L.; Harek, Y. Electrochemical, theoretical and XPS studies of 2-mercapto-1-methylimidazole adsorption on carbon steel in 1m $HClO_4$. *Appl. Sure. Sci.* **2007**, *253*, 6130–6139. [CrossRef]
45. Hussin, M.H.; Rahim, A.A.; Mohamad Ibrahim, M.N.; Brosse, N. The capability of ultrafiltrated alkaline and organosolv oil palm (elaeis guineensis) fronds lignin as green corrosion inhibitor for mild steel in 0.5 M HCl solution. *Measurement* **2016**, *78*, 90–103. [CrossRef]
46. Feng, R.; Beck, J.; Ziomek-Moroz, M.; Lvov, S.N. Electrochemical corrosion of ultra-high strength carbon steel in alkaline brines containing hydrogen sulfide. *Electrochim. Acta* **2016**, *212*, 998–1009. [CrossRef]
47. Yan, Y.; Li, W.; Cai, L.; Hou, B. Electrochemical and quantum chemical study of purines as corrosion inhibitors for mild steel in 1 M HCl solution. *Electrochim. Acta* **2008**, *53*, 5953–5960. [CrossRef]
48. Xia, G.; Wan, J.; Zhang, J.; Zhang, X.; Xu, L.; Wu, J.; He, J.; Zhang, J. Cellulose-based films prepared directly from waste newspapers via an ionic liquid. *Carbohyd polym.* **2016**, *151*, 223–229. [CrossRef]
49. Yahya, S.; Othman, N.K.; Daud, A.R.; Jalar, A.; Ismail, R. The influence of temperature on the inhibition of carbon steel corrosion in acidic lignin. *Anti-Corros. Method. Mater.* **2015**, *62*, 301–306. [CrossRef]
50. Cheng, S.; Chen, S.; Liu, T.; Chang, X.; Yin, Y. Carboxymenthyl chitosan as an ecofriendly inhibitor for mild steel in 1 M HCl. *Mater. Lett.* **2007**, *61*, 3276–3280. [CrossRef]

© 2019 by the authors. Licensee MDPI, Basel, Switzerland. This article is an open access article distributed under the terms and conditions of the Creative Commons Attribution (CC BY) license (http://creativecommons.org/licenses/by/4.0/).

Article

Investigation of the Corrosion Behavior of Atomic Layer Deposited Al_2O_3/TiO_2 Nanolaminate Thin Films on Copper in 0.1 M NaCl

Michael A. Fusco, Christopher J. Oldham and Gregory N. Parsons *

Department of Chemical and Biomolecular Engineering, North Carolina State University, Raleigh, NC 27695, USA; mfusco@ncsu.edu (M.A.F.); cjoldham@ncsu.edu (C.J.O.)

* Correspondence: gnp@ncsu.edu

Received: 31 January 2019; Accepted: 19 February 2019; Published: 24 February 2019

Abstract: Fifty nanometers of Al_2O_3 and TiO_2 nanolaminate thin films deposited by atomic layer deposition (ALD) were investigated for protection of copper in 0.1 M NaCl using electrochemical techniques. Coated samples showed increases in polarization resistance over uncoated copper, up to 12 MΩ-cm^2, as measured by impedance spectroscopy. Over a 72-h immersion period, impedance of the titania-heavy films was found to be the most stable, as the alumina films experienced degradation after less than 24 h, regardless of the presence of dissolved oxygen. A film comprised of alternating Al_2O_3 and TiO_2 layers of 5 nm each (referenced as ATx5), was determined to be the best corrosion barrier of the films tested based on impedance spectroscopy measurements over 72 h and equivalent circuit modeling. Dissolved oxygen had a minimal effect on ALD film stability, and increasing the deposition temperature from 150 °C to 250 °C, although useful for increasing film quality, was found to be counterproductive for long-term corrosion protection. Implications of ALD film aging and copper-based surface film formation during immersion and testing are also discussed briefly. The results presented here demonstrate the potential for ultra-thin corrosion barrier coatings, especially for high aspect ratios and component interiors, for which ALD is uniquely suited.

Keywords: atomic layer deposition; corrosion protection; copper; aluminum oxide; titanium oxide; nanolaminate; electrochemical impedance spectroscopy; barrier coatings

1. Introduction

Copper is ubiquitous in numerous industries due to its relatively low cost coupled with its malleability and high electrical and thermal conductivities. It has uses in everything from residential and commercial plumbing [1–4] and electrical wiring to industrial heat exchangers [5–7] and high-powered electronics [8–11]. Copper possesses adequate aqueous corrosion resistance due to the formation of a semi-protective native oxide film. Still, it corrodes at a finite rate and is susceptible to pitting dependent on solution constituents, pH, and temperature [2–4,12–16].

One specific application of interest here is the use of copper in radio frequency (RF) devices (i.e., traveling-wave tubes and crossed-field amplifiers). These devices typically contain fluid-cooled copper collectors and/or copper cooling channels [17–24]. Often, the cooling fluid is specified as deionized (DI) water with low dissolved oxygen content, in which case corrosion of the wetted copper components is insignificant within the useful lifetime of the device. However, the presence of contaminants—such as chlorides, additional oxygen, and carbon dioxide—in the cooling water supply can accelerate the corrosion to unacceptable levels. Chlorides are known to be particularly aggressive toward copper [12,14,25], as with many other metals, and are common contaminants in water supplies. Copper readily forms salt compounds in the presence of chlorides and can have its protective oxide film locally disrupted, allowing for pitting to occur.

The corrosion of copper in chloride media has been thoroughly studied [13–15,26–30], from which it has been concluded that the anodic and cathodic half reactions are given by Equations (1) and (2), respectively. The anodic reaction proceeds in two steps with the mostly insoluble CuCl produced in the first step and soluble $CuCl_2^-$ produced in the second. CuCl builds up as a film on the Cu surface, leading to passivation, or, more appropriately, pseudo-passivation. Higher chloride concentrations tend to shift the equilibrium of Equation (1) toward $CuCl_2^-$ and can even produce higher chloride complexes (i.e., $CuCl_3^{2-}$ and $CuCl_4^{3-}$) [27,31]. This increases dissolution of the CuCl surface film and accounts for the more aggressive corrosion seen in solutions with higher chloride content. Additionally, cuprous oxide (Cu_2O) may be produced from $CuCl_2^-$, but its stability also decreases as the chloride content increases [31].

Oxygen reduction and water reduction (Equation (2)) dominate the cathodic current, though there are complications in the presence of copper corrosion products and surface films [31], including the reduction of CuCl on the copper surface [30]. Water reduction dominates when dissolved oxygen levels are low, whereas oxygen reduction produces high cathodic currents in oxygenated solutions [32]. Although the corrosion of copper in pure water via water reduction and hydrogen evolution in the absence of oxygen is still under debate [33–37], it is reasonable to expect that this reaction could play a role in the presence of chlorides and at sufficient overpotential.

$$\begin{gathered} Cu + Cl^- \rightarrow CuCl + e^- \\ CuCl + Cl^- \rightarrow CuCl_2^- \end{gathered} \tag{1}$$

$$\begin{gathered} O_2 + 2H_2O + 4e^- \rightarrow 4OH^- \\ 2H_2O + 2e^- \rightarrow H_2 + 2OH^- \end{gathered} \tag{2}$$

A common approach to improve the corrosion resistance of copper is to alloy it with other metals, such as aluminum and nickel. However, alloying is not always an option, as it can significantly reduce the electrical (and thermal) conductivity [38]. Another option is the application of barrier coatings to the copper surface. To preserve the desirable bulk properties of the copper, the barrier coatings should be as thin as possible. Additionally, copper components of RF devices are often of complex geometry or high aspect ratio, which narrows the available thin film deposition techniques.

Atomic layer deposition (ALD) is a vapor-phase technique that is uniquely suited for deposition of conformal thin films over complex, non-planar surfaces and through high-aspect-ratio structures, like tubing. ALD has no line of sight requirement with sub-nanometer thickness control [39–43]. ALD can also proceed at low temperatures (<150 °C) [44,45], avoiding temperature limits of fabricated components and even enabling deposition on thermally-sensitive substrates, such as polymers [46–50] and microelectronics [51,52].

Alumina (Al_2O_3) and titania (TiO_2) are two of the most widely-studied ALD processes. Deposition of alumina via trimethylaluminum (TMA, $Al(CH_3)_3$) and H_2O is thermodynamically favorable and nucleates well on most surfaces [39,40,50,53]. Al_2O_3 has excellent sealing properties and has shown outstanding versatility as a barrier layer on metals [54–58], polymers [48,59–62], and electronics [63–67]. However, the use of Al_2O_3 as a corrosion barrier is limited by its chemical stability and dissolution in alkaline media [68,69]. On the other hand, titania is lauded for its chemical stability [68,70–72] and has shown favorable corrosion resistance [73–78]. Titania ALD, however, exhibits film nucleation issues using titanium tetrachloride ($TiCl_4$) as a precursor, especially on copper substrates, and tends to deposit with high roughness, leading to increased porosity and poor corrosion performance over time [73,75,79].

Previous studies of ALD films for corrosion protection of copper have mostly focused on aluminum oxide. Mirhashemihaghighi et al. [80] reported a 7x increase in polarization resistance over uncoated, polished copper with a 10 nm ALD Al_2O_3 film and a three order of magnitude increase with a 50 nm Al_2O_3 film in deoxygenated 0.5 M NaCl. Chai et al. [81,82] found better corrosion protection with increasing alumina film thickness in aerated 0.1 M NaCl, as did Daubert et al. [79]. Daubert et al. [79]

also performed electrochemical impedance spectroscopy (EIS) over 90 h for 5 different metal oxide ALD films, showing a decrease in high-frequency impedance for the Al_2O_3 film and better stability for the other four. A study by Abdulagatov et al. utilized an alumina base layer and titania capping layer to overcome the limitations of the individual layers, resulting in significantly improved corrosion protection of copper over 900 h in water at 25° and 90 °C [73]. Although not specifically related to aqueous corrosion, it is worth noting that Chang et al. successfully applied 100 nm thick ALD alumina films to copper for protection against oxidation in air at 200 °C [54]. Whereas ALD alumina possesses excellent sealing properties and can provide adequate initial corrosion protection with films less than 50 nm thick, its longer-term stability on copper is in question. Preliminary results of combining alumina and titania prove promising as a corrosion barrier, and further investigation into nanolaminate alumina/titania film structures is warranted. Reports of Al_2O_3/TiO_2 nanolaminated ALD thin films for corrosion protection of steel are available [75,83,84], showing increased corrosion and delamination resistance over single-layer films.

Here we investigate the corrosion behavior of ALD-coated copper and use Al_2O_3/TiO_2 nanolaminate films to enhance corrosion protection over the single-layer materials. The effect of dissolved oxygen on the corrosion protection of these ALD thin films has previously not been reported. Oxygen is always present in some amount, even when low oxygen levels are maintained in contact with the copper surfaces. Determining its impact on the protective performance of ALD films is essential for their use in RF devices and other copper-containing components. Also of interest is the effect of ALD film layer structure, copper-based interfacial films, and deposition temperature on the corrosion behavior of ALD alumina- and titania-coated copper. In this work, we utilize DC voltammetry and electrochemical impedance spectroscopy (EIS) to investigate corrosion behavior and probe the stability of Al_2O_3/TiO_2 ALD films on copper in a sodium chloride (NaCl) solution with and without dissolved oxygen.

2. Materials and Methods

2.1. Material Preparation

2.1.1. Substrate Preparation

Grade 110 copper sheets (99.9%, McMaster-Carr, Douglasville, GA, USA) were cut into 15 mm × 25 mm × 1 mm coupons. Coupons had a No. 8 mirror finish with average surface roughness of 0.1–0.3 μm. Prior to loading into the ALD reactor, all coupons were ultrasonically rinsed in acetone for 5 min, rinsed with isopropanol and DI water, then dipped in 35% H_3PO_4 for 30 s to reduce the native copper oxide, and finally rinsed in DI water and dried thoroughly with ultra-high-purity nitrogen.

2.1.2. Thin Film Deposition and Characterization

Films were deposited in a home-built, viscous-flow, hot-wall ALD reactor using ultra-high-purity nitrogen (N_2—99.999%, Arc3 Gases, Richmond, VA, USA) as the carrier and purge gas. Nitrogen was constantly flowing during deposition, maintaining a reactor pressure of roughly 1 torr. Alumina (Al_2O_3) was deposited using trimethylaluminum (TMA, 98% Strem Chemicals, Newburyport, MA, USA), and titania (TiO_2) was deposited using titanium tetrachloride ($TiCl_4$, 99% Strem Chemicals, Newburyport, MA, USA). DI water served as the co-reactant for both precursors. Each ALD cycle consisted of a 0.1 s precursor (TMA or $TiCl_4$) dose followed by a 45 second N_2 purge and a 0.1 s H_2O dose followed by a 45 s N_2 purge. Unless otherwise specified, Al_2O_3 and TiO_2 were deposited at 150 °C with nominal growth rates of 1.25 Å/cycle and 0.45 Å/cycle, respectively, measured using spectroscopic ellipsometry (J.A. Wollam Co., Lincoln, NE, USA) on the copper coupons and on silicon monitor wafers coated simultaneously with the copper coupons. A Keyence VKx1100 confocal laser scanning microscope (Itasca, IL, USA) equipped with a 404 nm violet laser was used for imaging, and energy dispersive X-ray spectroscopy (EDX) was performed using a JEOL 6010LA scanning electron microscope (Peabody, MA, USA, results not shown).

2.2. Electrochemistry

Coated and uncoated copper coupons were placed in a 3-electrode corrosion cell (Princeton Applied Research, model K0235, Oak Ridge, TN, USA) with a platinum mesh counter electrode, a double junction Ag/AgCl reference electrode filled with 3.8 M KCl (F-600, Broadley-James, Irvine, CA, USA), and a 1 cm^2 nominal working electrode area. All potentials given are measured against this Ag/AgCl reference. Measurements were performed at room temperature in neutral 0.1 M NaCl solution, prepared with crystalline sodium chloride (Reagent grade, Fisher Scientific, Hampton, NH, USA) and DI water, using a BioLogic VMP3 potentiostat (Seyssinet-Pariset, France). Approximately 200 mL of electrolyte was used for each test with an average initial pH of 6.7 $\pm$ 0.3. For experiments where deoxygenation was performed, nitrogen was bubbled through the electrolyte for at least 15 min prior to and continuously during testing. The nitrogen was passed through a gas washing bottle of 0.1 M NaCl to pre-saturate the dry gas before its introduction into the test cell. 'Oxygenated' here refers to cases in which no nitrogen purging of the electrolyte was performed.

All samples are cathodically polarized at -1.5 V for 2.5 min upon immersion in the electrolyte and prior to testing to reduce ambient or aqueously formed surface films, and all electrochemical measurements were performed at least twice to ensure repeatability.

2.2.1. Electrochemical Impedance Spectroscopy

Before impedance testing, the open circuit potential was typically measured for 2 h to ensure the cell was at equilibrium, after which impedance measurements were collected at the open circuit potential over a frequency range of 100 kHz to 5 mHz at 10 frequencies per decade with two measurements per frequency and a 10 mV rms perturbation signal. Fitting impedance spectra to equivalent circuits was performed using the EC-Lab software (BioLogic, Seyssinet-Pariset, France), and all fits produced a relative error of less than 5%.

The equivalent electrical circuits used to fit impedance spectra are displayed in Figure 1 and are often used to model impedance data for inert coatings on metals [76,77,79,85]. The model assumes that corrosion occurs at the base of coating defects or pores, where electrolyte encounters the substrate. In this simplified model, there are parallel current paths: one through the intact coating, which is a dielectric, and the other through pores in the coating structure, represented by the film (or pore) resistance. This resistance may be viewed as the sum of the resistance of individual pores filled with electrolyte. The path through coating pores results in the substrate in contact with electrolyte, and the interfacial impedance consists of a parallel R-C circuit, where the resistance (R_{int}) is typically the polarization or charge transfer resistance, and the capacitance is often the double layer capacitance. Capacitors have been replaced with constant phase elements (CPEs) to account for the frequency dispersion of the capacitance encountered in electrode-electrolyte systems [86–88]. The typical expression for the impedance of a CPE is given by [87,88]:

$$Z_{CPE} = \frac{1}{Q(j\omega)^{\alpha}}, \quad (3)$$

where Z_{CPE} is the impedance of the constant phase element [Ω cm^2], Q is the CPE coefficient [Ω^{-1} cm^{-2} s^{α}], j is the imaginary unit [j = $\sqrt{-1}$], ω is the angular frequency [rad/s], and α is the CPE exponent [unitless].

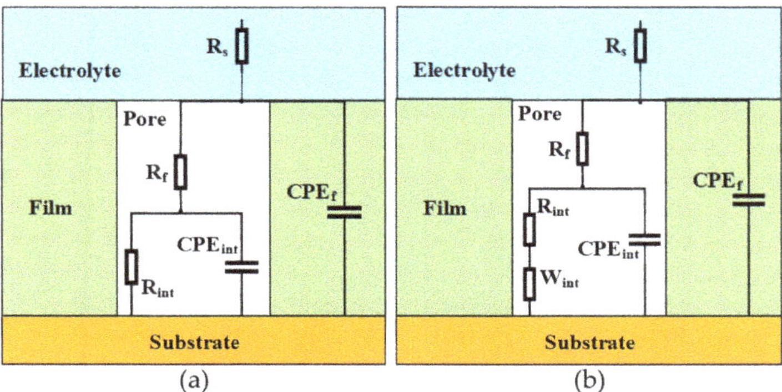

Figure 1. Equivalent electrical circuit used to fit impedance spectra measured in (**a**) deoxygenated and (**b**) oxygenated electrolyte. R_s is the solution (electrolyte) resistance, R_f is the film (or pore) resistance, CPE_f is the constant phase element representing film capacitance, CPE_{int} is the constant phase element representing interfacial capacitance (metal-solution interface), R_{int} is the interfacial resistance, and W_{int} is the interfacial Warburg diffusion element.

2.2.2. DC Voltammetry

Cyclic voltammetry was performed immediately following the reduction step (−1.5 V, 2.5 min) with potential scanned from −1.5 V to 0.8 V and back to −1.5 V at 10 mV/s. Additional polarization curves were measured from −1 V to 0.8 V and back to −1 V with a scan rate of 0.5 mV/s after the 2.5-min cathodic reduction with a 30-min rest period at open circuit.

Polarization resistance is determined as the slope of the V-I curve at ± 10 mV from the corrosion potential, and corrosion current is estimated using [89]:

$$\left.\frac{d\eta}{dI}\right|_{\eta\to 0} = R_p = \frac{\beta_a\beta_c}{2.3I_{corr}(\beta_a+\beta_c)}, \tag{4}$$

where η is the overpotential [V], I is the current [A cm^{-2}], R_p is the polarization resistance [Ω cm^2], I_{corr} is the corrosion current [A cm^{-2}], and β_a and β_c are the anodic and cathodic Tafel coefficients [V/decade], respectively.

3. Results and Discussion

3.1. Spectroscopic Ellipsometry

Al_2O_3 and TiO_2 ALD film thicknesses were measured using spectroscopic ellipsometry (SE) on polished copper coupons as well as silicon wafers coated simultaneously with the copper. The five coatings tested here are detailed in Table 1, along with their naming conventions and film thicknesses. Note that for all nanolaminate films (ATx1, ATx5, and ATx10), the alumina layer is deposited first (adjacent to the substrate), followed by titania. Thicknesses presented on copper were measured by spectroscopic ellipsometry (SE) after deposition and were averaged over six coupons coated together. Film thicknesses measured on copper are approximate based on a finite surface roughness and limitations in modeling of SE data, though they mostly agree with thicknesses measured on the silicon wafers. Increases in thickness above those measured on silicon may be indicative of a slightly higher growth rate for the $TiCl_4$-H_2O ALD process on copper or increased growth based on the proximity of the coupons to each other. The relative ratio of titania to alumina in the nanolaminate films is apparent in the refractive index, which was calculated from films grown on silicon. Despite differences in the number of layers, ATx10 and ATx5 have nearly identical refractive indices, confirming similar titania to alumina ratios.

Table 1. Samples tested in this work—naming convention, film type, and ellipsometric data. The Si Thickness and Cu Thickness columns indicate fitted film thicknesses on the respective substrates using spectroscopic ellipsometry. Refractive indices are determined from film measurements on silicon substrates.

Sample Name	Nominal Film Structure	Si Thickness (nm)	Cu Thickness (nm)	Refractive Index
Cu	None—uncoated copper	-	-	-
Al_2O_3	Single-layer Al_2O_3	52.9 ± 0.6	49.1 ± 0.7	1.64
TiO_2	Single-layer TiO_2	54.0 ± 1.6	63.8 ± 2.8	2.43
ATx10	Nanolaminate—(Al_2O_3 + TiO_2) × 10	56.0 ± 0.6	65.8 ± 2.5	2.1
ATx5	Nanolaminate—(Al_2O_3 + TiO_2) × 5	52.5 ± 0.8	56.6 ± 1.9	2.12
ATx1	Double layer—10 nm Al_2O_3 + 40 nm TiO_2	52.6 ± 0.9	60.9 ± 1.1	2.33

3.2. DC Voltammetry

ALD films were first investigated using DC voltammetry techniques. Figures 2 and 3 present cyclic polarization curves measured in deoxygenated and oxygenated 0.1 M NaCl, respectively, after a 30-min rest period at open circuit following cathodic polarization. Relevant parameters determined from these curves may be found in Table 2. Coated and uncoated copper in Figures 2 and 3 display similarly shaped polarization curves, suggesting analogous corrosion mechanisms consistent with electrochemically inert films. Moreover, both curves exhibit behavior that is typical of unalloyed copper in chloride media [26]. The anodic behavior includes: (i) an apparent Tafel region at low overpotentials, where Equation (1) accounts for most of the measured current; (ii) a peak current and a current reduction region, where CuCl (and possibly Cu_2O or $Cu(OH)_2$) builds up on the surface; (iii) a limited current, or passive, region where the formation and dissolution of surface films are somewhat at equilibrium; and (iv) a higher potential region where dissolution increases and oxidation of $CuCl_2^-$ to form Cu^{++} becomes significant.

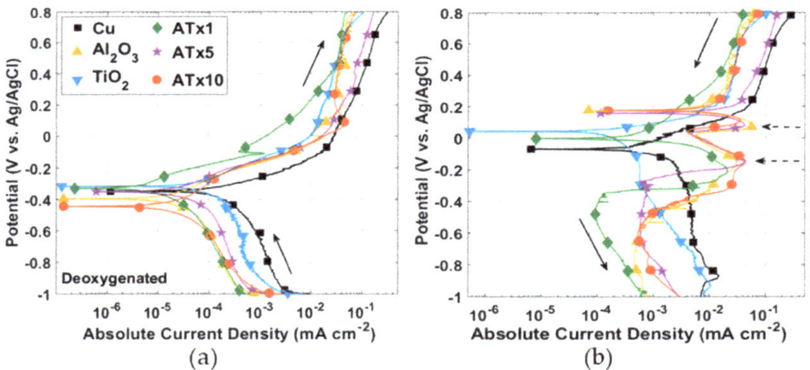

Figure 2. DC polarization curves: (**a**) forward scan [−1, 0.8] V and (**b**) reverse scan [0.8, −1] V measured at a scan rate of 0.5 mV/s in deoxygenated 0.1 M NaCl. Solid arrows specify scan direction and dashed arrows indicate features of interest. Refer to Table 1 for sample descriptions.

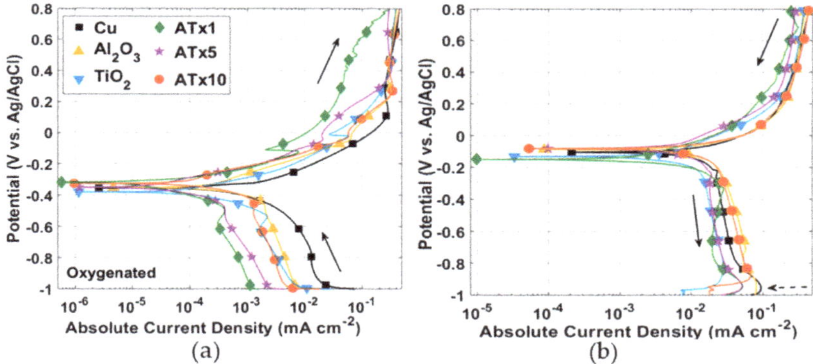

Figure 3. DC polarization curves: (**a**) forward scan [−1, 0.8] V and (**b**) reverse scan [0.8, −1] V measured at a scan rate of 0.5 mV/s in oxygenated 0.1 M NaCl. Solid arrows specify scan direction and dashed arrows indicate features of interest. Refer to Table 1 for sample descriptions.

Table 2. Corrosion potential (E_{corr}), polarization resistance (R_p), and corrosion current (I_{corr}) determined from polarization curves found in deoxygenated (Figure 2) and oxygenated (Figure 3) 0.1 M NaCl.

Sample	Deoxygenated			Oxygenated		
	E_{corr} (V vs. Ag/AgCl)	R_p (Ω cm^2) $\times 10^6$	I_{corr} (A cm^{-2}) $\times 10^{-7}$	E_{corr} (V vs. Ag/AgCl)	R_p (Ω cm^2) $\times 10^6$	I_{corr} (A cm^{-2}) $\times 10^{-7}$
Cu	−0.347	0.191	1.51	−0.355	0.066	3.77
Al_2O_3	−0.395	1.67	0.179	−0.314	0.204	1.07
TiO_2	−0.314	0.373	0.915	−0.379	0.193	0.992
ATx1	−0.328	3.37	0.103	−0.318	0.948	0.167
ATx5	−0.350	2.60	0.115	−0.349	0.596	0.340
ATx10	−0.444	1.85	0.197	−0.325	0.167	1.09

3.2.1. Deoxygenated Electrolyte

TiO_2 and ATx1 exhibit an anodic shift in corrosion potential from the uncoated copper. This has been attributed to modification of the substrate exposed through coating defects [55], as well as native oxide formation during ALD growth and aging of the oxide during storage [80]. The remaining corrosion potentials are shifted cathodically, a trend that has also been reported for ALD coatings on stainless steel [75,84], copper [79], and a Mg-Al alloy [83]. All coated coupons show an increase in polarization resistance and decrease in corrosion current compared to uncoated copper. Polarization resistance is better than an order of magnitude higher for the nanolaminate Al_2O_3/TiO_2 films.

The current peak associated with CuCl film buildup is apparent in Figure 4a, measured with a voltage scan rate of 10 mV/s. This suggests that the kinetics of the reaction leading to the precipitation of CuCl (Equation (1)) are faster than can be observed with a 0.5 mV/s scan rate. This is at least true for the uncoated copper. A current peak is visible in Figure 2a for several of the ALD-coated samples, likely corresponding to CuCl buildup at exposed copper surface sites. The peaks in the forward scan of Figure 4a are not as pronounced for coated copper as for the uncoated copper, showing that the coatings restrict the amount of surface area that is exposed to electrolyte.

Current measured over the forward scan is as much as an order of magnitude smaller for ALD-coated compared to uncoated copper. The curves never quite merge, though the difference in current decreases above ~0.4 V, as the CuCl surface coverage reaches a maximum at the Cu-ALD film interface. The reverse potential scans in Figure 2b show two large current peaks for Al_2O_3, ATx5, and ATx10 at ~ 0.1 V and -0.15 V. Only one of these peaks is seen for ATx1, and they do not appear at all in the curves for TiO_2 and uncoated copper. These current peaks are likely associated with copper-based film reduction at the Cu-ALD coating interface. Figure 4b shows that the presence of this reduction peak for uncoated copper is dependent on potential scan rate. This feature appears when the potential is scanned at 50 mV/s and becomes more prominent at 100 and 200 mV/s. The resistance of pores in some of the ALD films clearly slows this reduction to the point that the peaks are visible at low scan rates.

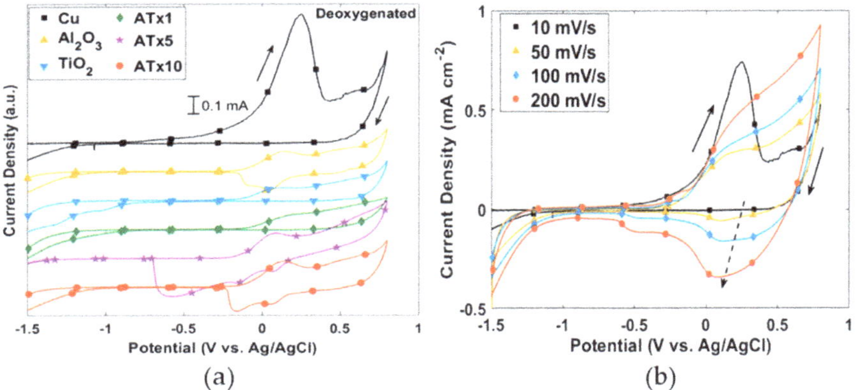

Figure 4. I–V curves in deoxygenated 0.1 M NaCl for: (**a**) uncoated and ALD-coated copper measured at 10 mV/s and (**b**) uncoated copper measured at scan rates from 10 mV/s to 200 mV/s. Solid arrows indicate scan direction and dashed arrow in (**b**) points in the direction of increasing scan rate. Refer to Table 1 for sample descriptions.

3.2.2. Oxygenated Electrolyte

Again, all coated samples demonstrate higher polarization resistance and lower corrosion current, and most have anodically-shifted corrosion potentials (see Figure 3). Several samples have similar corrosion potentials to the deoxygenated electrolyte, though all polarization resistances are smaller than their deoxygenated counterparts. Because dissolved oxygen alone is not typically enough to passivate the copper surface, the oxygen content should have little effect on the equilibrium potential. The larger differences in corrosion potential between oxygenated and deoxygenated conditions for Al_2O_3, TiO_2, and ATx10 may be attributed to slight differences in surface conditions, post-deposition aging, or film quality.

There is evidence of diffusion-limited current in the cathodic portion of the curves in Figure 3. In this case, the polarization resistance cannot be taken as a direct measurement of the charge transfer resistance, and thus the corrosion current should be considered approximate. Regardless of the diffusion-limited cathodic current, anodic dissolution of copper in chloride media is known to be under mixed charge transfer and mass transport control [31], so care should always be taken in interpreting corrosion parameters from polarization curves. In this case, the polarization resistance and corrosion current serve as adequate comparison points between samples.

Cathodic currents measured during the forward potential scan are all smaller for the ALD-coated copper, indicating a smaller exposed area for oxygen reduction at the copper surface. Measured current for most samples converges once passivation occurs above 0.1–0.3 V. The shape of the reverse scans (Figure 3b) is nearly identical for all samples, and a current peak appears close to −1 V associated with reduction of the Cu_2O surface film formed during the forward scan.

3.3. Electrochemical Impedance Spectroscopy

Figures 5 and 6 show impedance spectra of coated and uncoated copper in deoxygenated and oxygenated 0.1 M NaCl, respectively. The graphs are split into Bode plots of impedance modulus and phase angle versus frequency for clarity. All impedance spectra were collected at open circuit potential, and the results of equivalent circuit modeling of impedance data are provided in Table 3.

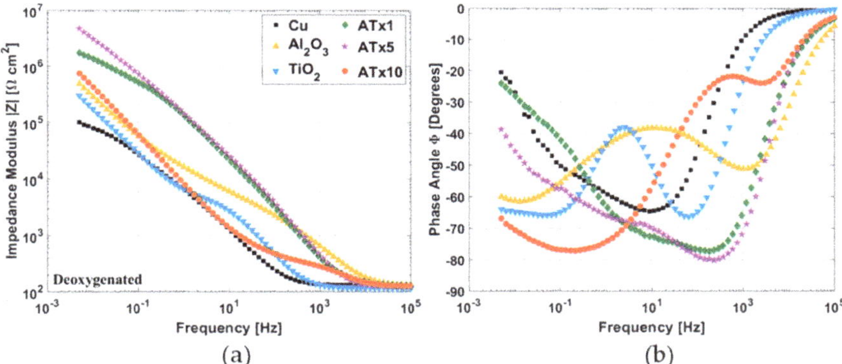

Figure 5. Impedance spectra measured from 100 kHz to 5 mHz in deoxygenated 0.1 M NaCl at OCP: (**a**) impedance modulus and (**b**) phase angle versus frequency. Refer to Table 1 for sample descriptions.

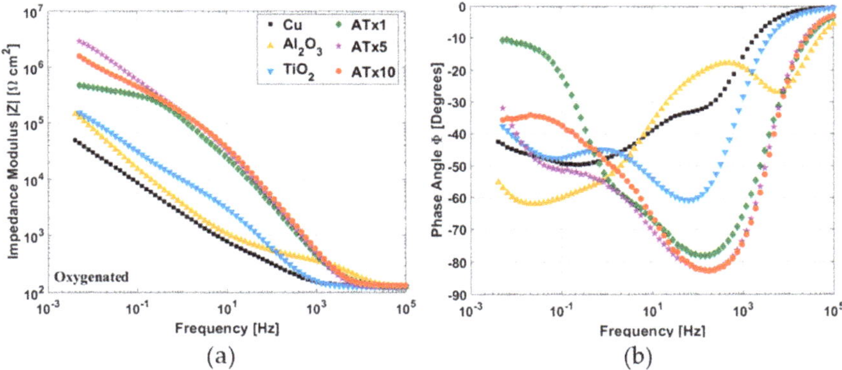

Figure 6. Impedance spectra measured from 100 kHz to 5 mHz in oxygenated 0.1 M NaCl at OCP: (**a**) impedance modulus and (**b**) phase angle versus frequency. Refer to Table 1 for sample descriptions.

3.3.1. Deoxygenated Electrolyte

ALD films exhibit increases in impedance over uncoated copper, generally 1–2 orders of magnitude at low frequency. The phase angle plot depicts differences in the high-frequency impedance response associated with the ALD films. Nanolaminate films have the largest impedance increase, though all the ALD films have distinct impedance response from the uncoated copper sample.

There are several things of note in Table 3. The film resistance of ATx1 and ATx5 is much higher than the other ALD coatings. Also, the interfacial CPE exponent is quite low for several of the samples, indicating significant capacitive dispersion at the underlying copper surface. This may partly explain the disagreement between interfacial resistance values from EIS and polarization resistances from DC voltammetry (Table 2), as well as the effect of having an insufficiently low cutoff frequency. However, the fact that the EIS-derived polarization resistances are higher is most likely influenced by the longer rest period following reduction before impedance spectra are recorded. The impact of rest time and aging is explored in Section 3.7. Equivalent circuit modeling reveals that several of the ALD films exhibit CPE exponents above 0.9, indicating high-quality coatings, and fitted interfacial resistances for all coatings are higher than the uncoated copper.

Table 3. Fitted parameters from equivalent electrical circuit modeling of impedance spectra in deoxygenated (Figure 5) and oxygenated (Figure 6) 0.1 M NaCl. Refer to Figure 1 for a depiction of the equivalent circuit and parameter definitions. Q and α are the coefficient and exponent of the constant phase element (see Equation (3)).

Deoxygenated								
	R_s (Ω cm^2)	Q_{film} (Ω^{-1} s^a cm^{-2})	α_{film}	R_{film} (Ω cm^2)	Q_{int} (Ω^{-1} s^a cm^{-2})	α_{int}	R_{int} (Ω cm^2)	
Cu	124.7	2.08×10^{-5}	0.87	6831	3.14×10^{-5}	0.62	1.23×10^{5}	
Al_2O_3	120.1	2.65×10^{-6}	0.74	5320	2.04×10^{-5}	0.65	6.30×10^{6}	
TiO_2	116.5	5.63×10^{-6}	0.93	4759	4.83×10^{-5}	0.80	1.48×10^{6}	
ATx1	130.2	9.05×10^{-7}	0.91	11,642	1.54×10^{-6}	0.4	3.97×10^{6}	
ATx5	126.2	4.43×10^{-7}	0.98	18,858	1.45×10^{-6}	0.63	1.11×10^{7}	
ATx10	126.3	3.75×10^{v6}	0.79	343.5	2.08×10^{-5}	0.86	5.19×10^{6}	
Oxygenated								
	R_s (Ω cm^2)	Q_{film} (Ω^{-1} s^a cm^{-2})	α_{film}	R_{film} (Ω cm^2)	Q_{int} (Ω^{-1} s^a cm^{-2})	α_{int}	R_{int} (Ω cm^2)	W_{int} (Ω s$^{-0.5}$ cm^2)
Cu	135.6	8.08×10^{-5}	0.90	272.5	1.46×10^{-4}	0.58	2.18×10^{5}	2.2×10^{-9}
Al_2O_3	116.8	3.02×10^{-6}	0.74	385	8.96×10^{-5}	0.73	7.60×10^{5}	4.8×10^{-10}
TiO_2	126	7.16×10^{-6}	0.86	5324	3.81×10^{-5}	0.58	4.85×10^{5}	9.2×10^{-10}
ATx1	128	8.03×10^{-7}	0.91	65,445	1.26×10^{-6}	0.77	3.02×10^{5}	1.6×10^{4}
ATx5	123.9	4.58×10^{-7}	0.97	72,013	1.96×10^{-6}	0.64	5.50×10^{6}	1.6×10^{-3}
ATx10	134.3	3.35×10^{-7}	0.99	24,467	2.63×10^{-6}	0.42	4.58×10^{6}	4.4×10^{5}

3.3.2. Oxygenated Electrolyte

The nanolaminate films show increases in impedance as high as two orders of magnitude. The equivalent circuit for oxygenated measurements (Figure 1b) contains an additional Warburg diffusion element at the copper-ALD film interface to account for enhanced mass transport control in the presence of dissolved oxygen. Data were only fit well with the Warburg element in two cases, ATx1 and ATx10. For other samples the fitted Warburg coefficient was negligible. Film resistances for the nanolaminate ALD films are much higher than the single layers and the deoxygenated measurements, indicating passivation of pores in the more complex film structures. Interfacial resistances for all coated samples are higher than the bare copper, although they are lower than those determined from deoxygenated measurements, as would be expected.

3.4. ALD Film Stability

Impedance spectra measured over a 72-h immersion in deoxygenated and oxygenated electrolyte are presented in Figures 7–10. Again, the graphs are split into Bode plots of impedance modulus and phase angle versus frequency, and measurements were made at open circuit conditions. Dashed arrows are provided in some of the graphs to indicate an increase in immersion time. Laser scanning microscope images of coated and uncoated copper after 72-h EIS testing are displayed in Figure 11.

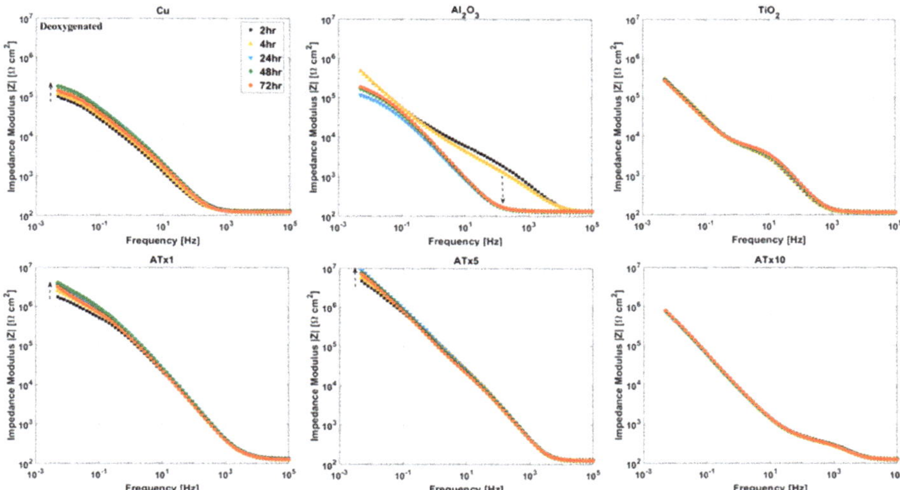

Figure 7. Impedance modulus versus frequency at selected time intervals over 72-h immersion in deoxygenated 0.1 M NaCl. Frequency range is 100 kHz to 5 mHz and sample type is given above each plot. Dashed arrows indicate increasing immersion time. Refer to Table 1 for sample descriptions.

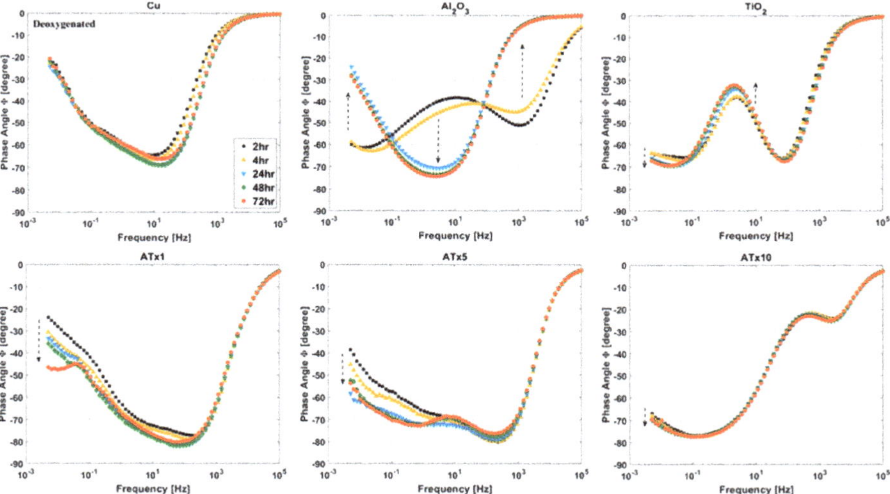

Figure 8. Impedance phase angle versus frequency at selected time intervals over 72-h immersion in deoxygenated 0.1 M NaCl. Frequency range is 100 kHz to 5 mHz and sample type is given above each plot. Dashed arrows indicate increasing immersion time. Refer to Table 1 for sample descriptions.

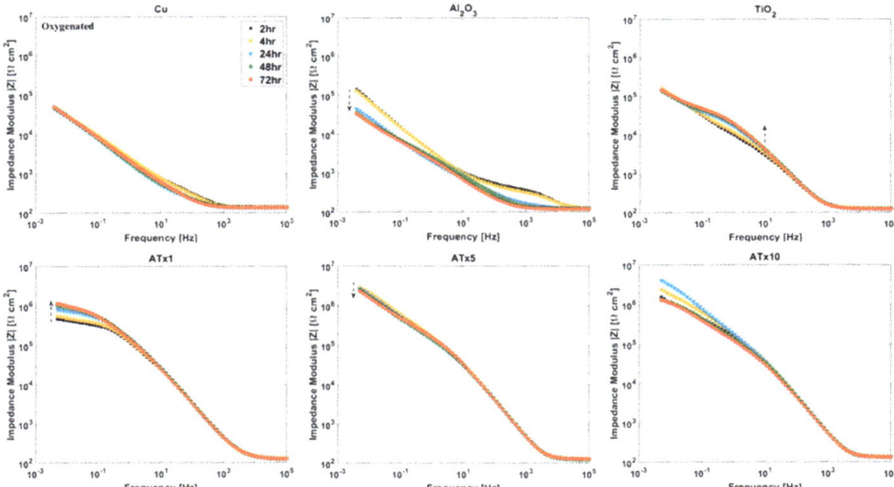

Figure 9. Impedance modulus versus frequency at selected immersion times over 72-h immersion in oxygenated 0.1 M NaCl. Frequency range is 100 kHz to 5 mHz and sample type is given above each plot. Dashed arrows indicate increasing immersion time. Refer to Table 1 for sample descriptions.

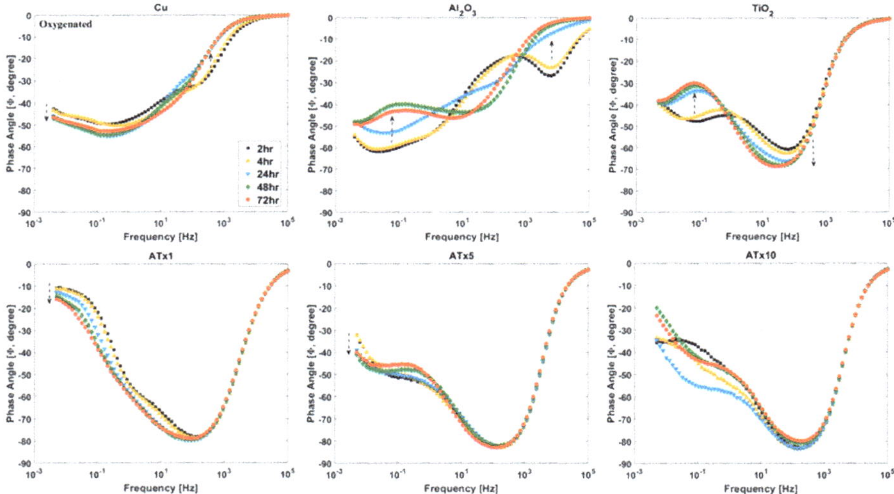

Figure 10. Impedance phase angle versus frequency at selected immersion times over 72-h immersion in oxygenated 0.1 M NaCl. Frequency range is 100 kHz to 5 mHz and sample type is given above each plot. Dashed arrows indicate increasing immersion time. Refer to Table 1 for sample descriptions.

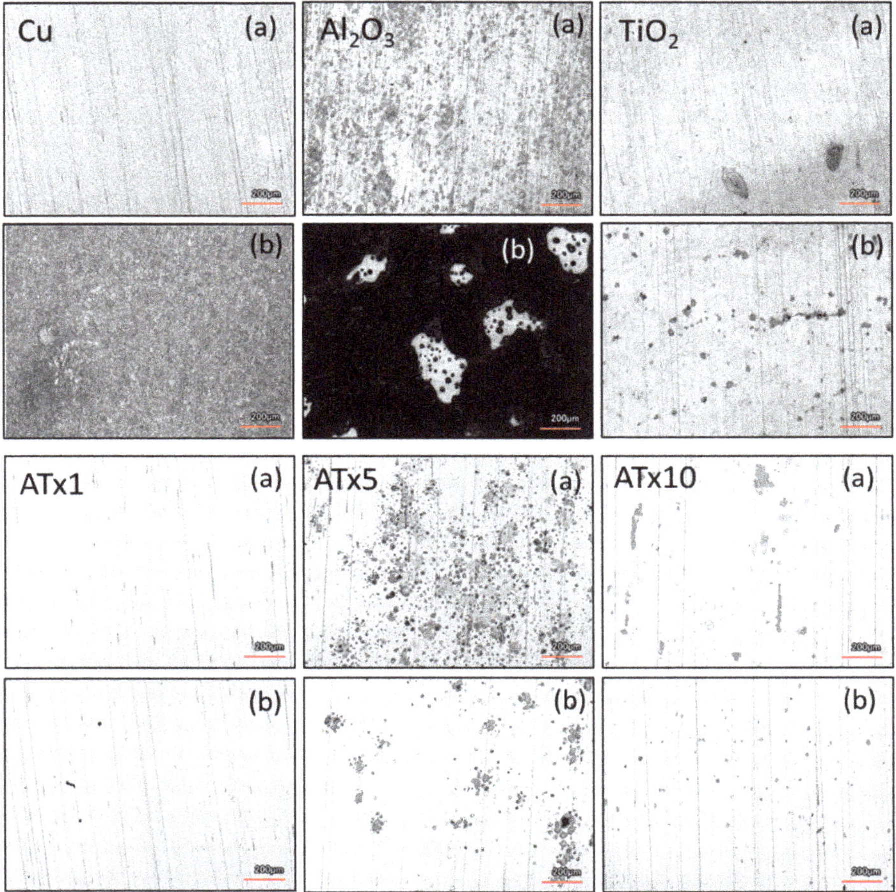

Figure 11. Confocal laser scanning microscope images of uncoated and ALD-coated copper samples after 72-h immersion and EIS testing in (**a**) deoxygenated and (**b**) oxygenated 0.1 M NaCl. Refer to Table 1 for sample descriptions.

3.4.1. Deoxygenated Electrolyte

Phase angle plots show differences in impedance response among the ALD films and highlight changes in the samples over time. Uncoated copper undergoes an increase in impedance, peaking at 24–48 h. This is due to the buildup of a protective surface film (i.e., CuCl). ATx5 and ATx1 also exhibit similar behavior. Variations in phase angle at low frequency for these two films indicate changes at the ALD film-substrate interface, rather than much alteration in the film itself. TiO_2 and ATx10 display good stability over 72 h, with minimal deviation in modulus or phase angle visible. A change in the Al_2O_3 sample over time is apparent in Figures 7 and 8. The change in shape of both the modulus and phase angle spectra indicate degradation of the ALD alumina film. The increase in phase angle and decrease in impedance at high frequency seen after 24 h point to a stark decrease in the dielectric quality of the film. Indeed, the shape of these curves for Al_2O_3 after 24 h closely resemble those of the bare copper, and the impedance of the Al_2O_3 sample drops below that of uncoated copper due to severe ALD film damage.

Laser scanning microscope images after 72 h of EIS testing reveal that the uncoated copper surface remains smooth and generally featureless after 72 h, having undergone uniform corrosion along with Cu-based surface film buildup and dissolution during the immersion period. TiO_2 and ATx1 remain mostly unchanged, in good agreement with the impedance spectra. ATx5 and ATx10 show signs of corrosion spots, though the ALD films are still intact, which also agrees with EIS measurements over 72 h. In the case of ATx5, these appear to be superficial spots of discoloration rather than corrosion of the underlying copper surface. Al_2O_3 seems to be somewhat degraded, as the smoothness of the substrate is apparently gone. Only a trace amount of aluminum is detected using EDX (not shown) after immersion and is approximately 2% of what was detected before exposure to the NaCl solution. An explanation of the degradation of the alumina ALD film is presented in Section 3.5.

3.4.2. Oxygenated Electrolyte

The titania-containing films have good stability over the immersion time, with TiO_2, ATx1, and ATx10 having increases in low-frequency impedance over time due to passivation of the substrate. Al_2O_3 again undergoes degradation, evidenced by decreasing impedance modulus in a frequency range corresponding to film response, which is mirrored by variations in shape of the phase angle over time. The change in impedance of the Al_2O_3 film makes the degradation seem less severe than in deoxygenated conditions, though the laser scanning microscope image in Figure 11 shows what appears to be worse damage. However, the aluminum content detected by EDX (not shown) after immersion is approximately 25% of what is detected before testing, which is 10 times more than that detected after exposure to deoxygenated NaCl. The high contrast in the Al_2O_3 image in Figure 11 shows regions of exposed copper surrounded by dark, rough patches. The substrate has clearly been oxidized in most places, likely slowing the deterioration of the remaining Al_2O_3 ALD film. Other ALD samples appear to be in good shape, and the laser images are similar to those taken after immersion in deoxygenated electrolyte.

3.5. *Dissolution of Al_2O_3*

According to thermodynamics, alumina should be stable in neutral aqueous electrolyte [69]. However, it has been shown in several instances that ALD alumina films experience dissolution in neutral NaCl solutions [76,90]. Diaz et al. proposed that this behavior for ALD Al_2O_3 on carbon steel resulted from cathodic oxygen reduction at the substrate surface, accessible through defects in the ALD film [90]. Recall that oxygen reduction proceeds through the first reaction in Equation (2). This reaction generates hydroxide ions, leading to a localized increase in pH and facilitating dissolution according to [90]:

$$Al_2O_3 + 2OH^- \rightarrow 2AlO_2^- + H_2O \tag{5}$$

The degradation of ALD Al_2O_3 on copper over 72 h in 0.1 M NaCl seen in this work supports the dissolution of alumina in neutral electrolyte. As in the work of Diaz et al. [90], deterioration of the alumina film took place in deoxygenated electrolyte as well as oxygenated. This suggests either incomplete removal of dissolved oxygen, coupled with a very low threshold for oxygen reduction to occur, or the generation of hydroxide ions through other means, possibly water reduction (see Equation (2)). Both scenarios likely contribute: incomplete removal of dissolved oxygen has been proposed in the past [80] and some water reduction should occur at the high cathodic overpotentials to which the samples are exposed. However, even in the absence of the reduction step (at -1.5 V) the ALD alumina film degrades similarly according to EIS spectra, displayed in Figure 12. Thus, hydroxide ions generated at equilibrium conditions must be sufficient to cause dissolution, assuming Equation (5) is the primary reaction pathway for alumina ALD films.

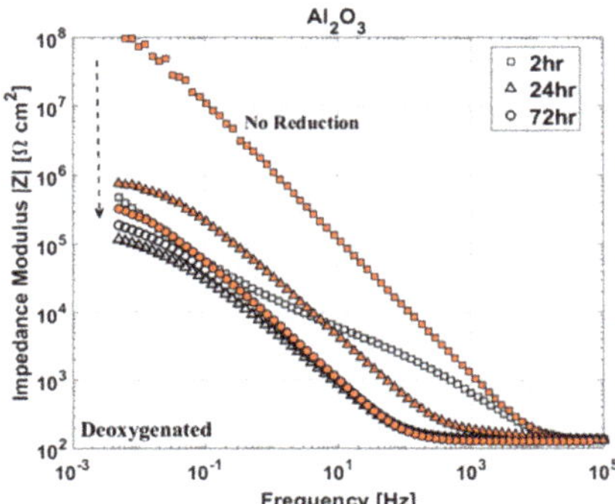

Figure 12. Impedance modulus versus frequency at selected immersion times over 72 h in deoxygenated 0.1 M NaCl for ALD Al_2O_3 on copper. Unfilled shapes (black) are with and filled shapes (red) are without an initial 2.5-min reduction at −1.5 V. Dashed arrow points in the direction of increasing immersion time.

A recent study investigated the stability of ALD alumina films on silicon substrates in aqueous media at elevated temperatures [68]. This study did not report dissolution of alumina in neutral electrolyte, which is not surprising since oxygen reduction would likely not occur at the Si surface to generate the required hydroxide ions, according to Equation (5). However, the authors did show hydration and swelling, followed by film restructuring and roughening during exposure to neutral DI water at room temperature with and without chloride ions. Alumina is known to exist in its hydrated form when in contact with water [69]. Swelling and roughening would increase porosity and exposed surface area, which could enhance the overall degradation of the alumina-coated copper. Note that the hydration and dissolution mechanisms discussed here are not mutually exclusive and could both contribute to the observed phenomena.

3.6. Effect of Dissolved Oxygen

The presence of dissolved oxygen is known to shift the equilibrium of Equation (1) toward $CuCl_2^-$, leading to a higher dissolution rate with higher oxygen content [28]. The enhanced dissolution of CuCl does not allow it to reach maximum coverage on the surface and inhibits passivation until a critical current of 0.3–0.4 mA cm^{-2} is reached. The passivation current is more variable in deoxygenated electrolyte, though it is roughly an order of magnitude smaller than in the presence of oxygen. Reverse potential scans with oxygen present (Figure 3b) are significantly less variable than in deoxygenated solutions. This, coupled with the single reduction peak for all samples, indicates a distinct mechanism from deoxygenated media, which is presumed to be the formation of a Cu_2O surface film. The reduction of this film appears to be a slower reaction, as this peak is evident at a scan rate of 0.5 mV/s, whereas the CuCl reduction on the bare copper was not visible until scan rates of 50–100 mV/s were used.

There is a current peak and subsequent reduction in the anodic portion of the polarization curves for some of the ALD samples in oxygenated NaCl. This additional current peak is more obvious in the CV scans provided in Figure 13 and generally results in a broadening of the passivation peak. This may be indicative of maximum CuCl coverage at the Cu surface sites exposed through coating defects. The much smaller substrate area exposed through pores in the ALD films compared to the uncoated copper could promote full CuCl surface coverage before full oxide film formation occurs.

ALD film stability is rather unaffected by dissolved oxygen in the electrolyte. Formation of a copper oxide film at the substrate-ALD film interface likely counteracts the more aggressive dissolution of copper with oxygen present. This is an important point for implementation of ALD coatings in RF devices for corrosion protection, as it is common for dissolved oxygen levels to be above specification for coolant supplies.

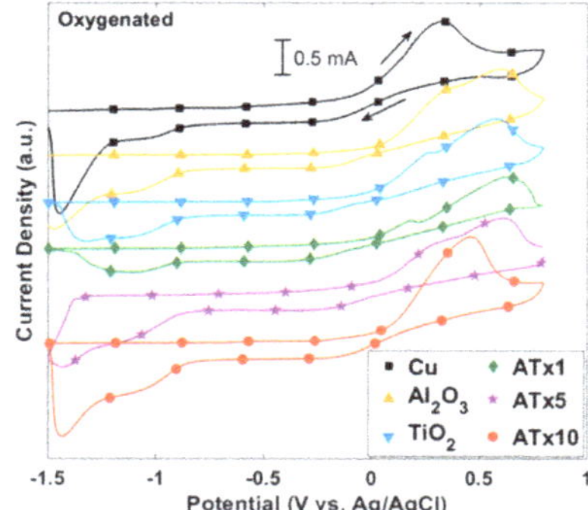

Figure 13. I–V curves in oxygenated 0.1 M NaCl for coated and uncoated copper, measured at a scan rate of 10 mV/s. Arrows indicate the scan direction.

3.7. Copper-Based Surface Films and Sample Aging

Voltammetric curves were measured with and without (results not shown) a 30-min rest period to investigate the effect of copper-based surface films on electrode polarization. Samples were held at −1.5 V for 2.5 min to reduce surface films, after which either the polarization scan immediately begins or there is a 30-min rest period at the open circuit potential. In both cases the scan starts at −1 V vs. Ag/AgCl. Uncoated copper exhibits higher polarization resistance (as much as 5×) without the rest period regardless of dissolved oxygen content in the electrolyte. Both CuCl and Cu_2O are thought to precipitate on the Cu surface during immersion in Cl-containing electrolyte [30,31,91]. The presence of these corrosion products formed during the open circuit measurement could account for the increase in cathodic currents, as copper corrosion products are known to complicate (and perhaps enhance) oxygen reduction [92]. Differences in polarization resistance and anodic current (deoxygenated electrolyte) may be indicative of patchy CuCl/Cu_2O film formation at OCP, resulting in enhanced dissolution, rather than passivation.

The effects of a rest period after electroreduction are mixed for the ALD-coated copper samples. Polarization resistance generally decreases for ALD samples without the 30-min OCP period. This is due to a likely (partial) passivation of the copper surface at ALD film defect sites during the OCP measurement. The much smaller exposed area through pores in the ALD film compared with the uncoated copper allows surface films to precipitate faster. This is supported by the fact that the TiO_2 sample sees an increase in polarization resistance comparable to uncoated copper. The TiO_2 possesses inferior sealing and nucleation properties compared with Al_2O_3, resulting in increased porosity and allowing for a larger area of exposed Cu. In some cases, the polarization behavior is largely unchanged with or without the rest period.

Assuming we may treat the ALD films as electrochemically inert simplifies this analysis considerably. In that case, changes in polarization behavior must be ascribed to changes in the state of the copper surface or to blocking effects by the ALD films. However, the dissolution of Al_2O_3 would complicate this process (refer to Section 3.5). A lack of passivation of surface sites during the immediate transition from −1.5 V to −1 V for the polarization scan allows for enhanced oxygen reduction and local increases in pH. This would lead to dissolution of alumina via Equation (5), contributing to the measured current and the apparent decrease in corrosion protection.

Aging of the ALD-coated samples plays an important role in parameters derived from electrochemical testing, as is apparent from the discussion of the effect of copper surface films formed during testing. After the coupons are coated (and during deposition [93]), copper oxide growth is expected at the base of coating pores and defects. Passivation in this manner can greatly increase initial corrosion resistance. This is evident in the work of Mirhashemihaghighi et al. [80,93], in which polarization resistances on the order of 100 MΩ cm^2 are measured without an electroreduction step during immersion in 0.5 M NaCl. Also, this is apparent in Figure 12, where the lack of a reduction step leads to a two order of magnitude increase in impedance of Al_2O_3-coated copper. It does not, however, reduce or prevent the dissolution of alumina over time. Further investigation is needed to determine the effect of aging of ALD films on copper on the long-term stability of the films.

3.8. Effect of Deposition Temperature

It is well known that many thermal ALD processes can proceed over a wide temperature range. For example, Al_2O_3 has been deposited by ALD from room temperature [45,94] up to 400 °C [76]. The substrate temperature during deposition can have a substantial impact on film properties, though its effect on the corrosion resistance of the film is not immediately apparent. Increasing temperature has two competing consequences for thin film growth: densification and crystallization. Higher deposition temperature for inorganic films almost always leads to an increase in density, which is desirable for corrosion protection. Along with an increase in density, certain materials begin to crystallize. Though crystallinity is preferred in many applications, amorphous corrosion barrier films are generally more effective. Crystalline films tend to have aligned pores and thus provide inferior separation between substrate and environment.

ALD Al_2O_3 is known to be amorphous up to temperatures well beyond the upper limit of atomic layer growth [95,96], which is important for its use as a corrosion barrier coating. Higher deposition temperature leads to a denser film and decreases impurities such as hydrogen and carbon [76,97]. Figure 14a shows that this does not translate into an increase in the stability of ALD alumina on copper. Despite the higher starting film quality of the 250 °C deposition, resulting in impedance that is as much as an order of magnitude higher than the film deposited at 150°C, the stability is still poor. The 250 °C Al_2O_3 film experiences a 100x decrease in low-frequency impedance after 72 h, which is ~10x that of the 150 °C film. This result agrees with a previous study of ALD alumina films, in which increased chemical stability was not observed until annealing at 900 °C was performed [68].

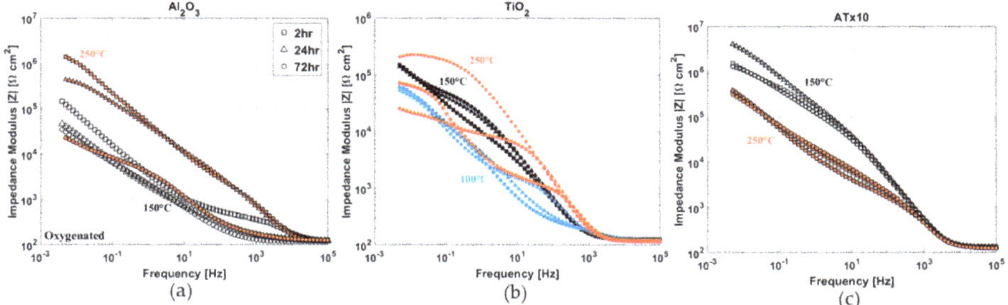

Figure 14. Impedance modulus versus frequency at selected immersion times over 72 h in oxygenated 0.1 M NaCl for various ALD substrate temperatures: (**a**) Al_2O_3-coated Cu; (**b**) TiO_2-coated Cu; and (**c**) ATx10 [(Al_2O_3+TiO_2)x10]-coated Cu. In (**a**) and (**c**) unfilled shapes (black) represent a substrate temperature of 150 °C during deposition and filled shapes (red) indicate 250 °C substrate temperature. In (**b**) substrate temperatures are: 250 °C—red, 150 °C—black, and 100 °C—blue. Deposition temperatures are also provided on each plot, and all films are nominally 50 nm thick.

The temperature dependence of TiO_2 deposition is a bit more complicated. Refractive index and impurity concentrations generally increase and decrease, respectively, with increasing deposition temperature [76,98,99]. TiO_2 ALD films deposited using the $TiCl_4$-H_2O process have been shown to be amorphous at temperatures $\leq$ 150 °C and crystalline above roughly 165 °C [76,98,100]. Studies have also presented film thickness [100,101] and substrate [102] dependence of TiO_2 crystallinity by ALD. There are competing mechanisms with TiO_2 ALD growth: higher deposition temperatures produce better film quality but also result in varying degrees of crystallization. This effect is demonstrated in Figure 14b, which shows impedance spectra for three different deposition temperatures of TiO_2 ALD growth on copper. The films deposited at 100 °C and 150 °C both show good stability over 72 h, though the higher film quality of the 150 °C film is apparent based on the larger measured impedance from ~100 Hz to 5 mHz. The titania film deposited at 250 °C exhibits vastly different impedance spectra after 2, 24, and 72 h immersion. Unlike alumina, the chemical stability of titania in neutral salts is not in question, so this behavior is indicative of changes in the copper substrate and may be attributed to the higher porosity of this coating due to its crystallinity.

A deposition temperature of 250 °C for the 10x nanolaminate (ATx10) results in lower overall impedance (Figure 14c), though there is very little change in impedance over 72 h immersion. This contrasts the poor stability seen for alumina and titania individually at 250 °C. It is presumed that the multilayered structure generates discontinuous pores and can reduce the negative effects of the crystallization of TiO_2 layers. It is also likely that the TiO_2 layer (~2.5 nm) is not thick enough to become crystalline in the ATx10 film structure [100,101]. Further investigation is needed to fully understand the effects of deposition conditions on ultra-thin ALD nanolaminate film layers.

3.9. Nanolaminate Coatings

The effects of having a conformal sealing layer are apparent in the increased initial corrosion protection afforded by the alumina-containing coatings over the single-layer TiO_2 film. However, due to the dissolution of alumina, this beneficial effect may be partially overshadowed. Each of the three nanolaminate coatings was a superior corrosion barrier to either of the individual materials. It was expected that increasing the number of film layers would lead to lower porosity and a better corrosion barrier, though ATx5 [(5 nm Al_2O_3 + 5 nm TiO_2)x5], rather than ATx10 [(2.5 nm Al_2O_3 + 2.5 nm TiO_2)x10], provided the best overall corrosion protection, as determined by EIS measurements over 72 h with and without deoxygenation. We attribute this to achieving a balance between the number of layers (and interfaces) and better film quality as thickness increases. Having multiple film layers likely facilitates the filling of pores in previous layers and the misalignment of pores across several interfaces. On the other hand, film density and uniformity generally increase with thickness for ultra-thin films. Thus, having fewer interfaces in ATx5 is balanced by generating higher-quality individual layers at 5 nm compared to 2.5 nm for ATx10. Nonetheless, the nanolaminate films successfully incorporate desirable properties of both film materials. The possibility of including additional materials, metal oxides or otherwise, allows these films to be tailorable to specific situations or environments. The precise thickness control of ALD enables these types of nanolaminate films with many individual layers, even when the total film thickness is only 50 nm or less.

4. Conclusions

Atomic layer deposition presents the opportunity to apply nanometer-scale corrosion barrier coatings to very high aspect ratio structures and component interiors, which is not possible with many other coating techniques. The results presented here show the promise of ultra-thin ALD barrier coatings in chloride environments. DC voltammetry measurements produced 10–15x increases in polarization resistance, up to 3.4 MΩ cm^2, with 50 nm nanolaminate films of Al_2O_3 and TiO_2, two materials that are readily available and widely deposited by ALD. With longer equilibration time than for the DC measurements, the nanolaminate films saw nearly a 100x increase in the polarization resistance determined from equivalent circuit fitting of impedance spectra, up to 12 MΩ cm^2. This was even higher, well above 100 MΩ cm^2, when the initial electroreduction step was removed and could be further increased using thicker films or different materials.

The stability of single-layer Al_2O_3 ALD films on copper was found to be poor during immersion in neutral 0.1 M NaCl, with severe degradation in measured impedance response occurring within 24 h. This agrees with previous reports of the dissolution of ALD alumina on active substrates [76,90] and has been attributed to hydroxide generation during oxygen reduction at the substrate surface exposed through pores or defects in the ALD film [90]. TiO_2 was found to be very stable, although it showed the smallest initial increase in corrosion resistance because of its inferior nucleation and sealing capabilities compared to alumina. Nanolaminate film structures, including ATx1 [10 nm Al_2O_3 + 40 nm TiO_2], ATx5 [(5 nm Al_2O_3 + 5 nm TiO_2)x5], and ATx10 [(2.5 nm Al_2O_3 + 2.5 nm TiO_2)x10], were successful at combining the superior sealing properties of alumina with the excellent chemical stability of titania. ATx5 exhibited the best overall performance according to impedance spectroscopy over 72 h. We attribute this to the beneficial effects of multiple thin film layers without individual layer quality suffering from being too thin to form a dense and effective barrier. It remains to be seen how these samples perform over longer immersion times and in flowing electrolyte conditions.

The effect of dissolved oxygen in the electrolyte was probed through various electrochemical measurements. As expected, the presence of oxygen enhances corrosion in chloride-containing media. However, perhaps unexpectedly, dissolved oxygen was found to have little effect on the stability of ALD films. The ability of copper exposed through coating pores to passivate caused there to be minimal variation in impedance over 72 h, comparable to deoxygenated conditions. Laser microscope imaging revealed little difference in surface morphology between nanolaminate samples exposed to oxygenated or deoxygenated electrolyte.

Also explored was the use of different deposition temperatures to determine its effect on film quality and corrosion barrier efficacy. An obviously higher initial film quality existed for Al_2O_3 single layers deposited at 250 °C, though degradation over time was found to be at least as, if not more, severe than for the films deposited at 150 °C, according to impedance spectra. Substrate temperature during deposition of TiO_2 is more complicated in that it impacts the crystallinity as well as the dielectric quality and contaminant concentrations. As with alumina, the initial dielectric quality of the titania films was found to increase with deposition temperature, although crystallization and increased porosity at 250 °C resulted in poor performance over the 72-h immersion. The stability of TiO_2 at 100 °C was good, but it was still outperformed by the 150 °C deposition, which appears to be in the range of an optimal temperature for titania ALD for corrosion protection. ATx10 was also deposited at 250 °C and showed good stability over 72 h, unlike both the single-layer alumina and titania deposited at this temperature.

Author Contributions: Conceptualization, M.A.F., C.J.O., G.N.P.; methodology, M.A.F.; software, M.A.F.; formal analysis, M.A.F.; investigation, M.A.F.; resources, G.N.P.; data curation, M.A.F.; writing—original draft preparation, M.A.F.; writing—review and editing, C.J.O., G.N.P., M.A.F.; visualization, M.A.F., C.J.O., G.N.P.; supervision, C.J.O. and G.N.P.; project administration, C.J.O. and G.N.P.; funding acquisition, C.J.O. and G.N.P.

Funding: This research was funded by the U.S. Navy under contract N00253-16-C-0002.

Acknowledgments: This work was performed in part at the Analytical Instrumentation Facility (AIF) at North Carolina State University, which is supported by the State of North Carolina and the National Science Foundation (award number ECCS-1542015). The AIF is a member of the North Carolina Research Triangle Nanotechnology Network (RTNN), a site in the National Nanotechnology Coordinated Infrastructure (NNCI).

Conflicts of Interest: The authors declare no conflict of interest.

References

1. Al-Kharafi, F.M.; Shalaby, H.M. Corrosion Behavior of Annealed and Hard-Drawn Copper in Soft Tap Water. *Corrosion* **1995**, *51*, 469–481. [CrossRef]
2. Sobue, K.; Sugahara, A.; Nakata, T.; Imai, H.; Magaino, S. Effect of free carbon dioxide on corrosion behavior of copper in simulated water. *Surf. Coat. Technol.* **2003**, *169–170*, 662–665. [CrossRef]
3. Lee, S.-H.; Kim, J.-G.; Koo, J.-Y. Investigation of pitting corrosion of a copper tube in a heating system. *Eng. Fail. Anal.* **2010**, *17*, 1424–1435. [CrossRef]
4. Suh, S.H.; Suh, Y.; Yoon, H.G.; Oh, J.H.; Kim, Y.; Jung, K.; Kwon, H. Analysis of pitting corrosion failure of copper tubes in an apartment fire sprinkler system. *Eng. Fail. Anal.* **2016**, *64*, 111–125. [CrossRef]
5. Lu, B.; Meng, W.J.; Mei, F. Experimental investigation of Cu-based, double-layered, microchannel heat exchangers. *J. Micromech. Microeng.* **2013**, *23*, 035017. [CrossRef]
6. Mei, F.; Phillips, W.A.; Lu, B.; Meng, W.J.; Guo, S. Fabrication of copper-based microchannel devices and analysis of their flow and heat transfer characteristics. *J. Micromech. Microeng.* **2009**, *19*, 035009. [CrossRef]

7. Medrano, M.; Yilmaz, M.O.; Nogués, M.; Martorell, I.; Roca, J.; Cabeza, L.F. Experimental evaluation of commercial heat exchangers for use as PCM thermal storage systems. *Appl. Energy* **2009**, *86*, 2047–2055. [CrossRef]
8. Wang, X.J.; Qiu, X.; Ben-Zvi, I. Experimental observation of high-brightness microbunching in a photocathode rf electron gun. *Phys. Rev. E* **1996**, *54*, R3121–R3124. [CrossRef]
9. Tooker, J.F.; Huynh, P.; Felch, K.; Blank, M.; Borchardt, P.; Cauffman, S. Gyrotron and power supply development for upgrading the electron cyclotron heating system on DIII-D. *Fusion Eng. Des.* **2013**, *88*, 521–524. [CrossRef]
10. Musumeci, P.; Cultrera, L.; Ferrario, M.; Filippetto, D.; Gatti, G.; Gutierrez, M.S.; Moody, J.T.; Moore, N.; Rosenzweig, J.B.; Scoby, C.M.; et al. Multiphoton Photoemission from a Copper Cathode Illuminated by Ultrashort Laser Pulses in an rf Photoinjector. *Phys. Rev. Lett.* **2010**, *104*, 084801. [CrossRef] [PubMed]
11. Li, R.K.; To, H.; Andonian, G.; Feng, J.; Polyakov, A.; Scoby, C.M.; Thompson, K.; Wan, W.; Padmore, H.A.; Musumeci, P. Surface-Plasmon Resonance-Enhanced Multiphoton Emission of High-Brightness Electron Beams from a Nanostructured Copper Cathode. *Phys. Rev. Lett.* **2013**, *110*, 074801. [CrossRef] [PubMed]
12. Edwards, M.; Ferguson, J.F.; Reiber, S.H. The pitting corrosion of copper. *J. Am. Water Works Assoc.* **1994**, *86*, 74–90. [CrossRef]
13. Brusic, V. Copper Corrosion With and Without Inhibitors. *J. Electrochem. Soc.* **1991**, *138*, 2253. [CrossRef]
14. Jeon, B.; Sankaranarayanan, S.K.R.S.; van Duin, A.C.T.; Ramanathan, S. Atomistic insights into aqueous corrosion of copper. *J. Chem. Phys.* **2011**, *134*, 234706. [CrossRef] [PubMed]
15. Feng, Y.; Siow, K.-S.; Teo, W.-K.; Tan, K.-L.; Hsieh, A.-K. Corrosion Mechanisms and Products of Copper in Aqueous Solutions at Various pH Values. *Corrosion* **1997**, *53*, 389–398. [CrossRef]
16. Pehkonen, S.O.; Palit, A.; Zhang, X. Effect of Specific Water Quality Parameters on Copper Corrosion. *Corrosion* **2002**, *58*, 156–165. [CrossRef]
17. Kompfner, R. The Traveling-Wave Tube as Amplifier at Microwaves. *Proc. IRE* **1947**, *35*, 124–127. [CrossRef]
18. Gill, C.; Wilczek, A. Crossed-Field Microwave Tube Having a Fluid Cooled Cathode and Control Electrode. U.S. Patent No. 3,612,932, 12 October 1971.
19. Boyd, R.; Hendry, F.; Mannette, R. Fluid Cooled Traveling Wave Tube. U.S. Patent No. 3,246,190, 14 March 1966.
20. Garven, M.; Calame, J.P.; Danly, B.G.; Nguyen, K.T.; Levush, B.; Wood, F.N.; Pershing, D.E. A gyrotron-traveling-wave tube amplifier experiment with a ceramic loaded interaction region. *IEEE Trans. Plasma Sci.* **2002**, *30*, 885–893. [CrossRef]
21. Calame, J.P.; Garven, M.; Danly, B.G.; Levush, B.; Nguyen, K.T. Gyrotron-traveling wave-tube circuits based on lossy ceramics. *IEEE Trans. Electron Devices* **2002**, *49*, 1469–1477. [CrossRef]
22. Skowron, J.F. The continuous-cathode (emitting-sole) crossed-field amplifier. *Proc. IEEE* **1973**, *61*, 330–356. [CrossRef]
23. MacPhail, G.R. Crossed-Field Amplifier. U.S. Patent No. 4,700,109, 13 October 1987.
24. Okress, E.C. Crossed-Field Amplifier. U.S. Patent No. 3,082,351, 19 March 1963.
25. McDougall, J.L.; McCall, L.; Stevenson, M.E. Water Chemistry and Processing Effects on the Corrosion Degradation of Copper Tubing in Cooling Water Systems. *Pract. Fail. Anal.* **2003**, *3*, 81–88. [CrossRef]
26. Lee, H.P. Kinetics and Mechanisms of Cu Electrodissolution in Chloride Media. *J. Electrochem. Soc.* **1986**, *133*, 2035. [CrossRef]
27. Braun, M.; Nobe, K. Electrodissolution Kinetics of Copper in Acidic Chloride Solutions. *J. Electrochem. Soc.* **1979**, *126*, 1666. [CrossRef]
28. Bjorndahl, W.D.; Nobe, K. Copper Corrosion in Chloride Media. Effect of Oxygen. *Corrosion* **1984**, *40*, 82–87. [CrossRef]
29. King, F.; Litke, C.D.; Quinn, M.J.; LeNeveu, D.M. The measurement and prediction of the corrosion potential of copper in chloride solutions as a function of oxygen concentration and mass-transfer coefficient. *Corros. Sci.* **1995**, *37*, 833–851. [CrossRef]
30. Deslouis, C.; Tribollet, B.; Mengoli, G.; Musiani, M.M. Electrochemical behaviour of copper in neutral aerated chloride solution. I. Steady-state investigation. *J. Appl. Electrochem.* **1988**, *18*, 374–383. [CrossRef]
31. Kear, G.; Barker, B.D.; Walsh, F.C. Electrochemical corrosion of unalloyed copper in chloride media—A critical review. *Corros. Sci.* **2004**, *46*, 109–135. [CrossRef]

32. Sherif, E.M.; Park, S.-M. Inhibition of Copper Corrosion in 3.0% NaCl Solution by *N*-Phenyl-1,4-phenylenediamine. *J. Electrochem. Soc.* **2005**, *152*, B428. [CrossRef]
33. Eriksen, T.E.; Ndalamba, P.; Grenthe, I. On the corrosion of copper in pure water. *Corros. Sci.* **1989**, *29*, 1241–1250. [CrossRef]
34. Hultquist, G. Why copper may be able to corrode in pure water. *Corros. Sci.* **2015**, *93*, 327–329. [CrossRef]
35. Ottosson, M.; Boman, M.; Berastegui, P.; Andersson, Y.; Hahlin, M.; Korvela, M.; Berger, R. Copper in ultrapure water, a scientific issue under debate. *Corros. Sci.* **2017**, *122*, 53–60. [CrossRef]
36. Hedin, A.; Johansson, A.J.; Lilja, C.; Boman, M.; Berastegui, P.; Berger, R.; Ottosson, M. Corrosion of copper in pure O_2-free water? *Corros. Sci.* **2018**, *137*, 1–12. [CrossRef]
37. Szakálos, P.; Hultquist, G.; Wikmark, G. Corrosion of Copper by Water. *Electrochem. Solid-State Lett.* **2007**, *10*, C63–C67. [CrossRef]
38. Lu, L.; Shen, Y.; Chen, X.; Qian, L.; Lu, K. Ultrahigh Strength and High Electrical Conductivity in Copper. *Science* **2004**, *304*, 422–426. [CrossRef] [PubMed]
39. Parsons, G.N.; George, S.M.; Knez, M. Progress and future directions for atomic layer deposition and ALD-based chemistry. *MRS Bull.* **2011**, *36*, 865–871. [CrossRef]
40. George, S.M. Atomic layer deposition: An overview. *Chem. Rev.* **2010**, *110*, 111–131. [CrossRef] [PubMed]
41. Leskelä, M.; Ritala, M. Atomic Layer Deposition Chemistry: Recent Developments and Future Challenges. *Angew. Chem. Int. Ed.* **2003**, *42*, 5548–5554. [CrossRef] [PubMed]
42. Leskelä, M.; Ritala, M. Atomic layer deposition (ALD): From precursors to thin film structures. *Thin Solid Films* **2002**, *409*, 138–146. [CrossRef]
43. Johnson, R.W.; Hultqvist, A.; Bent, S.F. A brief review of atomic layer deposition: From fundamentals to applications. *Mater. Today* **2014**, *17*, 236–246. [CrossRef]
44. Meyer, J.; Riedl, T. Low-Temperature Atomic Layer Deposition. In *Atomic Layer Deposition of Nanostructured Materials*; Pinna, N., Knez, M., Eds.; Wiley-VCH Verlag GmbH & Co. KGaA: Weinheim, Germany, 2012; pp. 109–130. ISBN 9783527639915.
45. Groner, M.D.; Fabreguette, F.H.; Elam, J.W.; George, S.M. Low-Temperature Al_2O_3 Atomic Layer Deposition. *Chem. Mater.* **2004**, *16*, 639–645. [CrossRef]
46. Peng, Q.; Sun, X.-Y.; Spagnola, J.C.; Hyde, G.K.; Spontak, R.J.; Parsons, G.N. Atomic Layer Deposition on Electrospun Polymer Fibers as a Direct Route to Al_2O_3 Microtubes with Precise Wall Thickness Control. *Nano Lett.* **2007**, *7*, 719–722. [CrossRef] [PubMed]
47. Mundy, J.Z.; Shafiefarhood, A.; Li, F.; Khan, S.A.; Parsons, G.N. Low temperature platinum atomic layer deposition on nylon-6 for highly conductive and catalytic fiber mats. *J. Vac. Sci. Technol. A* **2016**, *34*, 01A152. [CrossRef]
48. Brozena, A.H.; Oldham, C.J.; Parsons, G.N. Atomic layer deposition on polymer fibers and fabrics for multifunctional and electronic textiles. *J. Vac. Sci. Technol. A* **2016**, *34*, 010801. [CrossRef]
49. Daubert, J.S.; Mundy, J.Z.; Parsons, G.N. Kevlar-Based Supercapacitor Fibers with Conformal Pseudocapacitive Metal Oxide and Metal Formed by ALD. *Adv. Mater. Interfaces* **2016**, *3*, 1600355. [CrossRef]
50. Parsons, G.N.; Atanasov, S.E.; Dandley, E.C.; Devine, C.K.; Gong, B.; Jur, J.S.; Lee, K.; Oldham, C.J.; Peng, Q.; Spagnola, J.C.; et al. Mechanisms and reactions during atomic layer deposition on polymers. *Coord. Chem. Rev.* **2013**, *257*, 3323–3331. [CrossRef]
51. Biercuk, M.J.; Monsma, D.J.; Marcus, C.M.; Becker, J.S.; Gordon, R.G. Low-temperature atomic-layer-deposition lift-off method for microelectronic and nanoelectronic applications. *Appl. Phys. Lett.* **2003**, *83*, 2405–2407. [CrossRef]
52. Lim, S.J.; Kwon, S.; Kim, H.; Park, J.-S. High performance thin film transistor with low temperature atomic layer deposition nitrogen-doped ZnO. *Appl. Phys. Lett.* **2007**, *91*, 183517. [CrossRef]
53. Puurunen, R.L. Surface chemistry of atomic layer deposition: A case study for the trimethylaluminum/water process. *J. Appl. Phys.* **2005**, *97*. [CrossRef]
54. Chang, M.L.; Cheng, T.C.; Lin, M.C.; Lin, H.C.; Chen, M.J. Improvement of oxidation resistance of copper by atomic layer deposition. *Appl. Surf. Sci.* **2012**, *258*, 10128–10134. [CrossRef]
55. Díaz, B.; Härkönen, E.; Światowska, J.; Maurice, V.; Seyeux, A.; Marcus, P.; Ritala, M. Low-temperature atomic layer deposition of Al_2O_3thin coatings for corrosion protection of steel: Surface and electrochemical analysis. *Corros. Sci.* **2011**, *53*, 2168–2175. [CrossRef]

56. Fedel, M.; Deflorian, F. Electrochemical characterization of atomic layer deposited Al_2O_3 coatings on AISI 316L stainless steel. *Electrochim. Acta* **2016**, *203*, 404–415. [CrossRef]
57. Giorleo, L.; Ceretti, E.; Giardini, C. ALD coated tools in micro drilling of Ti sheet. *CIRP Ann.* **2011**, *60*, 595–598. [CrossRef]
58. Mirhashemihaghighi, S.; Światowska, J.; Maurice, V.; Seyeux, A.; Zanna, S.; Salmi, E.; Ritala, M.; Marcus, P. Corrosion protection of aluminium by ultra-thin atomic layer deposited alumina coatings. *Corros. Sci.* **2016**, *106*, 16–24. [CrossRef]
59. Hirvikorpi, T.; Vähä-Nissi, M.; Mustonen, T.; Iiskola, E.; Karppinen, M. Atomic layer deposited aluminum oxide barrier coatings for packaging materials. *Thin Solid Films* **2010**, *518*, 2654–2658. [CrossRef]
60. Hirvikorpi, T.; Vähä-Nissi, M.; Harlin, A.; Karppinen, M. Comparison of some coating techniques to fabricate barrier layers on packaging materials. *Thin Solid Films* **2010**, *518*, 5463–5466. [CrossRef]
61. Lahtinen, K.; Maydannik, P.; Johansson, P.; Kääriäinen, T.; Cameron, D.C.; Kuusipalo, J. Utilisation of continuous atomic layer deposition process for barrier enhancement of extrusion-coated paper. *Surf. Coat. Technol.* **2011**, *205*, 3916–3922. [CrossRef]
62. Kääriäinen, T.O.; Maydannik, P.; Cameron, D.C.; Lahtinen, K.; Johansson, P.; Kuusipalo, J. Atomic layer deposition on polymer based flexible packaging materials: Growth characteristics and diffusion barrier properties. *Thin Solid Films* **2011**, *519*, 3146–3154. [CrossRef]
63. Bae, D.; Kwon, S.; Oh, J.; Kim, W.K.; Park, H. Investigation of Al_2O_3 diffusion barrier layer fabricated by atomic layer deposition for flexible Cu(In,Ga)Se_2 solar cells. *Renew. Energy* **2013**, *55*, 62–68. [CrossRef]
64. Baba Heidary, D.S.; Qu, W.; Randall, C.A. Evaluating the merit of ALD coating as a barrier against hydrogen degradation in capacitor components. *RSC Adv.* **2015**, *5*, 50869–50877. [CrossRef]
65. Langereis, E.; Creatore, M.; Heil, S.B.S.; van de Sanden, M.C.M.; Kessels, W.M.M. Plasma-assisted atomic layer deposition of Al_2O_3 moisture permeation barriers on polymers. *Appl. Phys. Lett.* **2006**, *89*, 081915. [CrossRef]
66. Park, S.K.; Oh, J.; Hwang, C.; Lee, J.; Yang, Y.S.; Chu, H.Y. Ultrathin Film Encapsulation of an OLED by ALD. *Electrochem. Solid-State Lett.* **2005**, *8*, H21. [CrossRef]
67. Jarvis, K.L.; Evans, P.J. Growth of thin barrier films on flexible polymer substrates by atomic layer deposition. *Thin Solid Films* **2017**, *624*, 111–135. [CrossRef]
68. Correa, G.C.; Bao, B.; Strandwitz, N.C. Chemical Stability of Titania and Alumina Thin Films Formed by Atomic Layer Deposition. *ACS Appl. Mater. Interfaces* **2015**, *7*, 14816–14821. [CrossRef] [PubMed]
69. Pourbaix, M. *Atlas of Electrochemical Equilibria in Aqueous Solutions*; NACE: Houston, TX, USA, 1974; ISBN 978-0915567980.
70. Sun, Q.; Yu, H.; Liu, Y.; Li, J.; Lu, Y.; Hunt, J.F. Improvement of water resistance and dimensional stability of wood through titanium dioxide coating. *Holzforschung* **2010**, *64*, 757–761. [CrossRef]
71. Taruta, S.; Watanabe, K.; Kitajima, K.; Takusagawa, N. Effect of titania addition on crystallization process and some properties of calcium mica–apatite glass-ceramics. *J. Non. Cryst. Solids* **2003**, *321*, 96–102. [CrossRef]
72. Karpagavalli, R.; Zhou, A.; Chellamuthu, P.; Nguyen, K. Corrosion behavior and biocompatibility of nanostructured TiO_2 film on Ti6Al4V. *J. Biomed. Mater. Res. Part A* **2007**, *83A*, 1087–1095. [CrossRef] [PubMed]
73. Abdulagatov, A.I.; Yan, Y.; Cooper, J.R.; Zhang, Y.; Gibbs, Z.M.; Cavanagh, A.S.; Yang, R.G.; Lee, Y.C.; George, S.M. Al_2O_3 and TiO_2 Atomic Layer Deposition on Copper for Water Corrosion Resistance. *ACS Appl. Mater. Interfaces* **2011**, *3*, 4593–4601. [CrossRef] [PubMed]
74. Pan, J.; Leygraf, C.; Thierry, D.; Ektessabi, A.M. Corrosion resistance for biomaterial applications of TiO_2 films deposited on titanium and stainless steel by ion-beam-assisted sputtering. *J. Biomed. Mater. Res.* **1997**, *35*, 309–318. [CrossRef]
75. Marin, E.; Guzman, L.; Lanzutti, A.; Ensinger, W.; Fedrizzi, L. Multilayer Al_2O_3/TiO_2 Atomic Layer Deposition coatings for the corrosion protection of stainless steel. *Thin Solid Films* **2012**, *522*, 283–288. [CrossRef]
76. Matero, R.; Ritala, M.; Leskelä, M.; Salo, T.; Aromaa, J.; Forsén, O. Atomic layer deposited thin films for corrosion protection. *Le J. Phys. IV* **1999**, *9*, 493–499. [CrossRef]
77. Ćurković, L.; Ćurković, H.O.; Salopek, S.; Renjo, M.M.; Šegota, S. Enhancement of corrosion protection of AISI 304 stainless steel by nanostructured sol–gel TiO_2 films. *Corros. Sci.* **2013**, *77*, 176–184. [CrossRef]

78. Li, S.; Fu, J. Improvement in corrosion protection properties of TiO_2 coatings by chromium doping. *Corros. Sci.* **2013**, *68*, 101–110. [CrossRef]
79. Daubert, J.S.; Hill, G.T.; Gotsch, H.N.; Gremaud, A.P.; Ovental, J.S.; Williams, P.S.; Oldham, C.J.; Parsons, G.N. Corrosion Protection of Copper Using Al_2O_3, TiO_2, ZnO, HfO_2, and ZrO_2 Atomic Layer Deposition. *ACS Appl. Mater. Interfaces* **2017**, *9*, 4192–4201. [CrossRef] [PubMed]
80. Mirhashemihaghighi, S.; Światowska, J.; Maurice, V.; Seyeux, A.; Klein, L.H.; Härkönen, E.; Ritala, M.; Marcus, P. Electrochemical and Surface Analysis of the Corrosion Protection of Copper by Nanometer-Thick Alumina Coatings Prepared by Atomic Layer Deposition. *J. Electrochem. Soc.* **2015**, *162*, C377–C384. [CrossRef]
81. Chai, Z.; Li, J.; Lu, X.; He, D. Use of electrochemical measurements to investigate the porosity of ultra-thin Al_2O_3 films prepared by atomic layer deposition. *RSC Adv.* **2014**, *4*, 39365–39371. [CrossRef]
82. Chai, Z.; Liu, Y.; Li, J.; Lu, X.; He, D. Ultra-thin Al_2O_3 films grown by atomic layer deposition for corrosion protection of copper. *RSC Adv.* **2014**, *4*, 50503–50509. [CrossRef]
83. Marin, E.; Lanzutti, A.; Guzman, L.; Fedrizzi, L. Chemical and electrochemical characterization of TiO_2/Al_2O_3 atomic layer depositions on AZ-31 magnesium alloy. *J. Coat. Technol. Res.* **2012**, *9*, 347–355. [CrossRef]
84. Marin, E.; Lanzutti, A.; Lekka, M.; Guzman, L.; Ensinger, W.; Fedrizzi, L. Chemical and mechanical characterization of TiO_2/Al_2O_3 atomic layer depositions on AISI 316 L stainless steel. *Surf. Coat. Technol.* **2012**, *211*, 84–88. [CrossRef]
85. Liu, C.; Bi, Q.; Leyland, A.; Matthews, A. An electrochemical impedance spectroscopy study of the corrosion behaviour of PVD coated steels in 0.5 N NaCl aqueous solution: Part I. Establishment of equivalent circuits for EIS data modelling. *Corros. Sci.* **2003**, *45*, 1243–1256. [CrossRef]
86. Cole, K.S.; Cole, R.H. Dispersion and Absorption in Dielectrics I. Alternating Current Characteristics. *J. Chem. Phys.* **1941**, *9*, 341–351. [CrossRef]
87. Brug, G.J.; van den Eeden, A.L.G.; Sluyters-Rehbach, M.; Sluyters, J.H. The analysis of electrode impedances complicated by the presence of a constant phase element. *J. Electroanal. Chem. Interfacial Electrochem.* **1984**, *176*, 275–295. [CrossRef]
88. Láng, G.; Heusler, K.E. Remarks on the energetics of interfaces exhibiting constant phase element behaviour. *J. Electroanal. Chem.* **1998**, *457*, 257–260. [CrossRef]
89. Stern, M.; Geary, A.L. Electrochemical Polarization I. A Theoretical Analysis of the Shape of Polarization Curves. *J. Electrochem. Soc.* **1957**, *104*, 56–63. [CrossRef]
90. Díaz, B.; Härkönen, E.; Maurice, V.; Światowska, J.; Seyeux, A.; Ritala, M.; Marcus, P. Failure mechanism of thin Al_2O_3 coatings grown by atomic layer deposition for corrosion protection of carbon steel. *Electrochim. Acta* **2011**, *56*, 9609–9618. [CrossRef]
91. Bianchi, G.; Fiori, G.; Longhi, P.; Mazza, F. "Horse Shoe" Corrosion of Copper Alloys in Flowing Sea Water: Mechanism, and Possibility of Cathodic Protection of Condenser Tubes in Power Stations. *Corrosion* **1978**, *34*, 396–406. [CrossRef]
92. Vazquez, M.V.; de Sanchez, S.R.; Calvo, E.J.; Schiffrin, D.J. The electrochemical reduction of oxygen on polycrystalline copper in borax buffer. *J. Electroanal. Chem.* **1994**, *374*, 189–197. [CrossRef]
93. Mirhashemihaghighi, S.; Światowska, J.; Maurice, V.; Seyeux, A.; Klein, L.H.; Salmi, E.; Ritala, M.; Marcus, P. Interfacial native oxide effects on the corrosion protection of copper coated with ALD alumina. *Electrochim. Acta* **2016**, *193*, 7–15. [CrossRef]
94. Van Hemmen, J.L.; Heil, S.B.S.; Klootwijk, J.H.; Roozeboom, F.; Hodson, C.J.; van de Sanden, M.C.M.; Kessels, W.M.M. Plasma and Thermal ALD of Al_2O_3 in a Commercial 200 mm ALD Reactor. *J. Electrochem. Soc.* **2007**, *154*, G165. [CrossRef]
95. Miikkulainen, V.; Leskelä, M.; Ritala, M.; Puurunen, R.L. Crystallinity of inorganic films grown by atomic layer deposition: Overview and general trends. *J. Appl. Phys.* **2013**, *113*, 021301. [CrossRef]
96. Jakschik, S.; Schroeder, U.; Hecht, T.; Gutsche, M.; Seidl, H.; Bartha, J.W. Crystallization behavior of thin ALD- Al_2O_3 films. *Thin Solid Films* **2003**, *425*, 216–220. [CrossRef]
97. Härkönen, E.; Díaz, B.; Światowska, J.; Maurice, V.; Seyeux, A.; Vehkamäki, M.; Sajavaara, T.; Fenker, M.; Marcus, P.; Ritala, M. Corrosion Protection of Steel with Oxide Nanolaminates Grown by Atomic Layer Deposition. *J. Electrochem. Soc.* **2011**, *158*, C369. [CrossRef]

98. Aarik, J.; Aidla, A.; Mändar, H.; Uustare, T.; Schuisky, M.; Hårsta, A. Atomic layer growth of epitaxial TiO_2 thin films from $TiCl_4$ and H_2O on α-Al_2O_3 substrates. *J. Cryst. Growth* **2002**, *242*, 189–198. [CrossRef]
99. Aarik, J.; Aidla, A.; Kiisler, A.; Uustare, T.; Sammelselg, V. Effect of crystal structure on optical properties of TiO_2 films grown by atomic layer deposition. *Thin Solid Films* **1997**, *305*, 270–273. [CrossRef]
100. Aarik, J.; Aidla, A.; Uustare, T.; Sammelselg, V. Morphology and structure of TiO_2 thin films grown by atomic layer deposition. *J. Cryst. Growth* **1995**, *148*, 268–275. [CrossRef]
101. Aarik, J.; Aidla, A.; Mändar, H.; Uustare, T. Atomic layer deposition of titanium dioxide from $TiCl_4$ and H_2O: Investigation of growth mechanism. *Appl. Surf. Sci.* **2001**, *172*, 148–158. [CrossRef]
102. Ritala, M.; Leskelä, M.; Nykänen, E.; Soininen, P.; Niinistö, L. Growth of titanium dioxide thin films by atomic layer epitaxy. *Thin Solid Films* **1993**, *225*, 288–295. [CrossRef]

© 2019 by the authors. Licensee MDPI, Basel, Switzerland. This article is an open access article distributed under the terms and conditions of the Creative Commons Attribution (CC BY) license (http://creativecommons.org/licenses/by/4.0/).

Article

Corrosion Resistance of Heat-Treated Ni-W Alloy Coatings

Magdalena Popczyk [1], Julian Kubisztal [1], Andrzej Szymon Swinarew [1,2], Zbigniew Waśkiewicz [2,3], Arkadiusz Stanula [2] and Beat Knechtle [4,*]

1 Faculty of Science and Technology, University of Silesia in Katowice, 41-500 Chorzów, Poland; magdalena.popczyk@us.edu.pl (M.P.); julian.kubisztal@us.edu.pl (J.K.); andrzej.swinarew@us.edu.pl (A.S.S.)
2 Institute of Sport Science, The Jerzy Kukuczka Academy of Physical Education, 40-065 Katowice, Poland; z.waskiewicz@awf.katowice.pl (Z.W.); a.stanula@awf.katowice.pl (A.S.)
3 Department of Sports Medicine and Medical Rehabilitation, Sechenov University, 119991 Moscow, Russia
4 Institute of Primary Care, University of Zurich, 8091 Zurich, Switzerland
* Correspondence: beat.knechtle@hispeed.ch

Received: 30 January 2020; Accepted: 3 March 2020; Published: 6 March 2020

Abstract: The paper presents research on evaluation of corrosion resistance of Ni-W alloy coatings subjected to heat treatment. The corrosion resistance was tested in 5% NaCl solution by the use of potentiodynamic polarization technique and electrochemical impedance spectroscopy. Characteristics of the Ni-W coatings after heat treatment were carried out using scanning electron microscopy, scanning Kelvin probe technique and X-ray diffraction. Suggested reasons for the improvement of properties of the heat treated Ni-W coating, obtained at the lowest current density value (125 mA·cm^{-2}), are the highest tungsten content (c.a. 25 at.%) as well as the smallest and the most homogeneous electrochemically active surface area.

Keywords: Ni-W alloy coating; heat treatment; corrosion resistance

---

1. Introduction

The electroplating technique is increasingly used to obtain new materials with specific functional properties. This is due to the fact that by controlling the deposition parameters, i.e. voltage, current, bath composition, temperature, it is possible to influence the structure of the obtained material, and hence its properties. The advantage of this method is the possibility of simultaneously co-depositing several metals as well as incorporation powders of metals, non-metals or chemical compounds into the coating [1–44]. Thus, the electroplating technique allows obtaining alloy and composite coatings (amorphous or crystalline) with a specific chemical and phase composition, as well as modelled surface morphology. Many metals are currently used as electrode materials in various electrochemical processes. Among them are the metals from the group of irons, especially nickel, which is characterized by good corrosion resistance and high catalytic activity in the process of hydrogen evolution. In order to improve the utilization of nickel coatings, various methods of their modifications could be applied, such as the use of alloys instead of pure elements. The interest in electrodeposited nickel - tungsten alloys is due to their specific magnetic, electrical, mechanical, thermal and corrosion properties [19–39,44]. These alloy coatings are widely used in the elements of machines operating under high mechanical load, at high temperatures, as well as in aggressive environments. Ni-W coatings are also used as electrode materials for hydrogen evolution reaction (HER) [2,19,40]. It should be noted that nickel - tungsten alloys can only be obtained from aqueous solutions through an induced code position, that is, tungsten is code posited with nickel. Sulphate, sulfamine and citrate baths with the addition of sodium tungstate are usually used [19–39,44].

Generally, heat treatment of electrolytic coatings should increase their corrosion resistance what was confirmed in earlier studies e.q. in [10,13,15,17,41]. The formation of new phases and the reduction of the active surface after heat treatment are the main reasons for improving the corrosion resistance of these materials. Thus, we expect that heat treatment of investigated Ni-W alloy coatings can also significantly slow down corrosion processes occurring on its surface.

According to our knowledge there is lack of information about corrosion resistance of Ni-W coatings subjected to heat treatment in the air. Thus, the aim of this work is to study corrosion properties of heat-treated Ni-W coatings in 5% NaCl solution especially with respect to surface morphology, chemical and phase composition. The Ni-W coatings were deposited under galvanostatic conditions at the following cathodic current densities: 125, 150, 175 and 200 mA·cm^{-2}. The heat treatment of all coatings was carried out at a temperature of 1173 K. Therefore, the coatings discussed in the article were marked as follows: C125/1173, C150/1173, C175/1173 and C200/1173.

2. Materials and Methods

The Ni-W alloy coatings were obtained by electroplating from the electrolyte of the following composition (concentrations in g·dm^{-3}): $NiSO_4 \cdot 7H_2O$–13, $Na_2WO_4 \cdot 2H_2O$–68, $C_6H_5O_7Na_3 \cdot 2H_2O$–200 and NH_4Cl–50. For preparation of the bath ultrapure water (Millipore, 18.2 MΩ cm) and 'analytical grade' reagents (Avantor Performance Materials Poland S.A.) were used. The coatings were deposited galvanostatically at the current densities 125, 150, 175 and 200 mA·cm^{-2} and temperature of 343 K. The coatings were deposited on the steel (S235) plate of 1.0 cm^2 geometric surface area. A platinum mesh served as an auxiliary electrode. The chemical composition of the as-deposited Ni-W alloy coatings is presented in Table 1.

Table 1. Chemical composition of the as-deposited Ni-W alloy coatings determined by energy dispersive spectroscopy, in dependence on deposition current density.

Type of As-Deposited Coatings	At.% Ni	At.% W
Ni-W (j_{dep} = 125 mA·cm^{-2})	75.4 ± 0.4%	24.6 ± 0.4%
Ni-W (j_{dep} = 150 mA·cm^{-2})	77.3 ± 0.2%	22.7 ± 0.2%
Ni-W (j_{dep} = 175 mA·cm^{-2})	78.5 ± 0.7%	21.5 ± 0.7%
Ni-W (j_{dep} = 200 mA·cm^{-2})	80.2 ± 0.1%	19.8 ± 0.1%

Heat treatment of Ni-W alloy coatings was carried out in a muffle stove of the type FCF 2.5 SHMgO (Czylok Company, Jastrzębie-Zdrój, Poland) at 1173 K for 1 h in the air.

The surface morphology and chemical composition of the heat-treated coatings was studied using a scanning electron microscope (SEM, JEOL JSM–6480, JEOL Ltd., Tokyo, Japan) equipped with an energy dispersive spectroscopy (EDS) detector (JEOL Ltd., Tokyo, Japan). The phase composition was determined by means of X-ray diffraction method using Philips X'Pert PW 3040/60 X-ray diffractometer (U = 40 kV, I = 30 mA, Panalytical, Almelo, Netherlands) with copper radiation (λ (Cu K_α) = 1.54178 Å). The data collection was over the 2-theta range of 20° to 120° in steps of 0.02°.

Corrosion resistance of the heat-treated coatings was determined, using potentiodynamic polarization technique and electrochemical impedance spectroscopy (EIS). These measurements were carried out in a 5 wt.% NaCl solution, using three-electrode cell and an AUTOLAB® electrochemical system (PGSTAT30, Metrohm Autolab B.V., Utrecht, Netherlands). The auxiliary electrode was a platinum mesh and the reference electrode was a saturated calomel electrode (SCE). Potentiodynamic curves were recorded in the potential range ± 100 mV versus open circuit potential with rate v = 1 mV·s^{-1}.

The electrochemical impedance spectroscopy was performed at the corrosion potential. In these measurements, the amplitude of the ac signal was 10 mV. A frequency range from 10 kHz to 0.1 Hz was covered with 10 points per decade. All electrochemical investigations were made at 298 K.

Contact potential difference (*CPD*) maps and surface topography maps of the heat-treated coatings were recorded by means of Scanning Kelvin Probe (SKP) technique using PAR Model 370 Scanning Electrochemical Workstation (Princeton Applied Research, Oak Ridge, USA) equipped with a tungsten Kelvin probe (KP). The scanning area was 4000 × 4000 μm^2 and the distance between the probe and the sample was ca. 100 μm.

3. Results and Discussion

The heat-treated Ni-W coatings are characterized by grey, smooth and uniform surface. The surface morphology of the coatings differs, which means it depends on the deposition current density (Figure 1). The surface of C125/1173 coating shows small, separately located globules changing into larger ones with increasing of deposition current density. Coatings obtained at low current density values have a poorly developed surface. It can be explained by that low current densities favor the slow discharge of ions at electrodes, and therefore the growth rate of the resulting grains exceeds the speed of forming of new ones. As the current density increases, the rate of formation of new grains also increases what result in more developed surface. The increase in the density of the deposition current causes intense hydrogen evolution, which in turn can cause the formation of porous coatings.

Figure 1. Surface morphology of Ni-W coatings after heat treatment in the air, in dependence on deposition current density: (**a**) C125/1173, (**b**) C150/1173, (**c**) C175/1173 and (**d**) C200/1173.

The phase composition of the as-deposited Ni-W alloy coatings is independent of applied current conditions. All X-ray diffraction patterns show the presence of reflexes coming from solid solution of W in Ni. An example of X-ray diffraction pattern obtained for Ni-W coating deposited at current density of 175 mA·cm^{-2} is shown in the Figure 2a. The phase composition of the Ni-W alloy coatings after heat treatment is also independent of applied current conditions. During the heat treatment in the air the solid solution of tungsten in nickel breaks down and chemical reaction with oxygen proceeds leading to a formation of new phases. X-ray diffraction patterns shown in Figure 2b–e indicate that the C125/1173, C150/1173, C175/1173 and C200/1173 coatings consist of three phases, i.e., Ni_4W, WO_2 and WO_3.

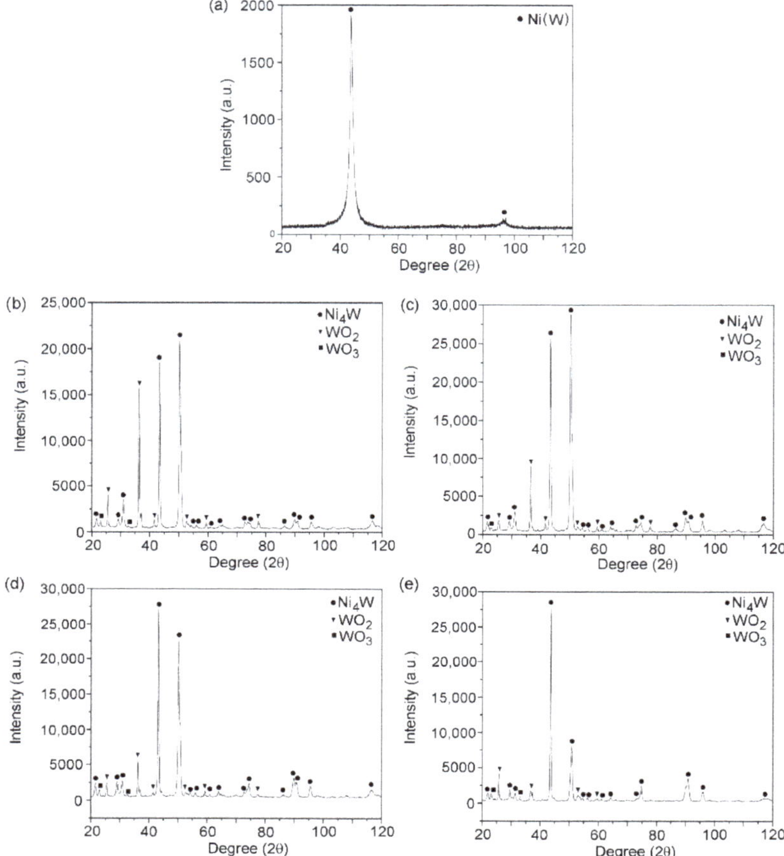

Figure 2. X-ray diffraction patterns for the as-deposited (**a**) C175/- and after heat treatment (**b**) C125/1173, (**c**) C150/1173, (**d**) C175/1173, (**e**) C200/1173 Ni-W coatings.

Values of the corrosion parameters i.e. corrosion potential E_{corr} and corrosion current density j_{corr} were determined from measured dependencies $j = f(E)$. It was found that the value of the corrosion potential for the C125/1173 coating is the highest compared to the E_{corr} obtained for the coatings deposited at larger current densities i.e. C150/1173, C175/1173 and C200/1173 (Figure 3, Table 2). It was also noted that, for the C125/1173 coating, the value of corrosion current density is lower compared to the other coatings (Table 2). This suggests that the C125/1173 coating, is more corrosion resistant in 5 wt.% NaCl solution than the other investigated coatings. It should be added that all heat-treated Ni-W coatings are characterized by the definitely higher corrosion resistance compared to the substrate (corrosion potential of S235 steel is −739 mV) [4].

The results of the EIS investigations presented in the form of Nyquist plots ($-Z'' = f(Z')$) were shown in Figure 4. For all investigated coatings one semicircle in the whole range of frequencies is observed. It has been found that this behavior of the heat-treated Ni-W coatings could be described by one-CPE electrode model (Figure 5). This is typical model for rough or porous materials. Such equivalent circuit is characteristic for materials composed of cylindrical pores of radius r and length l. As was shown in a paper [45] for short and wide pores l^2/r is very small and only one semicircle on the complex plane plot (Nyquist plot) was observed. The one-CPE model consists of the solution resistance R_s in series with a parallel connection of the CPE element ($Z_{CPE} = 1/[(j\omega)^{\phi} T]$ where T is the capacitive parameter, ϕ is a dimensionless parameter and ω is the angular frequency of ac voltage) and the polarization resistance R_p [42].

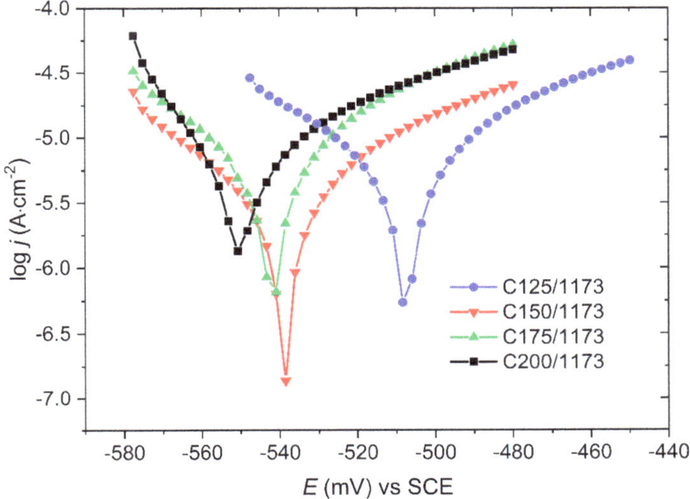

Figure 3. Potentiodynamic curves registered in 5 wt.% NaCl solution for the Ni-W coatings after heat treatment in the air, in dependence on deposition current density.

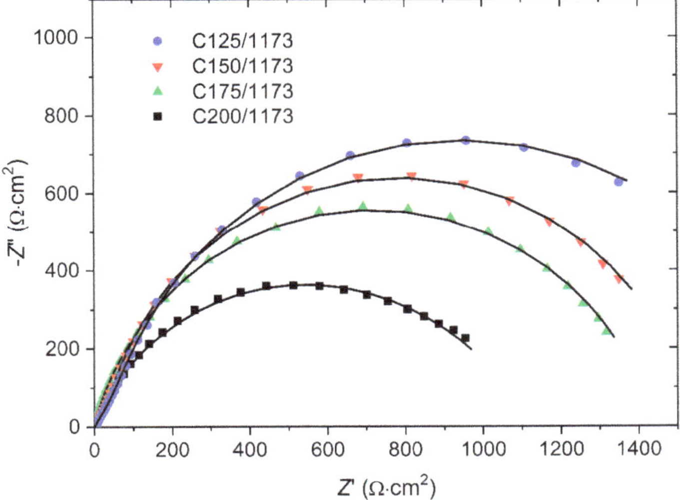

Figure 4. Nyquist plots registered in 5 wt.% NaCl solution for the Ni-W coatings after heat treatment in the air, in dependence on deposition current density; symbols–experimental points, solid lines–approximations using the one-CPE model.

Table 2. Corrosion potential E_{corr} and corrosion current density j_{corr} determined for Ni-W coatings after heat treatment in the air, in dependence on deposition current density.

Ni-W Coating	E_{corr} (mV)	j_{corr} ($\mu A \cdot cm^{-2}$)
C125/1173	−508	5.9
C150/1173	−538	9.2
C175/1173	−539	10.9
C200/1173	−550	11.7

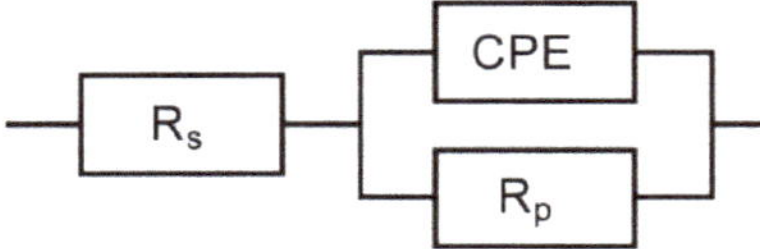

Figure 5. Equivalent circuit scheme, R_s–solution resistance, *CPE*–constant phase element, R_p–polarization resistance.

Approximations of the experimental impedances using the one-CPE model allowed to determine the following parameters: R_p, R_s, T, ϕ (Table 3). Note that lower values of polarization resistance R_p indicate that a material is more susceptible to corrosion.

Table 3. EIS parameters determined for Ni-W coatings after heat treatment in the air, in dependence on deposition current density.

Ni-W Coating	R_p ($k\Omega \cdot cm^2$)	T	ϕ	R_s ($\Omega \cdot cm^2$)	R_f
C125/1173	1.845	0.000146	0.87	1.19	2.00
C150/1173	1.535	0.000274	0.89	1.79	5.33
C175/1173	1.451	0.000328	0.89	1.51	6.39
C200/1173	1.122	0.000488	0.86	1.27	7.33

R_p is the polarization resistance, T is the capacitive parameter, ϕ is the parameter related to the rotation of the complex plane plot, R_s is the solution resistance, R_f is the factor of electrochemically active surface area.

The double-layer capacitance, C_{dl}, was calculated according to [43]:

$$T = C_{dl}{}^{\phi} (1/R_s + 1/R_p)^{1-\phi} \quad (1)$$

The ratio of capacitances C_{dl} determined for Ni-W coating and ideally smooth nickel electrode (20 $\mu F \cdot cm^{-2}$ [43]) gives factor of electrochemically active surface area, R_f (Table 3). Larger values of this parameter indicate larger interfacial surface, and hence deterioration of material corrosion resistance. The smallest electrochemically active surface area and the highest polarization resistance obtained for C125/1173 sample clearly indicate that this coating exhibit the best anticorrosion properties compared with the other coatings.

Figure 6 shows *CPD* maps registered for the studied Ni-W coatings. Statistical analysis of the obtained maps allows determining parameters describing quantitatively the surface properties i.e. average (CPD_{av}) and root mean square (CPD_q) of contact potential difference [46–48]. It was stated that the C125/1173 coating (Figure 6a, Table 4) is characterized by the highest value of CPD_{av} which equals c.a. −1060 mV_{KP} (mV_{KP} is the voltage measured in relation to the Kelvin probe). Increasing of the deposition current density to 200 $mA \cdot cm^{-2}$ (Figure 6d, Table 4) causes that the CPD_{av} decreases by about 140 mV_{KP}. Deviation of the *CPD* values from the mean (represented by CPD_q) is the smallest for C125/1173 and equals c.a. 16 mV_{KP}. It means that this coating shows the most homogeneous surface of all the coatings tested. It should be noted that in the case of C200/1173 coating, obtained at the highest current density, CPD_q increases more than two times in comparison with C125/1173. Figure 7 shows the tungsten content (at.%) and corrosion current density (j_{corr}) plotted versus the average contact potential difference (CPD_{av}). It has been found that the increase of tungsten content in the Ni-W coating causes linear increase of CPD_{av}. What is more, the corrosion rate (represented by j_{corr}) of Ni-W coatings linearly decreases with increasing CPD_{av}. Thus, CPD_{av} value allows estimating the corrosion rate of Ni-W coatings after heat treatment in air.

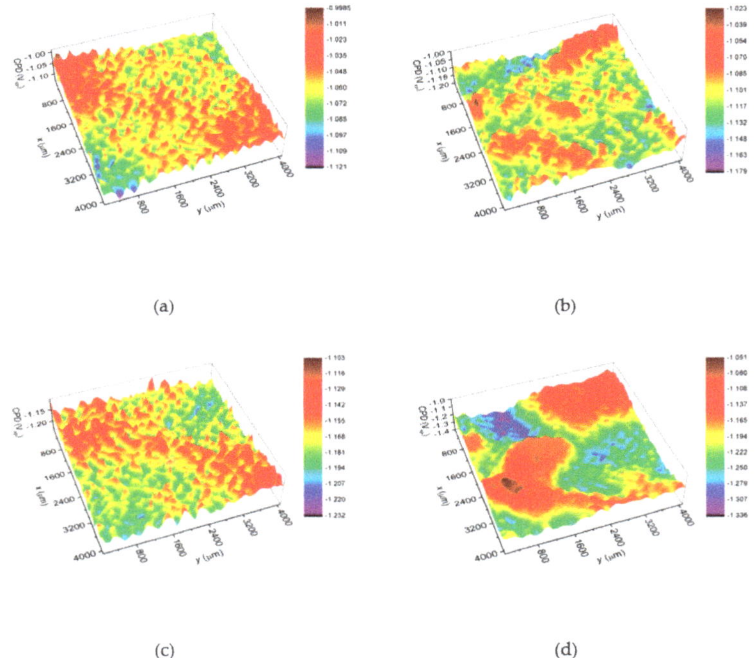

(a) (b)

(c) (d)

Figure 6. *CPD* maps determined for the Ni-W coatings after heat treatment in the air: (**a**) C125/1173, (**b**) C150/1173, (**c**) C175/1173 and (**d**) C200/1173.

Table 4. Statistical parameters obtained using *CPD* maps of the heat-treated Ni-W coatings.

Ni-W Coating	C125/1173	C150/1173	C175/1173	C200/1173
CPD_{av} (mV_{KP})	−1058	−1104	−1169	−1194
CPD_{q} (mV_{KP})	16	22	17	55

CPD_{av}—average value, CPD_{q}—root mean square, mV_{KP} is the voltage measured in relation to the Kelvin probe.

Figure 8 shows surface topography maps of the heat-treated Ni-W coatings obtained at deposition current density 125 mA·cm^{-2} (a) and 200 mA·cm^{-2} (b). Maps allow determining parameters describing quantitatively the surface roughness i.e. root mean square roughness (S_q), maximum peak height (S_p) and maximum pit depth (S_v). It was found that for the C125/1173 coating S_q = 0.8 μm, S_p = 2.9 μm, S_v = 2.6 μm and for C200/1173 coating S_q = 9.8 μm, S_p = 20.9 μm, S_v = 21.1 μm. It can be concluded that both coatings are characterized by a uniform distribution of peaks and valleys heights around the mean. However, it should be noted that for C200/1173 coating S_p and S_v parameters are 7-8 times higher in comparison with C125/1173. This fact can be explained by that as the deposition current density increases, the small globules visible on the C125/1173 surface (see Figure 1) change into larger ones. It was also stated that the deviation of peaks and valleys heights around the mean (S_q parameter) for C200/1173 is higher. This is due to the fact that the S_q parameter is directly related to the heights of peaks and valleys on the material surface.

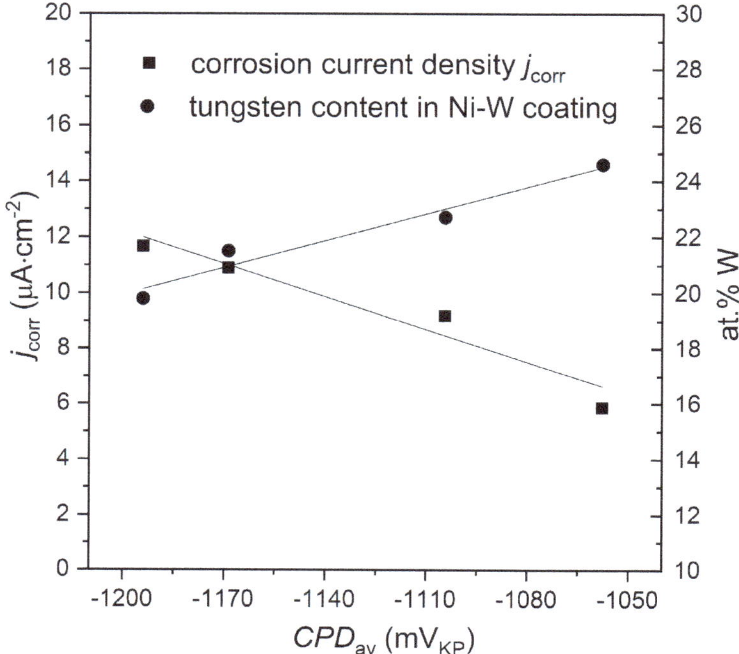

Figure 7. Tungsten content (at.%) and corrosion current density (j_{corr}) versus average contact potential difference (CPD_{av}) determined for the Ni-W coatings; mV_{KP} is the voltage measured in relation to the Kelvin probe.

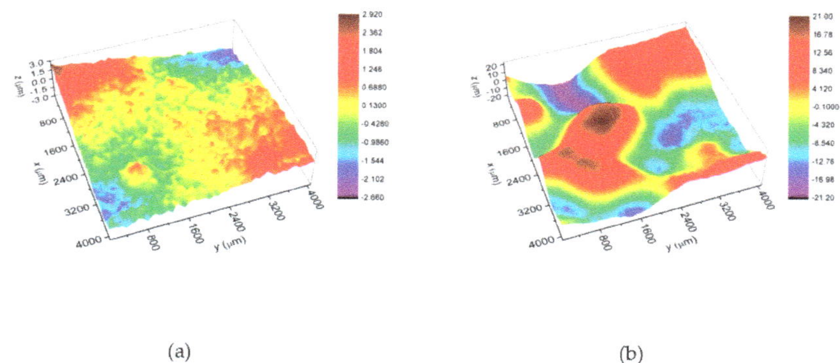

(a) (b)

Figure 8. Topography maps determined for the heat-treated Ni-W coatings: (**a**) C125/1173 and (**b**) C200/1173.

4. Conclusions

It was found that C125/1173 coating is the most resistant to corrosion in 5 wt.% NaCl solution of all the coatings tested. This is evidenced by the highest values of the corrosion potential, average contact potential difference and polarization resistance as well as the lowest value of the corrosion current density. The reason for this is the highest tungsten content in C125/1173 and the smallest surface area of this coating. Analysis of contact potential difference distribution shows also that the C125/1173 coating is characterized by the most homogeneous surface of all the coatings tested.

Author Contributions: Conceptualization, M.P. and J.K.; Methodology, M.P., J.K. and A.S.S.; Investigation, M.P. and J.K.; Resources, M.P. and J.K.; Formal Analysis, M.P., J.K. and A.S.S.; Writing—Original Draft Preparation, M.P., J.K. and A.S.S.; Data Curation, M.P., J.K., Z.W. and A.S.; writing—review and editing, Z.W., B.K. and A.S.; visualization, Z.W. and A.S.; All authors have read and agreed to the published version of the manuscript.

Funding: This research received no external funding.

Conflicts of Interest: The authors declare no conflict of interest.

References

1. Wang, Y.; Zhou, Q.; Li, K.; Zhong, Q.; Bui, Q.B. Preparation of Ni-W-SiO_2 nanocomposite coating and evaluation of its hardness and corrosion resistance. *Ceram. Int.* **2015**, *41*, 79–84. [CrossRef]
2. Popczyk, M. The influence of molybdenum and silicon on activity of Ni+W composite coatings in the hydrogen evolution reaction. *Surf. Interface Anal.* **2008**, *40*, 246–249. [CrossRef]
3. Allahyarzadeh, M.H.; Aliofkhazraei, M.; Sabour Rouhaghdam, A.R.; Torabinejad, V. Electrodeposition of Ni-W-Al_2O_3 nanocomposite coating with functionally graded microstructure. *J. Alloys Compd.* **2016**, *666*, 217–226. [CrossRef]
4. Popczyk, M.; Zacharz, M.; Osak, P.; Dercz, G.; Łosiewicz, B. Structure and corrosion resistance of nickel-molybdenum alloy coatings. *Acta Phys. Pol. A* **2016**, *130*, 1072–1074. [CrossRef]
5. Beltowska-Lehman, E.; Indyka, P.; Bigos, A.; Szczerba, M.; Kot, M. Ni-W/ZrO_2 nanocomposites obtained by ultrasonic DC electrodeposition. *Mater. Des.* **2015**, *80*, 1–11. [CrossRef]
6. Hou, K.-H.; Sheu, H.-H.; Ger, M.-D. Preparation and wear resistance of electrodeposited Ni-W/diamond composite coatings. *Appl. Surf. Sci.* **2014**, *308*, 372–379. [CrossRef]
7. Hosseini, M.; Teymorinia, H.; Farzaneh, A.; Khameneh-asl, S. Evaluation of corrosion, mechanical and structural properties of new Ni-W-PCTFE nanocomposite coating. *Surf. Coat. Technol.* **2016**, *298*, 114–120. [CrossRef]
8. Popczyk, M. The hydrogen evolution reaction on electrolytic nickel - based coatings containing metallic molybdenum. *Mater. Sci. Forum* **2010**, *636–637*, 1036–1041. [CrossRef]
9. Beltowska-Lehman, E.; Indyka, P.; Bigos, A.; Kot, M.; Tarkowski, L. Electrodeposition of nanocrystalline Ni-W coatings strengthened by ultrafine alumina particles. *Surf. Coat. Technol.* **2012**, *211*, 62–66. [CrossRef]
10. Popczyk, M.; Kubisztal, J.; Budniok, A. Structure and electrochemical characterization of electrolytic Ni+Mo+Si composite coatings in an alkaline solution. *Electrochim. Acta* **2006**, *51*, 6140–6144. [CrossRef]
11. Wykpis, K.; Popczyk, M.; Budniok, A. Electrolytic deposition and corrosion resistance of Zn-Ni coatings obtained from sulphate-chloride bath. *Bull. Mater. Sci.* **2011**, *34*, 997–1001. [CrossRef]
12. Wykpis, K.; Popczyk, M.; Niedbała, J.; Budniok, A.; Łągiewka, E. Influence of the current density of deposition on the properties of Zn-Ni coatings. *Mater. Sci.* **2012**, *47*, 838–847. [CrossRef]
13. Wykpis, K.; Popczyk, M.; Niedbała, J.; Budniok, A.; Łągiewka, E.; Bierska-Piech, B. Influence of thermal treatment on the corrosion resistance of electrolytic Zn-Ni coatings. *Mater. Sci.-Poland* **2012**, *29*, 177–183. [CrossRef]
14. Hashemi, M.; Mirdamadi, S.; Rezaie, H. Effect of SiC nanoparticles on microstructure and wear behavior of Cu-Ni-W nanocrystalline coating. *Electrochim. Acta* **2014**, *138*, 224–231. [CrossRef]
15. Popczyk, M.; Budniok, A.; Łągiewka, E. Structure and corrosion resistance of nickel coatings containing tungsten and silicon powders. *Mater. Charact.* **2007**, *58*, 371–375. [CrossRef]
16. Wykpis, K.; Niedbała, J.; Popczyk, M.; Budniok, A.; Łągiewka, E. The electrodeposition and properties of Zn-Ni+Ni composite coatings. *Russ. J. Electrochem.* **2012**, *48*, 1123–1129. [CrossRef]
17. Wykpis, K.; Popczyk, M.; Niedbała, J.; Bierska-Piech, B.; Budniok, A.; Łągiewka, E. Influence of thermal treatment on the corrosion resistance of electrolytic Zn-Ni+Ni composite coatings. *Adv. Compos. Mater.* **2015**, *24*, 431–438. [CrossRef]

18. Popczyk, M.; Serek, A.; Budniok, A. Production and properties of composite layers based on an Ni-P amorphous matrix. *Nanotechnology* **2003**, *14*, 341–346. [CrossRef]
19. Zemanová, M.; Druga, J.; Szúnyogh, J.; Dobročka, E. Ni-W alloys for hydrogen evolution. *Mater. Sci. Forum* **2016**, *844*, 167–171. [CrossRef]
20. Yang, F.-Z.; Guo, Y.F.; Huang, L.; Xu, S.K.; Zhou, S.M. Electrodeposition, structure and corrosion resistance of nanocrystalline Ni-W alloy. *Chinese J. Chem.* **2004**, *22*, 228–231.
21. Obradović, M.; Stevanović, J.; Despić, A.; Stevanović, R.; Stoch, J. Characterization and corrosion properties of electrodeposited Ni-W alloys. *J. Serb. Chem. Soc.* **2001**, *66*, 899–912. [CrossRef]
22. Allahyarzadeh, M.H.; Aliofkhazraei, M.; Rezvanian, A.R.; Torabinejad, V.; Sabour Rouhaghdam, A.R. Ni-W electrodeposited coatings: Characterization, properties and applications. *Surf. Coat. Tech.* **2016**, *307*, 978–1010. [CrossRef]
23. Quiroga Argañaraz, M.P.; Ribotta, S.B.; Folquer, M.E.; Gassa, L.M.; Benítez, G.; Vela, M.E.; Salvarezza, R.C. Ni-W coatings electrodeposited on carbon steel: Chemical composition, mechanical properties and corrosion resistance. *Electrochim. Acta* **2011**, *56*, 5898–5903. [CrossRef]
24. Alimadadi, H.; Ahmadi, M.; Aliofkhazraei, M.; Younesi, S.R. Corrosion properties of electrodeposited nanocrystalline and amorphous patterned Ni-W alloy. *Mater. Des.* **2009**, *30*, 1356–1361. [CrossRef]
25. Mroz, K.P.; Bigos, A.; Kucharski, S.; Dolinski, K.; Bełtowska-Lehman, E. Ni-W electrodeposited coatings on low carbon steel substrate: Fatigue observations. *J. Mater. Eng. Perform.* **2014**, *23*, 3459–3466. [CrossRef]
26. Indyka, P.; Bełtowska-Lehman, E.; Tarkowski, L.; Bigos, A.; García-Lecina, E. Structure characterization of nanocrystalline Ni-W alloys obtained by electrodeposition. *J. Alloy Compd.* **2014**, *590*, 75–79. [CrossRef]
27. Kirihara, S.; Umeda, Y.; Tashiro, K.; Honma, H.; Takai, O. Development of Ni-W alloy plating as a substitution of hard chromium plating. *Trans. Mater. Res. Soc. Jpn.* **2016**, *41*, 35–39. [CrossRef]
28. Jones, A.R.; Hamann, J.; Lund, A.C.; Schuh, C.A. Nanocrystalline Ni-W alloys coating for engineering applications. *Plat. Surf. Finish.* **2010**, *97*, 52–60.
29. Wasekar, N.P.; Sundararajan, G. Sliding wear behavior of electrodeposited Ni-W alloy and hard chrome coatings. *Wear* **2015**, *342–343*, 340–348. [CrossRef]
30. Lee, H.B. Synergy between corrosion and wear of electrodeposited Ni-W coating. *Tribol. Lett.* **2013**, *50*, 407–419. [CrossRef]
31. Chianpairot, A.; Lothongkum, G.; Schuh, C.A.; Boonyongmaneerat, Y. Corrosion of nanocrystalline Ni-W alloys in alkaline and acidic 3.5 wt.% NaCl solutions. *Corros. Sci.* **2011**, *53*, 1066–1071. [CrossRef]
32. Chen, H.; Ren, X.R.; Zhang, X.H.; Li, J.H. Wear and corrosion properties of crystalline Ni-W alloy coatings prepared by electrodeposition. *Mater. Sci. Forum* **2016**, *849*, 671–676. [CrossRef]
33. de Lima-Neto, P.; Correia, A.N.; Santana, R.A.C.; Colares, R.P.; Barros, E.B.; Casciano, P.N.S.; Vaz, G.L. Morphological, structural, microhardness and electrochemical characterizations of electrodeposited Cr and Ni-W coatings. *Electrochim. Acta* **2010**, *55*, 2078–2086. [CrossRef]
34. Kaninski, M.P.M.; Saponjic, D.P.; Perovic, I.M.; Maksic, A.D.; Nikolic, V.M. Electrochemical characterization of the Ni-W catalyst formed in situ during alkaline electrolytic hydrogen production—Part II. *Appl. Catal. A Gen.* **2011**, *405*, 29–35. [CrossRef]
35. Tasić, G.S.; Lačnjevac, U.; Tasić, M.M.; Kaninski, M.M.; Nikolić, V.M.; Žugić, D.L.; Jović, V.D. Influence of electrodeposition parameters of Ni-W on Ni cathode for alkaline water electrolyser. *Int. J. Hydrog. Energy* **2013**, *38*, 4291–4297.
36. Chen, S.H.; Lai, J.H.; Wu, M.Y.; Lee, H.B.; Lee, C.Y. A study on the corrosion and wear behavior of electrodeposited Ni-W coatings. *J. Chin. Corros. Eng.* **2012**, *26*, 1–8.
37. Sriraman, K.; Raman, S.G.S.; Seshadri, S. Corrosion behaviour of electrodeposited nanocrystalline Ni-W and Ni-Fe-W alloys. *Mater. Sci. Eng. A* **2007**, *460*, 39–45. [CrossRef]
38. Sunwang, N.; Wangyao, P.; Boonyongmaneerat, Y. The effects of heat treatments on hardness and wear resistance in Ni- W alloy coatings. *Surf. Coat. Technol.* **2011**, *206*, 1096–1101. [CrossRef]
39. Ko, Y.-K.; Chang, G.-H.; Lee, J.-H. Nickel tungsten alloy electroplating for the high wear resistant materials applications. *Solid State Phenom.* **2007**, *124–126*, 1589–1592.
40. Popczyk, M.; Łosiewicz, B. Influence of surface development of Ni/W coatings on the kinetics of the electrolytic hydrogen evolution. *Solid State Phenom.* **2015**, *228*, 293–298. [CrossRef]

41. Popczyk, M.; Łosiewicz, B.; Łągiewka, E.; Budniok, A. Influence of thermal treatment on the electrochemical properties of Ni+Mo composite coatings in an alkaline solution. *Solid State Phenom.* **2015**, *228*, 231–236. [CrossRef]
42. Karimi-Shervedani, R.; Lasia, A. Studies of the hydrogen evolution reaction on Ni-P electrodes. *J. Electrochem. Soc.* **1997**, *144*, 511–519. [CrossRef]
43. Karimi-Shervedani, R.; Lasia, A. Evaluation of the surface roughness of microporous Ni-Zn-P electrodes by in situ methods. *J. Appl. Electrochem.* **1999**, *29*, 979–986. [CrossRef]
44. Hou, K.-H.; Chang, Y.-F.; Chang, S.-M.; Chang, C.-H. The heat treatment effect on the structure and mechanical properties of electrodeposited nano grain size Ni-W alloy coatings. *Thin Solid Films* **2010**, *518*, 7535–7540. [CrossRef]
45. Lasia, A. Nature of the two semi-circles observed on the complex plane plots on porous electrodes in the presence of a concentration gradient. *J. Electroanal. Chem.* **2001**, *500*, 30–35. [CrossRef]
46. Kubisztal, J.; Kubisztal, M.; Haneczok, G. Quantitative characterization of material surface–application to Ni + Mo electrolytic composite coatings. *Mater. Charact.* **2016**, *122*, 45–53. [CrossRef]
47. Kubisztal, J.; Kubisztal, M.; Stach, S.; Haneczok, G. Corrosion resistance of anodic coatings studied by scanning microscopy and electrochemical methods. Surf. *Coat. Tech.* **2018**, *350*, 419–427. [CrossRef]
48. Bak, A.; Losiewicz, B.; Kozik, V.; Kubisztal, J.; Dybal, P.; Swietlicka, A.; Barbusinski, K.; Kus, S.; Howaniec, N.; Jampilek, J. Real-time corrosion monitoring of AISI 1010 carbon steel with metal surface mapping in sulfolane. *Materials* **2019**, *12*, 3276. [CrossRef]

© 2020 by the authors. Licensee MDPI, Basel, Switzerland. This article is an open access article distributed under the terms and conditions of the Creative Commons Attribution (CC BY) license (http://creativecommons.org/licenses/by/4.0/).

Article

Effects of the Combined Addition of Zn and Mg on Corrosion Behaviors of Electropainted AlSi-Based Metallic Coatings Used for Hot-Stamping Steel Sheets

Si On Kim [1], Won Seog Yang [2] and Sung Jin Kim [1,*]

1 Department of Advanced Materials Engineering, Sunchon National University, Jungang-ro, Suncheon, Jeonnam 57922, Korea; kzo1102@scnu.ac.kr

2 R&D Division, Hyundai Steel, 1480 Bukbusaneop-ro, Songak-eup, Dangjin, Chungnam 31719, Korea; wsyang@hyundai-steel.com

* Correspondence: sjkim56@sunchon.ac.kr; Tel.: +82-61-750-3557

Received: 8 July 2020; Accepted: 27 July 2020; Published: 30 July 2020

Abstract: The effects of the combined addition of Zn and Mg on the corrosion resistance of AlSi-based coating for automotive steel sheets were investigated using a variety of analytical and electrochemical techniques. The preferential dissolution of Mg and Zn from $MgZn_2/Mg_2Si$ phases occurred on the AlSi-based coating that had been alloyed with a smaller portion of Zn and Mg, which contributed to the rapid surface coverage by corrosion products with a protective nature, reducing the corrosion current density. On the other hand, localized corrosion attacks caused by the selective dissolution of Mg were also observed in the AlSi-based coating with a smaller portion of Zn and Mg. Such alloying can also worsen its corrosion resistance when coated additionally with electrodeposited paint. The mechanistic reasons for these conflicting results are also discussed.

Keywords: hot stamping steel; AlSi-based coating; electrodeposited paint; Zn; Mg; corrosion

---

1. Introduction

Metallic coatings have been applied widely to steel substrates used in the automotive industry, because of their anti-corrosion performance [1–3]. In general, coating materials are more electrochemically active than protected substrates, meaning they offer sacrificial protection. A recent technical issue in a coating system is the formation of suitable corrosion products and new phases that are effective in delaying the coating consumption and reducing the overall corrosion rate [4–6]. Regarding Al-based coatings for steel sheets, a passive film (Al_2O_3) forms on the coating surface, providing superior barrier protection [7–9]. In the case of AlSi-based coatings, the addition of Si decreases the thickness of the intermetallic phase at the inter-diffusion layer close to the steel substrates [10,11], which also provide good anti-corrosion performance [12]. On the other hand, Al_2O_3 films are unstable when exposed to environments containing chloride ions (Cl^-), which can decrease the stability of the film, leading to a decrease in coating efficiency by localized corrosion [13,14]. To overcome these drawbacks, other alloying elements, such as Zn and Mg, which can modify the coating potential, are added to the Al-based coating, so that they can provide a supplementary self-healing effect by the sacrificial dissolution of alloying elements [4,5]. At the same time, several corrosion products acting as a protective barrier can be formed on the outer surface [15,16]. Hence, AlSi-based coatings with Zn and Mg are promising candidates for metallic coatings on hot-stamping steels used for auto-parts [17]. Recently, Nicard et al. [18] examined the anti-corrosion mechanism of AlSi-based coatings with Zn (2~30 wt %) and Mg (1~10 wt %). They reported that the addition of Zn and Mg to AlSi-based coatings increases the corrosion resistance in chloride-containing environments. From a practical and an industrial point of view, however, lower concentrations of Zn and Mg, which have low melting temperatures, are favored to avoid the liquid metal embrittlement (LME) phenomenon [19] and localized corrosion attack caused by the selective dissolution of these alloying elements [6,20].

Furthermore, considering the applicability of the coating to hot-stamped, high strength steel sheets used for auto-body parts, an evaluation of the corrosion resistance should proceed after the electropainting process on the metallic coating. In this regard, this study examined on the corrosion behaviors not only of AlSi-based metallic coating with 10 wt % Zn and 0.5 wt % Mg (refer to AlSiZnMg-MC, here-in-after), but also of electropainted AlSiZnMg-MC. Field-emission scanning electron microscopy (FE-SEM), X-ray diffraction (XRD), X-ray photoelectron spectroscopy (XPS), and potentiostat measurements were used to analyze the corrosion behaviors of metallic coated steels. Moreover, an accelerated corrosion test [12] was conducted on electropainted samples that had been damaged by scratching the coating with an "×" incision, and by colliding the coating surface with fine stone.

2. Experimental

2.1. Materials and Specimen Preparation

The steel substrate, produced by Hyundai Steel Corp.(Dang Jin, Korea), was classified as 22MnB5 steel with 0.27–0.3 wt % C, 1.3–1.5 wt % Mn, 0.2–0.25 wt % Cr, 0.18–0.2 wt % Si, 0.0025–0.0035 wt % B, 0.02–0.035 wt % Ti. The S and P contents were kept as low as possible to avoid high-temperature cracking during the subsequent stamping process. Two types of metallic coated steel were used: AlSi and AlSi with Zn and Mg (AlSiZnMg) coated steels, which were manufactured using a hot-dip simulator. For the coating process, two types of molten bath were prepared: Al-Si (7 wt %) and Al-Si (7 wt %)-Zn (10 wt %)–Mg (0.5 wt %). The substrates were then dipped in each molten bath, and two types of metallic coating were produced: AlSi-MC and AlSiZnMg-MC, at 50/50 g/m^2. For the following stamping process, the two metallic coated steels were heated to 930 °C for 300 s and cooled to room temperature by die quenching.

Some of the two types of metallic coated steel sample were also coated with electrodeposited (ED) paint. For this ED coating, the coated steel samples underwent the following processes: degreasing, surface conditioning, and phosphating in a pretreatment simulator. The ED coating was then conducted at 300 V for 180 s while the voltage was increased uniformly for 30 s. The ED coated samples were dried in a circulating oven at 170 °C for 20 min. Detailed information on the ED coating process can be found elsewhere [21].

2.2. Microstructure Examination

The microstructure and composition of the coated steel samples before and after the corrosion test (potentiostatic (PS) polarization measurement) were examined by FE-SEM and energy dispersive spectroscopy (EDS). Before the corrosion test, the coated steel samples were cleaned ultrasonically in ethanol and dried in air. In particular, for the cross-sectional observations, the samples were mounted with the cut-edge sides facing the surface, and they were polished with a final polishing step of 0.04 μm. After the corrosion test, the sample surface was washed with distilled water and dried in air. The sample was then stored in a vacuum chamber. The various phases in the coating layers were also characterized by XRD.

2.3. Depth Profile Analysis

The compositional distributions of the coating layers as a function of depth were examined by glow discharge spectroscopy (GDS) analysis with a Leco (St. Joseph, MI, USA) GDS-850A spectrometer using argon plasma equipped with an RF lamp. The diameter of the analysis area in a sampling area of 20×20 mm^2 was 4 mm.

2.4. X-ray Photoelectron Spectroscopy Analysis

XPS was used to analyze the chemical state of the corrosion products formed on the metallic-coated layers after the corrosion test (PS polarization measurement). Prior to analysis, the two types of coated steel sheets that had been corroded by the polarization test were cut into 7 × 8 mm^2 samples and cleaned using ethanol. XPS (VG Scientific Escalab 250 (Waltham, MA, USA)) was conducted using mono chromatic Al Kα radiation (1486.7 eV) with a 500 μm diameter spot size. A constant analyzer energy mode with 200 and 50 eV for the survey and high-resolution spectra, respectively, were used. Data processing of the spectra was performed using a spectral data processor (SDP) v 3.0 software. The adventitious C 1s peak at 284.8 eV was used as a reference for charge correction [22].

2.5. Electrochemical Measurements

Potentiodynamic and potentiostatic polarization tests were performed after the open circuit potential (OCP) evolution measurements for 1 h in a 3.5% NaCl using a GAMRY (Philadelphia, PA, USA) reference 600. A three-electrode system was used, which consisted of the metallic-coated sample as the working electrode (tested area of 1 cm^2), a platinum grid as the counter electrode, and a saturated calomel electrode (SCE) as the reference electrode.

For the potentiodynamic polarization scan, the working electrode was polarized from −0.5 V to 0.25 V vs. OCP at a scan rate of 0.16 mV s^{-1}. The corrosion current densities were determined by curve-fitting using the Wagner-Traud equation (Equation (1)), described below, to the potentiodynamic polarization curves.

$$i = i_{corr}\left[\exp\left(\frac{2.303(E - E_{corr})}{\beta_a}\right) - exp\left(\frac{-2.303(E - E_{corr})}{\beta_c}\right)\right] \quad (1)$$

where i_{corr} is the corrosion current density (A/cm^2), E_{corr} and E represent the corrosion potential (V) and the measured potential (V), respectively. β_a and β_c are Tafel slopes (V/decade), and i is the total corrosion density (A/cm^2).

Based on the polarization curves, the applied potentials for the anodic and cathodic polarization during the potentiostatic polarization measurements were determined to be 50 mV above the OCP and 100 mV below the OCP, respectively. A preliminary test showed that under an applied potential of 100 mV above the OCP in the potentiostatic polarization, the coating dissolution was too severe to analyze the corrosion behaviors of the coating surface covered with some corrosion products. Therefore, the anodic potential was set to 50 mV lower than the cathodic potential in the polarization measurements.

Three repetitive tests were conducted for each experiment to ensure reproducibility.

2.6. Cyclic Corrosion Test for ED Coated Samples

To evaluate the corrosion resistance of the ED coated AlSi and AlSiZnMg samples, a VDA (Berlin, Germany) 233-102 cyclic corrosion test was conducted according to VDA 233-102 [12], which is a standard for the accelerated corrosion testing of coatings/paintings within the automotive industry. Before the corrosion test, the samples were cut to a size of 150 × 70 mm, and damaged artificially using the following two methods. One set of samples was scratched with an "×" incision in reference to ASTM D-1654 [23], and the other set of samples abraded with a spray of fine stone chips, as reported elsewhere [24]. Salt spraying with a 1% NaCl solution (pH 6.5) at a rate of 2.0 to 4.0 mL/h was then conducted on the samples positioned 65° to the horizontal in an enclosed chamber. A full test cycle lasted for seven days, and the multi-step cycles were composed of varying temperatures ranging from −15 °C to 50 °C and humidity between 50% and 95%. Further information on test conditions can be found in VDA 233-102 [12].

3. Results and Discussion

3.1. Microstructure Characterizations of Metallic Coatings

Figure 1 presents the surface and cross-sectional morphologies and EDS mapping of AlSi-MC (Figure 1a,b) and AlSiZnMg-MC (Figure 1c,d). EDS mapping showed that the two types of coating were composed mainly of Al matrix with two (Al,Fe,Si)-rich intermetallic phases. One of the intermetallic phases formed at the interface between the coating layer and Fe substrate during the hot-dip aluminizing process. The other phases formed in the coating layer during the austenitizing process. The formation of the former phase was attributed to the inter-diffusion of Fe atoms from the steel substrate into the coating and of Al atoms from the coating into the steel substrate [25]. At the same time, Si, known to be concentrated at the outer surface of the steel substrate [26], was also enriched at the inter-diffusion layer. According to previous studies [27,28], the composition of the intermetallic phase is close to $\tau 5$ or $\tau 6$ shown in the ternary phase diagram of Al-Fe-Si [27,28]. The latter phase was formed by the diffusion of Fe and Si atoms from the steel substrate into the coating during austenitizing for the hot-stamping process, as reported elsewhere [29]. A noticeable feature in the coating morphologies was the presence of micro-cracks and pores. These may be closely associated with the very high hardness (900~1100 HV0.05) of Al-rich Al-Fe intermetallic compounds, such as $Al_{13}Fe_4$ and Al_5Fe_2 [30,31], and the difference in thermal shrinkage among the intermetallic phases during cooling in the hot-stamping process [32].

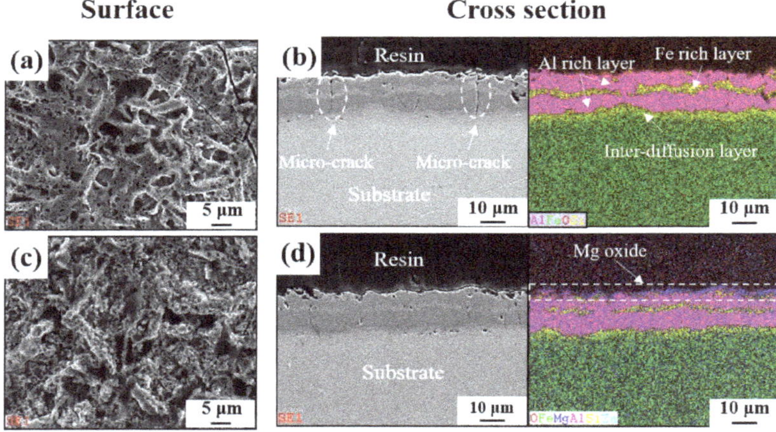

Figure 1. Surface and cross-sectional observations with eds mappings of (**a**,**b**) AlSi-MC and (**c**,**d**) AlSiZnMg-MC.

Although Figure 1 indicated that there were more pores and micro-cracks on the AlSi-MC, the microscopic observations could not be representative of the entire coating layers. On the other hand, the composition features were different apparently in that the outer-surface of AlSiZnMg-MC was covered with a thin (Mg, Zn)—based phase. GDS analysis was conducted for further clarification; Figure 2 presents the depth profile results. In contrast to the surface of AlSi-MC, consisting only of a thin Al-based oxide, the Mg and Zn-based oxide may cover the coating layer of AlSiZnMg-MC. In addition, based on the AlSiZnMg quaternary system [33], Zn-Mg intermetallic phases may exist on the coating surface of AlSiZnMg. For closer analysis, phase characterization was carried out using XRD; the results are shown in Figure 3. As expected, one of the major phases in both coating systems was an Al-Fe intermetallic compound.

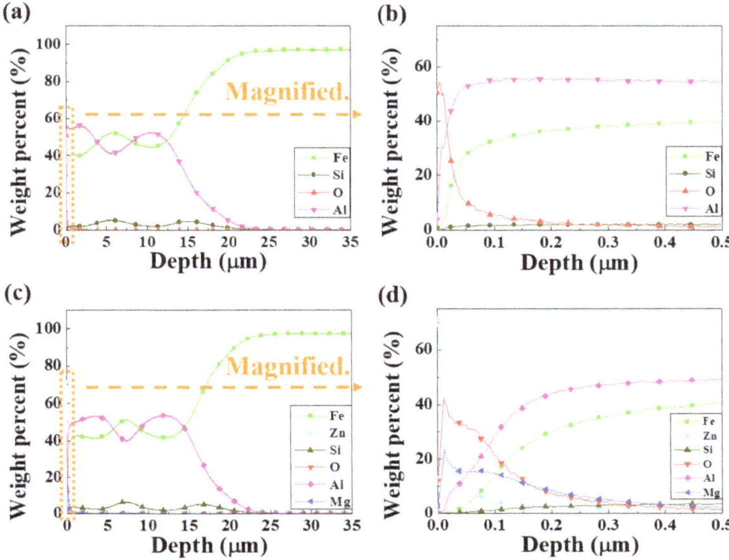

Figure 2. Glow discharge spectroscopy (GDS) depth profile of the coating layers: (**a**,**b**) AlSi-MC, (**c**,**d**) AlSiZnMg-MC.

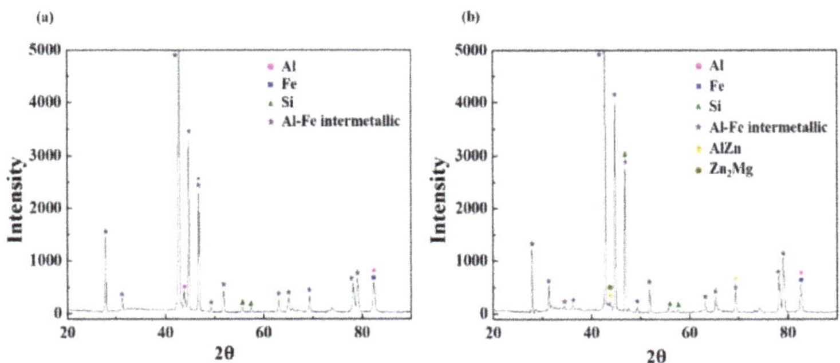

Figure 3. X-ray diffraction patterns of (**a**) AlSi-MC and (**b**) AlSiZnMg-MC.

Al-Fe, Fe-Si, and Fe-Al-Si intermetallic phases are generally difficult to distinguish by XRD due to overlap of the XRD peaks [12]. The phase existing only in AlSiZnMg-MC was $MgZn_2$. The other possible phases in the sample may be Mg_2Si and MgO. However, the initial fraction of Mg (0.5 wt %) was too low to give prominent peaks of Mg-containing phases in the XRD measurements. Even if some Mg-containing phases were formed, minor fractions that are below the detection limit of the XRD instrument, might be present. Instead of Mg-containing phases, some peaks for the Al-Fe intermetallic phase and Si were detected with low intensity. Considering that the phases, such as $MgZn_2$ and Mg_2Si, have different electrochemical potentials of their own, they can affect the corrosion kinetics and resulting coating life. From an electrochemistry point of view, the phases of $MgZn_2$ and Mg_2Si can act as an anode from the coating matrix because of their lower corrosion potential (−1538 and −1029 mV_{SCE}, respectively) [18]. On the other hand, unlike the case of $MgZn_2$, which provides sacrificial protection at an early stage of corrosion, Mg_2Si can form Si clusters after the selective dissolution of Mg, which will be cathodic and cause the localized corrosion attack of an Al-based coating layer. [6,20]. This will be discussed in more detail in the following section.

Another possible phase in AlSiZnMg-MC is MgO, which may be present at the outermost part of the coating surface. Although MgO with low electrical conductivity [34] could contribute in part to the suppression of the cathodic reduction reaction during the early corrosion stage, it is generally non-uniform and, in most cases, increases local corrosion attack. Under this coating composition with a minor fraction of Mg (0.5 wt %), it is extremely difficult to form a stable and uniform MgO film over the coating surface. These discussions were based primarily on the theoretical studies reported elsewhere, and they should be clarified experimentally, which will be discussed in the following section.

3.2. Electrochemical Characterizations

The evolution of the OCP (Figure 4) was measured for 1 h before the electrochemical polarization tests. The coating potentials of AlSi-MC and AlSiZnMg-MC after 1 h of immersion were approximately −0.59 V_{SCE}, and −0.62 V_{SCE}, respectively. In contrast to the Al coating covered by a thin Al oxide film (Al_2O_3), Zn and Mg, which were enriched at the coating surface, rarely form an oxide/hydroxide film with a high passivity coefficient in a neutral solution, which could result in a lower corrosion potential. As mentioned previously, the presence of active phases, such as Mg_2Si and $MgZn_2$, can contribute to the decrease in corrosion potential. On the other hand, the preferential dissolution of Mg from the active phases can provide sacrificial protection. Previous studies [16,35] have shown that Mg and Zn cations, which are supplied by the selective dissolution of $MgZn_2$ (reaction (2)), can also lead to rapid coverage of the coating surface by the precipitation of corrosion products with a protective nature, such as simonkolleite ($Zn_5(OH)_8Cl_2{\cdot}H_2O$), which can prevent further corrosion.

$$MgZn_2(s) \rightarrow Mg^{2+}(aq) + 2Zn^{2+}(aq) + 6e^- \quad (2)$$

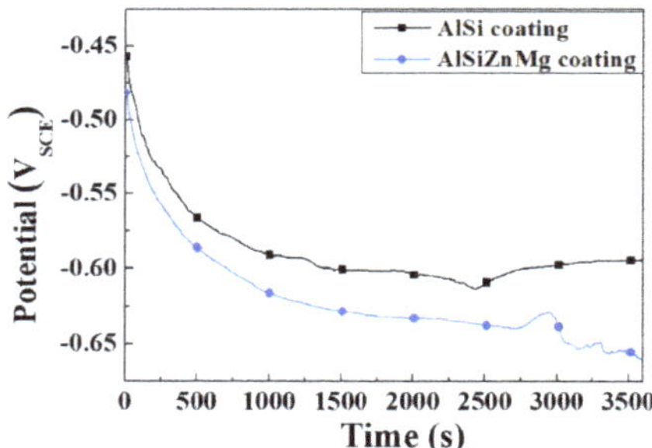

Figure 4. Open circuit potential (OCP) evolution measurement of AlSi-MC and AlSiZnMg-MC for 1 h in a 3.5% NaCl solution [20].

According to a previous study, the selective dissolution of Mg from the Mg_2Si phase can provide cathodic protection. They proposed that the formation of less conductive SiO_2 can decrease the galvanic current between Al/SiO_2 according to reaction (3) [18,36]:

$$Mg_2Si(s) + 2H_2O(l) \rightarrow 2Mg^{2+}(aq) + SiO_2(s) + 4H^+(aq) + 8e^- \quad (3)$$

On the other hand, the formation of a stable SiO_2 film uniformly formed on the coating surface cannot be guaranteed. Under this condition, an unstable film may create active local galvanic coupling, increasing the dissolution of the coating layer.

These beneficial and harmful effects on the corrosion resistance can be examined by polarization measurements, which are shown in Figure 5; Figure 6. Potentiodynamic and potentiostatic polarizations showed that the anodic and cathodic reaction rates of AlSiZnMg-MC were much smaller than those of AlSi-MC. The corrosion current densities (i_{corr}) determined by curve-fitting with the Wagner–Traud equation to the polarization curves of AlSiZnMg-MC and AlSi-MC were 3.5 and 8.5 μA cm^{-2}, respectively. The lower anodic and cathodic current densities of AlSiZnMg-MC suggest that the liberated Mg^{2+} and Zn^{2+} ions, supplied by the selective dissolution of Mg and Zn, provided an environment for the formation of corrosion products contributing to the effective reduction of the anodic and cathodic current densities. On the other hand, the polarization curve of AlSiZnMg-MC has many fluctuations, which may be closely associated with localized corrosion caused by the non-uniform distribution of MgO or the selective dissolution of Mg from Mg_2Si phases in the coating. This can be supported in part by several local corrosion attacks after the corrosion test, as shown in Figure 7. From morphological observations before (Figure 1) and after (Figure 7) the corrosion test (PS polarization), there was little difference except that the outermost surfaces were less uniform after the corrosion test. Compared to the case of AlSi-MC, the surface morphology of AlSiZnMg-MC after the corrosion test appeared to be denser (Figure 7c). On the other hand, AlSiZnMg-MC exhibited a more uneven cross-sectional view, which may be associated with the formation of corrosion products containing Mg, Zn, and Si, and localized corrosion attacks mentioned previously. Owing to the low resolution of EDS or GDS, XPS was performed to characterize the corrosion products formed on the two types of coating after the corrosion test.

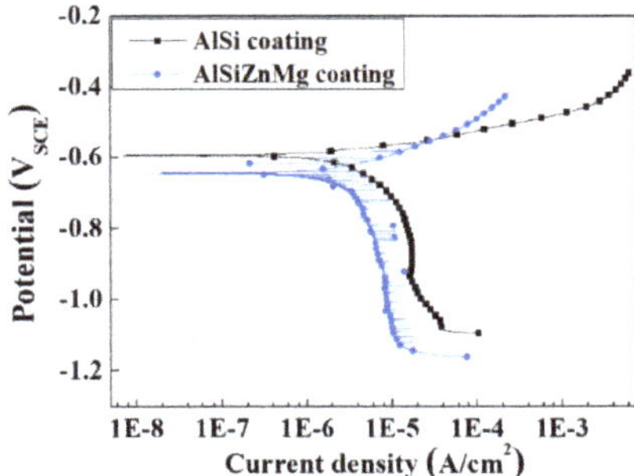

Figure 5. Potentiodynamic polarization curves of AlSi-MC and AlSiZnMg-MC, measured after 1 h immersion in a 3.5% NaCl solution [20].

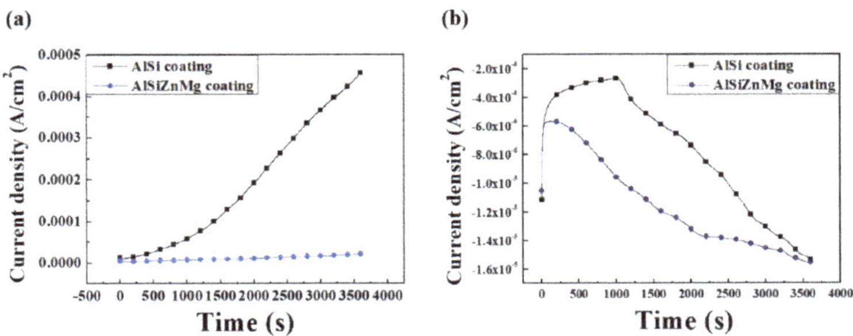

Figure 6. Potentiostatic polarization curves of AlSi-MC and AlSiZnMg-MC, measured after 1 h immersion in a 3.5% NaCl solution under (**a**) anodic (50 mV above OCP), and (**b**) cathodic (100 mV below OCP) potentials [20].

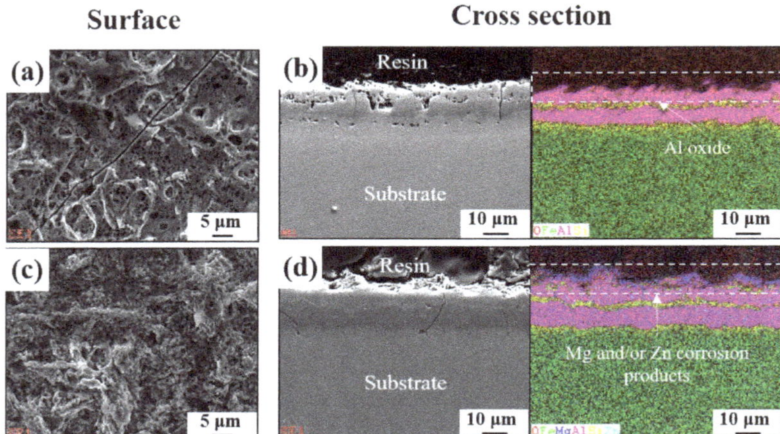

Figure 7. Surface and cross-sectional observations with EDS mappings of (**a**,**b**) AlSi-MC and (**c**,**d**) AlSiZnMg-MC after the potentiostatic (PS) test.

3.3. Characterization of the Corrosion Products by XPS

Figure 8 presents the XPS survey spectrum of the surface of AlSi-MC. The spectrum of the coating surface revealed C 1s, Al 2p, O 1s, and Si 2p primarily. The high-resolution spectra, shown in Figure 8c–e, indicated that the corrosion products formed on the surface were composed mainly of Al_2O_3 with a smaller portion of SiO_2. On the other hand, the survey spectrum of the surface on AlSiZnMg-MC (Figure 9) showed that, in addition to C 1s, Al 2p, O 1s, and Si 2p, Mg 2p and Zn 2p were observed on the coating surface, and the oxygen intensity was higher. This suggests that Zn and Mg-based oxide/hydroxide are major components in the coating surface. From the high-resolution spectra of the surface on AlSiZnMg-MC, as shown in Figure 9c–g, the coating surface was composed of a wider variety of corrosion products including Al_2O_3, $Mg(OH)_2$, SiO_2, $Zn(OH)_2$, $Zn_5(OH)_8Cl_2{\cdot}H_2O$, and Mg-Al layered double hydroxide (LDH) with the general formula, $M_xAl_y(A)_m(OH)_n{\cdot}_zH_2O$ (where A and M represent an anion and a di-valent cation, respectively) [16,18,37]. As mentioned previously, the addition of Zn and Mg to the coating can help increase the corrosion resistance in a neutral aqueous solution, in such a way that several corrosion products with an inhibiting nature are formed preferentially on the surface at the early stages of corrosion. Among the products, Mg-Al LDH and $Zn_5(OH)_8Cl_2{\cdot}H_2O$ can act as effective barriers for oxygen diffusion [38,39] and provide superior corrosion resistance in neutral aqueous solutions [38,39]. On the other hand, the formation or dissolution of other products can also help increase the corrosion resistance by stabilizing Mg-Al LDH and $Zn_5(OH)_8Cl_2{\cdot}H_2O$. First, the formation of Mg-Al LDH, as shown in Figure 9d,f involves reactions (2), (4), (5), and (6).

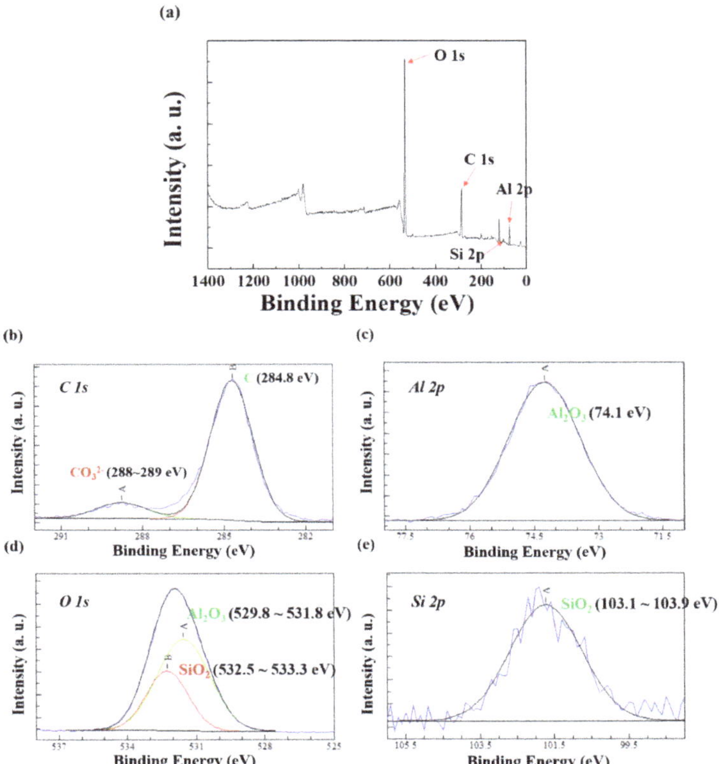

Figure 8. (**a**) XPS survey spectrum and (**b–e**) high-resolution spectra of the surface on AlSi-MC after the PS test.

Oxygen reduction:

$$O_2(g) + 2H_2O(l) + 4e^- \rightarrow 4OH^-(aq) \tag{4}$$

Dissolution of alumina:

$$Al_2O_3(s) + 3H_2O(l) + 2OH^-(aq) \rightarrow 2Al(OH)^{4-}(aq) \tag{5}$$

Dissolution of metallic Al:

$$Al(s) + 4OH^-(aq) \rightarrow Al(OH)^{4-}(aq) + 3e^- \tag{6}$$

Reactions (5) and (6) can occur locally at the cathode area because of the instability of Al/Al_2O_3 at alkaline pH [16]. As a result, the formation reaction of LDH can proceed, as described below:

$$2Al(OH)^{4-}(aq) + 6Mg^{2+}(aq) + 8OH^-(aq) + CO_3^{2-}(aq) \rightarrow Mg_6Al_2(OH)_{16}CO_3(s) \tag{7}$$

The stability of LDH is dependent on the pH [40,41], and LDH can dissolve under weakly alkaline condition. On the other hand, the dissolution of LDH can be suppressed by the formation of a thin $Mg(OH)_2$ layer (reaction (8)), called the skin effect [15,37]:

$$Mg^+(aq) + 2OH^-(aq) \rightarrow Mg(OH)_2(s) \tag{8}$$

Second, the formation of $Zn_5(OH)_8Cl_2{\cdot}H_2O$, as shown in Figure 9g, can be described below:

$$4ZnO(s) + Zn^{2+}(aq) + 5H_2O(l) + 2Cl^-(aq) \rightarrow Zn_5(OH)_8Cl_2{\cdot}H_2O(s) \quad (9)$$

The presence of Mg^{2+} supplied by reaction (2) can also delay the following reaction, which leads to slower dissolution kinetics of $Zn_5(OH)_8Cl_2{\cdot}H_2O$. The formations of Mg-Al LDH and $Zn_5(OH)_8Cl_2{\cdot}H_2O$ with high stability are the major mechanistic reasons for the higher corrosion resistance of AlSiZnMg in a neutral environment containing Cl^-. Nevertheless, the increased possibility of local corrosion of the AlSiZnMg sample cannot be excluded. In addition, its superiority and utility as a coated steel should also be evaluated after the electropainting process, which will be discussed in the following section.

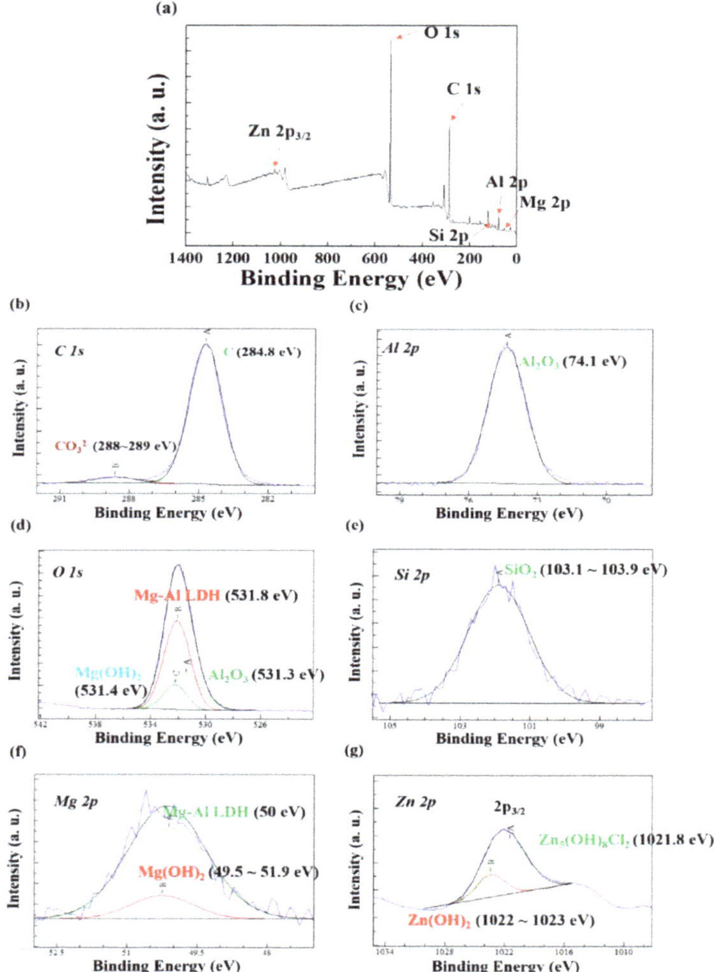

Figure 9. (**a**) XPS survey spectrum and (**b–g**) high-resolution spectra of the surface on AlSiZnMg-MC after PS test.

3.4. Surface Characteristics after Electropainting

Figure 10 presents the GDS depth profiles of AlSi-MC and AlSiZnMg-MC after the ED coating. In contrast to AlSi-MC, Zn(Mg)—based oxides were present on the outer surface of the ED-coated AlSiZnMg-MC. This suggests that the oxides remained on the outer surface, even after surface cleaning before the ED coating. As mentioned in the previous section, a stable thick MgO film cannot be formed in a minor fraction of Mg (0.5 wt %) in this coating system. An unstable oxide film on the surface may have weakened the adhesion between the inner metallic coating/outer ED coating and reduced the corrosion resistance.

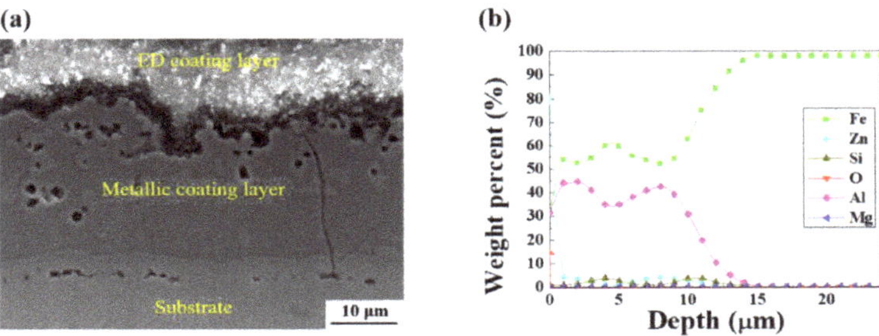

Figure 10. (**a**) Cross-sectional observation and (**b**) GDS depth profile of ED-treated AlSiZnMg-MC.

3.5. Corrosion Resistance Evaluations after Electropainting

Figure 11 shows the surface appearance of the ED-treated AlSi-MC (Figure 11a) and AlSiZnMg-MC (Figure 11b), after the accelerated corrosion test in reference to VDA 233-102 [12]. Considering the many test results presented above, it was expected that ED-treated AlSiZnMg-MC showed higher corrosion resistance when evaluated by an accelerated corrosion test. After the ED coating, however, its higher corrosion resistance was not observed clearly in that the damaged areas and extent of red rust formation on the scribed regions with an X between the two coated steels sheets became similar as the number of cycles increased. In addition, the surface morphologies of the two ED-coated steel samples that had been subjected to stone chipping and a subsequent corrosion test revealed even more severe surface degradation of ED-coated AlSiZnMg-MC. As shown in Figure 12, the size and number of surface blisters were larger on the ED-coated AlSiZnMg-MC. These blisters resulted from volumetric expansion caused by the formation of corrosion products on the steel substrate covered with a metallic coating layer. This suggests that the corrosive species are more able to permeate through the ED coating on AlSiZnMg-MC, and the electrochemical corrosion reactions occurred rapidly. The higher degradation by the corrosion reactions of ED coated AlSiZnMg-MC may be associated closely with the lower adhesion between the inner metallic coating/outer ED coating. As mentioned previously, the (Zn, Mg)-based oxide formed on AlSiZnMg can weaken the adhesion between the inner/outer coating layers, leading to delamination of the outer coating layer when attacked mechanically and chemically. This is supported by the lower coefficient of friction for the ED coating on AlSiZnMg-MC, which was measured by surface and interfacial cutting analysis, and was reported elsewhere [21]. Hence, the addition of Zn and Mg to the AlSi-based metallic coating does not always have beneficial effects on the corrosion resistance when the metallic-coated steels are coated additionally with ED paint. These results provide useful insights into the development of coated steel sheets used in automotive industries. Nevertheless, further technical research on optimizing the compositions of Zn and/or Mg alloying that provides sacrificial corrosion protection to the AlSi coated steel while reducing the possibility of local corrosion and securing superior corrosion resistance even after additional ED coating will be needed.

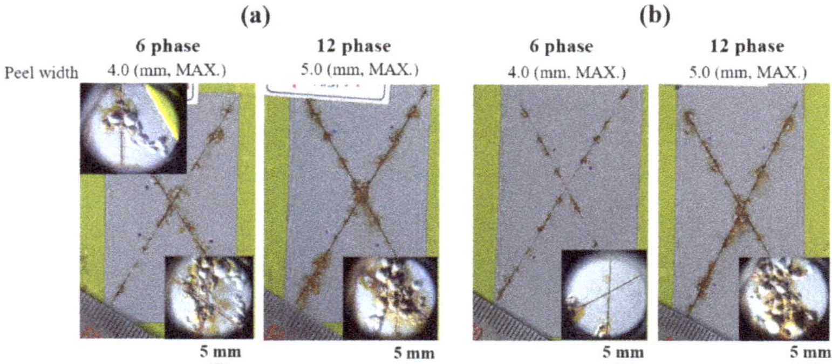

Figure 11. Surface appearance of electrodeposited (ED)-treated (**a**) AlSi-MC and (**b**) AlSiZnMg-MC, which had been scratched with an "×" incision, after the accelerated corrosion test.

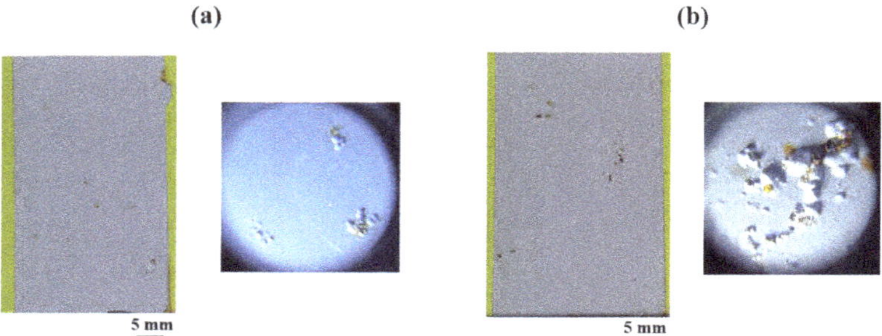

Figure 12. Surface appearance of ED-treated (**a**) AlSi-MC and (**b**) AlSiZnMg-MC, which had been collided with fine stone chips, after the accelerated corrosion test.

4. Conclusions

The influence of the combined addition of Mg and Zn on the microstructure and corrosion resistance of AlSi-based coatings for steel sheets was investigated using a range of analytical and experimental techniques. In particular, the corrosion resistance was also evaluated after ED coating on the metallic coated steel sheets. The major findings can be summarized as follows:

1. The two types of AlSi-based coating (AlSi-MC and AlSiZnMg-MC) were composed mainly of an Al matrix with two (Al,Fe,Si)-rich intermetallic phases. The compositional features were different in that the intermetallic phases of $MgZn_2$ and Mg_2Si existed only in AlSiZnMg-MC covered with a thin outermost layer of (Zn, Mg)-based oxide.
2. The active phases formed in AlSiZnMg-MC, such as Mg_2Si and $MgZn_2$, can decrease the corrosion potential. The preferential dissolution of Mg and Zn from the active phases can lead to rapid coverage of the coating surface by the precipitation of corrosion products with a protective nature, resulting in a much smaller current density for both anodic and cathodic reactions. On the other hand, the polarization curve of AlSiZnMg-MC showed many fluctuations, which may be closely associated with localized corrosion caused by the non-uniform distribution of MgO or the selective dissolution of Mg from Mg_2Si phases in the coating.
3. Compared to the case of the AlSi-MC, XPS showed that the surface of AlSiZnMg-MC was composed of a wider variety of corrosion products, including Al_2O_3, $Mg(OH)_2$, SiO_2, $Zn(OH)_2$, $Zn_5(OH)_8Cl_2{\cdot}H_2O$, and Mg-Al LDH. Among the products, the presence of $Zn_5(OH)_8Cl_2{\cdot}H_2O$

and Mg-Al LDH, stabilized by $Mg(OH)_2$, is the major mechanistic reason for the higher corrosion resistance of AlSiZnMg-MC.

4. In contrast to the corrosion resistance of metallic coatings, ED-coated AlSiZnMg-MC exhibited more severe surface degradation after the accelerated corrosion tests. The non-uniform formation of Mg-based oxide over the AlSiZnMg-MC can weaken the adhesion between the inner metallic coating/outer ED coating. Hence, the corrosive species are more able to permeate through the ED coating on AlSiZnMg-MC, and the electrochemical corrosion reactions occurred rapidly. Therefore, further optimization of the alloy contents in AlSi-based coatings needs to be investigated.

Author Contributions: Conceptualization, W.S.Y. and S.J.K.; methodology, S.O.K., W.S.Y., and S.J.K.; investigation, S.O.K., W.S.Y., and S.J.K.; data curation, S.O.K., W.S.Y., and S.J.K., writing—original draft preparation, S.O.K. and S.J.K. All authors have agreed to the published version of the manuscript.

Funding: This research was supported in part by the National Research Foundation of Korea (NRF) grant funded by the Korea government (MSIT), grant number No. 2019R1C1C1005007. In addition, this work was partly funded and conducted under the Competency Development Program for Industry Specialists of the Korean Ministry of Trade, Industry and Energy (MOTIE), operated by the Korea Institute for Advancement of Technology (KIAT) (No. P0002019, HRD Program for High Value-Added Metallic Materials Expert).

Conflicts of Interest: The authors declare no conflict of interest.

References

1. Ahmad, Z. *Principles of Corrosion Engineering and Corrosion Control*, 2nd ed.; Butterworth-Heinemann: London, UK, 2006.
2. Forsyth, M.; Hinton, B. *Rare Earth-Based Corrosion Inhibitors*; Woodhead Publishing: Cambridge, UK, 2014.
3. Schürz, S.; Luckeneder, G.H.; Fleischanderl, M.; Mack, P.; Gsaller, H.; Kneissl, A.C.; Mori, G. Chemistry of corrosion products on Zn-Al-Mg alloy coated steel. *Corros. Sci.* **2010**, *52*, 3271–3279. [CrossRef]
4. Thebault, F.; Vuillemin, B.; Oltra, R.; Ogle, K.; Allely, C. Investigation of self-healing mechanism on galvanized steels cut edges by coupling SVET and numerical modeling. *Electrochim. Acta* **2008**, *53*, 5226–5234. [CrossRef]
5. Ogle, K.; Morel, S.; Jacquet, D. Observation of self-healing functions on the cut edge of galvanized steel using SVET and pH microscopy. *J. Electrochem. Soc.* **2006**, *153*, B1–B5. [CrossRef]
6. Zeng, F.; Wei, Z.; Li, C.; Tan, X.; Zhang, Z. Corrosion mechanism associated with Mg_2Si and Si particles in Al–Mg–Si alloys. *Trans. Nonferrous Met. Soc. China* **2011**, *21*, 2559–2567. [CrossRef]
7. Zhang, Y.S.; Zhu, X.M. Electrochemical polarization and passive film analysis of austenitic Fe–Mn–Al steels in aqueous solutions. *Corros. Sci.* **1999**, *41*, 1817–1833. [CrossRef]
8. Bobby Kannan, M.; Singh Raman, R.K.; Khoddam, S. Comparative studies on the corrosion properties of a Fe–Mn–Al–Si steel and an interstitial-free steel. *Corros. Sci.* **2008**, *50*, 2879–2884. [CrossRef]
9. Panossian, Z.; Mariaca, L.; Morcillo, M.; Flores, S.; Rocha, J.; Peña, J.J.; Herrera, F.; Corvo, F.; Sanchez, M.; Rincon, O.T.; et al. Steel cathodic protection afforded by zinc, aluminum and zinc/aluminium alloy coatings in the atmosphere. *Surf. Coat. Technol.* **2005**, *190*, 244–248. [CrossRef]
10. Han, S.; Li, H.; Wang, S.; Jiang, L.; Liu, X. Influence of silicon on hot-dip aluminizing process and subsequent oxidation for preparing hydrogen/tritium permeation barrier. *Int. J. Hydrog. Energy* **2010**, *35*, 2689–2693. [CrossRef]
11. Cheng, W.J.; Wang, C.J. Microstructural evolution of intermetallic layer in hot-dipped aluminide mild steel with silicon addition. *Surf. Coat. Technol.* **2011**, *205*, 4726–4731. [CrossRef]
12. Allely, C.; Dosdat, L.; Clauzeau, O.; Ogle, K.; Volovitch, P. Anticorrosion mechanisms of aluminized steel for hot stamping. *Surf. Coat. Technol.* **2014**, *238*, 188–196. [CrossRef]
13. Szklarska-Smialowska, Z. Pitting corrosion of Al. *Corros. Sci.* **1999**, *41*, 1743–1767. [CrossRef]
14. Vargel, C. *Corrosion of Aluminium*; Elsevier: Oxford, UK, 2004.
15. Salgueiro Azevedo, M.; Allely, C.; Ogle, K.; Volovitch, P. Corrosion mechanisms of Zn (Mg,Al) coated steel: The effect of HCO_3^- and NH_4^+ ions on the intrinsic reactivity of the coating. *Electrochim. Acta* **2015**, *153*, 159–169. [CrossRef]

16. Volovitch, P.; Vu, T.N.; Allély, C.; Abdel Aal, A.; Ogle, K. Understanding corrosion via corrosion product characterization: II. Role of alloying elements in improving the corrosion resistance of Zn–Al–Mg coatings on steel. *Corros. Sci.* **2011**, *53*, 2437–2445. [CrossRef]
17. Prosek, T.; Larché, N.; Vlot, M.; Goodwin, F.; Thierry, D. Corrosion performance of Zn–Al–Mg coatings in open and confined zones in conditions simulating automotive applications. *Mater. Corros.* **2010**, *61*, 412–420. [CrossRef]
18. Nicarda, C.; Allély, C.; Volovitch, P. Effect of Zn and Mg alloying on microstructure and anticorrosion mechanisms of Al-Si based coatings for high strength steel. *Corros. Sci.* **2019**, *146*, 192–201. [CrossRef]
19. Hwang, Y.; Lee, C.W.; Shin, G.Y.; Yoo, J.H.; Choi, M. Study of the corrosion behavior, liquid metal embrittlement and resistance spot weldability of galvannealed hot stamping steel. *Korean J. Met. Mater.* **2019**, *57*, 193–201. [CrossRef]
20. Kharitonov, D.S.; Örnek, C.; Claesson, P.M.; Sommertune, J.; Zharskii, I.M.; Kurilo, I.I.; Pan, J. Corrosion inhibition of aluminum alloy AA6063-T5 by vanadates, microstructure characterization and corrosion analysis. *J. Electrochem. Soc.* **2018**, *165*, C116–C126. [CrossRef]
21. Yang, W.S.; Ahn, S.H.; Kim, S.J. Effects of Zn and Mg addition to AlSi metallic coating for hot stamping steel. In *Hyundai Steel Technical Report*; Hyundai Steel Company: Chungnam, Korea, 2019; Volume 7, pp. 12–17. ISSN 2288-6990.
22. Susi, T.; Pichler, T.; Ayala, P. X-ray photoelectron spectroscopy of graphitic carbon nanomaterials doped with heteroatoms. *Beilstein J. Nanotechnol.* **2015**, *6*, 177–192. [CrossRef]
23. *Standard Test Method for Evaluation of Painted or Coated Specimens Subjected to Corrosive Environments*; ASTM D1654–08; ASTM International: West Conshohocken, PA, USA, 2016.
24. *Standard Test Method for Chipping Resistance of Coatings*; ASTM D3170; ASTM International: West Conshohocken, PA, USA, 2014.
25. Wang, C.J.; Chen, S.M. The high-temperature oxidation behavior of hot-dipping Al–Si coating on low carbon steel. *Surf. Coat. Technol.* **2006**, *200*, 6601–6605. [CrossRef]
26. Jo, K.R.; Cho, L.; Sulistiyou, D.H.; Seo, E.J.; Kim, S.W.; De Cooman, B.C. Effects of Al-Si coating and Zn coating on the hydrogen uptake and embrittlement of ultra-high strength press-hardened steel. *Surf. Coat. Technol.* **2019**, *374*, 1108–1119. [CrossRef]
27. Grigorieva, R.; Drillet, P.; Mataigne, J.M.; Redjaïmia, A. Phase transformations in the Al-Si coating during the austenitization step. *Solid State Phenom.* **2011**, *172–174*, 784–790. [CrossRef]
28. Maitra, T.; Gupta, S.P. Intermetallic compound formation in Fe–Al–Si ternary system: Part II. *Mater. Charact.* **2003**, *49*, 293–311. [CrossRef]
29. Bruschi, S.; Ghiotti, A. *Comprehensive Materials Processing*; Elsevier: Amsterdam, The Netherlands, 2014.
30. Windmann, M.; Röttger, A.; Theisen, W. Phase formation at the interface between a boron alloyed steel substrate and an Al-rich coating. *Surf. Coat. Technol.* **2013**, *226*, 130–139. [CrossRef]
31. Kobayashi, S.; Yakou, T. Control of intermetallic compound layers at interface between steel and aluminum by diffusion-treatment. *Mater. Sci. Eng. A* **2002**, *338*, 44–53. [CrossRef]
32. Cho, Y.R.; Kim, T.H.; Park, S.H.; Baek, E.R. Aluminum-Plated Steel Sheet Having Superior Corrosion Resistance, Hot Press Formed Product Using the Same, and Method for Production Thereof. European Patent WO/2010/079995, 15 July 2010.
33. Li, Q.; Zhao, Y.Z.; Luo, Q.; Chen, S.L.; Zhang, J.Y.; Chou, K.C. Experimental study and phase diagram calculation in Al–Zn–Mg–Si quaternary system. *J. Alloy. Compd.* **2010**, *501*, 282–290. [CrossRef]
34. Samsonov, G.V. *The Oxide Handbook*, 2nd ed.; IFI/Plenum: New York, NY, USA, 1973.
35. Lee, J.W.; Park, B.R.; Oh, S.Y.; Yun, D.W.; Hwang, J.K.; Oh, M.S.; Kim, S.J. Mechanistic study on the cut-edge corrosion behaviors of Zn-Al-Mg alloy coated steel sheets in chloride containing environments. *Corros. Sci.* **2019**, *160*, 108170. [CrossRef]
36. Volovitch, P.; Barrallier, L.; Saikaly, W. Microstructure and corrosion resistance of Mg alloy ZE41 with laser surface cladding by Al–Si powder. *Surf. Coat. Technol.* **2008**, *202*, 4901–4914. [CrossRef]
37. Salgueiro Azevedo, M.; Allely, C.; Ogle, K.; Volovitch, P. Corrosion mechanisms of Zn (Mg,Al) coated steel: 2. The effect of Mg and Al alloying on the formation and properties of corrosion products in different electrolytes. *Corros. Sci.* **2015**, *90*, 482–490. [CrossRef]
38. Hosking, N.C.; StrÖm, M.A.; Shipway, P.H.; Rudd, C.D. Corrosion resistance of zinc–magnesium coated steel. *Corros. Sci.* **2007**, *49*, 3669–3695. [CrossRef]

39. Ishizaki, T.; Kamiyama, N.; Watanabe, K.; Serizawa, A. Corrosion resistance of $Mg(OH)_2$/Mg–Al layered double hydroxide composite film formed directly on combustion-resistant magnesium alloy AMCa602 by steam coating. *Corros. Sci.* **2015**, *92*, 76–84. [CrossRef]
40. Wu, L.; Pan, F.; Liu, Y.; Zhang, G.; Tang, A.; Atrens, A. Influence of pH on the growth behavior of Mg–Al LDH films. *Surf. Eng.* **2017**, *34*, 674–681. [CrossRef]
41. Zhang, F.; Zhang, C.L.; Song, L.; Zeng, R.C.; Liu, Z.G.; Cui, H.Z. Corrosion of in-situ grown MgAl-LDH coating on aluminum alloy. *Trans. Nonferrous Met. Soc. China* **2015**, *25*, 3498–3504. [CrossRef]

© 2020 by the authors. Licensee MDPI, Basel, Switzerland. This article is an open access article distributed under the terms and conditions of the Creative Commons Attribution (CC BY) license (http://creativecommons.org/licenses/by/4.0/).

Article

Effect of Spraying Power on Oxidation Resistance of $MoSi_2$-ZrB_2 Coating for Nb-Si Based Alloy Prepared by Atmospheric Plasma

Guanqun Zhuo [1,2], Linfen Su [1,2,*], Kaiyong Jiang [1,2] and Jianyong Yang [1,2]

1 College of Mechanical Engineering and Automation, Huaqiao University, Xiamen 361021, China; zhuoguanqun@hqu.edu.cn (G.Z.); kyjiang@hqu.edu.cn (K.J.); yangjianyong@hqu.edu.cn (J.Y.)
2 Fujian Key Laboratory of Special Energy Manufacturing, Huaqiao University, Xiamen 361021, China
* Correspondence: linfensu@hqu.edu.cn

Received: 29 September 2020; Accepted: 6 November 2020; Published: 10 November 2020

Abstract: The $MoSi_2$-ZrB_2 coatings were prepared on Nb-Si based alloy by atmospheric plasma spraying with the spraying power 40, 43 and 45 kW. The effect of spraying power on the microstructure and oxidation resistance of $MoSi_2$-ZrB_2 coating at 1250 °C were studied. The results showed that the main constituent phases of coatings were $MoSi_2$ at all spraying power. The coating became more compact as the spraying power increased. The coating prepared at 45 kW was dense and uniform, which exhibited the best oxidation resistance due to the formation of a dense and uniform glass layer consisting of SiO_2 and $ZrSiO_4$.

Keywords: Nb-Si based alloy; $MoSi_2$-ZrB_2 coating; atmospheric plasma; oxidation

1. Introduction

Nb-Si based alloys are considered to be one of the most promising high-temperature structural material, owing to the high melting points (1750 °C), medium density and excellent high-temperature strength [1–7]. However, the high-temperature oxidation resistance of Nb-Si based alloys are poor, which limits their application [8–10]. Adding elements such as Cr, B, Ta, Ti, Si in Nb-Si based alloys or preparing coatings on the Nb-Si based alloys can improve the oxidation resistance; however, alloying would compromise mechanical properties of alloys [11,12]. Therefore, more attention is paid to preparing coatings on Nb-Si based alloys.

$MoSi_2$ is a promising coating material for Nb-Si based alloy [13–16]. It can produce SiO_2 oxide film with excellent oxidation resistance at high temperatures. Besides, its thermal expansion coefficient (8.1×10^{-6} °C^{-1}) is close to that of Nb-Si based alloys (8.4×10^{-6} °C^{-1}) [17]. However, $MoSi_2$ would suffer pest oxidation at 400–600 °C. The addition of B can effectively avoid this disadvantage [18]. The formation of borosilicate glass at high temperatures can effectively protect $MoSi_2$ against pest oxidation at 400–600 °C [18]. Furthermore, borosilicate glass coating has a better self-repair ability than SiO_2 due to its lower viscosity [19–21]. Fu et al. prepared B_2O_3 modified SiC-$MoSi_2$ coating on C/C composites by a two-step pack cementation [22]. The coating could protect C/C composites from oxidation at 1500 °C in air for more than 242 h. Pang et al. prepared a Mo-Si-B coating on Nb-Si-based alloys by spraying Mo first and then co-deposition of Si and B [23]. The mass gain was 0.92 mg/cm^2 after oxidation at 1250 °C for 100 h. However, stresses caused by CTE mismatch between the MoSi2 coating and silica can produce cracks in the oxide film if they exceed the strength of the SiO_2.

Atmospheric plasma spraying has attracted widespread attention due to the advantages such as high spraying temperature, high deposition efficiency and precise control of the composition and thickness of the coating [24,25]. Le et al. directly deposited the oxidation-resistant coating $MoSi_2$ on Nb alloy substrate by supersonic air plasma spraying with pure agglomerated $MoSi_2$ powder [25].

After oxidation at 1500 °C in air for 43 h, it showed excellent oxidation resistance with mass loss of 5.31 mg cm^2. Some of the literature has mentioned that the addition of Zr to Mo-Si-B coating could effectively improve the mechanical properties of $MoSi_2$ at a high temperature. Furthermore, the $ZrSiO_4$ produced by the reaction of dispersive ZrO_2 and SiO_2 could minimize the CTE difference between silica and $MoSi_2$, as well as the consumption of SiO_2 at a high temperature [26–28]. In this study, the $MoSi_2$-ZrB_2 coatings were prepared on Nb-Si based alloy by the atmospheric plasma spraying technology. The effects of spraying power on coating structure and the oxidation mechanism of $MoSi_2$-ZrB_2 coating were investigated.

2. Materials and Methods

2.1. Preparation of $MoSi_2$-ZrB_2 Coating

Substrates (Nb-15Si-24Ti-13Cr-2Al-2Hf (at.%) were fabricated by non-consumable arc-melting. The ingots were re-melted and inverted at least four times to guarantee the uniformity of the composition. Samples with a size of 10 × 10 × 8 mm^3 were cut from the ingots. All surfaces were mechanically ground on wet SiC paper to 800 grit, then cleaned ultrasonically with ethanol and dried at about 80 °C for 1 h. Commercially available $MoSi_2$ with 95 wt.%, ZrB_2 with 5 wt.% powders were selected as raw materials with the purity of 99.9 wt.% and the particle size between 45 and 65 μm. The powders were ground in a planetary ball mill for 2 h, to ensure their uniformity. The $MoSi_2$-ZrB_2 coatings were prepared by atmospheric plasma spraying, at the power of 40, 43 and 45 kW, respectively. The samples were designated as mz40, mz43 and mz45, according to the spraying power. The spraying distance was set as 100 mm. Argon was used as primary gas and carrier gas, and hydrogen was used as secondary gas. The detailed parameters are listed in Table 1.

Table 1. Detailed parameters of the sprayed $MoSi_2$-ZrB_2 coating.

Content	Parameters
Spraying power (kW)	40–45
Primary gas Ar (L/min)	58
Carrier gas Ar (L/min)	10
Second gas H_2 (L/min)	9
Powder feed rate (g/min)	17
Spraying distance (mm)	100
Nozzle diameter (mm)	5.5

2.2. Isothermal Oxidation

An isothermal oxidation test was carried out in an open tube furnace, in air, at 1250 °C. Each sample was placed in a separate alumina crucible. Samples were taken from the furnace at intervals of 10, 20, 40 and 60 h, and weighed with a crucible, using a precision analytical balance (model CPA225D, Sartorius, Göttingen, Germany) with an accuracy of 0.00001 g.

2.3. Coating Characterization

Phase composition of the coating and oxidation specimens were analyzed by X-ray diffraction (XRD, CuKα-radiation, X'Pert Pro, Panalytical, Almelo, Holland) with Cu radiation. Morphology details and elemental distribution characteristics of the coated specimens were investigated by scanning electron microscope combined with energy dispersive spectroscopy (EDS) (Sigma 500, Zeiss, Oberkochen, Germany).

3. Results

3.1. Microstructure of $MoSi_2$-ZrB_2 Coating

Figure 1 shows the XRD patterns of the surface of as-prepared $MoSi_2$-ZrB_2 coatings. It could be seen that the constituent phases of coating at different spraying powers are $MoSi_2$, Mo_5Si_3, Mo and ZrB_2. In the process of plasma spraying, the temperature of the plasma arc was about 10,000 °C [28], which is much higher than the oxidation temperature of $MoSi_2$. Therefore, the raw materials are oxidized to form Mo_5Si_3, SiO_2 and Mo according to the Equations (1) and (2) [29–32]. SiO_2 is an amorphous phase, and its amount is relatively small; therefore, no SiO_2 phase is detected in XRD patterns.

$$5MoSi_2 + 7O_2 \rightarrow Mo_5Si_3 + 7SiO_2 \quad (1)$$

$$MoSi_2 + 2O_2 \rightarrow Mo + 2SiO_2 \quad (2)$$

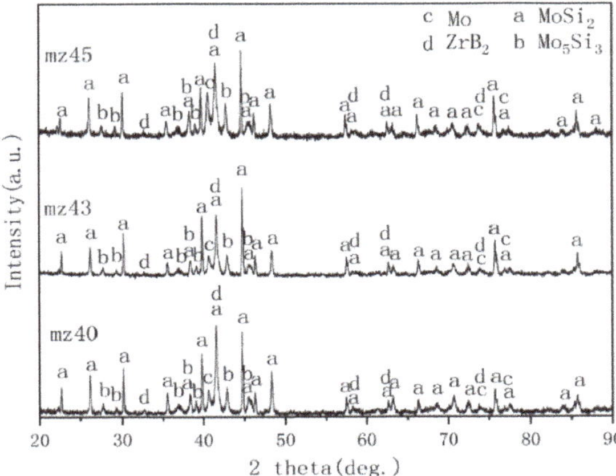

Figure 1. XRD patterns of $MoSi_2$-ZrB_2 coatings.

Figure 2 shows the surface morphology of $MoSi_2$-ZrB_2 coatings. The surfaces of all spraying samples are rough. In addition, the molten zone is interwoven with the incompletely molten particles, which is a typical structure feature of plasma sprayed coatings. In the process of coating preparation, high-speed particles are heated by plasma flame, and then the molten particles impinge on the substrate to form a flat structure. The temperature of the plasma arc elevates as the spraying power increases. Therefore, the full molten area of the mz43 sample is larger than that of the mz40 sample, leading to a much smoother surface of the mz43 sample. As for the mz45 sample, the completely molten area of the mz45 sample is the largest (as shown in Figure 2c). Therefore, the mz45 sample shows a more compact surface as compared with that of the mz40 and mz43 samples.

Figure 2. Surface morphology of $MoSi_2$-ZrB_2 coatings: (**a**) mz40, (**b**) mz43 and (**c**) mz45.

Figure 3 shows cross-sectional morphologies of $MoSi_2$-ZrB_2 coatings. The mean thickness of the coatings of the mz40, mz43 and mz45 samples are about 122, 138 and 158 μm, respectively. As shown in Figure 3, the interface between the coating and the substrate becomes denser and more uniform as the power increases. For the mz45 sample, the interface is more compact, indicating that the mz45 sample has a better combination of the coating and substrate as compared to that of the mz40 and mz43 samples. As shown in Table 2, the main constituent phase of the mz45 sample is confirmed to be $MoSi_2$. The elements mapping, as shown in Figure 4, reveals that the coating mainly consists of Mo, Si and O. The existence of O element may be induced from the spraying in oxygen atmosphere. Furthermore, the Vickers hardness of the coating prepared by 40, 43 and 45 Kw are measured to be 850, 924 and 979, respectively. This may be due to the better combination of the substrate and coating as the increase of the spraying power.

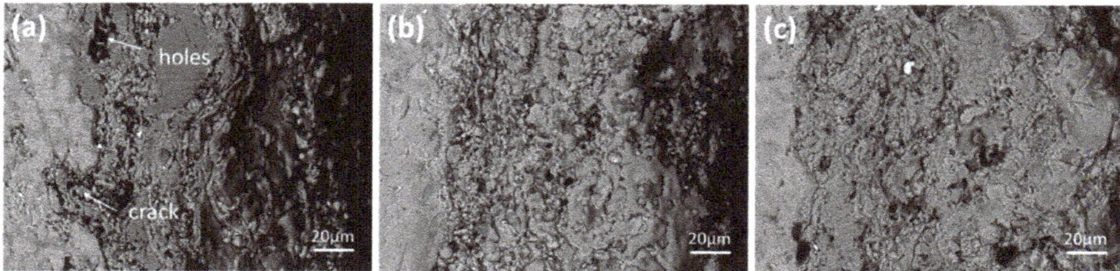

Figure 3. Cross-section morphology of $MoSi_2$-ZrB_2 coatings: (**a**) mz40, (**b**) mz43 and (**c**) mz45.

Table 2. The EDS results of constituent phase of mz45 sample (at.%).

Elements	Mo	Si	O	B
$MoSi_2$	34.29	65.71	0	0

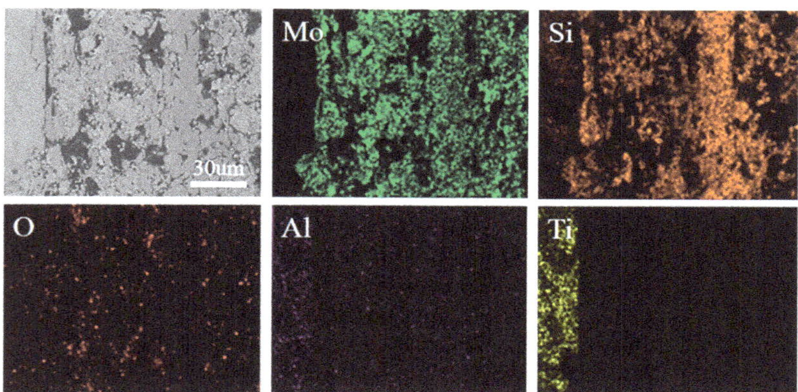

Figure 4. Elements mapping for the $MoSi_2$-ZrB_2 coating of the mz45 sample.

3.2. High Temperature Oxidation Resistance

Figure 5a shows the weight gain per unit area as a function of the exposure time at 1250 °C. The Nb-Si based alloy suffers serious oxidation with the mass gain of 205.24 mg cm^{-2} after oxidation at 1250 °C for 60 h and follows a linear oxidation behavior. The mass gains of the mz40, mz43 and mz45 samples were 11.81, 5.32 and 1.66 mg/cm^2, respectively. Therefore, $MoSi_2$-ZrB_2 coatings could effectively improve the oxidation resistance of Nb-Si based alloy. As shown in Figure 5b, the mz40 and mz43 samples follow linear oxidation behavior, and the linear kinetic constants (g^2/cm^4s) of the mz40 and mz43 sample are calculated to be 1.89×10^{-5} and 7.8×10^{-6}, respectively, according to Equation (3) [4], where Δm is the weight change of the sample, *A* is the surface area and t is the exposure time. During oxidation, the edges of the coating are the place where stress is easily concentrated, leading to the failure of the coating. As shown in Figure 5d, the coating edges of the mz40 and mz43 samples have peeled off after oxidation, while the coating edge of mz45 sample is compact. The mz45 sample shows excellent high temperature oxidation resistance and conforms to the parabolic oxidation behavior. The parabolic kinetic constant (g^2/cm^4s) of the mz45 sample is calculated to be 1.27×10^{-11} according to Equation (4) [17], where Δm is the weight change of the sample, and *A* is the surface area and t is the exposure time.

$$\frac{\Delta m}{A} = k_l t \quad (3)$$

$$\left(\frac{\Delta m}{A}\right)^2 = k_p t \quad (4)$$

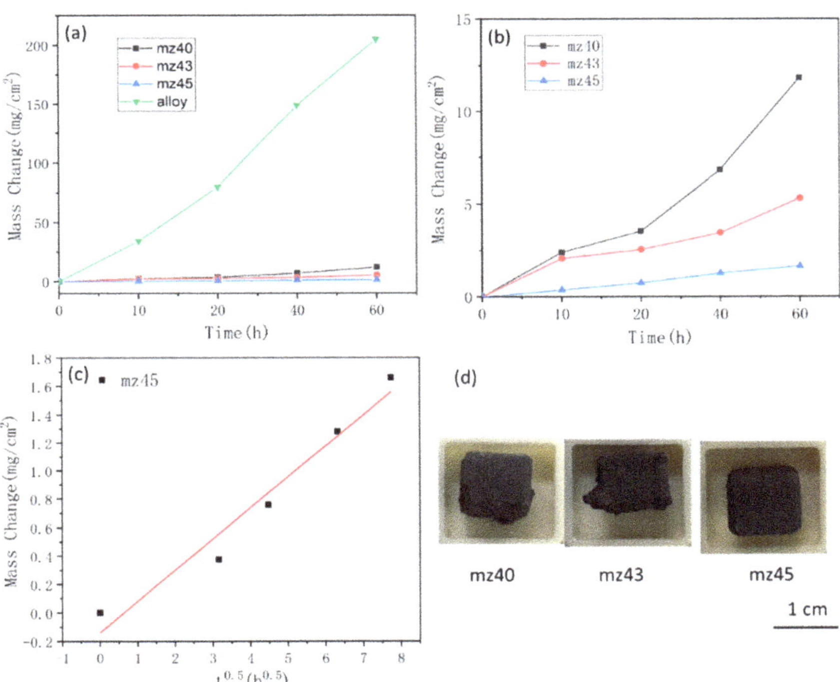

Figure 5. (**a**) Oxidation weight gain curve of Nb-Si based alloy and coatings, (**b**) oxidation weight gain curve of coatings, (**c**) representation of the weight gain versus the square root of time for mz45 oxidized in air, and (**d**) the photograph of oxidized samples.

Figure 6 shows the XRD patterns of $MoSi_2$-ZrB_2 coatings after oxidation at 1250 °C for 60 h. As shown in Figure 6, the oxidized $MoSi_2$-ZrB_2 coatings mainly consist of $MoSi_2$, Mo_5Si_3, SiO_2 and $ZrSiO_4$. Figure 7 demonstrates the surface morphologies of $MoSi_2$-ZrB_2 coatings after oxidation at 1250 °C for 60 h. The surface of the mz40 sample is loose and undulate, as shown in Figure 7a. It can be observed from Figure 7b that the surface of mz43 sample is much denser. As shown in Figure 7c, the mz45 sample displays a uniform, dense and integrated surface.

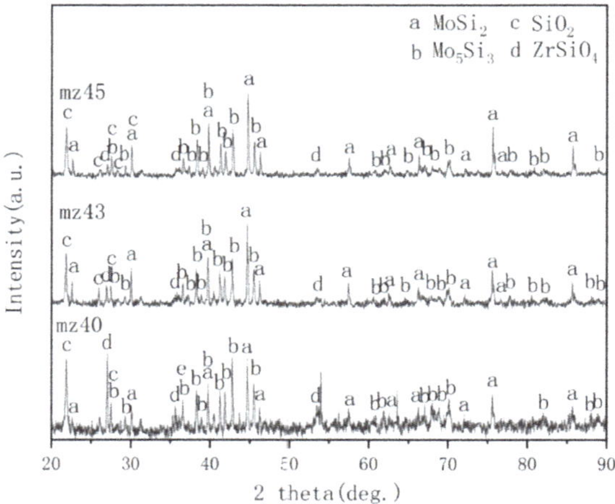

Figure 6. XRD patterns of $MoSi_2$-ZrB_2 coatings after oxidation at 1250 °C for 60 h.

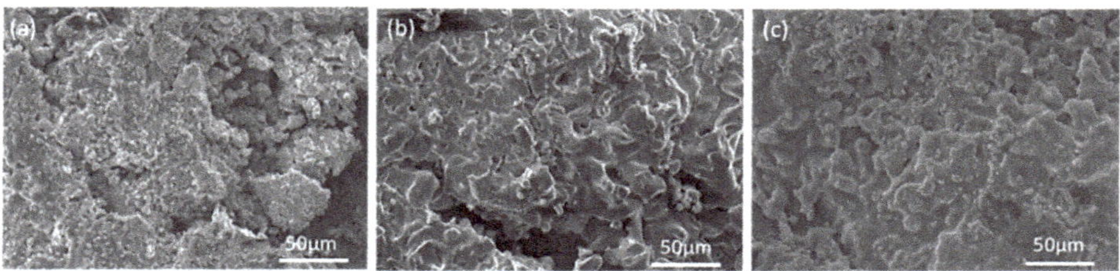

Figure 7. Surface morphology of $MoSi_2$-ZrB_2 coatings after oxidation in air at 1250 °C for 60 h: (**a**) mz40, (**b**) mz43 and (**c**) mz45.

Figure 8 shows the cross-sectional morphologies of $MoSi_2$-ZrB_2 coatings after oxidation at 1250 °C for 60 h. According to the XRD results and EDS analysis (as listed in Table 3), the coating consists of the black SiO_2, the black-gray $MoSi_2$, the gray-white Mo_5Si_3 and the white $ZrSiO_4$. Moreover, some cracks and holes are observed in all coatings, and they are filled with black SiO_2 phase. As shown in Figure 5d, the coating edges of the mz40 and mz43 samples are peeling off during oxidation. Therefore, the edge of the substrate (Nb-Si based alloy) exposes to the oxygen environment, leading to the worse oxidation performance of these two samples. Figure 8d shows the microstructure of the failure edge of the mz40 sample after oxidation. The oxides of the edge of mz40 sample are confirmed to be $TiNbO_4$ and SiO_2 according to EDS analysis, which is the typical oxides of Nb-Si based alloy at high temperature [17].

Table 3. The EDS results of constituent phase of the oxidized mz45 sample (at.%).

Element	Mo	Si	Zr	B	Nb	O
SiO_2	0	38.26	0	0	0	61.74
$MoSi_2$	32.99	67.01	0	0	0	0
Mo_5Si_3	52.02	35.4	0	2.93	3.08	6.57
$ZrSiO_4$	0	15.01	14.91	0	0	70.08

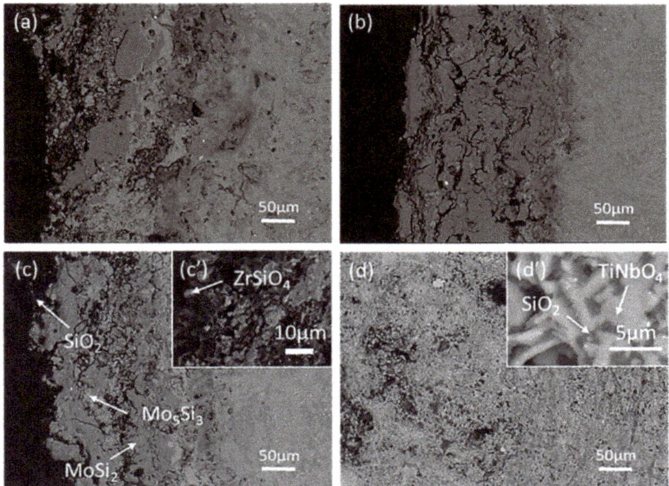

Figure 8. Cross-section morphology of $MoSi_2$-ZrB_2 coatings after oxidation in air at 1250 °C for 60 h: (**a**) mz40, (**b**) mz43, (**c**) mz45 and (**d**) the failure edge of the mz40 sample.

4. Discussion

4.1. Effect of Spraying Powder on the Microstructure of Coatings

As shown in Figure 2, the surface of the mz45 sample is more compact due to the completely molten area of the mz45 sample is the largest. In the process of spraying, the temperature of the plasma arc elevates as the spraying power increases. The plasma arc of 40 kW possesses a lower temperature. Therefore, the surface of the mz40 sample is rough consisting of many incompletely molten particles and holes, which is mainly due to the accumulation of incompletely molten particles and the escape of gas by-products during the spraying process. In addition, the incompletely molten powders are difficult to adhere to the substrate surface, which reduces the spraying efficiency, resulting in the thinnest thickness of the coating.

When the spraying power increases to 43 kW, the increasing temperature increases the melting ratio of spraying powder. These spraying powders can be riveted together with the substrate when they hit the surface of the substrate. Thus, the completely molten area increases in this sample. However, a few holes still exist at the interface between the coating and the substrate, suggesting that the spraying powders cannot be fully spread at 43 kW. When the power increases to 45 kW, completely molten particles can be uniformly distributed on the surface of the substrate, and the thickness of the coating increases. In addition, no obvious cracks and holes were observed at the interface between the coating and the substrate, indicating the good bonding of the mz45 sample. It could be concluded that the higher spraying power could produce a much more compact coating.

4.2. Oxidation Mechanism of $MoSi_2$-ZrB_2 Coatings

As shown in Figure 8, the oxidized $MoSi_2$-ZrB_2 coatings mainly consist of $MoSi_2$, Mo_5Si_3, SiO_2 and $ZrSiO_4$. The excellent oxidation resistance of the mz45 sample is due to the formation of dense SiO_2 glass layer on the surface, leading to a lower diffusion rate of oxygen. The existence of Mo_5Si_3 phase is due to the oxidation of $MoSi_2$ phase according to the oxidation reaction (Equation (1)) [30,31]; the formation of SiO_2 phase is due to the oxidation of silicides, such as $MoSi_2$ and Mo_5Si_3 according to Equations (5) and (6) [30–33]; ZrB_2 are oxidized to form ZrO_2 and B_2O_3 according to Equation (7). The $ZrSiO_4$ phase is the result of reaction between the SiO_2 and ZrO_2 according to Equation (8). Owing to the volatilization of MoO_3 and B_2O_3 at high temperature, the MoO_3 phase and B_2O_3 phase cannot be observed in the coating. Although the oxidation protective phase SiO_2 is formed in the mz40 and mz43 samples, these two samples exhibit worse oxidation due to the bad combination between coating and substrate at lower spraying power.

$$MoSi_2 + 7O_2 \rightarrow 2MoO_3 + 4SiO_2 \quad (5)$$

$$2Mo_5Si_3 + 21O_2 \rightarrow 10MoO_3 + 6SiO_2 \quad (6)$$

$$ZrB_2 + 5O_2 \rightarrow 2ZrO_2 + 2B_2O_3 \quad (7)$$

$$ZrO_2 + SiO_2 \rightarrow ZrSiO_4 \quad (8)$$

Moreover, it can be found that $ZrSiO_4$ distributed in the coating, as shown in Figure 8c. Dissolving a certain amount of zirconium oxide in amorphous silica scale could enhance its oxidation resistance [27]. Zr-based oxides have higher melting temperature, of which ZrO_2 is 2715 °C, $ZrSiO_4$ is 2550 °C, and pure SiO_2 is 1650 °C [34–36]. Therefore, the dispersion of $ZrSiO_4$ in SiO_2 glass could increase the melting temperature of the silica. Furthermore, the CTEs of ZrO_2 (10.5×10^{-6} °C^{-1}) and $ZrSiO_4$ (4.9×10^{-6} °C^{-1}) are larger than that of SiO_2 (0.55×10^{-6} °C^{-1}) [37–40]. Therefore, the formation of $ZrSiO_4$ could increase the CTE of silica. As a result, the difference of CTE between silica and $MoSi_2$ could minimize, reducing the internal stress of the coating.

In order to explain the oxidation mechanism of $MoSi_2$-ZrB_2 coating, the oxidation process is shown in Figure 9, which is similar to that of $MoSi_2$-based composite coating on Nb alloy at 1500 °C [25]. $MoSi_2$ and ZrB_2 are oxidized to form SiO_2 and ZrO_2, respectively. After that, the silica glass covers the surface of the coating and heals the cracks and holes. As oxidation continued, the $ZrSiO_4$ is produced by the reaction of dispersive ZrO_2 and SiO_2, which could minimize the CTE difference between silica and $MoSi_2$.

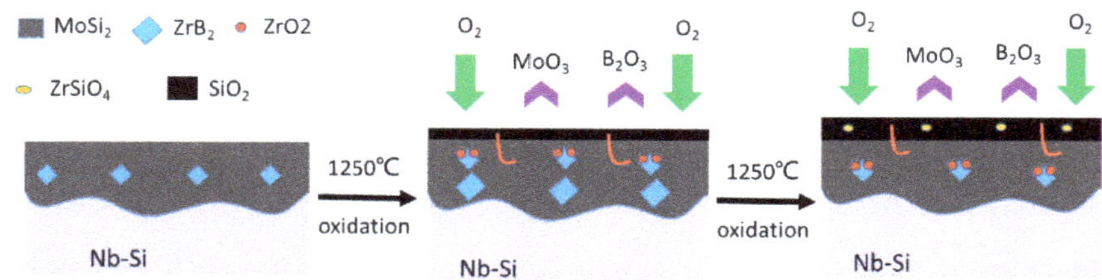

Figure 9. Oxidation process of $MoSi_2$-ZrB_2 coating at 1250 °C in air.

5. Conclusions

The effect of different spraying power on the microstructure and oxidation resistance of the $MoSi_2$-ZrB_2 coatings was investigated.

1. The microstructure of $MoSi_2$-ZrB_2 coatings prepared by atmospheric plasma spraying mainly consisted of $MoSi_2$. The higher spraying power could produce a much more compact coating.
2. The $MoSi_2$-ZrB_2 coating prepared under 45 Kw showed the best oxidation resistance with the mass gain of 1.66 mg cm^{-2} after oxidation at 1250 °C for 60 h. However, the $MoSi_2$-ZrB_2 coatings prepared under 40 and 43 Kw showed worse oxidation resistance, due to the bad combination between coating and substrate at the lower spraying power.
3. The excellent anti-oxidation protection of mz45 sample was mainly due to the formation of a silica glass layer, leading to a low diffusion rate of oxygen.

Author Contributions: Investigation, G.Z. and J.Y.; writing—original draft preparation, G.Z.; writing—review and editing, L.S. and K.J. All authors have read and agreed to the published version of the manuscript.

Funding: This research was funded by National Nature Science Foundation of P.R. China, grant number 51701077, Fujian Nature Science Foundation, grant number 2017J05082, Promotion Program for Young and Middle-Aged Teacher in Science and Technology Research of Huaqiao University, grant number ZQN-PY505 Foundation of Huaqiao University, grant number 605-50Y15065, and Subsidized Project for Postgraduates' Innovative Fund in Scientific Research of Huaqiao University.

Conflicts of Interest: The authors declare no conflict of interest.

References

1. Bewlay, B.P.; Jackson, M.R.; Subramanian, P.R.; Zhao, J.-C. A review of very-high-temperature Nb-silicide-based composites. *Met. Mater. Trans. A* **2003**, *34*, 2043–2052. [CrossRef]
2. Tang, Y.; Guo, X. High temperature deformation behavior of an optimized Nb–Si based ultrahigh temperature alloy. *Scr. Mater.* **2016**, *116*, 16–20. [CrossRef]
3. Guan, K.; Jia, L.; Kong, B.; Yuan, S.; Zhang, H. Study of the fracture mechanism of NbSS/Nb_5Si_3 in situ composite: Based on a mechanical characterization of interfacial strength. *Mater. Sci. Eng. A* **2016**, *663*, 98–107. [CrossRef]
4. Geng, J.; Tsakiropoulos, P.; Shao, G. The effects of Ti and Mo additions on the microstructure of Nb-silicide based in situ composites. *Intermetalic* **2006**, *14*, 227–235. [CrossRef]
5. Zhang, S.-N.; Guo, Y.; Kong, B.; Zhang, F.-X.; Zhang, H. High-temperature oxidation behavior of Nb-Si-based alloy with separate vanadium, tantalum, tungsten and zirconium addition. *Rare Met.* **2017**, *36*, 1–9. [CrossRef]
6. Bewlay, B.P.; Jackson, M.R.; Zhao, J.-C.; Subramanian, P.R.; Mendiratta, M.G.; Lewandowski, J.J. Ultrahigh-Temperature Nb-Silicide-Based Composites. *MRS Bull.* **2003**, *28*, 646–653. [CrossRef]
7. Bewlay, B.P.; Jackson, M.R.; Lipsitt, H.A. The balance of mechanical and environmental properties of a multielement niobium-niobium silicide-basedIn Situ composite. *Met. Mater. Trans. A* **1996**, *27*, 3801–3808. [CrossRef]
8. Wang, L.; Jia, L.; Cui, R.; Zheng, L.; Zhang, H. Microstructure, Mechanical Properties and Oxidation Resistance of Nb-22Ti-14Si-2Hf-2Al-xCr Alloys. *Chin. J. Aeronaut.* **2012**, *25*, 292–296. [CrossRef]
9. Li, X.; Chen, H.; Sha, J.; Zhang, H. The effects of melting technologies on the microstructures and properties of Nb–16Si–22Ti–2Al–2Hf–17Cr alloy. *Mater. Sci. Eng. A* **2010**, *527*, 6140–6152. [CrossRef]
10. Yao, D.; Cai, R.; Zhou, C.; Sha, J.; Jiang, H. Experimental study and modeling of high temperature oxidation of Nb-base in situ composites. *Corros. Sci.* **2009**, *51*, 364–370. [CrossRef]
11. Kang, Y.; Qu, S.; Song, J.; Huang, Q.; Han, Y. Microstructure and mechanical properties of Nb–Ti–Si–Al–Hf–xCr–yV multi-element in situ composite. *Mater. Sci. Eng. A* **2012**, *534*, 323–328. [CrossRef]
12. Geng, J.; Tsakiropoulos, P. A study of the microstructures and oxidation of Nb–Si–Cr–Al–Mo in situ composites alloyed with Ti, Hf and Sn. *Intermetalic* **2007**, *15*, 382–395. [CrossRef]
13. Xiao, L.-R.; Cai, Z.-G.; Yi, D.-Q.; Ying, L.; Liu, H.-Q.; Huang, D.-Y. Morphology, structure and formation mechanism of silicide coating by pack cementation process. *Trans. Nonferrous Met. Soc. China* **2006**, *16*, s239–s244. [CrossRef]
14. Yan, J.; Wang, Y.; Liu, L.; Wang, Y. Oxidation and interdiffusion behavior of Niobium substrate coated $MoSi_2$ coating prepared by spark plasma sintering. *Appl. Surf. Sci.* **2014**, *320*, 791–797. [CrossRef]
15. Hidouci, A.; Pelletier, J. Microstructure and mechanical properties of $MoSi_2$ coatings produced by laser processing. *Mater. Sci. Eng. A* **1998**, *252*, 17–26. [CrossRef]
16. Sun, S.; Li, X.; Wang, H.; Jiang, Y.; Yi, D. Prediction on anisotropic elasticity, sound velocity, and thermodynamic properties of $MoSi_2$ under pressure. *J. Alloy. Compd.* **2015**, *652*, 106–115. [CrossRef]
17. Su, L.; Jia, L.; Weng, J.; Hong, Z.; Zhou, C.; Zhang, H. Improvement in the oxidation resistance of Nb–Ti–Si–Cr–Al–Hf alloys containing alloyed Ge and B. *Corros. Sci.* **2014**, *88*, 460–465. [CrossRef]
18. Yokota, H.; Kudoh, T.; Suzuki, T. Oxidation resistance of boronized $MoSi_2$. *Surf. Coat. Technol.* **2003**, *169*, 171–173. [CrossRef]
19. Lange, A.; Braun, R. Magnetron-sputtered oxidation protection coatings for Mo–Si–B alloys. *Corros. Sci.* **2014**, *84*, 74–84. [CrossRef]
20. Majumdar, S.; Dönges, B.; Gorr, B.; Christ, H.-J.; Schliephake, D.; Heilmaier, M. Mechanisms of oxide scale formation on yttrium-alloyed Mo–Si–B containing fine-grained microstructure. *Corros. Sci.* **2015**, *90*, 76–88. [CrossRef]

21. Tian, X.; Guo, X.; Sun, Z.; Yin, Z.; Wang, L. Formation of B-modified $MoSi_2$ coating on pure Mo prepared through HAPC process. *Int. J. Refract. Met. Hard Mater.* **2014**, *45*, 8–14. [CrossRef]
22. Fu, Q.-G.; Li, H.; Wang, Y.-J.; Li, K.; Shi, X.-H. B_2O_3 modified SiC–$MoSi_2$ oxidation resistant coating for carbon/carbon composites by a two-step pack cementation. *Corros. Sci.* **2009**, *51*, 2450–2454. [CrossRef]
23. Pang, J.; Wang, W.; Zhou, C. Microstructure evolution and oxidation behavior of B modified $MoSi_2$ coating on Nb–Si based alloys. *Corros. Sci.* **2016**, *105*, 1–7. [CrossRef]
24. Wang, C.; Li, K.; Huo, C.; He, Q.; Shi, X. Oxidation behavior and microstructural evolution of plasma sprayed La_2O_3-$MoSi_2$-SiC coating on carbon/carbon composites. *Surf. Coat. Technol.* **2018**, *348*, 81–90. [CrossRef]
25. Sun, L.; Fu, Q.; Fang, X.-Q.; Sun, J. A $MoSi_2$-based composite coating by supersonic atmospheric plasma spraying to protect Nb alloy against oxidation at 1500 °C. *Surf. Coat. Technol.* **2018**, *352*, 182–190. [CrossRef]
26. Yanagihara, K.; Maruyama, T.; Nagata, K. Effect of third elements on the pesting suppression of Mo-Si-X intermetallics (X = Al, Ta, Ti, Zr and Y). *Intermetalics* **1996**, *4*, S133–S139. [CrossRef]
27. Wang, C.; Li, K.; He, Q.; He, D.; Huo, C.; Su, Y.; Shi, X.; Li, H. Oxidation and ablation protection of plasma sprayed LaB_6-$MoSi_2$-ZrB_2 coating for carbon/carbon composites. *Corros. Sci.* **2019**, *151*, 57–68. [CrossRef]
28. Wang, L.; Fu, Q.; Liu, N.; Shan, Y. Supersonic plasma sprayed $MoSi_2$-ZrB_2 antioxidation coating for SiC–C/C composites. *Surf. Eng.* **2016**, *32*, 508–513. [CrossRef]
29. Mao, J.; Ding, S.; Li, Y.; Li, S.; Liu, F.; Zeng, X.; Cheng, X.-D. Preparation and investigation of $MoSi_2$/SiC coating with high infrared emissivity at high temperature. *Surf. Coat. Technol.* **2019**, *358*, 873–878. [CrossRef]
30. Yan, J.; Liu, L.; Mao, Z.; Xu, H.; Wang, Y. Effect of Spraying Powders Size on the Microstructure, Bonding Strength, and Microhardness of MoSi2 Coating Prepared by Air Plasma Spraying. *J. Therm. Spray Technol.* **2014**, *23*, 934–939. [CrossRef]
31. Sun, J.; Li, T.; Zhang, G.-P. Effect of thermodynamically metastable components on mechanical and oxidation properties of the thermal-sprayed MoSi2 based composite coating. *Corros. Sci.* **2019**, *155*, 146–154. [CrossRef]
32. Zhang, G.; Sun, J.; Fu, Q. Effect of mullite on the microstructure and oxidation behavior of thermal-sprayed $MoSi_2$ coating at 1500 °C. *Ceram. Int.* **2020**, *46*, 10058–10066. [CrossRef]
33. Wu, H.; Li, H.; Lei, Q.; Fu, Q.-G.; Ma, C.; Yao, D.-J.; Wang, Y.-J.; Sun, C.; Wei, J.-F.; Han, Z.-H. Effect of spraying power on microstructure and bonding strength of $MoSi_2$-based coatings prepared by supersonic plasma spraying. *Appl. Surf. Sci.* **2011**, *257*, 5566–5570. [CrossRef]
34. Jia, Y.; Li, H.; Fu, Q.; Zhao, Z.; Sun, J. Ablation resistance of supersonic-atmosphere-plasma-spraying ZrC coating doped with ZrO_2 for SiC-coated carbon/carbon composites. *Corros. Sci.* **2017**, *123*, 40–54. [CrossRef]
35. Veytizou, C. Zircon formation from amorphous silica and tetragonal zirconia: Kinetic study and modelling. *Solid State Ionics* **2001**, *139*, 315–323. [CrossRef]
36. Chu, Y.; Li, H.-J.; Luo, H.; Li, L.; Qi, L. Oxidation protection of carbon/carbon composites by a novel SiC nanoribbon-reinforced SiC–Si ceramic coating. *Corros. Sci.* **2015**, *92*, 272–279. [CrossRef]
37. Fu, Q.-G.; Jing, J.-Y.; Tan, B.-Y.; Yuan, R.-M.; Zhuang, L.; Li, L. Nanowire-toughened transition layer to improve the oxidation resistance of SiC–$MoSi_2$–ZrB_2 coating for C/C composites. *Corros. Sci.* **2016**, *111*, 259–266. [CrossRef]
38. Lee, W.Y.; More, K.L.; Lara-Curzio, E. Multilayered Oxide Interphase Concept for Ceramic-Matrix Composites. *J. Am. Ceram. Soc.* **2005**, *81*, 717–720. [CrossRef]
39. Sun, C.; Li, H.; Luo, H.; Xie, J.; Zhang, J.; Fu, Q. Effect of Y_2O_3 on the oxidation resistant of $ZrSiO_4$/SiC coating prepared by supersonic plasma spraying technique for carbon/carbon composites. *Surf. Coat. Technol.* **2013**, *235*, 127–133. [CrossRef]
40. Chu, Y.; Fu, Q.; Li, H.; Li, K.; Zou, X.; Gu, C. Influence of SiC nanowires on the properties of SiC coating for C/C composites between room temperature and 1500 °C. *Corros. Sci.* **2011**, *53*, 3048–3053. [CrossRef]

Publisher's Note: MDPI stays neutral with regard to jurisdictional claims in published maps and institutional affiliations.

© 2020 by the authors. Licensee MDPI, Basel, Switzerland. This article is an open access article distributed under the terms and conditions of the Creative Commons Attribution (CC BY) license (http://creativecommons.org/licenses/by/4.0/).

Article

Kinetic Model of Incipient Hydride Formation in Zr Clad under Dynamic Oxide Growth Conditions

Qianran Yu [1], Michael Reyes [1], Nachiket Shah [2] and Jaime Marian [1,2,*]

1 Department of Mechanical and Aerospace Engineering, University of California Los Angeles, Los Angeles, CA 90095, USA; yuqianran0709@gmail.com (Q.Y.); mpreyes1974@yahoo.com (M.R.)
2 Department of Materials Science and Engineering, University of California Los Angeles, Los Angeles, CA 90095, USA; nachiketshah@outlook.com
* Correspondence: jmarian@ucla.edu; Tel.: +1-310-206-9161

Received: 21 January 2020; Accepted: 25 February 2020; Published: 29 February 2020

Abstract: The formation of elongated zirconium hydride platelets during corrosion of nuclear fuel clad is linked to its premature failure due to embrittlement and delayed hydride cracking. Despite their importance, however, most existing models of hydride nucleation and growth in Zr alloys are phenomenological and lack sufficient physical detail to become predictive under the variety of conditions found in nuclear reactors during operation. Moreover, most models ignore the dynamic nature of clad oxidation, which requires that hydrogen transport and precipitation be considered in a scenario where the oxide layer is continuously growing at the expense of the metal substrate. In this paper, we perform simulations of hydride formation in Zr clads with a moving oxide/metal boundary using a stochastic kinetic diffusion/reaction model parameterized with state-of-the-art defect and solute energetics. Our model uses the solutions of the hydrogen diffusion problem across an increasingly-coarse oxide layer to define boundary conditions for the kinetic simulations of hydrogen penetration, precipitation, and dissolution in the metal clad. Our method captures the spatial dependence of the problem by discretizing all spatial derivatives using a stochastic finite difference scheme. Our results include hydride number densities and size distributions along the radial coordinate of the clad for the first 1.6 h of evolution, providing a quantitative picture of hydride incipient nucleation and growth under clad service conditions.

Keywords: zirconium corrosion; zircalloy clad; Zr hydride; hydrogen; nuclear reactor

---

1. Introduction

Corrosion of metallic structural materials is a pervasive phenomenon in industry and technology [1–4]. In nuclear reactors, understanding the kinetics of corrosion of metallic components is grand materials science challenge due to the synergistic combination of high temperature, mechanical stresses, complex coolant and fuel chemistry, and irradiation [5–7]. In light-water nuclear reactors (LWR) zirconium alloys are used as cladding material in fuel elements to provide mechanical integrity between the coolant (water) and the fuel while keeping low levels of neutron absorption [8–11]. In principle, Zr clad is subjected to corrosion from the coolant (water) and fuel sides, both by way of oxygen and hydrogen penetration. The oxidation and hydrogenation of zirconium fuel components in LWR may affect reactor safety and efficiency, which makes corrosion a critical design aspect of Zr materials response in nuclear environments [12–18].

While the majority of the focus of corrosion studies has centered on oxidation and oxygen transport and chemistry in the clad, in the corresponding temperature range, zirconium is known to absorb hydrogen and form hydrides once the critical concentration is reached in the interior of the clad. The accumulation of hydrides during operation plays an important role in fuel performance and safety during steady-state operation and transients, accident conditions, and temporary and permanent fuel storage [19,20]. Examination of the Zr-H phase diagram below 810 K [21–24] indicates that the first stable compound that appears after the metal solid solution (α-Zr) is a cubic phase with a nominal stoichiometry of 1:1.5 (atomic) known as δ-hydride. Generally, a range of stoichiometries between 1.52 and 1.66 is accepted experimentally as corresponding to this phase [25,26]. While the presence of other metastable hydrides has been reported depending on temperature, aging time, or alloy composition [27], it is now well accepted that the needle-shaped structures that form in the metal region beneath the oxide layer are δ-hydride precipitates. Precipitation first starts when the hydrogen concentration reaches the terminal solubility limit, which ranges from zero at 523 K to approximately 7% at. at 810 K [21]. Although δ Zr_2H_3 displays good thermo-mechanical stability, it is also an exceedingly brittle phase [28–31] that can compromise the clad's mechanical integrity [18,31–33].

A key observation of the hydride microstructure is the elongated shape of the precipitates up to a few microns in length [34–37], typically aligned along directions consistent with the stress distribution within the clad. Calculations and experiments point to the large misfit strains between the cubic δ-hydride and the host α-Zr as the reason behind such preferential alignment [35,38–42], which may also impact the mechanical response of the clad. Indeed, the formation of brittle hydride phases is a principal cause of delayed hydride cracking (a subcritical crack growth mechanism facilitated by precipitation of hydride platelets at the crack tips in Zr clad [19,43–45]).

The phenomenology of corrosion is such that oxidation and hydrogenation are typically treated separately, despite some evidence suggesting that there might exist synergisms between oxygen and hydrogen pickup and transport that must be considered jointly in corrosion of Zr [46–50]. This is partially due to the formation of a clearly distinguishable outer oxide scale and and inner region where hydride platelets accumulate. In keeping with this distinction, existing models of hydrogen pickup and precipitation have been developed assuming no cooperative effects from oxygen on hydrogen transport and reaction [35,51]. Models based on hydrogen supersaturation of the α-Zr metal [52–54] assume a binary partition of hydrogen in the clad, either as solid solution or as part of precipitates without specification of their size, number, or orientation. Detailed cluster dynamics (CD) modeling offers a more accurate alternative to obtain hydride size distributions and number densities by solving the complete set of differential balance equations with one-dimensional spatial resolution [55,56]. Phase field methods can capture extra detail by furnishing the shape and orientation of hydrides in addition to concentrations and sizes [27,57,58].

An important aspect often overlooked in the models when studying hydrogen transport and hydride formation in the clad is that it occurs in a dynamic setting, with the oxide scale growing in time and hydrogen traversing an increasingly thicker layer before it can reach the interface. This is rationalized in terms of sluggish H diffusion through the oxide in the relevant temperature range (<300 °C), suggesting that this then would be the rate limiting step [34,36]. This is the accepted picture during the *pre-transition* regime, as, after that, fast H transport then occurs through percolated crack networks formed in the oxide layer [59]. However, there is contradicting evidence in the literature about this [60,61], and it is not clear what effect a dynamic boundary condition might have on hydrogen precipitation in the metal substrate at higher temperatures, and what the evolution of the hydride microstructure will be in those conditions. With the objective of shedding new light on these and other issues by using new computational and experimental understanding, in this paper, we present an comprehensive hydrogen transport and precipitation model in Zr formulated from first principles reaction kinetics and fundamental thermodynamics and mechanics. The model is parameterized using electronic structure calculations and experiments and captures both transport across the oxide layer growth and precipitation in the clad under dynamic hydrogen concentration profiles at the oxide/metal interface. First, we describe the fundamental chemistry and phenomenology of the hydrogen evolution in the clad followed by a mathematical formulation of the model. We then provide numerical results under a number of conditions relevant to LWR operation. We finalize with a discussion of the results and the implications of our modeling approach for zircalloy behavior.

2. Chemical Reaction Kinetics Model

2.1. Zr-Clad Hydrogen Chemistry

The formation of hydrides in the clad is predicated on exposure of its outer surface to bi-molecular hydrogen. This can occur as a consequence of exposure to water or steam, from the reduction of water molecules as:

$$2HO_2 + 4e^- \rightarrow 2O^{2-} + 2H_2 \quad (1)$$

or directly from exposure to hydrogen gas. It is well known that only a fraction of the hydrogen produced in this way is absorbed by the clad, ranging between 5 and 20% of the total hydrogen uptake (the total amount of hydrogen obtained stoichiometrically from reaction (1) [35,50,62–64]. This, known as the *pickup* fraction, sets the boundary condition for the adsorption of hydrogen at the clad's surface. Adsorbed H_2 molecules can split into atomic hydrogen by a number of processes [65,66], although whether this atomic H appears in a neutral or charged state in the metal is still an issue under debate [61,65]. Hydrogen atoms diffuse through the oxide layer and reach the oxide/metal interface, from which they can enter the α-Zr substrate and undergo a number of processes depending on temperature and concentration. Above the terminal solubility limit, hydrogen and zirconium react to form a hydride:

$$\mathrm{Zr} + x\mathrm{H} \rightarrow \mathrm{ZrH}_x \quad (2)$$

where x is the atomic hydrogen concentration.

By way of illustration, Figure 1 shows representative hydridized microstructures in Zirc-4 and Zirconioum.

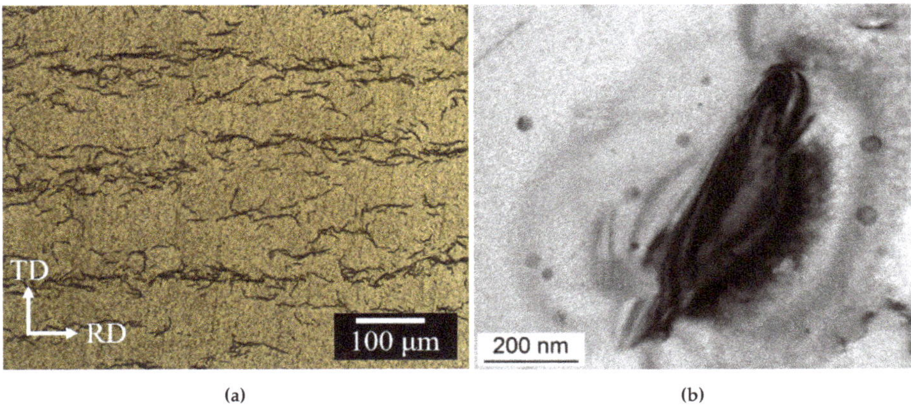

(a) (b)

Figure 1. (**a**) Optical micrographs of hydride morphologies in Zircaloy-4. 'TD' and 'RD' indicate the tangential and radial directions in the clad (from Ref. [67], reproduced with permission of the International Union of Crystallography). (**b**) Electron micrograph detail of a needle-like Zr hydride (reproduced with permission from Ref. [68]).

In view of this picture, and to be consistent with our recent work on oxide layer growth modeling [69], we split our model into two connected elements: (i) a transport part involving H diffusion through an evolving oxide layer, and (ii) a kinetic model of hydride formation and growth in the metal with a dynamic boundary condition set by the first part (i). Figure 2 shows a schematic diagram of the geometry considered for this study and the principal chemical processes taking place in the material. Although it is well known that the Zr oxide layer is not monolithic, containing various Zr-O phases depending on the external conditions and alloy composition [69–72], here we consider a single phase (monoclinic) ZrO_2 with thickness defined by the variable $s(t)$. Hydrogen's diffusion through this layer is thought to occur mostly along grain boundaries, in microstructures ranging from columnar in out-of-pile [71] tests to roughly equiaxed for in-pile conditions [70]. This phenomenon takes place during the pre-transition regime, before the oxide layer cracks and/or develops porosity due to Pilling-Bedworth stresses developed during the metal-to-oxide transformation [73,74]. Once cracking occurs, new diffusion avenues open up for hydrogen to reach the interface and diffusion is no longer seen as a rate limiting step. Our model applies only up to this transition point but not beyond.

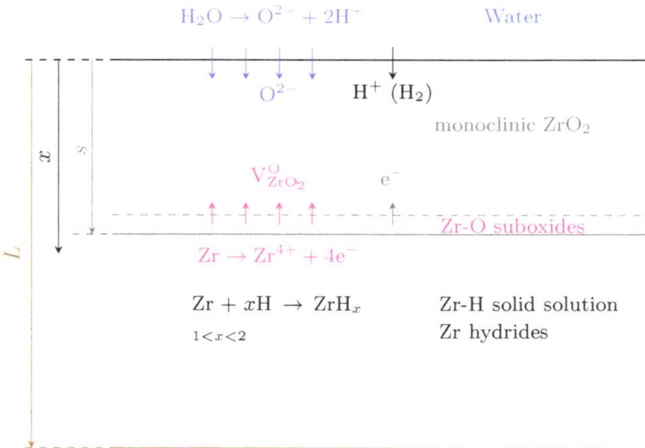

Figure 2. Schematic diagram (not to scale) of the geometry considered for the hydrogen penetration and hydride model developed in this work. x is the depth variable, s is the thickness of the oxide scale, and L is the total thickness of the clad. The chemical processes occurring at each interface are shown for reference.

2.2. *Diffusion Model of Hydrogen in ZrO_2*

The goal of this part of the model is to determine the hydrogen concentration at the metal/oxide interface as a function of time. For this, a generalized drift-diffusion equation is solved:

$$\frac{\partial c_{\mathrm{H}}}{\partial t} = \nabla\left(D_{\mathrm{H}}\nabla c_{\mathrm{H}}\right) - \frac{U_{\mathrm{H}}D_{\mathrm{H}}}{kT^2}\nabla c_i \nabla T + \frac{qD_{\mathrm{H}}}{kT}\nabla\left(c_{\mathrm{H}}\nabla\phi\right) \tag{3}$$

This equation includes the following contributions:

- The first term is standard *Fickian* diffusion in the presence of a concentration gradient.
- The second term is the so-called *thermo-migration* contribution, which depends on the temperature gradient and where U_{H} is the activation energy for diffusion. The convention is for interstitial solutes to move in the direction opposing the gradient, i.e., a 'negative' drift contribution in the equation.
- The third term represents *electro-migration*, where q is the charge of the diffusing species (+1 for protons), and ϕ is the electrical potential, which can be determined by solving Poisson's equation:

$$\nabla^2\phi = -\frac{\rho}{\varepsilon} \tag{4}$$

where ρ is the charge density and ε is the dielectric permittivity.

Consistent with our previous work [69] and other studies [61], we assume the existence of a charge gradient across the oxide layer that originates from the onset of an electron density profile [75]. As well, in this work we consider autoclave conditions and thus neglect the thermomigration contribution.

Equation (3) is solved in one dimension (x) using the finite difference model with the following dynamic boundary conditions:

$$\begin{aligned} c_{\mathrm{H}}(x,0) &= 0 \\ J_{\mathrm{H}}(0,t) &= D_{\mathrm{H}}\frac{\partial c_{\mathrm{H}}(0,t)}{\partial x} = 2f_{\mathrm{H}}C_0\frac{\partial s}{\partial t} \end{aligned}$$

where C_0 is the amount of oxygen (per unit volume) absorbed into the clad to form Zr oxide.The first condition trivially states that the hydrogen content in the clad at the beginning of time is equal to zero, while the second one prescribes the flux of hydrogen at the water/oxide interface. This condition is time-varying as indicated by the growth rate of the oxide layer, $\dot{s}$. As well, it depends on the H pickup fraction, f_{H}, which albeit may also be time dependent [62,64], we fix at 15% for the remainder of this work. The factor of '2' represents the fact that there are two atoms of hydrogen per oxygen atom available to penetrate the clad. Under homogeneous oxide formation conditions, $C_0 \approx 2\rho_{\mathrm{Zr}}$, with ρ_{Zr} the Zr atomic density.

Expressions for $s(t)$ have been provided in our previous study for a number of nuclear-grade Zr alloys [69]. In general, $s(t) = at^n$ such that the growth rate can be directly expressed as

$$\frac{\partial s}{\partial t} = ant^{n-1} \tag{5}$$

a values range between 0.33 (pure Zr) and 0.37 (Zirc-4), while $n = 0.34$ in both cases. These values give s in microns when t is entered in days. With this, $\dot{s} \approx 0.11t^{-0.66}$ (microns per day).

2.3. Stochastic Cluster Dynamics Model with Spatial Resolution

Here we use the stochastic cluster dynamics method (SCD) [76] to perform all simulations. SCD is a stochastic variant of the mean-field rate theory technique, alternative to the standard implementations based on ordinary differential equation (ODE) systems, that eliminates the need to solve exceedingly large sets of ODEs and relies instead on sparse stochastic sampling from the underlying kinetic master equation [76]. Rather than dealing with continuously varying defect concentrations C_i in an infinite volume, SCD evolves an integer-valued defect population N_i in a finite material volume V, thus limiting the number of 'active' ODEs at any given moment. Mathematically, SCD recasts the standard ODE system:

$$\frac{dC_i}{dt} = g_i - \sum_i s_{ij}C_i + \sum_i s_{ji}C_j - \sum_{i,j} k_{ij}C_jC_i + \sum_{j,k} k_{jk}C_iC_k \tag{6}$$

into stochastic equations of the form:

$$\frac{dN_i}{dt} = \tilde{g}_i - \sum_i \tilde{s}_{ij}N_i + \sum_i \tilde{s}_{ji}N_j - \sum_{i,j} \tilde{k}_{ij}N_jN_i + \sum_{j,k} \tilde{k}_{jk}N_iN_k \tag{7}$$

The set $\{\tilde{g}, \tilde{s}, \tilde{k}\}$ represents the reaction rates of 0th (insertion), 1st (thermal dissociation, annihilation at sinks), and 2nd (binary reactions) order kinetic processes taking place inside V, and is obtained directly from the standard coefficients $\{g, s, k\}$ as:

$$\tilde{g} \equiv gV,\ \tilde{s} \equiv s,\ \tilde{k} \equiv kV^{-1}$$

The value of V chosen must satisfy

$$V^{\frac{1}{3}} > \ell$$

where ℓ is the maximum diffusion length l_i of any species i in the system, defined as:

$$\ell = \max_i \{l_i\} \tag{8}$$

$$l_i = \sqrt{\frac{D_i}{R_i}} \tag{9}$$

Here, D_i and R_i are the diffusivity and the lifetime of a mobile species within V. The above expression is akin to the stability criterion in explicit finite difference models. From Equation (7), $R_i = \tilde{s} + \sum_j \tilde{k}_{ij} N_j$. The system of Equation (7) is then solved using the kinetic Monte Carlo (residence-time) algorithm by sampling from the set $\{\tilde{g}, \tilde{s}, \tilde{k}\}$ with the correct probability and executing the selected events. Details of the microstructure such as dislocation densities and grain size are captured within SCD in the mean-field sense, i.e., in the form of effective sink strengths for hydrogen atoms. For example, the value of s_{ij} in Equation (6) for dislocation and grain boundary defect sinks is:

$$s_{ij} = s_d + s_g b = \rho + \frac{6\sqrt{\rho}}{d} \tag{10}$$

where ρ is the dislocation density and d is the grain size.

SCD has been applied in a variety of scenarios not involving concentration gradients [76,77]. The SCD code has been developed in house and is available at: http://jmarian.bol.ucla.edu/packages/packages.html. However, Equation (6) must be expanded into a transport equation (i.e., a *partial differential equation*, or PDE) by adding a *Fickian term* of the type:

$$\frac{dC_i}{dt} = \nabla (D_i \cdot \nabla C_i) + f(t; C_1, C_2, C_3 \dots) \tag{11}$$

where D_i is the diffusivity of species i, and $f(t; C_1, C_2, C_3 \dots)$ is used for simplicity to represent all of the terms in the r.h.s. of Equation (6). To cast Equation (11) into a stochastic form, the transport term must be converted to a reaction rate in the finite volume V. As several authors have shown, this can be readily done by applying the divergence theorem and approximating the gradient term in terms of the numbers of species in neighboring elements [78,79]. For a one-dimensional geometry such as that schematically shown in Figure 3, the Fickian term simply reduces to:

$$D_i \frac{N_i^\alpha - N_i^\beta}{l^2}$$

where Greek superindices refer to neighboring elements, i is the cluster species, and l is the element size. When summed over all neighboring elements, this term then represents the rate of migration of species i from volume element α to β, which can be now added to the r.h.s. of Equation (7) and sampled stochastically as any other event using the residence-time algorithm.

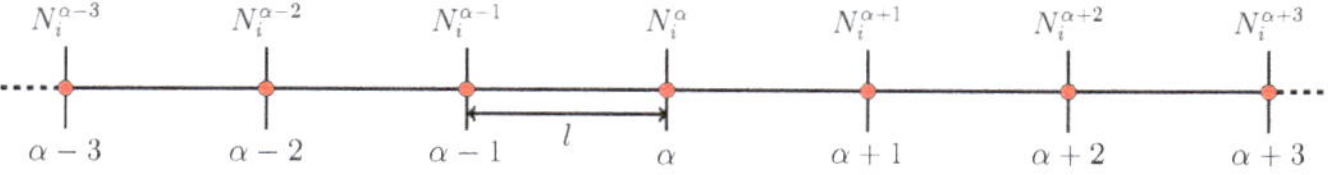

Figure 3. Schematic diagram of two volume elements of a 1D space discretization used to calculate spatial gradients within the stochastic cluster dynamics method (SCD). The superindex α refers to the physical element, while the subindex i refers to the cluster species.

The full stochastic PDE system, written for a generic species i in volume element α, takes then the following form:

$$\frac{dN_i^\alpha}{dt} = D_i \sum_\beta \frac{N_i^\alpha - N_i^\beta}{l^2} + \tilde{g}_i - \sum_i \tilde{s}_{ij} N_i^\alpha + \sum_i \tilde{s}_{ji} N_j^\alpha - \sum_{i,j} \tilde{k}_{ij} N_j^\alpha N_i^\alpha + \sum_{j,k} \tilde{k}_{jk} N_i^\alpha N_k^\alpha \tag{12}$$

Further, the model assumes the following:

(i) The only mobile species considered are hydrogen atoms.
(ii) The source term $\tilde{g}_i$ only applies to element 0 (oxide/metal boundary) and is calculated from the hydrogen arrival flux calculated from the model in Section 2.2.
(iii) The only processes considered in the metal are:

(a) H diffusion
(b) Immobilization of H atoms through formation of Zr_2H_3 molecules (equivalent to nucleation of hydride platelets).
(c) Growth of Zr_2H_3 clusters.
(d) Thermal dissolution of Zr_2H_3 clusters.

Next, we provide suitable expressions for each of the kinetic processes just listed.

2.3.1. H Atom Diffusion

The hydrogen diffusivity in both the oxide and the metal is assumed to follow an Arrhenius temperature dependence:

$$D_H^\alpha(T) = D_0^\alpha \exp\left\{\left(-\frac{e_m^\alpha}{kT}\right)\right\} \tag{13}$$

where D_0 is the exponential pre-factor, e_m is the migration energy, k is Boltzmann's constant, and the superscript α can refer to the oxide ('ox') or the metal ('m'). The diffusivity of H in Zircaloy-4 oxides has been recently measured by Tupin et al. [80], which give values of 2.5×10^{-14} m·s^{-1} and 0.41 eV for D_0^{ox} and e_m^{ox}, respectively. Alloy composition, however, has been shown to have a significant impact on diffusion parameters. For example, values of e_m^{ox} = 0.55 and 1.0 eV have been reported for Zr-2.5%Nb and pure Zr, respectively, with D_0 numbers in as high as 1.1×10^{-12} m·s^{-1} [81,82]. Here, we use the parameters for Zirc-4 given by Tupin et al.

Similarly, the only mobile species in the metal is monoatomic hydrogen. The most widely used parameters for hydrogen diffusion in metal Zr, and Zircaloy-2 and -4 (D_i in Equation (12)) are those by Kearns [83] in the 200–700 °C temperature range, with values of $D_0^m = 7.90 \times 10^{-7}$ m·s^{-1} and $e_m^m = 0.46$ eV. Earlier literature on these measurements [25,84,85] reveals pre-factors ranging from 7.00×10^{-8} to 4.15×10^{-7} m·s^{-1} and migration energies between 0.3 and 0.5 eV, all in a similar temperature range. More recent experiments and molecular dynamics simulations are also consistent with these values [86,87].

The values chosen here for each case (diffusion in the oxide and in the metal) are given un Table 1.

2.3.2. Nucleation of Zr_2H_3 Hydride

As shown in Figure 2, once hydrogen penetrates into the metal clad, the hydration reaction $Zr + xH \rightarrow ZrH_x$ starts occurring. Although the formation of the δ-hydride is seen for a range of x values, here we assume a perfect stoichiometry of x=1.5. Consequently, the governing equilibrium constant for the reaction can be expressed as:

$$K_\delta = \frac{[ZrH_{1.5}]}{[Zr]\,[H]^{1.5}}$$

However, it is more convenient to use an expression that is linear in the hydrogen concentration. From this, one can write the reaction rate as:

$$k_\delta = 4\pi \left(r_{\rm H} + r_{\rm Zr}\right) \left(V^{-\frac{1}{3}} \rho_{\rm Zr}^{\frac{2}{3}}\right) D_{\rm H} N_{\rm H} p(x) \exp\left(-\frac{\Delta E_\delta}{kT}\right) \tag{14}$$

which is simply a coagulation rate for two species –H and Zr– in the proportions indicated by the exponents of $\rho_{\rm Zr}$ and $N_{\rm H}$. ΔE_δ is the formation energy of a molecule of δ hydride (≈ 0.52 eV at 350 °C according to Blomqvist et al. [88]) and $p(x)$ represents the thermodynamic probability for this reaction to occur, which can be directly extracted from the Zr-H phase diagram using the lever rule:

$$p(x) = \frac{x - x_{\rm TTS}}{x_\delta - x_{\rm TTS}} \tag{15}$$

where $x_{\rm TTS}$ is the *terminal thermal solubility* at the temperature of interest, and x_δ is the phase boundary. A phase diagram of the Zr-H system in the temperature and concentration region relevant to the present study is shown in Figure 4. By way of example, at 660 K (horizontal dashed line in the figure) $x_{\rm TTS}$ is approximately 1.6% at. and $x_\delta \approx 60.0\%$ at.

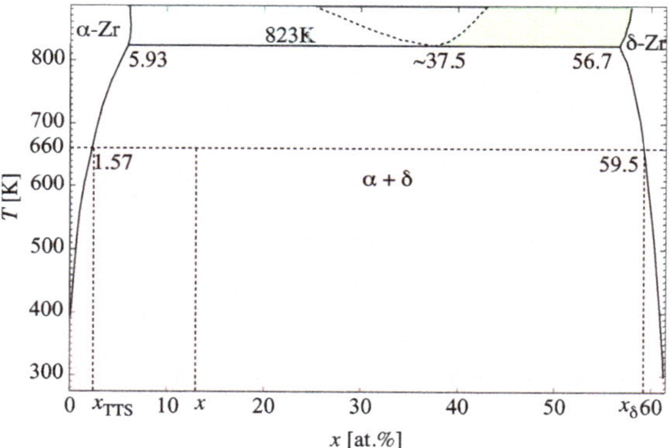

Figure 4. Phase diagram of the Zr-H system in the temperature and concentration region relevant to the present study (adapted from several sources [21,89,90]).

Equation (15) ensures that $p(x_{\rm TTS}) = 0$ and $p(x_\delta) = 1$, i.e. the nucleation probability is zero at the phase boundary between the (α) and $(\alpha + \delta)$ regions, and unity at the $(\alpha + \delta)$ and (δ) boundary. This simple factor captures the thermodynamic propensity for the hydride reaction to take place, and thus ties the thermodynamics and kinetics of hydration together. $r_{\rm H}$ and $r_{\rm Zr}$ in Equation (14) are the interaction radii of H and Zr, respectively, their values given in Table 2. In the SCD calculations, the atomic fraction x is simply defined at any instant in time as:

$$x = \frac{N_{\rm H}}{N_{\rm H} + \rho_{\rm Zr} V}$$

2.3.3. Growth of Zr_2H_3 Hydride

Once hydride nuclei appear in the clad, their growth is treated as a standard coagulation process in 3D with rate constant:

$$k_n = 4\pi V^{-1} \left(r_{\rm H} + r_\delta(n)\right) D_{\rm H} N_{\rm H} N_{({\rm Zr}_{0.66n}{\rm H}_n)} \exp\left(-\frac{\Delta E_\delta}{kT}\right) \tag{16}$$

where r_δ is the interaction radius of the hydride clusters, and $N_{(\mathrm{Zr}_{0.66n}\mathrm{H}_n)}$ is the concentration of a hydride cluster containing n H atoms (which implies having $0.66n$ Zr atoms). It is assumed that hydride clusters are immobile. In accordance with previous works, hydrides grow as circular platelets whose size is directly related to the number of hydrogen monomers contained in it [56]:

$$r_\delta(n) = \sqrt{\frac{n\Omega_\mathrm{H}}{\pi d}}$$

with Ω_H and d the formation volume of hydrogen and the thickness of the platelet, respectively. As given in Table 2, here we use $\Omega_\mathrm{H} = 2.8 \times 10^{-3}$ nm^3 per atom [91], and $d \approx 0.28$ nm [37].

The growth of hydride platelets is known to be highly directional, and influenced by stress and microstructure. Typically hydrides align themselves along the direction of the dominant axial stress components and grow preferentially in-plane on grain boundaries [27,36,37,53]. These details are not captured in our model at present.

2.3.4. Dissolution of Zr_2H_3 Hydride

The last process considered in our model is the thermal dissolution of the hydrides, as Figure 4 shows, strictly speaking, hydrides are stable up to 550 °C (eutectoid temperature), although there is ample evidence of their decomposition at much lower temperatures, as well as the observation of thermal hysteresis during heating/cooling cycles [37,92–94]. The dissociation rate is a first-order process that can be expressed as:

$$s_n = 4\pi r_\delta(n) D_\mathrm{H} N_{(\mathrm{Zr}_{0.66n}\mathrm{H}_n)} \exp\left\{\left(-\frac{e_b(n)}{kT}\right)\right\} \tag{17}$$

where e_b is the binding energy between a H monomer and a cluster containing n hydrogen atoms. Here, we assume a capillary approximation for e_b [55,56]:

$$e_b(n) = e_s - 0.44\left[n^{\frac{2}{3}} - (n-1)^{\frac{2}{3}}\right]$$

with e_s being the heat of solution of H in the α-Zr matrix. This parameter has been found to be approximately 0.45 eV in electronic structure calculations [95,96], compared to 0.66 eV in experiments [97].

2.3.5. Metal/oxide Interface Motion

Finally, the motion of the interface must also be considered as a viable stochastic event. To turn the interface velocity, Equation (5), into an event rate, r_i, one simply normalizes it by the interface thickness, s.

$$r_i = \left(\frac{1}{s}\right)\frac{ds}{dt} = \frac{ant^{n-1}}{at^n} = \frac{n}{t}$$

which results in the following expression for r_i:

$$r_i = 0.34t^{-1} \tag{18}$$

This is added to the event catalog and sampled with the corresponding probability as given by Equation (18). As the equation shows, this is a time-dependent rate that reflects the nonlinear growth of the oxide layer with time. In the context of the SCD model, it implies that the one-dimensional mesh shown in Figure 3 must be dynamically updated with time because the physical dimensions of the simulation domain are dynamically changed. To our knowledge, this has not been attempted in any prior models of hydride formation and buildup.

2.4. Parameterization, Physical Dimensions, and Boundary Conditions

All the material constants used in the present model are given in Tables 1 and 2. External parameters representing the geometry and the boundary conditions are given in Table 3.

Table 1. Zr-H energetics used in the model with the respective source.

Parameter	Unit	Symbol	Value	Source
Hydrogen diffusivity prefactor in Zr oxide	$m \cdot s^{-1}$	D_H^{ox}	2.50×10^{-14}	[80]
Hydrogen migration energy in Zr oxide	eV	e_m^{ox}	0.41	[80]
Hydrogen diffusivity prefactor in Zr metal	$m \cdot s^{-1}$	D_H^{m}	7.90×10^{-7}	[83]
Hydrogen migration energy in Zr metal	eV	e_m^{m}	0.46	[83]
δ-hydride formation energy	eV	ΔE_δ	0.88	[88]
H solution energy in Zr metal	eV	e_s	0.66	[97]

Table 2. Physical constants for the Zr-H system employed here. In actuality, the interaction radii of Zr and H atoms are extended by a distance equal to the Burgers vector $\langle a \rangle$ in α-Zr, which is equal to 3.23 Å.

Physical Constant	Symbol	Unit	Value	Source
Zr atomic density	ρ_{Zr}	m^{-3}	4.31×10^{28}	-
H-atom interaction radius	r_H	Å	0.31	[98]
Zr-atom interaction radius	r_{Zr}	Å	1.75	[98]
H-atom formation volume	Ω_H	nm^3 per atom	2.8×10^{-3}	[91]
δ-hydride platelet thickness	d	nm	0.28	[37]

Table 3. Numerical parameters used in the model.

f_H	x_{TTS} [%]	x_δ [%]	T [K]	V $[m^{-3}]$	l [nm]	L [nm]
0.15	1.6	59.5	660	10^{-18}	100	900

3. Results

The first two figures show results intended to set the stage for the SCD calculations.

We begin with the time evolution of the hydrogen concentration at the oxide/metal interface. This results from solving the diffusion equation in the the oxide layer subjected to a moving boundary as explained in Section 2.2. Figure 5 shows the buildup of hydrogen up to the first 580 h. This represents a *dynamic* Dirichlet boundary condition for the spatially-resolved SCD calculations of hydride nucleation and growth in the metal substrate ($\tilde{g}$ term in Equation (12)). Second, we track the sampling rate r_i defined in Section 2.3.5 to confirm that it matches Equation (18). Figure 6 shows a comparison between both, indeed demonstrating their equivalency and confirming the correctness of its implementation in the code. The effect of this interface motion is that, over the course of the time scale covered in the SCD simulations, the oxide layer effectively sweeps over the first mesh element of the metal depth profile (recall that we assume that such sweep results in dissolution of the hydrides existing within that element at that point, and re-solution of the immobilized hydrogen in the metal). In practice, this allows us to subsequently discard the first spatial element of the 1D mesh. That is the reason why in the figures shown next the spatial range shown spans 800 (as opposed to the original 900) nm.

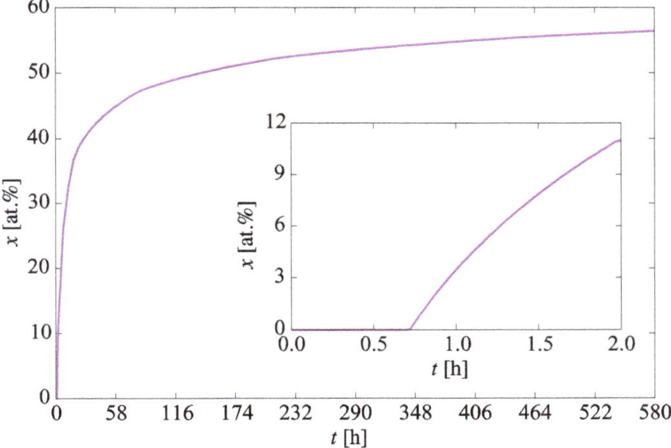

Figure 5. Evolution with time of the hydrogen concentration at the oxide/metal interface. This represents the boundary condition for the spatially-resolved SCD calculations of hydride nucleation and buildup.

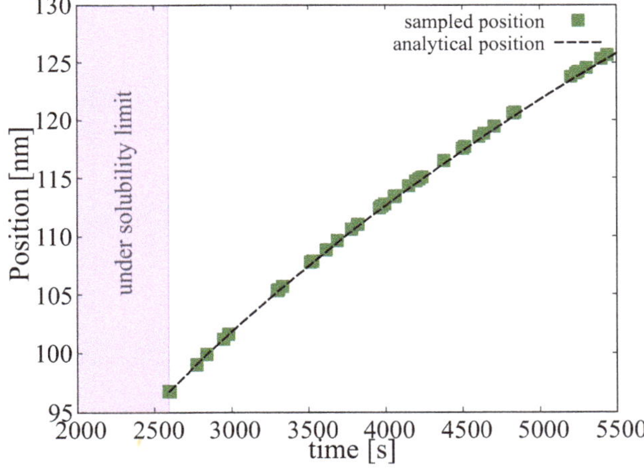

Figure 6. Comparison between the predictions of r_i and Equation (18) of the position of the oxide/metal interface as a function of time. We track the interface position only after the concentration of hydrogen has reached the solubility limit.

Next, we study the generation of hydride molecules in the metal layer as a function of time and depth. The results are shown in Figure 7a, which shows a histogram with the concentration of hydride molecules at several instants in time for each of the mesh elements of the metal region. As discussed in Section 2.3.2, the probability that a new hydride molecule will form depends primarily on the relative H concentration at the interface and the heat of formation of δ-hydride. With a probability per unit time k_δ (Equation (14)), freely-diffusing H atoms are immobilized to form $Zr_{2/3}H$ molecules that act as incipient hydride nuclei. The concentrations of such nuclei are strongly depth-dependent, as shown in the figure, ranging over two orders of magnitude over the entire specimen thickness L of 900 nm. As well, the nucleation rate, i.e., the derivative of the evolution curves shown in Figure 7b (which display the same data as Figure 7a but plotted as a function of time), can be seen to decrease gradually in time across the entire depth profile.

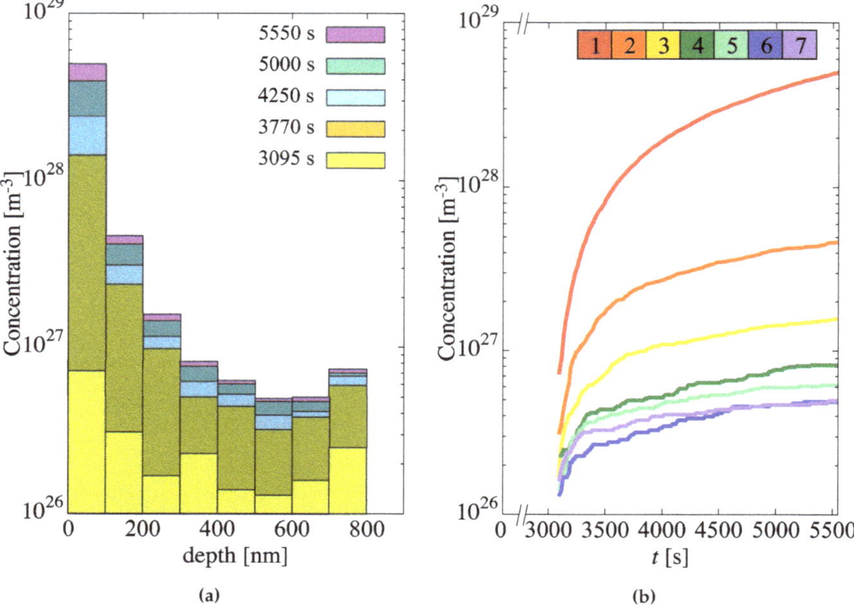

Figure 7. (**a**) Concentration of incipient hydride nuclei in the metal layer as a function of depth for several time snapshots. (**b**) Hydride concentration buildup as a function of time for each depth element. Each curve is colored according to the key at the top of the figure (element 1 is closest to the oxide/metal interface). Per Table 3, each element is 100-nm thick.

Subsequent growth of these embryos occurs at a rate given by the combination of the rates of H-atom absorption (Equation (16)) and dissolution (Equation (17)), i.e., $(k_n - s_n)$, as shown in Figure 8. Rapid net growth is seen in the initial stages of hydridization close to the oxide/metal interface. However, these rates gradually abate both in time and with increasing depth until almost no net growth is observed, particularly at depths greater than 700 nm after 1.4 h of evolution.

The resulting hydride concentrations across the 900-nm metal layer at the end of the simulated time can be found in Figure 9a. As the graph indicates, the hydride number densities suffer almost a 100-fold decrease through the metal layer studied. In relative terms, these are large concentrations of small clusters, so it is to be expected that further time evolution of the hydride subpopulations will be dominated by growth, perhaps by way of some type of coarsening or ripening mechanism. The associated size distributions of the hydride clusters are shown in Figure 9b, where both the average and maximum cluster sizes are shown. We emphasize that, during the incipient nucleation of the hydrides, they grow as circular discs, and so the sizes simulated ($\approx$ 50 nm or less), correspond to the regime prior to the acicular growth of the hydrides.

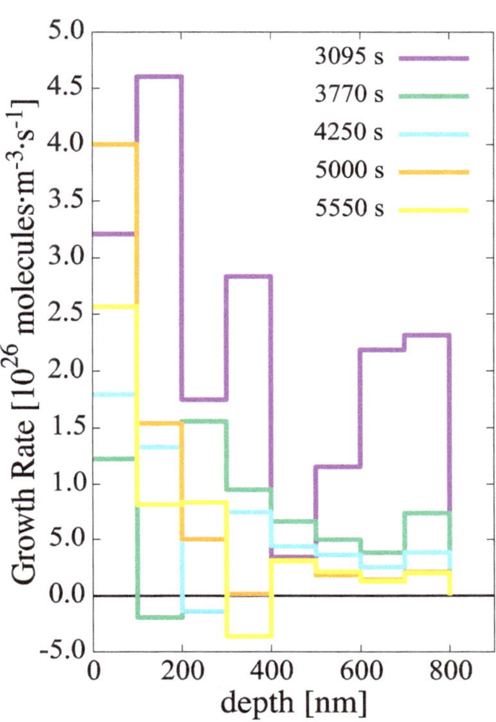

Figure 8. Growth rate of hydride clusters in the metal layer as a function of time and depth.

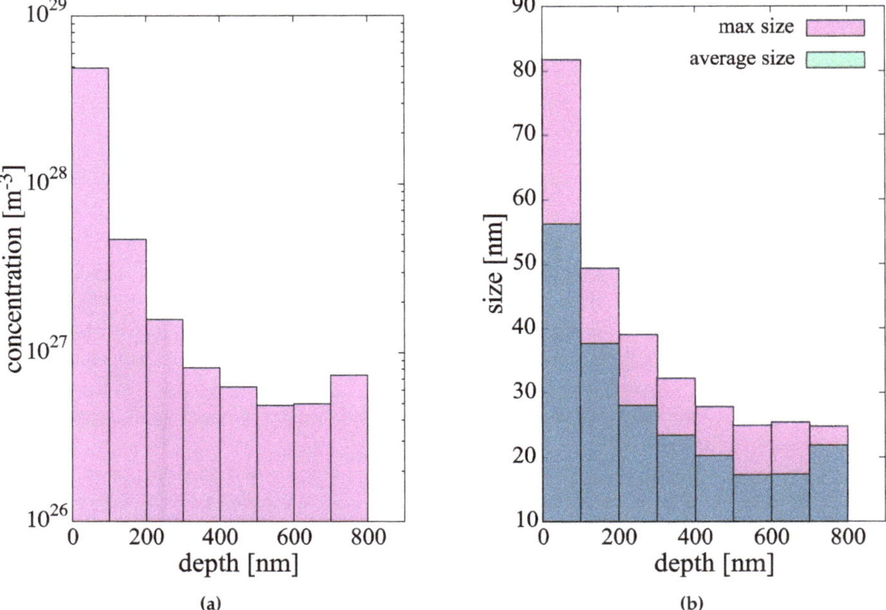

Figure 9. (**a**) Concentration profile and (**b**) size distributions of the hydride cluster population after 1.53 h of simulated evolution.

4. Discussion

Several of the most important features of the model presented here are: (i) consideration of a moving interface representing the growth of the oxide scale during operation in corrosive conditions; (ii) using a hydride nucleation criterion that is consistent with the thermodynamics of the Zr-H system; (iii) using a mean-field growth/dissolution model that respects; (iv) a completely physics-based parameterization based on calculated atomistic data. Some of these features were part of a comparable study [56], to which new ones have been added and existing ones augmented. All these features combined are the basis of a model that has been developed as an attempt to break the phenomenological vicious cycle in which models of materials degradation in nuclear environments are often found.

To study the nucleation of the hydride clusters, our method samples discrete kinetic processes defined by the corresponding energetics and thermodynamics. For example, hydride nucleation is simulated by considering the interplay between (i) aggregation, (ii) growth, and (iii) dissolution processes, which together determine the net nucleation and growth rates. Processes (i), (ii), and (iii) are embodied in Equations (14), (16) and (17), respectively. Each one of these processes is treated as a stochastic event sampled with the probabilities given by each respective rate. If the conditions are such that dissolution would dominate over nucleation, the clusters would never form. If growth dominates over nucleation, the clusters would grow bigger, etc. All the energetics are given by the parameters in each of those equations.

As is often the case, the price paid for an increased physical fidelity in the simulations is computational efficiency. For this reason, our simulations can only extend to times of several thousand seconds (<2 h), which is of course only representative of the initial stages of hydridation in Zr clad (and, of course, part of the pre-transition corrosion regime)). In these relatively short time scales (which are still orders of magnitude higher than what direct atomistic methods can cover), one can only claim to faithfully study the incipient nucleation phase of the hydride microstructure. In this sense, our results do not include important features of the Zr hydride particles such as their elongated shape and/or their orientation. Excellent recent examples of experimental characterization displaying all of these structure complexities exist now in the literature [99–101]. They can act, however, as a good springboard from which to connect to other methods such as phase field simulations [27,102,103], or orientation-dependent precipitation models [104,105]. Therefore, it is reasonable to assume that the time scale of the next phase of hydride formation/growth kinetics would be one dominated by coarsening/ripening, where population densities suffer a gradual decline at the expense of an increased average precipitate size. Thus, it is important to emphasize this aspect of the work: our results correspond to the incipient hydride nucleation and growth phase, before steady state populations are established. Steady state sizes and concentrations in corroded Zr specimens range from 100 nm to 1 μm [101] and $\sim 10^{24}$ m^{-3}. On this aspect, it is also difficult to reconcile calculated H-atom diffusivities in the clad with almost cross-clad uniform hydride distributions observed experimentally [106]. Calculated migration energies suggest a much more sluggish diffusion in the metal, and screening of the clad interior by hydrides formed near the oxide metal interface, as seen in this study, compared to experimental results. While validation on the time and length scales covered in this work is always difficult, it is encouraging to see reasonable qualitative agreement with experimental studies, e.g., hydride precipitation completion fractions in ref. [36] (Figure 3) vs. Figure 7b in this paper. As well, our predictions for the size (long axis) of the precipitates in Figure 9b are in good agreement with *in situ* SEM observations [107].

As reviewed in the Introduction section, the formation of Zr hydrides in the metal clad is considered to be highly detrimental to reactor performance due to their embrittling effect. However, the high thermal stability of these hydride phases also makes them a matter of concern for reactor safety due to the potential for hydrogen storage and release during loss-of-coolant conditions and core meltdown. Palliative measures such as increasing the enthalpy of formation of δ-ZrH by microstructure tailoring [108], or by hindering H diffusion in Zr oxide by selective alloying in the clad [109,110], have been proposed for future candidate materials in novel nuclear fuel designs.

5. Conclusions

We end this paper with a list of the most important conclusions:

- We have developed a spatially-resolved kinetic model of hydrogen transport/accumulation in Zr-metal clad. The model includes state-of-the-art hydride energetics data from atomistic calculations and is formulated as a stochastic version of the cluster dynamics method. Notably, boundary conditions are dynamically updated in time during the simulations, by accounting for oxide/metal interface motion due to the time-dependent growth of the oxide scale.
- In doing so, our model is consistent with the oxidation in the clad, as well as with the equilibrium thermodynamics of the Zr-H system.
- As most cluster dynamics models based on mean-field rate theory, our model does not capture the orientation dependence of elongated hydride platelets observed experimentally, and microstructural information such as grain sizes and dislocation densities is included only in an effective way. As such, our results are representative of the 'average' structure along the depth direction.
- Our results show that high concentrations of small hydride nuclei form across the entire metal clad. This results in a very fine microstructure that sets the stage for the next kinetic phase, likely to be one of ripening and coarsening.
- Gaps in our knowledge identified in this work include, among others: (i) how to model the H dissolved from hydrides swept by the growing oxide layer, (ii) how to reconcile existing H-atom diffusion energies with almost cross-clad uniform hydride distributions, and (iii) the reasons for the acicular (or capsular) growth of the precipitates are still not clear and, while such geometries can be adopted in the models, a physical approach that yields these geometric features is still lacking.

Author Contributions: J.M. designed the research; Q.Y. performed the S.C.D. calculations and made all the figures; M.R. developed the early version of the hydrogen diffusion code; N.S. assisted with some of the calculations and figures; J.M. and Q.Y. discussed the data and participated in writing the manuscript. All authors have read and agreed to the published version of the manuscript.

Funding: This research was supported by the Consortium for Advanced Simulation of Light Water Reactors (CASL), an Energy Innovation Hub for Modeling and Simulation of Nuclear Reactors under U.S. Department of Energy Contract No. DE-AC05-00OR22725.

Acknowledgments: This work was done using computer time allocations at UCLA's IDRE Hoffman2 supercomputer. Helpful discussions with A. Couet, A. Motta, B. D. Wirth, and A. Van der Ven are gratefully acknowledged.

Conflicts of Interest: The authors declare no conflict of interest.

References

1. Scully, J.C. *The Fundamentals of Corrosion*, 2nd ed.; Pergamon: Oxford, UK, 1978.
2. North, N.; MacLeod, I.D.; Pearson, C. *Corrosion of Metals*; Butterworth-Heinemann: Oxford, UK, 1987; pp. 68–98.
3. Young, D.J. *High Temperature Oxidation and Corrosion of Metals*; Elsevier: Amsterdam, The Netherlands, 2008; Volume 1.
4. Comstock, R.; Motta, A.T. *Zirconium in the Nuclear Industry: 18th International Symposium*; ASTM International: West Conshohocken, PA, USA, 2018.
5. Clayton, J.C. Out-of-pile nickel alloy-induced accelerated hydriding of zircaloy fasteners. In *Zirconium in the Nuclear Industry*; ASTM International: West Conshohocken, PA, USA, 1984.
6. Jacques, P.; Lefebvre, F.; Lemaignan, C. Deformation–corrosion interactions for Zr alloys during I-SCC crack initiation: part I: chemical contributions. *J. Nucl. Mater.* **1999**, *264*, 239–248. [CrossRef]
7. Sabol, G.P.; Moan, G.D. *Zirconium in the Nuclear Industry: Twelfth International Symposium*; ASTM: West Conshohocken, PA, USA, 2000.
8. Féron, D. *Nuclear Corrosion Science and Engineering*, 1st ed.; Woodhead Publishing: Sawston, UK, 2012.

9. Cattant, F.; Crusset, D.; Féron, D. Corrosion issues in nuclear industry today. *Mater. Today* **2008**, *11*, 32–37. doi:10.1016/S1369-7021(08)70205-0. [CrossRef]
10. Allen, T.; Konings, R.; Motta, A. 5.03 corrosion of zirconium alloys. *Compr. Nuclear Mater.* **2012**, *5*, 49–68.
11. Preuss, M. Zirconium cladding-the long way towards a mechanistic understanding of processing and performance. In Proceedings of the Second International Conference on Advances in Nuclear Materials: Abstract Booklet and Souvenir, Mumbai, India, 9–11 February 2011.
12. Hillner, E. Corrosion of zirconium-base alloys? An overview. In *Zirconium in the Nuclear Industry*; ASTM International: West Conshohocken, PA, USA, 1977.
13. Zaimovskii, A. Zirconium alloys in nuclear power. *At. Energy* **1978**, *45*, 1165–1168. [CrossRef]
14. Cox, B. Some thoughts on the mechanisms of in-reactor corrosion of zirconium alloys. *J. Nucl. Mater.* **2005**, *336*, 331–368. [CrossRef]
15. Causey, R.A.; Cowgill, D.F.; Nilson, R.H. *Review of the Oxidation Rate of Zirconium Alloys*; Technical Report SAND2005-6006; Sandia National Laboratories: Albuquerque, NM, USA, 2005.
16. Motta, A.T.; Yilmazbayhan, A.; da Silva, M.J.G.; Comstock, R.J.; Was, G.S.; Busby, J.T.; Gartner, E.; Peng, Q.; Jeong, Y.H.; Park, J.Y. Zirconium alloys for supercritical water reactor applications: Challenges and possibilities. *J. Nucl. Mater.* **2007**, *371*, 61–75. [CrossRef]
17. Bossis, P.; Pecheur, D.; Hanifi, K.; Thomazet, J.; Blat, M. Comparison of the high burn-up corrosion on M5 and low tin Zircaloy-4. In *14th International Symposium on Zirconium in the Nuclear Industry*; ASTM Special Technical Publication: West Conshohocken, PA, USA, 2006; Volume 3, pp. 494–525.
18. Motta, A.T.; Capolungo, L.; Chen, L.Q.; Cinbiz, M.N.; Daymond, M.R.; Koss, D.A.; Lacroix, E.; Pastore, G.; Simon, P.C.A.; Tonks, M.R.; et al. Hydrogen in zirconium alloys: A review. *J. Nucl. Mater.* **2019**, *518*, 440–460. [CrossRef]
19. McRae, G.; Coleman, C.; Leitch, B. The first step for delayed hydride cracking in zirconium alloys. *J. Nucl. Mater.* **2010**, *396*, 130–143. [CrossRef]
20. Zieliński, A.; Sobieszczyk, S. Hydrogen-enhanced degradation and oxide effects in zirconium alloys for nuclear applications. *Int. J. Hydrog. Energy* **2011**, *36*, 8619–8629. [CrossRef]
21. Zuzek, E.; Abriata, J.; San-Martin, A.; Manchester, F. The H-Zr (hydrogen-zirconium) system. *Bull. Alloy Phase Diagr.* **1990**, *11*, 385–395. [CrossRef]
22. Dupin, N.; Ansara, I.; Servant, C.; Toffolon, C.; Lemaignan, C.; Brachet, J. A thermodynamic database for zirconium alloys. *J. Nucl. Mater.* **1999**, *275*, 287–295. doi:10.1016/S0022-3115(99)00125-7. [CrossRef]
23. Steinbrück, M. Hydrogen absorption by zirconium alloys at high temperatures. *J. Nucl. Mater.* **2004**, *334*, 58–64.
24. Grosse, M.; Steinbrueck, M.; Lehmann, E.; Vontobel, P. Kinetics of Hydrogen Absorption and Release in Zirconium Alloys During Steam Oxidation. *Oxid. Met.* **2008**, *70*, 149–162. [CrossRef]
25. Gulbransen, E.A.; Andrew, K.F. Diffusion of hydrogen and deuterium in high purity zirconium. *J. Electrochem. Soc.* **1954**, *101*, 560–566. [CrossRef]
26. Root, J.; Small, W.; Khatamian, D.; Woo, O. Kinetics of the δ to γ zirconium hydride transformation in Zr-2.5Nb. *Acta Mater.* **2003**, *51*, 2041–2053. [CrossRef]
27. Zhao, Z.; Blat-Yrieix, M.; Morniroli, J.; Legris, A.; Thuinet, L.; Kihn, Y.; Ambard, A.; Legras, L. Characterization of zirconium hydrides and phase field approach to a mesoscopic-scale modeling of their precipitation. In *Zirconium in the Nuclear Industry: 15th International Symposium*; ASTM International: West Conshohocken, PA, USA, 2009.
28. Ackland, G. Embrittlement and the bistable crystal structure of zirconium hydride. *Phys. Rev. Lett.* **1998**, *80*, 2233. [CrossRef]
29. Olsson, P.; Massih, A.; Blomqvist, J.; Holston, A.M.A.; Bjerkén, C. Ab initio thermodynamics of zirconium hydrides and deuterides. *Comput. Mater. Sci.* **2014**, *86*, 211–222. [CrossRef]
30. Zhu, W.; Wang, R.; Shu, G.; Wu, P.; Xiao, H. First-principles study of different polymorphs of crystalline zirconium hydride. *J. Phys. Chem. C* **2010**, *114*, 22361–22368. [CrossRef]
31. Chernov, I.I.; Staltsov, M.S.; Kalin, B.A.; Guseva, L.Y. Some problems of hydrogen in reactor structural materials: A review. *Inorg. Mater. Appl. Res.* **2017**, *8*, 643–650. [CrossRef]
32. Coleman, C.; Hardie, D. The hydrogen embrittlement of α-zirconium? A review. *J. Less Common Met.* **1966**, *11*, 168–185. [CrossRef]

33. Tummala, H.; Capolungo, L.; Tome, C.N. *Quantifying the Stress Fields Due to a Delta-Hydride Precipitate in Alpha-Zr Matrix*; Technical Report; Los Alamos National Lab. (LANL): Los Alamos, NM, USA, 2017.
34. Bloch, J. The temperature-dependent changes of the kinetics and morphology of hydride formation in zirconium. *J. Alloys Compd.* **1995**, *216*, 187–195. [CrossRef]
35. Motta, A.T.; Chen, L.Q. Hydride formation in zirconium alloys. *JOM* **2012**, *64*, 1403–1408. [CrossRef]
36. Blackmur, M.S.; Robson, J.; Preuss, M.; Zanellato, O.; Cernik, R.; Shi, S.Q.; Ribeiro, F.; Andrieux, J. Zirconium hydride precipitation kinetics in Zircaloy-4 observed with synchrotron X-ray diffraction. *J. Nucl. Mater.* **2015**, *464*, 160–169. [CrossRef]
37. Cinbiz, M.N.; Koss, D.A.; Motta, A.T.; Park, J.S.; Almer, J.D. In situ synchrotron X-ray diffraction study of hydrides in Zircaloy-4 during thermomechanical cycling. *J. Nucl. Mater.* **2017**, *487*, 247–259. [CrossRef]
38. Ells, C. Hydride precipitates in zirconium alloys (A review). *J. Nucl. Mater.* **1968**, *28*, 129–151. [CrossRef]
39. Carpenter, G. The dilatational misfit of zirconium hydrides precipitated in zirconium. *J. Nucl. Mater.* **1973**, *48*, 264–266. [CrossRef]
40. Singh, R.N.; Ståhle, P.; Massih, A.R.; Shmakov, A. Temperature dependence of misfit strains of δ-hydrides of zirconium. *J. Alloys Comp.* **2007**, *436*, 150–154. [CrossRef]
41. Barrow, A.; Korinek, A.; Daymond, M. Evaluating zirconium–zirconium hydride interfacial strains by nano-beam electron diffraction. *J. Nucl. Mater.* **2013**, *432*, 366–370. [CrossRef]
42. Lumley, S.; Grimes, R.; Murphy, S.; Burr, P.; Chroneos, A.; Chard-Tuckey, P.; Wenman, M. The thermodynamics of hydride precipitation: The importance of entropy, enthalpy and disorder. *Acta Mater.* **2014**, *79*, 351–362. [CrossRef]
43. Chan, K.S. An assessment of delayed hydride cracking in zirconium alloy cladding tubes under stress transients. *Int. Mater. Rev.* **2013**, *58*, 349–373. [CrossRef]
44. Markelov, V.A. Delayed hydride cracking of zirconium alloys: Appearance conditions and basic laws. *Russ. Metall. (Met.)* **2011**, *2011*, 326. [CrossRef]
45. Lee, H.; min Kim, K.; Kim, J.S.; Kim, Y.S. Effects of hydride precipitation on the mechanical property of cold worked zirconium alloys in fully recrystallized condition. *Nuclear Eng. Technol.* **2019**. [CrossRef]
46. Likhanskii, V.; Evdokimov, I. Review of theoretical conceptions on regimes of oxidation and hydrogen pickup in Zr-alloys. In Proceedings of the International Conference on WWER Fuel Performance, Modelling and Experimental Eupport, Albena, Bulgaria, 17–21 September 2007.
47. Steinbrück, M.; Birchley, J.; Goryachev, A.; Grosse, M.; Haste, T.; Hozer, Z.; Kisselev, A.; Nalivaev, V.; Semishkin, V.; Sepold, L.; et al. Status of studies on high-temperature oxidation and quench behaviour of Zircaloy-4 and E110 cladding alloys. In Proceedings of 3rd European Review Meeting on Severe Accident Research (ERMSAR-2008), Nesseber, Bulgaria, 23–25 September 2008.
48. Lindgren, M.; Panas, I. On the fate of hydrogen during zirconium oxidation by water: effect of oxygen dissolution in [small alpha]-Zr. *RSC Adv.* **2014**, *4*, 11050–11058. doi:10.1039/C4RA00020J. [CrossRef]
49. Chen, W.; Wang, L.; Lu, S. Influence of oxide layer on hydrogen desorption from zirconium hydride. *J. Alloys Comp.* **2009**, *469*, 142–145. [CrossRef]
50. Couet, A.; Motta, A.T.; Ambard, A.; Comstock, R. Oxide electronic conductivity and hydrogen pickup fraction in Zr alloys. In Proceedings of the 2014 Annual Meeting on Transactions of the American Nuclear Society and Embedded Topical Meeting: Nuclear Fuels and Structural Materials for the Next Generation Nuclear Reactors, NSFM, Reno, NV, USA, 15–19 June 2014; pp. 845–848.
51. Puls, M.P. Review of the thermodynamic basis for models of delayed hydride cracking rate in zirconium alloys. *J. Nucl. Mater.* **2009**, *393*, 350–367. [CrossRef]
52. Marino, G. Hydrogen supercharging in Zircaloy. *Mater. Sci. Eng.* **1971**, *7*, 335–341. [CrossRef]
53. Tikare, V. *Simulation of Hydride Reorientation in Zr-Based Claddings During Dry Storage*; Technical Report; Sandia National Lab. (SNL-NM): Albuquerque, NM, USA, 2013.
54. Courty, O.; Motta, A.T.; Hales, J.D. Modeling and simulation of hydrogen behavior in Zircaloy-4 fuel cladding. *J. Nucl. Mater.* **2014**, *452*, 311–320. [CrossRef]
55. Aryanfar, A.; Thomas, J.; Van der Ven, A.; Xu, D.; Youssef, M.; Yang, J.; Yildiz, B.; Marian, J. Integrated computational modeling of water side corrosion in zirconium metal clad under nominal LWR operating conditions. *JOM* **2016**, *68*, 2900–2911. [CrossRef]
56. Xu, D.; Xiao, H. Cluster Dynamics Model for the Hydride Precipitation Kinetics in Zirconium Cladding. In *Proceedings of the 18th International Conference on Environmental Degradation of Materials in Nuclear Power*

Systems–Water Reactors; Jackson, J.H., Paraventi, D., Wright, M., Eds.; Springer International Publishing: Cham, Switzerlands, 2019; pp. 1759–1768.

57. Ma, X.; Shi, S.; Woo, C.; Chen, L. The phase field model for hydrogen diffusion and γ-hydride precipitation in zirconium under non-uniformly applied stress. *Mech. Mater.* **2006**, *38*, 3–10. [CrossRef]
58. Guo, X.; Shi, S.; Zhang, Q.; Ma, X. An elastoplastic phase-field model for the evolution of hydride precipitation in zirconium. Part I: Smooth specimen. *J. Nucl. Mater.* **2008**, *378*, 110–119. [CrossRef]
59. Aryanfar, A.; Goddard, W., III; Marian, J. Constriction Percolation Model for Coupled Diffusion-Reaction Corrosion of Zirconium in PWR. *Corros. Sci.* **2019**, *158*, 108058. [CrossRef]
60. Une, K. Kinetics of reaction of Zirconium alloy with hydrogen. *J. Less Common Met.* **1978**, *57*, 93–101. doi:10.1016/0022-5088(78)90165-0. [CrossRef]
61. Wang, X.; Zheng, M.J.; Szlufarska, I.; Morgan, D. Continuum model for hydrogen pickup in zirconium alloys of LWR fuel cladding. *J. Appl. Phys.* **2017**, *121*, 135101. [CrossRef]
62. Lim, B.H.; Hong, H.S.; Lee, K.S. Measurements of hydrogen permeation and absorption in zirconium oxide scales. *J. Nucl. Mater.* **2003**, *312*, 134–140. [CrossRef]
63. Geelhood, K.; Beyer, C. Hydrogen Pickup Models for Zircaloy-2, Zircaloy-4, M5^{TM}, and ZIRL^{TM}. In Proceedings of the 2011 Water Reactor Fuel Performance Meeting, Chengdu, China, 11–14 September 2011.
64. Couet, A.; Motta, A.T.; Comstock, R.J. Hydrogen pickup measurements in zirconium alloys: Relation to oxidation kinetics. *J. Nucl. Mater.* **2014**, *451*, 1–13. [CrossRef]
65. Chernyayeva, T.P.; Ostapov, A. Hydrogen in zirconium part 1. *Probl. At. Sci. Technol.* **2013**, *87*, 16–32.
66. Hu, J.; Liu, J.; Lozano-Perez, S.; Grovenor, C.R.; Christensen, M.; Wolf, W.; Wimmer, E.; Mader, E.V. Hydrogen pickup during oxidation in aqueous environments: The role of nano-pores and nano-pipes in zirconium oxide films. *Acta Mater.* **2019**, *180*, 105–115. [CrossRef]
67. Heuser, B.J.; Lin, J.L.; Do, C.; He, L. Small-angle neutron scattering measurements of δ-phase deuteride (hydride) precipitates in Zircaloy 4. *J. Appl. Crystallogr.* **2018**, *51*, 768–780. [CrossRef]
68. Zhao, Z.; Morniroli, J.P.; Legris, A.; Ambard, A.; Khin, Y.; Legras, L.; Blat-Yrieix, M. Identification and characterization of a new zirconium hydride. *J. Microsc.* **2008**, *232*, 410–421. [CrossRef]
69. Reyes, M.; Aryanfar, A.; Baek, S.W.; Marian, J. Multilayer interface tracking model of zirconium clad oxidation. *J. Nucl. Mater.* **2018**, *509*, 550–565. doi:10.1016/j.jnucmat.2018.07.025. [CrossRef]
70. Garzarolli, F.; Seidel, H.; Tricot, R.; Gros, J. Oxide growth mechanism on zirconium alloys. In *Zirconium in the Nuclear Industry: Ninth International Symposium*; ASTM International: West Conshohocken, PA, USA, 1991.
71. Billot, P.; Cox, B.; Ishigure, K.; Johnson, A.; Lemaignan, C.; Nechaev, A.; Petrik, N.; Reznichenko, E.; Ritchie, I.G.; Sukhanov, G.I. *Corrosion of Zirconium Alloys in Nuclear Power Plants*; Technical Report IAEA-TECDOC-684; International Atomic Energy Agency: Vienna, Austria, 1993.
72. Motta, A.T.; Couet, A.; Comstock, R.J. Corrosion of Zirconium Alloys Used for Nuclear Fuel Cladding. *Ann. Rev. Mater. Res.* **2015**, *45*, 311–343. [CrossRef]
73. Chevalier, J.; Gremillard, L.; Virkar, A.V.; Clarke, D.R. The Tetragonal-Monoclinic Transformation in Zirconia: Lessons Learned and Future Trends. *J. Am. Ceram. Soc.* **2009**, *92*, 1901–1920. [CrossRef]
74. Whitney, E.D. Kinetics and mechanism of the transition of metastable tetragonal to monoclinic zirconia. *Trans. Faraday Soc.* **1965**, *61*, 1991–2000. [CrossRef]
75. Couet, A.; Motta, A.T.; Ambard, A. The coupled current charge compensation model for zirconium alloy fuel cladding oxidation: I. Parabolic oxidation of zirconium alloys. *Corros. Sci.* **2015**, *100*, 73–84. doi:10.1016/j.corsci.2015.07.003. [CrossRef]
76. Marian, J.; Bulatov, V.V. Stochastic cluster dynamics method for simulations of multispecies irradiation damage accumulation. *J. Nucl. Mater.* **2011**, *415*, 84–95. [CrossRef]
77. Marian, J.; Hoang, T.L. Modeling fast neutron irradiation damage accumulation in tungsten. *J. Nucl. Mater.* **2012**, *429*, 293–297. [CrossRef]
78. Dunn, A.Y.; Capolungo, L.; Martinez, E.; Cherkaoui, M. Spatially resolved stochastic cluster dynamics for radiation damage evolution in nanostructured metals. *J. Nucl. Mater.* **2013**, *443*, 128–139. [CrossRef]
79. Dunn, A.; Capolungo, L. Simulating radiation damage accumulation in α-Fe: a spatially resolved stochastic cluster dynamics approach. *Comput. Mater. Sci.* **2015**, *102*, 314–326. [CrossRef]
80. Tupin, M.; Martin, F.; Bisor, C.; Verlet, R.; Bossis, P.; Chêne, J.; Jomard, F.; Berger, P.; Pascal, S.; Nuns, N. Hydrogen diffusion process in the oxides formed on zirconium alloys during corrosion in pressurized water reactor conditions. *Corros. Sci.* **2017**, *116*, 1–13. [CrossRef]

81. Cox, B. *Mechanisms of Hydrogen Absorption by Zirconium Alloys*; Technical Report; Atomic Energy of Canada Ltd.: Laurentian Hills, ON, Canada, 1985.
82. Khatamian, D.; Manchester, F. An ion beam study of hydrogen diffusion in oxides of Zr and Zr-Nb (2.5 wt%): I. Diffusion parameters for dense oxide. *J. Nucl. Mater.* **1989**, *166*, 300–306. [CrossRef]
83. Kearns, J. Diffusion coefficient of hydrogen in alpha zirconium, Zircaloy-2 and Zircaloy-4. *J. Nucl. Mater.* **1972**, *43*, 330–338. [CrossRef]
84. Sawatzky, A. The diffusion and solubility of hydrogen in the alpha phase of Zircaloy-2. *J. Nucl. Mater.* **1960**, *2*, 62–68. [CrossRef]
85. Someno, M. Solubility and diffusion of hydrogen in zirconium. *Nippon Kinzoku Gakkaishi Jpn.* **1960**, *24*, 003131.
86. Grosse, M.; Van Den Berg, M.; Goulet, C.; Kaestner, A. In-situ investigation of hydrogen diffusion in Zircaloy-4 by means of neutron radiography. *J. Phys. Conf. Ser.* **2012**, *340*, 012106. [CrossRef]
87. Siripurapu, R.K.; Szpunar, B.; Szpunar, J.A. Molecular Dynamics Study of Hydrogen in α-Zirconium. *Int. J. Nuclear Energy* **2014**, *2014*. [CrossRef]
88. Blomqvist, J.; Olofsson, J.; Alvarez, A.M.; Bjerkén, C. Structure and Thermodynamical Properties of Zirconium Hydrides from First-Principle. In *Proceedings of the 15th International Conference on Environmental Degradation of Materials in Nuclear Power Systems — Water Reactors*; Busby, J.T., Ilevbare, G., Andresen, P.L., Eds.; Springer International Publishing: Cham, Switzerlands, 2016; pp. 671–681.
89. Miyake, M.; Uno, M.; Yamanaka, S. On the zirconium–oxygen–hydrogen ternary system. *J. Nucl. Mater.* **1999**, *270*, 233–241. [CrossRef]
90. LaGrange, L.D.; Dykstra, L.; Dixon, J.M.; Merten, U. A Study of the Zirconium-Hydrogen and Zirconium-Hydrogen–Uranium Systems between 600 and 800°. *J. Phys. Chem.* **1959**, *63*, 2035–2041. [CrossRef]
91. Weekes, H.; Dye, D.; Proctor, J.; Smith, D.; Simionescu, C.; Prior, T.; Wenman, M. The Effect of Pressure on Hydrogen Solubility in Zircaloy-4. *arXiv* **2018**, arXiv:1806.09657.
92. Northwood, D.; Kosasih, U. Hydrides and delayed hydrogen cracking in zirconium and its alloys. *Int. Met. Rev.* **1983**, *28*, 92–121. [CrossRef]
93. Une, K.; Ishimoto, S. Dissolution and precipitation behavior of hydrides in Zircaloy-2 and high Fe Zircaloy. *J. Nucl. Mater.* **2003**, *322*, 66–72. [CrossRef]
94. Zanellato, O.; Preuss, M.; Buffiere, J.Y.; Ribeiro, F.; Steuwer, A.; Desquines, J.; Andrieux, J.; Krebs, B. Synchrotron diffraction study of dissolution and precipitation kinetics of hydrides in Zircaloy-4. *J. Nucl. Mater.* **2012**, *420*, 537–547. [CrossRef]
95. Domain, C.; Besson, R.; Legris, A. Atomic-scale Ab-initio study of the Zr-H system: I. Bulk properties. *Acta Mater.* **2002**, *50*, 3513–3526. doi:10.1016/S1359-6454(02)00173-8. [CrossRef]
96. Nazarov, R.; Majevadia, J.S.; Patel, M.; Wenman, M.R.; Balint, D.S.; Neugebauer, J.; Sutton, A.P. First-principles calculation of the elastic dipole tensor of a point defect: Application to hydrogen in α-zirconium. *Phys. Rev. B* **2016**, *94*, 241112. doi:10.1103/PhysRevB.94.241112. [CrossRef]
97. Fukai, Y. *The Metal-Hydrogen System: Basic Bulk Properties*; Springer Science & Business Media: Cham, Switzerlands, 2006; Volume 21.
98. Cordero, B.; Gómez, V.; Platero-Prats, A.E.; Revés, M.; Echeverría, J.; Cremades, E.; Barragán, F.; Alvarez, S. Covalent radii revisited. *Dalton Trans.* **2008**, *21*, 2832–2838. [CrossRef] [PubMed]
99. Blackmur, M.S.; Preuss, M.; Robson, J.D.; Zanellato, O.; Cernik, R.J.; Ribeiro, F.; Andrieux, J. Strain evolution during hydride precipitation in Zircaloy-4 observed with synchrotron X-ray diffraction. *J. Nucl. Mater.* **2016**, *474*, 45–61. [CrossRef]
100. Weekes, H.; Jones, N.; Lindley, T.; Dye, D. Hydride reorientation in Zircaloy-4 examined by in situ synchrotron X-ray diffraction. *J. Nucl. Mater.* **2016**, *478*, 32–41. [CrossRef]
101. Wang, S.; Giuliani, F.; Britton, T.B. Microstructure and formation mechanisms of δ-hydrides in variable grain size Zircaloy-4 studied by electron backscatter diffraction. *Acta Mater.* **2019**, *169*, 76–87. [CrossRef]
102. Bair, J.; Zaeem, M.A.; Tonks, M. A review on hydride precipitation in zirconium alloys. *J. Nucl. Mater.* **2015**, *466*, 12–20. [CrossRef]
103. Heo, T.W.; Colas, K.B.; Motta, A.T.; Chen, L.Q. A phase-field model for hydride formation in polycrystalline metals: Application to δ-hydride in zirconium alloys. *Acta Mater.* **2019**, *181*, 262–277. [CrossRef]

104. Vizcaíno, P.; Santisteban, J.; Alvarez, M.V.; Banchik, A.; Almer, J. Effect of crystallite orientation and external stress on hydride precipitation and dissolution in Zr2.5. *J. Nucl. Mater.* **2014**, *447*, 82–93.
105. Tikare, V.; Weck, P.F.; Mitchell, J.A. *Modeling of Hydride Precipitation and re-Orientation*; Technical Report SAND2015-8260R; Sandia National Laboratories (SNL-NM): Albuquerque, NW, USA, 2015.
106. Blat, M.; Noel, D. Detrimental role of hydrogen on the corrosion rate of zirconium alloys. In *Zirconium in the Nuclear Industry: Eleventh International Symposium*; ASTM International: West Conshohocken, PA, USA, 1996.
107. Shinohara, Y.; Abe, H.; Iwai, T.; Sekimura, N.; Kido, T.; Yamamoto, H.; Taguchi, T. In situ TEM observation of growth process of zirconium hydride in Zircaloy-4 during hydrogen ion implantation. *J. Nuclear Sci. Technol.* **2009**, *46*, 564–571. [CrossRef]
108. Krishna, K.M.; Sain, A.; Samajdar, I.; Dey, G.; Srivastava, D.; Neogy, S.; Tewari, R.; Banerjee, S. Resistance to hydride formation in zirconium: An emerging possibility. *Acta Mater.* **2006**, *54*, 4665–4675.
109. Škarohlíd, J.; Ashcheulov, P.; Škoda, R.; Taylor, A.; Čtvrtlík, R.; Tomáštík, J.; Fendrych, F.; Kopeček, J.; Cháb, V.; Cichoň, S.; et al. Nanocrystalline diamond protects Zr cladding surface against oxygen and hydrogen uptake: Nuclear fuel durability enhancement. *Sci. Rep.* **2017**, *7*, 6469. [CrossRef] [PubMed]
110. Youssef, M.; Yang, M.; Yildiz, B. Doping in the Valley of Hydrogen Solubility: A Route to Designing Hydrogen-Resistant Zirconium Alloys. *Phys. Rev. Appl.* **2016**, *5*, 014008. doi:10.1103/PhysRevApplied.5.014008. [CrossRef]

© 2020 by the authors. Licensee MDPI, Basel, Switzerland. This article is an open access article distributed under the terms and conditions of the Creative Commons Attribution (CC BY) license (http://creativecommons.org/licenses/by/4.0/).

Article

The Characterization of Stress Corrosion Cracking in the AE44 Magnesium Casting Alloy Using Quantitative Fractography Methods

Maria Sozańska [1,*], Adrian Mościcki [2] and Tomasz Czujko [3,*]

1 Faculty of Materials Engineering and Metallurgy, Silesian University of Technology, Krasińskiego 8, 40-019 Katowice, Poland

2 BGH Polska Sp. z o.o., ul. Żelazna 9, 40-851 Katowice, Poland; adrian.moscicki@bgh.pl

3 Faculty of Advanced Technology and Chemistry, Military University of Technology, Gen. S. Kaliskiego 2, 00-908 Warsaw, Poland

* Correspondence: maria.sozanska@polsl.pl (M.S.); tomasz.czujko@wat.edu.pl (T.C.)

Received: 23 October 2019; Accepted: 3 December 2019; Published: 9 December 2019

Abstract: In this work an assessment of the susceptibility of the AE44 magnesium alloy to stress corrosion cracking in a 0.1M Na_2SO_4 environment is presented. The basic assumed criterion for assessing the alloy behavior under complex mechanical and corrosive loads is deterioration in mechanical properties (elongation, reduction in area, tensile strength and time to failure). The AE44 magnesium alloy was subjected to the slow strain rate test (SSR) in air and in a corrosive environment under open circuit potential (OCP) conditions. In each variant, the content of hydrogen in the alloy was determined. The obtained fractures were subjected to a quantitative evaluation by original fractography methods. It was found that under stress corrosion cracking (SCC) conditions and in the presence of hydrogen the mechanical properties of AE44 deteriorated. The change in the mechanical properties under SCC conditions in a corrosive environment was accompanied by the presence of numerous cracks, both on fracture surfaces and in the alloy microstructure. The developed method for the quantitative evaluation of cracks on the fracture surface turned out to be a more sensitive method, enabling the assessment of the susceptibility of AE44 under complex mechanical and corrosive loads in comparison with deterioration in mechanical properties. Mechanical tests showed a decrease in properties after SSRT tests in corrosive environments (UTS ≈ 153 MPa, ε = 11.2%, Z = 4.0%) compared to the properties after air tests (UTS ≈ 166 MPa, ε = 11.9%, Z = 7.8%) but it was not as visible as the results of quantitative assessment of cracks at fractures (number of cracks, length of cracks): after tests in corrosive environment (900; 21.3 μm), after tests in air (141; 34.5 μm). These results indicate that the proposed new proprietary test methodology can be used to quantify the SSC phenomenon in cases of slight changes in mechanical properties after SSRT tests in a corrosive environment in relation to the test results in air.

Keywords: magnesium alloy; stress corrosion cracking; hydrogen; fractography methods

1. Introduction

Magnesium-based casting alloys, with their low density and good technological (workability, castability, machinability, and full recyclability) and mechanical properties (considerable specific rigidity and specific strength), are a very attractive construction material in those industries where it is important to reduce the mass of a structure and at the same time retain its mechanical properties [1,2]. The main examples are the motor and aerospace industries, where magnesium alloys are used for manufacturing various components, such as engine parts, transmission cases, pumps, oil sumps, and steering wheels [3–6]. Over the last 20 years, there has been a steady tendency to use increasingly strong magnesium alloys, often working at elevated temperatures (above 150 °C) and in a corrosive environment. Such properties of magnesium alloys could be achieved by reducing the content of aluminum (in alloys from the Mg–Al group, such as AZ91, AM50, and AM60) and introducing alloying agents such as Ca, Si, Sr, and rare earth (RE) elements. Thermodynamically stable Al-RE intermetallic phases that form at grain boundaries during solidification (Al_4RE or $Al_{11}(RE)_3$ and Al_2RE and $Mg_{17}Al_{12}$) enhanced the mechanical properties (in particular creep resistance). However, their effect on corrosion resistance under SCC conditions in hydrogen-containing environment has not been described.

Literature data [7,8] concerning this problem suggest that the phenomenon may be caused by stresses that are less than 50% of the yield point of particular magnesium alloys being in service, in environment causing only negligible corrosion in those alloys [8–10]. Hydrogen-assisted stress corrosion cracking is particularly dangerous for the group of high-strength magnesium alloys containing rare earth elements. Literature data concerning hydrogen degradation and SCC are very scarce for high-strength Mg–Al–RE alloys. This concerns in particular the AE44 casting alloy.

The authors of many publications [1,11–13] agree that the development of stress corrosion cracking in magnesium alloys can take place according to two major destruction mechanisms, which are determined both by the alloy's microstructure and its operating environment:

- continuous crack propagation caused by anodic dissolution of the alloy at the top of cracks,
- discontinuous crack propagation caused by a series of cracks occurring at the crack tip.

In the case of the mechanism associated with the continuous dissolution of the alloy at the apex of the crack, several models describing this mechanism can be found in the literature. The most common applies to magnesium alloys characterized by a continuous or almost continuous mesh of phase separation occurring at grain boundaries. The model assumes that material destruction is carried out by galvanic corrosion leading to inter-crystalline stress corrosion cracking [15,16].

The potential difference between the α-Mg solution, which is the anode in this system, and the non-metallic phase precipitates located at the alloy grains, which are the cathode, causes the formation of galvanic cells, which consequently leads to anodic dissolution of the solid solution adjacent to the phase separations. Stresses affecting the material cause continuous tearing of the crack thus created, providing the corrosive environment with constant access to the metallic substrate at the tip of the crack, enabling it to spread continuously.

Another model assumes that in alloys where there is no continuous or almost continuous mesh of phase separation at grain boundaries, a similar process may also occur that causes anodic dissolution of the α-Mg solid solution. However, unlike the model described earlier, it assumes that a galvanic cell can be formed between the passive layer formed on the surface of a magnesium alloy and the place where this layer has been damaged as a result of mechanical stress or an aggressive environment revealing a metallic substrate [16]. In this arrangement, the exposed metallic substrate acts as an anode, while the passive layer is a cathode. As a result of the corrosive cell, a pitting is formed which is also a stress concentrator. Their presence at the crack tip prevents reproduction passive layer, which leads to crack growth. It should be noted that the operation of this model is possible only for alloys and environments in which the repassivation speed is insufficient to restore the protective layer on the alloy surface.

The mechanism of discontinuous crack propagation caused by a series of microcracks at the tip of the crack is quite different. The result of this mechanism is the formation of both trans-crystalline and inter-crystalline stress corrosion cracking. The authors of numerous studies [1,11–15] agree that the impact of hydrogen on the alloy plays a key role in the functioning of this type of destruction, while it is difficult to precisely determine its role. Studies devoted to the phenomenon of stress corrosion cracking suggest that in the case of magnesium alloys, material cracking may occur according to the following models taking into account hydrogen as the driving force of the cracking process [1]:

- local decohesion of the crystal lattice caused by hydrogen (hydrogen enhanced decohesion—HEDE),

- local hydrogen-induced plasticization (hydrogen enhanced localized plasticity—HELP),
- dislocation emission caused by adsorption (adsorption-induced dislocation-emission—AIDE),
- delayed hydride cracking (DHC).

Material destruction according to HEDE, HELP, DHC, and AIDE models also takes place in the case of hydrogen embrittlement phenomenon of magnesium alloys. The difference that separates the phenomenon of hydrogen embrittlement and stress cracking is the presence of external mechanical stress in the case of stress corrosion cracking, while the main mechanism of destruction is the interaction of hydrogen deformation [17].

Stress corrosion cracking processes can be analyzed using various criteria and testing techniques, including mechanical tests, electrochemical methods and examinations of the microstructure and fracture surfaces. The most frequently used test is the slow strain rate test (SSRT) [10,18]. Mechanical and corrosive tests related to stress corrosion cracking are often complemented by fractographic examination—a qualitative analysis of fracture surfaces or profiles, including the determination of the nature of the fracture, most frequently using scanning electron microscopy (SEM) methods [19].

In this study, it is attempted to analyze stress corrosion cracking in a modern magnesium-based alloy containing rare earth elements—AE44. The RE mixture in the AE44 alloy contains, apart from Al and Si, also La, Ce and Nd. It is commonly believed that precipitates enriched with rare earth metals may enhance not only creep resistance but also stress corrosion cracking resistance. The susceptibility of AE44 to stress corrosion cracking was assessed by determining changes in its mechanical properties under SSRT tests [5] in a 0.1 M Na_2SO_4 solution. The mechanical tests were complemented by an analysis of the alloy's microstructure and by a quantitative evaluation of changes in the morphology of the fracture surfaces being the result of stress corrosion cracking by means of quantitative fractography methods. The developed method for the quantitative evaluation of cracks on fracture surfaces turned out to be a more sensitive method enabling the assessment of the susceptibility of AE44 under complex mechanical and corrosive loads in comparison with mechanical property degradation.

2. Materials and Methods

The AE44 magnesium-based alloy containing rare earth elements (in wt.%: 4.2 Al, 0.2 Mn, 0.1 Si, 4.2 RE (La, Ce and Nd), balance—Mg) was manufactured by die casting in the form of a 30 mm long rods with a diameter of 12 mm. The microstructure of raw material was analyzed using a Hitachi S-3400N (Hitachi, Tokyo, Japan) scanning electron microscope. All samples before structural examinations were subjected to a following metallographic preparation:

- placing the cut samples in a conductive resin using a press;
- grinding of sample surfaces on abrasive papers with SiC particles with gradation of grains: 120, 320, 600, and 1200;
- rinsing of the ground surface of the samples under running water and polishing with the use of diamond suspensions with the following particle sizes: 6 μm, 3 μm, and 1 μm;
- the last stage of sample preparation was polishing in a Al_2O_3 suspension with a grain size of 0.05 μm.

Due to the high sensitivity of magnesium alloys to corrosion in water environments, special care was taken to ensure that the time between polishing and drying of samples was as short as possible and not exceed 10 seconds.

The slow strain rate test (SSR) was carried out in accordance with ASTM G129-00 (2013) "Standard Practice for Slow Strain Rate Testing to Evaluate the Susceptibility of Metallic Materials to Environmentally Assisted Cracking" [5]. Cylindrical samples with a length of 20 mm and a diameter of 5 mm have been used. Before the test, samples were cleaned with acetone. The SRT test was performed both in air and in a corrosive environment (0.1M Na_2SO_4 solution) under open circuit potential (OCP) conditions, at ambient temperature and at a strain rate of $\varepsilon = 9{\cdot}10^{-7}$ s^{-1}. Following parameters were determined: elongation at failure ε (%), reduction in area at failure Z (%), ultimate tensile strength UTS (MPa) and time to failure t (h).

After SSR test, corrosion products were removed from the fracture surfaces by bathing in a 200 g/L CrO_3 + 10 g/L $AgNO_3$ solution for 1 ÷ 3 minutes (as necessary). The samples were then rinsed three times in acetone using an ultrasonic cleaner. The fracture surfaces were analyzed using a Hitachi S-3400N (Hitachi, Tokyo, Japan) scanning electron microscope. A quantitative characterization of cracks observed on the fracture surfaces was performed using an image analysis program—Met-Ilo® (version 15.3, Instytut Inżynierii Materiałowej, Katowice, Poland) [20]. The analysis comprised an assessment of the morphology and distribution of cracks present on the fracture surfaces and the determination of number and length of cracks.

The quantitative fractography analysis covered entire fracture surfaces. The use of the systemic scanning method (a series of photographs showing the entire fracture surface) and the subsequent arrangement of the images into one high-resolution photograph (Figure 1) enabled the assessment of details on a micrometer scale.

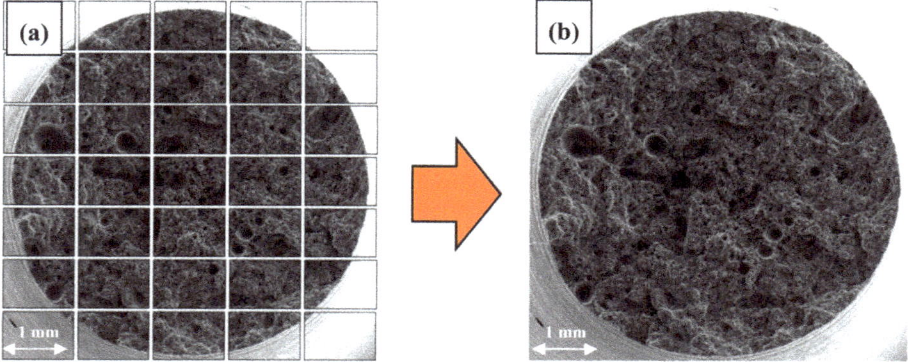

Figure 1. Scheme of the image merging operation (SE) into the image of high resolution (HR) of the fracture surface of the sample using the scanning electron microscope (SEM): (**a**) scheme of systematic scanning operations of the real surface of the fracture - C_{ij} matrix of images, (**b**) image of the total area surface of the fracture after combining images from the C_{ij} matrix.

The original authors' own procedure for the quantitative evaluation of cracks in AE44 comprised:

- **Stage 1**—the preparation of a high-resolution image of the entire fracture surface, made up of a series of images with marked cracks (cracks had to be marked manually in the GIMP image manipulation program as the greyness differences between cracks and some elements of the structure were insufficient for automatic detection),
- **Stage 2**—the use of image analysis methods (the correction of brightness and contrast and the creation of a binary image of cracks) to determine the quantitative crack parameters.

Apart from the qualitative and quantitative analyses of the fracture surfaces, a qualitative characterization of fracture profiles was performed. Fracture profiles were obtained by grinding one side of a fracture surface with abrasive paper. Due to the considerable susceptibility of magnesium-based alloys to corrosion in water environments, the time between the polishing and drying of specimens was maximally shortened and did not exceed 10 seconds.

The hydrogen concentration in the specimens was analyzed using a LECO TCHEN 600 (LECO, St. Joseph, MI, USA) elemental analyzer. Hydrogen concentration was analyzed for specimens that were subjected to the SSR tests both in air and in a corrosive environment.

3. Results and Discussion

The microstructure of as-cast AE44 is characterized by the presence of fine grains with phase precipitates present mainly at grain boundaries (Figure 2). Moreover, few pores of various sizes are also observed in the structure (Figure 2b).

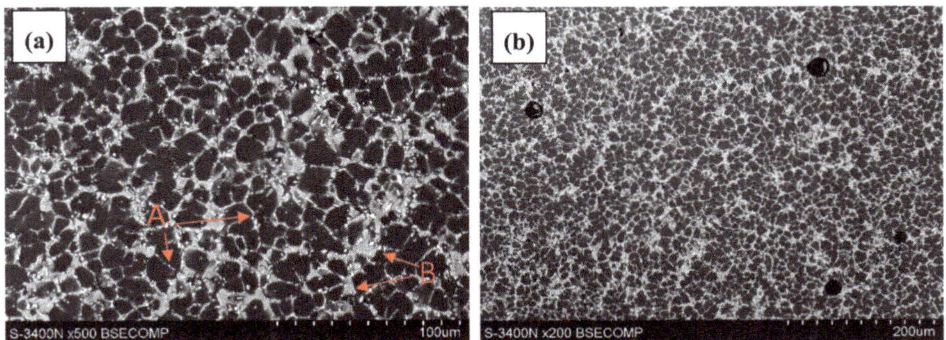

Figure 2. Microstructure of magnesium alloy AE44 in initial state: (**a**) cross-section perpendicular to the axis of the sample; (**b**) casting pores.

SEM observations (Figure 2) revealed typical microstructure of diecast AE44 alloy consisted of fine grains of α-Mg matrix and two sorts of tiny precipitates [21]. The first one was with lamellar morphology (phase type A, Figure 2). The second was coarse with globular shape (phase type B, Figure 2). Micro-zone compositions analysis was performed on these white phases by energy dispersive spectrometer (EDS), and the results show that they contain aluminum and rare earth elements, cerium, lanthanum and neodymium. In our research, it was confirmed, according to [21], that these particles are precipitates of $Al_{11}RE_3$ and Al_2RE, respectively.

3.1. The Slow Strain Rate (SSR) Test Results

AE44 magnesium alloy was subjected to slow strain rate tests in order to determine its susceptibility to stress corrosion cracking in the applied environment (0.1 M Na_2SO_4 solution) and under external mechanical loads. The samples were subjected to strain in air as well as exposure to the corrosive environment under OCP conditions. Such conditions were selected in order to determine the influence of a corrosive environment, corresponding to the working conditions of AE44 alloy, on its mechanical properties [22].

The stress–strain curves obtained as a result of the two SSR test variants for the AE44 samples are shown in Figure 3. They have a normal shape, with an elastic and plastic strain zone, but with no distinct yield point. The highest ultimate tensile strength (UTS ≈ 166 MPa) is recorded for the material subjected to strain in air.

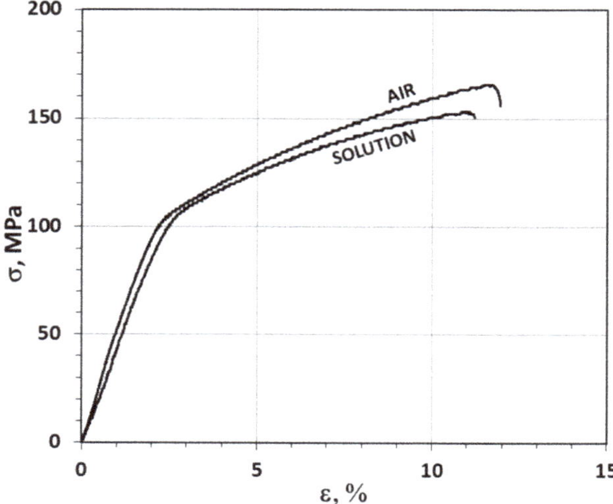

Figure 3. Stress–strain curves ($\varepsilon = 9 \times 10^{-7}$ s^{-1}) for samples of magnesium alloy AE44 deformed in air and in a corrosive environment under open circuit potential (OCP) conditions.

The samples subjected to strain with simultaneous exposure to the corrosive solution under OCP conditions were characterized by lower strength (UTS ≈ 153 MPa). The best plasticity (ε, Z) was displayed by the samples tested in air (Figure 3, Table 1). The elongation recorded for these samples was ε = 11.9% and the reduction in area was Z = 7.8%. For samples subjected to strain in a corrosive environment under OCP conditions, a slightly lower elongation (ε = 11.2%) and a noticeably smaller reduction in area (Z = 4.0%) were recorded (Figure 3, Table 1).

Table 1. Mechanical properties, hydrogen concentration C_H and hydrogen embrittlement index EI_ε for AE44 magnesium alloy AE44 obtained as a result of slow strain rate (SSR) test ($\varepsilon = 9 \times 10^{-7}$ s^{-1}) in air and in a corrosive solution.

Test Environment	Elongation ε [%]	Necking Z [%]	Ultimate Tensile Strength UTS [MPa]	Fracture Time [h]	Hydrogen Concentration C_H [ppm]	Hydrogen Embrittlement Index	
						EI_ε [%]	EI_Z [%]
Strained in air	11.9	7.8	166	39.5	99.7	-	-
Strained in solution	11.2	4.0	153	37.7	98.7	5.9	49

It can be concluded that the mechanical properties of the AE44 alloy after the test in the air were close to the results obtained by Rzychoń and Kiełbus [23]: UTS – 146 MPa, ε – 7.1%. It should be taken into account that the magnesium alloy AE44 is a casting alloy and its properties strongly depend on the volume fraction of gas pores. Literature data indicate that the greatest deterioration of the mechanical properties of magnesium alloys is observed during their deformation in a corrosive environment, under cathodic polarization conditions, as well as in open potential conditions. This effect is observed for many magnesium alloys (AZ31, Mg-8.6Al, ZK60, AZ91D, and Mg-7%Gd-5%Y-1%Nd-0.5%Zr) in various corrosive environments (sensitivity of Mg alloys in 0.1 M neutral salt solution decreases in the following order: Na_2SO_4 > $NaNO_3$ > Na_2CO_3 > NaCl > CH_3 COONa) [13,14,24–27].

The results of hydrogen concentration analyses for the AE44 alloy after the SSR tests shows that the hydrogen concentrations for both test variants were comparable (Table 1) and that the hydrogen embrittlement index $EI\varepsilon$, determined using Equation (1), was approx. 5.9% for elongation ε and approx. 49% for reduction in area EI_Z.

$$EI = |X_H - Xair|/Xair \times 100\%, \quad (1)$$

where: X_H is the value of a parameter (e.g., reduction in area Z or elongation ε) in the presence of hydrogen, Xair is the value of a parameter (e.g., reduction in area Z or elongation ε) without hydrogen, $i = \varepsilon$ or Z.

The role of hydrogen in the destruction of the AE44 alloy is extremely complex and requires a lot of further research regarding both its penetration process into the alloy, but also subsequent diffusion in the alloy and recombination on the internal surfaces of the pores. In practice, it is possible to have several destruction mechanisms simultaneously with any combination thereof. However, there is no doubt that the hydrogen entering the material as a result of electrochemical corrosion processes is in atomic or proton form. In this form it diffuses in the alloy microstructure, and the microstructure of the alloy and existing structural defects—separation of intermetallic phases, grain boundaries, dislocations and pores—play an important role in the processes of hydrogen diffusion in magnesium alloys. In addition, in rare-earth magnesium alloys, hydrogen can form hydrides with RE elements, and pores also play an important role in magnesium foundry alloys. In the pores, hydrogen may recombine to molecular form, accumulate and cause increased stress in these areas [28].

When analyzing the mechanical properties of the AE44 magnesium-based casting alloy, it is necessary to take into account the relationship between the porosity of AE44 alloy and its mechanical properties. Lee et al. [29] demonstrated that the mechanical properties of this alloy greatly depended on the number of pores present in the microstructure. Moreover, the results that they obtained indicated a relationship between the number of pores present on the fracture surface and the plastic and strength properties. The distribution of hydrogen in AE44 material is also affected by porosity. In their work [30], Volkova and Morozova analyzed the chemical composition of gases present in magnesium-based casting alloys containing rare earth elements, trapped at the stage of their manufacture. Their research revealed that hydrogen accounted for as much as 93–95% of the gases' volume. Bearing in mind that pores present in AE44 alloy may contain hydrogen in the form of molecules, trapped in there during the alloy crystallization, the assessment of hydrogen concentration in the AE44 samples seems to involve a potential for major errors. The presence of hydrogen in pores following the crystallization process in AE44 alloy may also be the reason why there is hardly any difference in hydrogen concentration for the specimens after the SSR tests in air (99.7 ppm) and in the corrosive environment (98.7 ppm) under OCP conditions.

3.2. Qualitative and Quantitative Evaluation of the Fracture Surface after the SSR Tests Using The Oryginaly Developed Quantitative Fractography Method

The analysis of the AE44 sample surfaces and fracture surfaces after the SSR tests in air and in the corrosive environment showed major differences. There were a few cracks on the side surfaces of the AE44 sample subjected to strain in air (Figure 4a,b). Their propagation direction was perpendicular to the force applied during the strain test.

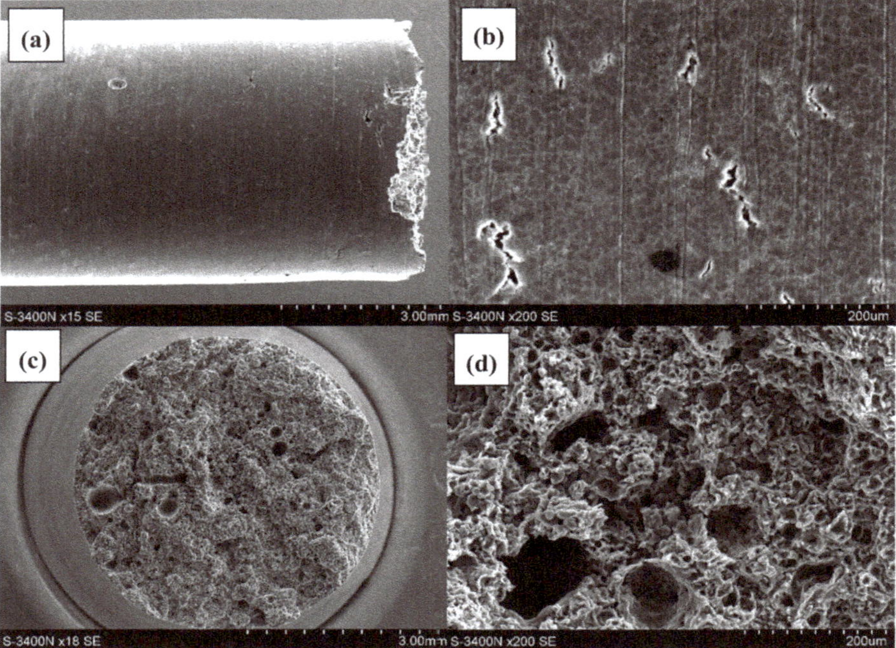

Figure 4. Lateral and fracture surface of the sample after SSR test in air: (**a**) macroscopic image of lateral surface; (**b**) magnified image of the lateral surface; (**c**) macroscopic image of fracture surface; (**d**) magnified image of fracture surface.

On the main fracture surface (Figure 4c,d), there are numerous pores that formed during the casting process, of various sizes reaching even several hundred micrometers (Figure 4c).

In the case of the specimen subjected to strain in the corrosive environment under OCP conditions, there are cracks and very numerous but shallow corrosion pits on its surface side (Figure 5a,b). On the main fracture surface (Figure 5c,d), there are pores of various sizes (Figure 5c) and numerous cracks (Figure 5d). However, there are also areas free of cracks. The morphology of the fracture surface of the AE44 alloy after the air test is typical—a ductile fracture with numerous pores, the elongated shrinkage porosity and spherical gas (air) porosity. It is comparable with the results of other authors [24].

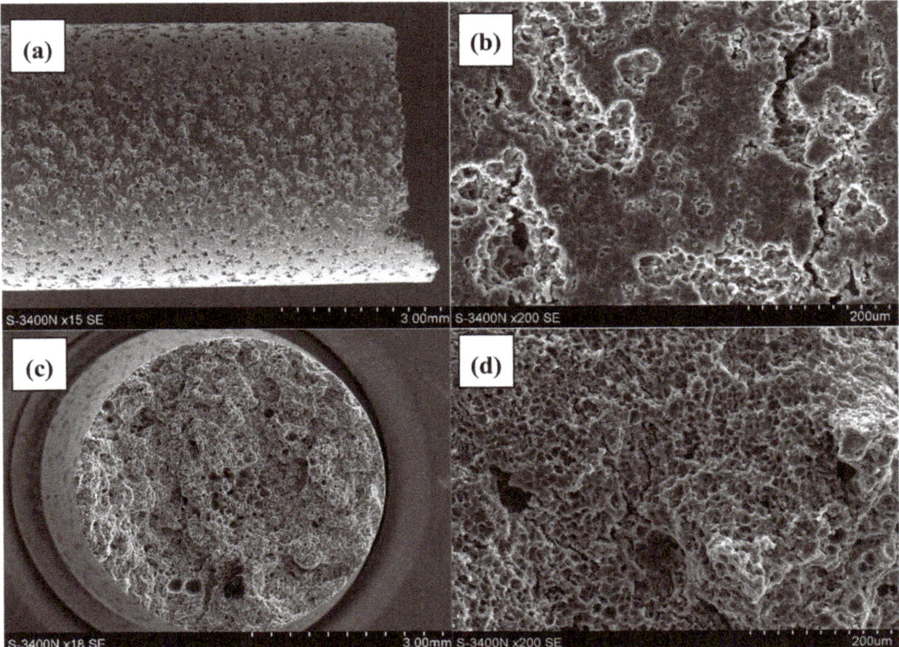

Figure 5. Lateral and fracture surface of the sample after SSR test in corrosive solution under OCP conditions: (**a**)—macroscopic image of lateral surface; (**b**)—magnified image of the lateral surface; (**c**) macroscopic image of fracture surface; (**d**) magnified image of fracture surface.

A detailed analysis of the fracture surface of the AE44 specimens subjected to strain in air revealed that the fracture was intercrystalline. There were also many intermetallic phase precipitates on the fracture surfaces (Figure 6a), which were located along grain boundaries in AE44. Cracking along grain boundaries was also observed on the fracture profiles (Figure 6a,b). The analysis of the fracture surfaces of the AE44 specimens subjected to strain in the corrosive solution under OCP conditions revealed a significantly larger number of cracks. Moreover, areas of a transcrystalline nature were observed on the specimen surfaces (Figure 6c,d). According to Atrens et al. [31] fracture surface after SSC of Mg alloy is intergranular (IG-SSC) or transgranular (TG-SSC). A continuous or nearly continuous second phase, typically along grain boundaries, causes IG-SSC by microgalvanic corrosion of the adjacent Mg-matrix. IG-SSC is expected in all such alloys. TG-SSC is most likely caused by interaction of hydrogen with the microstructure. Hydrogen strongly interacts with rare earth elements (RE) in magnesium alloy AE44 to form hydrides with them. It is commonly recognized that transgranular SCC (TGSCC) of Mg alloys is a type of HE. HE models involve hydrogen-enhanced decohesion, hydrogen-enhanced localized plasticity, adsorption-induced dislocation emission, and delayed hydride cracking. The presence of hydride (MgH_2) phase on the fracture surface designates the DHC mechanism for a friction-stir welded magnesium alloy AZ31 [32].

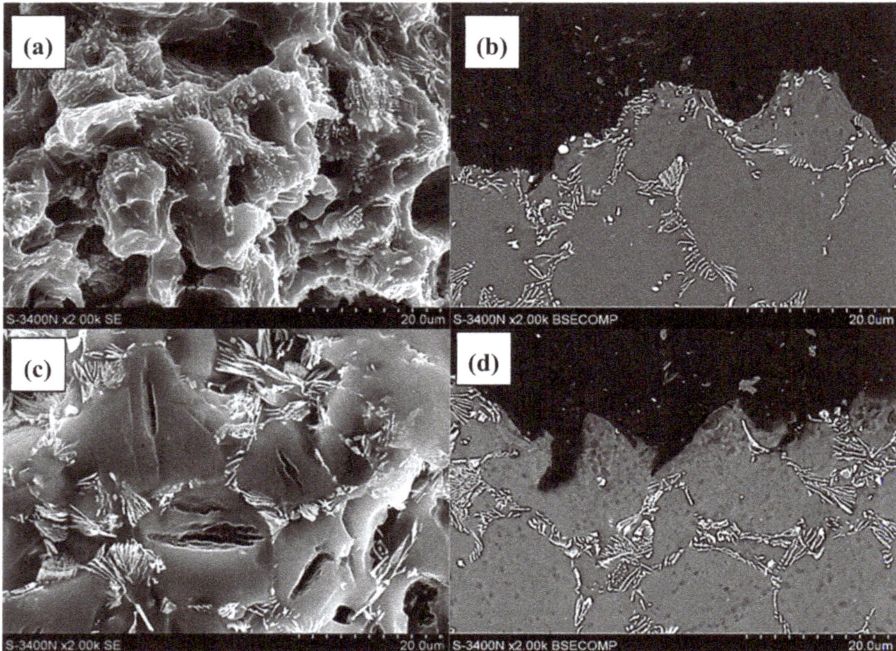

Figure 6. Surface and profiles of AE44 magnesium alloys obtained as a result of the slow strain rate test (SSRT): (**a**,**b**)—air, intergranular fracture; (**c**,**d**)—solution, transgranular fracture, and secondary cracks in the volume of the solid solution α-Mg.

Cracks were also identified in the alloy matrix (α-Mg), where very often they propagated on parallel crystallographic planes (Figure 7). At the same time, it was observed that in the corrosive environment, the cracking occurred not only through the growth of the main crack, but also through the growth of secondary cracks, which together formed more complex crack configurations (Figure 7a).

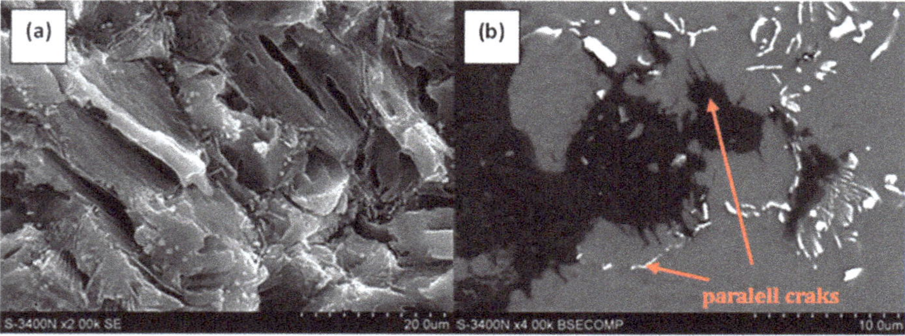

Figure 7. Series of parallel cracks in the AE44 magnesium alloy as a result of SSRT in a corrosive environment under OCP conditions: (**a**) on the fracture surface; (**b**) on the profile of fracture.

Analysis of the places of occurrence of cracks on fracture surfaces indicates that in the AE44 alloy these cracks occurred in the areas of α-Mg solid solution after SSRT tests in corrosive environment and they were not accompanied by the separation of intermetallic phases (Figure 6). No such cracks were found after testing SSRT in the air. At the same time, it should be remembered that in the presence of a corrosive environment, the nature of the fracture changed from inter-crystalline after SSRT tests in the air to trans-crystalline after tests in a corrosive environment. This fact and the presence of numerous cracks in the alloy matrix after SSRT tests in a corrosive environment may indicate the presence of hydrogen, which caused the matrix brittleness in this place.

On fracture profiles, porosity is observed in the entire volume of the specimens, with the pores being located mainly near the specimen axis (Figure 8). This is typical behavior for pores forming during the casting process.

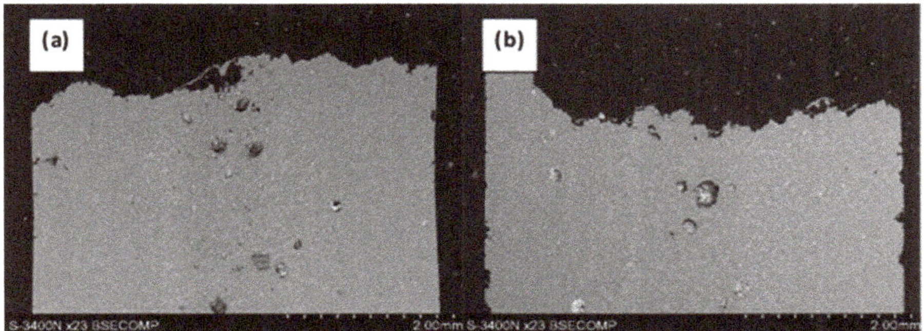

Figure 8. Profile of the fracture in magnesium alloy AE44 after SSRT: (**a**) in the air; (**b**) in solution.

The presence of local cracks propagating from the side surface deeper into the specimen was detected on the side surfaces of all the AE44 specimens (Figure 9b,d).

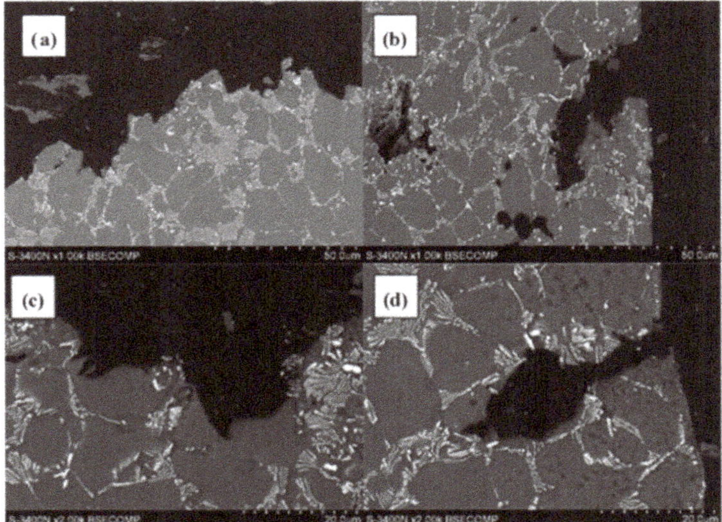

Figure 9. Profile of the fracture in AE44 magnesium alloy after SSRT in the air: (**a**)—fracture surface; (**b**)—lateral surface of the sample and after SSRT in solution: (**c**)—fracture surface; (**d**)—lateral surface of the sample.

There are many research techniques allow quantitative assessment of surface fractures [18]. However, each of these techniques has its limitations. The origins of quantitative fractography can be found in the works Undervood [19], Wrigth and Karlsson [33], Coster and Chermant [34], and

Undervood and Banerji [35]. The authors discuss in some detail the possibilities and limitations in the methods of quantitative description of surface fracture in these works. Generally, there are two groups of quantitative fractography techniques using:

- methods based on metallographic cross-sections of the fracture surface, i.e., profilometric analysis,
- methods based on images of the fracture surface, i.e., quantitative description of the fracture surface.

The authors of this publication have used both techniques—quantitative description of the profile fracture [7,8] or a fracture surface [9,10,36]. However, the most important thing, regardless of the chosen research technique, is the selection of detail of surface fracture for the quantitative description, e.g., a fraction of the ductile fracture, cracks.

The basis for the quantitative evaluation of cracks on the AE44 fracture surfaces were binary images of total fracture surfaces with marked cracks (Figure 10).

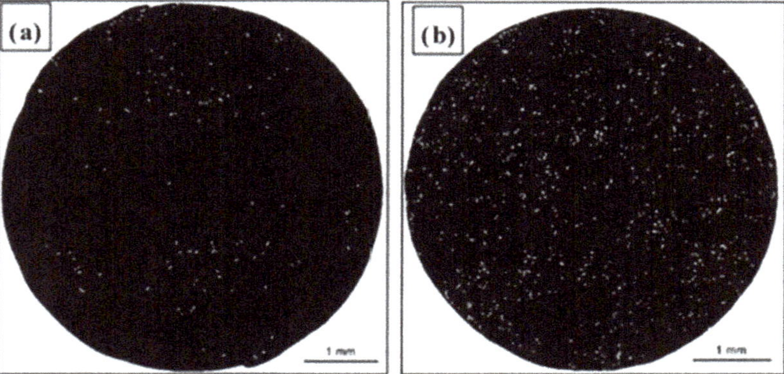

Figure 10. Binary image of cracks on the fracture surfaces of AE44 magnesium alloy after SSRT tests: (**a**) in the air and (**b**) in solution.

The results of the quantitative evaluation of cracks on the fracture surfaces (Figure 10, Table 2) show major differences in the numbers and lengths of cracks depending on the SSR test conditions—in air and in the corrosive solution under OCP conditions.

Table 2. Results of the quantitative analysis of cracks on the surfaces of AE44 magnesium alloy fracture after SSRT tests in air and in a corrosive solution in OCP conditions.

SSRT Environment	Number of Cracks	Length of Cracks [μm]	Width of Cracks [μm]	Distance between Adjacent Cracks [μm]
strained in air	141	34.5 ± 43.8%	1.02 ± 14.1%	118 ± 99%
strained in solution	900	21.3 ± 48.7%	1.04 ± 18.9%	58 ± 76%

The fracture surface of the specimen subjected to strain in air is characterized by the smallest number of cracks (141). Their distribution was uneven – the cracks formed small clusters (Figure 10a). Moreover, the cracks are characterized by the greatest average length (34.5 μm). The most numerous group of cracks present on the surfaces of the specimens tested in air were cracks that were 20–30 μm long—they accounted for 37.6% of all cracks. No cracks shorter than 14 μm were found. The longest crack found in these specimens was 114 μm long. Detailed information concerning the crack length distribution on the fracture surface of the AE44 specimen subjected to strain in air is shown in the form of a histogram (Figure 11a). The quantitative analysis of the cracks present on the fracture surface of the specimen tested in the corrosive environment under OCP conditions revealed that they were much more numerous than in the case of the specimen tested in air—there were 900 cracks. They were evenly distributed on the fracture surface (Figure 10b). Their average length was 21.3 µm. The most numerous group of cracks was in the rage of length of 10–20 µm. They accounted for 54.3% of all cracks present on the fracture surface. No shorter cracks were found. The longest crack was 139 µm long and this value was greater than that recorded for the specimens tested in air.

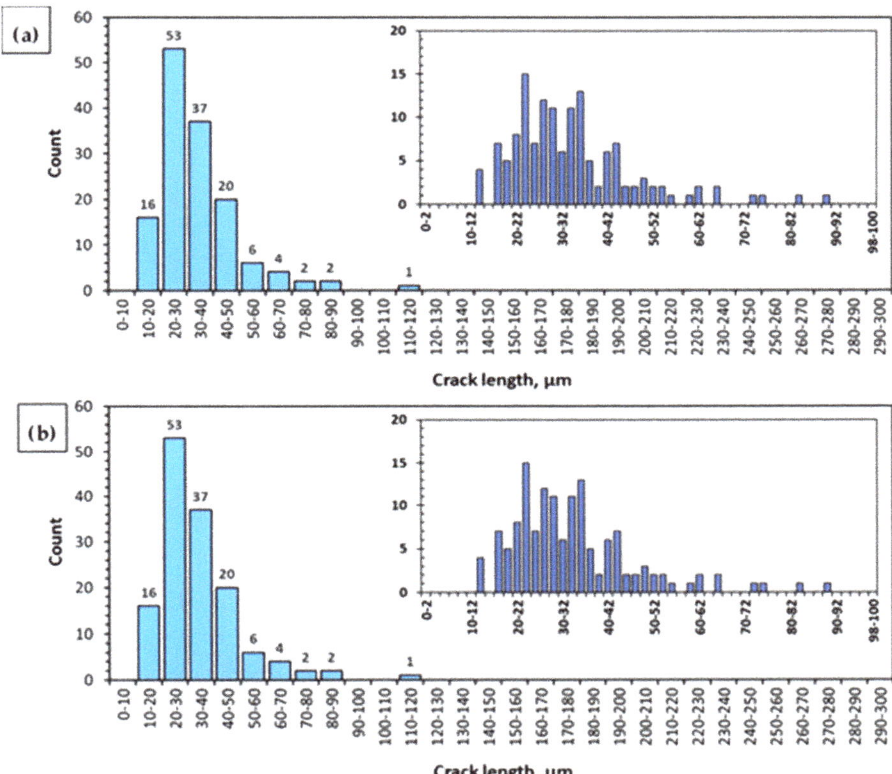

Figure 11. Histogram of the length of cracks occurring in the binary image of cracks present on the fracture surface of the AE44 magnesium alloy after SSRT in the air (**a**) and in solution (**b**).

The distribution of cracks on the fracture surface is shown Figure 11b. These results indicate that the presence of the corrosive environment resulted in a greater number of cracks, however, the cracks were shorter and thus their average length was also smaller. An important feature of the method presented in this work is that it covers the entire fracture area (resulting from the final specimen failure) and all cracks (taking into account the image resolution) of very different sizes, and the results obtained can be used to assess the susceptibility of an alloy to the action of a corrosive environment.

Analysis of the SSC fracture surface in the Na_2SO_4 solution demonstrates the active role of hydrogen in the process of initiating and developing microcracks. This can be demonstrated by the presence of an intercrystalline surface fracture (TG-SSC) on SSRT samples in a corrosive environment. The presence of TG-SSC in magnesium alloys (in particular with rare earth elements RE) after the SSRT test in a hydrogenation environment, is considered according to Atrens et al. [31], for the effect of hydrogen interaction with the alloy microstructure.

Based on the research carried out, it was proposed that the model description of stress corrosion cracking in magnesium alloys AE44 consists of the following stages:

- Stage one: as a result of exposure of the alloy to the environment containing hydrogen, due to the difference in its concentration in both media, it diffuses from the environment to the surface layer of the alloy. Hydrogen diffusion is gradually weakened by the $Mg(OH)_2$ layer formed on the surface of the alloy, which arises as a result of its exposure to the corrosive environment,
- Stage two: as a result of the synergic interaction of the external mechanical loads to which the alloy is subjected, and internal stresses induced by the presence of hydrogen in the material microstructure, cracking occurs in the hydrogen-saturated surface layer of the alloy,
- Stage three: the cracks created in the microstructure of the alloy reveal previously covered with $Mg(OH)_2$ layer, a metal substrate enabling easier and faster diffusion of hydrogen to the inside of the alloy through the surfaces of freshly formed cracks,
- Stage four: the cracks in the material propagate into the material, followed by decohesion of the alloy in the areas of the microstructure located in front of the crack tip, where there is a high concentration of stress caused by external mechanical loads and internal stresses induced by the diffusion of hydrogen. In addition, secondary cracks also develop in these areas.

4. Conclusions

In the summary of mechanical test results, obtained under complex mechanical and corrosive loads (SSR) tests in air and in a 0.1M Na_2SO_4 corrosive environment and of the qualitative and quantitative evaluation of cracks on the fracture surfaces after the SSR tests, the following conclusions were formulated for AE44 alloy:

1. The impact of mechanical loads and the corrosive environment (OCP) under SSC conditions on the properties of AE44 took the form of deterioration in mechanical properties.
2. The influence of hydrogen on the mechanical properties of AE44 was difficult to evaluate as the measured content of hydrogen in AE44 was the sum of hydrogen that penetrated into the alloy from the solution during the SSR test and hydrogen contained in pores that formed in the alloy during the manufacture process.
3. The qualitative analysis of the fracture surfaces of the AE44 specimens after the SSR tests in air and in the corrosive solution under OCP conditions revealed a significantly greater presence of cracks in the case of the specimens tested in the corrosive environment.
4. The quantitative analysis of the cracks (number, length, and size distribution) in the AE44 specimens after the SSR tests indicated a significant difference between the alloy's susceptibility to SCC in a corrosive environment and in air. Such clear differences could not be observed when the adopted criterion of the alloy's susceptibility was the changes in its mechanical properties in SSR tests.
5. The analysis of the results concerning the susceptibility of the AE44 magnesium-based alloy to the action of a corrosive environment under SSC conditions can lead to the conclusion that the results obtained from the quantitative fractography methods are an important complement to the SSR tests (changes in mechanical properties) being the basic criterion for the assessment of an alloy's susceptibility to the combined action of stresses and a corrosive environment, but also that the developed quantitative fractography method is more unambiguous and more sensitive to changes in the corrosive environment and the mechanical load than the change in mechanical properties during SSR tests.

Author Contributions: Conceptualization, M.S.; methodology, M.S.; formal analysis, T.C.; investigation, A.M.; resources, M.S.; writing—original draft preparation, M.S.; writing—review and editing, T.C.; visualization, A.M.; supervision, A.M.

Funding: This work was supported by the National Science Centre in Poland under the research grant No. 2011/03/B/ST8/06387.

Conflicts of Interest: The authors declare no conflict of interest.

References

1. Raja, V.S.; Shoji, T. *Stress Corrosion Cracking: Theory and Practice*; Woodhead Publishing Limited: Cambridge, UK, 2011; pp. 1–89.
2. Song, R.G.; Blawert, C.; Dietzel, W.; Atrens, A. A study on stress corrosion cracking and hydrogen embrittlement of AZ31 magnesium alloy. *Mater. Sci. Eng. A* **2005**, *399*, 308–317. [CrossRef]
3. Winzer, N.; Atrens, A.; Dietzel, W.; Song, G.; Kainer, K.U. Comparison of the linearly increasing stress test and the constant extension rate test in the evaluation of transgranular stress corrosion cracking of magnesium. *Mater. Sci. Eng. A* **2008**, *472*, 97–106. [CrossRef]
4. Chen, J.; Wang, J.; Han, E.; Dong, J.; Ke, W. States and transport of hydrogen in the corrosion process of an AZ91 magnesium alloy in aqueous solution. *Corros. Sci.* **2008**, *50*, 1292–1305. [CrossRef]
5. ASTM G129. In *Standard Practice for Slow Strain Rate Testing to Evaluate the Susceptibility of Metallic Materials to Environmentally Assisted Cracking*; ASTM International: West Conshohocken, PA, USA, 2013.
6. ISO 7539-7. In *Corrosion of metals and alloys—Stress corrosion testing—Part 7: Method for slow strain rate testing*; ISO: Genewa, Switzerland, 1987.
7. Kłyk-Spyra, K.; Sozańska, M. Quantitative fractography of 2205 duplex stainless steel after a sulfide stress cracking test. *Mater. Charact.* **2006**, *56*, 384–388.
8. Michalska, J.; Sozańska, M.; Hetmańczyk, M. Application of quantitative fractography in the assessment of hydrogen damage of duplex stainless steel. *Mater. Charact.* **2009**, *60*, 1100–1106. [CrossRef]
9. Sozańska, M.; Sojka, J.; Beťáková, P.; Dagbert, C.; Hyspecká, L.; Galland, J.; Tvrdý, M. Examination of hydrogen interaction in carbon steel by means of quantitative microstructure and fracture description. *Mater. Charact.* **2001**, *46*, 239–243.
10. Sozańska, M.; Sojka, J. Quantitative description of the "fish eyes" type hydrogen embrittlement in steels for power industry. *Inż. Mater.* **2008**, *4*, 247–249.
11. Wang, J.; Chen, J.; Han, E.; Ke, W. Investigation of Stress Corrosion Cracking Behaviors of an AZ91 Magnesium Alloy in 0.1kmol/m3 Na_2SO4 Solution Using Slow Strain Rate Test. *Mater. Trans.* **2008**, *5*, 1052–1056. [CrossRef]
12. Chen, J.; Wang, J.; Han, E.; Ke, W. Electrochemical corrosion and mechanical behaviors of the charged magnesium. *Mater. Sci. Eng. A* **2008**, *494*, 257–262. [CrossRef]
13. Zhou, L.F.; Liu, Z.Y.; Wu, W.; Li, X.G.; Du, C.W.; Jiang, B. Stress corrosion cracking behavior of ZK60 magnesium alloy under different conditions. *Int. J. Hydrogen Energy* **2017**, *42*, 26162–26174. [CrossRef]
14. Kannan, M.B.; Dietzel, W.; Raman, R.K.S.; Lyon, P. Hydrogen-induced-cracking in magnesium alloy under cathodic polarization. *Scripta Mater.* **2007**, *57*, 579–581. [CrossRef]
15. Narayanan, T.S.; Park, I.S.; Lee, M.H. Volume 1: Biological interactions, mechanical properties and testing. In *Surface Modification of Magnesium and Its Alloys for Biomedical Applications*; Elsevier: Amsterdam, The Netherlands, 2015; pp. 1–25.
16. Song, G.L. *Corrosion of Magnesium Alloys*; Woodhead Publishing: Cambridge, UK, 2011; pp. 1–656.
17. Zieliński, A. *Niszczenie Wodorowe Metali Nieżelaznych i Ich Stopów*; Gdańskie Towarzystwo Naukowe: Gdańsk, Poland, 1999; pp. 1–35.
18. Antolovich, S.; Gokhale, A.; Bathias, C. Applications of Quantitative Fractography and Computed Tomography to Fracture Processes in Materials. In *Quantitative Methods in Fractography*; Strauss, B.M., Putatunda, S.K., Eds.; ASTM International: West Conshohocken, PA, USA, 1990; pp. 3–25.
19. Undervood, E.E. *Quantitative Stereology*; Addison-Wesley: Reading, MA, USA, 1970; pp. 1–42.
20. *Met-Ilo®*; Version 15.03; Computer program; Instytut Inżynierii Materiałowej: Katowice, Poland, 2015.
21. Rzychoń, T.; Kiełbus, A.; Dercz, G. Structural and quantitative analysis of die cast AE44 magnesium alloy, JAMME. *J. Achiev. Mater. Manuf. Eng.* **2007**, *22*, 43–46.
22. Revie, R.W.; Uhlig, H.H. *Corrosion and Corrosion Control*; John Wiley & Sons, Inc: Hoboken, NJ, USA, 2008; pp. 399–406.
23. Rzychoń, T.; Kiełbus, A. Microstructure and tensile properties of sand cast and die cast AE44 magnesium alloy. *Arch. Metall. Mater.* **2008**, *58*, 901–907.
24. Zeng, R.C.; Yin, Z.Z.; Chen, X.B.; Xu, D.K. Corrosion Types of Magnesium Alloys. In *Magnesium Alloys: Selected Issue*; Tański, T., Ed.; IntechOpen: London, UK, 2018; pp. 29–52.

25. Choudhary, L.; Raman, R.K.S.; Hofstetter, J.; Uggowitzer, P.J. In-vitro characterization of stress corrosion cracking of aluminium-free magnesium alloys for temporary bio-implant applications. *Mater. Sci. Eng. C* **2014**, *42*, 629–636. [CrossRef] [PubMed]
26. Wang, S.D.; Xu, D.K.; Wang, B.J.; Sheng, L.Y.; Han, E.H.; Dong, C. Effect of solution treatment on stress corrosion cracking behavior of an as-forged Mg-Zn-Y-Zr alloy. *Sci. Rep.* **2016**, *6*, 1–12.
27. Choudhary, L.; Raman, R.K.S. Magnesium alloys as body implants: Fracture mechanism under dynamic and static loadings in a physiological environment. *Acta Biomater.* **2012**, *8*, 916–923. [CrossRef]
28. Song, G.L. Corrosion behavior and prevention strategies for magnesium (Mg) alloys. *Woodhead Publ. Ser. Met. Surf. Eng.* **2013**, *1*, 3–37.
29. Lee, S.G.; Patel, G.R.; Gokhale, A.M.; Sreeranganathan, A.; Horstemeyer, M.F. Quantitative fractographic analysis of variability in the tensile ductility of high-pressure die-cast AE44 Mg-alloy. *Mater. Sci. Eng. A* **2006**, *427*, 255–262. [CrossRef]
30. Volkova, E.F.; Morozova, G.I. Role of hydrogen in deformed magnesium alloys of the Mg-Zn-Zr-REM system. *Met. Sci. Heat Treat.* **2008**, *50*, 105–109. [CrossRef]
31. Atrens, A.; Dietzel, W.; Bala, S.P.; Bobby Kannan, M. Stress corrosion cracking (SSC) of magnesium alloy. In *Stress Corrosion Cracking: Theory and Practice*; Raja, V.S., Shoji, T., Eds.; Woodhead Publishing Ltd: Cambridge, UK, 2011; Chapter 9; pp. 341–380.
32. Zeng, R.C.; Dietzel, W.; Zettler, R.; Gan, W.M.; Sun, X.X. Microstructural evolution and delayed hydride cracking of FSW-AZ31 magnesium alloy during SSRT. *Trans. Nonferr. Mrtal. Soc.* **2014**, *24*, 3060–3069. [CrossRef]
33. Wrigth, K.; Karlsson, B. Topograpic Quantification of Nonplanar Localized Surfaces. In Proceedings of the Third European Symposium for Stereology, Ljubljana, Yugoslavia, 22–27 June 1981; pp. 247–253.
34. Coster, M.; Chermant, J.L. *Précis d'Analyse d'Images*; Editions du CNRS: Paris, France, 1985; pp. 13–60.
35. Undervood, E.E.; Banerji, K. Quantitative Fractography. In *ASM Handbook Volume 12: Fractography*, 9th ed.; ASM International: Materials Park, OH, USA, 1987; pp. 193–210.
36. Sozańska, M.; Mościcki, A.; Chmiela, B. Investigation of stress corrosion cracking in magnesium alloys by quantitative fractography methods. *Arch. Metall. Mater.* **2017**, *62*, 557–562. [CrossRef]

© 2019 by the authors. Licensee MDPI, Basel, Switzerland. This article is an open access article distributed under the terms and conditions of the Creative Commons Attribution (CC BY) license (http://creativecommons.org/licenses/by/4.0/).

Article

Effect of Load on the Corrosion Behavior of Friction Stir Welded AA 7075-T6 Aluminum Alloy

Marina Cabrini [1,2,3,*], Sara Bocchi [4], Gianluca D'Urso [4], Claudio Giardini [4], Sergio Lorenzi [1,2,3], Cristian Testa [1,2,3] and Tommaso Pastore [1,2,3]

1 Department of Engineering and Applied Sciences, University of Bergamo, 24044 Dalmine (BG), Italy; sergio.lorenzi@unibg.it (S.L.); cristian.testa@unibg.it (C.T.); tommaso.pastore@unibg.it (T.P.)
2 National Interuniversity Consortium of Materials Science and Technology (INSTM) Research unit of Bergamo, 24044 Dalmine (BG), Italy
3 Center for Colloid and Surface Science (CSGI) Research unit of Bergamo, 24044 Dalmine (BG), Italy
4 Department of Management, Information and Production Engineering, University of Bergamo, 24044 Dalmine (BG), Italy; sara.bocchi@unibg.it (S.B.); gianluca.d-urso@unibg.it (G.D.); claudio.giardini@unibg.it (C.G.)
* Correspondence: marina.cabrini@unibg.it

Received: 21 April 2020; Accepted: 4 June 2020; Published: 7 June 2020

Abstract: The paper focuses on the corrosion behavior of aluminum joints made by friction stir welding as a function of loading conditions. A four-points bend-beam test, constant loading test, and slow strain-rate test were carried out on AA 7075-T6 alloy in aerated NaCl 35g/L solution at room temperature monitoring the free corrosion potential. The penetration depth of the intergranular attack was deeper after the four-point bent-beam tests compared to all the other testing techniques. Preferential dissolution along the grain boundaries was found in the heat-affected zone and the attack follows the elongated grains structure along the rolling direction. However, no stress-corrosion cracking phenomena were detected. No relevant stress corrosion cracking (SCC) crack embryos propagation was noticed under uniaxial tensile tests—both constant loading and slow strain-rate tests—manly due to the high dissolution rate occurring at the crack tip which promoted premature shear ruptures.

Keywords: corrosion; stress-corrosion cracking; friction stir welding (FSW); AA7075

---

1. Introduction

Friction stir welding (FSW) is a recent solid-state joining technique developed by the Welding Institute [1,2]. This technique has promising applications on difficult-to-weld alloys, i.e., age-hardening aluminum alloys [3,4]. FSW technique has been successfully applied to Al-Cu 2XXX series [5,6], Al-Si-Mg 6XXX [7,8], and Al-Zn-Mg 7XXX series alloys [9,10].

The 7XXX series aluminum alloys are widely used in aircraft and automotive industry due to the high tensile strength on weight ratio. The corrosion resistance of the alloy is significantly affected by the presence of Zn and Mg, which could induce several corrosion phenomena, such as localized corrosion such as pitting [11–13], intergranular [14,15], and stress-corrosion cracking [16,17].

The corrosion behavior of such alloys is strictly related to the temper. Several authors confirm that the 7XXX alloy exhibits SCC susceptibility in peak-aged conditions (T6 temper). However, over-aging (T7) or retrogression and re-aging (RRA) treatments are not critical compared to SCC [18,19]. Even if FSW is a solid-state joining technique, it is well known that it implies deep modifications to the alloy's microstructure and precipitates' distribution in the welding zones [20–23]. Therefore, this process can significantly affect the corrosion resistance and SCC [24,25].

The aim of this work is the study of the SCC susceptibility of AA 7075 T6 FSW butt joints by means of constant deformation tests with a four-point bending device (4PBD), uniaxial constant loading (CL) and slow strain-rate (SSR) tests in NaCl 35g/L aerated solutions at room temperature. During the tests, the open circuit potential (OCP) was monitored. The corrosion morphology was observed after exposure by means of the optical microscope and the scanning electron microscope (SEM) equipped with an energy dispersive X-ray spectroscopy device (EDX).

2. Materials and Methods

2.1. Welds

Sheets with a thickness equal to 4 mm were considered. Table 1 shows the chemical composition of the AA 7075 T6 alloy. Sheets of 200 mm × 80 mm were welded by using a custom-made tool designed with a smooth plane shoulder (16 mm diameter) coupled with a frustum cone-shaped pin (maximum and minimum pin diameters were 6 mm and 4 mm, pin height was 3.8 mm). The rotational speed of the tool (S) was 1500 rpm and feed rate (F) was 10 mm/min. The set-up of the welding procedure was reported in a previous work [26]. The mechanical properties of the base materials and the welded joints are reported in Table 2. Good reproducibility was observed in terms of tensile strength of the FSW joints. Owing to non-uniform strain distribution, it was not possible to calculate the yield strength of the welded specimens.

Table 1. Chemical composition (% weight) of the alloy 7075-T6.

Al	Si	Fe	Cu	Mn	Mg	Zn	Ti	Cr
bulk	0.05	0.10	1.53	0.008	2.54%	5.72%	0.04	0.20

Table 2. Mechanical properties of the base materials and the welded joint.

Base Material			FWS Joint		
YS (MPa)	**UTS (MPa)**	**Max Strain (%)**	**YS (MPa)**	**UTS (MPa)**	**Max Strain (%)**
512	576	13.5	-	329	11

2.2. Metallographic Analysis

The microstructure was revealed through Keller's etchant after grinding by means of emery papers up to 4000 grit, then polishing with 0.1 μm alumina aqueous suspension and finally observed with Nikon (Eclipse MA100N, Tokyo, Japan) optical microscope and SEM (Zeiss EVO 50, Zeiss, Oberkochen, Germany) Oxford x-act sdd (silicon drift detector, nitrogen free) EDX equipment type model (Oxford Instruments, High Wycombe, UK).

2.3. OCP Monitoring

OCP was measured on the three different zones of FSW butt welds, i.e., the nugget, the thermomechanical affected zone/heat-affected zone (TMAZ/HAZ—it was not possible to separate the two zones) and the base metal. The specimens were cold-mounted in epoxy resin with an embedded wire to allow electrochemical measurements. A 20 mm × 20 mm area was grounded with emery paper and then polished with colloidal alumina up to 0.1 μm. After polishing, the specimens were stored at room temperature (23 °C) with relative humidity in the range of 25–30% for 48 h before the immersion, to allow the formation of the natural protective film. The specimens were dipped in different testing cells filled with aerated 35 g/L NaCl solution (Carlo Erba RPA reagent, Cornaredo, Italy). The OCP was measured by using a saturated calomel electrode (SCE E = +0.240 V vs Standard Hydrogen Electrode, Amel Instrument, Milano, Italy) positioned close to the surface of the specimens to reduce the ohmic drop in the electrolyte during electrochemical tests.

2.4. Four-Point Bending Tests

Four-point bending (4PBB) tests have been performed according to ASTM G39 on 4 mm × 20 mm × 160 mm prismatic specimens obtained by FSW joints (Figure 1). The specimens were polished by emery papers up to 1000 grit and degreased in acetone in ultrasonic bath for 5 min. After polishing, the specimens were left in still air for 48 h to allow the formation of the protective film.

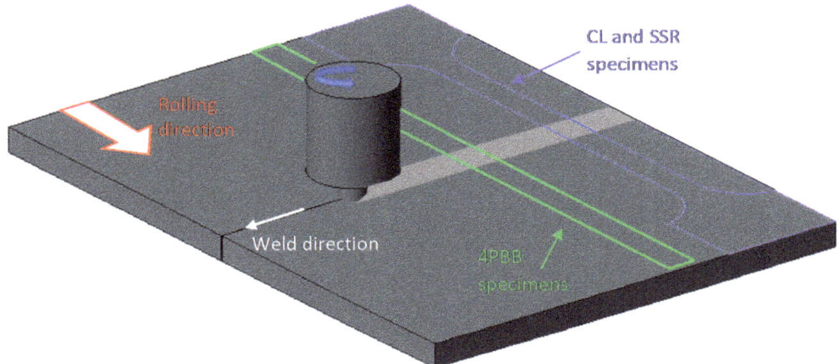

Figure 1. Schematic representation of specimens obtained by means of friction stir welding (FSW) welded sheet (draw not in scale).

The specimens were positioned inside the bending device to attain a uniform tensile strain distribution over the welded surface. Four glass cylinders were used to avoid galvanic coupling between the stainless-steel specimen holder and the aluminum alloy. Four specimens were loaded up to 80% of the tensile strength of FSW joints derived by tensile tests (refer to Table 2). The other four specimens were dipped without loading (unloaded condition). Tests have been performed on four specimens for each condition (loaded and unloaded). All the specimens were exposed in a cell filled with about 30 L water, and 35 g/L sodium chloride (Carlo Erba RPS analytical grade reagents, Cornaredo, Italy) for 1500 h. To avoid the contamination of the testing solution by corrosion products of aluminum during long-term exposures, the test solution was refreshed after 150 h and 500 h. The amount of water that was substituted was equal to 10 liters. During the procedure, the specimens were always left well below the waterline. The OCP was measured during the exposure. More details about the experimental set-up are reported in a previous work [27].

At the end of the long-term exposure tests, the specimens were unloaded, washed in distilled water, and then rinsed with acetone in an ultrasonic bath. The surfaces were then observed by means of optical microscopy up to 600 times magnification and SEM equipped with EDX.

2.5. Constant Load Tests

CL tests were carried out on a specimen of 4 mm × 8 mm × 82 mm gage length, with the weld positioned at the center of each specimen (Figure 1). The size of the specimens is reported in Figure 2. The specimen was settled in a double compartment cell (Figure 3a). All the compartments were maintained at room temperature filled with aerated NaCl 35 g/L solution (Carlo Erba RPS analytical grade reagents, Cornaredo, Italy). Water recirculation in the cell was granted during the tests. The OCP of the specimens was continuously monitored using a SCE positioned close to the metal surface by means of a Lugging capillary.

The specimens in the cell were positioned into the CL machine (no loading), then they were dipped in the test solution. After 120 h of OCP monitoring, the specimens were loaded at 190 MPa, corresponding to 54% of the ultimate tensile strength (UTS) of the weld. This value is slightly lower than the elastic limit of the weld obtained by means of tensile tests. The specimens were left at this value for 350 h and the OCP was measured during all the period. Then, the applied load increased up to 279 MPa, corresponding to 80% of the UTS of the weld (homogeneous plastic deformation field). The OCP was measured until 720 h of exposure (one month).

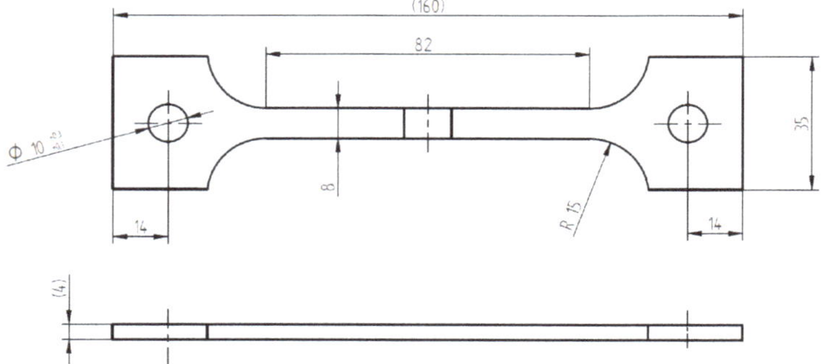

Figure 2. CL and SSR specimen: the weld is in the center of the gouge length, the thermomechanical affected zone/heat-affected zone (TMAZ/HAZ) extension is 40 mm for each side.

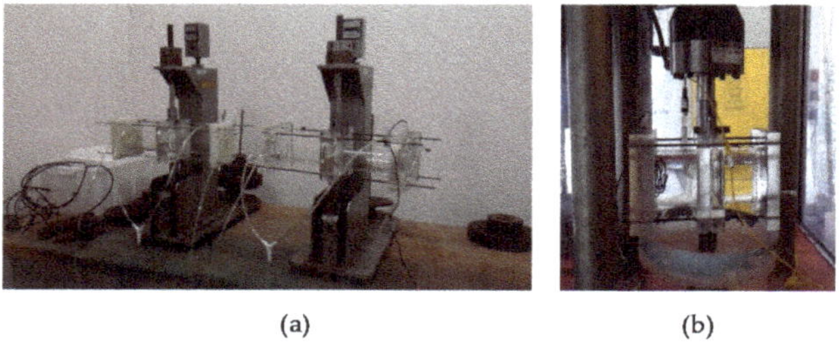

(a) (b)

Figure 3. CL test device (**a**) and SSR test device (**b**).

At the end of the test, the specimens were unloaded and observed under the optical microscope and under the SEM to analyze the corrosion morphology. Metallographic cross-section was also taken along the longitudinal direction of the specimen.

2.6. Slow Strain-Rate Tests

The SSR tests were carried out by using specimens of the same dimensions as the CL tests. The set-up of the tests is shown in Figure 3b. The tests were carried out on a 30 kN testing machine at displacement rates varying from 5×10^{-7} to 5×10^{-3} mm/s. A displacement rate of 8.2×10^{-4} mm/s was used in order to have a strain rate of 10^{-6} s^{-1} as average value on all the gage length. During the tests, the load and OCP were measured continuously. At the end of the tests the specimens were washed in distilled water and rinsed in acetone to permit the observation of the fracture surface under the SEM. Then the specimens were longitudinally sectioned for the metallographic observation.

3. Results

3.1. Metallographic Analysis

The microstructure of the AA 7075 T6 alloy has already been presented in [26]. Coarse precipitates along the grain boundaries (Figure 4) and micrometric precipitates within the grains of α-Al, oriented in the rolling direction [28], can be well demonstrated. The strengthening $MgZn_2$ nanoprecipitates are not visible under the optic or SEM microscope, while macro-precipitates are copper and iron-rich phases as Al_2CuMg and Al_7Cu_2Fe or zinc and magnesium rich phases, i.e., undissolved $MgZn_2$ during alloy solution annealing treatment, or MgSi.

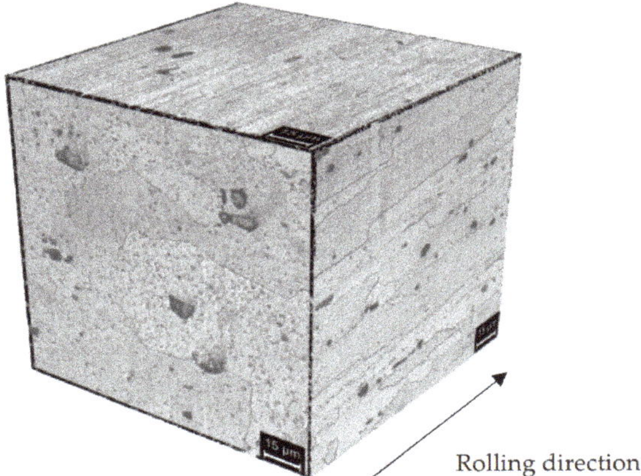

Figure 4. Optical image (Keller attack) of the alloy AA 7075 T6 (base material). Elongated grains along the rolling direction and the macro-precipitates are visible.

In the nugget, re-crystalized fine equiaxed grain structure can be noticed. In addition, $MgZn_2$ strengthening precipitates are dissolved (T6 artificial aging temper) into the supersaturated solid solution and reprecipitated after cooling (Figure 5). Venugopol et al. affirmed that the temperature obtained during the process is below the melting temperature of the alloy, but above the solutioning temperature, as demonstrated by the re-crystallization of the weld nugget and the redistribution of the precipitates [21]. For the same authors, the absence of fine precipitates in the weld nugget indicates that the cooling rates are such that larger precipitates could nucleate and grow but not the finer ones. In the TMAZ, the precipitates are quite random, and the coarsening of finer precipitates observed in base material can occur during welding [21].

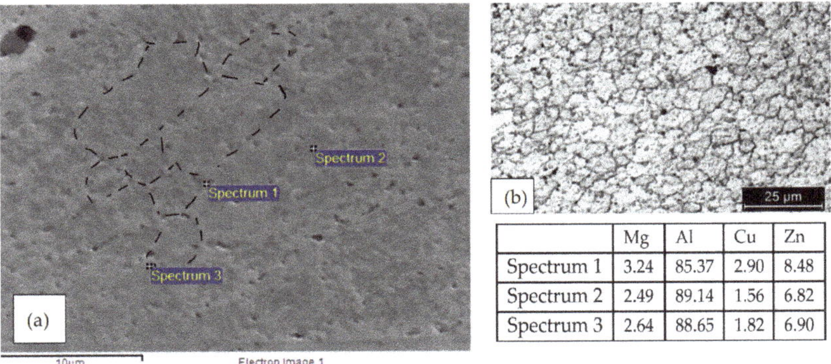

	Mg	Al	Cu	Zn
Spectrum 1	3.24	85.37	2.90	8.48
Spectrum 2	2.49	89.14	1.56	6.82
Spectrum 3	2.64	88.65	1.82	6.90

Figure 5. (**a**) SEM image of the nugget of the FWS 7075 alloy (no metallographic attack) the rounded recrystallized grains are visible owing to the precipitates of $MgZn_2$ on their grain borders and EDX spectra in which is visible an enrichment of zinc and copper in the correspondence of the precipitates. In the image the grain border is demonstrated; (**b**) optical image of the nugget with metallographic attack (Keller's reagent) that better demonstrated the rounded recrystallized grains.

The size of the precipitates at the grain boundaries of the HAZ (Figure 6) is unfortunately too small to permit their observation under the SEM, but many transmission electron microscope (TEM) observations were reported in the literature and they demonstrated that these smaller precipitates play a fundamental role in the intergranular corrosion and the SCC of these alloy series [18].

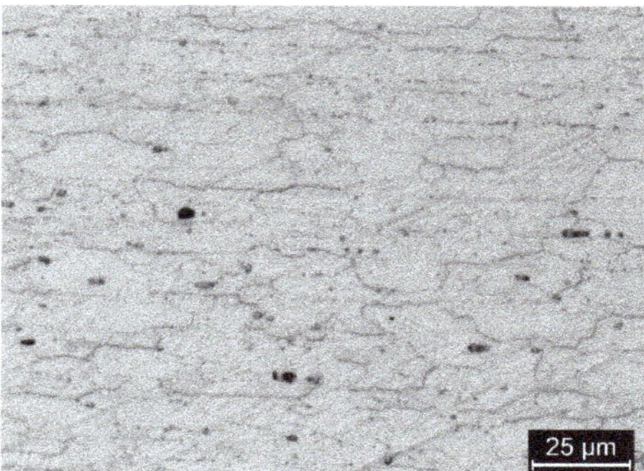

Figure 6. Optical image (Keller attack) of the HAZ of the FWS AA 7075 alloy. The grains appear elongated along the rolling direction; presence of macro-precipitates is evident.

3.2. 4PBB Specimens

The OCP measurements (average values of four specimens) did not show any evident differences between 4PBB loaded and unloaded specimens (Figure 7). The corrosion potentials generally were in the range of −870–−840 mV vs SCE; the corrosion potential showed a sharp decrease during the solution refresh owing to the partial removal of non-adherent scale of corrosion products that exposes the corroded surface on the fresh solution; after this, the corrosion potential increases to the same potential values detected prior to the partial substitution of the testing solution. This effect was more marked for specimens without loading compared to loaded specimens. The corrosion potential of the loaded specimens decreases after 1300 h, then it increases again. Compared to the values obtained by the analysis of the three different zones (nugget, TMAZ/HAZ and base materials—Figure 8) similar results can be outlined in both the TMAZ/HAZ and the nugget. The base metal showed slightly nobler potential values.

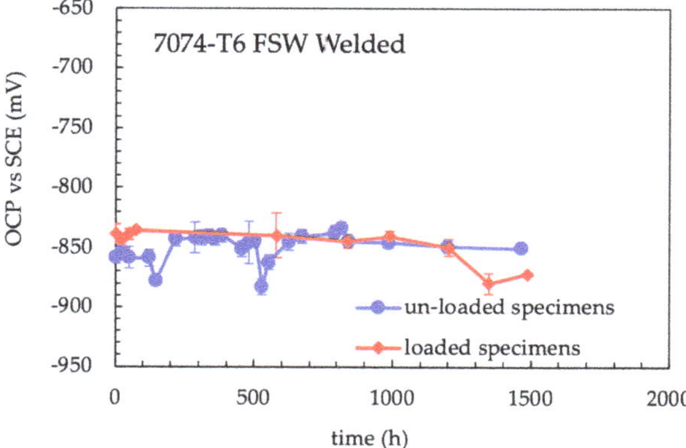

Figure 7. Open circuit potential (OCP) vs time for loaded and unloaded 4PBB specimens. The decreasing of the OCP of the specimens in the correspondence of the partial refresh of the test solution that breaks the aluminum oxide scale and exposes fresh metal surface to the aggressive solution.

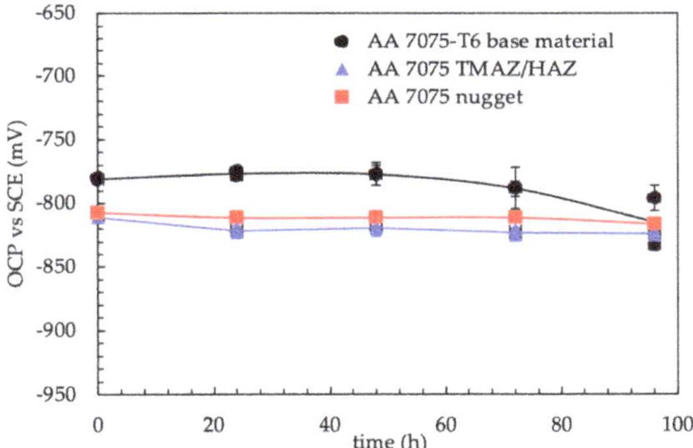

Figure 8. OCP vs time of the different zones (base materials, TMAZ and nugget) of the alloy AA7075 T6.

Figure 9a shows the corrosion morphology of the specimens at the end of the 4PBB tests. No differences in the corrosion morphology were observed between the loaded and unloaded specimens. The HAZ is preferentially corroded. The attack propagates along the grain boundaries, following the rolling direction (Figure 9b). The EDX analysis (Figure 9c) demonstrated the depletion of zinc and copper in the corroded zone (Spectrum 1) with respect to the un-corroded zones (Spectrum 2).

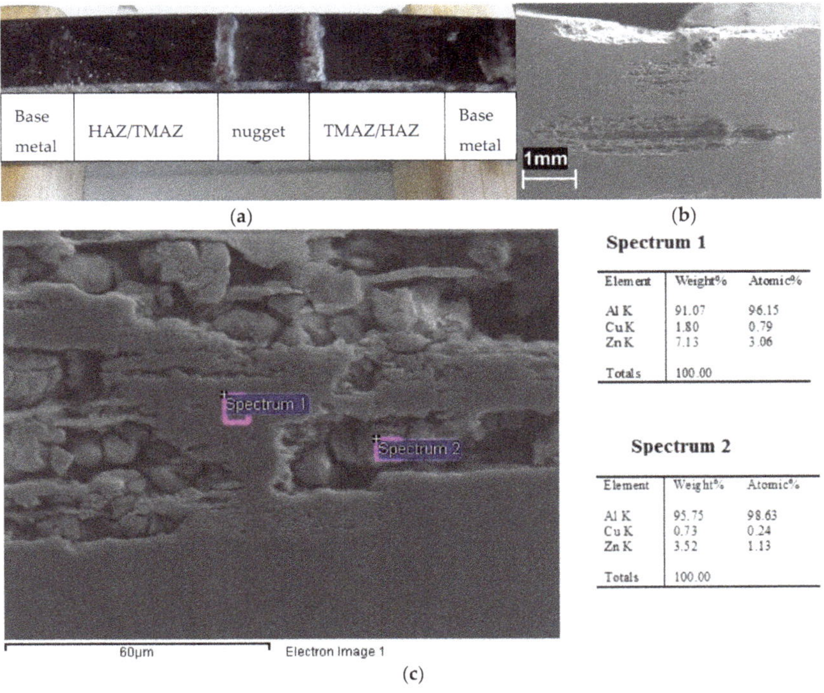

Spectrum 1

Element	Weight%	Atomic%
Al K	91.07	96.15
Cu K	1.80	0.79
Zn K	7.13	3.06
Totals	100.00	

Spectrum 2

Element	Weight%	Atomic%
Al K	95.75	98.63
Cu K	0.73	0.24
Zn K	3.52	1.13
Totals	100.00	

Figure 9. (**a**) Image of a 4PBB specimens after the test: an intense localized attack in the correspondence of the TMAZ and (**b**) SEM image of the metallographic longitudinal section of the loaded specimen in which is visible the deviation of the attack along the rolling direction, (**c**) particular of the attack and EDX analysis.

3.3. Constant Loading (CL) Tests

Figure 10 shows the effect of the applied load on the OCP of the CL specimen. The OCP of the unloaded specimens rapidly stabilized between −850 and −800 mV vs SCE. The fluctuations are mainly ascribable to the removal of corrosion products of aluminum formed at very early exposures, due to the recirculation of the testing solution. When the specimen is loaded in elastic field, a decrease in the OCP was noticed but the potential values stayed in the same range measured in unloaded condition. Conversely, at strain level exceeding the yield stress—i.e., in the plastic field—100 mV decrease in the OCP occurred due both to the rupture of the thick corrosion product scale of aluminum and the plastic straining exposes the very active metal to the aggressive environment. The OCP came back to the initial value after about 24 h. This value remained almost constant, until the end of the tests. The specimen did not break during the tests but showed a corrosion attack mainly localized at the HAZ (Figure 11). Differently from the 4PBB specimens, the attack appears wider, and positioned perpendicularly to the loading direction coinciding with the rolling direction (Figures 11c and 12a). Some small ramifications along the grain boundaries were observed (Figure 12b).

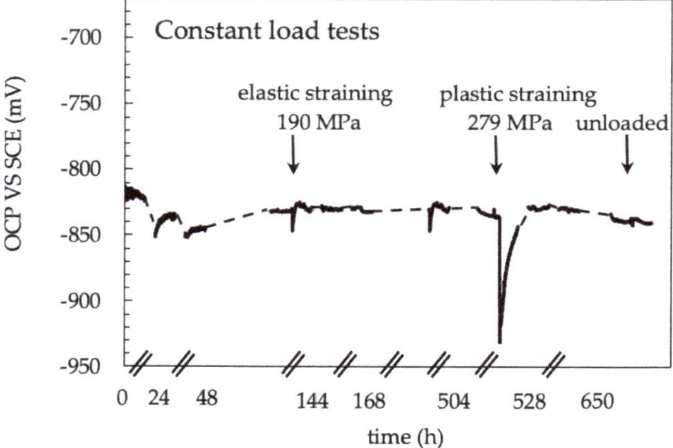

Figure 10. Open circuit potential measurements of the specimen loaded at different values during CL test. The decrease in the OCP mainly occurs as loading increase due to aluminum corrosion products cracking.

3.4. Slow Strain-Rate (SSR) Tests

Figure 13 reports the load and OCP measurements performed during the SSR tests of the AA 7075-T6 base materials and the FSW joints. The tests were twice repeated, but only one curve for each condition is reported in the graph for simplicity. The tensile curves of the base material and the weld obtained in air or in NaCl 0.6 M are practically overlapped. The time to failure of the base material is close to the mean value with a very small deviation of 7%. 25% deviation was observed for the welded specimens. No stress-corrosion crack occurrence was noticed for the base metal and the time to failure in 0.6 M NaCl solution is even longer compared to the value measured at air. The role of active corrosion in NaCl solution can be hypothesized to enhance dislocation mobility and thus the plastic deformation, as reported by Jones et al. [28,29]. The authors named this phenomenon anodic attenuation of strain hardening.

Figure 11. Aspect of the specimen at the end of the constant load test. (**a**) Macro image: the dark zone in the central part exposed to the NaCl 0.6 M solution; (**b**) close-up of the red circle with the selective attack of the TMAZ; (**c**) metallographic section of localized attack.

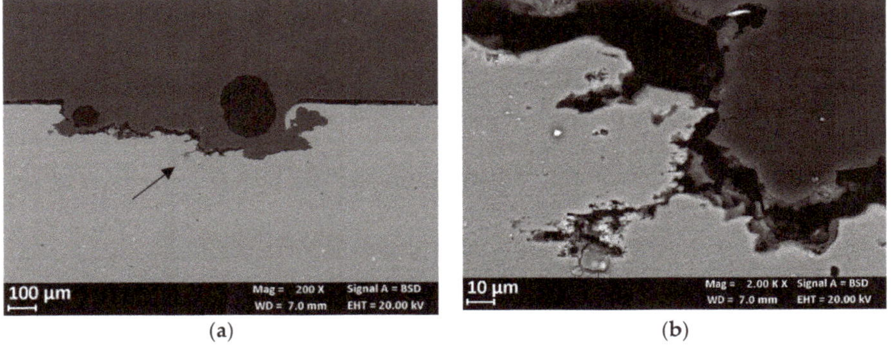

Figure 12. (**a**) Metallographic section of the enlarged localized attack in the TMAZ zone of the CL specimen. Close-up of the arrow in (**b**) in which is evident the change the morphology of attack that allow the rolling direction.

Figure 14 compares the fracture surface of the SSR specimens after the test in air and in 0.6 M NaCl solution. The fracture surface at air (Figure 14a) showed a shearing failure, typical of the prismatic specimens, while the fracture surface of the specimen after the SSR test in 0.6 M NaCl solution exhibits an initial flat area (Figure 14b) due to the presence of microdefects able to trigger localized corrosion initiation. These microcracks are mainly at macro-precipitates (black arrows in Figure 15a) and have depth less than 100 μm. In the correspondence of these microcracks, the fracture surface is heavily corroded (Figure 15b) and it is possible to observe the presence of several micro-precipitates. The flat zone of the fracture surface (Figure 16b) is mixed brittle/ductile as some small dimples and quasi cleavage areas were observed, similarly to the fracture surface at air (Figure 16a). The presence of very small zones with typical SCC morphology, indicated with the arrows in Figure 16b [29] was also confirmed; the final shearing fracture of these specimens was similar to that in air.

No macroscopic SCC phenomena were observed probably due to the strain rate (10^{-6} s^{-1}) value adopted and very aggressive NaCl solution. Results presented in other works [30] confirmed the absence of stress corrosion on this alloy during the SSR tests at the OCP and strain rate of 10^{-6} s^{-1}. The OCP remains constant in the elastic field, but it increases with the plastic deformation, probably due to the enrichment in iron and copper due to the dissolution of the surrounding aluminum matrix. This observation is supported by back-scattered electron (BSE) image and EDX spectrum (Figure 17).

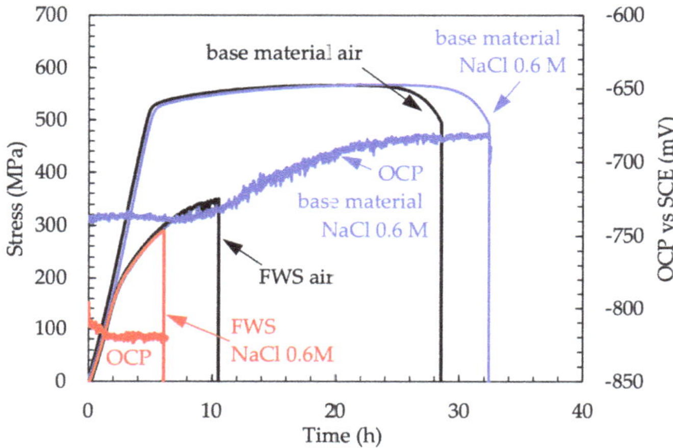

Figure 13. Stress vs time and OCP of the specimens during the slow strain-rate tests.

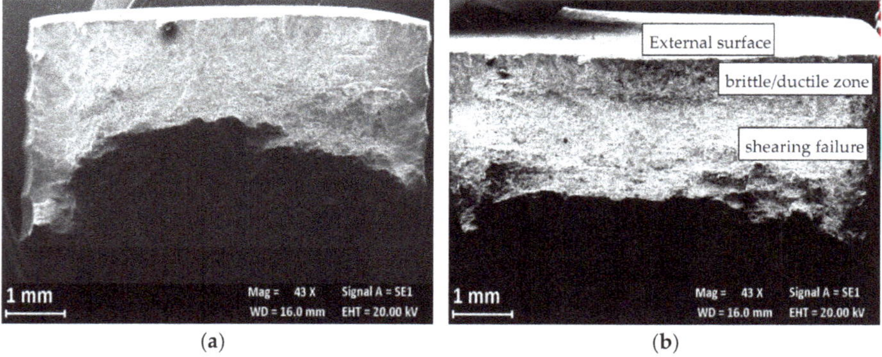

Figure 14. SEM image of the fracture surface of the AA7075 T6 specimen after the SSR test (**a**) macro at air, (**b**) macro in 0.6 M NaCl solution.

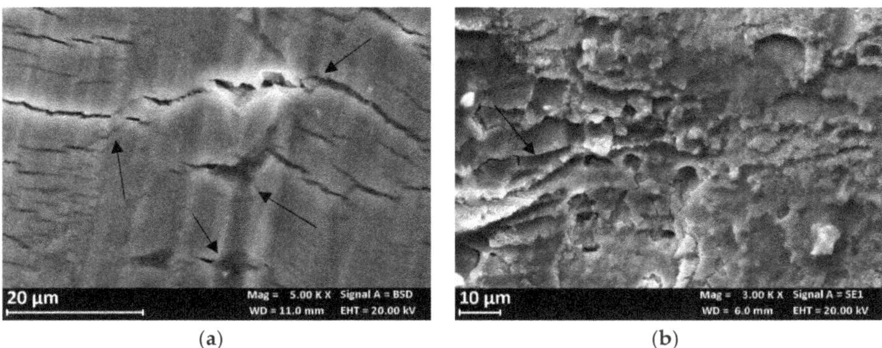

(a) (b)

Figure 15. SEM image of the fracture surface of the AA7075 T6 specimen after the SSR test (**a**) specimen surface with corrosion products of aluminum and several microcracks at macro-precipitates (black arrows); (**b**) particular of fracture initiation zone after SSR test in 0.6 M NaCl solution.

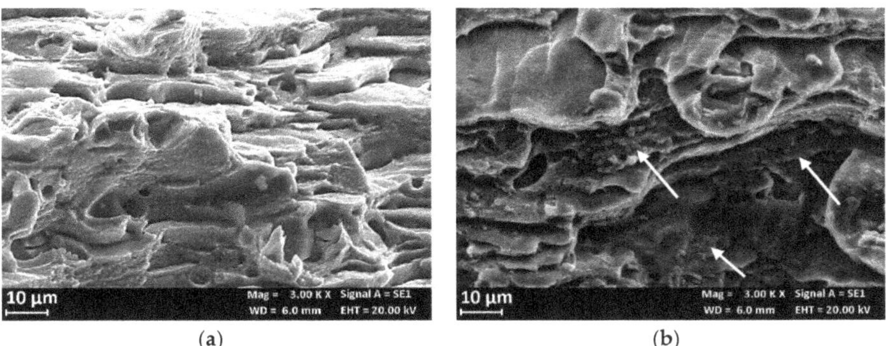

(a) (b)

Figure 16. SEM image of the fracture surface of the AA7075 T6 specimen after the SSR test: (**a**) particular of the fracture surface at air and (**b**) in 0.6M NaCl solution (the arrows indicate small areas with typical SCC morphology).

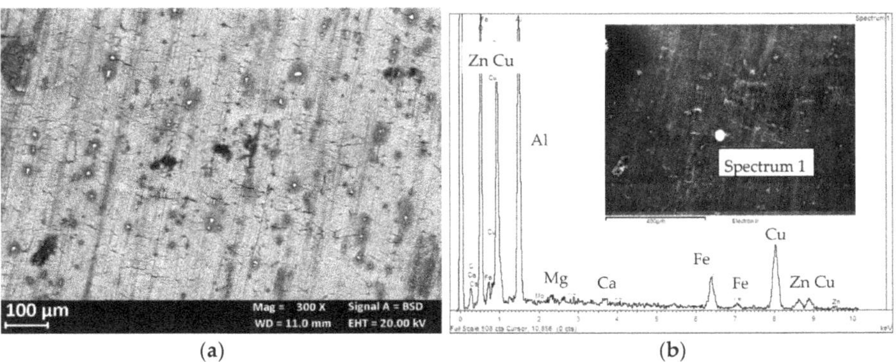

(a) (b)

Figure 17. (**a**) Backscatter electron detector (BSD) image of the external surface of a base material specimen after the SSR test in NaCl 0.6 M, the bright zone indicates precipitates with higher atomic weight than the aluminum matrix; (**b**) EDX spectrum of one of these particles.

The curves of the FSW joints exhibit time-to-failure values lower than the base material. This can be mainly ascribable to the decrease in the tensile properties at the joint TMAZ/HAZ zone, as observed also elsewhere [26,31–33]. Consequently, the hardness and then the UTS of the alloy decrease due to microstructural modifications [34,35].

Figure 18a reports the hardness profile of the longitudinal section of the weld. The hardness sharply decreases at the TMAZ/HAZ and, therefore, the plastic deformation mainly occurs at this point, thus decreasing the total elongation at break and the time of failure. Despite the high strain rate due to plastic strain localization at TMAZ/HAZ, slight decrease in the time to failure in 0.6 M NaCl solution was noticed compared to air—i.e., 10 ± 0.5 h—thus denoting SCC occurrence. Bala Srinivasana et al. reported about SCC phenomena in the HAZ of the AA7075 of mixed FWS joints of AA7075 and AA6056, when the specimens were tested at strain rate 10^{-7} s^{-1} in NaCl solutions, but only localized corrosion in the tests carried out at 10^{-6} s^{-1} was detected [36]. The SEM observation of the surfaces demonstrate a flat zone at fracture initiation with maximum depth of 1 mm (Figure 18b,c). The presence of non-coherent corrosion products of aluminum and several precipitates was also noticed (Figure 19a). EDX analysis detected the presence of zinc, aluminum, and copper, as well as magnesium depletion (Figure 20). The final fracture is by shearing (Figure 18b) and its morphology is characterized by elongated dimples and a few brittle areas Figure 19b), similarly to the base material (Figure 16).

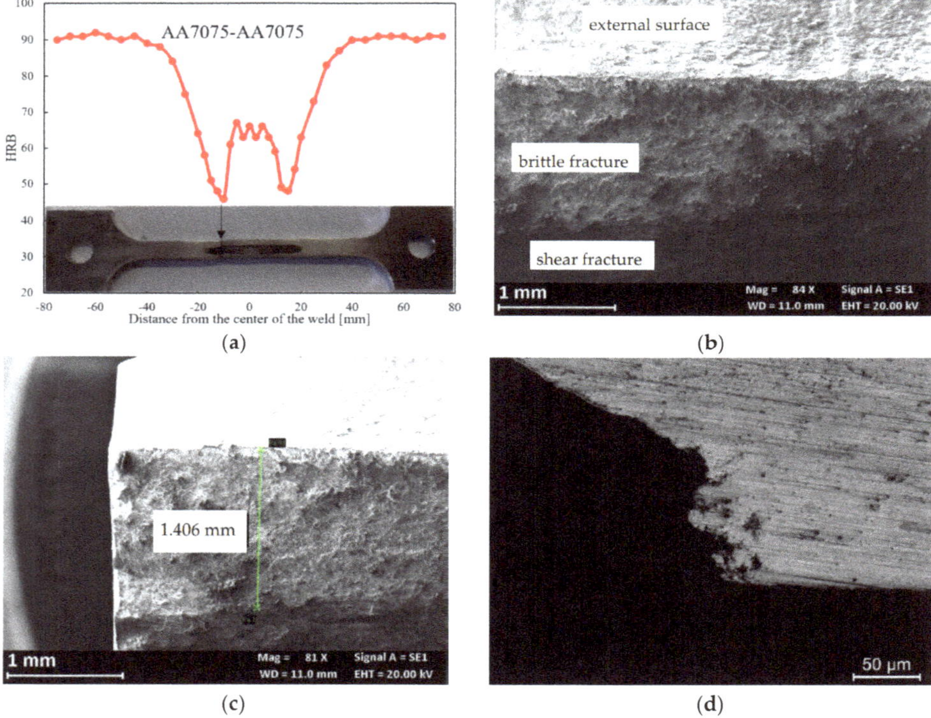

(a) (b) (c) (d)

Figure 18. (**a**) Hardness profile along the FSW butt joint; (**b**) SEM image of the fracture surface, a flat zone of about 1 mm of growth (close-up in (**c**) is highlighted in (**d**)) metallographic section of the fracture surface in which is visible the flat zone and the small ramifications.

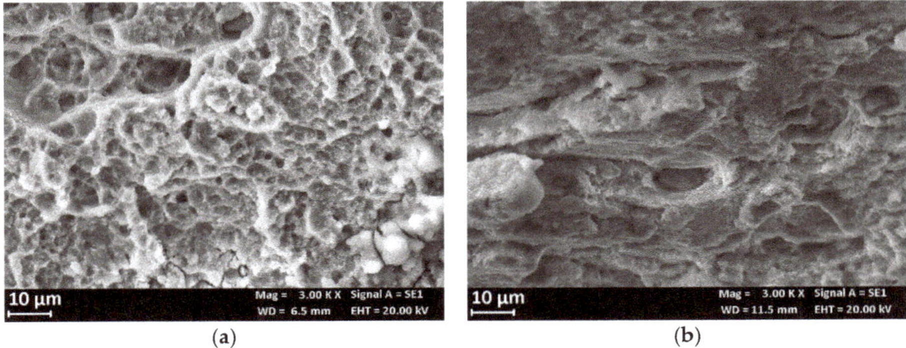

Figure 19. (**a**) Particular of the mixed ductile/brittle fracture zone and (**b**) of the ductile shearing final rupture.

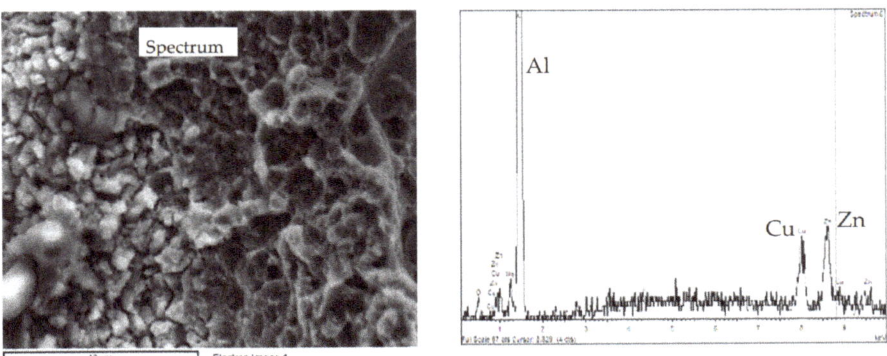

Figure 20. EDS spectrum of the precipitates present on the mixed ductile/brittle zone of the fracture surface of the FWS specimen after the SSR test in 0.6 M NaCl solution.

The OCP of the FSW butt joints lies in the range of −820–−810 mV vs SCE and it is less noble than the base material, whose potential was in the range between −725 and −670 mV vs SCE. The OCP of the FSW alloy decreases during the elastic deformation. Conversely, constant values were detected once the specimen was strained in the plastic field. The values of the OCP of the FSW specimens during SSR tests are close to that measured during 4PBB tests and CL tests (Figures 7 and 10), corresponding to the OCP of the TMAZ/HAZ and the nugget (Figure 8). The very low OCP potential indicates that these zones are more susceptible to corrosion than the base material, as confirmed by the intense attack observed on the specimens at the end of the test.

4. Discussion

These results confirmed the higher corrosion susceptibility of the TMAZ/HAZ of the AA 7075 FSW joints with respect to the nugget and the base material, but there is no agreement between the different geometries of the specimens.

No SCC occurrence was noticed after 4PBB tests, but severe localized attack along the rolling direction at TMAZ/HAZ. Shallow defects at TMAZ/HAZ were observed after CL and SSR tests, which direction was perpendicular respect to the applied load.

The higher corrosion susceptibility of the TMAZ/HAZ can be attributed to the microstructure modification produced by the FSW process. The presence of precipitates at the grain boundaries less noble than the matrix promotes the intergranular corrosion of the alloy [37]. Different mechanisms were proposed to explain their effect, Maitra and English affirmed that the grain boundaries have a more active breakdown than the matrix and preferentially initiated the pitting [38], while other authors consider the effect of acidification generated by the preferential dissolution of the anodic particles as the cause of the continuous grain-boundary attack [11,14]. The intergranular attack through thickness direction (short transverse—S) is much lower than that in either longitudinal (L) and transverse (T) directions for rolled plates owing to the microstructural anisotropy while the breakdown potentials were independent of sample orientations [39].

Similar mechanisms have been formulated to explain the SCC mechanism of aluminum 7XXX alloys, mainly hypothesizing hydrogen embrittlement [18,40], but also slip-dissolution mechanism [41]. Naijar et al. [42] proposed a mixed mechanism, in which the SCC phenomena were initiated from localized attacks at macro-precipitates present on the alloy surface (i.e., Al_7Cu_2Fe or Al_2CuMg) or on grain boundaries. The hydrolysis of the produced cations induces the local acidification of the occluded cell and, as consequence, the possibility to electrochemical evolution of hydrogen. These authors suggested that localized dissolution at precipitates or at the grain boundary correspond to the formation of critical defects, which is the first step of the SCC. The deterioration of the passive film and the anodic dissolution of the metal promotes hydrogen uptake and diffusion inside the metal lattice. As the defects act as local stress intensifier, intergranular fracture are promoted, especially in the short-transverse direction for which grain boundaries are favorably oriented to the applied tensile stress [43,44]. Vasudevan and Sadananda [45] reported the decrease in the crack propagation rate in NaCl solution with the increasing of the precipitates size at the grain boundaries, and the depletion in copper of the aluminum matrix. The positive effect of precipitates size was also underlined by other authors [46].

On the contrary, the results of Ashai et al. [47] on smooth specimens of Al-4.3Zn-1.7Mg alloys in 3.5% NaCl + 0.2% H_2O_2 solution, demonstrated an inverse correlation between the time to failure (failure rate) and the size of precipitates at the grain boundaries. This can be ascribed to the preferential dissolution of the $MgZn_2$ precipitates at the grain boundaries [48].

Talianker and Cina demonstrated a clear relationship between the presence of dislocations aside to grain boundaries and the susceptibility to stress corrosion of AA7XXX. [48].

The fractographic analysis of the specimens demonstrated the microcracks initiation preferentially at larger precipitates. The nugget and the TMAZ/HAZ showed lower OCP, thus indicating higher susceptibility to corrosion than the base material. No SCC microcracks propagation was noticed at the nugget due to the large size of the strengthening $MgZn_2$ micro-precipitates—that decreased the SCC susceptibility—and the higher tensile strength compared to the TMAZ/HAZ zones. Under constant deformation tests, such as 4PBB tests, the applied stress naturally decreases during the exposure as localized attack propagate toward the specimen thickness. In addition, the tensile stress acts on these specimens in the direction of the elongated grains, and therefore the SCC cracks should propagate in the perpendicular direction. For these reasons, corrosion takes place in form of localized attacks at the elongated grain boundaries along the rolling direction (Figure 9).

In SSR and CL tests, the localized attacks and the plastic strain mainly occurred at TMAZ/HAZ, thus triggering the formation of SCC crack embryos. Based on the abovementioned mechanism, hydrogen embrittlement should promote SCC microcracks propagation. Hydrogen supplied by both the cathodic reaction and the anodic dissolution of the matrix/precipitate interphase can diffuse in the high stressed zone and promotes brittle crack propagation. The SEM analysis confirms the presence of several submicrometric precipitates in the correspondence of the brittle areas of the fracture surface. The iron-rich phases can consequently enhance the hydrogen evolution mechanism, because the hydrogen overpotential on these phases is lower than that on aluminum. Conversely, the applied plastic strain enhanced the corrosion rate by means of the chemomechanical effect underlined by Gutman. This effect is higher at the interface between the matrix and second phases, which acts as an obstacle to the movement of the dislocations [28,49]. In these zones, the very high dissolution rate of the alloy is prevalent, and hydrogen cannot reach the critical concentration to produce cracking. In the CL tests the attack becomes enlarged and no failure of the specimens were observed. On the other hand, the only small propagation of microcracks noticed after SSR tests suggests that the final rupture quickly occurred with a shear mechanism because of the high strain rate at the TMAZ/HAZ. For this reason, no evident stress-corrosion cracking phenomena were observed.

5. Conclusions

The paper studied the stress-corrosion cracking behavior of AA7075 T6 aluminum joints made by FSW as a function of loading conditions. SCC tests were performed by considering constant deformation (4PBB and CL) conditions and SSR conditions in aerated NaCl 35g/L solution at room temperature.

Certain susceptibility to SCC was demonstrated for AA 7075 FSW butt joints tested in this work mainly at the TMAZ/HAZ as nucleation of crack embryos occurred during both CL and SSR tests. However, high dissolution rate of the metal inside the localized attack prevents the achievement of critical crack size to promote its propagation up to rupture.

Slight decrease in the free corrosion potential was noticed because of the modification of the microstructure introduced by the welding process mainly at the TMAZ/HAZ and the nugget with respect the base material. As the TMAZ/HAZ zone is characterized by low tensile strength compared to both the nugget and the base material, it is preferentially strained under loading conditions. The plastic strain then enhances the active dissolution rate of the metal matrix and severe attack takes place, thus hindering SCC crack propagation.

The penetration depth of the intergranular attack was deeper after the four-point bent-beam tests compared to all the other testing techniques. Preferential dissolution along the grain boundaries was found in heat-affected zones and the attack follows the elongated grain structure along the rolling direction.

Author Contributions: Methodology, S.L., C.T. and S.B.; Investigation; S.L., C.T. and S.B.; Data Curation, G.D. and M.C.; Writing—original draft of the paper, M.C., S.B., G.D., C.G., S.L., C.T. and T.P. Writing—Review & Editing T.P., C.G. and M.C. Supervision, T.P. and C.G.; Project Administration, T.P. All authors have read and agreed to the published version of the manuscript.

Funding: This research received no external funding

Conflicts of Interest: The authors declare no conflict of interest.

References

1. Thomas, W.M.; Nicholas, E.D. Friction stir welding for the transportation industries. *Mater. Des.* **1997**, *18*, 269–273. [CrossRef]
2. Thomas, W.M.; Nicholas, E.D.; Needham, J.C.; Murch, M.G.; Temple-Smith, P.; Dawes, C.J. Friction Welding. U.S. Patent 5460317, 24 Octorber 1995.
3. Threadgill, P.L.; Leonard, A.J.; Shercliff, H.R.; Withers, P.J. Friction stir welding of aluminium alloys. *Int. Mater. Rev.* **2009**, *54*, 49–93. [CrossRef]
4. Çam, G.; Mistikoglu, S. Recent developments in friction stir welding of Al-Alloys. *J. Mater. Eng. Perform.* **2014**, *23*, 1936–1953. [CrossRef]
5. Aydin, H.; Bayram, A.; Durgun, I. The effect of post-weld heat treatment on the mechanical properties of 2024-T4 friction stir-welded joints. *Mater. Des.* **2010**, *31*, 2568–2577. [CrossRef]
6. Bitondo, C.; Prisco, U.; Squillace, A.; Giorleo, G.; Buonadonna, P.; Dionoro, G.; Campanile, G. Friction stir welding of AA2198-T3 butt joints for aeronautical applications. *Int. J. Mater. Form.* **2010**, *3*, 1079–1082. [CrossRef]
7. Rodrigues, D.M.; Loureiro, A.; Leitao, C.; Leal, R.M.; Chaparro, B.M.; Vilaça, P. Influence of friction stir welding parameters on the microstructural and mechanical properties of AA 6016-T4 thin welds. *Mater. Des.* **2009**, *30*, 1913–1921. [CrossRef]
8. Tongne, A.; Desrayaud, C.; Jahazi, M.; Feulvarch, E. On material flow in Friction Stir Welded Al alloys. *J. Mater. Process. Technol.* **2017**, *239*, 284–296. [CrossRef]

9. Rajakumar, S.; Muralidharan, C.; Balasubramanian, V. Influence of friction stir welding process and tool parameters on strength properties of AA7075-T6 aluminium alloy joints. *Mater. Des.* **2011**, *32*, 535–549. [CrossRef]
10. Bayazid, S.M.; Farhangi, H.; Ghahramani, A. Effect of Pin Profile on Defects of Friction Stir Welded 7075 Aluminum Alloy. *Procedia Mater. Sci.* **2015**, *11*, 12–16. [CrossRef]
11. Muller, I.L.; Galvele, J.R. Pitting potential of high purity binary aluminium alloys-I. AlCu alloys. Pitting and intergranular corrosion. *Corros. Sci.* **1977**, *17*, 179–193. [CrossRef]
12. Birbilis, N.; Cavanaugh, M.K.; Buchheit, R.G. Electrochemical behavior and localized corrosion associated with Al7Cu2Fe particles in aluminum alloy 7075-T651. *Corros. Sci.* **2006**, *48*, 4202–4215. [CrossRef]
13. Na, K.H.; Pyun, S., II. Comparison of susceptibility to pitting corrosion of AA2024-T4, AA7075-T651 and AA7475-T761 aluminium alloys in neutral chloride solutions using electrochemical noise analysis. *Corros. Sci.* **2008**, *50*, 248–258. [CrossRef]
14. Ramgopal, T.; Gouma, P.I.; Frankel, G.S. Role of grain-boundary precipitates and solute-depleted zone on the intergranular corrosion of aluminum alloy 7150. *Corrosion* **2002**, *58*, 687–697. [CrossRef]
15. Andreatta, F.; Terryn, H.; De Wit, J.H.W. Corrosion behaviour of different tempers of AA7075 aluminium alloy. *Electrochim. Acta* **2004**, *49*, 2851–2862. [CrossRef]
16. Holroyd, N.J.H.; Scamans, G.M. Stress corrosion cracking in Al-Zn-Mg-Cu aluminum alloys in saline environments. *Metall. Mater. Trans. A* **2013**, *44*, 1230–1253. [CrossRef]
17. Loto, C.A.; Cottis, R.A. Electrochemical noise generation during stress corrosion cracking of the high-strength aluminum AA 7075-T6 alloy. *Corrosion* **1989**, *45*, 136–141. [CrossRef]
18. Knight, S.P.; Birbilis, N.; Muddle, B.C.; Trueman, A.R.; Lynch, S.P. Correlations between intergranular stress corrosion cracking, grain-boundary microchemistry, and grain-boundary electrochemistry for Al-Zn-Mg-Cu alloys. *Corros. Sci.* **2010**, *52*, 4073–4080. [CrossRef]
19. Park, J.K.; Ardell, A.J. Effect of retrogression and reaging treatments on the microstructure of Ai-7075-T651. *Metall. Mater. Trans. A* **1984**, *15*, 1531–1543. [CrossRef]
20. Nandan, R.; DebRoy, T.; Bhadeshia, H.K.D.H. Recent advances in friction-stir welding-Process, weldment structure and properties. *Prog. Mater. Sci.* **2008**, *53*, 980–1023. [CrossRef]
21. Venugopal, T.; Rao, K.S.; Rao, K.P. Studies on Friction Stir Welded Aa 7075 Aluminum Alloy. *Trans. Indian Inst. Met.* **2004**, *57*, 659–663.
22. Khodir, S.A.; Shibayanagi, T. Friction stir welding of dissimilar AA2024 and AA7075 aluminum alloys. *Mater. Sci. Eng. B* **2008**, *148*, 82–87. [CrossRef]
23. Azimzadegan, T.; Serajzadeh, S. An investigation into microstructures and mechanical properties of AA7075-T6 during friction stir welding at relatively high rotational speeds. *J. Mater. Eng. Perform.* **2010**, *19*, 1256–1263. [CrossRef]
24. Navaser, M.; Atapour, M. Effect of Friction Stir Processing on Pitting Corrosion and Intergranular Attack of 7075 Aluminum Alloy. *J. Mater. Sci. Technol.* **2017**, *33*, 155–165. [CrossRef]
25. Lumsden, J.B.; Mahoney, M.W.; Rhodes, C.G.; Pollock, G.A. Corrosion behavior of friction-stir-welded AA7050-T7651. *Corrosion* **2003**, *59*, 212–219. [CrossRef]
26. Bocchi, S.; Cabrini, M.; D'Urso, G.; Giardini, C.; Lorenzi, S.; Pastore, T. The influence of process parameters on mechanical properties and corrosion behavior of friction stir welded aluminum joints. *J. Manuf. Process.* **2018**, *35*, 1–15. [CrossRef]
27. Cabrini, M.; Lorenzi, S.; Bocchi, S.; Pastore, T.; D'Urso, G.; Giardini, C. Evaluation of corrosion behavior of AA2024 T3 welded by means of FSW. In Proceedings of the EUROCORR 2017-The Annual Congress of the European Federation of Corrosion; 20th International Corrosion Congress and Process Safety Congress 2017, Congress Centre, Prague, Czech Republic, 3–7 September 2017.
28. Vijaya Kumar, P.; Madhusudhan Reddy, G.; Srinivasa Rao, K. Microstructure and pitting corrosion of armor grade AA7075 aluminum alloy friction stir weld nugget zone—Effect of post weld heat treatment and addition of boron carbide. *Def. Technol.* **2015**, *11*, 166–173. [CrossRef]
29. Fooladfar, H.; Hashemi, B.; Younesi, M. The effect of the surface treating and high-temperature aging on the strength and SCC susceptibility of 7075 aluminum alloy. *J. Mater. Eng. Perform.* **2010**, *19*, 852–859. [CrossRef]
30. Hatamleh, O.; Singh, P.M.; Garmestani, H. Corrosion susceptibility of peened friction stir welded 7075 aluminum alloy joints. *Corros. Sci.* **2009**, *51*, 135–143. [CrossRef]

31. D'Urso, G.; Giardini, C.; Lorenzi, S.; Cabrini, M.; Pastore, T. The Effects of Process Parameters on Mechanical Properties and Corrosion Behavior in Friction Stir Welding of Aluminum Alloys. *Procedia Eng.* **2017**, *183*, 270–276. [CrossRef]
32. D'Urso, G.; Giardini, C.; Lorenzi, S.; Cabrini, M.; Pastore, T. The influence of process parameters on mechanical properties and corrosion behaviour of friction stir welded aluminum joints. *Procedia Eng.* **2017**, *207*, 591–596. [CrossRef]
33. Bocchi, S.; Cabrini, M.; D'Urso, G.; Giardini, C.; Lorenzi, S.; Pastore, T. Stress enhanced intergranular corrosion of friction stir welded AA2024-T3. *Eng. Fail. Anal.* **2020**, *111*, 104483. [CrossRef]
34. Rhodes, C.G.; Mahoney, M.W.; Bingel, W.H.; Spurling, R.A.; Bampton, C.C. Effects of friction stir welding on microstructure of 7075 aluminum. *Scr. Mater.* **1997**, *36*, 69–75. [CrossRef]
35. Mahoney, M.W.; Rhodes, C.G.; Flintoff, J.G.; Spurling, R.A.; Bingel, W.H. Properties of friction-stir-welded 7075 T651 aluminum. *Metall. Mater. Trans. A* **1998**, *29*, 1955–1964. [CrossRef]
36. Srinivasan, P.B.; Dietzel, W.; Zettler, R.; dos Santos, J.F.; Sivan, V. Stress corrosion cracking susceptibility of friction stir welded AA7075-AA6056 dissimilar joint. *Mater. Sci. Eng. A* **2005**, *392*, 292–300. [CrossRef]
37. Maitra, S.; English, G.C. Environmental factors affecting localized corrosion of 7075-t7351 aluminum alloy plate. *Metall. Trans. A* **1982**, *13*, 161–166. [CrossRef]
38. Huang, T.S.; Frankel, G.S. Influence of grain structure on anisotropic localised corrosion kinetics of AA7XXX-T6 alloys. *Corros. Eng. Sci. Technol.* **2006**, *41*, 192–199. [CrossRef]
39. Goswami, R.; Lynch, S.; Holroyd, N.J.H.; Knight, S.P.; Holtz, R.L. Evolution of grain boundary precipitates in Al 7075 upon aging and correlation with stress corrosion cracking behavior. *Metall. Mater. Trans. A* **2013**, *44*, 1268–1278. [CrossRef]
40. Burleigh, T.D. Postulated mechanisms for stress corrosion cracking of aluminum alloys. A review of the literature 1980–1989. *Corrosion* **1991**, *47*, 89–98. [CrossRef]
41. Najjar, D.; Magnin, T.; Warner, T.J. Influence of critical surface defects and localized competition between anodic dissolution and hydrogen effects during stress corrosion cracking of a 7050 aluminium alloy. *Mater. Sci. Eng. A* **1997**, *238*, 293–302. [CrossRef]
42. Gruhl, W. Stress Corrosion Cracking of High Strength Aluminium Alloys. *Zeitschrift fuer Metallkunde* **1984**, *75*, 819–826.
43. Scamans, G.M.; Alani, R.; Swann, P.R. Pre-exposure embrittlement and stress corrosion failure in AlZnMg Alloys. *Corros. Sci.* **1976**, *16*, 443–459. [CrossRef]
44. Vasudevan, A.K.; Sadananda, K. Role of slip mode on stress corrosion cracking behavior. *Metall. Mater. Trans. A* **2011**, *42*, 405–414. [CrossRef]
45. Rajan, K.; Wallace, W.; Beddoes, J.C. Microstructural study of a high-strength stress-corrosion resistant 7075 aluminium alloy. *J. Mater. Sci.* **1982**, *17*, 2817–2824. [CrossRef]
46. Asahi, T.; Yabusaki, F.; Osamura, K.; Murakami, Y. Studies on Microstructure and Stress-Corrosion Cracking Behavior of Al-Zn-Mg Alloys. *Int. Conf. Light Met.* **1975**, 64–67.
47. Talianker, M.; Cina, B. Retrogression and reaging and the role of dislocations in the stress corrosion of 7000-type aluminum alloys. *Metall. Trans. A* **1989**, *20*, 2087–2092. [CrossRef]
48. Gutman, E.M. *Mechanochemistry of Materials*; Cambridge International Science Publishing: Cambridge, UK, 1998; ISBN 1898326320.
49. Ferri, M.; Trueba, M.; Trasatti, S.P.; Cabrini, M.; Lo Conte, A. Electrochemical investigation of corrosion and repassivation of structural aluminum alloys under permanent load in bending. *Corros. Rev.* **2017**, *35*, 225–239. [CrossRef]

© 2020 by the authors. Licensee MDPI, Basel, Switzerland. This article is an open access article distributed under the terms and conditions of the Creative Commons Attribution (CC BY) license (http://creativecommons.org/licenses/by/4.0/).

Article

Effects of Alloying Elements (C, Mo) on Hydrogen Assisted Cracking Behaviors of A516-65 Steels in Sour Environments

Jin Sung Park [1], Jin Woo Lee [2], Joong Ki Hwang [3] and Sung Jin Kim [1,*]

[1] Department of Advanced Materials Engineering, Sunchon National University, Jungang-ro, Suncheon, Jeonnam 57922, Korea; pjs1352@naver.com

[2] POSCO Steel Solution Center, Pohang, Gyungbuk 790-704, Korea; jwlee36@posco.co.kr

[3] Department of Mechanical Engineering, Tongmyong University, Busan 48520, Korea; jkhwang@tu.ac.kr

* Correspondence: sjkim56@scnu.ac.kr; Tel.: +82-61-750-3557

Received: 23 August 2020; Accepted: 15 September 2020; Published: 21 September 2020

Abstract: This study examined the effects of alloying elements (C, Mo) on hydrogen-induced cracking (HIC) and sulfide stress cracking (SSC) behaviors of A516-65 grade pressure vessel steel in sour environments. A range of experimental and analytical methods of HIC, SSC, electrochemical permeation, and immersion experiments were used. The steel with a higher C content had a larger fraction of banded pearlite, which acted as a reversible trap for hydrogen, and slower diffusion kinetics of hydrogen was obtained. In addition, a higher hardness in the mid-thickness regions of the steel, due to center segregation, resulted in easier HIC propagation. On the other hand, the steel with a higher Mo content showed more dispersed banded pearlite and a larger amount of irreversibly trapped hydrogen. Nevertheless, the addition of Mo to the steel can deteriorate the surface properties through localized pitting and the local detachment of corrosion products with uneven interfaces, increasing the vulnerability to SSC. The mechanistic reasons for the results are discussed, and a desirable alloy design for ensuring an enhanced resistance to hydrogen assisted cracking (HAC) is proposed.

Keywords: steel; hydrogen induced cracking; sulfide stress cracking; sour; hydrogen permeation; diffusion; corrosion

---

1. Introduction

Hydrogen assisted cracking (HAC), which is classified into hydrogen induced cracking (HIC) and sulfide stress cracking (SSC) by the presence and absence of applied stress, is one of the most important issues in the petrochemical industries [1–5]. In a recent refining system for low-grade crude oil, the refining temperature and pressure required have increased remarkably. The sensitivity of material deterioration, caused by the inflow of hydrogen atoms in the steel structures, is becoming higher than in the past [6,7]. Accordingly, considerable efforts have been made to develop the steel used as pressure vessel facilities with superior resistance to HAC [7–10]. For this purpose, a variety of technical solutions, such as de-phospho/sulfu-rization of molten steel [11,12], Ca-treatment [13], cast steel soft reduction [14], and thermal mechanical controlled process (TMCP) [15–17], have been proposed. Microstructural modifications by reducing the number of second phases, such as retained austenite (RA), martensite-austenite constituents (MA), and other hard phases, and their role in improving the HAC resistance, have been reported [18,19]. Nevertheless, there are uncertainties regarding the relationship between the alloying elements in the steel and the HAC behaviors, primarily because of the lack of systematic efforts to understand the HAC behaviors, in terms of the hydrogen diffusion/trapping behaviors in the steel matrix and corrosion behaviors on the steel surface.

Theoretically, hydrogen atoms, formed by corrosion reactions on the steel surface, can diffuse easily into the steel matrix with a body-centered cubic (BCC) structure, and become trapped at several metallurgical defects with high binding energy [20–22]. According to the internal pressure and de-cohesion theories, proposed by Zapffe et al. [23], and Troiano [24], respectively, cracks were initiated by the continued accumulation of hydrogen atoms into local areas and their recombination reaction ($H + H \rightarrow H_2$), and propagated easily through the embrittled regions formed by a weakening of the binding force between Fe-Fe atoms. Therefore, clarification of the physical nature of hydrogen diffusion and trapping phenomena is critically important. Moreover, the other important feature of HAC, occurring in sour environments, is the corrosion behaviors on the steel surface. In particular, the amount of hydrogen infusion and its infusion kinetics can also be controlled by the characteristics of the corrosion products formed on the steel surface [25,26]. Kim et al. [26] investigated the combined addition of Cu and Ni on sour corrosion and the HIC resistance of A516 steel in sour environments. On the other hand, some alloying elements can have both beneficial and detrimental effects on sour corrosion and HAC resistance of steels in sour environments. Hence, a further in-depth study is required.

In this study, the HAC resistance of A516 steel, used in pressure vessel facilities, was investigated systematically, in terms of microstructural modifications, hydrogen diffusion/trapping behaviors, and surface characteristics, which were dependent on the addition of alloying element of C and Mo to the steel. Based on the experimental and analytical results, a desirable alloy design concept for improving the resistance to HAC is suggested.

2. Experimental

2.1. Specimen Preparation and Microstructure Observation

The test material used in this study was a low carbon steel equivalent to an ASTM A516-65 grade pressure vessel steel plate. Table 1 lists the chemical composition and mechanical properties of the three tested steel samples. The major differences in the chemical composition among the three samples are the C and Mo contents. The steel was austenitized by heating to 910 °C for 10 min and cooled to room temperature in air.

Table 1. Chemical compositions and mechanical properties of the three tested steel samples.

Specimens	Chemical Composition (wt.%)								YS (MPa)	TS (MPa)	Strain (%)
	C	Mn	Si	Mo	Cr	Nb	Ti	Fe			
Steel A	0.16	1–1.5	0.3–0.4	<0.003	<0.05	<0.01	<0.01	Bal.	349.11	512.18	16.22
Steel B	0.13	1–1.5	0.3–0.4	0.01–0.015	<0.05	<0.01	<0.01	Bal.	345.1	487.21	17.55
Steel C	0.13	1–1.5	0.3–0.4	0.05–0.055	<0.05	<0.01	<0.01	Bal.	347.33	492.61	17.22

For microstructural analysis, the samples were polished to 1 μm and etched with a 5% nital solution (a mixture of 5% nitric acid and ethanol). They were then observed by field emission-scanning electron microscopy (FE-SEM) (Hitachi, Tokyo, Japan). The fractions of pearlite in the samples were measured by image analysis via optical microscopy (OM) (Zeiss, Jena, Germany). The morphological features of the precipitates in the microstructures were observed by transmission electron microscopy (TEM) (SELMI, Sumy, Ukraine) using the extraction replica method.

2.2. Hydrogen Induced Cracking Test (HIC)

HIC sensitivity of three samples was evaluated by the HIC experiment in reference to the National Association of Corrosion Engineers (NACE) TM 0284-96A standard [27]. Prior to conducting the experiment, the samples with 100 mm × 20 mm × 15 mm were cleaned ultrasonically in acetone. The samples were then immersed in a NACE solution (5% NaCl + 0.5% CH_3COOH) saturated fully with H_2S gas for 96 h. Subsequently, the level of HIC in the samples was evaluated, using an ultrasonic testing (UT) method, and the crack area ratio (CAR) was derived using Equation (1):

$$\text{CAR} = \frac{\sum \textit{Crack area}}{\textit{Total area}} \times 100\ (\%). \tag{1}$$

An ultrasonic inspection, a hardness test and microstructure observation were performed according to the thickness direction, in order to determine the location of cracks and microstructural properties along the thickness direction.

2.3. Electrochemical Hydrogen Permeation Test

The hydrogen diffusion behaviors in the steel samples were evaluated using an electrochemical permeation experiment in reference to the ISO 17081 [28] standard. The samples of the sheet-type steel membrane with a 1 mm thickness were prepared by mechanical polishing with SiC paper (P2000 grit sand paper). The polished samples were then cleaned ultrasonically in ethanol and distilled water to remove the impurities on their surfaces. It is generally considered that there is no change in the structure of the ferritic steel during sample preparation. The prepared sample was placed in the center of the permeation cell. Prior to the permeation measurement, a thin palladium (Pd) coating layer, approximately 100 nm thick, was deposited electrochemically on the sample surface in the detection side of the cell. The sample in the detection side was polarized at a constant potential of 270 mV_{SCE} in a deaerated 0.1 M NaOH solution. After reaching a background current density below 1 $\mu m/cm^2$, a constant cathodic current density of 1 mA/cm^2 was applied to the sample in the charging side filled with a charging solution (NACE + 0.05 M Na_2S + 0.3 wt.% NH_4SCN). Na_2S was added to simulate the H_2S atmosphere at the laboratory level because of the toxic risk of H_2S. As the measured permeation current (rising transient) reached a steady-state value, the application of a cathodic charging was stopped and the charging solution was drained immediately for the desorption process. Information about the test procedure is reported elsewhere [29–31].

Preliminary tests showed that the reproducibility of the measured permeation flux in the rising transients is seldom obtained. Instead, the permeation flux in the decay transients was much more reproducible. For this reason, the apparent hydrogen diffusivity of the three samples was obtained by curve fitting with the theoretical Equation (2) [32] to the experimental permeation curves in the decay transients.

The hydrogen current decay phase,

$$\frac{i_t - i_0}{i_0 - i_\infty} = 1 - \frac{2L}{\sqrt{\pi D t}} \sum_{n=0}^{\infty} exp\left[-\frac{(2n+1)^2 L^2}{4Dt}\right] \tag{2}$$

where D, i_t, i_0, and i_∞ are the apparent hydrogen diffusion coefficient, measured hydrogen permeation current density at time t, initial hydrogen permeation current density, and steady-state hydrogen permeation current density for each phase, respectively. While, L is the thickness of the steel membrane, and the number n of 0, 1, 2, 3 could be generally taken for accurate curve fitting.

Two consecutive permeation experiments, involving 1st permeation—Desorption—2nd permeation, were normally conducted to measure the irreversibly trapped hydrogen concentration in the sample [28]. On the other hand, it was difficult to completely exclude the effect of corrosion products with a low electrical conductivity [33], which can be formed on the surface during desorption, on subsequent hydrogen reduction, ad/absorption, and diffusion in the steel matrix during the second permeation. Therefore, the authors attempted to obtain the irreversibly trapped hydrogen concentration in the sample using only a first permeation curve (rising and decay transients) based on the assumption that the irreversible trapping phenomenon can contribute only to the rising transient. Therefore, the irreversibly trapped hydrogen concentrations (C_{irr}) in the samples were estimated quantitatively by measuring the area between the two curves of the fitted decay transients with the two D_{app} values, obtained using Equation (3), describing the rising transient and experimental decay transient.

The hydrogen permeation phase:

$$\frac{i_t - i_0}{i_\infty - i_0} = \frac{2L}{\sqrt{\pi D t}} \sum_{n=0}^{\infty} exp\left[-\frac{(2n+1)^2 L^2}{4Dt}\right]. \quad (3)$$

The total hydrogen trap density values (# of trap site/unit volume) of the samples were determined using Equation (4) [34],

$$\frac{D_L}{D_\tau} - 1 = \frac{3N_T}{\langle C \rangle} \quad (4)$$

where D_L, D_τ, N_T, and <C> are the lattice diffusion coefficient of hydrogen, diffusion coefficient of hydrogen in the presence of traps, the number of trap site per unit volume, and average hydrogen concentration (=$J_{ss}L/D_{app}$) obtained from rising transient, respectively. While, D_L was determined to be 3.25×10^{-9} m^2/s, which was obtained from American Petroleum Institute (API) grade carbon steel [35]. This value was slightly smaller than that (7.5×10^{-9} m^2/s) obtained from pure iron [36]. Such differences could be caused by the fact that C atoms at the interstitial lattice sites and some substitutional atoms (C, Cr, Mo, Cu, and Ni), which may form a stress field inside the lattice, can delay hydrogen diffusion in the steel [34,37].

2.4. SSC Test and Fracture Surface Observation

An SSC experiment was conducted based on NACE TM 0177 [38], and tensile stresses of 85 and 95% vs. yield strength of the steel were applied using the dead-weight method. At the same time, the samples were exposed to a NACE solution fully saturated with H_2S for 720 h. Subsequently, the SSC susceptibility was evaluated by measuring the time to rupture. Field emission scanning electron microscopy (FE-SEM) was performed to observe the fracture surfaces of the samples.

2.5. Corrosion Characteristic Analysis

Immersion tests were also conducte4d in the NACE solution with 0.05 M Na_2S for four weeks to analyze the effects of corrosion and corrosion products on the surface of the samples on the SSC properties. After conducting the tests, the cross-sectional images of the samples were observed by FE-SEM. X-ray diffraction (XRD) (Bruker, Karlsruhe, Germany) was also performed to characterize the phases in the corrosion product layer formed on the steel surface.

The nature of corrosion products formed on the sample with the lowest resistance to SSC was characterized by glow discharge spectroscopy (GDS) and X-ray photoelectron spectroscopy (XPS). GDS analysis was conducted using a Leco (St. Joseph, MI, USA) GDS-850A spectrometer with Ar plasma equipped with an RF lamp. The diameter of the analysis area was 4 mm. XPS (VG Scientific Escalab 250 (Waltham, MA, USA)) was performed using monochromatic Al K-alpha radiation (1486.6 eV) with a 500 μm diameter spot size. A constant analyzer energy mode with 150 eV for the survey and high-resolution spectra was used.

3. Results and Discussion

3.1. Microstructure Observation

Figure 1 presents FE-SEM images of the microstructures of the three samples. They were composed of typical ferrite and band-shaped pearlite, but the noticeable differences among the samples were the fraction, thickness, and distribution of pearlite. As expected from the C contents in the samples, the fraction of pearlite in Steel A was higher than that in the other two samples (Figure 2). There were also differences between Steel B and C with relatively low fractions of pearlite. Compared to Steel B, Steel C showed a high degree of pearlite dispersion in the microstructure, meaning that the pearlite banding index, defined previously [39,40], of Steel C was lower. A high degree of pearlite dispersion in Steel C is closely associated with the addition of Mo in the steel. According to previous studies [41], the addition of Mo in the low C steel lowers the transformation temperature of $\gamma \rightarrow \alpha$, leading to larger nucleation sites, and suppresses the diffusion of C during the cooling process due to the increased super-cooling. A more dispersed banded structure in steel with increasing Mo content was also reported [42]. The dispersion of pearlite in the microstructure has two opposite effects on hydrogen assisted cracking (HAC) of steels. The larger interface between ferrite/banded pearlite, which acts as a reversible trap for hydrogen atoms, leads to increased diffusible hydrogen contents and has a detrimental effect. The other is a beneficial effect that arises through the limited propagation path of hydrogen induced crack (HIC). In addition to these two factors, there are a variety of metallurgical parameters that affect the HAC of steels [42,43]. Further mechanistic discussion, based on the test results of HAC is needed, which will be presented in the following sections.

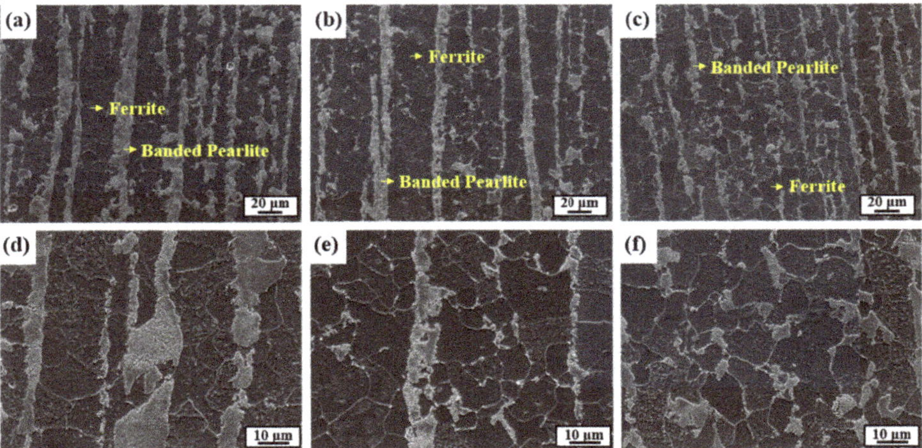

Figure 1. Microstructures observed by FE-SEM: (**a**,**d**) Steel A, (**b**,**e**) Steel B, and (**c**,**f**) Steel.

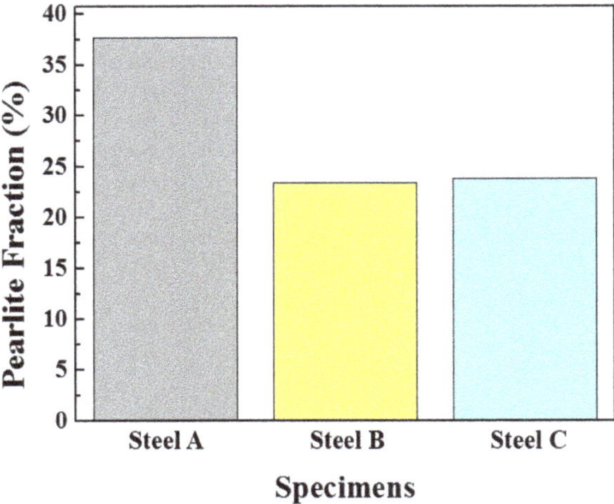

Figure 2. Fractions of pearlite in the microstructures of the three samples, which were measured using an image analyzer in OM.

3.2. Hydrogen Induced Cracking (HIC) Test

Figure 3 presents the distribution and level of the occurrence of HIC of three samples, which were inspected ultrasonically after the HIC standard test [27]. In terms of the crack formation patterns among the samples, Steel B showed somewhat different characteristics, compared to the other two samples. A large number of fine cracks formed mostly in a quarter of Steel B in the thickness direction, which is known as the region where inclusions are mainly formed [44]. In contrast to the sample, much coarser cracks were formed mostly in the center of Steels A and C, presuming that Steels A and C had much lower resistance to the propagation of HIC. The crack sensitivity index, expressed as the CAR, increased in the order of Steel B, Steel C, and Steel A. The susceptibility of Steel A to HIC can be understood partly by the higher fraction of pearlite, with high banding properties in its microstructure, as the banded pearlite can act as a main crack propagation path. Moreover, considering that the interface between ferrite/cementite can act as a reversible trap for hydrogen (11–17 kJ/mol [45,46]), a higher fraction of pearlite in Steel A can provide a larger interfacial area for hydrogen trapping, facilitating embrittlement and cracking. Contrary to expectations, Steel C, with a high degree of pearlite dispersion, was more susceptible to HIC than Steel B. This suggests that the other factors may have been involved in the occurrence of HIC. It is notable that the locations of crack formation in the two samples were different. Similar to Steel A, Steel C had coarse cracks formed at the center region in the thickness direction. From a metallurgical point of view, the microstructure of the center region of the steel plates is somewhat different from that of surface region owing to the centerline segregation phenomenon, and some low-temperature transformation phases with high hardness can be more segregated in the center region of the steel plates. For a more precise understanding, the Vickers hardness of the three samples was measured in the thickness direction, and the morphological features of pearlite formed in surface and center regions of the samples were observed (Figure 4). In the case of Steel A and C, the mid-thickness regions had higher hardness than the top and bottom regions. On the other hand, no high hardness values were particularly measured in the center regions of Steel B. Considering the inverse relationship between the hardness level and HIC resistance [47], the cause of the higher CAR and much more prominent crack propagation in the mid-thickness region of Steel C, compared to Steel B, is better understood. The detrimental effects by local areas with high hardness in the mid-thickness region overtake the beneficial effect by a high degree of pearlite dispersion in the microstructure of Steel C, resulting in lower resistance to HIC. The presence of local areas with high hardness in the mid-thickness region of Steel A or C can also be understood by the degenerated pearlite formed in the mid-thickness region of Steel A or C, as shown in Figure 5. The degenerated pearlite, which is generally referred to as pseudopearlite [48], can be formed under the condition of limited C diffusion during the cooling process. Lee et al. [49] reported that the transformation temperature of the 2nd phase (pearlite) decreased sharply from approximately 600 °C to 500 °C by the addition of Mo in low alloy steel. Hence, C diffusion was not quick enough to bring about the complete growth of pearlite. Based on these facts, it is expected that an increase in C and Mo in these low C steels can form the degenerated pearlite with a higher hardness in the mid-thickness region due the center-segregation, thereby resulting in a lower HIC propagation resistance. These HIC properties are also dependent on hydrogen diffusion and trapping behaviors in steels. These behaviors are discussed in more detail in the following section.

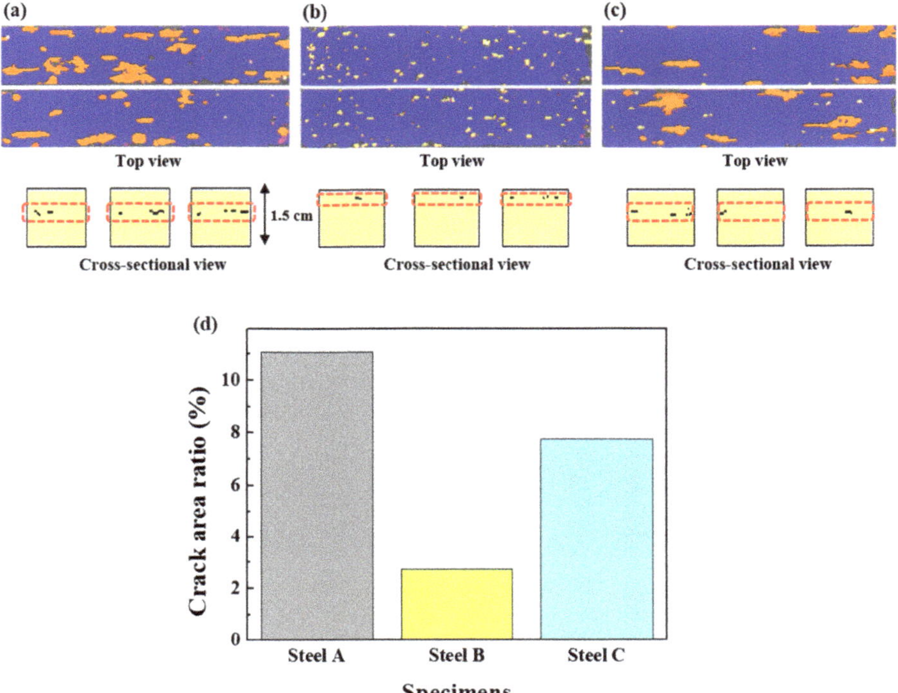

Figure 3. HIC test results obtained by ultrasonic detection: (**a**) Steel A, (**b**) Steel B, (**c**) Steel C, and (**d**) CAR values of the three samples.

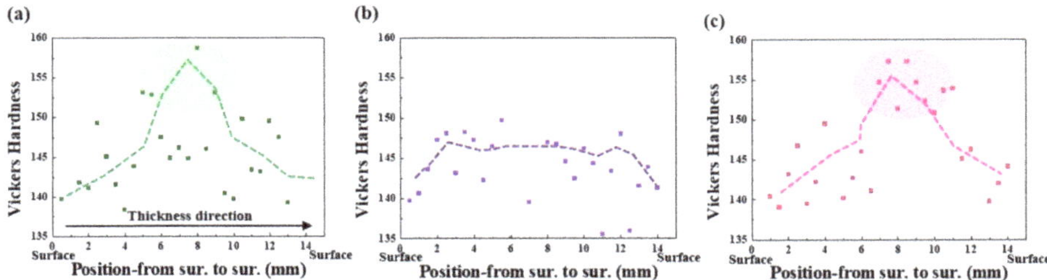

Figure 4. Through-thickness hardness distribution: (**a**) Steel A, (**b**) Steel B, and (**c**) Steel C.

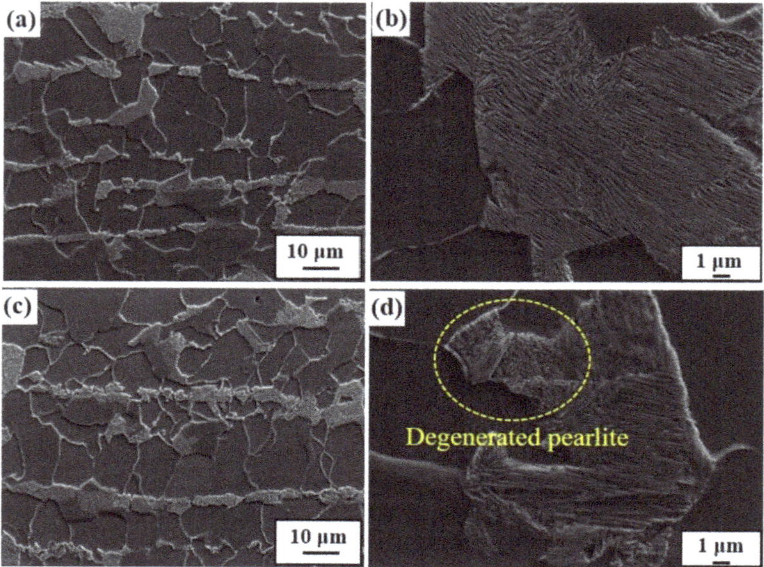

Figure 5. Microstructure observation of Steel C using FE-SEM: (**a**) surface region, (**b**) magnified image of (**a**), (**c**) center region, and (**d**) magnified image of (**c**).

3.3. Electrochemical Hydrogen Permeation Test

Figure 6 presents the hydrogen diffusion coefficients of three samples, obtained by curve fitting to the experimental decay transients. The diffusion coefficients increased in the order of Steel A, Steel C, and Steel B. As mentioned above, the more extensive interface between ferrite/pearlite acting as a reversible trap [45,46] of Steel A can decrease the diffusion kinetics of hydrogen in the steel. The highest diffusion coefficient of Steel B can be understood in this context. On the other hand, the diffusion coefficient of Steel C with a similar pearlite fraction to that of Steel B was lower than expected. Here, the mechanistic reasons for this phenomenon are proposed as follows. A comparatively lower diffusivity in Steel C may be closely associated with the distribution of the banded structure. More dispersed banded pearlite can provide a larger interfacial area for hydrogen trapping, leading to slower hydrogen diffusion kinetics. The formation of a stress field in the lattice structure of Steel C could also be one of the plausible reasons. Hydrogen can diffuse into areas where the stress field is set up [34,37]. The stress field in the lattice structure, which was formed by the difference in atomic size between Fe and Mo, can also delay hydrogen diffusion/transport with the hydrogen binding energy of 26–27 kJ/mol [37,45]. These factors are somewhat speculative, and their impacts on the hydrogen diffusion kinetics may be insignificant. The other factors may also be involved in the diffusion kinetics. In this context, the irreversible trapping phenomenon was also evaluated, as shown in Figure 7. The irreversibly trapped hydrogen contents (C_{irr}) in Steel A and B showed similar levels within the error range, whereas a much higher level of C_{irr} was obtained in Steel C with a higher Mo content. This suggests that additional traps associated with Mo could act as an irreversible trap for hydrogen, leading to slower hydrogen diffusion kinetics. For a more precise understanding, TEM analysis was conducted on Steel B and C, and the results are shown in Figure 8. The fine-sized precipitates of (Ti, Nb) C, classified as an irreversible trap for hydrogen [50], were identified in both steels. Although, it appears that the precipitates observed in Steel C were slightly smaller than those observed in Steel B, their density and distribution cannot be discussed from these limited data. On the other hand, no Mo-based precipitate was observed. Considering that it is generally difficult for fine precipitates, several nanometers in size, to be extracted using the replica method, Mo-based precipitates are too fine to be detected easily. In this case, it would be better for the TEM observation to be conducted with the thin film sample. Although, the nature of the irreversible traps was not clearly characterized in this study, a few comments are noteworthy. Steel C had the largest amount of irreversibly trapped hydrogen among the steels, which was closely associated with the presence of fine precipitates, resulting in slower diffusion kinetics of hydrogen.

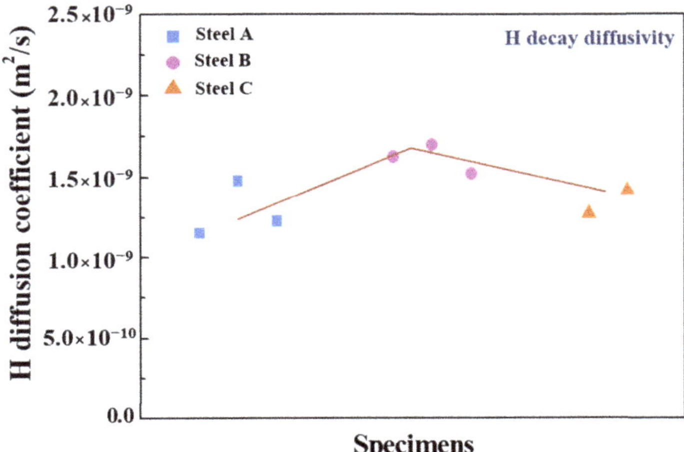

Figure 6. Hydrogen diffusion coefficients obtained by curve fitting to decay permeation curves.

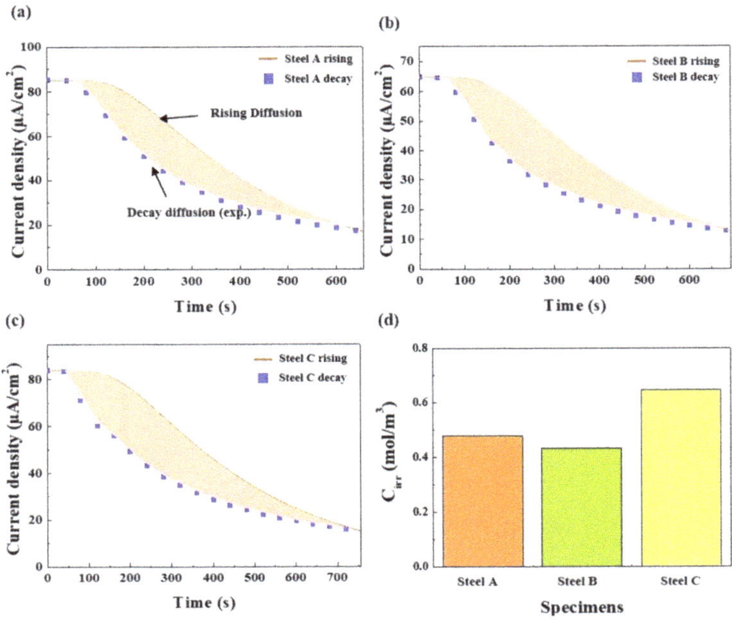

Figure 7. Irreversibly trapped hydrogen contents measured by the area between the two fitted decay transients: (**a**) Steel A, (**b**) Steel B, and (**c**) Steel C, and (**d**) bar chart of the irreversibly trapped hydrogen contents of the three steel samples.

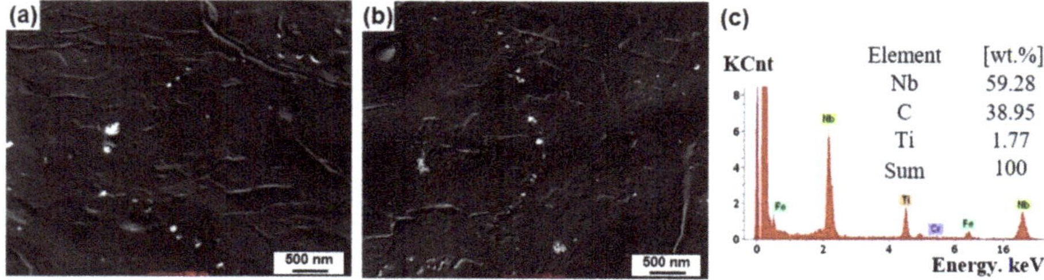

Figure 8. TEM images showing fine-sized precipitates in the steel matrix: (**a**) Steel B and (**b**) Steel C, and (**c**) EDS analysis of fine-sized precipitates.

Figure 9 shows the overall hydrogen trap density (N_T), considering both reversible and irreversible trapping, which was derived from Equation (4). The highest N_T of Steel A can be understood by the higher fraction of pearlite, which corresponds to the lowest diffusivity of hydrogen in Steel A, suggesting that the C content in steels can be a crucial element increasing the hydrogen trap density and reducing the hydrogen diffusivity. On the other hand, the N_T of Steel C was between those of Steel A and B. The N_T of Steel C, which was higher than expected, was attributed to be due to the high level of irreversibly trapped hydrogen in the steel, as discussed above. In general, the presence of a higher density of irreversible traps in steel is considered to be due to a higher resistance to HIC as the traps with a high hydrogen binding energy immobilize the diffusible hydrogen and suppress the local hydrogen concentrations around potential crack areas [51]. As shown above, however, the HIC resistance of Steel C was much lower than that of Steel B. In relation to the effects of irreversible traps on the resistance to HAC in steels, there are additional factors that need to be considered, such as size [52], shape [53], and coherency with the matrix [54]. These also suggest that the HAC resistance cannot be evaluated simply by the hydrogen diffusion/trapping parameters obtained from the electrochemical permeation experiment. Hence, further in-depth analysis is required.

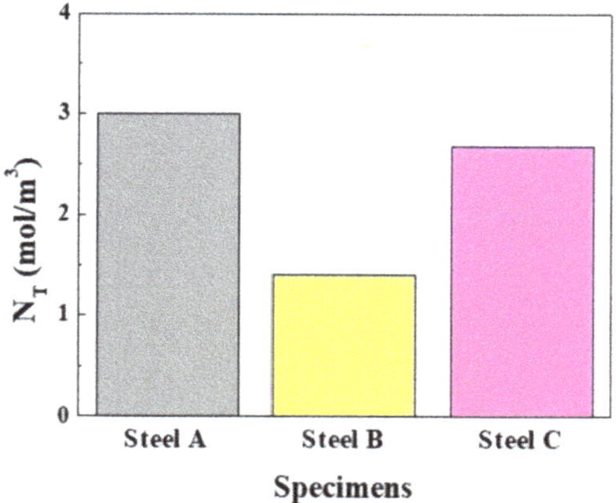

Figure 9. Total hydrogen trap density in the steel.

3.4. *Sulfide Stress Corrosion Cracking (SSC) Test*

The SSC resistance was also evaluated, and Figure 10 presents the time to rupture of the samples after the SSC test which was conducted in reference to NACE TM0177 [38]. In the three samples, the time to rupture decreased with increasing applied stress. Under applied stress of 95% vs. YS of steel, there was no significant difference in the time to rupture among the samples. On the other hand, the time to rupture under an applied stress of 85% versus YS of steel decreased in the order of Steel

B, Steel A, and Steel C. The highest SSC resistance of Steel B is understandable based on the HIC resistance, Vickers hardness distribution, and hydrogen diffusivity/trap density, as shown above.

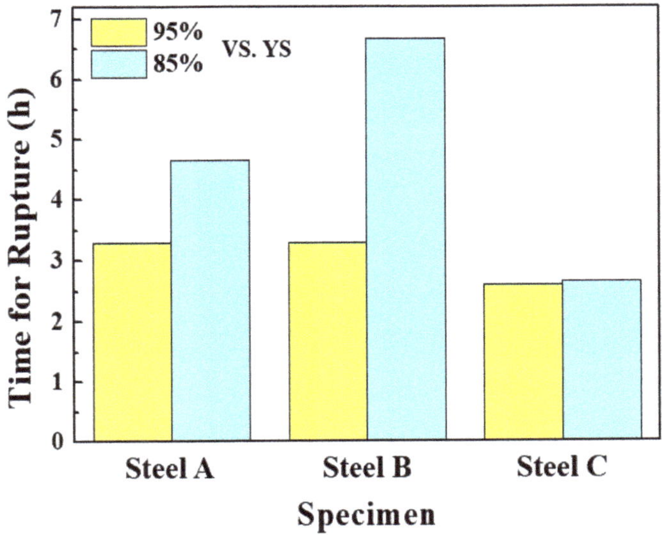

Figure 10. Time for rupture of the three samples, obtained by the SSC test.

Regarding the fracture surface showing the micro-dimple (MD) and quasi-cleavage (QC) patterns (Figure 11), in Steel B, the MD/QC ratio was the largest and the size of the hydrogen induced blister crack (HIBC) in the fracture surface was the smallest. On the other hand, the time to rupture of Steel C was much shorter than expected. From the fracture surface observations, the MD/QC ratio of Steel C was larger than that of Steel A, suggesting that the embrittlement index of Steel C was slightly lower than that of Steel A. On the other hand, it is notable that surface degradation as a form of local-pit like corrosion was observed in Steel C. The major difference between the HIC and SSC experiments lies in the presence of applied stress. Considering the relationship between the applied stress and corrosion behaviors of steels in sour environments [55], the reason for the lowest SSC resistance of Steel C could be closely related to the surface properties. Based on the observations of the fracture surface of Steel C, it appears that pit-like corrosion on the surface acting as a stress intensifier under the applied stress conditions was connected to the internal HIBC, facilitating rupture. Therefore, it is surface defects, such as local-pit like corrosion, can make Steel C more susceptible to SSC, and the corrosion behaviors of Steel C containing Mo should be analyzed further.

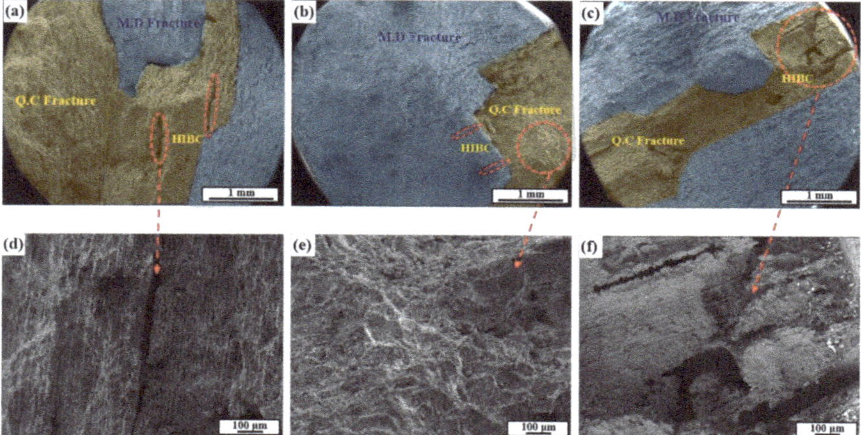

Figure 11. Fracture surface morphologies of the three samples after the SSC test: (**a**) Steel A, (**b**) Steel B, and (**c**) Steel C, and (**d**–**f**) magnified images of the circle marked on (**a**–**c**), respectively.

3.5. Corrosion Product Analysis

Figure 12 showed the cross-sectional images of the samples after immersing in a NACE solution containing 0.05 M Na_2S for four weeks. The corrosion product formed on Steel A was thicker than Steel B and C. The highest growth rate of corrosion products formed on Steel A can be attributed to the highest rate of anodic dissolution, considering the positive relationship between the rate of precipitation and the amount of Fe^{2+} supplied by the anodic dissolution of steel [56]. From the viewpoint of electrochemical corrosion, the Fe_3C contained in steels acts as a cathode, and a higher fraction and larger size of the cathode lead to a higher dissolution rate of the anode (Fe matrix) [57]. On the other hand, it appears from the cross-sectional observation of Steel C that the corrosion products were formed unevenly, presuming that some portions of products were detached locally, and localized pitting occurred below the corrosion products. Preliminary tests showed that the corrosion product formed on Steel C had poor adhesion to the steel substrate, and the weight loss rate measured after three weeks of immersion was the highest in Steel C.

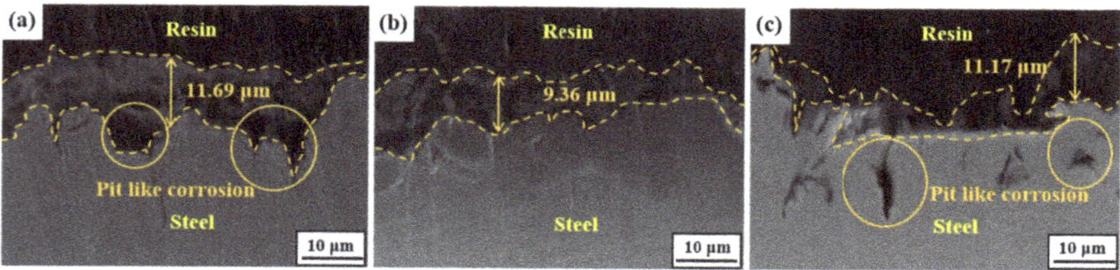

Figure 12. Cross-sectional images after immersion in a NACE solution with 0.05 M Na_2S for four weeks: (**a**) Steel A, (**b**) Steel B, and (**c**) Steel C.

Although XRD analysis (Figure 13) showed that the corrosion products that formed on all three samples consisted mainly of Fe_3O_4 and FeS_{1-x}, there is some doubt as to whether the Mo in Steel C can adversely affect the stability of the corrosion products on the surface. Therefore, additional analyses of the corrosion products formed on Steel C were conducted. GDS and XPS analyses (Figure 14) showed that the corrosion products were composed of a mixture of oxides and sulfides. Based on XPS analyses, the binding energies and relative quantity of compounds for S 2p3/2 spectra of the corrosion product formed on Steel C were obtained, and they are listed in Table 2. The sulfides were composed mainly of

FeS_{1-x}, CuS, NiS, and MoS_2. A previous study [26] reported that the small addition of Cu and Ni to steels leads to the formation of thin and dense corrosion products in sour environments, and contributes to the enhanced corrosion resistance. Accordingly, local detachment of corrosion products with uneven interfaces, and localized pitting of Steel C may have resulted from the formation of MoS_2. Koh et al. [58] pointed out that the formation of MoS_2 in iron sulfide based corrosion products formed on the low C steel exposed to a sour environment weakens the stability of the sulfide products, due to the differences in their crystal structures and lattice parameters. Based on these facts, the desirable alloy design strategy to enhance the HAC of A516-65 steel is proposed as follows. It is preferable that the C content should be reduced, in order to decrease the fraction of pearlite. Moreover, the optimal Mo content providing only a beneficial effect (i.e., dispersion of banded structure) on the HAC needs to be determined.

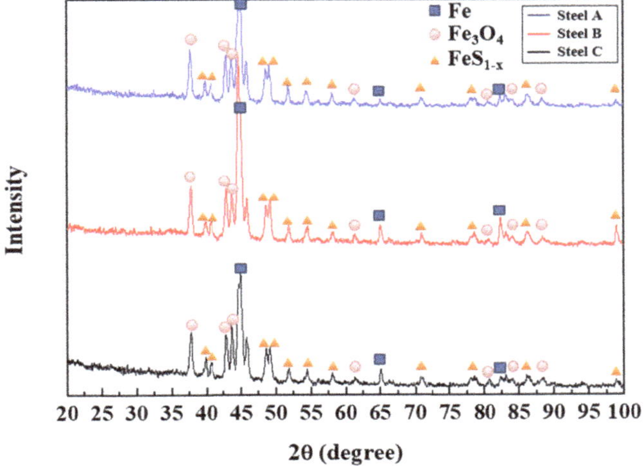

Figure 13. XRD analysis on the corrosion product layer formed on the steel surface after immersion test in a NACE solution with 0.05 M Na_2S for four weeks.

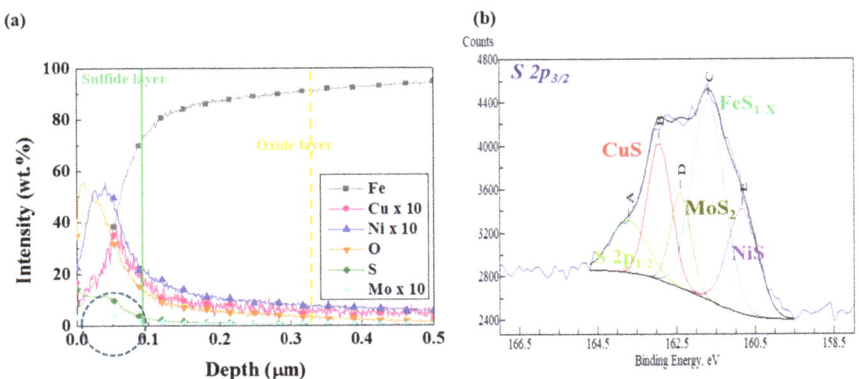

Figure 14. (**a**) GDS analysis, and (**b**) XPS analysis of the corrosion product layer formed on Steel C after the immersion test in a NACE solution with 0.05 M Na_2S for four weeks.

Table 2. Binding energies and relative quantity of compounds for S $2p_{3/2}$ spectra of Steel C after immersion for four weeks.

Phase	Binding Energy (eV)	Relative Quantity
S $2p_{1/2}$	163.7	0.095 (±0.009)
CuS	162.8	0.207 (±0.021)
MoS_2	162.3	0.100 (±0.009)
FeS_{1-X}	161.66	0.374 (±0.037)
NiS	160.79	0.224 (±0.022)

4. Conclusions

The effects of alloying elements (C and Mo) on the HAC properties of ASTM A516-65 grade steel use, as pressure vessel facilities, were investigated using a range of experimental and analytical methods. The main conclusions are as follows:

1. The microstructures of the three steel samples (Steel A, B and C) were composed of ferrite and band-shaped pearlite. A higher C content in the steel resulted in a higher fraction of banded pearlite. On the other hand, the addition of Mo contributed to the dispersion of the banded structure.
2. The sample with lower C and Mo contents (Steel B) showed the highest resistance to HIC, with fine cracks initiated mostly in a quarter of the sample in the thickness direction. In contrast, much coarser cracks were formed mostly in the center of the other samples with higher C, and Mo contents (Steel A and C), respectively, and they showed higher susceptibility to HIC. This is closely associated with the difference in hardness distribution in the through- thickness direction caused by the center segregation phenomenon.
3. The sample with lower C and Mo contents and a smaller fraction of pearlite (Steel B) had the highest diffusion coefficient of hydrogen. On the other hand, the diffusion coefficient of the sample with a higher Mo content (Steel C) was rather high and similar to the case of the sample with a higher C content (Steel A). Therefore, more dispersed banded pearlite, which provides a larger interfacial area for hydrogen trapping, the formation of a stress field in the lattice structure by the difference in atomic size between Fe and Mo, and the presence of irreversible trap site for hydrogen, are the proposed mechanistic reasons.
4. The sample with a higher Mo content (Steel C), however, showed the shortest rupture time by the SSC experiment. In contrast to the fracture surfaces of the other samples (Steel A and B), pit-like corrosion occurred on the surface of Steel C. This was connected to the internal HIBC, which facilitates rupture under the applied stress conditions. The corrosion products formed on Steel C, which were composed of FeS_{1-x}, MoS_2, CuS, and NiS have an uneven interface with the steel substrate and were locally detached. These results suggest that the addition of Mo in the steel should be optimized further to improve the resistance to HAC in sour environments.

Author Contributions: Conceptualization, J.W.L., J.K.H., and S.J.K.; methodology, J.S.P., J.W.L., and S.J.K.; investigation, J.S.P. and S.J.K.; data curation, J.S.P. and S.J.K.; writing-original draft preparation, J.S.P. and S.J.K.; writing-review and editing; S.J.K.; Supervision. All authors have read and agreed to the published version of the manuscript.

Funding: This research was supported in part by the National Research Foundation of Korea (NRF) grant funded by the Korea government (MSIT), grant number No. 2019R1C1C1005007. In addition, this work was partly funded and conducted under the Competency Development Program for Industry Specialists of the Korean Ministry of Trade, Industry and Energy (MOTIE), operated by the Korea Institute for Advancement of Technology (KIAT) (No. P0002019, HRD Program for High Value-Added Metallic Materials Expert).

Conflicts of Interest: The authors declare no conflict of interest.

References

1. Shi, X.B.; Yan, W.; Wang, W.; Zhao, L.Y.; Shan, Y.Y.; Yang, K. HIC and SSC behavior of High-strength pipeline steels. *Acta Met. Sin.* **2015**, *28*, 799–808. [CrossRef]
2. Ghosh, G.; Rostron, P.; Garg, R.; Panday, A. Hydrogen induced cracking of pipeline and pressure vessel steel: A review. *Eng. Fract. Mech.* **2018**, *199*, 609–618. [CrossRef]
3. Homrossukon, S.; Mostovoy, S.; Todd, J.A. Investigation of Hydrogen assisted cracking in high and low strength steels. *J. Press. Vessel. Technol.* **2009**, *131*, 041405. [CrossRef]
4. Hobson, J.D.; Sykes, C. Effect of hydrogen on the properties of low-alloy steel. *J. Iron Steel Inst.* **1951**, *169*, 209–215.
5. Merrick, R.D. An overview of hydrogen damage to steels at low temperature. *Mater. Perform.* **1989**, *28*, 53–55.
6. Gabetta, G.; Pagliari, F.; Rezgui, N. Hydrogen embrittlement in pipeline transporting sour hydrocarbons. *Proc. Struct. Integr.* **2018**, *13*, 746–752. [CrossRef]
7. Huang, F.; Cheng, P.; Zhao, X.Y.; Liu, J.; Hu, Q.; Cheng, Y.F. Effect of sulfide films formed on X65 steel surface on hydrogen permeation in H2S environment. *Int. J. Hydrog. Energy* **2017**, *42*, 4561–4570. [CrossRef]
8. Chen, Y.; Zheng, S.; Zhou, J.; Wang, P.; Chen, L.; Qi, Y. Influence of H_2S interaction with prestrain on the mechanical properties of high-strength X80 steel. *Int. J. Hydrog. Energy* **2016**, *41*, 10412–10420. [CrossRef]
9. Kittel, J.; Smanio, V.; Fregonese, M.; Garnier, L.; Lefebvre, X. Hydrogen induced cracking (HIC) testing of low alloy steel in sour environment: Impact of time of exposure on the extent of damage. *Corros. Sci.* **2010**, *52*, 1386–1392. [CrossRef]
10. Domizzi, G.; Anteri, G.; Ovejero-García, J. Influence of sulphur content and inclusion distribution on the hydrogen induced blister cracking in pressure vessel and pipeline steels. *J. Corros. Sci.* **2001**, *43*, 325–339. [CrossRef]
11. Jiang, X.F.; Chen, Z.P.; Lu, Z.X.; Zhang, G.; Zhong, Z.M. Development and application of BRP technology in Baosteel. *Rev. Met. Paris* **2007**, *104*, 29–34. [CrossRef]
12. Schrama, F.N.H.; Beunder, E.M.; Berg, B.V.D.; Yang, Y.; Boom, R. Sulphur removal in ironmaking and oxygen steelmaking. *Ironmak. Steelmak.* **2017**, *44*, 333–343. [CrossRef]
13. Chattoraj, I. The effect of hydrogen induced cracking on the integrity of steel components. *Sadhana* **1995**, *20*, 199–211. [CrossRef]
14. Nieto, J.; Elías, T.; López, G.; Campos, G.; López, F.; Garcia, R.; De, A.K. Effective process design for the production of HIC-resistance linepipe steels. *J. Mater. Eng. Perform.* **2013**, *22*, 2493–2499. [CrossRef]
15. Yoon, B.H. Characteristics of sulfide stress cracking of high strength pipeline steel weld by heat input. *J. Weld. Join.* **2018**, *36*, 38–44. [CrossRef]
16. Tamehiro, H.; Yamada, N.; Matsuda, H. Effect of the thermo-mechanical control process on the properties of high-strength low alloy steel. *Trans. ISIJ* **1985**, *25*, 54–61. [CrossRef]
17. Tomić, T.; Kožuh, Z.; Garašič, I.; Samardžič, I. Effect of hydrogen upon the properties of thermo mechanical controlled process (TMCP) steel. *Metalurgija* **2016**, *55*, 99–102.
18. Liao, C.M.; Lee, J.L. Effect of molybdenum on sulfide stress cracking resistance of low-alloy steels. *Corrosion* **1994**, *50*, 695–704. [CrossRef]
19. Shi, X.B.; Yan, W.; Wang, W.; Zhao, L.Y.; Shan, Y.Y.; Yang, K. Effect of microstructure on hydrogen induced cracking behavior of a high deformability pipeline steel. *J. Iron Steel Res.* **2015**, *22*, 937–942. [CrossRef]
20. Hong, G.W.; Lee, J.Y. The interaction of hydrogen and the cementite-ferrite interface in carbon steel. *J. Mater. Sci.* **1983**, *18*, 271–277. [CrossRef]
21. Jiang, Y.F.; Zhang, B.; Zhou, Y.; Wang, J.Q.; Han, E.H.; Ke, W. Atom probe tomographic observation of hydrogen trapping at carbides/ferrite interfaces for a high strength steel. *J. Mater. Sci. Tech.* **2018**, *34*, 1344–1348. [CrossRef]
22. Park, J.S.; Seong, H.G.; Kim, S.J. Effect of heat treatment conditions on corrosion and hydrogen diffusion behaviors of ultra-strong steel used for automotive applications. *Corros. Sci. Tech.* **2019**, *18*, 267–276.
23. Zapffe, C.A.; Sims, C.E. Hydrogen embrittlement, internal stress and defects in steel. *Trans. AIME* **1941**, *145*, 225–261.
24. Troiano, A.R. The role of hydrogen and other interstitials in the mechanical behavior of metals. *Trans ASM* **1960**, *52*, 54–80. [CrossRef]
25. Carneiro, R.A.; Ratnapuli, R.C.; Lins, V.D.F.C. The influence of chemical composition and microstructure of API linepipe steels on hydrogen induced cracking and sulfide stress corrosion cracking. *Mater. Sci. Eng.* **2003**, *357*, 104–110. [CrossRef]

26. Kim, S.J.; Jung, H.G.; Park, G.T.; Kim, K.Y. Effect of Cu and Ni on sulfide film formation and corrosion behavior of pressure vessel steel in acid sour environment. *Appl. Surf. Sci.* **2014**, *313*, 396–404. [CrossRef]
27. *Standard Test Method for Evaluation of Pipeline and Pressure Vessel Steels for Resistance to Hydrogen Induced Cracking*; NACE TM 0284-96A; NACE International: Houston, TX, USA, 2016.
28. *Method of Measurement of Hydrogen Permeation and Determination of Hydrogen Uptake and Transport in Metals by an Electrochemical Technique*; ISO 17081; ISO: Geneva, Switzerland, 2004.
29. Hwang, E.H.; Seong, H.G.; Kim, S.J. Effect of carbon contents on corrosion and hydrogen diffusion behaviors of ultra-strong steels for automotive applications. *Korean J. Met. Mater.* **2018**, *56*, 570–579. [CrossRef]
30. Kim, S.J.; Hwang, E.H.; Park, J.S.; Ryu, S.M.; Yun, D.W.; Seong, H.G. Inhibiting hydrogen embrittlement in ultra-strong steels for automotive applications by Ni-alloying. *NPJ Mater. Degrad.* **2019**, *3*, 12. [CrossRef]
31. Park, J.S.; Hwang, E.H.; Lee, M.J.; Kim, S.J. Effect of tempering condition on hydrogen diffusion behavior of martensitic high-strength steel. *Corros. Sci. Tech.* **2018**, *17*, 242–248.
32. Gan, L.; Huang, F.; Zhao, X.; Liu, J.; Cheng, Y.F. Hydrogen trapping and hydrogen induced cracking of welded X100 pipeline steel in H_2S environment. *Int. J. Hydrog. Energy* **2018**, *43*, 2293–2306. [CrossRef]
33. Rao, C.N.R.; Subba Rao, G.V. Electrical conduction in metal oxides. *Phys. Status Solidi* **1970**, *1*, 579–652. [CrossRef]
34. Frappart, S.; Feaugas, X.; Creus, J.; Thebault, F.; Delattre, L.; Marchebois, H. Study on the hydrogen diffusion and segregation into Fe-C-Mo martensitic HSLA steel using electrochemical permeation test. *J. Phys. Chem. Solids* **2010**, *71*, 1467–1479. [CrossRef]
35. Bolzon, G.; Boukharouba, T.; Gabetta, G.; Elboujdaini, M.; Mellas, M. *Integrity of Pipelines Transporting Hydrocarbons*; Springer Science & Business Media: Dordrecht, The Netherlands, 2011.
36. Zakroczymski, T. Adaptation of the electrochemical permeation technique for studying entry, transport and trapping of hydrogen in metals. *Electrochim. Acta* **2006**, *51*, 2261–2266. [CrossRef]
37. Shirley, A.I.; Hall, C.K. Trapping of hydrogen by substitutional and interstitial impurities in α-iron. *Scr. Met.* **1983**, *17*, 1003–1008. [CrossRef]
38. *Standard Test Method for Laboratory Testing of Metals for Resistance to Sulfide Stress Cracking and Stress Corrosion Cracking in H2S Environments*; NACE TM 0177; NACE International: Houston, TX, USA, 2016.
39. Kop, T.A.; Sietsma, J.; Zwaag, S.V.D. Anisotropic dilatation behavior during transformation of hot rolled steels showing banded structure. *Mater. Sci. Tech.* **2001**, *17*, 1569–1574. [CrossRef]
40. Farahani, H.; Xu, W.; Zwaag, S.V.D. A novel approach for controlling the band formation in medium Mn steels. *Metall. Mater. Trans.* **2018**, *49A*, 1998–2010. [CrossRef]
41. De, A.C.G.; Capdevila, C.; Caballero, F.G.; San, M.D. Effect of molybdenum on continuous cooling transformations in two medium carbon forging steels. *J. Mater. Sci.* **2001**, *36*, 565–571.
42. Hu, H.; Xu, G.; Zhou, M.; Yuan, Q. Effect of Mo content on microstructure and property of low-carbon bainitic steels. *Metals* **2016**, *6*, 173. [CrossRef]
43. Komenda, J.; Sandstrom, R. Assessment of pearlite banding using automatic image analysis: Application to hydrogen-induced cracking. *Mater. Charact.* **1993**, *31*, 143–153. [CrossRef]
44. Cheng, R.; Zhang, J.; Wang, B. Deformation behavior of inclusion system CaO-Al_2O_3-SiO_2 with different compositions during hot rolling processes. *Trans. Indian Inst. Met.* **2018**, *71*, 705–713. [CrossRef]
45. Bhadeshia, H.K.D.H. Prevention of hydrogen embrittlement in steels. *ISIJ Int.* **2016**, *56*, 24–36. [CrossRef]
46. Choo, W.Y.; Lee, J.Y. Thermal analysis of trapped hydrogen in pure iron. *Met. Trans.* **1982**, *13*, 135–140. [CrossRef]
47. Fragiel, A.; Serna, S.; Malo-Tamayo, J.; Silva, P.; Campillo, B.; Martinez-Martinez, E.; Cota, L.; Staia, M.H.; Puchi-Cabrera, E.S.; Perez, R. Effect of microstructure and temperature on the stress corrosion cracking of two microalloyed pipeline steels in H_2S environment for gas transport. *Eng. Fail. Anal.* **2019**, *105*, 1055–1068. [CrossRef]
48. Jeng, H.W.; Chiu, L.H.; Johnson, D.L.; Wu, J.K. Effect of pearlite morphology on hydrogen permeation, diffusion, and solubility in carbon steels. *Mater. Trans.* **1990**, *21*, 1990–3257. [CrossRef]
49. Lee, W.B.; Hong, S.G.; Park, C.G.; Kim, K.H.; Park, S.H. Influence of Mo on precipitation hardening in hot rolled HSLA steels containing Nb. *Scr. Mater.* **2000**, *43*, 319–324. [CrossRef]
50. Wei, F.G.; Tsuzaki, K. Quantitative analysis on hydrogen trapping of TiC particles in steel. *Metall. Mater. Trans.* **2006**, *37A*, 331–353. [CrossRef]
51. Eeckhout, E.V.D.; Depover, T.; Verbeken, K. The effect of microstructural characteristics on the hydrogen permeation transient in quenched and tempered martensitic alloys. *Metals* **2018**, *8*, 779. [CrossRef]
52. Lee, J.; Lee, T.; Kwon, Y.J.; Mun, D.J.; Yoo, J.Y.; Lee, C.S. Role of Mo/V carbides in hydrogen embrittlement of tempered martensitic steel. *Corros. Rev.* **2015**, *33*, 433–441. [CrossRef]

53. Nagao, A.; Hayashi, K.; Oi, K.; Mitao, S. Effect of uniform distribution of fine cementite on hydrogen embrittlement of low carbon martensitic steel plates. *ISIJ Int.* **2012**, *52*, 213–221. [CrossRef]
54. Takahashi, J.; Kawakami, K.; Kobayashi, Y.; Tarui, T. The first direct observation of hydrogen trapping sites in TiC precipitation-hardening steel through atom probe tomography. *Scr. Mater.* **2010**, *63*, 261–264. [CrossRef]
55. Kim, S.J. Effect of the elastic tensile load on the electrochemical corrosion behavior and diffusible hydrogen content of ferritic steel in acidic environment. *Int. J. Hydrog. Energy* **2017**, *42*, 19367–19375. [CrossRef]
56. Mullin, J.W. *Crystallization*, 3rd Ed. ed; Butterworth-Heinemann: London, UK, 1993.
57. Farelas, F.; Galicia, M.; Brown, B.; Nesis, S.; Castaneda, H. Evolution of dissolution processes at the interface of carbon steel corroding in a CO_2 environment studied by EIS. *Corros. Sci.* **2010**, *52*, 509–517. [CrossRef]
58. Koh, S.U.; Lee, J.M.; Yang, B.Y.; Kim, K.Y. Effect of molybdenum and chromium addition on the susceptibility to sulfide stress cracking of high-strength, low-alloy steels. *Corrosion* **2007**, *63*, 220–230. [CrossRef]

© 2020 by the authors. Licensee MDPI, Basel, Switzerland. This article is an open access article distributed under the terms and conditions of the Creative Commons Attribution (CC BY) license (http://creativecommons.org/licenses/by/4.0/).

Article

Hydrogen Accumulation and Distribution in Pipeline Steel in Intensified Corrosion Conditions

Anatolii I. Titov [1], Aleksandr V. Lun-Fu [1], Aleksandr V. Gayvaronskiy [2], Mikhail A. Bubenchikov [1,*], Aleksei M. Bubenchikov [3], Andrey M. Lider [4], Maxim S. Syrtanov [4] and Viktor N. Kudiiarov [4]

1 PAO 'Gazprom Transgaz Tomsk', 9, Frunze Ave., 634029 Tomsk, Russia; office@gtt.gazprom.ru (A.I.T.); office@gtt.gazprom.ru (A.V.L.-F.)
2 OOO 'Gazprom Transgaz Ukhta', 10/1, Naberezhnaya Gazovikov, 169300 Ukhta, Komi Republic; sgp@sgp.gazprom.ru
3 National Research Tomsk State University, 36, Lenin Ave., 634050 Tomsk, Russia; Aleksy121@mail.ru
4 National Research Tomsk Polytechnic University, 30, Lenin Ave., 634050 Tomsk, Russia; lider@tpu.ru (A.M.L.); maxim-syrtanov@mail.ru (M.S.S.); viktor.kudiiarov@gmail.com (V.N.K.)
* Correspondence: M.Bubenchikov@gtt.gazprom.ru

Received: 26 March 2019; Accepted: 28 April 2019; Published: 30 April 2019

Abstract: Hydrogen accumulation and distribution in pipeline steel under conditions of enhanced corrosion has been studied. The XRD analysis, optical spectrometry and uniaxial tension tests reveal that the corrosion environment affects the parameters of the inner and outer surface of the steel pipeline as well as the steel pipeline bulk. The steel surface becomes saturated with hydrogen released as a reaction product during insignificant methane dissociation. Measurements of the adsorbed hydrogen concentration throughout the steel pipe bulk were carried out. The pendulum impact testing of Charpy specimens was performed at room temperature in compliance with national standards. The mechanical properties of the steel specimens were found to be considerably lower, and analogous to the properties values caused by hydrogen embrittlement.

Keywords: trunk gas pipelines; hydrogen embrittlement; hydrogen adsorption; pipeline inner surface; cathodic protection current; hydrogen distribution

---

1. Introduction

Metal corrosion is currently one of the most important issues of oil and gas transportation [1]. Cumulative corrosion damage in the oil and gas industry is estimated to be more than $1 billion annually [2]. At the same time, revenue losses due to repair operations exceed $10 billion annually [3]. The United States Department of Transportation reported one of the major causes of pipeline failure to be stress corrosion cracking (SCC) [4]. In accordance with the pipeline classification (by soil type and structure, pressure, temperature, and cathodic protection) accepted by the expert community, SCC is caused by different mechanisms [5]. Propagation of stress corrosion cracking in a soil with pH higher than 9 is induced by cathodic protection currents and the coating condition, usually SCC occurs in carbonate-based electrolytes. The latter represent carbonic acid (H_2CO_3) which dissociates in water and forms -CO_3 and -HCO_3 which also appear in soil due to dissolution of carbon dioxide (CO_2) [6]. This SCC mechanism includes intercrystalline fracture caused by the changes in chemical properties along the grain boundaries. In this case, primary dissolution and decrease in cohesive energy of the grain boundary inclusions are observed. This SCC mechanism occurs in Australia, Argentina, Iran, Saudi Arabia, the United States, and also in regions of Central Asia and Kazakhstan [7–9]. In Russia this mechanism is not observed.

Propagation of stress corrosion cracking in a soil with pH varying from 6–8 is caused by the mechanism of transcrystalline fracture that propagates through the grain, regardless of the grain boundary. Most of the theories explain this SCC mechanism by the synergetic effect caused by hydrogen dissolution in steel, corrosive medium, strain and adsorption of other substances on the pipeline inner surface. This results in reduction in the atomic binding energy of the crystal lattice. This SCC mechanism is observed in Canada, Italy and some regions of Russia and other countries [7–9].

According to [10,11], propagation of stress corrosion cracking in a soil with pH lower than 6 occurs due to the mechanism of hydrogen permeation in the pipe steel and its hydrogen embrittlement. Saturation of the steel pipe with hydrogen is possible in acid soil due to the free migration or electrophoretic mobility of H ions from corrosive environment to the pipe metal provided by the excessively intensive cathodic protection [12]. This SCC mechanism occurs mostly in China and Russia [13].

Atomic hydrogen absorbed into steel and diffused in it by the interstitial mechanism, is pressed from all sides of the crystal lattice except for defects (dislocations, vacancies, grain boundaries), and tends to be displaced therein. With the increase in the hydrogen amount by hundreds of times, its molecularization occurs, thereby creating the local pressure greater than 100 atmospheres [14–16]. Moreover, when hydrogen reacts with some of alloying elements, it forms hydrides that bring these elements out of the functional state of substitutional or interstitial solutions [14–16]. Hydride, carbide and hydride segregations, which form in dislocations or along the grain boundaries, lower their mobility and cohesive energy resulting in formation of coalescences with defects [14–16]. In addition, hydrogen causes reduction in the material hardness changing, the surface energy during the simple interpolation, i.e., the Rehbinder effect [17]. At the macro-scale level, hydrogen modifies the elastoplastic properties of steel and facilitates both reduction in its plasticity via hydrogen embrittlement, and localized plasticization of crack nuclei; in its turn, the localized plasticization facilitates cracks growth [7].

Hydrogen embrittlement is a very serious problem widely discussed by researchers worldwide. Researchers consider hydrogen embrittlement in relation to development of hydrogen energy technologies and pipeline transport of hydrogen rather than natural gas [18,19]. Hydrogen permeates a steel pipe not only from the outside but also from the inside due to dissociation of hydrocarbons (methane). This is because of either chemisorption of methane, the main gas component [20], on the steel surface and its transformation on the inner, slightly rusted surface at evaporation or CO_2 conversion of methane [21]. Since VIII group metals Fe, Ni, Co can catalyze dehydrogenation reaction of methane [22], it is possible for hydrogen to adsorb on the inner surface as a reaction product. The rate of chemical conversion of methane is not zero even at 100 °C, although in industrial conditions it usually occurs at the temperature above 700 °C. This is due to the presence of molecules having higher kinetic energy as a result of the velocity distribution of gas molecules [23].

It is worth noting that when the products of the dehydrogenation reaction of methane are withdrawn from the reaction site which absorbs hydrogen atoms, the chemical equilibrium shifts to the right and, therefore, adsorption is the limiting factor of the inverse reaction [23].

Methane Dissociation

The degree of dissociation of methane (CH_4) molecules is determined by their chemisorption on Ni catalyst. The Gibbs free energy (Table 1) of CH_4 decomposition reaction is indicated as:

$$H_4 \leftrightarrow C + 2H_2 \tag{1}$$

Table 1. Standard values of the Gibbs free energy, enthalpy and entropy of formation of substances involved in Reaction (1) [24].

Substances	CH_4	C	H_2
Standard Gibbs free energy, ΔG^0, kJ/mole	50	0	0
Standard enthalpy of formation, ΔH^0, kJ/mole	74.6	0	0
Standard entropy of formation, ΔS^0, kJ/mole	186.19	5.7	130.5

The change in the Gibbs free energy is $\Delta G = \Delta H - T\Delta S$, where H is the enthalpy, T is the absolute temperature, S is the entropy. This change describes the permissible process of the chemical reaction [23]. Thus, under the given thermodynamic conditions, the reaction occurs at $\Delta G < 0$ and does not occur at $\Delta G > 0$. As for the decomposition Reaction (1), $\Delta H < 0$, $\Delta S < 0$ and $\Delta G < 0$ at $T < 929$ K. Hence, methane dissociation occurs due to the enthalpy of formation only and is typical for low temperatures. The literature presents disaggregated data [25] on methane dissociation on iron oxides and hydroxides. Nevertheless, it can be measured by the upper level of its efficiency using the data on methane dissociation on Ni catalysts because they are more efficient than Fe [26]. According to [27], the reaction equilibrium constant of methane dissociation at T = 373 K is written as

$$K_p^1 = 0.606 \times 10^{-7} \tag{2}$$

This indicates the absolutely permissible methane dissociation on the inner surface of the pipeline under the conditions approaching to normal.

As for other permissible transformations of methane, the reactions of combination of methane and water vapour, carbon dioxide, and oxygen can occur by the following mechanisms [28,29]:

$$CH_4 + H_2O \leftrightarrow CO + 3H_2 \tag{3}$$

$$CH_4 + 2H_2O \leftrightarrow CO_2 + 4H_2 \tag{4}$$

$$CH_4 + CO_2 \leftrightarrow 2CO + 2H_2 \tag{5}$$

$$CH_4 + O_2 \leftrightarrow C + 2H_2O \tag{6}$$

$$2CH_4 + O_2 \leftrightarrow 2CO + 4H_2 \tag{7}$$

$$CH_4 + 2O_2 \leftrightarrow CO_2 + 2H_2O \tag{8}$$

Reactions (2) and (3) are steam methane reforming; Reaction (4) is CO_2 conversion; Reactions (5) and (6) are partial oxidation, and Reaction (7) is complete oxidation or burning of methane. Reactions (2) and (3) describe the same process, provided that Reaction (3) describes further CO_2 conversion. For Reactions (5–7) oxygen is required, however, its content in natural gas is very low [30].

The reactions of methane conversion (2–4) are usually performed at the temperature over 400 °C and 1–4 MPa pressure, in the presence of catalysts [31,32]. This is because of their high activation barrier due to relatively high energy of methane dissociation or its C–H chemical bonds, such as the first 435 kJ/mole bond, the second and third bonds 444 kJ/mole each, and the fourth 335 kJ/mole bond [26]. According to [31,32], Pt, Pd, Co, Ni, Fe and Cu and their oxides are more efficient catalysts for such reactions.

It is assumed that steam methane reforming consists of two stages:

$$CH_4 + H_2O \leftrightarrow 3H_2 + CO\ (-206\ \text{kJ}) \tag{9}$$

$$CO + H_2O \leftrightarrow H_2 + CO_2\ (+42\ \text{kJ}) \tag{10}$$

The first stage is strongly endothermal and critical to steam methane reforming. Both stages are reversible reactions.

Measurements of the Gibbs free energy for steam methane reforming demonstrate that $\Delta G > 0$ at $T = 373$ K. According to [29], at this temperature, the reaction equilibrium constant for steam methane reforming is

$$K_p^1 = 0.269 \times 10^{-19} \tag{11}$$

Therefore, hydrogen formation on the inner surface and in the depth of the pipe can be explained by methane chemisorption. This is proven by the data about carbon forming on the inner pipe surface [33].

In compliance with the national standards [30], the concentration of water vapour in the transported natural gas composition should not exceed 30 mg/m^3. Hydrostatic testing of gas pipelines demonstrates the water remains on the inner surface of the pipeline in the amount of 30 tons per 100 km owing to its non-zero wettability and roughness [13]. Despite the traditional purification of natural gas from CO_2 with the view to prevent crystallohydrate formation, it may contain up to 30 g/m^3 CO_2 [30].

The possibility of low-temperature methane transformations is supported by the nature. Thus, iron is used as a catalyst not only in chemical industry technologies. Obligate methanotrophs are methane loving microorganisms which receive energy and raw material for cellular structures necessary for life from methane [34] and oxidize methane such that

$$CH_4 + O_2 + NAD(P)H + H^+ \leftrightarrow CH_3OH + NAD(P)^+ + H_2O \tag{12}$$

Methanotrophs use methane monooxygenase (MMO), an enzyme that exists in two forms, namely the soluble form (sMMO) and the particulate form (pMMO). The active site in sMMO contains an iron atom, whereas the active site in pMMO utilizes copper [34]. There are several thousand species of methanotrophs that effectively develop at the temperatures ranging from 10–70 °C [34]. Some species of methanotrophs live deep underwater in the arctic region. Therefore thermodynamic conditions inside gas pipelines with the temperature range of 5–15 °C and 25–70 atm pressure and iron oxides and hydroxides on the inner surface are comparable to methanotrophs thermodynamics. Most of methanotrophs are aerobic and cannot live inside gas pipelines because of the insufficient oxygen percentage of 0.001% in transport gas. There are, however, anaerobic methanotrophs which are currently being investigated [34]. Consequently, the possibility of reaction of methane decomposition at low temperatures is confirmed by measurements of the Gibbs free energy and metabolic capabilities of methanotrophs.

The goal of this research is to prove the possibility of hydrogenation of the inner surface of gas pipelines caused by dissociation of hydrocarbons. A gas reactor is designed to simulate the operating conditions for the real gas pipeline. Hydrogen accumulation and distribution in steel specimens are studied in the proposed gas reactor under the conditions of the accelerated corrosion test.

2. Materials and Methods

On the territory of the Russian Federation for the construction of trunk gas pipelines, structural low-alloy steels of grades 09G2S and X70 are commonly used, since their characteristics allow for operating under the pressure within the wide temperature range (−70, +450 C), being durable and resistant to dynamic loads. Therefore, a pipe made of 09G2S steel was chosen as a sample for the experiment. The production technology of this steel does not imply extensive use of hydrogen at the stage of dispersal, therefore its initial content in steel after smelting remains negligible. By its composition, steel 09G2S most closely relates to international analogues—steel grades A 516-55, A 516-60, A 516-65, A 561 Gr70, produced in the USA. The group of these steels demonstrates resistance to sulfide stress corrosion cracking (SSCC) and is recommended for the use with acid gases. At the same time, there is sufficient data on the susceptibility of these steels to stress corrosion cracking [6–9], which in some cases is also explained by hydrogen damage. Hydrogen-induced cracking (HIC) test results demonstrate that high-strength steels are most susceptible to it, and in the series of X52, X60, X70, and 100XF steels, the latter is most vulnerable. However, X70 also proved to be sufficiently exposed to HIC, which makes it possible to speak of the possibility of its corrosion stress cracking, which develops with the help of HIC.

The gas reactor supplying methane-based mixture was designed to simulate the corrosive environment matching the operating conditions for the real gas pipeline that enabled intensification of the studied processes increasing the temperature and humidity. The simulation system for hydrogenation of the pipe inner surface is illustrated in Figure 1.

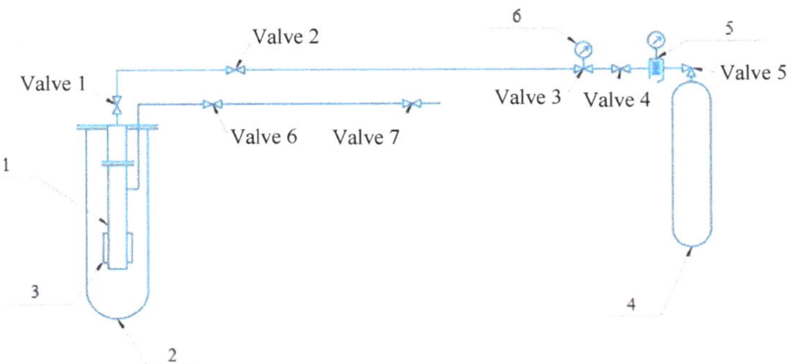

Figure 1. Simulation system for the inner surface hydrogenation in the intensified corrosion conditions: 1—reactor, 2—casing, 3—heater element, 4—methane gas canister, 5—pressure controller, 6—manometer.

Table 2. summarizes the gas mixture components supplied to the gas reactor.

Table 2. Percentage of gas mixture components.

Substance	CH_4	C_2H_6	C_3H_8	C_4H_{10}	CO_2	H_2S
Weight content, %	97	1.5	0.5	10^{-2}	10^{-3}	10^{-5}

The proposed gas reactor is a closed steel pipe prepared for operating under pressure. The steel pipe has two valves. Valve 1 supplies gas in the reactor, whereas valve 6 serves for gas sampling. Reactor 1 has a removable pipe section just above heater element 3 which is withdrawn for examination from the reactor after the experiment. Steel specimens (templates) are then cut from the removable pipe section and analyzed.

The gas mixture is supplied to the reactor under 40 atm pressure which is measured with manometer 5. The composition of the gas mixture is given in Table 2. The installation was designed so that it would be possible to conduct studies of steel samples subjected to aggressive impact both from the outside and from the inside. However, the goal of this study was to investigate the concentration and distribution of hydrogen adsorbed in the process of methane chemisorption inside the pipe, and therefore there was no aggressive action outside.

The 3 kW nickel chromium clamp heater is connected to the temperature controller and two thermocouples mounted inside the reactor. The thermocouples are accurate to within 0.1 °C, with 1° accuracy of the controller, and 100 °C heating stop temperature. The heater element allows the corrosion tests to be intensified at various temperatures, because, in accordance with the main statements of chemical kinetics, it affects the rate of chemical reactions.

In order to conduct the uniaxial tension tests and determine the elemental composition of the pipe, the electro discharge machining was used to cut off specimens from the pipe center parallel and normal to the axial direction. The X-ray diffraction (XRD) analysis was carried out on templates 2 mm thick cut off from the pipe at various depth. For the uniaxial tension tests, the dog bone specimens were cut off in compliance with the ISO 6892-84 [29–32], as presented in Figure 2.

Figure 2. Specimens dimensions in millimeters: (**a**)—for elemental analysis, (**b**)—for uniaxial tension tests.

The steel specimens were studied in two states, before and after 24-h exposure to the humid and heating environment. The templates and dog bone specimens were obtained from the treated specimens for further investigations.

The phase composition of the templates was investigated on the XRD-7000S X-ray diffractometer (Shimadzu, Japan), and the obtained diffraction patterns were analyzed. Measurements were conducted using copper radiation ($K_{\alpha 1}$, $K_{\alpha 2}$). The operating parameters for the XRD-7000S included 10–90° scan range; 0.0143° step angle; 2.149 s exposure time. The XRD patterns were recorded on the OneSight wide-range array detector with 1280 channels (Shimadzu, Japan).

The qualitative and quantitative elemental analysis of the treated steel specimens was performed on the GD-Profiler 2 spectrometer (Horiba Scientific, Japan) which combines glow discharge powered by the radio frequency source with the optical emission spectrometer (RF-GD-OES).

The test machine Com-Ten Industries DFM-5000 was used in this experiment for tensile testing under the dead load. Metal parts with curved square section 5 × 5 mm^2 were cut from pipe and then were mechanically grinded. Then flat specimens with square section 4.5 × 4.5 mm^2 were tested on the impact pendulum Instron 450MPX impact tester (UK) at room temperature. The impact testing was carried out in compliance with national standards.

3. Results and Discussion

3.1. XRD Analysis

The XRD analysis of the specimens was performed before and after 24-h exposure to humid and heating environment. Five specimens were used for each test. The diffraction patterns obtained for these specimens are given in Figure 3.

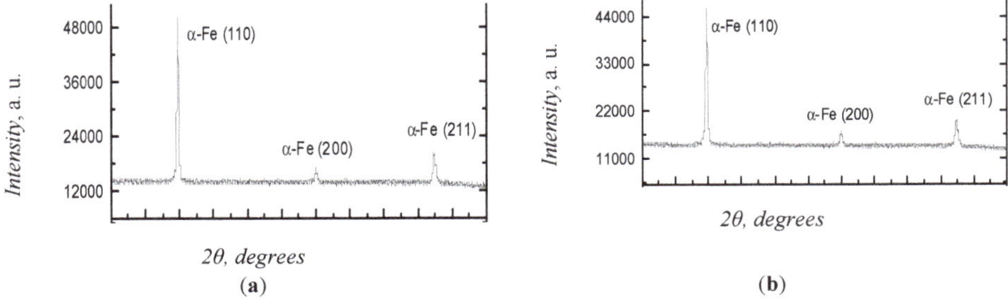

Figure 3. XRD patterns of the steel specimens: (**a**)—before exposure to; (**b**)—after exposure.

The analysis of the diffraction patterns demonstrates that all the specimens contain only the α-Fe phase with the body-centered cubic (BCC) lattice. Table 3 presents the lattice parameters, the average grain size estimated by the XRD data-based calculations of the size of the coherent scattering region, and the microstress values.

Table 3. XRD data.

Treatment	Phases	Phase Composition, vol.%	Lattice Parameters	Microstress
Before	BCC α-Fe phase	100	$a = 2.8697$	0.001065
After	BCC α-Fe phase	100	$a = 2.8692$	0.000425

As can be seen from Table 3, long-term exposure of steel specimens to the corrosive environment leads to the microstress relaxation without any changes in the phase composition and lattice parameters.

3.2. Elemental Composition

The elemental composition of the steel templates was examined both on the inner and outer surfaces of the pipeline. Table 4 summarizes the elemental composition detected by the RF-GD-OES for the outer and inner surfaces of the untreated specimens. The weight content of the chemical elements is presented as a result of ten measurements. The relative uncertainty of measurements is about 1%.

Table 4. Elemental composition of the outer and inner surfaces in the untreated specimens.

Chemical Elements	Weight Content, % (Outer Surface)	Weight Content, % (Inner Surface)
C	0.078	0.072
Si	0.321	0.352
Mn	1.331	1.266
S	0.002	0.003
Fe	98.268	98.307

The elemental analysis reveals that both surfaces of the untreated specimens are similar. The relative difference in the weight content is 1%. The elemental composition matches AISI 1513 carbon steel grade.

The RHEN 602 gas analyzer (LECO, USA) is used to measure the absolute hydrogen concentration in the untreated steel specimens cut off from the inner and outer surfaces and from the bulk. The obtained results are presented in Table 5. The hydrogen concentration values were obtained after three experiments. In calculating the standard deviation of the mean, the confidence figure was assumed to be 0.95; the Student's coefficient was 4.3.

Table 5. H content in the untreated steel specimens.

Weight Content, wt.%	Outer Surface	Bulk	Inner Surface
H content	0.00069	0.00067	0.00065
Direct measurement error	0.00005	0.00004	0.00003

Within the measurement uncertainties, the hydrogen concentration in the specimens before the treatment is similar in all the pipe sections. Table 6 presents the RF-GD-OES data for the elementary composition of the outer and inner surfaces of the treated specimens. The weight content of the chemical elements is presented as a result of ten measurements. The relative uncertainty of measurements is about 1%.

Table 6. Elemental composition of the outer and inner surfaces in the treated specimens.

Chemical Elements	Weight Content, % (Outer Surface)	Weight Content, % (Inner Surface)
C	0.075	0.071
Si	0.318	0.339
Mn	1.315	1.331
S	0.004	0.004
Fe	98.288	98.255

According to the elemental analysis, the outer and inner surfaces of the steel specimens exposed to the corrosive environment are similar. The relative difference in the weight content ranges within 1%. The elemental composition matches AISI 1513 carbon steel grade.

The results of the gas analysis of the absolute hydrogen concentration in the treated steel specimens cut off from the inner and outer surfaces and from the bulk are presented in Table 7. The hydrogen concentration values are obtained after three experiments. In calculating the standard deviation of the mean, the confidence figure was assumed to be 0.95; the Student's coefficient was 4.3.

Table 7. H content in the treated steel specimens.

Weight Content, wt.%	Outer Surface	Bulk	Inner Surface
H content	0.00020	0.00083	0.00086
Direct measurement error	0.00005	0.00006	0.00004

The gradient hydrogen distribution is observed in steel specimens after their treatment in the gas environment, humid and heated. On the inner surface and in the bulk of the material the H concentration is 4 times higher than on the outer surface. The decrease in the hydrogen content on the outer surface of the pipe after treatment may be due both to the diffusion of hydrogen into the pipe bulk and to the partial desorption of hydrogen as a result of heating.

The H concentration on the inner surface in the initial steel state is 0.00065 wt.%, whereas after the gas treatment it is 0.00086 wt.% i.e., hydrogen permeates the steel when subjected to the gas treatment during heating.

3.3. Uniaxial Tension Tests

The uniaxial tension tests were performed at room temperature. The dog bone specimens were cut off from the pipe steel in its initial state and after treatment in the corrosive environment. Such cases are presented in Figure 4.

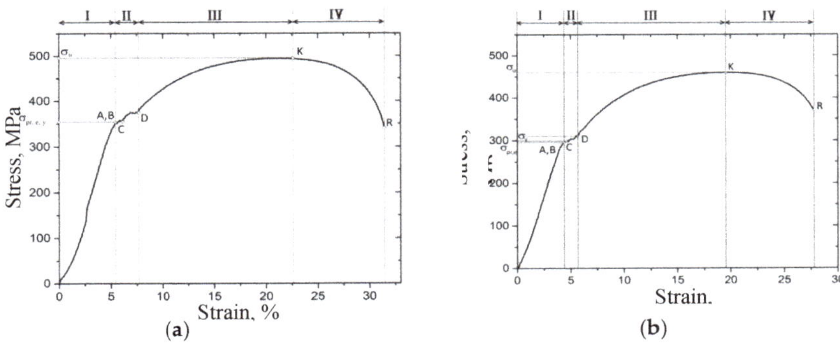

Figure 4. The stress-strain curves: (**a**)—before treatment, (**b**)—after treatment.

Table 8 presents the main parameters of steel specimens before and after treatment.

Table 8. Main parameters of steel specimens in different states.

	Before Treatment	After Treatment
Proportional limit strength ± 20, MPa	352	296
Elastic limit strength ± 20, MPa	352	296
Yield strength ± 20, MPa	352	300
Tensile strength ± 20, MPa	494	462
Maximum percent elongation ± 0.1, %	32	27

According to Table 8, the effect from the humid gas environment on the heated specimen results in significant degradation of its mechanical properties. Thus, the values of the proportional limit strength, elastic limit strength and the yield strength are reduced by 20%; tensile strength decreases by 7% and the maximum percent elongation decreases by 18%.

3.4. Pendulum Impact Testing

The pendulum impact testing of Charpy specimens was carried out at room temperature in accordance with the national standards. The impact testing results are summarized in Table 9.

Table 9. Fracture toughness of steel specimens in different states.

Resilience KCV, J/cm^2	Initial State	State after Treatment
KCV ± 20	355	345

The impact tests reveal that the resilience of the untreated and treated specimens is 355 and 345 J/cm^2 respectively. These values are averaged by three dimensions for each specimen. There is no serious change in both states of specimens due to the low H concentration which does not exceed the solubility limit of hydrogen in steel. The impact of aggressive environment and the accumulation of hydrogen influence on the behavior of the material in the process of uniaxial stretching. However, these factors did not affect the behavior of the material when tested for impact strength. The reasons for this will be investigated further.

4. Conclusions

The effect the corrosive environment produces on the structure, elemental composition, mechanical properties and the impact toughness of the pipe steel have been studied. The XRD analysis, optical spectrometry and uniaxial tension tests revealed that the corrosion environment affects the parameters of the inner and outer surfaces of the steel specimens as well as the steel pipeline bulk.

In the initial state, on the inner surface of the specimens H content was 0.00062 wt.%, whereas after exposure to the corrosive heating environment it reached 0.00086 wt.%. That indicated hydrogen permeation in the steel during the experiment. Hydrogen accumulation resulted in steel embrittlement proven by the mechanical tests.

Author Contributions: A.I.T. conceived the paper; all the authors contributed to the data collection, analysis and comments; all authors wrote the paper.

Funding: This work was carried out within the framework of the Competitiveness Enhancement Program of National Research Tomsk Polytechnic University (VIU-OEF-66/2019).

Conflicts of Interest: The authors declare no conflict of interest.

References

1. Popoola, L.T.; Grema, A.S.; Latinwo, G.K.; Gutti, B.; Balogun, A.S. Corrosion Problems During Oil and Gas Production and its Mitigation. *Int. J. Ind. Chem* **2013**, 4–35. [CrossRef]
2. Walsh, M.R.; Hancock, S.H.; Wilson, S.J.; Patil, S.L.; Moridis, G.J.; Boswell, R.; Sloan, E.D. Preliminary report on the commercial viability of gas production from natural gas hydrates. *Energy Econ.* **2009**, *31*, 815–823. [CrossRef]
3. Popoola, L.T.; Grema, A.S.; Latinwo, G.K.; Gutti, B.; Balogun, A.S. *Opened Report: Corrosion Mitigation for Complex Environments*; Champion Technologies: Houston, TX, USA, 2012.
4. US Department of Transportation Pipeline and Hazardous Materials Safety Administration. *Opened Report: Development of Guidelines for Identification of SCC Sites and Estimation of Re-Inspection Intervals for SCC Direct Assessment*; US Department of Transportation Pipeline and Hazardous Materials Safety Administration: New Jersey Ave, Washington, USA, 2010.
5. CEPA Pipeline Integrity Working Group. *Recommended Practices for Managing Near-neutral pH Stress Corrosion Cracking*, 3rd ed.; CEPA Pipeline Integrity Working Group: Calgary, AB, Canada, 2015; p. 162.
6. Sandana, D. Stress Corrosion Cracking of Pipeline Steels in Contaminated Aqueous CO_2 Environments. Ph.D. Thesis, School of Chemical Engineering and Advanced Materials, Newcastle University, Newcastle, UK, July 2016.
7. Cheng, Y.F. (Ed.) *Stress Corrosion Cracking of Pipelines*; John Wiley & Sons, Inc.: Hoboken, TX, Canada, 2013; p. 275.
8. Raja, V.S.; Shoji, T. (Eds.) *Stress Corrosion Cracking, Theory and Practice*; Woodhead Publishing Ltd.: Cambridge, UK, 2011; p. 816.
9. Jones, R.H. (Ed.) *Stress-Corrosion Cracking: Materials Performance and Evaluation*; ASM International: Materials Park, OH, USA, 2017.
10. Liu, Y.; Li, Q.; Cui, Z.Y.; Wua, W.; Li, Z.; Du, C.W.; Li, X.G. Field Experiment of Stress Corrosion Cracking Behavior of High Strength Pipeline Steels in Typical Soil Environments. *Constr. Build. Mater.* **2017**, *148*, 131–139. [CrossRef]
11. Gao, S.-J.; Dong, C.-F.; Fu, A.-Q.; Xiao, K.; Li, X.-G. Corrosion Behavior of the Expandable Tubular in Formation Water. *Int. J. Min. Met. Mater.* **2015**, *22*, 149–156. [CrossRef]
12. Zoski, C.G. *Handbook of Electrochemistry*, 1st ed.; Elsevier: Amsterdam, The Netherlands; Kidlington, UK, 2007; p. 935.
13. Mazur, I.I.; Ivantsev, O.M. *Pipeline System Safety [in Russian]*; ELIMA: Moscow, Russia, 2004; p. 1104.
14. Barrera, O.; Bombac, D.; Chen, Y.; Daff, T.D.; Galindo-Nava, E.; Gong, P.; Haley, D.; Horton, R.; Katzarov, I.; Kermode, J.R.; Liverani, C.; Stopher, M.; Sweeney, F. Understanding and Mitigating Hydrogen Embrittlement of Steels: A Review of Experimental, Modelling and Design Progress From Atomistic to Continuum. *J. Mater. Sci.* **2018**, *53*, 6251–6290. [CrossRef]

15. Nechaev, Y.S. *NATO Advanced Research Workshop on Hydrogen Materials Science and Chemistry of Metal Hydrides*; Ser. II, Kluwer Acad. Publ.: Dordrecht, The Netherlands, 2002; p. 161.
16. Malkin, A.I. Regularities and Mechanisms of the Rehbinder's Effect. *Colloid J.* **2012**, *74*, 223–238. [CrossRef]
17. Nagumo, M. *Fundamentals of Hydrogen Embrittlement*; Springer: Berlin, Germany, 2016; p. 239.
18. Gangloff, R.P.; Somerday, B.P. *Gaseous Hydrogen Embrittlement of Materials in Energy Technologies: The Problem, its Characterisation and Effects on Particular Alloy Classes*; Elsevier, Woodhead Publishing: Sawston, Cambridge, UK, 2012; p. 964.
19. Briottet, L.; Moro, I.; Lemoine, P. Quantifying the Hydrogen Embrittlement of Pipeline Steels for Safety Considerations. *Int. J. Hydrogen Energy* **2012**, *37*, 616–623. [CrossRef]
20. Ashwani, S.N.; Tiwari, K.; Jackson, B. Dissociative Chemisorption of Methane on Ni and Pt Surfaces: Mode-Specific Chemistry and the Effects of Lattice Motion. *J. Phys. Chem. A* **2014**, *118*, 9615–9631.
21. Guo, H. The Dissociative Chemisorption of Methane and its Isotopologues on Metal Surfaces. Doctoral Dissertation, University of Massachusetts Amherst, Amherst, MA, USA, 2018.
22. Germain, J.E. *Catalytic Conversion of Hydrocarbons*; Academic Press Inc.: London, UK, 1969; p. 322.
23. Kauzmann, W. *Kinetic Theory of Gases*; Dover Publications: Mineola, NY, USA, 2013; p. 272.
24. Dean, J.A. *Lange's Handbook of Chemistry*, 15th ed.; McGraw Hill, Inc.: New York, NY, USA, 1999; p. 1291.
25. Bao, S.; Fan, C.; Li, H.; Xu, Y. The Adsorption and Decomposition of Methane on Fe/Cu(110) Bimetallic Surface. *Sci. China Math. Phys. Astron. Technol. Sci.* **1995**, *38*, 813. [CrossRef]
26. Galaktionova, L.V.; Arcatova, L.A.; Kurina, L.N.; Gorbunova, E.I.; Belousova, V.N.; Nyborodenko, Yu.S.; Kasatsky, N.G.; Golobokov, N.N. Fe-Containing Intermetallic Compounds as Catalysts for Carbon Dioxide Reforming of Methane. *J. Phys. Chem. A* **2008**, *82*, 271–275.
27. Anderson, J.R.; Boudart, M. (Eds.) *Catalysis Science and Technology*; Springer-Verlag: Berlin/Heidelberg, Germany, 1984; p. 281.
28. Xu, J.; Froment, G.F. Methane Steam Reforming, Methanation and Water-Gas Shift: 1. Intrinsic Kinetics. *AICHE J.* **1989**, *35*, 88–96. [CrossRef]
29. Arutyunov, V.C.; Krylov, O.V. *Oxidative Transformations of Methane*; Nauka: Moscow, Russia, 1998; p. 362.
30. Tikhomirov, D.V.; Krashennikov, S.V.; Don, B.D.; Makinsky, A.A. *Combustible Natural Gas Supplied and Transported through Gas Pipelines*; STO Gazprom 089-2010: Moscow, Russia, 2011.
31. Pismen, M.K. *Production of Hydrogen in Oil Industry*; Khimiya: Moscow, Russia, 1976; p. 208.
32. Bargin, O.V.; Liberman, A.L. *Hydrocarbons Incremented by Metal Catalysts*; Khimiya: Moscow, Russia, 1981; p. 264.
33. Nechaev, Y.S. Physical Complex Problems of Aging, Embrittlement and Destruction of Metallic Materials of Hydrogen Power Engineering and Gas Mains. *Phys. Usp.* **2008**, *178*, 709–726.
34. Galchenko, V.F. *Methanotrophic Bacteria*; GEOS: Moscow, Russia, 2001; p. 500.

© 2019 by the authors. Licensee MDPI, Basel, Switzerland. This article is an open access article distributed under the terms and conditions of the Creative Commons Attribution (CC BY) license (http://creativecommons.org/licenses/by/4.0/).

Article

Corrosion of High-Strength Steel Wires under Tensile Stress

Shanglin Lv [1,2], Kefei Li [1,*], Jie Chen [2] and Xiaobin Li [2]

[1] Department of Civil Engineering, Tsinghua University, Beijing 100084, China; lvshanglin@126.com
[2] National Construction Steel Quality Supervision and Test Centre, Central Research Institute of Building and Construction, China Metallurgical Group Cooperation, Co., Ltd., Beijing 100088, China; xjj_xjj@163.com (J.C.); lxb2001230@126.com (X.L.)
* Correspondence: likefei@tsinghua.edu.cn

Received: 22 September 2020; Accepted: 22 October 2020; Published: 27 October 2020

Abstract: The stress corrosion cracking is the central issue for high-strength wires under high tensile stress used in civil engineering. This paper explores the resistance of stress corrosion cracking of three typical steel wires of high-strength carbon through a laboratory test, combining the actions of tensile stress and corrosive solution. Besides, the impact of tensile stress and immersion time are also investigated. During the tests, the wires were subject to electrochemical measurements of potentiodynamic polarization and electrochemical impedance spectroscopy, and the microstructure analysis was performed on the fractured cross sections. The obtained results show the following: the high-strength wire, conforming to GB/T 5224, has higher resistance to the combined actions of tensile stress and corrosive solution; tensile stress of 70% fracture strength and longer loading-immersion time make the film of corrosion products on steel surface unstable and weaken the corrosion resistance; the surface film consisted of the iron oxide film and the corrosion products film whose components are mainly iron thiocyanate and iron sulphide.

Keywords: high-strength wires; stress corrosion cracking; potentiodynamic polarization; electrochemical impedance spectroscopy; metallography; fracture morphology

---

1. Introduction

Due to the high bearing capacity and stiffness, the prestressed concrete (PC) structures are widely used in large-span building, bridge and hydraulic structures [1]. The PC structures adopt either bonded or unbounded systems and are usually considered to acquire sufficient durability due to the limitation of concrete cracking and appropriate protection measures for the high-strength wires and strands [2,3]. However, prestress failures in PC structures have been reported in literature. Schupack and Suarez [4] showed, through an investigation from 1978 to 1982 in USA, about 50 PC structures had undergone different degrees of corrosion, among which 10 cases were caused by stress corrosion or hydrogen brittleness. Hydrogen is introduced by the improperly pickling process in general. As for the hydrogen-induced stress corrosion cracking mechanism, Vehovar et al. [5] reported its cause as the embrittlement of the prestressing tendons due to the penetration of hydrogen atoms, generated by hydrogen reduction at the alloy's surface. Enos and Scully et al. [6] used a simulated steel-concrete interface and laboratory-scale prestressed concrete pilings to study the safe cathodic protection limits for prestressing steel in concrete about hydrogen embrittlement. Moreover, the dislocations in the prestressing tendons relative to the production process and heat treatment are sensitive to hydrogen embrittlement. On this basis, Gertsman [7] and Jaka Kovac et al. [8] studied and characterized the intergranular SCC (IGSCC) processes. The results show that geometrical parameters and chemical parameters play an important role in the intergranular stress corrosion cracking of materials.

Walter [9] reported 242 cases of prestress failure, from 1951 to 1979, among which a large portion were attributed to the environmental actions. Among these failure causes, the stress corrosion cracking (SCC) is always a main engineering concern as it reduces the fracture strength of high-strength steel wires, leading to unexpected brittle failures of PC structures [10,11]. Moreover, the SCC can be sensitive to the aggressive agents present in the corrosive environments [12].

The SCC is caused by the combined action of stress, environmental agents and metal components, and the widely accepted mechanisms include the surface (depassivation) film of wires is broken jointly by the tensile stress and the environmental actions; the freshly exposed steel surface serves as the anode dissolved at the crack tip, leading to the steel fracture [13,14]. The conventional test methods include the deformation, sustained loading and slow strain rate testing [15]. During the mechanical loading, the electrochemical methods have been used to characterize the SCC process, identifying the susceptibility potential range for prestress wires [8,16] and detecting the stress corrosion fracture of a stainless steel through the phase shift of electrochemical impedance spectroscopy (EIS) [17]. As for the corrosive environment, a SCC test method was developed by the International Prestressed Concrete Federation (FIP, now merged into *fib*), using ammonium thiocyanate (NH_4SCN) solution as the corrosive environment for SCC test. This method has been standardized by ISO 15630 [18] and GB/T 21839 [19]. Perrin et al. [20] used this method to investigate the SCC mechanisms together with detection of EIS, electrochemical noise and metallographic analysis, and confirmed the validity of this method for SCC investigation.

Nowadays, high-strength steel wires for civil engineering use are usually classified following the mechanical properties including the rupture strength and the elongation, with a rough account for chemical compositions [21–23]. However, the SCC risk of high-strength steel is rather sensitive to the chemical composition. Thus, the SCC performance should always be investigated with respect to the environmental actions and steel chemical compositions. Accordingly, it is of interest to compare the SCC performance for streel wires with similar chemical compositions from different standards. To this purpose, this study retains three steel wires, conforming to BS 5896 [21] and GB/T 5224 [22], and investigates the SCC resistance of these wires through stress-immersion tests and the role of stress on the electrochemical behaviors.

2. Experiments

2.1. Material and Composition

The samples used in this investigation were central wires of prestress strand. Three kinds of samples were chosen: Sample A from Jiangyin, China, Walsin steel cable Co., Ltd., conforming to BS 5896 [21] and using C86D2 grade steel [24] and Samples B and C, according to GB/T 5224 [22] and using YL87B grade steel [25] but from different manufacturers (Xinhua Metal Products Co., Ltd., Xinyu, China and, Walsin steel cable Co., Ltd., Jiangyin, China respectively). The chemical compositions of steel samples were determined by GB/T 4336 [26], and the results are given in Table 1. Besides, the measurement uncertainty for chemical composition is given in Table 2.

Table 1. Chemical composition of steel wires (mass fraction)/%.

Wire	C	Si	Mn	P	S	Cr	Ni	Cu	V
A	0.870	0.235	0.694	0.0184	0.0121	0.007	0.009	0.012	0.007
B	0.875	0.253	0.795	0.0188	0.0113	0.190	0.010	0.024	0.005
C	0.886	0.267	0.837	0.0202	0.0109	0.280	0.014	0.031	0.004

Table 2. Measurement uncertainty for chemical composition (%).

Wire	C	Si	Mn	P	S	Cr	Ni	Cu	V
A	0.0229	0.0116	0.0167	0.0017	0.0016	0.0031	0.0012	0.0015	0.0052
B	0.0230	0.0121	0.0185	0.0017	0.0015	0.0069	0.0013	0.0024	0.0089
C	0.0233	0.0125	0.0194	0.0018	0.0016	0.0087	0.0016	0.0028	0.0107

2.2. Specimen Preparation

The rupture strength f_t of wires, A, B and C, were measured through axial tension loading tests on three specimens for each wire in lab-air environment, and the average values were retained as its fracture strength: 1971 MPa, 2106 MPa and 2088 MPa for Wires A, B and C, respectively. The relative composite measurement uncertainty for fracture strength values 1971 MPa, 2106 MPa and 2088 MPa is 0.699%, 0.722% and 0.718% respectively.

For the immersion tests, the sample length is 1280 mm, twice as long as the immersion part in the solution, and the diameter of sample is 5 mm. The NH_4SCN solution was prepared by dissolving 200 g of analytically pure NH_4SCN in 800 mL of distilled water. The stress corrosion testing device is composed of a loading frame and an immersion pool; see Figure 1. During the immersion tests, electrochemical measurements were performed for the steel wires in immersion.

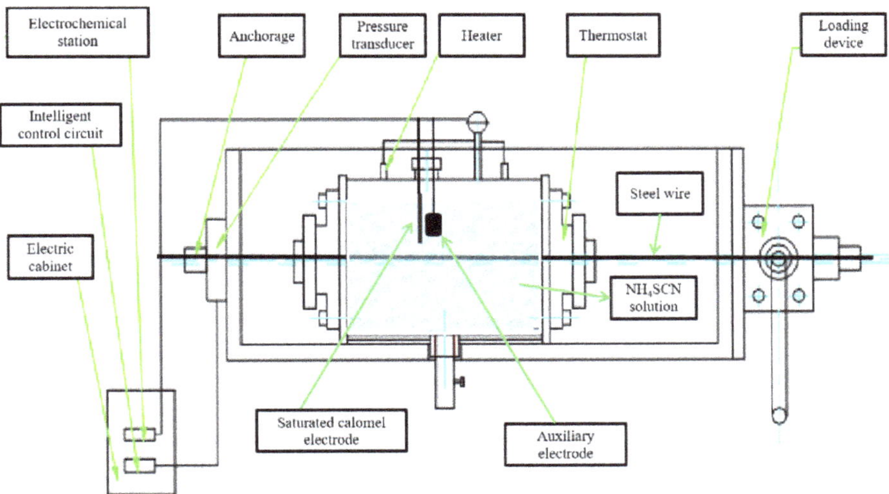

Figure 1. Experimental setup for accelerated corrosion tests for prestressed wires.

2.3. Test Procedure

The samples were wiped, degreased with acetone (CH_3COCH_3), and air-dried. At least 50 mm in length parts at both ends of the samples were coated with sealant to prevent corrosion. The samples were loaded to 70% of the rupture strength and the loading was sustained, with variation controlled within +2%, during the entire test duration. After loading, the container was sealed to prevent leakage, and the solution was replaced after each test. The NH_4SCN solution (Tengtai Chemical Technology Co., Ltd., Suzhou, China) was first deoxidized by introducing nitrogen gas during 2 h, then pre-heated to 50–55 °C, and injected into the container and kept at a constant temperature. The solution was filled within 1 min, covering the surface of sample and kept stagnant during the stress corrosion tests. The following cases were tested: (1) Samples A, B and C under stress level 70% f_t with loading-immersion duration 10 min; (2) Samples A under free stress (0% f_t) with loading-immersion duration 10 min, and stress level (70% f_t) with loading-immersion duration 30 min.

During all these tests, the potentiodynamic polarization and EIS (Chenhua Instrument Co., Ltd., Chi600E, Shanghai, China) were used to characterize the corrosion behavior of steel wires. The electrochemical measurements were performed via a three-electrode system containing a stainless-steel auxiliary electrode and a saturated calomel electrode (SCE) as reference, and the potentials hereafter refer to the SCE. The prestressed steel wire, acting as working electrode, was immersed in NH_4SCN solution with an exposed working area of 100 cm^2. The electrochemical tests were conducted when the open circuit potential (OCP) of the working electrode became stable. The scanning potential of the anodic polarization curve was −1.5–1.0 V vs. OCP, and the scanning rate was 5 mV/s. The sinusoidal voltage excitation signal with disturbance amplitude of 5 mV vs. OCP was used for EIS testing, and the frequency range was 10^5–10^{-2} Hz. All the electrochemical measurements were repeated three times for good reproducibility, and a typical group of data were selected to study for clarity.

3. Electrochemical Analysis

3.1. Potentiodynamic Polarization Analysis

The potentiodynamic polarization results are given in Figure 2 and Table 3. For the cases of loading-immersion duration of 10 min, the potentiodynamic polarization curves of wires B and C were on the left side of Wire A, and the curves of B and C were rather close. For the case of Wire A of loading-immersion duration 30 min, the polarization curve of wire A moves towards the bottom right, and the stable range becomes narrower, indicating that longer exposure time promotes the anodic dissolution of steel and the corrosion resistance is weakened [27,28]. Furthermore, compared to the A—10 min case, the case of A-without stress presents a left-upward shift in the polarization curves, showing wider range of the stable interval and the degree of corrosion was minimized [29].

In Table 3, the corrosion potential (E_{corr}), corrosion current density (i_{corr}), anodic Tafel slope (b_a) and cathodic Tafel slope (b_c) values were taken from the polarization curves in Figure 2. Compared to the cases B, C—10 min, the case A—10 min has lower E_{corr}, b_a and b_c values but larger i_{corr} values. By extending the loading-immersion time from 10 min to 30 min (Case A—10 min to Case A—30 min), the value of i_{corr} increased from 1.28 × 10^{-2} A/cm^2 to 1.90 × 10^{-2} A/cm^2, due to the E_{corr} descended from −315.5 to −349.0 mV vs. SCE. Compared to Case A—10 min, the stress-free case, A-without stress, has the E_{corr} value increased from −315.5 mV to −290.3 mV and i_{corr} decreased from 1.28 × 10^{-2} A/cm^2 to 0.34 × 10^{-2} A/cm^2. From these values, the current density of A was higher than those of B and C under the same corrosion conditions, which indicates that wires B and C had lower corrosion rate via the immersion tests. Moreover, for wires A, the current density increased with the application of tensile stress loading and longer loading-immersion duration, indicating that the corrosion process was accelerated. Tang [30] studied the effect of stress on corrosion of X70 pipeline steel in neutral solution with microzone electrochemical method. They found that the corrosion rate increased significantly with the stress loading, which confirmed that stress promoted the occurrence of corrosion.

Table 3. Electrochemical parameters of polarization curves of prestressed steel wires.

Steel Wire	Cases	E_{corr} (mV vs. SCE)	i_{corr} (10^{-2} A/cm^2)	b_a (V/dec)	b_c (V/dec)
A	70% f_t, 10 min	−315.5	1.28	25.76	13.45
A	70% f_t, 30 min	−349.0	1.90	12.13	9.55
A	0% f_t, 10 min	−290.3	0.34	85.31	23.34
B	70% f_t, 10 min	−286.1	0.65	64.13	15.63
C	70% f_t, 10 min	−281.2	0.51	58.46	18.03

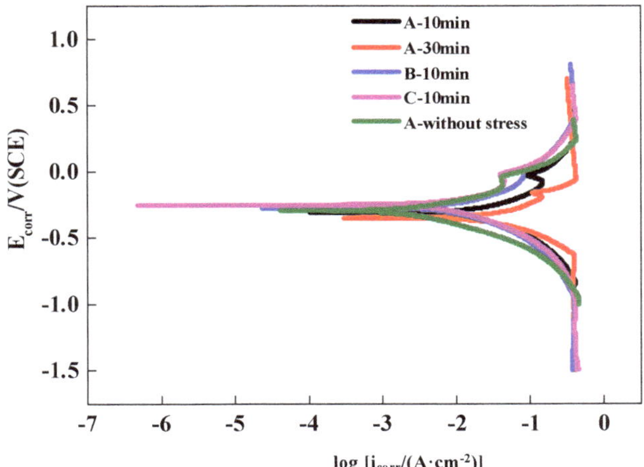

Figure 2. Potentiodynamic polarization curves of steel wires in NH_4SCN solution.

3.2. EIS Analysis

Figure 3 provides the experimentally obtained Nyquist curves and the EIS of response evolution of the corrosion products in different cases. Nyquist curves for the prestressed wire samples showed the similar characteristics. All of the Nyquist curves showed incomplete depressed semi-circular arcs that were affected by the frequency dispersion, and inductive shrinking occurred in the low-frequency zone, showing that such adsorbents as ferro-thiocyanate film existed on the steel surface and the corrosion was activated during this frequency range [26]. Under the same stress-loading duration (A, B, C—10 min cases), the Nyquist curves showed that the capacitance arc magnitude of B were larger than those of A and C, which suggests a more compact film of corrosion products was formed on the surface of B and the corrosion resistance of B wire is better than A and C wires. The polarization curves are extrapolated from the dynamic electrode behavior of strong polarization, while the EIS is measured under the stable equilibrium state. Therefore, the behavior of the electrode in static equilibrium can be better characterized by EIS results.

Further, as the loading-immersion time increases, from 10 min (A—10 min) to 30 min (A—30 min), the capacitance arc magnitudes are basically the same. By comparing the Nyquist curves of stressed wires (A, B, C under 70% f_t) and stress free wire (A, 0% f_t), it was found that the capacitance arc magnitude of stress free wire A is much larger than the wires (A, B, C) under 70% f_t tensile stress, indicating that the tensile stress promotes substantially the occurrence of corrosion. This observation is consistent with the anodic polarization curves in Figure 2. In Figure 3, the Z_{re} values, real part of the impedance, were not the same for different wires. This is due to the fact that the NH_4SCN solution was replaced and refilled into the accelerated corrosion container after each individual wire test, resulting in a slight change in the corrosive environment. However, this change in Z_{re} value does not change the judgement on the role of tensile stress.

The polarization resistance can be obtained from low frequency impedance, which determines the change in transfer resistance [31]. As shown in Figure 4a, when the influence caused by the change of solution impedance was ignored, the Bode curves of different wires revealed that the impedance modulus of Wire C in the low frequency region was slightly higher than that of Wire A under the same loading-immersion duration. Moreover, the impedance modulus of B was much higher than those of A and C, which indicates that the surface film resistance of B was higher and a significant charge transfer resistance of Wire B was created; therefore, the corrosion resistance of Wire B was higher than Wires A and C. Figure 4b shows the phase angle of three stressed wires (A, B, C—10 min) increased with the frequency under the same stress-corrosion time. The peak value of C was the highest, showing that the surface was smoother, and the pitting corrosion was inhibited. With corrosion time prolonged from 10 min to 30 min, the impedance modulus and phase angle of Wire A decreased, and the corrosion resistance decreased accordingly. When Wire A is under stress-free condition, the resistance of the surface film in the low frequency region was larger than the stressed case of A—10 min, and the resistance decreased also with frequency. Meanwhile, the peak value of stressed A—10 min was lower than the stress-free case, indicating that the formation of compact products was inhibited under tensile stress, and the product film was rougher.

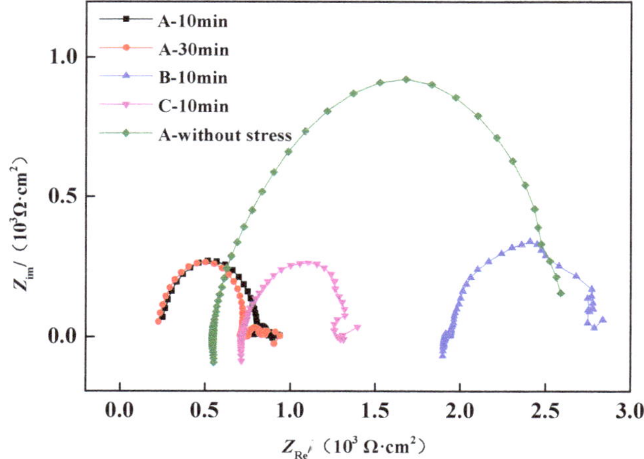

Figure 3. Nyquist curves of steel wires of different cases in NH_4SCN solution.

The models in literature [32–34] are used to describe the electrochemical processes of the surface films exposed to different conditions, and interpret the obtained EIS curves, see Figure 5. In the figure, R_{sol} represents the resistance of the solution, the constant phase angle element CPE_1 corresponds to the double-layer capacitance of the interface between the sample surface and solution, R_1 represented the charge transfer resistance, R_2 was the surface film resistance, the constant phase angle element CPE2 was used as a substitute for the surface film capacitance, and n represented the dispersion exponent. Through this model, the electrochemical parameters can be regressed to represent corrosion kinetics for different cases. In general, the electron transfer resistance determines the impedance at low frequencies, the solution resistance determines the impedance at high frequencies, and the electrochemical corrosion kinetics can be extrapolated by the low frequency impedance.

Due to the heterogeneity of the surface film [35], a constant phase element (CPE) is used to represent the non-ideal capacitance responses of the interface. The CPE was defined as,

$$Z_{CPE} = \frac{1}{Y_0(jw)^n} \tag{1}$$

where Y_0 is the admittance magnitude of CPE; ω is the angular frequency; j is the imaginary number ($j^2 = -1$) and n is the exponent ($-1 < n < 1$). Y_0 and n can be converted into CPE. The n value is interrelated to the heterogeneity and smoothness of the surfaces. When n values are close to 1, the CPE will approach an ideal capacitance.

The parameters are regressed and given in Table 4 from the EIS data for steel wires in different cases. Under the same stress-corrosion time (70% f_t, 10 min), the R_1 value of Wires B and C are higher than Wire A, indicating the transfer of electrons is more difficult in Wire B and C. Therefore, Wires B and C show better resistance to the redox reaction of corrosion relative to A. The R_2 value of Wire B is the highest among A, B, C—10 min cases, which means that Wire B has the largest surface film impedance and highest compactness. Compared to Wires B and C, the CPE_1 and CPE_2 values of Wire A are slightly higher, showing Wire A is more prone to corrode. The values from A—10 min and A—30 min cases show a general decrease of resistance and capacitance values, indicating the charge transfer processes were promoted with the immersion time from 10 min to 30 min.

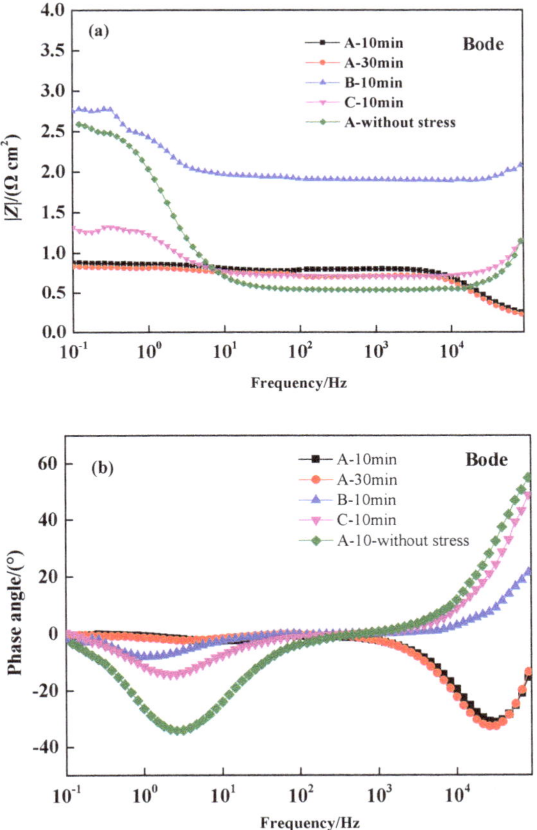

Figure 4. Bode curves of steel wires immersed in NH_4SCN solution: impedance modulus (**a**) and impedance angle (**b**).

The R_1 values of Sample A cases showed the longer loading-immersion duration, in immersion solution, tends to decrease R_1 value, in other terms, promote the electron transfer and weaken the corrosion resistance. The R_2 values decreased accordingly by longer loading-immersion duration and tensile stress, possibly attributed to the deterioration of surface film. The evolution of CPE_1 and CPE_2 increased with prolonged loading-immersion time, indicating that the surface film formed under prolonged loading duration leads to capacitive behaviors, and the corrosion resistance is deteriorated. In addition, with the increase in loading-immersion duration, n (n_1, n_2) values of the double layer capacitance and the surface film capacitance decreased, respectively. These observations confirm that the corrosion of steel wires was gradually intensified. Finally, when comparing the stress-free case of

Wire A with other stressed cases, one gets a similar R_{sol} value but a much higher R_1 value in Table 4, meaning the electrons transfer resistance in stress-free wire A is much higher than those stressed wires (A, B or C). When pitting corrosion occurred on the surface of the stressed wire, the potential of pitting corrosion area was lower than that of other parts, which resulted in the area that became active and provided crack core for stress corrosion. The concentrated stress made the crack tip and the surrounding area yield deformation, and then the micro-slip destroyed the surface film of the crack tip again, which accelerated the dissolution of the tip. At the same time, because of the existence of micro cracks, the corrosion resistance of the surface film deteriorated, and the electron transfer resistance decreased. This observation confirms further that the tensile stress, 70% f_t, substantially decreased the surface electron transfer resistance, thus promoting the SCC of the wires.

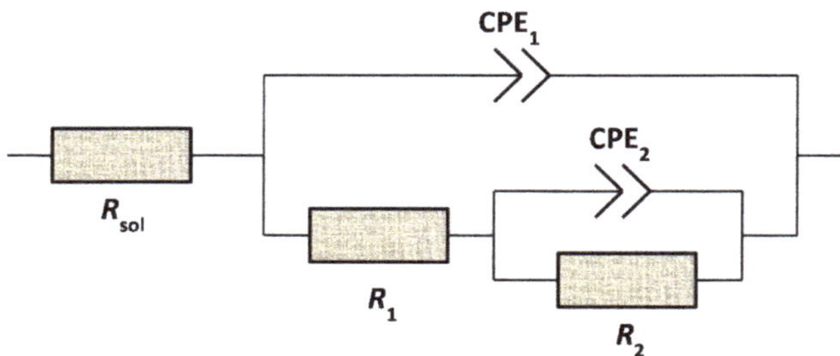

Figure 5. Equivalent electrical circuit used to represent the measured EIS data.

Table 4. Fitting parameters for experimental EIS of wires immersed in NH_4SCN solution.

Steel Wire	Cases	R_{sol} (10^3 Ω cm^2)	R_1 (10^3 Ω cm^2)	CPE$_1$		R_2 (10^3 Ω cm^2)	CPE$_2$	
				Y_0/(10^{-4} $Ω^{-1}$ cm^2 S^n)	n_1		Y_0/(10^{-4} $Ω^{-1}$ cm^2 S^n)	n_2
A	70% f_t, 10 min	0.26	0.32	1.035	0.95	0.47	1.29	0.91
A	70% f_t, 30 min	0.24	0.29	0.63	0.82	0.42	1.58	0.84
A	0% f_t, 10 min	0.53	1.23	0.11	0.97	0.82	1.19	0.95
B	70% f_t, 10 min	1.86	0.46	0.29	0.96	0.58	1.14	0.93
C	70% f_t, 10 min	0.67	0.42	0.23	0.95	0.52	1.08	0.92

4. Microstructure Analysis

4.1. Metallography

The microstructure of high strength wires was observed by metallography microscope (Olympus GX51, Tokyo, Japan) (Magnification factor: 50×–1000×, Light source: 6V30WHAL halogen lamp) before the corrosion test. Figure 6 shows the metallographic microstructures of Wires A, B and C: All the three wires are composed of sorbite, product of austenite isothermal transformation, having eutectoid structures of alternate thin layers of ferrite and cementite. The sorbite lamellae size of B and C is finer; the grain boundary defects are accordingly reduced, and the corrosion resistance is expected to enhance. The results are consistent with Ren's study [36].

The ferrite and cementite were grey-white in Figure 6, and the interface between them was black. The three wires all had sorbitization treatment applied to ensure a high degree of strength and toughness. When the wires are produced, chemical composition adjusting is often used to reduce the lower limit temperature of sorbite transformation and avoid the occurrence of bainite and martensite. Generally, the alloying elements Mn, Cr, Ni and Cu will improve the stability of austenite and delay the decomposition of austenite. As shown in Table 1, the carbon content of Samples A, B and C is similar while the Mn, Cr and Cu contents of Wires B and C are higher. Therefore, the completion of the sorbite transformation is prolonged, the precipitation of eutectoid ferrite and pearlite is reduced, and the termination temperature of the sorbite transformation is also significantly decreased; all these factors give Wires B and C higher tensile strength. In addition, the surface film of wires B and C formed by the corrosion was more compact due to the higher content of Cr and Ni. The corrosion resistance was improved, which was consistent with the anodic polarization curve and EIS conclusion.

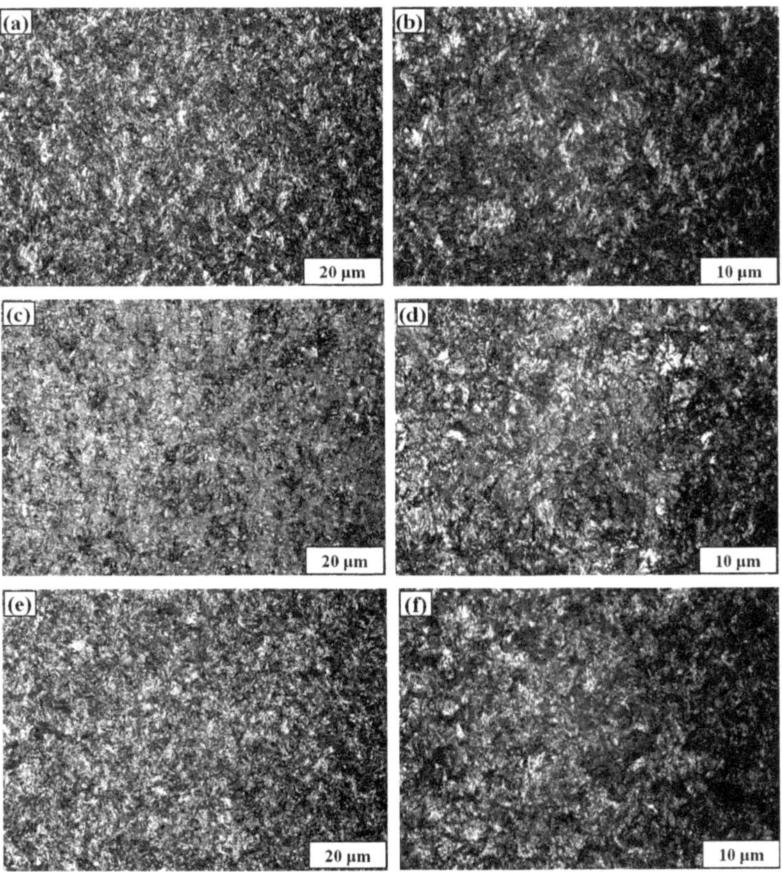

Figure 6. Metallographic microstructures of three high strength wires: Wire A ×500 (**a**) and ×1000 (**b**), Wire B ×500 (**c**) and ×1000 (**d**), Wire C ×500 (**e**) and ×1000 (**f**).

4.2. Fracture Surface

The fracture cross sections of the three wires (A, B, C—10 min) under 70% f_t tensile stress were observed by Scanning Electronic Microscopy (SEM, Quanta 250 FEG, Hillsboro, OR, USA). The acceleration voltage was 20 kV, and high vacuum mode was $<6 \times 10^{-4}$ Pa, resolution was 1.2 nm; see Figure 7.

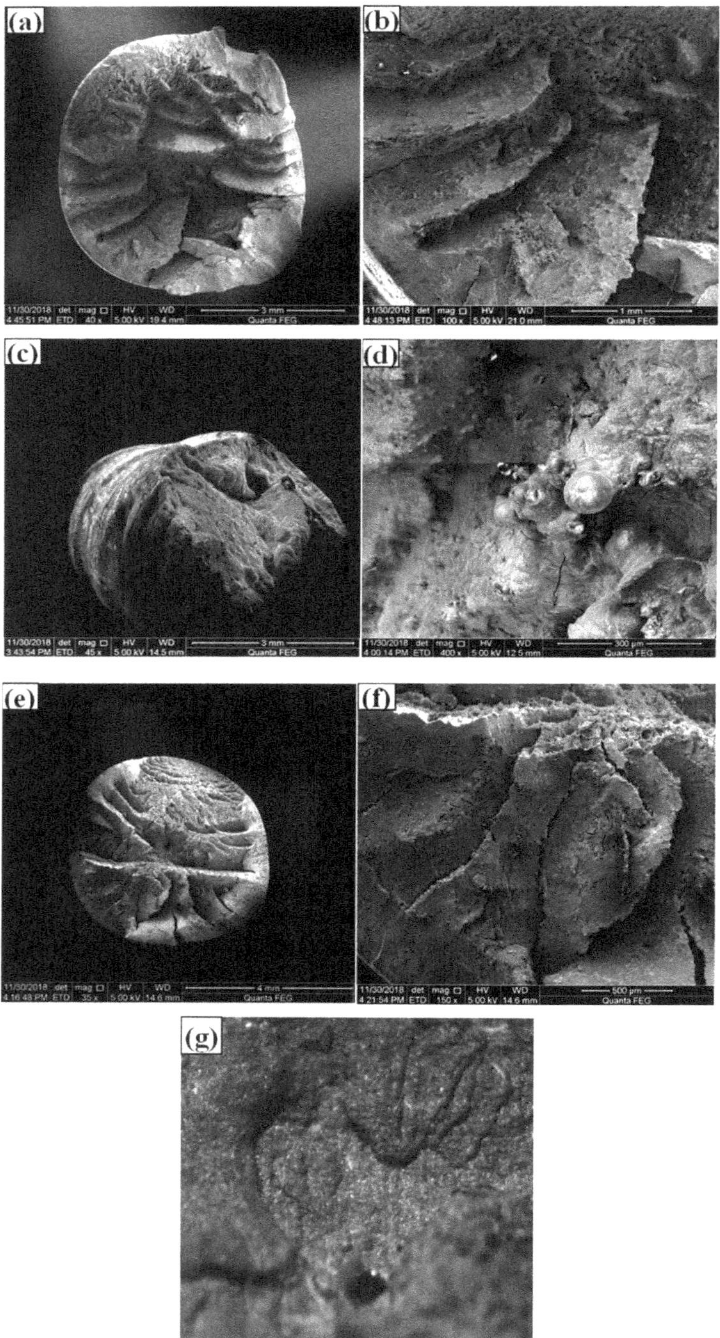

Figure 7. SEM images of rupture surface of wires: (**a**,**b**) wire A, (**c**,**d**) wire B, (**e**,**f**) wire C, (**g**) corrosion pits.

The fractographic and SEM images in Figure 7 confirmed that the macroscopic fracture of all three wires under stress corrosion is of brittle nature. These fracture surfaces present radial pattern without plastic deformation. The splitting surface is strip-like, and the end of the fracture is radial from the section center to the periphery, which was consistent with the literature observations [37,38]. Under higher magnifications, all sections showed more characteristics related to corrosion: the corrosion pits on the sections are generally deep and narrow, and the surface fractures develop into the solid matrix of steel. These fractures were developed from the microcracks on the wire surface extended along the wire length and tension, results of tensile stress and NH_4SCN solution. Some cracks, torn edges and quasi-cleavage fractures are found on the wire surface in the crack growth area. The fractography analysis of these fractures attributes the major pattern to be trans-granular.

Energy spectrum (EDX) was used to analyze the chemical composition of inclusions in the fractures in Figure 8: Wire A has carbides non-metallic inclusions while Wires B and C had oxide non-metallic inclusions. The existence of non-metallic inclusions in the samples destroyed the continuity of the metal matrix structure, which deteriorates the mechanical properties of all the wires. These inclusions can be the crack nucleation sites, which promotes the fracture propagation in solid matrix of granular nature, especially under the high level of tensile stress.

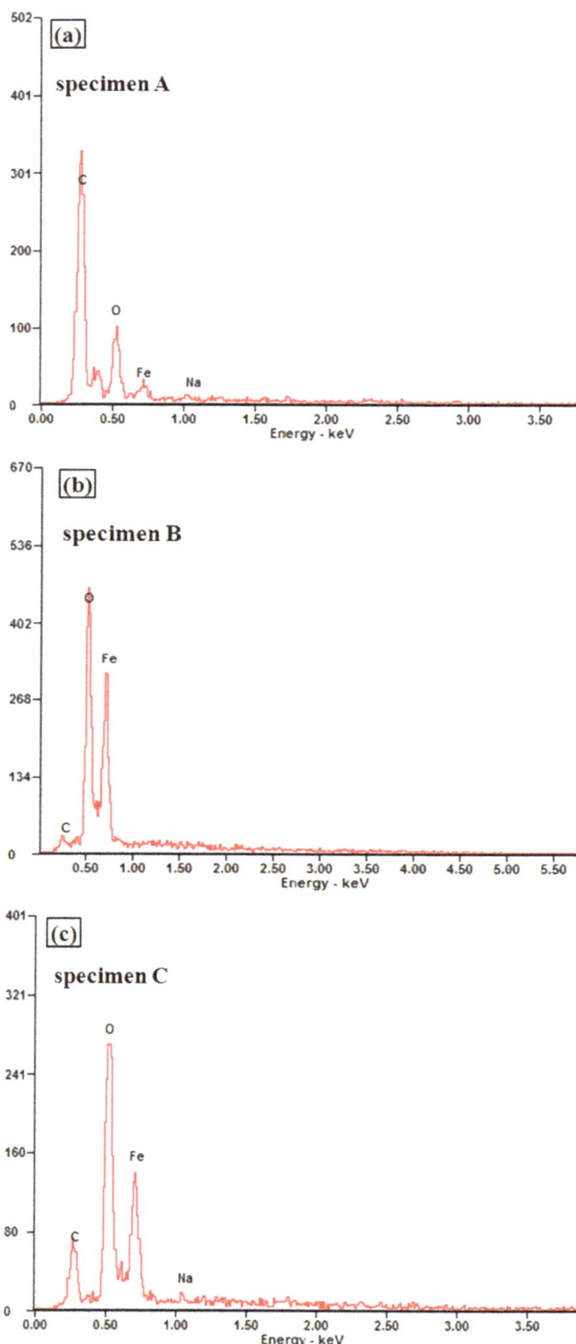

Figure 8. EDX analysis of the fracture inclusions on cross sections of high-strength wires: (**a**) A, (**b**) B and (**c**) C.

4.3. Corrosion Products

The original oxide film defects of steel wire were split to expose the fresh metal under tensile loading. The exposed substrate and the oxide film around the defects formed a corrosion source in NH_4SCN solution. The exposed fresh metal acted as anode while oxide film acted as cathode, which led to an increase in the corrosion rate. The corrosion products film is formed in the anodic dissolution process, and the defects trigger continuously under tensile loading leading the film to be deteriorated.

Some ferric iron formed in the anodic dissolution process, and the reaction of ferric iron in solution containing thiocyanate produces ferro-thiocyanate compound, i.e., $Fe^{3+} + xSCN^{-} \rightarrow [Fe(SCN)_x]^{3-x}$. The ferro-thiocyanate compounds form on the steel surface and will promote the anodic dissolution of steel [30]. From the available EIS and other results in this study, the film becomes loose, and the surface of the steel wire becomes rougher under tensile loading. The decreased transfer resistance and the sharp increase of corrosion current under tensile stress indicate that the film was rapidly deteriorated, which is confirmed also by potentiodynamic polarization curves.

Figures 9 and 10 showed the SEM microstructure and the energy spectrum analysis of prestressed steel wire surface after corrosion in NH_4SCN solution, respectively. The surface film consisted of the original iron oxide film and the corrosion products film whose components are mainly iron thiocyanate and iron sulphide. The original iron oxide film was dense while the corrosion products film was rough and loose.

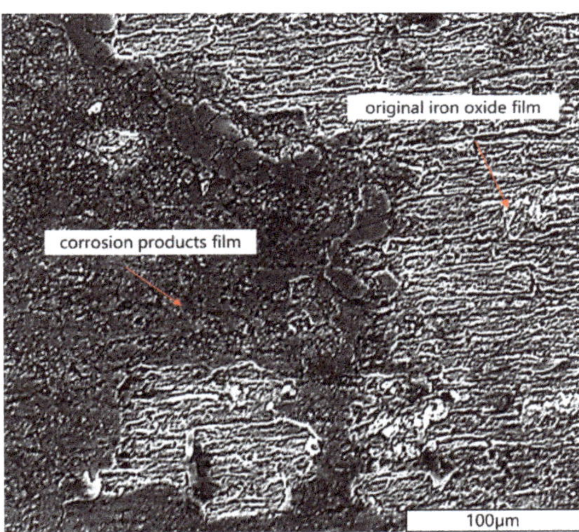

Figure 9. SEM microstructure of prestressed steel wire surface in NH_4SCN solution.

The steel wires with similar chemical composition were selected to study the stress corrosion behavior in this paper. The results showed that they had different corrosion resistance in the same corrosion environment. It indicated that even if the contents of Cr, Ni and Cu increased slightly, the performance of corrosion product film on the wires surface would be affected. Meanwhile, stress corrosion is the result of multiple factors synergy, such as composition and microstructure. Thus, in order to improve the stress corrosion resistance of steel, it is necessary to improve many aspects.

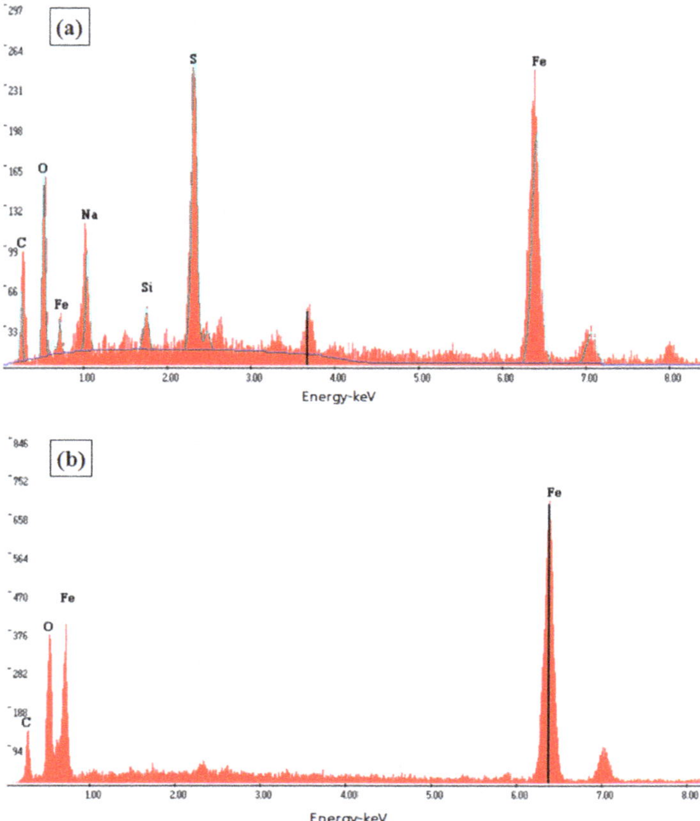

Figure 10. Energy spectrum of prestressed steel wire surface in NH_4SCN solution: (**a**) corrosion products film, (**b**) original iron oxide film.

5. Conclusions

1. The electrochemical measurements show that, with loading duration, the corrosion current density increases, and the corrosion resistance is weakened. The results from polarization curves and EIS are consistent with those from corrosion morphologies.
2. The tensile stress leads to a more unstable corrosion products film formed on the steel, which means worse flatness and compactness. The stress substantially decreased the surface electron transfer resistance, thus promoting the corrosion of the high strength wires.
3. The surface film consisted of the original iron oxide film and the corrosion products film, whose components are mainly iron thiocyanate and iron sulphide.
4. The corrosion-resistance performance of the wires, conforming to GB/T 5224, is better than the selected wire conforming to BS 5896.
5. The corrosion resistance of steel wire will be affected by the change of chemical composition. Stress corrosion is the result of multiple factors synergy, such as composition and microstructure.

Author Contributions: Conceptualization, K.L. and S.L.; methodology, K.L.; software, S.L.; validation, S.L., J.C. and X.L.; formal analysis, K.L.; investigation, S.L.; resources, J.C.; data curation, S.L.; writing—original draft preparation, S.L.; writing—review and editing, K.L.; visualization, S.L.; supervision, X.L.; project administration, K.L.; funding acquisition, K.L. All authors have read and agreed to the published version of the manuscript.

Funding: The research is supported by Zhejiang Transportation Science and Technology Project No. 2018035.

Conflicts of Interest: The authors declare no conflict of interest. The funders had no role in the design of the study; in the collection, analyses, or interpretation of data; in the writing of the manuscript, or in the decision to publish the results.

References

1. Nawy, E.G. *Prestressed Concrete. A Fundamental Approach*, 5th ed.; Prentice Hall: New York, NY, USA, 2006.
2. TR 47 Durable Bonded Post-Tensioned Concrete Bridges. 2nd edition. Available online: https://www.thenbs.com/PublicationIndex/documents/details?Pub=CS&DocID=265790 (accessed on 30 May 2019).
3. Li, K.F. *Durability Design of Concrete Structures: Phenomena, Modeling and Practice*; John Wiley & Sons: Singapore, 2016.
4. Schupack, M.; Suarez, M.G. Some recent corrosion embrittlement failures of prestressing systems in the United States. *PCI J.* **1982**, *27*, 38–55. [CrossRef]
5. Vehovar, L.; Kuhar, V.; Vehovar, A. Hydrogen-assisted stress-corrosion of prestressing wires in a motorway viaduct. *Eng. Fail. Anal.* **1998**, *5*, 21–27. [CrossRef]
6. Enos, D.G.; Williams, A.J.; Scully, J.R. Long-term effects of cathodic protection on prestressed concrete structures: Hydrogen embrittlement of prestressing steel. *Corrosion* **1997**, *11*, 891–908. [CrossRef]
7. Gertsman, V.Y.; Bruemmer, S.M. Study of grain boundary character along intergranular stress corrosion crack paths in austenitic alloys. *Acta Mater.* **2001**, *49*, 1589–1598. [CrossRef]
8. Kovac, J.; Alaux, C.; Marrow, T.J. Correlations of electrochemical noise, acoustic emission and complementary monitoring techniques during intergranular stress-corrosion cracking of austenitic stainless steel. *Corros. Sci.* **2010**, *52*, 2015–2025. [CrossRef]
9. Podolny, W. Corrosion of Prestressing Steels and Its Mitigation. *PCI J.* **1992**, *5*, 34–55. [CrossRef]
10. Isecke, B.; Mietz, J. The risk of hydrogen embrittlement in high-strength prestressing steels under cathodic protection. *Steel Res.* **1993**, *64*, 97–101. [CrossRef]
11. Toribio, J.; Lancha, A.M. Anisotropic stress corrosion cracking behaviour of prestressing steel. *Mater. Corros.* **1998**, *49*, 34–38. [CrossRef]
12. Spencer, D.T.; Edwards, M.N.; Wenman, M.R. The initiation and propagation of chloride-induced transgranular stress-corrosion cracking (TGSCC) of 304L austenitic stainless steel under atmospheric conditions. *Corros. Sci.* **2014**, *88*, 76–88. [CrossRef]
13. Barnes, A.; Senior, N.; Newman, R.C. *Environment-Induced Cracking of Materials*; Elsevier: Amsterdam, The Netherlands, 2008; pp. 47–57.
14. Galvele, J.R. Recent developments in the surface-mobility stress-corrosion-cracking mechanism. *Electrochim. Acta* **2000**, *45*, 3537–3541. [CrossRef]
15. Popov, B.N. *Corrosion Engineering. Principles and Solved Problems*; Elsevier: Amsterdam, The Netherlands, 2015.
16. Diaz, B.; Freire, L.; Novoa, X.R. Electrochemical behavior of high strength steel wires in the presence of chlorides. *Electrochim. Acta* **2009**, *54*, 5190–5198. [CrossRef]
17. Bosch, R.W. Electrochemical impedance spectroscopy for the detection of stress corrosion cracks in aqueous corrosion systems at ambient and high temperature. *Corros. Sci.* **2005**, *47*, 125–143. [CrossRef]
18. *ISO 15630-3: Steel for the Reinforcement and Prestressing of Concrete—Test Methods—Part 3. Prestressing Steel*; Ministry of Defence (MOD): London, UK, 2010.
19. *GB/T 21839 Steel for Prestressed Concrete-Test Methods*; Standardization Administration: Beijing, China, 2019.
20. Perrin, M.; Gaillet, L.; Tessier, C. Hydrogen embrittlement of prestressing cables. *Corros. Sci.* **2010**, *52*, 1915–1926. [CrossRef]
21. *BS 5896—High Tensile Steel Wire and Strand for the Prestressing of Concrete–Specification*; BSI: London, UK, 2012.
22. *GB/T 5224—Steel Strand for Prestressed Concrete*; Standardization Administration: Beijing, China, 2014.
23. *ISO 6934-4—Steel for the Prestressing of Concrete-Part 4*; ISO: Geneva, Switzerland, 1991.
24. *ISO 16120-4—Non-Alloy Steel Wire Rod for Conversion to Wire*; ISO: Geneva, Switzerland, 2017.

25. *GB/T 24238-2017—Hot-Rolled Wire Rod for Prestressed Steel Wire and Strand*; Standardization Administration: Beijing, China, 2017.
26. *GB/T 4336-2016—Carbon and Low-Alloy Steel-Determination of Multi-Element Contents-Spark Discharge Atomic Emission Spectrometric Method (Routine Method)*; Standardization Administration: Beijing, China, 2016.
27. Yang, Z.M.; Zhang, B. Corrosion and passive behaviour of duplex stainless steel 2205 at different cooling rates in a simulated marine-environment solution. *J. Iron Steel Res. Int.* **2018**, *25*, 943–953.
28. Bo, R.; Kosec, T.; Kranjc, A. Electrochemical impedance spectroscopy of pure copper exposed in bentonite under oxic conditions. *Electrochim. Acta* **2011**, *56*, 7862–7870.
29. Cao, C.N. *Principle of Corrosion Electrochemistry*; Chemical Industry Press: Beijing, China, 2008.
30. Tang, X.; Cheng, Y.E. Micro-electrochemical characterization of the effect of applied stress on local anodic dissolution behavior of pipeline steel under near-neutral pH condition. *Electrochim. Acta* **2009**, *54*, 1499–1505. [CrossRef]
31. Nishikata, A.; Ichihara, Y.; Tsuru, T. Electrochemical impedance spectroscopy of metals covered with a thin electrolyte layer. *Electrochim. Acta* **1996**, *41*, 1057–1062. [CrossRef]
32. Mohammadi, F.; Nickchi, T.; Attar, M.M.; Alfantazi, A. EIS study of potentiostatically formed passive film on 304 stainless steel. *Electrochim. Acta* **2011**, *56*, 8727–8733. [CrossRef]
33. Oskuie, A.A.; Shahrabi, T.; Shahriari, A. Electrochemical impedance spectroscopy analysis of X70 pipeline steel stress corrosion cracking in high pH carbonate solution. *Corros. Sci.* **2012**, *61*, 111–122. [CrossRef]
34. Baek, J.S.; Kim, J.G.; Hur, D.H. Anodic film properties determined by EIS and their relationship with caustic stress corrosion cracking of Alloy 600. *Corros. Sci.* **2003**, *45*, 983–994. [CrossRef]
35. Rammelt, U.; Reinhard, G. The influence of surface roughness on the impedance data for iron electrodes in acid solutions. *Corros. Sci.* **1987**, *27*, 373–382. [CrossRef]
36. Ren, A.C.; Zhu, M. Study on early corrosion mechanism of U75V and U68CrCu rail steels. *China Railw. Sci.* **2014**, *35*, 7–12.
37. Wu, S.; Chen, H.; Ramandi, H.L. Effects of environmental factors on stress corrosion cracking of cold-drawn high-carbon steel wires. *Corros. Sci.* **2018**, *132*, 234–243. [CrossRef]
38. Caballero, L. Discussion on "Failure mechanisms of high strength steels in bicarbonate solutions under anodic polarization" by E. Proverbio and P. Longo. *Corros. Sci.* **2004**, *46*, 1813–1820. [CrossRef]

Publisher's Note: MDPI stays neutral with regard to jurisdictional claims in published maps and institutional affiliations.

© 2020 by the authors. Licensee MDPI, Basel, Switzerland. This article is an open access article distributed under the terms and conditions of the Creative Commons Attribution (CC BY) license (http://creativecommons.org/licenses/by/4.0/).

Article

Influence of the Loading Speed on the Ductility Properties of Corroded Reinforcing Bars in Concrete

Angela Moreno Bazán [1], María de las Nieves González [2], Marcos G. Alberti [1] and Jaime C. Gálvez [1,*]

[1] Departamento de Ingeniería Civil: Construcción, E.T.S de Ingenieros de Caminos, Canales y Puertos, Universidad Politécnica de Madrid, C/Profesor Aranguren, s/n, 28040 Madrid, Spain; angela.moreno@upm.es (A.M.B.); marcos.garcia@upm.es (M.G.A.)

[2] Departamento de Construcciones Arquitectónicas y su Control, E.T.S de Edificación, Universidad Politécnica de Madrid, Avda/Juan de Herrera, 6, 28040 Madrid, Spain; mariadelasnieves.gonzalez@upm.es

* Correspondence: Jaime.galvez@upm.es; Tel.: +34-910-674-125

Received: 25 January 2019; Accepted: 19 March 2019; Published: 22 March 2019

Abstract: In this work 144 reinforcing bars of high-ductility steel named B500SD were subjected to an accelerated corrosion treatment and then tested under tension at different loading speeds in order to assess the effect of corrosion on the ductility properties of the rebars. Results showed that the bars with a corrosion level as low as the one reducing the steel mass by 1% gave rise to a significant degradation on the ductility properties when a high loading speed was applied in tensile tests. In that case, the equivalent steel concept is useful to reduce the destabilising effect. Thus, the research significance lies in the assessment of the influence of the loading speed at which the tensile test is performed for the reinforcement bars that largely depends of the ductility criteria used.

Keywords: corrosion; ductility; mechanical properties; reinforced concrete; tensile strength; equivalent steel

1. Introduction

Corrosion of steel rebars is one of the most common deterioration mechanisms identified in reinforced concrete structures. Such corrosion, whether induced by carbonation, chlorides or another attack, affects the overall serviceability and durability of the structure with consequences such as a reduction of the effective cross-section of the steel rebars, cracking and spalling of the concrete cover and the degradation of bond strength [1–3].

In the case of chloride corrosion, the attack is mostly local and commonly known as pit corrosion. These chlorides can be found in sea water, industrial wastewater, and, among others, deicing salts [4]. Corrosion occurs after the depassivation of the alkaline barrier when sufficient oxygen and moisture are available. Then the passive film is locally destroyed and a process of local corrosion is initiated. The crucial amount of chlorides needed to interrupt the passive cover is 0.4%–1% by mass of cement as an appropriate chloride threshold [5].

Over the last years more and more qualitative and quantitative research has been carried out about the reduction of the steel bars effective cross-section areas through tensile tests run by chloride corrosion. The Spanish Structural Concrete Code EHE-08 and the Eurocode EC-2 [6,7] require limited values in the mechanical properties of high-ductility steel in terms of both strength and strain. The rebar strength has a significant influence on the structural strength of concrete reinforced members and the codes require minimum values for the steel yield strength and maximum tensile strength. Additionally, due to the consideration of dynamic and seismic actions, consideration of properties in relation to the steel ductility is also required. It is essential that an effective method to reflect the relationship between the mechanical and ductility property of steel bars and the corrosion be found [8–11].

One way in which ductility can be considered is in relationship with the fracture energy that is the area covered by the strain-stress curve. This energy depends on the plastic deformation capacity of steel up to breaking point. The higher the area, the higher is the capacity of steel to dissipate energy under dynamic loads. In addition, for dynamic and impact loads, the speed at which the load is applied is important.

The Codes EHE-08 and EC-2 require that the steel bars meet ductility properties based on total elongation at maximum force (A_{gt}) and the ratio between tensile strength and yield strength (R_m/R_e). Table 1 shows the limits of those parameters as required by the EHE-08 code in order to qualify the steel as of high ductility according to standard UNE 36065:2011 [12]. In addition, although the percentage elongation after fracture ($A_{u,5}$) is included in this code, it is not considered in other codes.

Table 1. Requirements of the EHE-08 code for B500SD steel.

R_e (MPa)	R_m (MPa)	R_m/R_e	A_{gt} (%)	$A_{u,5}$ (%)
500	575	$1.15 \le R_m/R_e \le 1.35$	≥7.5	≥16

The conventional approach to steel rebar corrosion considers a reduction in the area of the bar section proportional to the degree of corrosion. Most of the published works [13–17] report systematic reduction of strength and strain at maximum load when the degree of corrosion increases. Recent studies [18] have shown that corrosion takes place in local spots of the bar surface (pitting), a weakening of strength occurs at these spots (a notch effect), and the bar strength falls under the minimum values required by the codes, even with very small degrees of corrosion. Nevertheless, the reduction of strain is greater than the loss of strength in the bar [19].

With low levels of corrosion the loss of strength is also low: this means that the structural elements can still meet their resistance function, though the reduction of strain may not meet the minimum values required in Table 1 to ensure enough ductility. Previous studies [20,21] have shown that the ratio R_m/R_e remains constant with the increase of the corrosion level. This means that the steel may amply meet the R_m/R_e requirement but not the requirement of A_{gt}.

In these cases, the use of the equivalent steel concept as a ductility criterion based on both R_m/R_e and A_{gt} may be highly useful [22,23]. Table 2 shows the minimum values obtained with the EHE-08 requirements in application of the equivalent steel formulas [24–28] as proposed by Cosenza (p), Creazza (A^*) and Ortega (Id).

Table 2. Values of the equivalent steel parameters obtained with EHE-08 high ductility steel requirements.

Equivalent Steel Concept	Cosenza (p)	Creazza (A^*) (MPa)	Ortega (Id)
Normative EHE-08	0.82	3.87	63.65

Moreover, although the standard test procedure of tensile tests of reinforcing bars ISO 15630-1:2010 [29] is remarkably complete, no recommendation regarding the loading speed for the tensile test is set. This might have a significant impact on the test results. For this reason, in this study the variations of the mechanical properties of steel rebars as a function of the degree of corrosion and the loading speed applied in the tensile tests are reported. For such a purpose, 144 12-mm diameter high-ductility steel rebars, named B500SD, were tested in tension after an accelerated corrosion treatment, embedded in NaCl contaminated concrete. The results showed that the influence of high-speed loading is significant if only the Codes EHE-08 and EC-2 are used. In order to establish a better relationship between the tensile test and the minimum effective cross-sectional area, the use of the equivalent concept is required.

2. Materials and Methods

2.1. Materials

Twelve concrete slabs of 30 × 40 × 10 cm^3 were fabricated, with each one having 12 reinforcing bars partially embedded, making a total of 144 bars to be corroded and tested in tension (Figure 1). Table 3 shows the concrete composition. The mean compression strength of the concrete was 26 MPa. The steel type was B500SD, used normally in structures in seismic areas due to the higher properties in terms of ductility.

Figure 1. Setup used for the fabrication of slabs (30 × 40 × 10 cm^3).

Table 3. Composition of concrete.

Material	Value
Cement (kg/m^3)	290
Water (kg/m^3)	174
Cement/water	0.6
Gravel (kg/m^3)	1110
Sand (kg/m^3)	760
$CaCl_2$ (% cement weight)	3

River silica, sand, round gravel, and cement CEM II/A-L 32.5, according to standard RC-16 [30] were the mix basic materials. The water/binder ratio was 0.6. $CaCl_2$ with a concentration of 3% relative to the weight of cement was diluted in the mixing tap water in order to destroy the passive state of the reinforcing bars. After casting and demoulding, the slabs were cured for 28 days in chamber under ambient conditions at 25 °C and 99% of relative humidity. In order to avoid corrosion initiation and propagation at the point where the bars protrude the concrete slab, each bar was wrapped with insulating tape in these points for lengths of 3 cm both outside and inside the concrete.

2.2. Accelerated Corrosion

Because natural corrosion of the steel bars is a time-consuming process, an artificial accelerated test by an electrochemical method was utilised to simulate the corrosion process [31]. Thus, corrosion was accelerated by imposing an electrical current between the reinforcement, the working-electrode and a counter-electrode, in a way that oxidation is enforced. The flow of the current was established by an external constant intensity. Nevertheless, a corrosion pit was found on the steel bars by electrochemical method that showed that non-uniform corrosion occurred too [32,33]. Therefore, the uniformity of corrosion was unconnected with the corroded method, but rather with the degree of corrosion. Many researchers have already used this technique to gather information about corrosion processes in reinforced concrete [34,35]. In this test, the accelerated corrosion process was activated by an electrical current by the application of a constant anodic current between the bars and a lead plate placed on top of the slabs, acting as the cathode. A soaked textile pad placed between the concrete slab and the lead plate ensured an even distribution of the electric current.

The value of the current in each bar was controlled by means of a digital multimeter, involving periodical recording of the voltage and adjustment of the electrical potential at the power source to ensure a constant current value of approximately 10 $\mu A/cm^2$ in each bar. This accelerated corrosion test is only valid if the intensity remains rather low (<200 $\mu A/cm^2$) with respect to Faraday's law. Otherwise a significant increase of strain response, and consequently the crack width, will occur. Densities can reach 200 $\mu A/cm^2$, with increasingly more appearing that will undermine accuracy of the experiment. These current values are broadly explained by Andrade [36], Maaddawy [37] or Suvash [38]. In order to obtain distinct corrosion levels, the current was disconnected at different ages after cracks appeared in the concrete slabs (Figure 2).

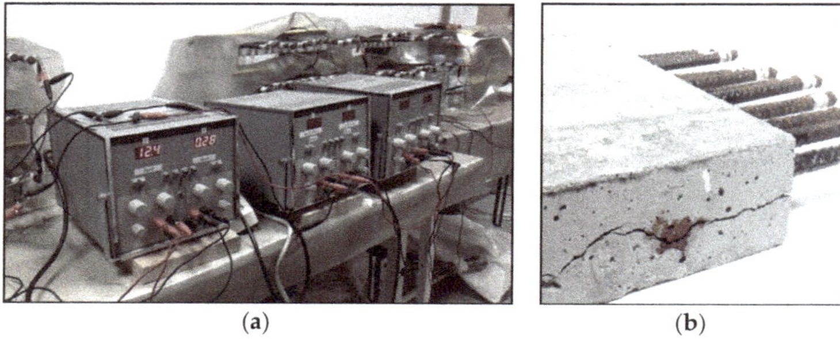

(**a**) (**b**)

Figure 2. Electrical connection of the bars for the accelerated corrosion process (**a**) and cracks in the concrete slabs after disconnecting and ready for demolition (**b**).

Once the corrosion process was over, the slabs were demolished, the oxide and cement of each bar surface were mechanically cleaned by a brush according to standard ASTM G1-03 (2017) [39]. Then, if such a treatment could not eliminate all the corrosion products, the oxide was cleaned by a chemical process by immersing the bars in a bath for 10 min. It was then rinsed with ethanol and water and dried according to standard ISO 8407 [40]. As it can be observed in Figure 3, the corrosion was extended throughout the whole surface of embedded bars.

Figure 3. Rebars after the first treatment according to standard ASTM G1-03 (2017) [39].

The degree of corrosion or corrosion level (Q_{corr}) was measured by the gravimetric procedure of weighing the bars after the full cleaning of all corrosion products. The gravimetric cross-sectional loss of the corroded steel reinforcement was deduced by Equation (1) [41].

$$Q_{corr} = \frac{m_0 - m_{res}}{m_0} \times 100 \tag{1}$$

where Q_{corr} is the corrosion degree of the steel reinforcement in percentage and m_0 and m_{res} are the mass of initial reinforcement corresponding to the portion of rebar that participates, respectively, in corrosion and the residual mass of the corroded steel reinforcement portion.

In order to calculate the residual diameter value of the corroded bar cross-section, the residual mass of the corroded steel reinforcement is used and determined following the equivalent section definition [14,42].

Thus, the residual diameter of the corroded bars was computed by Equation (2) as follows:

$$\varnothing_{res} = \sqrt{\frac{4m_{res}}{\pi\, 7.85\ L_c}} \tag{2}$$

where L_c is the corroded length of the bar (cm) and 7.85 is the specific weight of steel (g/cm^3).

2.3. Strength Tests

Once the corrosion process was finished and the bars free of all corrosion products, the bars were tested under tension according to standard EN ISO 7500-1:2018 [43] by a multitest IBERTEST press with a loading capacity of 100 kN, controlled by the WINTest 32 software program. The strain measurements were performed with an extensometer 2-IBER-25 with a 50 mm base.

Before the test, for the manual determination of the elongation after fracture (Equation (3), the entire bar received fine marks with multiples of 5mm following the UNE EN ISO 6892-1:2017 standard [44]. In addition, for the total elongation at maximum force a class two strain gauge, according to the ISO 9513:2012 standard, was used [45].

$$A_{u,5} = \frac{L_u - L_0}{L_0} \times 100 \tag{3}$$

where $A_{u,5}$ is the permanent elongation of the gauge length expressed as a percentage of the original gauge length, L_u is the final gauge length after fracture and L_0 is the original gauge length.

The tests were controlled in terms of load when the bar behaved in the elastic range and deformation when the test was running in the plastic zone at three different speeds. The standard speed V_m (medium loading speed) was that recommended by standard UNE-EN ISO 15630-1:2011 [29], that is to say, 3.7 kNs and 20.1 mm/min. A speed three times faster V_h (high loading speed) with 11.1 kN/s and 60.3 mm/m, and a speed V_l (low loading speed) three times slower with 1.23 kN/s and 6.7 mm/min were also used.

A designating code B-*XXX*-*Y* was used to identify each bar sample, with B meaning bar, *XX* the bar number (1 to 144) and *Y* the loading speed (low, medium or high). The (*) in the nomenclature shows that the rebar did not meet certain EHE-08 requirements for B500SD steel.

3. Results and Discussion

Based on the test results, the values of the equivalent steel ductility parameters as per Ortega (*Id*), Cosenza (*p*) and Creazza (*A**) were calculated for each bar sample (see Equations (4)–(6)). Tables A1–A3 in Appendix A show the results and the ductility parameters for, respectively, low, medium and high loading speed. The level of corrosion was quantified by Q_{corr} as per Equation (1) and used to order the table lists.

The results include the yield strength (R_e), tensile strength (R_m), the total elongation at maximum force (A_{gt}) and the permanent elongation of the gauge length $A_{u,5}$. All mechanical properties were calculated with respect to the residual diameter of the corroded bars ($\varnothing_{res}$ in Equation (2)). In addition, the equivalent steel ductility parameters were calculated with respect to Equations (4)–(6).

$$Id = 1 + \left(1 + \frac{R_m}{R_e}\right)\left(\frac{A_{gt}}{\varepsilon_y} - 1\right) \tag{4}$$

$$A^* = \frac{2}{3}(R_m - R_y)(A_{gt} - \varepsilon_{sh}) \quad (5)$$

$$p \approx A_{gt}^{0.75}\left(\frac{R_m}{R_e} - 1\right)^{0.9} \quad (6)$$

Those bars with lower values of mechanical parameters than those required for steel B500SD in EHE-08 have been highlighted with an asterisk in Tables A1–A3 (see Appendix A). The maximum and minimum values for the parameters can be seen in Table 4.

Table 4. Range of the mechanical and ductility parameters of tensile tests run at three loading speeds V_l (low loading speed), V_m (medium loading speed) and V_h (high loading speed).

Loading Speed	Values	R_e (R_m) MPa	R_m/R_e	A_{gt} (%)	$A_{u,5}$ (%)	*Id*	*P*	A^* (N/mm²)
V_l	Min.	429.6(518.8)	1.07	6.05	23	47.0	0.4	2.2
	Max.	651.8(726.4)	1.21	17.29	34	135.1	1.8	10.3
V_m	Min.	449.0(498.4)	1.10	3.7	7	27.8	0.4	1.3
	Max.	626.5(620.40)	1.21	14.8	35	369.0	1.5	8.6
V_h	Min.	206.0(283.6)	1.03	5.2	19	41.6	0.5	2.4
	Max.	618.6(688.9)	1.4	26.4	34	205.2	2.3	8.2

A general overview of the values in Table 4 reveals that the high loading speeds (V_h) cause a distorting of the results, with yield strength and tensile strength values being significantly lower than it expected. Thus, the equivalent steel ductility parameters at this loading speed would be highly recommended. As the values of total elongation at maximum force declined substantially, in the majority of the cases three times less than the total elongation recorded for the control, corrosion is more sensitive to strain than to stress.

Table 5 shows the yield strength and tensile strength for the three loading speeds of steel bars at four levels of corrosion and the percentage of strength loss of yield and tensile strength with 1%, 2%, 3% and 4% of corrosion degree and yield and tensile strength without corrosion, following Equation (7).

$$\text{Strength reduction rates} = \left(\frac{\left(\frac{R_{ei}}{R_{e0}}\right) + \left(\frac{R_{mi}}{R_{m0}}\right)}{2}\right) \times 100 \quad (7)$$

Table 5. Range of yield strength and tensile strength of tensile tests run at three loading speeds and the percentage of strength reduction.

Loading Speed	Q_{corr} (%)	R_e (R_m) MPa	Strength Reduction Rates (%)
V_l	0	531.7 (612.6)	–
	1	505.9 (587.3)	5%
	2	483.0 (570.3)	8%
	3	449.4 (534.9)	14%
	4	458.0 (541.9)	13%
V_m	0	544.4 (626 0)	–
	1	511.5 (594 7)	6%
	2	489.5 (577.1)	9%
	3	449.0 (542.7)	16%
	4	449.4 (498.4)	19%
V_h	0	547.3 (619.8)	–
	1	509.1 (591.0)	6%
	2	497.1 (574.6)	8%
	3	453.0 (544.8)	15%
	4	454.4 (533.3)	16%

With a corrosion degree of 1%, reduction rates of yield strength and tensile strength are around 5% and 6% for all the loading speeds. Similar strength reduction rates of approximately 8% and 9% are found in bars when the corrosion level increases to 2%. However, a corrosion level of more than 3% will induce a greater tensile-strength reduction of approximately 16% and 19% for medium and high levels of loading speeds and 13% for a low loading speed. This confirms that the loading speed has a greater influence when the rebars have higher degrees of corrosion.

Figure 4 shows how these deviations in the results can be studied based on the equivalent steel criteria and how they can be analysed in the results. The figure shows the comparison of the percentage of specimens that meet the various equivalent criteria, at the different loading speeds V_l, V_m and V_h.

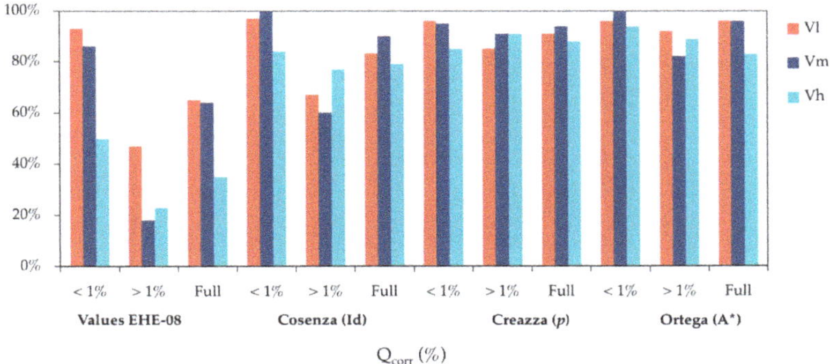

Figure 4. Comparison of the percentage of bar specimens that meet different ductility criteria in tensile strength tests run at different loading speeds: low V_l, medium V_m and high V_h.

As expected, when using EHE-08 criteria, the loading speed is important with high levels of corrosion. Even with corrosion rates of up to 1% there is a significant difference for high loading speed (V_h) as compared with low and medium speeds (V_l, V_m). In addition, for corrosion rates of up to 1% all equivalent steel criteria are met for low and medium loading speeds. For high-speed loading the criteria offered by Creazza are scarcely met, although fulfilment is frequent for Cosenza and Ortega criterion. With corrosion rates of higher than 1%, fulfilment of EHE-08 ductility criteria was low, less so for Cosenza criterion and high for the Creazza and Ortega criterion. Thus, in general, the three equivalent steel concepts offered by Cosenza, Creazza and Ortega serve as useful criteria for high loading speeds and corrosion rates under 1%. Given that more that 90% of the bar specimens meet the ductility criteria, the concept is quite advantageous in assessing structural ductility with corroded reinforcement. Regardless of the loading speed considered, the EHE-08 ductility requirements are met by more than 90% of the bar specimens for corrosion rates of up to 1%, though only 20% of the specimens meet such requirements when the corrosion rate is higher than 1%. The main reason is the systematic reduction of the total elongation at maximum load (A_{gt}) when it increases the corrosion rates [30] up to values that fail the minimum ones required by EHE-08.

Summaries of representative strain-stress curves are plotted in Figures 5–7 for, respectively, each low, medium and high loading speed. The numbers in % indicate the corrosion level of the rebar.

It can be seen that with the development of corrosion, the yield strength, tensile strength and total elongation at maximum load decreased under different strain rates. In addition, the yield plateau shortened or even disappeared. In contrast, given that with high-corrosion levels an anomalous elongation of the yield plateau is produced, the loading speed has a significant influence on this area of the curve. Compared with the uncorroded rebars, the decreased yield and tensile strength of the corroded rebars were mainly caused by the reduction of fracture cross-sectional areas. The decreased total elongation and the shortened yield plateau were due to intensified stress concentrations at the corrosion pits [46].

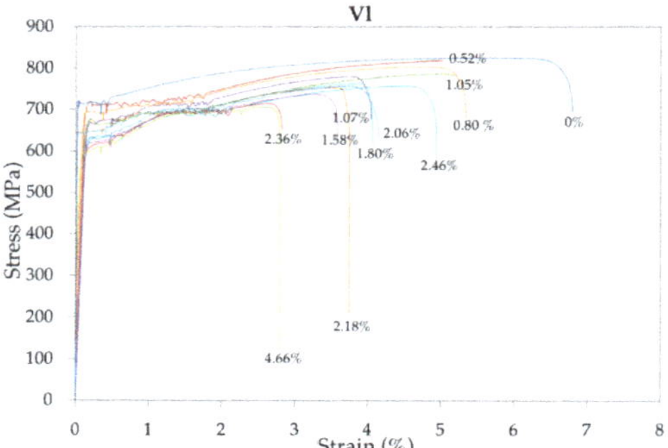

Figure 5. Representative summary of the strain-stress curves of the 48 bar specimens tested at low speed V_l for increasing corrosion rates.

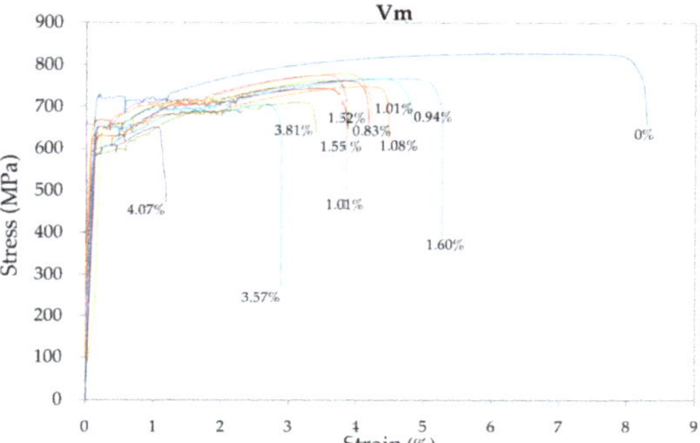

Figure 6. Representative summary of the strain-stress curves of the 48 bar specimens tested at medium (standard) speed V_m for increasing corrosion rates.

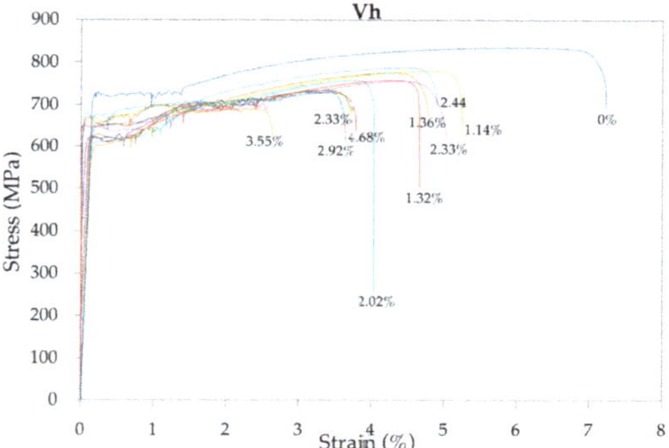

Figure 7. Representative summary of the strain-stress curves of the 48 bar specimens tested at high speed V_h for increasing corrosion rates.

Figure 8 shows the range of deformations (mean value) in the yield zone for corrosion rates of lower than 1% (a) and higher than 1% (b), for the three loading speeds used. It can be seen that deformation is similar for low and medium speeds regardless of the corrosion rate. At high corrosion rates, deformation is much greater for the high loading speed (as discussed above).

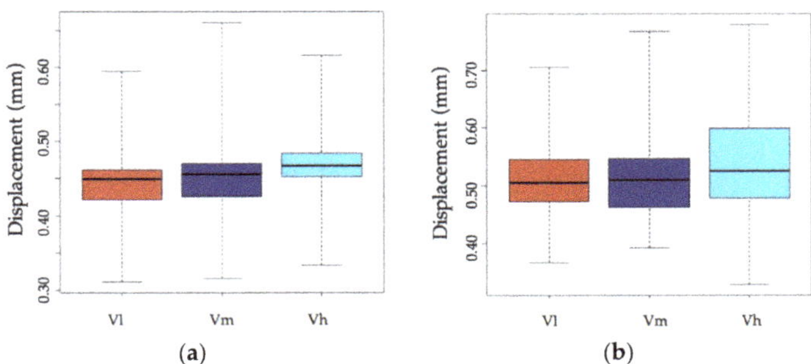

Figure 8. Deformation of bar specimens in the yield zone for corrosion rates under 1% (**a**) and over 1% (**b**) and for the three loading speeds V_l, V_m and V_h.

Regardless of the corrosion level, in Figure 9 it is possible to see that the deformation of bar specimens in the yield zone showed less dispersion for low and medium speeds in comparison with high speed.

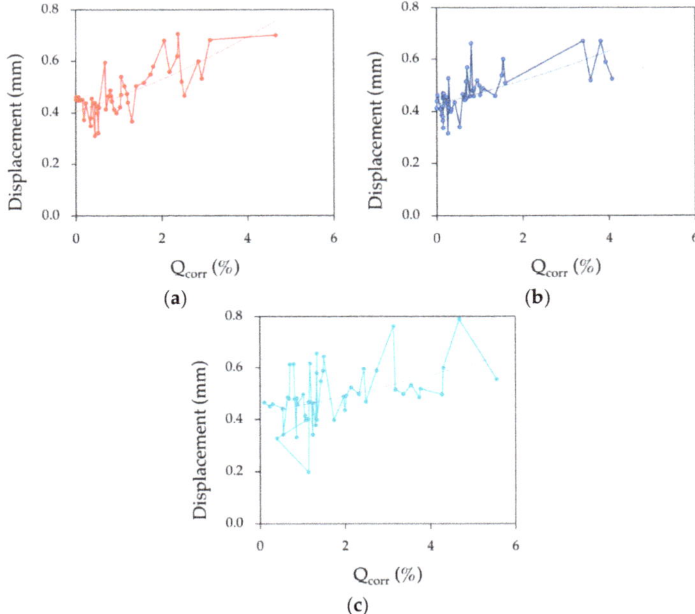

Figure 9. Deformation of bar specimens in the yield zone as a function of the loading speed V_l (**a**), V_m (**b**) and V_h (**c**).

The evolution of the mechanical properties obtained in the tensile tests as a function of the corrosion rate can be observed in Figure 10. It shows the average cross-section diameter of the specimen $\varnothing_{res}$ after the corrosion process. Each colour denotes the loading speed. The figure also shows that the yield strength, tensile strength and total elongation at maximum force, regardless of the loading speed, decrease with the increase in corrosion level. The dispersion of the results occurs only with high corrosion levels. Therefore, the loading speed in this test program seemed to have no significant effect on either strength at a low level of corrosion.

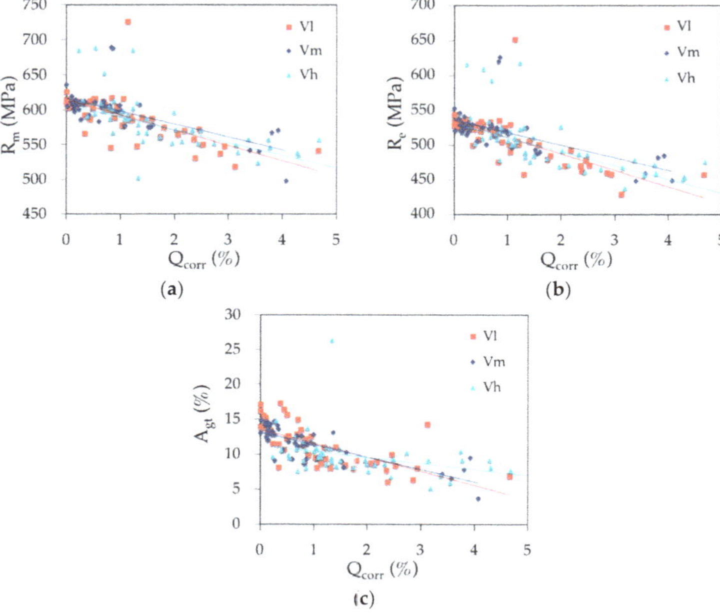

Figure 10. Effects of the corrosion rate and loading speed on the (**a**) tensile strength R_m, (**b**) yield strength R_e and (**c**) total elongation at maximum force A_{gt}.

If the corrosion had been uniform along the bar, the adjusting trend lines would have been horizontal. However, the lines decrease for all loading speeds [21]. This is because corrosion is not homogeneous in the bar surface, but occurs in a series of pitting spots typical for chloride corrosion of steel [46,47]. In these spots, the cross-section area of the bar is smaller than the average in the test results (see Figure 11). Additionally, corrosion takes place in the outer thickness of the bar surface composed by martensite, a metallographic material produced by the rolling mill when the bar was fabricated. Martensite has higher strength properties (R_m, R_e) than the ferrite composing the internal core of the bar. The destruction of part of this stronger outer layer explains the reduction of the average strength values in the bar cross-section.

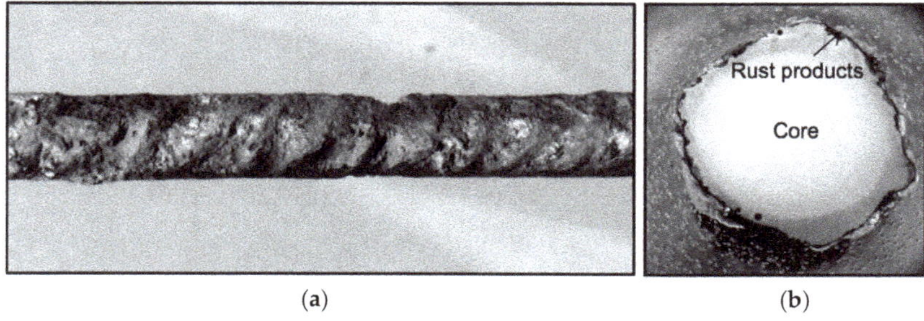

Figure 11. Microscopy image of the corroded surface (**a**) of bar specimen B087M with Q_{corr} = 1.07% and the cross-section (**b**).

The evolution of ratio R_m/R_e for the three loading speeds is shown in Figure 12. Regardless of the loading speed, the ratio has a small fluctuation near a fixed value while the corrosion ratio changed from zero to more than 4%. It is indicated that the ratio of R_m/R_e is irrelevant to the average corrosion ratio which shows that the decrease of yield strength is induced by the decrease of effective cross-section area of the steel bars [9,21].

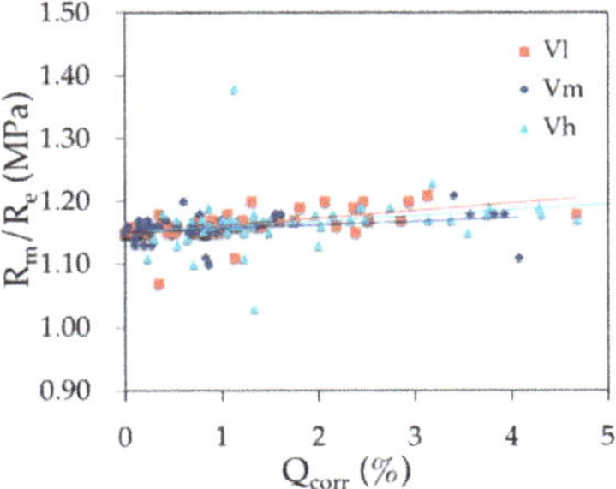

Figure 12. Effects of the corrosion rate and the loading speed on the ratio R_m/R_e.

Figure 13 shows the evolution of the three ductility parameters based on the steel equivalent concept as a function of the corrosion rate for the three loading speeds. All parameter values decrease when the corrosion rate increases regardless of the loading speed. Parameters evolve similarly for low and medium loading speeds (parallel lines). However, the degree of scatter with high-speed loading would not provide accurate conclusions.

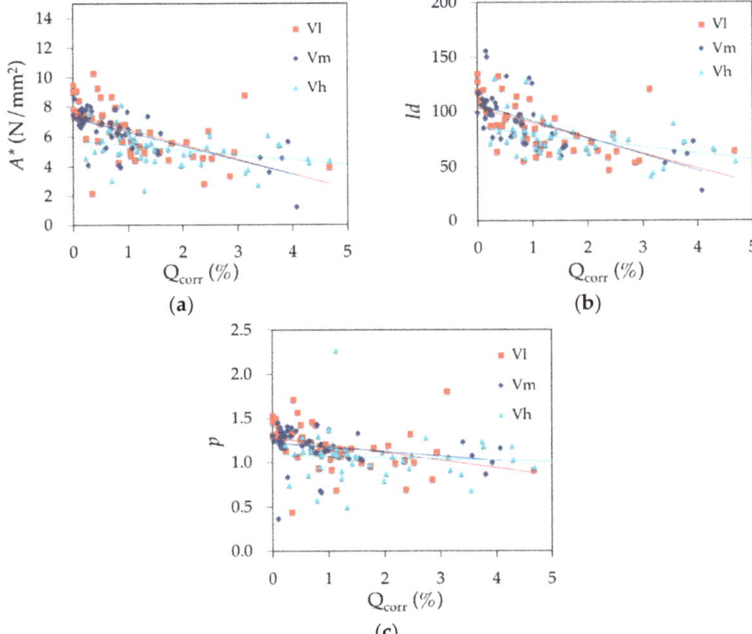

Figure 13. Effect of corrosion rate and loading speed on the equivalent steel concept parameters (**a**) A^*, (**b**) Id and (**c**) p.

4. Conclusions

The main conclusions can be summarised as follows:

- With the exception of the ratio R_m/R_e there is a systematic reduction of all strength and durability parameters (R_m, R_e, A_{gt}, p, A^* and Id) for increasing corrosion rates.
- The increasing of the corrosion level leads to modification of the stress-strain diagram, losing the yield plateau and showing cold-drawn behaviour.
- The loading speed of the tensile test is a variable which, along with the corrosion rate, governs the values and the evolution of all of the studied parameters (except the ratio R_m/R_e).
- The equivalent steel concept is useful to evaluate the ductility behaviour of corroded reinforcement bars in concrete regardless of the loading speed in the tensile test.
- The higher the tensile test loading speed, the higher is the yield zone in the strain–stress relationship curve.
- With corrosion rates as low as 1%, there is a change in the strain-stress curve which means that in some cases the yield plateau disappears and the steel behaves as a cold-formed steel.
- Finally, the research significance lies in the assessment of the influence of the loading speed at which the tensile test is performed for the reinforcement bars that largely depends of the ductility criteria used.

Author Contributions: Conceptualisation, A.M.B. and M.d.l.N.G.; Methodology, A.M.B. and M.d.l.N.G.; Formal Analysis, A.M.B. and M.d.l.N.G.; Investigation, A.M.B.; Resources, A.M.B.; Data Curation, A.M.B. and M.G.A.; Writing–Original Draft Preparation, A.M.B. and M.G.A.; Writing–Review and Editing, M.G.A. and J.C.G.; Supervision, J.C.G.; Project Administration, J.C.G.; Funding Acquisition, A.M.B. and J.C.G.

Funding: This research was funded by the Spanish Ministry of Economy, Industry and Competitiveness (No. BIA 2016 78742-C2-2-R). It was also supported by a university-lecturer training grant (formación de profesorado universitario, FPU) provided by the Spanish Ministry of Science and Innovation to Angela Moreno Bazan (No. FPU14-06362).

Conflicts of Interest: The authors declare no conflict of interest.

Appendix A

Table A1. Results of tensile tests on 12 mm diameter bars with increasing corrosion rates (Q_{corr}) using load increment speed (V_l) of 1.23 kN/s up to the yields strength (R_e) and a deformation speed of 6.7 mm/min the plastic zone. The three values of equivalent steel criteria for each bar have also been included. The (*) in the nomenclature shows that the rebar did not meet certain EHE-08 requirements for B500SD steel.

Rebars	Loading Speed	Q_{corr} (%)	R_e (MPa)	R_m (MPa)	R_m/R_e	A_{gt} (%)	$A_{u,5}$ (%)	Id	p	A^* (N/mm²)
B001L	1.23	0.00	534.02	614.83	1.15	17.11	27.00	135.10	1.53	9.51
B002L	1.23	0.00	529.28	610.33	1.15	16.19	27.00	127.77	1.46	9.03
B003L	1.23	0.03	526.54	603.69	1.15	13.89	25.00	118.30	1.30	7.37
B004L	1.23	0.06	526.99	612.10	1.16	15.55	30.00	110.80	1.50	9.11
B005L	1.23	0.08	526.89	606.71	1.15	14.06	28.00	106.81	1.32	7.72
B006L	1.23	0.10	529.58	609.96	1.15	15.24	27.00	120.21	1.40	8.43
B007L	1.23	0.15	535.34	612.49	1.15	13.17	28.00	99.98	1.25	6.99
B008L	1.23	0.19	526.24	607.37	1.15	13.40	28.00	101.74	1.27	7.48
B009L	1.23	0.23	537.76	612.43	1.15	11.49	28.00	87.08	1.13	5.90
B010L *	1.23	0.34	527.42	566.81	1.07	8.15	26.00	63.82	0.44	2.21
B011L	1.23	0.34	500.29	592.84	1.18	11.43	28.00	87.81	1.33	7.28
B012L	1.23	0.37	523.30	609.90	1.17	17.29	28.00	132.83	1.72	10.30
B014L	1.23	0.44	530.93	613.33	1.16	16.41	29.00	121.07	1.57	9.30
B015L	1.23	0.44	508.94	587.13	1.15	10.70	26.00	87.33	1.07	5.76
B016L	1.23	0.49	520.56	602.64	1.16	12.27	28.00	93.49	1.26	6.93
B017L	1.23	0.50	534.85	615.64	1.15	15.63	–	–	1.43	8.69
B018L	1.23	0.52	530.37	616.35	1.16	12.64	28.00	96.35	1.29	7.48
B019L	1.23	0.53	530.37	616.35	1.16	12.64	28.00	96.35	1.29	7.48
B020L	1.23	0.68	533.98	612.36	1.15	12.87	26.00	105.28	1.23	6.94

Table A1. *Cont.*

Rebars	Loading Speed	Q_{corr} (%)	R_e (MPa)	R_m (MPa)	R_m/R_e	A_{gt} (%)	$A_{u,5}$ (%)	Id	p	A^* (N/mm²)
B021L	1.23	0.70	524.22	608.98	1.16	14.94	28.00	114.09	1.46	8.71
B022L	1.23	0.76	516.12	601.71	1.17	13.51	30.00	96.55	1.43	7.96
B023L	1.23	0.80	520.95	601.50	1.15	11.34	27.00	89.15	1.12	6.28
B024L *	1.23	0.82	475.96	545.98	1.15	8.93	34.00	55.32	0.94	4.30
B025L	1.23	0.84	536.17	617.64	1.15	12.13	30.00	85.78	1.18	6.80
B026L	1.23	0.89	504.97	589.03	1.17	9.88	28.00	75.40	1.13	5.71
B027L	1.23	0.94	526.27	605.27	1.15	12.57	24.00	111.46	1.21	6.83
B028L	1.23	1.03	500.76	578.25	1.16	9.54	24.00	84.70	1.04	5.09
B029L *	1.23	1.05	490.76	578.25	1.18	9.54	29.00	70.53	1.16	5.74
B030L *	1.23	1.05	530.38	616.35	1.16	8.05	29.00	58.80	0.92	4.76
B031L *	1.23	1.13	651.71	726.36	1.11	8.64	27.00	66.41	0.69	4.44
B032L	1.23	1.19	506.27	589.77	1.16	11.02	26.00	90.39	1.16	6.33
B033L	1.23	1.21	502.03	585.89	1.17	9.34	28.00	71.22	1.08	5.39
B034L *	1.23	1.30	458.23	548.45	1.20	7.99	28.00	61.58	1.12	4.96
B035L	1.23	1.40	506.27	589.77	1.16	11.02	25.00	94.05	1.16	6.33
B036L	1.23	1.58	501.61	588.23	1.17	8.43	25.00	72.00	1.00	5.02
B037L *	1.23	1.74	479.76	561.71	1.17	7.92	26.00	64.93	0.96	4.47
B038L *	1.23	1.80	485.23	575.16	1.19	9.09	25.00	78.44	1.17	5.62
B039L *	1.23	2.06	470.60	565.26	1.20	8.71	26.00	72.50	1.19	5.67
B040L *	1.23	2.18	493.22	570.42	1.16	8.91	29.00	65.20	0.99	4.73
B041L *	1.23	2.36	470.79	558.30	1.19	7.67	28.00	58.80	1.03	4.62
B042L *	1.23	2.38	462.84	531.12	1.15	6.05	27.00	47.03	0.70	2.84
B043L *	1.23	2.46	478.80	572.32	1.20	9.94	27.00	79.79	1.32	6.40
B044L *	1.23	2.52	471.37	550.49	1.17	8.41	28.00	64.01	1.00	4.58
B045L *	1.23	2.85	460.34	537.38	1.17	6.35	25.00	53.95	0.81	3.37
B046L *	1.23	2.93	458.23	548.45	1.20	8.02	31.00	55.72	1.12	4.98
B047L *	1.23	3.12	429.57	518.82	1.21	14.30	26.00	120.34	1.81	8.78
B048L *	1.23	4.66	458.03	541.86	1.18	6.90	23.00	64.22	0.91	3.98

Table A2. Results of tensile tests on 12 mm diameter bars with increasing corrosion rates (Q_{corr}), using load increment speed (V_m) of 3.7 kN/s up to the yield strength (R_e) and a deformation speed of 20.1 mm/min in the plastic zone. The three values of equivalent steel criteria for each bar are also included. The (*) in the nomenclature shows that the rebar did not meet certain EHE-08 requirements for B500SD steel.

Rebars	Loading Speed	Q_{corr} (%)	R_e (MPa)	R_m (MPa)	R_m/R_e	A_{gt} (%)	$A_{u,5}$ (%)	Id	p	A^* (N/mm²)
B049M *	3.70	0.00	544.38	626.03	1.15	13.94	25.00	118.73	1.31	7.83
B050M	3.70	0.00	553.64	636.67	1.15	13.05	28.00	99.06	1.25	7.45
B051M	3.70	0.02	521.19	605.10	1.16	14.82	27.00	117.40	1.45	8.56
B052M *	3.70	0.09	538.32	612.13	1.14	14.18	27.00	111.25	1.25	7.20
B053M	3.70	0.11	529.48	608.73	1.15	12.48	31.00	85.40	1.20	6.80
B054M	3.70	0.14	518.72	604.69	1.17	13.14	30.00	93.88	1.40	7.77
B055M *	3.70	0.14	522.14	602.23	1.15	12.05	7.00	368.96	1.17	6.64
B056M	3.70	0.15	522.38	606.17	1.16	12.06	25.00	103.04	1.24	6.95
B057M	3.70	0.15	521.38	597.59	1.15	13.89	19.00	156.03	1.30	7.28
B058M	3.70	0.15	530.81	609.75	1.15	14.49	28.00	110.11	1.35	7.87
B059M	3.70	0.15	531.79	609.79	1.15	13.45	26.00	110.07	1.27	7.22
B060M	3.70	0.15	531.00	611.39	1.15	12.99	27.00	102.29	1.24	7.18
B061M	3.70	0.16	530.90	610.20	1.15	14.02	27.00	110.49	1.31	7.65
B062M	3.70	0.17	529.11	614.76	1.16	13.35	19.00	150.61	1.34	7.87
B063M *	3.70	0.19	541.84	612.54	1.13	14.31	–	–	1.17	6.96
B064M	3.70	0.22	525.96	608.02	1.16	12.79	28.00	97.51	1.30	7.22
B065M	3.70	0.23	515.56	601.22	1.17	13.21	28.00	101.21	1.41	7.79
B066M	3.70	0.26	532.26	612.23	1.15	14.75	28.00	112.11	1.36	8.12
B067M *	3.70	0.27	520.20	586.33	1.13	9.11	25.00	76.49	0.84	4.14
B068M	3.70	0.27	528.04	610.59	1.16	12.73	28.00	97.04	1.30	7.23
B069M	3.70	0.32	526.21	608.61	1.16	14.12	28.00	107.77	1.40	8.00
B070M	3.70	0.34	530.45	612.86	1.16	13.60	28.00	103.75	1.36	7.71
B071M	3.70	0.42	505.66	593.07	1.17	10.60	30.00	75.50	1.19	6.37
B072M	3.70	0.53	525.69	611.00	1.16	11.78	19.00	132.76	1.22	6.91

Table A2. *Cont.*

Rebars	Loading Speed	Q_{corr} (%)	R_e (MPa)	R_m (MPa)	R_m/R_e	A_{gt} (%)	$A_{u,5}$ (%)	Id	p	A^* (N/mm^2)
B073M	3.70	0.60	501.54	584.03	1.20	9.29	26.00	77.41	1.25	5.27
B074M	3.70	0.65	524.12	604.00	1.15	11.62	30.00	82.13	1.14	6.39
B075M	3.70	0.68	531.43	611.32	1.15	12.64	29.00	92.56	1.22	6.95
B076M *	3.70	0.69	527.09	613.29	1.16	11.11	9.00	265.48	1.17	6.59
B077M	3.70	0.70	528.74	606.55	1.15	11.18	26.00	91.30	1.11	5.99
B078M	3.70	0.77	514.27	604.54	1.18	12.63	28.00	97.15	1.43	7.84
B079M	3.70	0.80	525.60	606.32	1.15	11.26	27.00	88.51	1.11	6.25
B080M *	3.70	0.83	620.40	689.72	1.11	8.61	25.00	71.56	0.69	4.11
B081M *	3.70	0.85	518.02	600.39	1.16	11.67	10.00	250.91	1.21	6.61
B082M *	3.70	0.86	626.49	688.27	1.10	9.31	24.00	80.36	0.67	3.96
B083M	3.70	0.87	521.07	598.55	1.15	11.40	34.00	70.94	1.13	6.08
B084M	3.70	0.94	521.33	597.96	1.15	11.67	19.00	130.91	1.14	6.15
B085M	3.70	0.99	520.49	607.14	1.17	12.87	22.00	125.78	1.38	7.67
B086M	3.70	1.01	516.76	599.52	1.16	11.41	25.00	97.42	1.19	6.50
B087M *	3.70	1.08	497.26	577.39	1.16	9.55	31.00	65.38	1.04	5.26
B088M	3.70	1.36	525.72	607.42	1.16	13.13	35.00	79.87	1.33	7.38
B089M *	3.70	1.52	493.80	575.99	1.17	9.00	32.00	59.86	1.05	5.09
B090M *	3.70	1.55	488.25	575.28	1.18	8.14	26.00	67.07	1.03	4.87
B091M *	3.70	1.60	490.75	578.97	1.18	10.29	32.00	68.92	1.23	6.25
B092M*	3.70	3.40	449.01	542.67	1.21	7.19	29.00	53.58	1.08	4.63
B093M *	3.70	3.57	459.52	540.64	1.18	6.55	22.00	63.72	0.87	3.66
B094M *	3.70	3.81	482.57	567.84	1.18	7.81	27.00	61.88	1.00	4.58
B095M *	3.70	3.92	485.04	571.67	1.18	9.54	28.00	73.10	1.16	5.69
B096M *	3.70	4.07	449.41	498.42	1.11	3.70	27.00	27.80	0.37	1.25

Table A3. Results of tensile tests on 12 mm diameter bars with increasing corrosion rates (Q_{corr}), using load increment speed (V_h) of 11.1kN/s up to the yield strength (R_e) and a deformation speed of 60.3 mm/min in the plastic zone. The three values of equivalent steel criteria for each bar are also included. The (*) in the nomenclature shows that the rebar did not meet certain EHE-08 requirements for B500SD steel.

Rebars	Loading Speed	Q_{corr} (%)	R_e (MPa)	R_m (MPa)	R_m/R_e	A_{gt} (%)	$A_{u,5}$ (%)	Id	p	A^* (N/mm^2)
B097H *	11.1	0.1	547.27	619.76	1.13	14.62	29	106.25	1.19	7.29
B098H *	11.1	0.23	616.51	685.65	1.11	9.69	28	71.91	0.75	4.61
B099H *	11.1	0.3	540.17	613.85	1.14	14.78	24	130.65	1.28	7.49
B100H	11.1	0.53	510.04	594.97	1.17	9.91	29	72.98	1.13	5.79
B101H *	11.1	0.54	610.05	688.88	1.13	9.51	19	105.48	0.86	5.16
B102H *	11.1	0.64	534.57	609.86	1.14	12.41	31	84.53	1.13	6.43
B103H	11.1	0.68	513.14	599.51	1.17	10.77	29	79.42	1.21	6.40
B104H *	11.1	0.7	593.91	653.23	1.1	7.64	26	60.61	0.58	3.12
B105H	11.1	0.79	517.42	603.96	1.17	11.84	27	93.99	1.30	7.05
B106H	11.1	0.81	509.05	587.82	1.15	9.04	25	76.59	0.95	4.90
B107H	11.1	0.85	518.77	600.78	1.16	10.54	27	83.16	1.12	5.95
B108H *	11.1	0.86	478.64	567.93	1.19	8.08	30	57.79	1.08	4.96
B109H	11.1	0.88	526.21	612.61	1.16	13.76	23	128.06	1.37	8.18
B110H	11.1	1.01	506.82	594.51	1.17	10.48	28	80.05	1.18	6.32
B111H	11.1	1.05	519.43	598.94	1.15	10.87	34	67.59	1.09	5.95
B112H	11.1	1.08	501.11	587.48	1.17	9.49	27	75.10	1.10	5.64
B113H	11.1	1.13	510.5	592.25	1.16	10.18	34	63.51	1.10	5.73
B114H *	11.1	1.13	206.02	283.55	1.38	9.5	27	82.36	2.27	5.07
B115H	11.1	1.14	501.47	586.46	1.17	10.06	33	64.98	1.15	5.88
B116H	11.1	1.17	505.71	588.83	1.16	9.99	30	70.77	1.08	5.71
B117H *	11.1	1.23	618.59	685.66	1.11	11.03	24	95.86	0.83	5.09
B118H *	11.1	1.23	524.17	602.87	1.15	8.62	26	70.13	0.91	4.67
B119H	11.1	1.3	526.07	611.08	1.16	10.84	26	88.90	1.15	6.34
B120H *	11.1	1.33	490.11	503.35	1.03	26.41	26	205.17	0.50	2.41
B121H *	11.1	1.33	483.85	568.02	1.17	9.42	26	77.45	1.09	5.46
B122H *	11.1	1.33	490.15	579.27	1.18	8.99	25	77.21	1.11	5.51
B123H *	11.1	1.42	476.63	556.05	1.17	8.26	20	88.45	0.99	4.51

Table A3. *Cont.*

Rebars	Loading Speed	Q_{corr} (%)	R_e (MPa)	R_m (MPa)	R_m/R_e	A_{gt} (%)	$A_{u,5}$ (%)	Id	p	A^* (N/mm^2)
B124H	11.1	1.48	511.58	587.61	1.15	10.78	28	81.63	1.08	5.64
B125H *	11.1	1.49	485.18	567.64	1.17	9.1	30	64.65	1.06	5.16
B126H *	11.1	1.73	481.25	561.14	1.17	8.06	29	59.14	0.97	4.43
B127H *	11.1	1.96	469.95	552.83	1.18	9.13	24	81.75	1.12	5.21
B128H *	11.1	1.99	527.06	596.64	1.13	8.61	27	66.79	0.80	4.12
B129H *	11.1	2.02	492.4	571.87	1.16	7.61	24	67.33	0.88	4.16
B130H *	11.1	2.14	471.82	555.3	1.18	8.52	29	62.87	1.07	4.89
B131H *	11.1	2.33	487.94	570.91	1.17	8.66	27	68.43	1.02	4.94
B132H *	11.1	2.44	461.62	551.6	1.19	6.79	–	–	0.94	4.20
B133H *	11.1	2.49	487.02	570.88	1.17	9.31	25	79.64	1.08	5.37
B134H *	11.1	2.74	465.86	552.9	1.19	10.2	29	75.84	1.28	6.11
B135H *	11.1	3.14	468.11	549.96	1.17	9.13	26	75.03	1.07	5.14
B136H *	11.1	3.18	437.8	539.57	1.23	5.19	27	41.64	0.92	3.63
B137H *	11.1	3.37	479.02	558.81	1.17	6.97	33	44.66	0.87	3.83
B138H *	11.1	3.55	456.71	524.63	1.15	5.98	26	48.30	0.69	2.79
B139H *	11.1	3.74	472.56	557.92	1.18	10.44	25	89.86	1.24	6.13
B140H *	11.1	3.77	459.41	546.98	1.19	9.15	27	73.03	1.18	5.51
B141H *	11.1	4.28	451.88	538.98	1.19	9.14	27	72.95	1.18	5.48
B142H *	11.1	4.3	454.59	536.25	1.18	8.14	26	67.07	1.03	4.57
B143H *	11.1	4.68	476.94	557.9	1.17	7.7	25	65.67	0.94	4.29
B144H *	11.1	5.56	419.32	513.58	1.22	6.89	27	55.43	1.09	4.47

References

1. Soltani, A.; Harries, K.A.; Shahrooz, B.M. Crack opening behavior of concrete reinforced with high strength reinforcing steel. *Int. J. Concr. Struct. Mater.* **2013**, *7*, 253–264. [CrossRef]
2. Paul, S.C.; van Zijl, G.P.A.G. Corrosion deterioration of steel in cracked SHCC. *Int. J. Concr. Struct. Mater.* **2017**, *11*, 557–572. [CrossRef]
3. Broomfield, J.P. *Corrosion of Steel in Concrete Understanding, Investigating and Repair*, 2nd ed.; Taylor & Francis: New York, NY, USA, 2007.
4. Liu, Y.; Weyers, R.E. Time to cracking for chloride-induced corrosion in reinforced concrete. *Spec. Publ. R. Soc. Chem.* **1996**, *183*, 88–104.
5. Manera, M.; Vennesland, Ø.; Bertolini, L. Chloride threshold for rebar corrosion in concrete with addition of silica fume. *Corros. Sci.* **2008**, *50*, 554–560. [CrossRef]
6. *EHE-08. Code on Structural Concrete (EHE-08)*; Ministerio de la Presidencia: Madrid, Spain, 2008.
7. *EN 1992-1-1: Eurocode 2: Design of Concrete Structures: Part 1-1: General Rules and Rules for Buildings*; CEN European Committee for Standardization: Brussels, Belgium, 2004.
8. Cairns, J.; Plizzari, G.A.; Du, Y.; Law, D.W.; Franzoni, C. Mechanical properties of corrosion-damaged reinforcement. *ACI Mater. J.* **2005**, *102*, 256–264.
9. Zhang, W.; Song, X.; Gu, X.; Li, S. Tensile and fatigue behavior of corroded rebars. *Constr. Build. Mater.* **2012**, *34*, 409–417. [CrossRef]
10. Chen, A.; Pan, Z.; Ma, R. Mesoscopic simulation of steel rebar corrosion process in concrete and its damage to concrete cover. *Struct. Infrastruct. Eng.* **2017**, *13*, 478–493. [CrossRef]
11. Apostolopoulos, C.A. Mechanical behavior of corroded reinforcing steel bars S500s tempcore under low cycle fatigue. *Constr. Build. Mater.* **2007**, *21*, 1447–1456. [CrossRef]
12. *UNE 36065:2011 Ribbed Bars of Weldable Steel with Special Characteristics of Ductility for the Reinforcement of Concrete*; AENOR: Madrid, Spain, 2011.
13. Apostolopoulos, A.; Matikas, T.E. Corrosion of bare and embedded in concrete steel bar–impact on mechanical behavior. *Int. J. Struct. Integr.* **2016**, *7*, 240–259. [CrossRef]
14. Du, Y.; Clark, L.A.; Chan, A. Residual capacity of corroded reinforcing bars. *Mag. Concr. Res.* **2005**, *57*, 135–147. [CrossRef]
15. Fernández, I.; Bairán, J.M.; Marí, A.R. Corrosion effects on the mechanical properties of reinforcing steel bars. Fatigue and σ–ε behavior. *Constr. Build. Mater.* **2015**, *101*, 772–783. [CrossRef]

16. Maslehuddin, M.; Allam, I.A.; Al-Sulaimani, G.J.; Al-Mana, A.; Abduljauwad, S.N. Effect of rusting of reinforcing steel on its mechanical properties and bond with concrete. *ACI Mater. J.* **1990**, *87*, 496–502.
17. Apostolopoulos, C.A.; Papadakis, V.G. Consequences of steel corrosion on the ductility properties of reinforcement bar. *Constr. Build. Mater.* **2008**, *22*, 2316–2324. [CrossRef]
18. Imperatore, S.; Rinaldi, Z.; Drago, C. Degradation relationships for the mechanical properties of corroded steel rebars. *Constr. Build. Mater.* **2017**, *148*, 219–230. [CrossRef]
19. Fernandez, E.M.; Escamilla, A.C.; Cánovas, M.F. Ductility of reinforcing steel with different degrees of corrosion and the 'equivalent steel' criterion. *Materiales de Construcción* **2007**, *57*, 5–18.
20. Zhang, W.; Chen, H.; Gu, X. Tensile behaviour of corroded steel bars under different strain rates. *Mag. Concr. Res.* **2016**, *68*, 127–140. [CrossRef]
21. Shetty, A.; Venkataramana, K.; Gogoi, I.; Praveen, B.B. Performance enhancement of TMT rebar in accelerated corrosion. *J. Civ. Eng. Res.* **2012**, *2*, 14–17. [CrossRef]
22. Bazán, A.M.; Cobo, A.; Montero, J. Study of mechanical properties of corroded steels embedded concrete with the modified surface length. *Constr. Build. Mater.* **2016**, *117*, 80–87. [CrossRef]
23. Cobo, A.; Moreno, E.; Canovas, M.F. Mechanical properties variation of B500SD high ductility reinforcement regarding its corrosion degree. *Materiales de Construccion* **2011**, *61*, 517–532.
24. Comité Euro-International du Béton. *Durable Concrete Structures: Design Guide*, 2nd ed.; Thomas Telford Ltd.: London, UK, 1992.
25. Cosenza, E.; Greco, C.; Manfredi, G. The concept of equivalent steel. *CEB Bulletin d'information* **1993**, *218*, 163–184.
26. Cosenza, E.; Greco, C.; Manfredi, G. An equivalent steel index in the assessment of the ductility performances of the reinforcement. *Bulletin D Information-Comite Eurointernational Du Beton* **1998**, *242*, 157–170.
27. Creazza, G.; Russo, S. A new proposal for defining the ductility of concrete reinforcement steels by means of a single parameter. *Bulletin D Information-Comite Eurointernational Du Beton* **1998**, *242*, 171–182. [CrossRef]
28. Ortega, H. Estudio Experimental de la Influencia del tipo de Acero en la Capacidad de Redistribución en losas de Hormigón Armado. Ph.D. Thesis, Polytechnic University of Madrid, Madrid, Spain, 1998.
29. *UNE-EN ISO 15630-1:2011 Steel for the Reinforcement and Prestressing of Concrete–Test Methods–Part 1: Reinforcing Bars. Wire Rod and Wire*; AENOR: Madrid, Spain, 2011.
30. Real Decreto 256/2008. de 10 de junio Instrucción para la recepción de cementos (RC-16). Available online: https://portal.uah.es/portal/page/portal/epd2_profesores/prof121896/docencia/CEMENTOS.pdf (accessed on 25 January 2019).
31. Andrade, C.; Alonso, C.; Molina, F.J. Cover cracking as a function of bar corrosion: Part I-Experimental test. *Mater. Struct.* **1993**, *26*, 453–464. [CrossRef]
32. Caré, S.; Raharinaivo, A. Influence of impressed current on the initiation of damage in reinforced mortar due to corrosion of embedded steel. *Cem. Concr. Res.* **2007**, *37*, 1598–1612. [CrossRef]
33. Jin, J.; Zhao, Y. Effect of corrosion on bond behavior and bending strength of reinforced concrete beams. *J. Zhejiang Univ. Sci. A* **2001**, *2*, 298–308. [CrossRef]
34. van Zijl, G.P.A.G.; Paul, S.C. A novel link of the time scale in accelerated chloride-induced corrosion test in reinforced SHCC. *Constr. Build. Mater.* **2018**, *167*, 15–19. [CrossRef]
35. de Alcantara, N.P.; da Silva, F.M.; Guimarães, M.T.; Pereira, M.D. Corrosion assessment of steel bars used in reinforced concrete structures by means of eddy current testing. *Sensors* **2016**, *16*, 15. [CrossRef] [PubMed]
36. Andrade, C.; Alonso, C. Test methods for on-site corrosion rate measurement of steel reinforcement in concrete by means of the polarization resistance method. *Mater. Struct.* **2004**, *39*, 623–643. [CrossRef]
37. El Maaddawy, T.A.; Soudki, K.A. Effectiveness of impressed current technique to simulate corrosion of steel reinforcement in concrete. *J. Mater. Civ. Eng.* **2003**, *15*, 41–47. [CrossRef]
38. Paul, S.C.; van Zijl, G.P.A.G. Crack formation and chloride induced corrosion in reinforced strain hardening cement-based composite (R/SHCC). *J. Adv. Concr. Technol.* **2014**, *12*, 340–351. [CrossRef]
39. *ASTM G1-03. Standard Practice for Preparing, Cleaning, and Evaluating Corrosion Test Specimens*; ASTM: West Conshohocken, PA, USA, 2003.
40. *ISO 8407:1991 Corrosion of Metals and Alloys–Removal of Corrosion Products from Corrosion Test Specimens*; ISO: Geneva, Switzerland, 1991.

41. Moreno Fernández, E. Corrosión de Armaduras en Estructuras de Hormigón: Estudio Experimental de la Variación de la Ductilidad en Armaduras Corroídas Aplicando el Criterio de acero Equivalente. Ph.D Thesis, Universidad Carlos III de Madrid, Madrid, Spain, October 2008.
42. Wang, X.; Zhang, W.; Gu, X.; Dai, H. Determination of residual cross-sectional areas of corroded bars in reinforced concrete structures using easy-to-measure variables. *Constr. Build. Mater.* **2013**, *38*, 846–853. [CrossRef]
43. *ISO 7500-1:2018 Metallic Materials–Calibration and Verification of Static Uniaxial Testing Machines–Part 1: Tension/Compression Testing Machines–Calibration and Verification of the Force-Measuring System*; ISO: Geneva, Switzerland, 2018.
44. *UNE-EN ISO 6892-1:2017 Metallic Materials–Tensile Testing–Part 1: Method of Test at Room Temperature (ISO 6892-1:2016)*; AENOR: Madrid, Spain, 2017.
45. *ISO 9513:2012 Metallic Materials–Calibration of Extensometer Systems Used in Uniaxial Testing*; ISO: Geneva, Switzerland, 2012.
46. Zhu, W.; Dai, J.; Poon, C.-S. Prediction of the bond strength between non-uniformly corroded steel reinforcement and deteriorated concrete. *Constr. Build. Mater.* **2018**, *187*, 1267–1276. [CrossRef]
47. Moreno, E.; Cobo, A.; Palomo, G.; González, M.N. Mathematical models to predict the mechanical behaviour of reinforcements depending on their degree of corrosion and the diameter of the rebars. *Constr. Build. Mater.* **2014**, *61*, 156–163. [CrossRef]

© 2019 by the authors. Licensee MDPI, Basel, Switzerland. This article is an open access article distributed under the terms and conditions of the Creative Commons Attribution (CC BY) license (http://creativecommons.org/licenses/by/4.0/).

Article

On the Influence of Corrosion on the Load-Carrying Capacity of Old Riveted Bridges

Jozef Gocál and Jaroslav Odrobiňák *

Department of Structures and Bridges, Faculty of Civil Engineering, University of Žilina, Univerzitná 8215/1, 010 26 Žilina, Slovakia; jozef.gocal@uniza.sk
* Correspondence: jaroslav.odrobinak@uniza.sk; Tel.: +421-41-513-5650

Received: 31 December 2019; Accepted: 2 February 2020; Published: 5 February 2020

Abstract: Steel corrosion is one of the most dominant factors in the degradation of transport infrastructure. This article deals with the impact of the atmospheric corrosion of structural steel on the load-carrying capacity of old riveted bridge structures. A study on the impact of corrosion losses on the resistance and, thus, the load-carrying capacity of eight chosen bridge members with riveted I-sections from three different bridge substructures is presented. The load-carrying capacity calculation is carried out using modern procedures and on the basis of the diagnosed state of the structural elements. Within the analysis of the results, the need for long-term in situ corrosion measurements, as well as the need for regular inspections on the existing bridges are also discussed.

Keywords: corrosion of steel; riveted bridges; degradation; load-carrying capacity

---

1. Introduction

1.1. Degradation of Bridges due to Corrosion

Existing bridges are structures that reflect not only the level of the society in which they were built, but also the cultural and economic power of the present generations, as they reflect the care for these inherited engineering works. Therefore, professionals centered on bridges must be consistent in all activities related to the design, construction, and management of bridges from the initial idea to the end of their service life. Despite efforts during the design and construction of bridge structures, various damages and failures occur during their exploitation. Initially, small defects can gradually develop into failures significantly affecting load capacity and traffic safety.

Today, transport, energy, and environmental infrastructure structures (including bridge structures) account for approximately 70% of national assets in European countries. Their operation, maintenance, repair, and reconstruction consume around 35% of the total material and energy consumption and produce about 30% of all environmental burdens and waste. The above information highlights the impact of transport infrastructure on the economies of the countries and environments in which we live. The public usually perceives this fact only marginally, until the underestimated inspections and maintenance activities result in fatal consequences, such as the known collapses of footbridges or bridges in our and surrounding countries. Unfortunately, only then, the public discussion, supported by the media environment, immediately focuses on the management and maintenance of bridge structures.

The most powerful tool for the evaluation of bridge structures is diagnostics-supported determination of the load-carrying capacity of bridges and estimation of their residual life. These are extremely demanding and responsible tasks, in which the maximum permissible traffic load of the bridge is calculated in reverse form from the existing condition. A common problem is how to do these analyses with respect to environmental degradation processes and their future development.

The main factors influencing the condition of bridges, besides natural changes of the material, are their hidden structural defects and increasing traffic intensities, especially the degradation processes taking place in structural elements, which are caused by environmental load from the surrounding environment [1]. Polluted air has a significant impact on the degradation of all materials used in the transport infrastructure and throughout the construction industry. As the share of human-caused pollution is increasing, the polluted air becomes considerably more aggressive with a greater impact on the structural parts of bridges and footbridges. The damages caused by corrosion and related environmental degradation phenomena annually account for about 3%–4% of GDP in developed countries [2]. Several studies estimate that from 25% to 30% of annual corrosion costs could be saved if optimum corrosion management practices were employed [3]. Sophisticated expert assessment of damages and failures of bridge structures in terms of their impact on bridge reliability is an important task in determining the bearing capacity of bridges and designing their reconstruction.

1.2. Influence of Corrosion on Load-Carrying Capacity

As mentioned above, the corrosion degradation of structural steel has a significant impact on the bridge's reliability, in particular its safety and durability. Corrosion losses reduce the effective cross-sectional area of the load-bearing elements and thereby reduce their mechanical resistance to the effects of loads on the superstructure of the bridge. Depending on the level of safety of the individual load-bearing elements, it may happen during the service life of the bridge that the element reduced by progressive corrosion is no longer able to transmit the load effects, in particular the operational loads for which the bridge was primarily designed. The ability of the bridge structure to transmit the effects of the traffic load is quantified by the so-called "load-carrying capacity" (LCC), which is a basic quantification indicator for the evaluation of existing bridges. LCC represents a criterion that is valid not only for future planning, but is also used as the decision parameters for the evaluation of the passage of the actual railway service load. In recent years, several European-wide research projects have been completed [4] (such as, e.g., [5–8]). These projects have led to the development of guidelines that provide state-of-the-art methods for the safety assessment of existing bridges. Within the frame of this trend, the newest recommendation for the determination of the load-carrying capacity of metal railway bridges is being developed [9]. The newly elaborated guidelines for the determination of load-carrying capacity of railway bridges in the Slovak Republic [10] and Czech Republic [11] are also based on the latest knowledge combining the actual design codes and experiences from the area of evaluation.

Within transport infrastructure, there are relatively many bridge structures older than 50 years in Slovakia. For example, only in the case of railway bridges out of a total of 2300 bridge structures, up to 28% are over 75 years old, and almost one fifth of the bridges are even older than 100 years [12]. Thus, it is obvious that monitoring and consequently taking into account degradation due to environmental load is a very important factor in evaluating these structures and determining their load-bearing capacity. The LCC is generally defined as the ratio Z of the limiting effects of the vertical variable traffic load (in terms of the corresponding limit state) to the effects caused by the design load model in the member. This ratio represents the factor by which the multiplied effects of the load model (stresses, internal forces, deformations, etc.), in combination with other applied loads, cause the occurrence of corresponding limit states. In the case of railway bridges, this factor defines a multiple of the Load Model 71 (LM71) [13]; therefore, the ratio representing the LCC is referred to as Z_{LM71}. More details regarding LCC estimation can be found in [14,15].

In the case of riveted cross-sections of the old bridge structures, it is necessary to carry out several types of assessments in order to find a decisive check, which leads to the load-carrying capacity of the element itself. The impact of substantial imperfections of elements and parts of the steel structure should be taken into account in the global bridge analysis itself. Thus, major defects due to corrosion, in particular the significant reduction of the cross-section by corrosion, are to be included in the global analysis of the structural behavior already. Of course, any significant corrosion loss must also be taken into account in the verification of the cross-sections and members of the bridge structure.

2. Study Description

2.1. Inputs for Analysis

To point out how corrosion losses can reduce the resistance of a bridge member and its load-carrying capacity, the following study was executed. LCC of members with typical riveted I-sections of three old real railway bridges in service were calculated. The measured corrosion losses were taken into account in the process of bridge analysis and cross-section verifications. The main girders of two bridges were made of plate girders, while the last bridge had truss girders with both chords curved. All three bridges had a typical open member deck.

The first bridge, designated as "Bridge 1", is the smallest one, but it had been in the service for 142 years already. It bridges a railway line across a local road; thus, the span is only 10.92 m (Figure 1). For the presented study, the left stringer (outer side of the railway line directional curve), the second crossbeam, and the left (outer) main plate girders were chosen. These members emerged from the analysis as critical to the LCC of the bridge.

As mentioned before, the second railway bridge is also the plate girder bridge with an open member deck (Figure 2). "Bridge 2" was built across local stream in 1910. The span of the main girders is 22.90 m, and their mutual axial distance is 5.24 m. For the load-carrying capacity of this structure, the determining members were: the last right (outer one) stringer, the ninth cross-beam, and the right main girder on the outer side of the railway line directional curve.

The last bridge structure consisting of three simply supported superstructures was built on the main railway line. The main middle structure ("Bridge 3") with a span of 57.4 m is 76 years old. It was built as truss girder bridge with an intermediate open member deck (Figure 3). In order to estimate the effect of corrosion on the resistance of the I-shaped cross-sections of this bridge, the first left (inner) stringer and the sixth crossbeam of the bridge deck were chosen, as they showed the lowest LCC.

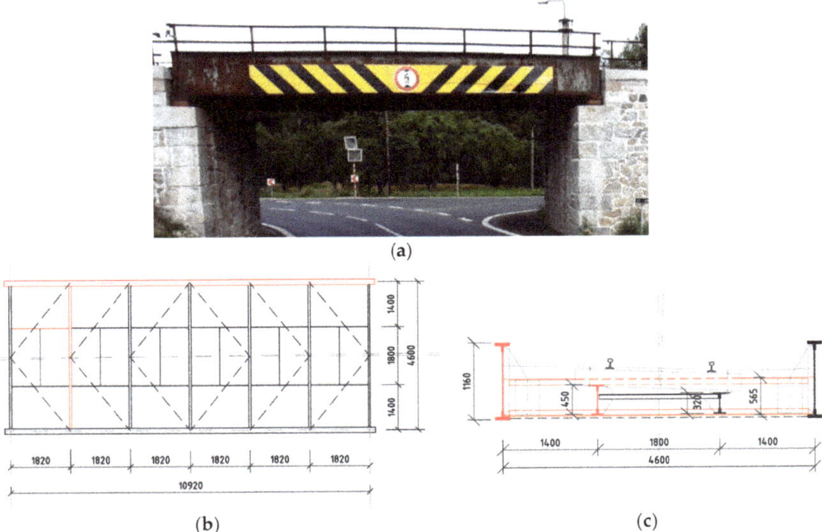

Figure 1. Bridge 1: the 142-year-old plate girder bridge with a bottom open member deck: (**a**) picture from the side; (**b**) schematic ground plan of the superstructure; (**c**) cross-section of the bridge; the chosen elements are drawn in red in (**b**) and (**c**).

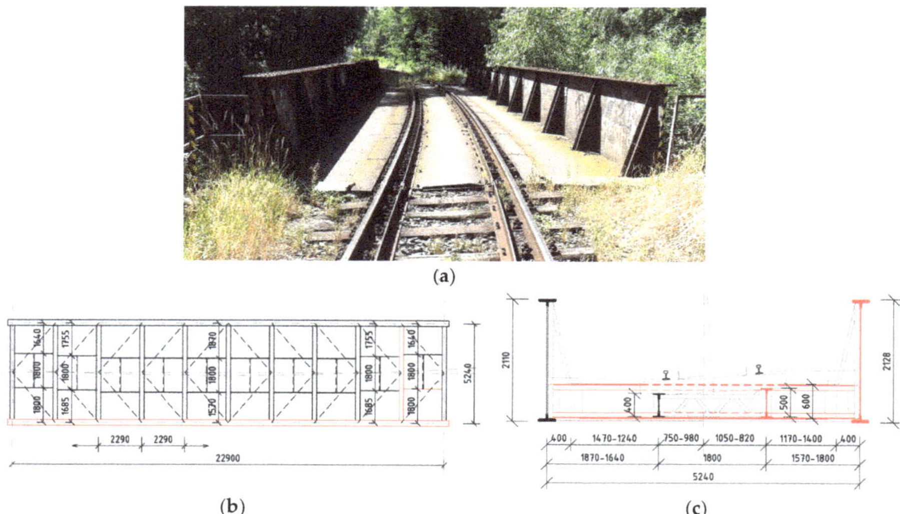

Figure 2. Bridge 2: the 109-year-old plate girder bridge with the bottom open member deck across a local stream: (a) picture of the bridge from the track; (b) schematic ground plan of the superstructure; (c) cross-section of the bridge; the chosen elements are drawn in red in (b) and (c).

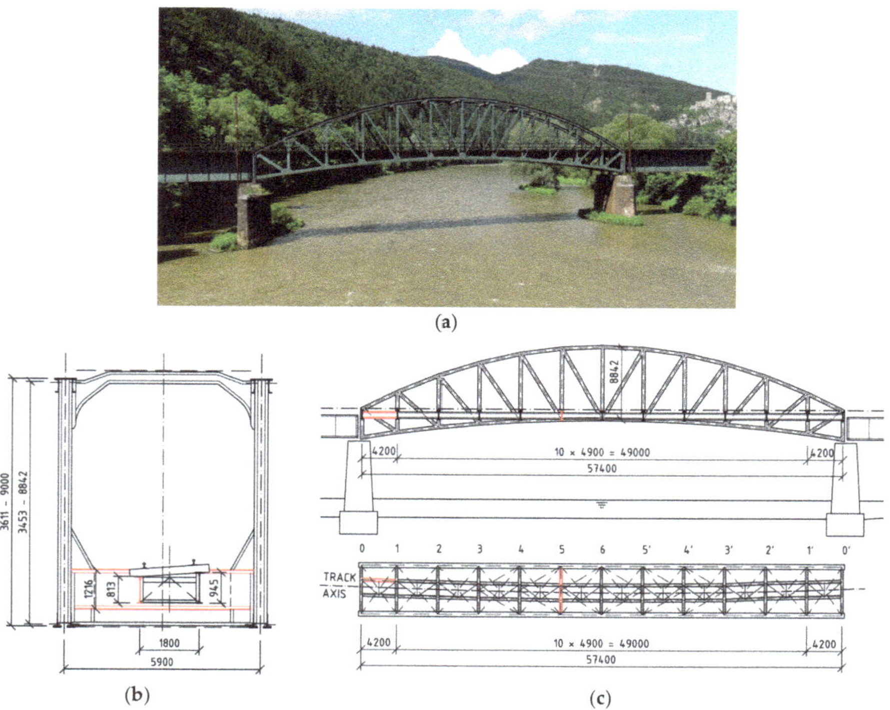

Figure 3. Bridge 3: the 76-year-old superstructure with the middle span built as the truss girder bridge with an intermediate open member deck: (a) picture from the side; (b) the cross-section; (c) schematic side view and ground plan of the superstructure; the chosen elements are drawn in red in (b) and (c).

The basic characteristic of each bridge and chosen member for the following study are summarized in Table 1.

Table 1. Basic data and designations of the chosen bridges and their members.

Bridge Superstructure				Chosen Member		
No./type	Year/age	Span	Figure	Designation	Length	Section
Bridge **1** plate girder	1877 142 years	10.92 m	Figure 1	Stringer 1 Crossbeam 1 Main girder 1	1.82 m 4.60 m 10.92 m	Figure 4a Figure 4b Figure 4c,d
Bridge **2** plate girder	1910 109 years	22.90 m	Figure 2	Stringer 2 Crossbeam 2 Main girder 2	2.29 m 5.24 m 22.90 m	Figure 5a Figure 5b Figure 5c,d
Bridge **3** truss girder	1943 76 years	57.40 m	Figure 3	Stringer 3 Crossbeam 3	4.20 m 5.90 m	Figure 6a Figure 6b

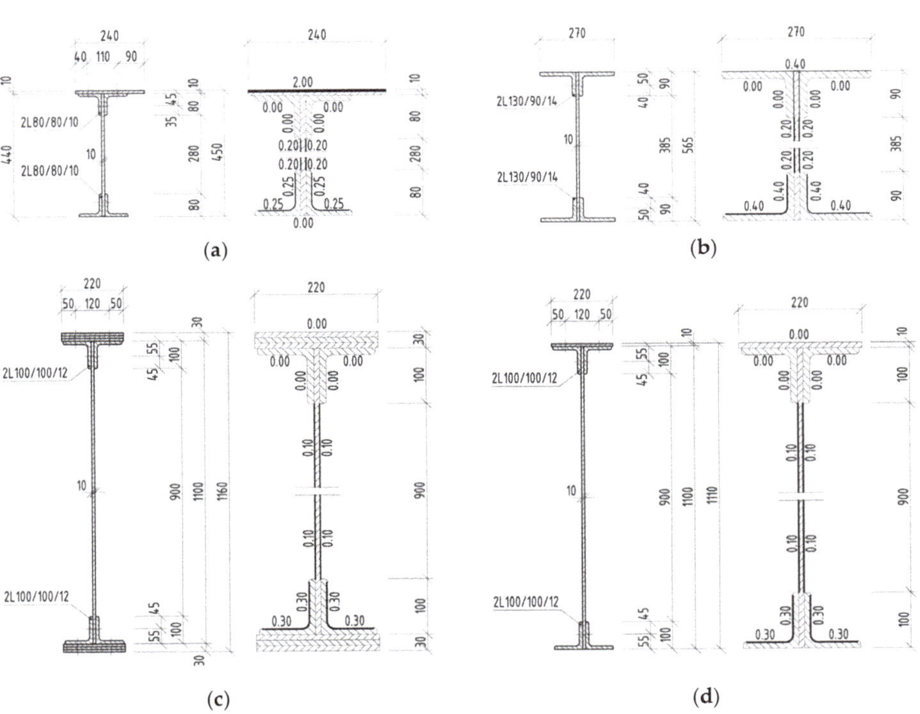

Figure 4. Riveted I-sections and average corrosion losses of elements in the case of Bridge 1: (**a**) left stringer; (**b**) crossbeam; (**c**) left main girder in the midspan; (**d**) left main girder in the support zone.

First, detailed inspections of the condition and diagnostics were carried out on all three bridge structures. The necessary geometry and imperfection data were measured and verified. In situ measurements to determine the material characteristics were also performed. In the case of degradation caused by corrosion, the main focus was on the current corrosion attack.

The in situ measurement of the corrosion attack of the elements took place most often after removal of the corrosion products using thickness gauges. Individual measured corrosion losses were recorded over the cross-section of the attacked element in several places. The number of measured places depended on the length of the element to be evaluated, the size of the cross-section of the element itself, as well as the cross-sectional structure and its corrosion damage. Subsequently, the data were statistically evaluated, and an effective cross-section of the riveted element was determined. In the case of the riveted I-sections listed above, the average corrosion losses are shown in Figures 4–6, in which all dimensions are given in millimeters.

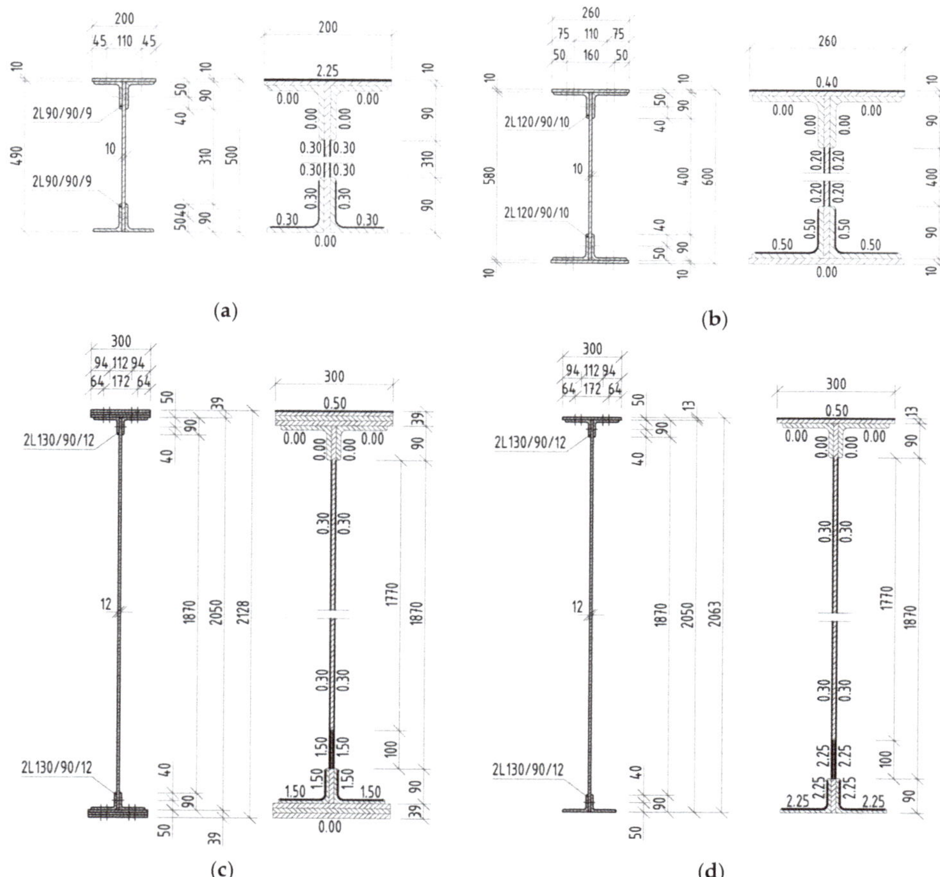

Figure 5. Riveted I-sections and average corrosion losses of elements in the case of Bridge 2: (**a**) right stringer; (**b**) crossbeam; (**c**) right main girder in the midspan; (**d**) right main girder near the support.

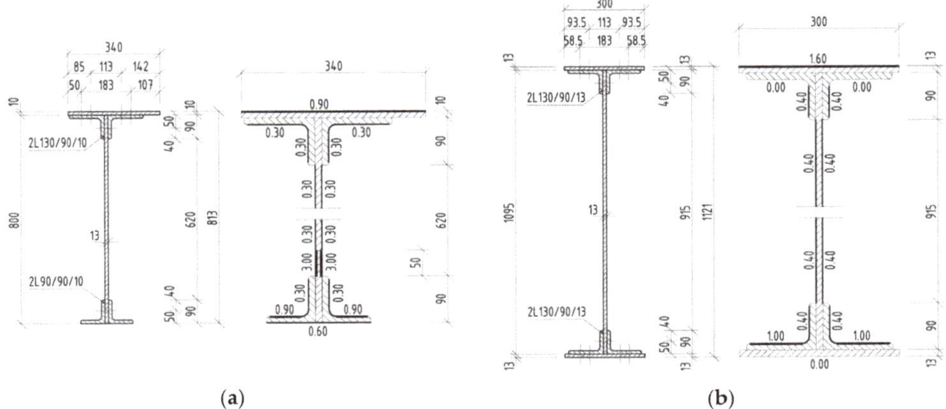

Figure 6. Riveted I-sections and average corrosion losses of deck elements in the case of Bridge 3: (**a**) first stringer; (**b**) crossbeam.

The photos in Figure 7 illustrate the conditions of some structural elements or the details of one of the bridges. Obviously, the corrosion attack was not always necessarily uniform and depended greatly on the position of a particular member or element in the structure, as well as on the position of the measured local point on the element itself.

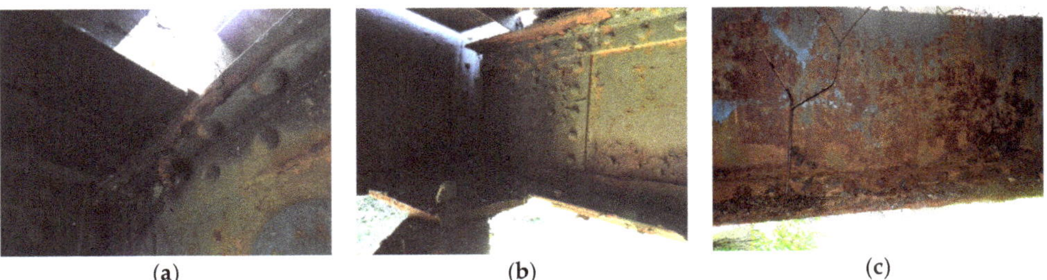

Figure 7. Examples of the degradation of the elements of Bridge 2: (**a**) local severe degradations of the top flange under the sleeper contact; (**b**) detail of the joint where the crossbeam, main girder, and bottom bracing are connected; (**c**) main girder; strong corrosion of the bottom flange and the bottom part of the girder web.

2.2. Load-Carrying Capacity

The guideline [10] presents general rules and a methodology for determining the load-carrying capacity of the railway bridges. As corrosion can be described as a random process, the best way to analyze such an effect in time is through stochastic approaches [16]. Moreover, other inputs such as material properties and of course actions are also of a stochastic nature. If the more sophisticated calculation was the issue, stochastic numerical approaches replacing traditional finite element method (FEM) analysis are also available [17].

Anyway, the aim of the paper is to show the importance of corrosion losses in load-carrying capacity calculation. The worldwide-used semi-probabilistic method of load and resistance factor design (LRFD) was applied in this research. As the existing structures were to be analyzed, partial safety factors were calibrated. The methodology in guidance [10] for the modification of reliability indexes for the evaluation of existing bridges was taken into account. The basic concept of how the reliability levels were transformed into the design values of the material properties and load effects could also be found in the paper [15].

For the classification of the riveted cross-sections, the widths of the respective parts of the cross-section are defined in Figure 8. In contrast to the welded cross-sections, it was also necessary to verify the classification in terms of the distance of the rivets parallel to the direction of the applied stresses in addition to the transverse direction.

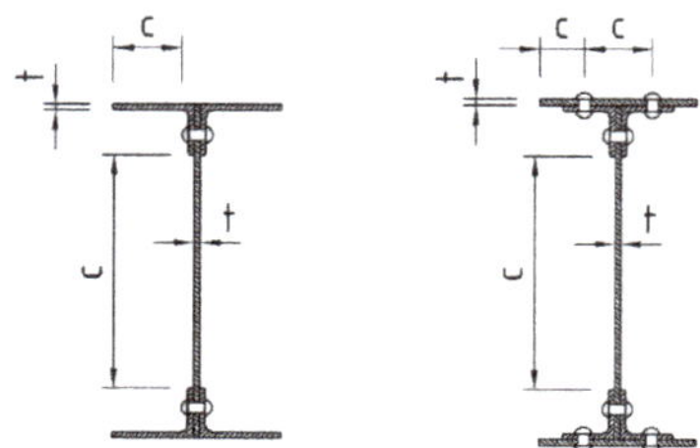

Figure 8. Definition of widths for the classification of riveted I-sections.

The determination of the LCC of cross-section under bending and tensile force or axial compression can be performed according to the Equation (1), in which the degradation due to corrosion is covered in cross-sectional parameters:

$$Z_{LM71} = (1 - \eta_{1,rs})/\eta_{1,LM71}, \tag{1}$$

where:

$$\eta_{1,rs} = N_{rs,Ed}/(A \times f_{yd}) + M_{y,rs,Ed}/(W_{el,y} \times f_{yd}) + M_{z,rs,Ed}/(W_{el,z} \times f_{yd}), \tag{2}$$

$$\eta_{1,LM71} = N_{LM71,Ed}/(A \times f_{yd}) + M_{y,LM71,Ed}/(W_{el,y} \times f_{yd}) + M_{z,LM71,Ed}/(W_{el,z} \times f_{yd}), \tag{3}$$

and designations $N_{LM71,Ed}$, $M_{y,LM71,Ed}$, and $M_{z,LM71,Ed}$ represent the design values of axial force and bending moments due to vertical variable rail traffic load effects including the dynamic factor, while $N_{rs,Ed}$, $M_{y,rs,Ed}$, and $M_{z,rs,Ed}$ are the design, combination, or group values of axial force and bending moments due to other load effects acting simultaneously with the vertical rail traffic load. In the cross-sectional characteristics, A (the area), $W_{el,y}$, and $W_{el,z}$ (the section modules) of the riveted cross-section, the holes for rivets were excluded in the tensile area of the cross-section. Moreover, the cross-sectional characteristics took into account the degradation due to corrosion by varying the thickness of the respective cross-sectional part. Finally, the design value of steel yield stress could be obtained from $f_{yd} = f_y/\gamma_{M0}$, where γ_{M0} is a partial factor for the material and resistance of cross-sections.

The value of LLC obtained from (1) is valid if the shear force meets the condition:

$$\eta_3 = V_{Ed}/V_{Rd} = (Z_{LM71} \times V_{LM71,Ed} + V_{rs,Ed})/V_{Rd} \leq 0.5, \tag{4}$$

where $V_{LM71,Ed}$ is the design value of shear force due to vertical variable rail traffic load effects represented by the LM71 including dynamic factors and $V_{rs,Ed}$ is the design, combination, or group value of shear force due to other load effects acting simultaneously with the vertical rail traffic load. The minimum from the design values of the shear resistance of the cross-section or design value of shear resistance of the web is designated as V_{Rd}.

If the above-mentioned Assumption (4) is not satisfied, the LCC in the form of the value Z_{LM71} should be determined from the quadratic equation:

$$\begin{aligned} Z_{LM71}^2 \times (4 \times k \times \eta_{3,LM71}^2) + Z_{LM71} \times (\eta_{1,LM71} + 8 \times k \times \eta_{3,LM71} \times \eta_{3,rs} - 4 \times k \times \eta_{3,LM71}) + \\ + (\eta_{1,rs} + 4 \times k \times \eta_{3,rs}^2 - 4 \times k \times \eta_{3,rs} + k - 1) = 0, \end{aligned} \tag{5}$$

where symbols $\eta_{1,rs}$ and $\eta_{1,LM71}$ were already defined in (2) and (3). For the other parameters, see the following three equations:

$$\eta_{3,rs} = V_{rs,Ed}/V_{pl,Rd}, \tag{6}$$

$$\eta_{3,LM71} = V_{LM71,Ed}/V_{pl,Rd}, \tag{7}$$

$$k = 1 - M_{f,N,Rd}/M_{pl,N,Rd}, \tag{8}$$

in which $M_{f,N,Rd}$ represents the design value of the plastic bending resistance of a cross-section consisting of the flanges only (i.e., without the contribution of the web) and $M_{pl,N,Rd}$ is the design value of the plastic bending resistance of the entire cross-section.

Since the value of the shear force V_{Ed} in Relation (4) is dependent on the investigated load-carrying capacity, the calculation of LCC should run in an iterative form.

Of course, the design value of shear force should be less than shear resistance. Thus, based on the above-mentioned equations and symbols defined below, the LCC of cross-section affected by pure shear can be derived from condition $\eta_3 \leq 1.0$ as follows:

$$Z_{LM71} = (1 - \eta_{3,rs})/\eta_{3,LM71}. \tag{9}$$

When verifying the resistance of cross-sections, it is also necessary to verify the biaxial stress state in the web. The LCC can then be derived from the next quadratic equation:

$$\begin{aligned} Z^2_{LM71} \times (\eta^2_{1,LM71} + \eta^2_{2,LM71} - \eta_{1,LM71} \times \eta_{2,LM71} + 3 \times \eta^2_{3,LM71}) + \\ + Z_{LM71} \times (2 \times \eta_{1,rs} \times \eta_{1,LM71} + 2 \times \eta_{2,rs} \times \eta_{2,LM71} - \eta_{1,rs} \times \eta_{2,LM71} - \eta_{2,rs} \times \eta_{1,LM71} + 2,2 \times \eta_{3,rs} \times \eta_{3,LM71}) + \\ + (\eta^2_{1,rs} + \eta^2_{2,rs} - \eta_{1,rs} \times \eta_{2,rs} + 3 \times \eta^2_{3,rs} - 1) = 0, \end{aligned} \tag{10}$$

where variables $\eta_{1,rs}$, $\eta_{1,LM71}$, $\eta_{3,rs}$, and $\eta_{3,LM71}$ were already defined before, while designations $\eta_{2,rs}$ and $\eta_{2,LM71}$ represent the influence of local vertical stress in the web if the local vertical force is present (e.g., a sleeper on the top flange of the riveted stringer). They can be calculated on the basis of equations:

$$\eta_{2,LM71} = \sigma_{z,LM71,Ed}/(f_y/\gamma_{M0}), \tag{11}$$

$$\eta_{2,rs} = \sigma_{z,rs,Ed}/(f_y/\gamma_{M0}), \tag{12}$$

where $\sigma_{z,LM71,Ed}$ represents the value of vertical stresses in the web due to vertical variable rail traffic load effects represented by the wheels of LM71 including dynamic factors and $\sigma_{z,rs,Ed}$ is the design, combination, or group value of vertical stresses in the web due to other load effects acting simultaneously with the vertical rail traffic load.

The verification of the resistance of slender cross-sections shall respect the shear lag effects and plate buckling effects, which may be calculated by means of effective cross-sectional characteristics. More information concerning the load-carrying capacity estimation, including techniques for how LCC should be calculated in the case of compressed member buckling and/or the loss of lateral and torsional stability due to bending, can be found in [10,11].

However, the process of calculating LCC has to be preceded by a very important task, which is global analysis. For the analysis of the behavior of each bridge structure, a spatial transformation numerical model was processed taking into account the real geometrical, stiffness, and material characteristics. That is why we did not put all details concerning FEM models. Two of the executed FEM models are shown in Figure 9.

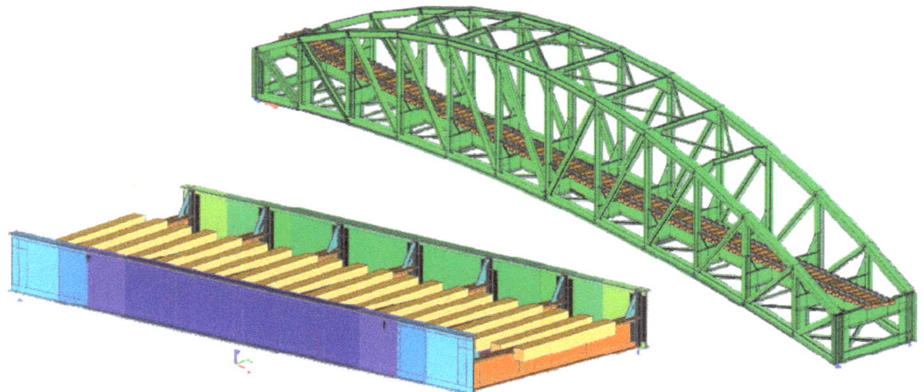

Figure 9. Global view at the FEM models of Bridge 1 (bottom left) and Bridge 3 (top right).

The computational models were created on the basis of long-term experience with the creation of FEM models of old and newly designed bridge structures. There is lack of space in this paper for a comprehensive description of each model and applied analysis. Therefore, only the basic features of the implemented models are given. Especially for the bridge elements, the beam finite elements were used, respecting their cross-sectional characteristics, shape variability, and mutual eccentricity. The interconnections of bridge deck members (stringer-to-crossbeam) were modeled as semi-rigid joints with stiffness on the basis of executed connections. The corner stiffeners of the main girders at the crossbeams' locations were incorporated into the models, thus helping to approximate the real rigidity of the crossbeam connection to the main girders. Increased attention was paid to this detail to obtain stiffness that was more accurate for the lateral-torsional stability analysis of the main girders. The stiffness of bridge bearings was also taken into account. Because of the better redistribution of traffic load to the stringers, the complete railway track (rails and sleepers) was included in all models so that the wheel forces from the traffic could be redistributed more correctly, but at the same time, the model of the rail track did not significantly affect the bridge deck behavior.

By utilization of the FEM, the necessary internal forces to determine the load-carrying capacities were determined by the elastic analysis. As riveted bridges are not supposed to behave in the plastic zone, the ultimate states were defined as the first occurrence of plastic strain (no matter if it was in the steel plates or in a rivet).

All relevant loads were included in the global analyses. Then, internal forces, stresses, and deformations produced by vertical variable rail traffic load were used from the design, combination, or group values caused by other load effects. After that, the process of the determination of relevant LCC could started.

3. Results of the Study

Baseline results of calculated actual LCC in the study are summarized in Table 2. Decreases of the cross-sectional area A, cross-section modulus W_y, and the load-carrying capacity Z_{LM71} due to the corrosion losses are given in the table. In addition, the relative values of load-carrying capacity decrement are also presented. For determination of the load-carrying capacity values, corresponding equations from (1–12) given in previous section were utilized.

Table 2. Results of the load-carrying capacity (LCC) of chosen members.

Element of the Bridge	Age of the Bridge	Cross-Sectional Area	Section Modulus	LCC from Bending and Axial Load: Equation (1) or (5)		LCC from Biaxial Stress State in Web: Equation (10)		LCC Derived from Pure Shear: Equation (9)	
	t (y)	A (mm^2)	W_y (mm^3)	Z_{LM71}	Z_t/Z_{t0}	Z_{LM71}	Z_t/Z_{t0}	Z_{LM71}	Z_t/Z_{t0}
Stringer 1	0	12,300.0	1,540,127.1	0.639	1.000	1.585	1.000	–	–
	142	11,733.0	1,477,247.9	0.582	0.911	1.505	0.950	–	–
Crossbeam 1	0	17,186.0	3,348,120.4	0.668	1.000	0.958	1.000	1.661	1.000
	142	16,768.8	3,223,218.9	0.642	0.961	0.918	0.958	1.590	0.957
Main girder 1	0	33,224.0	1,3455,418.9	0.818	1.000	0.875	1.000	1.224	1.000
	142	32,821.2	1,3280,169.9	0.806	0.985	0.861	0.984	1.179	0.963
Stringer 2	0	13,056.0	1,799,326.9	0.886	1.000	1.940	1.000	–	–
	109	12,317.4	1,701,602.0	0.812	0.916	1.814	0.935	–	–
Crossbeam 2	0	19,000.0	3,997,194.4	0.465	1.000	0.767	1.000	1.490	1.000
	109	18,536.0	3,870,092.9	0.448	0.963	0.739	0.963	1.427	0.958
Main girder 2	0	57,984.0	4,1565,670.1	0.908	1.000	1.109	1.000	1.436	1.000
	109	55,848.0	4,0166,758.5	0.868	0.956	1.049	0.946	1.295	0.902
Stringer 3	0	20,878.0	3,954,636.7	1.204	1.000	1.236	1.000	–	–
	76	19,567.0	3,572,712.3	1.040	0.864	1.102	0.892	–	-
Crossbeam 3	0	31,032.0	1,0593,420.7	1.404	1.000	1.043	1.000	1.736	1.000
	76	29,582.8	1,0216,187.5	1.344	0.957	1.016	0.974	1.622	0.934

In order to specify the development of reduction of LCC over time, it would be necessary to know the corrosion rate data at a given location. Moreover, it was evident that some parts of the cross-section corroded significantly more quickly, and others were slower. It depended on the position of the element in the structure and on the shape of the element itself. Thus, one of the dominant parameters seemed to be the so-called position coefficient in the structure [18].

Partial results of LCC estimation from the bending resistance of Stringer 3 in Table 2, were already a part of the study [19].

Other necessary data for correct time analysis and subsequent prediction of the development of corrosion and hence load-carrying capacity of the bridge were undoubtedly data about the renewals and repairs of the coating system in the past. For most of conventional bridges, this information was unobservable, and if so, there was no knowledge about the quality of the works and about the quality of the materials used nor their durability. Hence, it was evident that the time-corrosion relationship in the case of bridges had an irregular course similar to that shown in Figure 10.

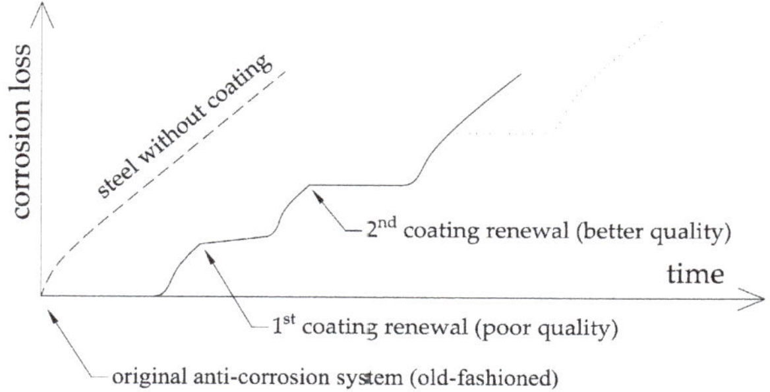

Figure 10. Possible time-corrosion loss relationship in the case of old bridge structure.

Generally, it is not possible to recover the whole history of anti-corrosion adjustments and renewals on a bridge structure. Moreover, in the case of the corrosion rate at a given site, only not very accurate estimates can be made usually. In such cases, there is no possibility to reliably estimate degradation development without longer corrosion in situ measurements.

Consequently, for comparison of the reduction of LCC and its prediction in the future, the percentage values of corrosion attack D′ to the measured actual values were simply used. The right graph in Figure 11 shows the dependence of the decrease of the cross-sectional area of riveted I-sections due to degradation caused by the corrosion process. At the same time, the decrease of the load-carrying capacity Z_{LM71} of the same cross-sections is presented in the left graph.

From Figure 11, it is evident that decrease process of both the cross-sectional area A and load-carrying capacity was almost in a perfect linear relationship with corrosion attack D′. Interestingly, this conclusion applied to each of the eight cases examined in the presented study, no matter which part of the cross-section was attacked by corrosion and how big the differences in corrosion attack observed within a cross-section were. However, the decrease of the load-carrying capacity was stronger than the decrease of the corresponding cross-sectional area, which was due to the fact that the LCC of the cross-section was reduced by the other load effects. The data shown in Figure 12 indicate that the decrease in calculated load-carrying capacity was indeed faster than the decrease in cross-sectional area. From the results in the graph, it could be assumed that the percentage decrease in the load-carrying capacity at the cross-sections solved in the study was at least 1.4 times and at most 2.8 times higher compared to the loss of cross-sectional area (i.e., corrosive loss throughout the cross-section).

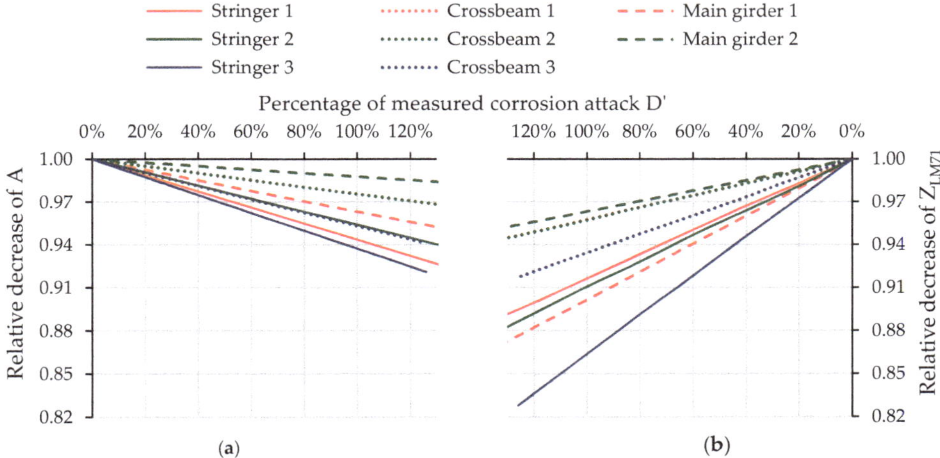

Figure 11. Relative decrease of the parameters of cross-sections as a function of the gradual increase of corrosion attack D′: (**a**) relative decrease of the cross-sectional area A; (**b**) relative decrease of the load-carrying capacity Z_{LM71} of the same cross-sections.

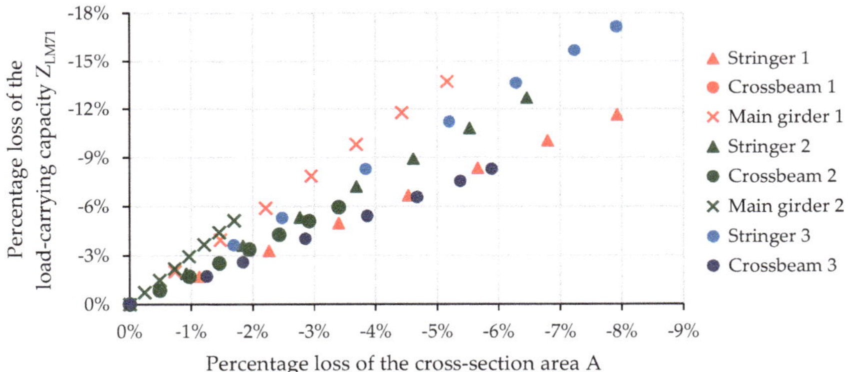

Figure 12. Relation between the losses of the cross-sectional area and load-carrying capacity due to corrosion.

4. Conclusions and Discussion

The results of the study focus on the effect of the corrosion of steel structures on the reliability of bridge structures showed a practically linear course of decreasing cross-sectional area and load-carrying capacity of observed elements of the bridge structures due to increasing corrosion losses during their service life. A similar dependence was observed when determining the load-carrying capacity of the riveted I-section from:

- Bending resistance or combined with axial force: Equation (1)
- Combination of bending, normal force, and shear force: Equation (5)
- Shear resistance of the web: Equation (9)
- Resistance of the cross-sectional web to the biaxial stress state caused by normal stresses in the longitudinal and vertical directions in combination with the shear stresses: Equation (10)

However, the governing criteria for LCC may change over time due to corrosion losses [20]. Thus, the linear dependence can be disrupted, especially in the case of very severe damages of the cross-section due to corrosion. It seemed that the speed of reduction of load-carrying capacity could be approximately 1.5–3.0 times faster than the corrosion process speed expressed by the corrosion loss within the cross-section. The faster reduction in LCC compared to the reduction of the corresponding cross-sectional area emphasized the importance of monitoring the effect of corrosion on the static safety of the bridge structure.

Neglecting inspections can lead to substantial degradation and a consequent decrease of the load-carrying capacity [21]. Thus, only perfect up-to date protection of steel together with regular periodic inspections and basic routine maintenance can ensure the required service life and save much money. Underestimating corrosion usually results in the poor condition of bridges, requiring major repairs and reconstruction [22]. Moreover, the careful inspection activities and records from them may in the future provide valuable data for a qualified estimation of the corrosion rate for a particular bridge object or its critical elements, and thus for LCC determination and remaining service life estimation.

As already mentioned, more relevant data for the evaluation of the influence of corrosion on any steel structural bridge element can be derived only from long-term measurements. Therefore, experimental investigation on real structures or on specimens located near or directly on bridges are needed. In many European countries, these data are processed for structural steel for many years, as in [23–25]. In some countries, extensive research has also been devoted to weathering of steel [26,27]. Similar measurements are running also in the Slovak Republic [28,29]. The data will also be used to determine inputs for refining the corrosion map of Slovakia, as many more data are needed to improve its accuracy in regions.

Author Contributions: J.G. and J.O. both participated in diagnostics of bridges and executed the global analyses of the bridges; moreover, J.G. executed the LLC calculations and prepared pictures for the paper, while J.O. prepared the graphs and wrote the paper. All authors have read and agreed to the published version of the manuscript.

Funding: This research was supported by the Slovak Research and Development Agency under Contract No. APVV-14-0772 and by Research Project No. 1/0413/18 of the Slovak Grant Agency.

Conflicts of Interest: The authors declare no conflict of interest.

References

1. Leygraf, C.; Wallinder, I.O.; Tidblad, J.; Graedel, T. *Atmospheric Corrosion*; Wiley & Sons: New York, NY, USA, 2016.
2. Hays, G.F. Now is the Time. The World Corrosion Organization, Editorial. 2010. Available online: https://www.scientific.net/AMR.95.-2.pdf (accessed on 31 December 2019).
3. Schmitt, G. Control global needs for knowledge dissemination, research, and development in materials deterioration and corrosion control. *World Corrosion Organ.* **2009**, *38*, 14. Available online: https://www.scribd.com/document/104293675 (accessed on 31 December 2019).
4. Wiśniewski, D.F.; Casas, J.R.; Ghosn, M. Codes for safety assessment of existing bridges—Current state and further development. *Struct. Eng. Int.* **2012**, *22*, 552–561. [CrossRef]
5. Adey, B.; Bailey, S.; Das, P.; O'Brien, E.J.; González, A. *Cost345—Procedures Required for Assessing Highway Structures, European Cooperation in the Field of Scientific and Technical Research*; Project Reports; COST: Brussels, Belgium, 2004.
6. Piau, J.M. *SAMARIS—Sustainable and Advanced Materials for Road Infrastructures*; Project Reports; V FP: Brussels, Belgium, 2006.
7. *Sustainable Bridges—Assessment for Future Traffic Demands and Longer Lives*; Project Reports; VI FP: Brussels, Belgium, 2007.
8. *ARCHES—Assessment and Rehabilitation of Central European Highway Structures*; Project Reports; VI FP: Brussels, Belgium, 2009.
9. *UIC 778-2—Recommendations for Determining the Carrying Capacity and Fatigue Risks of Existing Metal Bridges*; Guideline; UIC: Paris, France, 2018.

10. Vičan, J.; Gocál, J.; Hlinka, R.; Odrobiňák, J.; Moravč, M.; Koteš, P. *Určovanie zaťažiteľnosti Železničných mostov/Determination of Load Carrying Capacity of Railway Bridges*; Gudeline; Railways of the Slovak Republic: Bratislava, Slovakia, 2015.
11. Vičan, J.; Gocál, J.; Hlinka, R.; Odrobiňák, J.; Moravčík, M.; Koteš, P. *Metodický pokyn pro určování zatížitelnosti železničních mostních objektů/Methodological Guideline for Determination of Load Carrying Capacity of Railway Bridges*; Guideline; Czech Railway Infrastructure Administration: Prague, Czech Republic, 2015.
12. Vičan, J.; Koteš, P. *Hodnotenie Existujúcich Mostných Objektov/Evaluation of Existing Bridges*; EDIS: Žilina, Slovakia, 2019.
13. EN 1991-2. *Eurocode 1: Actions on Structures—Part 2: Traffic Loads on Bridges*; CEN: Brussels, Belgium, 2003.
14. Vičan, J.; Gocál, J.; Odrobiňák, J.; Koteš, P. Existing steel railway bridges evaluation. *Civ. Environ. Eng.* **2016**, *12*, 103–110. [CrossRef]
15. Vičan, J.; Odrobiňák, J.; Koteš, P. Determination of load-carrying capacity of railway steel and concrete composite bridges. *Key Eng. Mater.* **2016**, *691*, 172–184. [CrossRef]
16. Gomes, W.J.; Beck, A.T.; da Silva, C.R. Modeling Random Corrosion Processes via Polynomial Chaos Expansion. *J. Braz. Soc. Mech. Sci. Eng.* **2012**, *34*, 561–568. [CrossRef]
17. Kamiński, M.; Szafran, J. Bridges for pedestrians with random parameters using higher order Stochastic Finite Element Method. *Int. J. Appl. Mech. Eng.* **2017**, *22*, 175–197. [CrossRef]
18. Křivý, V.; Urban, V.; Fabian, L. Experimental investigation of corrosion processes on weathering steel structures. *Key Eng. Mater.* **2013**, *577*, 397–400. [CrossRef]
19. Odrobiňák, J.; Gocál, J. Experimental measurement of structural steel corrosion. *Proc. Struct. Integr.* **2018**, *13*, 1947–1954. [CrossRef]
20. Kayser, J.R.; Nowak, A.S. Capacity loss due to corrosion in steel-girder bridges. *J. Struct. Eng.* **1989**, *115*, 1525–1537. [CrossRef]
21. Odrobiňák, J.; Hlinka, R. Degradation of steel footbridges with neglected inspection and maintenance. *Proc. Eng.* **2016**, *156*, 304–311. [CrossRef]
22. Ágocs, Z.; Brodniansky, J.; Vičan, J.; Ziólko, J. *Assessment and Refurbishment of Steel Structures*; CRC Press: Boca Raton, FL, USA, 2004.
23. Morcillo, M.; Simancas, J.; Feliu, S. Long-Term Atmospheric Corrosion in Spain: Results after 13–16 Years of Exposure and Comparison with Worldwide Data. In *Atmospheric Corrosion*; Kirk, W., Lawson, H., Eds.; ASTM International: West Conshohocken, PA, USA, 1995; pp. 195–214. [CrossRef]
24. Kreislová, K.; Knotková, D. The results of 45 years of atmospheric corrosion study in the Czech Republic. *Materials* **2017**, *10*, 394. [CrossRef] [PubMed]
25. Tidblad, J. Atmospheric corrosion of metals in 2010-2039 and 2070-2099. *Atmos. Environ.* **2012**, *55*, 1–6. [CrossRef]
26. Morcillo, M.; Chico, B.; Díaz, I.; Cano, H.; Fuente, D.D.L. Atmospheric corrosion data of weathering steels. A review. *Corr. Sci.* **2017**, *77*, 6–24. [CrossRef]
27. Křivý, V.; Kubzová, M.; Konečný, P.; Kreislová, K. Corrosion Processes on Weathering Steel Bridges Influenced by Deposition of De-Icing Salts. *Materials* **2019**, *12*, 1089. [CrossRef] [PubMed]
28. Odrobiňák, J.; Gocál, J.; Jošt, J. NSS test of structural steel corrosion. *Roczniki Inżynierii Budowlanej* **2017**, *15*, 7–14.
29. Koteš, P.; Vičan, J.; Ivašková, M. Influence of reinforcement corrosion on reliability and remaining lifetime of RC bridges. *Mater. Sci. Forum* **2016**, *844*, 89–96. [CrossRef]

© 2020 by the authors. Licensee MDPI, Basel, Switzerland. This article is an open access article distributed under the terms and conditions of the Creative Commons Attribution (CC BY) license (http://creativecommons.org/licenses/by/4.0/).

MDPI
St. Alban-Anlage 66
4052 Basel
Switzerland
Tel. +41 61 683 77 34
Fax +41 61 302 89 18
www.mdpi.com

Materials Editorial Office
E-mail: materials@mdpi.com
www.mdpi.com/journal/materials

www.ingramcontent.com/pod-product-compliance
Lightning Source LLC
LaVergne TN
LVHW071625170726
843515LV00009B/2482
* 9 7 8 3 0 3 6 5 0 2 9 0 8 *